ODEN and REDDY—An Introduction to the Mathematical Theory of Finite Elements
PASSMAN—The Algebraic Structure of Group Rings
PETRICH—Inverse Semigroups
PIER—Amenable Locally Compact Groups
PRENTER—Splines and Variational Methods
RIBENBOIM—Algebraic Numbers
RICHTMYER and MORTON—Difference Methods for Initial-Value Problems, 2nd Edition
RIVLIN—The Chebyshev Polynomials
ROCKAFELLAR—Network Flows and Monotropic Optimization
RUDIN—Fourier Analysis on Groups
SAMELSON—An Introduction to Linear Algebra
SCHUMAKER—Spline Functions: Basic Theory
SHAPIRO—Introduction to the Theory of Numbers
SIEGEL—Topics in Complex Function Theory
    Volume 1—Elliptic Functions and Uniformization Theory
    Volume 2—Automorphic Functions and Abelian Integrals
    Volume 3—Abelian Functions and Modular Functions of Several Variables
STAKGOLD—Green's Functions and Boundary Value Problems
STOKER—Differential Geometry
STOKER—Nonlinear Vibrations in Mechanical and Electrical Systems
STOKER—Water Waves
TURÁN—On A New Method of Analysis and Its Applications
WHITHAM—Linear and Nonlinear Waves
WOUK—A Course of Applied Functional Analysis
ZAUDERER—Partial Differential Equations of Applied Mathematics

INTEGRATION THEORY

INTEGRATION THEORY

Volume 1: MEASURE AND INTEGRAL

Volume 2: REAL INTEGRATION THEORY

Volume 3: SELECTED TOPICS

INTEGRATION THEORY

Volume 1: MEASURE AND INTEGRAL

CORNELIU CONSTANTINESCU

KARL WEBER

IN COLLABORATION WITH

ALEXIA SONTAG

A Wiley-Interscience Publication

JOHN WILEY & SONS

New York • Chichester • Brisbane • Toronto • Singapore

Library of Congress Cataloging in Publication Data:

Constantinescu, Corneliu.
Integration theory.

(Pure and applied mathematics)
Translated from the German.
"A Wiley-Interscience publication."
Includes index.
Contents: v. 1. Measure and integral—
1. Integrals, Generalized. 2. Measure theory.
I. Weber, Karl, 1947 May 25- II. Sontag, Alexia.
III. Title. IV. Series: Pure and applied mathematics
(John Wiley & Sons)

QA312.C577 1984 515.4′2 84-15344
ISBN 0-471-04479-2 (v. 1)

Printed in the United States of America

10 9 8 7 6 5 4 3 2 1

PREFACE

Since about 1915 integration theory has consisted of two separate branches: the abstract theory required by probabilists and the theory, preferred by analysts, that combines integration and topology. As long as the underlying topological space is reasonably nice (e.g., locally compact with countable basis) the abstract theory and the topological theory yield the same results, but for more complicated spaces the topological theory gives stronger results than those provided by the abstract theory. The possibility of resolving this split fascinated us, and it was one of the reasons for writing this book.

The unification of the abstract theory and the topological theory is achieved by using new definitions in the abstract theory. The integral in this book is defined in such a way that it coincides in the case of Radon measures on Hausdorff spaces with the usual definition in the literature. As a consequence, our integral can differ in the classical case. Our integral, however, is more inclusive. In Volume 1 the chief goals are to define the integral, to provide a rationale for the definition that is given, to establish important properties of the integral, and to develop certain important tools.

It is our belief that certain important topics in integration theory are best understood in the context of vector lattices. These topics are studied in Volume 2, which accordingly begins with a thorough study of the theory of vector lattices. Volume 2 then presents the theory of L^p-spaces (for $0 < p \leq \infty$), including duality, and the theory of spaces of real-valued measures on a ring of sets, including the Radon-Nikodym Theorem. Vector lattices are not merely a passive backdrop for this presentation. Rather, the vector-lattice setting suggests important concepts and provides powerful techniques.

Volume 2 concludes with an extensive chapter on classical integration theory for $\mathbb{R}$. Although we have chosen a rather abstract and very general framework for integration theory, we are convinced that the classical theory represents a very important part of integration theory and should not be neglected in a book such as this. We also believe that the beauty of the classical theory is enhanced when this theory is viewed through the abstract theory.

Three main themes, followed by another return to classical integration theory, comprise Volume 3. The first chapter treats complex integration theory (i.e., complex-valued functions are integrated with respect to complex-valued measures). Then various operations with measures, including images, restrictions, and products, appear. Products are defined so they agree in the case of Radon measures with the generally used notion of product. Differences arise again in the classical case, and here too our notion is more inclusive. The third topic is the theory of integration on Hausdorff spaces (including Haar measures). True to our beliefs about the importance of the classical theory, we conclude Volume 3 with a chapter on integration in Euclidean space. All of the earlier themes appear again in this final chapter.

Appended to each of the three volumes is a brief sketch of the historical development of that portion of integration theory covered by the volume in question. Exercises provided for each section augment the treatment in the text. A number of exercises are new. Some others stem from the existing literature, although we do not provide references. We would like to make clear that in many exercises significant results have made their way into the book, results for which there was no room in the text. Many such exercises are of interest in themselves.

Certain points of view that influenced this book should be mentioned. One use we hope the book will find is as a text for a graduate course or seminar or for self-study. For this and other reasons the book is essentially self-contained: the only prerequisite is familiarity with elementary real analysis. In other words, we have tried to write a book so that someone who knows no integration theory (but has the prerequisite real analysis and is moderately perseverant) can read and learn integration theory by doing so. In particular, we have purposely chosen not to gloss over technical details. Such details may occasionally dim the appeal of the story, but they are nevertheless necessary. After all, what is trivial to one who knows and understands a theory is not always so to one who is just learning it. Such factors contribute to the length of the book.

It is also expected that many users of the book will use it as a reference in their own research, whether in integration theory or in other fields. It is our hope that such users not only will find the information they need but also will use the book as a conveniently citable, and readily decipherable, source.

The text is organized in the definition–theorem–proof format that is by now familiar. We find that this organizational format when strictly adhered to, simplifies the task both for the beginner who seeks to learn by reading the book and the expert who seeks to refresh his or her memory or to find an appropriate theorem to quote. We hope that our use of this format has not altogether hidden the fascination that prompted our writing of the book. Moreover it need not preclude narration and commentary. We have attempted to describe the ideas of the proofs, not just the formal details. Commentary has been used to signal important features of the definition–theorem–proof landscape. We hope that readers can thereby see in advance roughly where they are headed (and where possible the course to be pursued).

Finally, we have chosen to develop the abstract theory in full generality, which is only possible in a long book. An unfortunate consequence of this choice is that many important theorems of integration theory make their first appearance in the text very late in the book (i.e., in Volume 2 or even Volume 3). In many cases, however, special cases of the general result could appear earlier. Such special results have been included in the exercises, both to anticipate the general result and to fill out the abstract treatment in the text. Historically, of course, partial results and special cases also preceded the general theorems. To enable the reader to make effective use of the exercises, sketches of solutions are often included. For additional remarks describing the role played in this book by the exercises we ask the reader to consult the chapter entitled "Suggestions to the Reader."

CORNELIU CONSTANTINESCU
KARL WEBER
ALEXIA SONTAG

Zurich, Switzerland
Zurich, Switzerland
Wellesley, Massachusetts
May 1985

ACKNOWLEDGMENTS

The original version of this book was written in German. We are very much indebted to Alexia Sontag, who not only took care of the translation into English, but brought substantial contributions to the book by making important expository improvements such as changing the order of presentation of the material, writing more detailed proofs, and providing the whole text with many comments and explanations in order to make it more understandable to the reader. We wish to thank Imre Bokor very much for translating the exercises from German into English. We are much indebted to Wolfgang Filter for his conscientious reading of the manuscript. We are very grateful to Hannelore Aquilino, Marieluise Budde, Christina Frick, and Rose-Marie Grossmann for the skillful typing of the manuscript. We wish to express our sincere thanks to the staff of John Wiley & Sons for complying with our wishes with constant patience and understanding.

C. C.
K. W.

CONTENTS

INTRODUCTION

The integral defined in this book extends the integral of the usual abstract theory. Every function integrable according to the old definition is still integrable according to the new definition, and has the same integral, but there are more integrable functions. All important properties of the integral remain intact, and some new ones even appear.

In a sense the definition of integral used here is the best possible. An integral is always constructed by extending a small collection of "integrable" functions for each of which the value of the "integral" is known. It is desirable to obtain as many integrable functions as possible, but this goal must not be achieved at the expense of arbitrariness in choosing the extension. It is shown here that the class of all "reasonable" extensions, reasonable in the sense that they introduce no arbitrariness, forms a complete lattice. There is a smallest extension, which is in fact the integral of the classical abstract theory, and there is a largest extension, which is the extension used here to define the integral.

Both here and in the classical abstract theory, the collection of integrable functions that one constructs has the following property: if two functions differ only on some "insignificant" set and if one of them is integrable, then the other is also integrable and has the same integral. The difference in the two definitions is that here there are more "insignificant" sets, or, in the conventional terminology, there are more null sets. Loosely speaking, for the old abstract definition, null sets are somehow globally insignificant; for the definition used here, null sets are only locally insignificant. When σ-finiteness is hypothesized, the two definitions coincide.

The abstract definition of integral presented here is suggested, but not developed, in the book of I. E. Segal and R. A. Kunze (*Integrals and Operators*, McGraw-Hill, New York, 1968). As far as we know, however, nowhere is there developed a theory based on this definition.

In this book, as in the original papers of Lebesgue but in contrast to many others books on integration theory, positive measures are not allowed to take the value $+\infty$. Historically the value $+\infty$ came into play, on the one hand, out of the wish to assign (Lebesgue) measure to all Borel sets in Euclidean space and, on the other hand, via the Caratheodory theory of outer measure, which made possible extensions of measures on σ-rings. In any case, sets of infinite measure play no role at all in the definition of integral: they only appear as ornaments. These useless ornaments have a rather disturbing effect, namely the need for additional hypotheses. Later, when measures are allowed to take negative values, the values $+\infty$ and $-\infty$ have a devastating effect on the entire theory, so that many authors only allow the value $+\infty$. Even then, many properties of spaces of measures fail. Allowing infinite values becomes completely absurd when one moves to complex-valued measures, and books in which vector-valued or group-valued measures appear have dispensed with the value infinity. One small consequence of ruling out infinity was the replacement of σ-rings by δ-rings. Accordingly, σ-rings play only a subordinate role in this book.

Chapter 1 defines and discusses the objects and notions with which the theory begins. Chief among these is what is called a Daniell space. The two-step process by which one constructs an integral begins with a Daniell space. Exceptional sets and exceptional functions, an important tool in the construction of the integral, also receive attention in Chapter 1.

Chapter 2 presents two major examples of Daniell spaces: Daniell spaces associated with measure spaces and the Daniell spaces for the Stieltjes functionals.

The theoretical portion of Chapter 3 consists primarily of a discussion of one step in constructing the integral, namely the monotone-approximation process of Daniell. For a mere Daniell space, no completeness is required. If the Daniell space possesses a certain completeness property, it is said to be closed. The monotone-approximation process constructs, from a given Daniell space, an extension rightfully called the closure of the given Daniell space: it is the smallest, closed Daniell space extending the original. (This closure is the integral according to Daniell.)

The theory of summable families of real numbers, discussed and first applied in Sections 3.4 and 3.5, is an important example of the construction of the closure of a Daniell space and, at the same time, a tool in the theory of integration. A second major example, namely the closures of the Daniell spaces for the Stieltjes functionals, also merits a section of its own in Chapter 3.

Chapter 4 presents the definition of the integral for a Daniell space. The format of the chapter is to present a precise description of the sense in which the eventual definition is the best possible one. En route one obtains a concrete description of the second step in constructing the integral, including a description of the null sets, as well as important characterizations and properties of the integral that are used later.

Chapter 5 treats integrals that are derived from measures. The Daniell space associated with a measure space was already discussed in Chapter 2, and one need only apply the results of Chapters 3 and 4 to obtain the definitions and basic properties of integrals derived from measures (and of the completion of a measure space). Various concepts and theorems which serve as important tools in Volume 2 (upper and lower integral, inner and outer measure, measurability, approximation theorems) round out the theoretical portion of Chapter 5. The chapter concludes with two important examples: Stieltjes measure and integral, and then, in particular, Lebesgue measure and integral.

SUGGESTIONS TO THE READER

Each section of the book consists of two parts that have different goals. The first part, namely the text itself, is systematically developed. It consists of definitions and proven assertions assembled in an organized fashion and with no significant gaps for the reader to fill. All propositions and theorems, unless ready consequences of definitions and previously proven assertions, are proved in detail.

The second part, appearing in small print, consists of various kinds of exercises with accompanying suggestions and commentary. This part demands more independent work on the part of the reader and contains several kinds of information.

Some exercises provide straightforward practice using the material treated in the text and examples to illustrate the various concepts and techniques. They are designed in part to help the readers check whether they have understood the definitions and the theory and whether they can use the various notions themselves. Studying these exercises represents the minimum that readers should do on their own.

Other exercises extend the text in various directions. They generalize theorems from the text and present other possibilities for developing the theory, and they also serve to point out certain aspects that played an important role in the historical development of the material in the text. Counterexamples are included to show, in some instances, that various hypotheses of theorems in the main text cannot be dispensed with, or, in other instances, how certain hypotheses can be weakened. There are also exercises that point ahead to later chapters of the book; such exercises contain special cases of complete, general theorems proved later in the book.

The remaining exercises treat separate topics that are more or less closely related to integration theory. Various new results of the authors are included in these exercises.

The exercises are labeled E or C. We have used E for those exercises that are quite straightforward and are closely related to the text. Exercises marked C (for *colloquia*) are of two kinds. Some of them are closely related to the material in the text but are very extensive. The others treat separate topics having some connection to integration theory.

Most of the difficult exercises are accompanied by a suggested solution, the completeness of which increases with the difficulty of the exercise. In particular, for certain extensions of the theory and where problems are discussed that are far removed from the material in the text, the outlined solution is virtually complete. We hope to thereby enable even readers with little experience to work successfully, profitably through the exercises and the additional information they contain.

We want to emphasize that the texts of the successive sections form a unified whole independent of the exercises.

The requirements on the reader increase from one volume to the next. A book such as this is generally not read in a short time, and the qualifications of the reader increase in the duration, so we may safely assume that the preparation readers bring to the second or third volume is considerably greater than for Volume 1.

We want to stress at this point that a reader with sufficient background can easily read either the second or the third volume independently of the preceding volume(s). Volumes 2 and 3 both treat independent topics, and each contains a brief summary of the information it assumes. For a coverage of all important aspects of integration theory, however, the reader must consider all three volumes as one unit.

It is expected that many users of the book will treat it as a reference work. For such readers, especially, each volume contains a list of symbols both for the volume in question and for all volumes that precede it. The index is extensive, and we have tried to organize the text in such a way that readers using the book as a reference will readily find what they need.

One comment about our use of notation is in order here. We often work with mathematical objects Ω that depend on several previously defined objects $\Omega_1, \Omega_2, \ldots, \Omega_n$. A precise denotation for Ω requires mention of all of the objects $\Omega_1, \Omega_2, \ldots, \Omega_n$. We adhere to this requirement where notation is *defined*. In our *use* of such notation, where one or more of the objects $\Omega_1, \Omega_2, \ldots, \Omega_n$ is clear from context, we often delete the names of those objects from the notation. We follow the same practice for terminology. For example, $\mathscr{L}^1(X, \mu)$ may be abbreviated in any one of the forms: $\mathscr{L}^1(\mu)$, $\mathscr{L}^1(X)$, $\mathscr{L}^1$. Similarly the expression "μ-integrable functions on X" may be abbreviated as "μ-integrable functions," "integrable functions on X," or simply "integrable functions."

Two suggestions to the student conclude these remarks. Students of mathematics often assume that one must understand all the details of a subject the first time one studies it. It has long been evident that such a learning process is virtually impossible. Readers should not hesitate simply to skip over those points that even after considerable effort are not well understood and to study further, taking as mere facts the results not well understood. On returning to a point that was once confusing, the reader will often be amazed at how easy it has become to understand what was previously confusing or to see where the author has made an error. At other points readers studying the book will see another path that seems better. We encourage the reader to follow the other path. This holds even for the beginner, who often hesitates to embark on such a venture.

NOTATION AND TERMINOLOGY

In this preliminary section we assemble some notation and terminology used throughout the book. We assume that the items listed here are ones with which the reader is familiar.

All the investigations in the book are based on the notions of set theory, by which we mean naive set theory. We expect that the reader is familiar with set-theoretic terminology and notation [set, element, subset, union, intersection, (set-theoretic) difference, $\in$, $\notin$, etc.].

Throughout the book we assume the Axiom of Choice. Some of the exercises also hypothesize other nontrivial axioms (e.g., the Continuum Hypothesis). At such places we delve further into the foundations of set theory, but in the body of the text these more advanced set-theoretic considerations play no role at all. It goes without saying that the material in the text is developed in such a way that the reader who is familiar with axiomatic set theory will have no difficulty placing the material in axiomatic context.

We also assume that the reader is familiar with the natural numbers, including proofs by induction, and with the integers and the rational numbers. Although we do assume the reader to be familiar with the real number system, we begin Chapter 1 with a brief review of the real numbers. In traditional fashion we denote these various number systems by $\mathbb{N}$ (*natural numbers*), $\mathbb{Z}$ (*integers*), $\mathbb{Q}$ (*rational numbers*), and $\mathbb{R}$ (*real numbers*).

We want to mention explicitly that we do not include 0 among the natural numbers. For each natural number n, we denote by $\mathbb{N}_n$ the set whose elements are the natural numbers $1, 2, \ldots, n$, and we call $\mathbb{N}_n$ the *section* of $\mathbb{N}$ spanned by n.

Before proceeding any further, we want to describe our use of the symbols $:=$ and $:\Leftrightarrow$. When we write an expression of the form $P := Q$, we mean that Q is a term or a symbol whose meaning is already known and we are defining P to have the same meaning as Q. Similarly, when we write $P :\Leftrightarrow Q$, we mean that Q is a statement whose meaning is already known and we are defining the statement P to have the same meaning as Q. An expression of the form

$$P_n := P_{n-1} := \cdots := P_1 := Q$$

is understood to mean that Q is a term or symbol whose meaning is already known, that P_1 is being defined to have the same meaning as Q, and that P_k, for $k = 2, \dots, n$, is being defined to have the same meaning as P_{k-1}, hence inductively to have the same meaning as Q.

If P is a property that refers to elements of a set X (i.e., that might or might not be possessed by any particular element of X), we denote by

$$\{x \mid x \in X, P(x)\} \quad \text{or} \quad \{x \in X \mid P(x)\}$$

the set consisting of all elements of X that possess the property P. Similarly, if P is a property that refers to subsets of a set X, we denote by

$$\{A \mid A \subset X, P(A)\} \quad \text{or} \quad \{A \subset X \mid P(A)\}$$

the set consisting of all subsets of X that possess the property P. For a set X that has $x_1, x_2, \dots, x_n$ as its only elements, we write

$$X = \{x_1, x_2, \dots, x_n\}.$$

Thus $\{x\}$ denotes the set whose only element is x. Sets of the type $\{x\}$ are called *singletons* or *singleton sets*. The symbol $\varnothing$ always denotes the *empty set*.

For A and B sets, we write $A \cup B$, $A \cap B$, and $A \setminus B$, respectively, for the union, intersection, and difference of A and B. The *symmetric difference* of A and B is

$$A \triangle B := (A \setminus B) \cup (B \setminus A).$$

For $\mathfrak{R}$ an arbitrary set whose elements are sets,

$$\bigcup \{A \mid A \in \mathfrak{R}\} := \bigcup_{A \in \mathfrak{R}} A := \text{the union of the sets belonging to } \mathfrak{R}$$

and, provided $\mathfrak{R}$ is nonempty,

$$\bigcap \{A \mid A \in \mathfrak{R}\} := \bigcap_{A \in \mathfrak{R}} A := \text{the intersection of the sets belonging to } \mathfrak{R}.$$

Note that $\bigcup_{A \in \varnothing} A = \varnothing$, but $\bigcap_{A \in \varnothing} A$ is not defined.

To denote that B is a subset of a set A, we write $B \subset A$. If B is a subset of a set A, the set $A \setminus B$ is also called the *complement* of B in A (or the *complement* of B relative to A), or, provided A is clear from context, simply the *complement* of B. A subset of a set A is called a *proper* subset of A iff its complement in A is nonempty. To denote that B is a proper subset of A, we write $B \subsetneqq A$. For every set A, we denote by $\mathfrak{P}(A)$ the set whose elements are the various subsets of A, and we call $\mathfrak{P}(A)$ the *power set* of A.

Recall that

$$A \setminus \left(\bigcup_{B \in \mathfrak{R}} B \right) = \bigcap_{B \in \mathfrak{R}} (A \setminus B)$$

$$A \setminus \left(\bigcap_{B \in \mathfrak{R}} B \right) = \bigcup_{B \in \mathfrak{R}} (A \setminus B)$$

for every set A and every nonempty set of sets $\mathfrak{R}$ (*de Morgan laws*).

We call (x, y) the *ordered pair* consisting of x and y, and we call x the *first entry* or first *term* in the ordered pair (x, y) and y the *second* entry or term. For A and B sets, the *Cartesian product* of A and B is the set

$$A \times B := \{(x, y) \mid x \in A, y \in B\}.$$

A *map* or a *mapping* is a subset φ of $A \times B$, where A and B are sets and for which no two elements of φ have the same first entry. For A and B sets and $\varphi \subset A \times B$, if φ is a map with the additional property that, for each $x \in A$ there exists $y \in B$ such that $(x, y) \in \varphi$, then we call φ a *mapping from* A or *of* A *into* B. For φ a mapping from a set A into a set B, we write $y = \varphi(x)$ to mean that $(x, y) \in \varphi$. We employ various notations for mappings, most commonly the expression

$$\varphi\colon A \to B, \qquad x \mapsto \varphi(x). \tag{1}$$

This expression signifies that

$$\{(x, \varphi(x)) \mid x \in A\}$$

is a mapping from the set A into the set B, and we are naming this mapping φ. Shorthand versions of (1) include "$\varphi\colon A \to B$" and "the mapping $x \mapsto \varphi(x)$ from A into B."

Let φ be a mapping from a set A into a set B. For $x \in A$ and $y \in B$, if $y = \varphi(x)$, we call y the *image* of x under φ, and we say that φ *maps* x to y or φ *assigns* to x the *value* y. The set A is called the *domain* of φ, and the set $\{\varphi(x) \mid x \in A\}$, that is, the set $\{y \in B \mid y = \varphi(x)$ for some $x \in X\}$, is called the *range* of φ or the *image set* or the *set of values* of φ. The set B is sometimes called a *target space* for φ.

For arbitrary sets A and B we denote by A^B the set consisting of all mappings from B into A. For A an arbitrary set, we call the mapping

$$A \to A, \qquad x \mapsto x$$

the *identity map* on A, and we denote this map by id_A.

Let A, B, C be sets, and let φ: $A \to B$, ψ: $B \to C$ be mappings. Then the mapping from A into C defined by

$$A \to C, \qquad x \mapsto \psi(\varphi(x))$$

is called the *composition* of φ with ψ, and we denote this map by $\psi \circ \varphi$. The compositions of finitely many mappings, with appropriately matched domains, are defined by an obvious induction.

A mapping φ from a set A into a set B is said to be:

injective iff for all x, $y \in A$, $\varphi(x) = \varphi(y)$ implies $x = y$;
surjective (or a mapping from A *onto* B) iff the range of φ is the entire set B;
bijective iff it is both injective and surjective;
constant iff $\varphi(x) = \varphi(y)$ for all x, $y \in A$.

If φ is a bijective mapping from the set A onto the set B, then there is a uniquely determined bijective mapping from B onto A, which we denote by φ^{-1}, such that $\varphi^{-1} \circ \varphi = \mathrm{id}_A$ and $\varphi \circ \varphi^{-1} = \mathrm{id}_B$. The mapping φ^{-1} is called the *inverse* of φ or the mapping *inverse to* φ.

Let φ be a mapping from a set A into a set B, and let $C \subset A$, $D \subset B$. Then

$$\varphi(C) := \{\varphi(x) \mid x \in C\}$$

$$\varphi^{-1}(D) := \{x \in A \mid \varphi(x) \in D\}$$

and we call $\varphi(C)$ and $\varphi^{-1}(D)$, respectively, the *image* of C under φ and the *preimage* of D under φ. Note that if φ is bijective, the two possible interpretations of $\varphi^{-1}(D)$, for $D \subset B$, coincide (i.e., the image of D under φ^{-1} and the preimage of D under φ are identical).

Let φ be a mapping from a set A into a set B. Then for each subset C of A, the mapping

$$C \to B, \qquad x \mapsto \varphi(x)$$

is called the *restriction* to C of φ, and we denote this map by $\varphi|_C$.

Let A, B, and C be sets, and let φ: $A \to C$, ψ: $B \to C$ be mappings. We say that ψ is an *extension* or a *continuation* of φ iff A is a subset of B and φ is the restriction to A of ψ. In this situation we write $\varphi \preccurlyeq \psi$ or $(A, \varphi) \preccurlyeq (B, \psi)$. This notation is also used in a more general situation as follows. We say that

$$(A_1, A_2, \ldots, A_n, B, \psi)$$

is an *extension* or a *continuation* of $(A_1, A_2, \ldots, A_n, A, \varphi)$

and we write

$$(A_1, A_2, \ldots, A_n, A, \varphi) \preccurlyeq (A_1, A_2, \ldots, A_n, B, \psi)$$

iff $\varphi \preccurlyeq \psi$. For $\varphi \preccurlyeq \psi$ we also write $\psi \succcurlyeq \varphi$, and " $\succcurlyeq$ " is used similarly in the other possible usages of " $\preccurlyeq$." We consider "*extends*" and "is an extension of" synonymous.

For B a set, we say that $(x_\alpha)_{\alpha \in A}$ is a *family* from B or in B iff A is a set and the set $\{(\alpha, x_\alpha) \mid \alpha \in A\}$ is a mapping from A into B. In this case the family $(x_\alpha)_{\alpha \in A}$ is said to be *indexed* by the set A, A is called the *index set*, and its elements are called *indexes*, for the family $(x_\alpha)_{\alpha \in A}$. For $\alpha \in A$, x_α is called a *member* of the family $(x_\alpha)_{\alpha \in A}$.

Note that families and mappings are really two presentations of the same objects. To every mapping φ: $A \to B$ there corresponds the family $(\varphi(x))_{x \in A}$, a family from B, and to every family $(x_\alpha)_{\alpha \in A}$ from B there corresponds the mapping

$$\varphi\colon A \to B, \qquad \alpha \mapsto x_\alpha.$$

For $(x_\alpha)_{\alpha \in A}$ a family from a set B, we write

$$x(A) := \{x_\alpha \mid \alpha \in A\}$$

and we call $x(A)$ the *image* of the family $(x_\alpha)_{\alpha \in A}$.

Sequences are families for which the index set is $\mathbb{N}$. Families for which the index set is some section $\mathbb{N}_n$ of $\mathbb{N}$ are occassionally referred to as *finite sequences*. The members of a sequence are usually referred to as *terms* of the sequence, or occasionally as *entries* in the sequence.

When we call $(X_\alpha)_{\alpha \in A}$ a *family of* sets, for instance, we are indicating that X_α, for each $\alpha \in A$, is a set. Let $(A_\iota)_{\iota \in I}$ be a family of sets with I nonempty. Then we define

$$\bigcup_{\iota \in I} A_\iota := \text{the union of the sets belonging to } \{A_\iota \mid \iota \in I\}$$

$$\bigcap_{\iota \in I} A_\iota := \text{the intersection of the sets belonging to } \{A_\iota \mid \iota \in I\}$$

$$\prod_{\iota \in I} A_\iota := \left\{ (x_\iota)_{\iota \in I} \in \left(\bigcup_{\iota \in I} A_\iota \right)^I \mid x_\iota \in A_\iota \text{ for each } \iota \in I \right\}.$$

We set $\bigcup_{\iota \in \varnothing} A_\iota := \varnothing$ and $\prod_{\iota \in \varnothing} A_\iota := \{\varnothing\}$, but we leave $\bigcap_{\iota \in \varnothing} A_\iota$ undefined. The sets $\bigcup_{\iota \in I} A_\iota$, $\bigcap_{\iota \in I} A_\iota$, and $\prod_{\iota \in I} A_\iota$ are called the *union*, *intersection*, and *Cartesian product*, respectively, of the family $(A_\iota)_{\iota \in I}$. For a family $(A_k)_{k \in \mathbb{N}_n}$

of sets indexed by some section $\mathbb{N}_n$ of $\mathbb{N}$, we also write

$$\bigcup_{k\leq n} A_k := \bigcup_{k=1}^{n} A_k := A_1 \cup A_2 \cup \cdots \cup A_n := \bigcup_{k\in\mathbb{N}_n} A_k$$

$$\bigcap_{k\leq n} A_k := \bigcap_{k=1}^{n} A_k := A_1 \cap A_2 \cap \cdots \cap A_n := \bigcap_{k\in\mathbb{N}_n} A_k$$

$$\prod_{k\leq n} A_k := \prod_{k=1}^{n} A_k := A_1 \times A_2 \times \cdots \times A_n := \prod_{k\in\mathbb{N}_n} A_k$$

and we refer to the members of $\prod_{k\leq n} A_k$ as *n-tuples*. If $(A_\iota)_{\iota\in I}$ is a constant family of sets, say $A_\iota = A$ for every $\iota \in I$, there is a natural way to identify the sets A^I and $\prod_{\iota\in I} A_\iota$, and we consider these sets to be identical. If $(A_k)_{k\in\mathbb{N}_n}$ is a constant family of sets indexed by some section of $\mathbb{N}$, say $A_k = A$ for every $k \in \mathbb{N}_n$, we write A^n in place of $A^{\mathbb{N}_n}$. Elements of A^n are usually written in the form $(x_1, x_2, \ldots, x_n)$ and called (ordered) *n-tuples from A*. We identify $\prod_{k\leq 2} A_k$ with $A_1 \times A_2$, so ordered pairs are the same as 2-tuples. Instead of "3-tuple" we usually write "*triple*."

Two sets A and B are said to be *disjoint* iff their intersection is the empty set. A family $(A_\iota)_{\iota\in I}$ of sets is *disjoint* iff, for all ι, $\kappa \in I$ with $\iota \neq \kappa$, the sets A_ι and A_κ are disjoint. A set $\mathfrak{R}$ whose elements are sets is said to be *disjoint* iff, for all A, $B \in \mathfrak{R}$, either $A = B$ or A and B are disjoint.

Two sets A and B are said to be *equipotent* or to have the *same cardinality* iff there exists a bijective mapping of A onto B. A set A is said to be *finite* iff A is empty or A is equipotent with some section $\mathbb{N}_n$ of $\mathbb{N}$. If A is equipotent with $\mathbb{N}_n$, for some natural number n, we say that A has (exactly) n elements. A set A is said to be *countably infinite* iff it is equipotent with $\mathbb{N}$, and it is *countable* iff it is either finite or countably infinite. A set A is said to be *uncountable* or *uncountably infinite* iff it is not countable. A family $(x_\alpha)_{\alpha\in A}$ is *empty*, *finite*, *countably infinite*, *countable*, or *uncountable* according as its index set A is empty, finite, and so forth.

A *relation* on a family of sets is a subset of the Cartesian product of the family. For n a natural number, an *n-ary relation* on a set A is a subset of the Cartesian product A^n. We use the terms *unary relation* and *binary relation* as synonyms for 1-ary and 2-ary relations, respectively. For R a binary relation on a set A, and for x, $y \in A$, we write xRy to mean that the ordered pair (x, y) belongs to the relation R.

For R an n-ary relation on a set A, and B a subset of A, we define

$$R\restriction_B := R \cap B^n$$

and we call $R\restriction_B$ the relation *induced* on B by R or the relation on B *inherited* from A. In particular, if R is binary,

$$R\restriction_B := \{(x, y) \mid (x, y) \in B^2, xRy\}.$$

A binary relation R on a set A is said to be:

reflexive iff xRx for every $x \in A$;
symmetric iff, for all $x, y \in A$, xRy implies yRx;
transitive iff, for all $x, y, z \in A$, xRy and yRz imply xRz;
antisymmetric iff, for all $x, y \in A$, xRy and yRx imply $x = y$.

An *equivalence relation* on a set A is a binary relation on A that is reflexive, symmetric, and transitive. Let R be an equivalence relation on a set A. For $x \in A$ the set $\{y \in A \mid xRy\}$ is called the *equivalence class* (in A) of x *modulo* R. The set $\{\{y \in A \mid xRy\} \mid x \in A\}$ is disjoint, and its union is A. This situation can be described by saying that each equivalence relation on a set A *partitions* A into disjoint equivalence classes. These equivalence classes are also called *equivalence classes* of A *modulo* R, and A/R denotes the set whose elements are these equivalence classes. For R an equivalence relation on a set A and B a subset of A, the induced relation $R\restriction_B$ is an equivalence relation on B.

A *preorder* on a set X is a reflexive, transitive binary relation on X. An antisymmetric preorder on a set X is called an *order* or an *order relation* on X. An order relation R on a set X is called a *total ordering* of X iff, for all $x, y \in X$, either xRy or yRx. If $\leq$ is a preorder on a set X, then we say that X is a *preordered set* with preorder $\leq$, or $(X, \leq)$ is a *preordered set*. A preordered set X for which the preorder is an order relation is called an *ordered set*. An ordered set for which the order relation is a total ordering is called a *totally ordered set*.

For X an ordered set with order relation $\leq$, and for Y an arbitrary subset of X, the relation $\leq\restriction_Y$ is an order relation on Y. Similar remarks hold for preorders and total orderings. Note, however, that an induced preorder might be an order relation or a total ordering.

For X a preordered set with some preorder, unless explicitly stated otherwise, we use $\leq$ to denote the preorder on X. In particular, we do not distinguish notationally between the order relation on X and the order relations it induces on the subsets of X, unless it is necessary to do so.

For X a set and $\mathfrak{S}$ a subset of $\mathfrak{P}(X)$, $\mathfrak{S}$ can be viewed as an ordered set with set inclusion as the order relation, which we often tacitly do. We must explicitly state that in this situation we continue to denote the order relation on $\mathfrak{S}$ by $\subset$, not by $\leq$.

For X an ordered set and for $x, y \in X$, we write $x \geq y$ to mean that $y \leq x$, we write $x < y$ to mean that $x \leq y$ but $x \neq y$, and we write $x > y$ to mean that $y < x$. The statements "x is *less* than y" and "y is *greater* than x" are both taken to mean that $x \leq y$, whereas "x is *strictly less* than y" and "y is *strictly greater* than x" are taken to mean that $x < y$. For X an ordered set and for $x, y \in X$, we define

$$[x, y] := \{z \in X \mid x \leq z \leq y\},$$

$$[x, y[:= \{z \in X \mid x \leq z < y\},$$

$$]x, y[:= \{z \in X \mid x < z < y\},$$

$$]x, y] := \{z \in X \mid x < z \leq y\}.$$

Such sets are called *intervals* from X or in X, and provided $x < y$, x is called the *left endpoint* and y the *right endpoint* of the interval. The interval $[x, y]$ is said to be *closed*, $]x, y[$ is called *open*, and we call $]x, y]$ *left half-open* and $[x, y[$ *right half-open*. For each of the intervals $[x, y]$, $]x, y]$, $]x, y[$, and $[x, y[$, the *interior* of the interval is the open interval $]x, y[$.

Expressions such as $x \leq y$ and $y \geq x$ are called *inequalities*. Replacing $\leq$, $\geq$ by $<$, $>$, we have what are called *strict inequalities*. It should perhaps be stressed that when set inclusion is the order relation, we write $\subset$ and $\subsetneqq$ rather than $\leq$ and $<$.

Let A and B be ordered sets with order relations $\leq$ and $\preccurlyeq$, respectively, and let φ be a mapping from A into B. We say that φ *increases* (or φ *increases relative* to the order relations $\leq$ and $\preccurlyeq$) iff for all $x, y \in A$, if $x \leq y$ then $\varphi(x) \preccurlyeq \varphi(y)$. Similarly, we say that φ *decreases* iff for all $x, y \in A$, if $x \leq y$ then $\varphi(y) \preccurlyeq \varphi(x)$. We call φ an *order isomorphism* iff φ is bijective and both φ and φ^{-1} increase. If X and A are ordered sets and $(x_\alpha)_{\alpha \in A}$ is a family from X indexed by the set A, then the family $(x_\alpha)_{\alpha \in A}$ is called *increasing* (*decreasing*) iff the mapping

$$A \to X, \qquad \alpha \mapsto x_\alpha$$

is an increasing (decreasing) mapping. In particular, a sequence $(x_n)_{n \in \mathbb{N}}$ from an ordered set X is called *increasing* (*decreasing*) iff $x_k \leq x_{k+1}$ $(x_k \geq x_{k+1})$ for each $k \in \mathbb{N}$. A *monotone* sequence is a sequence that is either increasing or decreasing. We use the terms *strictly increasing*, *strictly decreasing*, or *strictly monotone* when strict inequality holds in the relevant defining inequality. For families of sets and sequences of sets the order relation (on the set of sets to which the members of the family belong) is set inclusion unless explicitly stated otherwise. For example, a sequence $(A_n)_{n \in \mathbb{N}}$ of sets *increases* iff $A_n \subset A_{n+1}$ for every $n \in \mathbb{N}$.

Let x be an element and A a subset of an ordered set X. Then we say that:

x is an *upper bound* (*lower bound*) in X of A or for A iff $z \leq x$ $(x \leq z)$ for every $z \in A$;

x is a *maximal* element (*minimal* element) of A iff $x \in A$ and the conditions $z \in A$ and $x \leq z$ $(z \leq x)$ imply $z = x$;

x is the *largest* or *greatest* element (*smallest* or *least* element) of A iff $x \in A$ and $z \leq x$ ($x \leq z$) for every $z \in A$;

A is *bounded above* (*bounded below*) in X *by* x iff x is an upper bound in X for A.

Now let A be a subset of an ordered set X. Then A is said to be:

bounded above (*bounded below*) in X iff there exists at least one upper bound (lower bound) in X for A;

bounded in X iff A is both bounded above in X and bounded below in X;

directed upward iff for all $x, y \in A$, there exists $z \in A$ such that $x \leq z$ and $y \leq z$;

directed downward iff for all $x, y \in A$, there exists $z \in A$ such that $z \leq x$ and $z \leq y$;

doubly directed iff it is directed both upward and downward.

If the set of upper bounds in X of A has a least element, we call this element the *supremum* in X of A. Similarly the greatest element of the set of lower bounds in X for A, if it exists, is called the *infimum* in X of A. The infimum of A in X, if it exists, is denoted by $\bigwedge^X A$ or $\bigwedge^X_{x \in A} x$, the supremum by $\bigvee^X A$ or $\bigvee^X_{x \in A} x$. For the infimum and supremum of a set $\{x, y\}$, if they exist, we write simply $x \overset{X}{\wedge} y$ and $x \overset{X}{\vee} y$.

A family $(x_\iota)_{\iota \in I}$ from an ordered set X is said to be *bounded above*, *bounded below*, *bounded*, *directed upward*, *directed downward*, *doubly directed*, and so forth, according as its image $x(I)$ has the corresponding property. The supremum, if it exists, of the image $x(I)$ is denoted by $\bigvee^X_{\iota \in I} x_\iota$, the infimum by $\bigwedge^X_{\iota \in I} x_\iota$. Such a supremum or infimum is called the supremum or infimum in X of the relevant family, rather than the supremum or infimum of the image of the family.

For $n \in \mathbb{N}$, we often write $m \leq n$ or $\overset{n}{m} = 1$ instead of $m \in \mathbb{N}_n$. Similarly, we write $m \geq n$ or $\overset{\infty}{m} = n$ to mean $m \in \{k \in \mathbb{N} \mid k \geq n\}$. This practice occurs, for instance, in connection with such symbols as $\bigcap$ and $\bigvee$ and should cause no confusion.

Many of the definitions given in the preceding paragraphs make sense when X is only a preordered set. We do not repeat these definitions here, but we shall use them as needed. Probably the most frequent usage will refer to a family from a preordered set being directed upward or downward.

In conclusion, we state without apology that we occasionally make such statements as "There is a natural map . . ." or "There is a natural isomorphism. . . ." We have tried to do so, however, only when we were confident that the reader could readily interpret the statement and that further explanation would hinder, not aid, the reader's understanding.

1

RIESZ LATTICES AND DANIELL SPACES

1.1. REAL AND EXTENDED-REAL NUMBERS AND ORDER-CONVERGENT SEQUENCES

The first part of this section summarizes the important properties of $\mathbb{R}$ and $\overline{\mathbb{R}}$. Although familiarity with the real numbers is a prerequisite for this book, there are good reasons for beginning with some remarks about this number system. The entire book is based on the axioms of the real numbers and the rules they imply, which alone compels us to display the axiom system and to point out any crucial facts.

Another reason may be even more important. It is the order structure of $\mathbb{R}$ and of $\overline{\mathbb{R}}$ that is most important for integration theory. Yet this aspect, especially the connection between the order structure and the algebraic operations, is often neglected in elementary courses, so many readers may welcome a review.

The second part of this section treats the question of convergence in $\overline{\mathbb{R}}$ and order convergence in general. By studying the relation between the familiar notion of convergence in $\mathbb{R}$ and the order structure of $\mathbb{R}$, we see how to define convergence in an arbitrary ordered set. The resulting definition provides us with both a notion of convergence in $\overline{\mathbb{R}}$ and the convergence that is most important for our treatment of real integration theory.

Definition 1.1.1. *A **real number system** is a set $\mathbb{R}$ for which there are defined a binary relation $\leq$ and mappings $(\alpha, \beta) \mapsto \alpha + \beta$ (called addition) and $(\alpha, \beta) \mapsto \alpha\beta$ (called multiplication) of $\mathbb{R} \times \mathbb{R}$ into $\mathbb{R}$ satisfying the following axioms.*

I. Algebraic axioms

(*F*1) *For all $\alpha, \beta, \gamma \in \mathbb{R}$, $(\alpha + \beta) + \gamma = \alpha + (\beta + \gamma)$.*

(*F*2) *There exists $0 \in \mathbb{R}$ such that $0 + \alpha = \alpha$ for every $\alpha \in \mathbb{R}$.*

(*F*3) *For each $\alpha \in \mathbb{R}$ there exists $-\alpha \in \mathbb{R}$ such that $\alpha + (-\alpha) = 0$.*

(*F4*) *For all* $\alpha, \beta \in \mathbb{R}$, $\alpha + \beta = \beta + \alpha$.
(*F5*) *For all* $\alpha, \beta, \gamma \in \mathbb{R}$, $(\alpha\beta)\gamma = \alpha(\beta\gamma)$.
(*F6*) *There exists* $1 \in \mathbb{R}$, $1 \neq 0$, *such that* $1\alpha = \alpha$ *for every* $\alpha \in \mathbb{R}$.
(*F7*) *For every* $\alpha \in \mathbb{R}$, $\alpha \neq 0$, *there exists* $\alpha^{-1} \in \mathbb{R}$ *such that* $\alpha\alpha^{-1} = 1$.
(*F8*) *For all* $\alpha, \beta \in \mathbb{R}$, $\alpha\beta = \beta\alpha$.
(*F9*) *For all* $\alpha, \beta, \gamma \in \mathbb{R}$, $\alpha(\beta + \gamma) = \alpha\beta + \alpha\gamma$.

II. Order axioms

(*O1*) *For every* $\alpha \in \mathbb{R}$, $\alpha \leq \alpha$.
(*O2*) *For all* $\alpha, \beta \in \mathbb{R}$, $\alpha \leq \beta$ *and* $\beta \leq \alpha$ *imply* $\alpha = \beta$.
(*O3*) *For all* $\alpha, \beta, \gamma \in \mathbb{R}$, $\alpha \leq \beta$ *and* $\beta \leq \gamma$ *imply* $\alpha \leq \gamma$.
(*O4*) *For all* $\alpha, \beta \in \mathbb{R}$, *either* $\alpha \leq \beta$ *or* $\beta \leq \alpha$.
(*O5*) *For all* $\alpha, \beta, \gamma \in \mathbb{R}$, $\alpha \leq \beta$ *implies* $\alpha + \gamma \leq \beta + \gamma$.
(*O6*) *For all* $\alpha, \beta \in \mathbb{R}$, $0 \leq \alpha$ *and* $0 \leq \beta$ *imply* $0 \leq \alpha\beta$.
(*O7*) *If A and B are nonempty subsets of* $\mathbb{R}$, *and if* $\alpha \leq \beta$ *for all* $\alpha \in A$ *and all* $\beta \in B$, *then there exists* $\gamma \in \mathbb{R}$ *such that* $\alpha \leq \gamma \leq \beta$ *for all* $\alpha \in A$ *and all* $\beta \in B$. □

As is well known, there are various ways of constructing a real number system, but they all give the same result: the axioms in Definition 1 determine $\mathbb{R}$ uniquely up to isomorphism. Thus we may freely use the expression "***the real number system.***" Throughout this book the letter $\mathbb{R}$ will denote the real number system.

The numbers 0 and 1 are uniquely determined by the axioms. Moreover $-\alpha$ is uniquely determined for each α in $\mathbb{R}$, as is α^{-1} for each α in $\mathbb{R} \setminus \{0\}$. We adopt the usual notation that $\alpha - \beta := \alpha + (-\beta)$ and, provided $\beta \neq 0$, $\frac{\alpha}{\beta} := \alpha/\beta := \alpha\beta^{-1}$. The numbers $\alpha + \beta$, $\alpha\beta$, $\alpha - \beta$, and α/β are called, respectively, the ***sum, product, difference,*** and ***quotient*** of α and β.

The number systems $\mathbb{N}$ (natural numbers), $\mathbb{Z}$ (integers), and $\mathbb{Q}$ (rational numbers) are embedded in $\mathbb{R}$. We set

$$\mathbb{R}_+ := \{\alpha \mid \alpha \in \mathbb{R}, \alpha \geq 0\} \qquad \mathbb{Q}_+ := \{\alpha \mid \alpha \in \mathbb{Q}, \alpha \geq 0\}.$$

A real number α is said to be ***positive*** (***strictly positive***) iff $\alpha \geq 0$ ($\alpha > 0$). Similarly we say that α is ***negative*** (***strictly negative***) iff $\alpha \leq 0$ ($\alpha < 0$). Note that for each natural number n, the *section* $\mathbb{N}_n$ of $\mathbb{N}$ spanned by n is

$$\mathbb{N}_n = \{m \mid m \in \mathbb{N}, m \leq n\}.$$

The axioms for $\mathbb{R}$ are not all equally significant for real integration theory. In particular, the order axioms turn out to be essentially more important than the algebraic ones. Of the rules that result from (F1)–(F9), we actually need only those that concern finite sums.

For a real number family $(\alpha_k)_{k \in \mathbb{N}_n}$ indexed by a section $\mathbb{N}_n$ of $\mathbb{N}$, it is clear how to define inductively the sum $\sum_{k=1}^{n} \alpha_k$ (or: $\alpha_1 + \alpha_2 + \cdots + \alpha_n$). The properties of such sums are familiar. For instance, the sum does not depend on grouping or order, and constants can be factored out. For integration theory one also needs sums of the form $\sum_{\iota \in I} \alpha_\iota$ where each α_ι is still a real number but where the index set I, except for being finite, is completely arbitrary. The reader will understand what is meant by such an expression and will have used the properties of finite sums many times. For the sake of completeness, however, we state a theorem describing these arbitrary sums of finitely many numbers. All the properties that we need are included in this theorem. Lest the reader fear that we will also, for completeness, prove the theorem, we state right away that we have no such intention. (See, however, Exercise 1.) Recall that for an arbitrary set I, $\mathbb{R}^I$ denotes the set of all real number families indexed by I (equivalently, the set of all mappings from I into $\mathbb{R}$).

Theorem 1.1.2. *Let I be a finite set. Then there is a uniquely determined, natural map*

$$\sum : \mathbb{R}^I \to \mathbb{R}, \qquad (\alpha_\iota)_{\iota \in I} \mapsto \sum_{\iota \in I} \alpha_\iota$$

satisfying the following conditions.

(*a*) *For all families $(\alpha_\iota)_{\iota \in I}, (\beta_\iota)_{\iota \in I} \in \mathbb{R}^I$,*

$$\sum_{\iota \in I} (\alpha_\iota + \beta_\iota) = \sum_{\iota \in I} \alpha_\iota + \sum_{\iota \in I} \beta_\iota.$$

(*b*) *For every family $(\alpha_\iota)_{\iota \in I} \in \mathbb{R}^I$ and every real number α,*

$$\sum_{\iota \in I} \alpha \alpha_\iota = \alpha \sum_{\iota \in I} \alpha_\iota.$$

(*c*) *If, for every ι in I and every κ in I, $\delta_{\iota\kappa}$ is defined by*

$$\delta_{\iota\kappa} := \begin{cases} 1 & \text{if } \iota = \kappa \\ 0 & \text{if } \iota \neq \kappa \end{cases}$$

then

$$\sum_{\iota \in I} \delta_{\iota\kappa} = 1$$

for every κ in I.

In addition, the map $\sum$: $\mathbb{R}^I \to \mathbb{R}$ has the following properties.

(*d*) *If I is empty, then $\sum_{\iota \in I} \alpha_\iota = 0$.*

(*e*) *If I contains exactly n elements ($n \in \mathbb{N}$), and if $\alpha_\iota = \beta$ for some $\beta \in \mathbb{R}$ and for every $\iota \in I$, then*

$$\sum_{\iota \in I} \alpha_\iota = n\beta.$$

Finally, let I, K, and P be finite sets. The map Σ, defined on each of the sets $\mathbb{R}^I$, $\mathbb{R}^K$, $\mathbb{R}^P$, and $\mathbb{R}^{I\times K}$, satisfies the following conditions.

(*f*) *If K and P are disjoint and $I = K \cup P$, then*

$$\sum_{\iota\in I} \alpha_\iota = \sum_{\kappa\in K} \alpha_\kappa + \sum_{\rho\in P} \alpha_\rho$$

for every family $(\alpha_\iota)_{\iota\in I} \in \mathbb{R}^I$.

(*g*) *If $\varphi\colon K \to I$ is a bijective mapping, then*

$$\sum_{\iota\in I} \alpha_\iota = \sum_{\kappa\in K} \alpha_{\varphi(\kappa)}$$

for every family $(\alpha_\iota)_{\iota\in I} \in \mathbb{R}^I$.

In particular, the following rule holds for calculating sums.

(*h*) *If I is nonempty and contains exactly n elements $(n \in \mathbb{N})$, and if $\varphi\colon \mathbb{N}_n \to I$ is an arbitrary bijective mapping, then*

$$\sum_{\iota\in I} \alpha_\iota = \sum_{k=1}^{n} \alpha_{\varphi(k)}.$$

(*i*) *For every family $(\alpha_{\iota\kappa})_{(\iota,\kappa)\in I\times K} \in \mathbb{R}^{I\times K}$,*

$$\sum_{(\iota,\kappa)\in I\times K} \alpha_{\iota\kappa} = \sum_{\iota\in I}\left(\sum_{\kappa\in K} \alpha_{\iota\kappa}\right) = \sum_{\kappa\in K}\left(\sum_{\iota\in I} \alpha_{\iota\kappa}\right).$$

(*j*) *For every family $(\alpha_\iota)_{\iota\in I} \in \mathbb{R}^I$ and every family $(\beta_\kappa)_{\kappa\in K} \in \mathbb{R}^K$,*

$$\left(\sum_{\iota\in I} \alpha_\iota\right)\left(\sum_{\kappa\in K} \beta_\kappa\right) = \sum_{(\iota,\kappa)\in I\times K} \alpha_\iota\beta_\kappa. \qquad \square$$

We call $\Sigma_{\iota\in I}\alpha_\iota$ the **sum** of the family $(\alpha_\iota)_{\iota\in I}$.

The crucial order properties of $\mathbb{R}$ are formulated in the following theorem.

Theorem 1.1.3. *$\mathbb{R}$ is a totally ordered set. Every nonempty set in $\mathbb{R}$ that is bounded above has a supremum. Similarly every nonempty set in $\mathbb{R}$ that is bounded below has an infimum. Every nonempty finite subset of $\mathbb{R}$ has a largest element and a smallest element.*

Proof. The first assertion merely restates Axioms (O1)–(O4) of Definition 1. Let A be a subset of $\mathbb{R}$ that is bounded above. Then the set B of all upper bounds of A is nonempty, and $\alpha \leq \beta$ for every α in A and every β in B. If A is also nonempty, then by Axiom (O7) in Definition 1 there exists a real number γ such that $\alpha \leq \gamma \leq \beta$ for every α in A and every β in B. This number γ possesses exactly the defining properties for the supremum of A. The statement about the existence of an infimum is proved analogously. The final assertion is easily proved by induction. □

Although very strong, the order properties of $\mathbb{R}$ are not quite so strong as is desirable for constructing integrals. $\mathbb{R}$ can easily be extended, however, to a system with stronger order properties, and such an extension is also natural from the point of view of applications. To be sure, a gain in order properties can only be achieved by a loss in algebraic properties. After all, the axioms in Definition 1 determine $\mathbb{R}$ uniquely. We accept the loss because the order structure provides the most important tools needed to construct the integral. The announced extension, achieved by adding two elements ∞ and $-\infty$ to $\mathbb{R}$, pays for itself in numerous simplifications.

Definition 1.1.4. *Let $\overline{\mathbb{R}} := \mathbb{R} \cup \{\infty, -\infty\}$. The algebraic operations and the order relation on $\mathbb{R}$ are extended to $\overline{\mathbb{R}}$ as follows. For every $\alpha \in \mathbb{R}$,*

$$-\infty < \alpha < \infty$$

$$\alpha\infty := \infty\alpha := \begin{cases} \infty & \text{if } \alpha > 0 \\ 0 & \text{if } \alpha = 0 \\ -\infty & \text{if } \alpha < 0 \end{cases}$$

$$\alpha(-\infty) := (-\infty)\alpha := \begin{cases} -\infty & \text{if } \alpha > 0 \\ 0 & \text{if } \alpha = 0 \\ \infty & \text{if } \alpha < 0 \end{cases}$$

$$\alpha + \infty := \infty + \alpha := \infty \qquad \alpha + (-\infty) := (-\infty) + \alpha := -\infty.$$

Furthermore

$$-\infty < \infty$$

$$\infty + \infty := \infty \qquad (-\infty) + (-\infty) := -\infty$$

$$\infty\infty := (-\infty)(-\infty) := \infty \qquad \infty(-\infty) := (-\infty)\infty := -\infty.$$

$\overline{\mathbb{R}}$, with the accompanying operations and order relation, is called the ***extended real number system****. As before, $\alpha - \beta := \alpha + (-\beta)$, provided now that the latter expression is defined. We set*

$$\overline{\mathbb{R}}_+ := \{\alpha \mid \alpha \in \overline{\mathbb{R}}, \alpha \geq 0\}.$$

For $\alpha \in \overline{\mathbb{R}}$ we define the ***absolute value*** *of α by*

$$|\alpha| := \begin{cases} \alpha & \text{if } \alpha \geq 0 \\ -\alpha & \text{if } \alpha < 0. \end{cases}$$ □

The laws of arithmetic for $\overline{\mathbb{R}}$ are defined in such a way that as many rules as possible still hold. The convention $0\infty = \infty 0 = 0$, which probably seems most questionable, proves very useful for integration theory and leads to no difficulties. The sums $\infty + (-\infty)$ and $(-\infty) + \infty$, on the other hand, cannot be defined in any meaningful way, so we have excluded them. In order to avoid a continual exclusion of these cases when statements are made about $\overline{\mathbb{R}}$, we adopt the following convention.

Convention 1.1.5 (***The*** $\infty - \infty$ ***convention for*** $\overline{\mathbb{R}}$). *If P is an assertion about extended-real numbers, then "P holds" is understood to mean "P is true provided that every sum appearing in P is defined."* □

This convention brings considerable simplification, but it is crucial that the reader bear it in mind. We illustrate with an example. Proposition 9(g), which appears later in this section, makes the following assertion:

"For every nonempty family $(\alpha_\iota)_{\iota \in I}$ from $\overline{\mathbb{R}}$ and for every extended-real number α, $\alpha + \sup_{\iota \in I}\alpha_\iota = \sup_{\iota \in I}(\alpha + \alpha_\iota)$."

According to the $\infty - \infty$ convention, this assertion is understood to state the following:

"For every nonempty family $(\alpha_\iota)_{\iota \in I}$ from $\overline{\mathbb{R}}$ and for every extended-real number α, if none of the situations

(a) $\alpha = \infty$ and $\alpha_\iota = -\infty$ for some $\iota \in I$
(b) $\alpha = -\infty$ and $\alpha_\iota = \infty$ for some $\iota \in I$
(c) $\alpha = \infty$ and $\sup_{\iota \in I}\alpha_\iota = -\infty$
(d) $\alpha = -\infty$ and $\sup_{\iota \in I}\alpha_\iota = \infty$

occurs, then

$$\alpha + \sup_{\iota \in I}\alpha_\iota = \sup_{\iota \in I}(\alpha + \alpha_\iota).$$

If any of the situations (a)–(d) occurs, then no assertion is made."

We examine the essential properties of $\overline{\mathbb{R}}$. The next theorem describes the order properties.

Theorem 1.1.6. *$\overline{\mathbb{R}}$ is a totally ordered set. The largest and smallest elements of $\overline{\mathbb{R}}$ are ∞ and $-\infty$, respectively. Every subset of $\overline{\mathbb{R}}$ has a supremum and an infimum. The supremum of the empty set is $-\infty$. The infimum of the empty set is ∞.*

The induced order relation $\leq \upharpoonright_{\mathbb{R}}$ is the same as the order relation originally given on $\mathbb{R}$. A subset of $\mathbb{R}$ that possesses a supremum (infimum) in $\mathbb{R}$ has the same supremum (infimum) when viewed as a subset of $\overline{\mathbb{R}}$. Every set in $\mathbb{R}$ that is *not bounded above has* ∞ *as supremum. Every set in* $\mathbb{R}$ *that is not bounded below has* $-\infty$ *as infimum.*

Proof. These statements are all trivial consequences of Theorem 3 and the definitions. □

Definition 1.1.7. *For* $A \subset \overline{\mathbb{R}}$ *the supremum in* $\overline{\mathbb{R}}$ *of* A *will be denoted by* $\sup A$. *Similarly* $\inf A$ *denotes the infimum in* $\overline{\mathbb{R}}$ *of* A. *For a family* $(\alpha_\iota)_{\iota \in I}$ *from* $\overline{\mathbb{R}}$ *we write*

$$\sup_{\iota \in I} \alpha_\iota := \sup\{\alpha_\iota \mid \iota \in I\} \quad \textit{and} \quad \inf_{\iota \in I} \alpha_\iota := \inf\{\alpha_\iota \mid \iota \in I\}.$$

For $\sup\{\alpha, \beta\}$ *and* $\inf\{\alpha, \beta\}$ *we also write* $\alpha \vee \beta$ *and* $\alpha \wedge \beta$, *respectively.* □

The elementary rules of computation for $\overline{\mathbb{R}}$ are given in the next proposition. These rules can easily be verified by using the axioms for $\mathbb{R}$ and the definition of $\overline{\mathbb{R}}$, its order relation, and its algebraic operations. We leave the verifications to the reader.

Proposition 1.1.8. *For all* $\alpha, \beta, \gamma \in \overline{\mathbb{R}}$ *the following statements hold (of course, with Convention 5 in force).*

(*a*) $(\alpha + \beta) + \gamma = \alpha + (\beta + \gamma)$ (*b*) $\alpha + \beta = \beta + \alpha$
(*c*) $\alpha(\beta\gamma) = (\alpha\beta)\gamma$ (*d*) $\alpha\beta = \beta\alpha$
(*e*) $\alpha(\beta + \gamma) = \alpha\beta + \alpha\gamma$ (*f*) $0 + \alpha = \alpha,\ \alpha - \alpha = 0$
(*g*) $1\alpha = \alpha,\ (-1)\alpha = -\alpha$ (*h*) $\alpha \le \beta \Rightarrow \alpha + \gamma \le \beta + \gamma$
(*i*) $0 \le \alpha,\ 0 \le \beta \Rightarrow 0 \le \alpha\beta$ (*j*) $\alpha + \beta = (\alpha \vee \beta) + (\alpha \wedge \beta)$
(*k*) $|\alpha - \beta| = (\alpha \vee \beta) - (\alpha \wedge \beta)$ (*l*) $\alpha \vee \beta = \frac{1}{2}(\alpha + \beta + |\alpha - \beta|)$
(*m*) $\alpha \wedge \beta = \frac{1}{2}(\alpha + \beta - |\alpha - \beta|)$ (*n*) $|\alpha + \beta| \le |\alpha| + |\beta|$
(*o*) $\big||\alpha| - |\beta|\big| \le |\alpha - \beta|$ (*p*) $|\alpha \vee \beta| \le |\alpha| \vee |\beta| \le |\alpha| + |\beta|$
(*q*) $|\alpha \wedge \beta| \le |\alpha| \vee |\beta| \le |\alpha| + |\beta|$ (*r*) $|\alpha\beta| = |\alpha|\,|\beta|$ □

Next we list the rules governing sup and inf for families from $\overline{\mathbb{R}}$.

Proposition 1.1.9. *Let* $(\alpha_\iota)_{\iota \in I}$ *and* $(\beta_\kappa)_{\kappa \in K}$ *be nonempty families from* $\overline{\mathbb{R}}$, *and let* α *belong to* $\overline{\mathbb{R}}$. *Then the following assertions hold.*

(*a*) $(\sup_{\iota \in I} \alpha_\iota) \vee (\sup_{\kappa \in K} \beta_\kappa) = \sup_{(\iota, \kappa) \in I \times K} (\alpha_\iota \vee \beta_\kappa)$
(*b*) $(\inf_{\iota \in I} \alpha_\iota) \wedge (\inf_{\kappa \in K} \beta_\kappa) = \inf_{(\iota, \kappa) \in I \times K} (\alpha_\iota \wedge \beta_\kappa)$
(*c*) $\alpha \wedge (\sup_{\iota \in I} \alpha_\iota) = \sup_{\iota \in I} (\alpha \wedge \alpha_\iota)$
(*d*) $\alpha \vee (\inf_{\iota \in I} \alpha_\iota) = \inf_{\iota \in I} (\alpha \vee \alpha_\iota)$

(*e*) $\inf_{\iota \in I} \alpha_\iota = -\sup_{\iota \in I}(-\alpha_\iota)$

(*f*) $\sup_{\iota \in I} \alpha_\iota = -\inf_{\iota \in I}(-\alpha_\iota)$

(*g*) $\alpha + \sup_{\iota \in I} \alpha_\iota = \sup_{\iota \in I}(\alpha + \alpha_\iota)$

(*h*) $\alpha + \inf_{\iota \in I} \alpha_\iota = \inf_{\iota \in I}(\alpha + \alpha_\iota)$

(*i*) $\alpha \sup_{\iota \in I} \alpha_\iota = \begin{cases} \sup_{\iota \in I}(\alpha\alpha_\iota) & \text{for } \alpha \geq 0,\ \alpha \neq \infty \\ \inf_{\iota \in I}(\alpha\alpha_\iota) & \text{for } \alpha \leq 0,\ \alpha \neq -\infty \end{cases}$

(*j*) $\alpha \inf_{\iota \in I} \alpha_\iota = \begin{cases} \inf_{\iota \in I}(\alpha\alpha_\iota) & \text{for } \alpha \geq 0,\ \alpha \neq \infty \\ \sup_{\iota \in I}(\alpha\alpha_\iota) & \text{for } \alpha \leq 0,\ \alpha \neq -\infty \end{cases}$

(*k*) *If* $(I_\rho)_{\rho \in P}$ *is a family of sets with* $\bigcup_{\rho \in P} I_\rho = I$, *then*

$$\sup_{\iota \in I} \alpha_\iota = \sup_{\rho \in P}\left(\sup_{\iota \in I_\rho} \alpha_\iota\right)$$

and

$$\inf_{\iota \in I} \alpha_\iota = \inf_{\rho \in P}\left(\inf_{\iota \in I_\rho} \alpha_\iota\right).$$

Also the following assertions hold, for all families $(\alpha_\iota)_{\iota \in I}$ *and* $(\beta_\iota)_{\iota \in I}$ *from* $\overline{\mathbb{R}}$.

(*l*) $\sup_{\iota \in I}(\alpha_\iota \vee \beta_\iota) = (\sup_{\iota \in I} \alpha_\iota) \vee (\sup_{\iota \in I} \beta_\iota)$

(*m*) $\inf_{\iota \in I}(\alpha_\iota \wedge \beta_\iota) = (\inf_{\iota \in I} \alpha_\iota) \wedge (\inf_{\iota \in I} \beta_\iota)$.

Proof. These statements are easy to prove. In each case one establishes equality by establishing inequality in both directions. The inequalities are obtained by using successively the two defining properties, for instance, of supremum: it is an upper bound, and there is no smaller upper bound. We illustrate by proving assertion (g).

The assertion to be proved is obviously true if $\alpha = \infty$ or $\alpha = -\infty$. (Remember that the $\infty - \infty$ convention is in force.) So let α be a real number. For every ι in I we have

$$\alpha + \alpha_\iota \leq \sup_{\iota \in I}(\alpha + \alpha_\iota)$$

so

$$\alpha_\iota \leq \sup_{\iota \in I}(\alpha + \alpha_\iota) - \alpha.$$

Hence

$$\sup_{\iota \in I} \alpha_\iota \leq \sup_{\iota \in I}(\alpha + \alpha_\iota) - \alpha$$

or

$$\alpha + \sup_{\iota \in I} \alpha_\iota \leq \sup_{\iota \in I} (\alpha + \alpha_\iota).$$

On the other hand, for every ι in I

$$\alpha_\iota \leq \sup_{\iota \in I} \alpha_\iota$$

so

$$\alpha + \alpha_\iota \leq \alpha + \sup_{\iota \in I} \alpha_\iota.$$

Therefore

$$\sup_{\iota \in I} (\alpha + \alpha_\iota) \leq \alpha + \sup_{\iota \in I} \alpha_\iota$$

and the equality is established. □

In the special case of monotone sequences from $\overline{\mathbb{R}}$, there are some additional rules governing sup and inf.

Proposition 1.1.10. *If* $(\alpha_n)_{n \in \mathbb{N}}$ *and* $(\beta_n)_{n \in \mathbb{N}}$ *are increasing sequences from* $\overline{\mathbb{R}}$, *then*:

(*a*) $\sup_{n \in \mathbb{N}}(\alpha_n + \beta_n) = (\sup_{n \in \mathbb{N}} \alpha_n) + (\sup_{n \in \mathbb{N}} \beta_n)$
(*b*) $\sup_{n \in \mathbb{N}}(\alpha_n \wedge \beta_n) = (\sup_{n \in \mathbb{N}} \alpha_n) \wedge (\sup_{n \in \mathbb{N}} \beta_n)$.

Similarly, if $(\alpha_n)_{n \in \mathbb{N}}$ *and* $(\beta_n)_{n \in \mathbb{N}}$ *are decreasing sequences from* $\overline{\mathbb{R}}$, *then*:

(*c*) $\inf_{n \in \mathbb{N}}(\alpha_n + \beta_n) = (\inf_{n \in \mathbb{N}} \alpha_n) + (\inf_{n \in \mathbb{N}} \beta_n)$
(*d*) $\inf_{n \in \mathbb{N}}(\alpha_n \vee \beta_n) = (\inf_{n \in \mathbb{N}} \alpha_n) \vee (\inf_{n \in \mathbb{N}} \beta_n)$.

Proof. We prove (a) as an example. From

$$\alpha_m + \beta_m \leq \left(\sup_{n \in \mathbb{N}} \alpha_n \right) + \left(\sup_{n \in \mathbb{N}} \beta_n \right)$$

for every m in $\mathbb{N}$, it follows that

$$\sup_{n \in \mathbb{N}} (\alpha_n + \beta_n) \leq \left(\sup_{n \in \mathbb{N}} \alpha_n \right) + \left(\sup_{n \in \mathbb{N}} \beta_n \right).$$

To get the inequality in the other direction, we use the monotonicity. Fix $k \in \mathbb{N}$. Then for every n in $\mathbb{N}$ we have

$$\alpha_k + \beta_n \leq \alpha_{k+n} + \beta_{k+n} \leq \sup_{m \in \mathbb{N}} (\alpha_m + \beta_m).$$

Taking the supremum over all n (using Proposition 9(g)), we conclude that

$$\alpha_k + \sup_{n\in\mathbb{N}} \beta_n = \sup_{n\in\mathbb{N}} (\alpha_k + \beta_n) \le \sup_{n\in\mathbb{N}} (\alpha_n + \beta_n).$$

This last relation holds for every k in $\mathbb{N}$. Thus taking the supremum over all k yields

$$\sup_{n\in\mathbb{N}} \alpha_n + \sup_{n\in\mathbb{N}} \beta_n \le \sup_{n\in\mathbb{N}} (\alpha_n + \beta_n). \qquad \square$$

We have formulated the rules for sup and inf as we shall generally use them, namely in the language of families from $\overline{\mathbb{R}}$. It should be easy for the reader to reformulate these rules for sets in $\overline{\mathbb{R}}$. After all, to every set $A \subset \overline{\mathbb{R}}$ is naturally associated the family $(\alpha)_{\alpha\in A}$, and we have $\{\alpha \mid \alpha \in A\} = A$.

The promised review is now complete. We have listed all the rules of computation and order properties for $\mathbb{R}$ and $\overline{\mathbb{R}}$ that we need. Ordered sets having the same order properties as $\mathbb{R}$ or $\overline{\mathbb{R}}$ will occur frequently and in many different contexts. It is useful to have a name for them.

Definition 1.1.11. *An ordered set X is called a **lattice** iff for every two elements x, y belonging to X, there exist both $x \vee y$ and $x \wedge y$.*

*A lattice in which every nonempty bounded set has both a supremum and an infimum is said to be **conditionally complete**.*

*A **complete** lattice is a lattice in which every subset has both a supremum and an infimum.*

*A lattice in which every countable, nonempty, bounded set has both a supremum and an infimum is said to be **conditionally σ-complete**.*

*A **σ-complete** lattice is a lattice in which every countable subset has both a supremum and an infimum.* $\square$

Obviously the implications

$$\begin{array}{ccccc} & \nearrow & \sigma\text{-complete} & \searrow & \\ \text{Complete} & & & & \text{Conditionally } \sigma\text{-complete} \\ & \searrow & \text{Conditionally complete} & \nearrow & \end{array}$$

all hold. According to Theorems 3 and 6, $\mathbb{R}$ is a conditionally complete lattice, whereas $\overline{\mathbb{R}}$ is a complete lattice. The difference between complete and conditionally complete points up what one gains by adjoining ∞ and $-\infty$. In $\overline{\mathbb{R}}$, with ∞ and $-\infty$ available, every set is bounded. One need no longer hypothesize that a set be bounded in order to conclude that its supremum and infimum exist.

Having summarized the basic properties of $\mathbb{R}$ and $\overline{\mathbb{R}}$, we turn to the problem of defining convergence in $\overline{\mathbb{R}}$. The reader knows what it means for a sequence of real numbers to converge (i.e., it satisfies condition (a) of Theorem 12, which follows). The ε-m definition of convergence in $\mathbb{R}$ cannot be used for $\overline{\mathbb{R}}$ since it is not always meaningful. Various alternatives are possible. Since order relations and their consequences play a prominent role in this book, it is appropriate to use the following (more general) formulation of the problem: given an ordered set X, what does it mean to say that a sequence from X converges? Since $\overline{\mathbb{R}}$ is also an ordered set, whatever definition we use will make sense in $\mathbb{R}$, and we insist that it give a notion equivalent to that given by the ε-m definition. What is needed is an alternate characterization of sequential convergence in $\mathbb{R}$, a characterization that uses the order relation on $\mathbb{R}$ but does not use the rich algebraic structure of $\mathbb{R}$. The next theorem provides just such a characterization.

Theorem 1.1.12. *For every sequence $(\alpha_n)_{n\in\mathbb{N}}$ of real numbers and for every real number α, the following assertions are equivalent.*

(*a*) *For every real number $\varepsilon > 0$ there exists a natural number m such that*

$$|\alpha_n - \alpha| < \varepsilon$$

for every integer $n \geq m$ (**ε-*m* definition**).

(*b*) *There exist in $\mathbb{R}$ an increasing sequence $(\alpha'_n)_{n\in\mathbb{N}}$ and a decreasing sequence $(\alpha''_n)_{n\in\mathbb{N}}$ such that*

$$\alpha'_n \leq \alpha_n \leq \alpha''_n$$

for every n in $\mathbb{N}$ and

$$\sup_{n\in\mathbb{N}} \alpha'_n = \inf_{n\in\mathbb{N}} \alpha''_n = \alpha. \tag{1}$$

Proof. (a) $\Rightarrow$ (b). Note first that the sequence $(\alpha_n)_{n\in\mathbb{N}}$ is bounded. [Choosing m so that $|\alpha_n - \alpha| < 1$ for $n \geq m$, we have

$$|\alpha_n| \leq (|\alpha|+1) \vee \sup_{k\in\mathbb{N}_m} |\alpha_k|$$

for every n.] For every n in $\mathbb{N}$, define

$$\alpha'_n := \inf_{m\geq n} \alpha_m \qquad \alpha''_n := \sup_{m\geq n} \alpha_m.$$

Since $(\alpha_n)_{n\in\mathbb{N}}$ is bounded in $\mathbb{R}$, every α'_n as well as every α''_n is a real number. Clearly $(\alpha'_n)_{n\in\mathbb{N}}$ increases, $(\alpha''_n)_{n\in\mathbb{N}}$ decreases, and

$$\alpha'_n \leq \alpha_n \leq \alpha''_n$$

for every n. Moreover for each $\varepsilon > 0$ there exists m such that the inequality

$$\alpha - \varepsilon \le \alpha'_n \le \alpha''_n \le \alpha + \varepsilon$$

holds for every $n \ge m$. We conclude, in light of the monotonicity, that

$$\alpha - \varepsilon \le \sup_{n \in \mathbb{N}} \alpha'_n \le \inf_{n \in \mathbb{N}} \alpha''_n \le \alpha + \varepsilon$$

for every $\varepsilon > 0$. Thus

$$\alpha = \sup_{n \in \mathbb{N}} \alpha'_n = \inf_{n \in \mathbb{N}} \alpha''_n$$

and (b) holds.

(b) $\Rightarrow$ (a). Let $\varepsilon > 0$ be given. By (1) and the monotonicity of the sequences $(\alpha'_n)_{n \in \mathbb{N}}$, $(\alpha''_n)_{n \in \mathbb{N}}$, there exists a natural number m such that

$$\alpha - \varepsilon < \alpha'_n \le \alpha \le \alpha''_n < \alpha + \varepsilon$$

for every $n \ge m$. Since

$$\alpha'_n \le \alpha_n \le \alpha''_n$$

for every n, we conclude that

$$|\alpha_n - \alpha| < \varepsilon$$

for every $n \ge m$. □

Definition 1.1.13. *A sequence $(x_n)_{n \in \mathbb{N}}$ from an ordered set X is said to be* ***order-convergent*** *toward $x \in X$, or to* ***order-converge*** *to x, if the following condition holds: there exist in X an increasing sequence $(x'_n)_{n \in \mathbb{N}}$ and a decreasing sequence $(x''_n)_{n \in \mathbb{N}}$ such that*

$$x'_n \le x_n \le x''_n$$

for every n in $\mathbb{N}$, and

$$\bigvee_{n \in \mathbb{N}} x'_n = \bigwedge_{n \in \mathbb{N}} x''_n = x. \tag{2}$$

A sequence $(x_n)_{n \in \mathbb{N}}$ that order-converges to x and is also monotone is said to be ***monotonely convergent****, or to* ***converge monotonely****, to x.*

Whenever the ordered set in question is $\mathbb{R}$ or $\overline{\mathbb{R}}$, we shall simply speak of sequences that converge or are convergent. □

We want to establish some facts about order-convergent sequences. We first verify that limits are unique.

Proposition 1.1.14. *Let $(x_n)_{n\in\mathbb{N}}$ be a sequence from an ordered set X that order-converges to both $x \in X$ and $y \in X$. Then $x = y$.*

Proof. By definition there exist in X two increasing sequences $(x_n')_{n\in\mathbb{N}}$ and $(y_n')_{n\in\mathbb{N}}$ and two decreasing sequences $(x_n'')_{n\in\mathbb{N}}$ and $(y_n'')_{n\in\mathbb{N}}$ such that

$$x_n' \le x_n \le x_n'' \quad \text{and} \quad y_n' \le x_n \le y_n'' \tag{3}$$

for every n and

$$\bigvee_{n\in\mathbb{N}} x_n' = \bigwedge_{n\in\mathbb{N}} x_n'' = x, \quad \bigvee_{n\in\mathbb{N}} y_n' = \bigwedge_{n\in\mathbb{N}} y_n'' = y.$$

We interlace the sequences. From (3) it follows that

$$x_n' \le x_n \le y_n'' \quad \text{and} \quad y_n' \le x_n \le x_n''$$

for every n. Thus

$$x = \bigvee_{n\in\mathbb{N}} x_n' \le \bigwedge_{n\in\mathbb{N}} y_n'' = y = \bigvee_{n\in\mathbb{N}} y_n' \le \bigwedge_{n\in\mathbb{N}} x_n'' = x. \tag{4}$$

Equality must hold everywhere in (4), so $x = y$. □

Definition 1.1.15. *Let $(x_n)_{n\in\mathbb{N}}$ be an order-convergent sequence from an ordered set X. The element x in X toward which $(x_n)_{n\in\mathbb{N}}$ converges (which the previous proposition shows to be uniquely determined) is called the* **order-limit**, *or simply the* **limit**, *of the sequence $(x_n)_{n\in\mathbb{N}}$. We write*

$${}^{X}\lim_{n\to\infty} x_n := x.$$

For ${}^{\mathbb{R}}\lim$ *and* ${}^{\overline{\mathbb{R}}}\lim$ *we generally write* $\lim$. □

Not unexpectedly, monotone sequences that are also order-convergent converge toward the supremum or the infimum.

Proposition 1.1.16. *If $(x_n)_{n\in\mathbb{N}}$ is a monotonely convergent sequence from an ordered set X, then*

$$\lim_{n\to\infty} x_n = \begin{cases} \bigvee_{n\in\mathbb{N}} x_n & \text{if } (x_n)_{n\in\mathbb{N}} \text{ increases} \\ \bigwedge_{n\in\mathbb{N}} x_n & \text{if } (x_n)_{n\in\mathbb{N}} \text{ decreases.} \end{cases}$$

Proof. Suppose that $(x_n)_{n\in\mathbb{N}}$ increases. We show first that $\lim_{n\to\infty}x_n$ is an upper bound for the sequence $(x_n)_{n\in\mathbb{N}}$ and then that it is the least upper bound.

Let $(x'_n)_{n\in\mathbb{N}}$ and $(x''_n)_{n\in\mathbb{N}}$ be increasing and decreasing sequences, respectively, such that $x'_n \le x_n \le x''_n$ for every n and

$$\bigvee_{n\in\mathbb{N}} x'_n = \bigwedge_{n\in\mathbb{N}} x''_n = \lim_{n\to\infty} x_n.$$

The monotonicity of the two sequences $(x_n)_{n\in\mathbb{N}}$ and $(x''_n)_{n\in\mathbb{N}}$, combined with the inequality $x_n \le x''_n$, guarantees that $x_m \le x''_n$ for all $m, n \in \mathbb{N}$. Thus for every m in $\mathbb{N}$

$$x_m \le \bigwedge_{n\in\mathbb{N}} x''_n = \lim_{n\to\infty} x_n.$$

Now if $x_n \le y$ for some $y \in X$ and for every n, then $x'_n \le y$ for every n, and so

$$\lim_{n\to\infty} x_n = \bigvee_{n\in\mathbb{N}} x'_n \le y.$$

This completes the proof in the case where $(x_n)_{n\in\mathbb{N}}$ increases; the case where $(x_n)_{n\in\mathbb{N}}$ decreases can be handled analogously. □

Proposition 1.1.17. *For every ordered set X, every order-convergent sequence from X is bounded.*

Proof. Let $(x_n)_{n\in\mathbb{N}}$ be an order-convergent sequence from an ordered set X. Denote by $(x'_n)_{n\in\mathbb{N}}$ and $(x''_n)_{n\in\mathbb{N}}$ a pair of sequences from X that satisfy, relative to $(x_n)_{n\in\mathbb{N}}$, the conditions described in the definition of order-convergence (Definition 13). Then, for every n in $\mathbb{N}$,

$$x'_1 \le x'_n \le x_n \le x''_n \le x''_1. \qquad \square$$

$\overline{\mathbb{R}}$ has a stronger order structure than that of an ordered set, and it has algebraic structure as well. Consequently convergent sequences from $\overline{\mathbb{R}}$ have additional properties, some of which we eventually use. These results will be very familiar to the reader, and we state them without proof.

Theorem 1.1.18. *Every increasing sequence from $\overline{\mathbb{R}}$ converges to its supremum. Every decreasing sequence from $\overline{\mathbb{R}}$ converges to its infimum.*

If $(\alpha_n)_{n\in\mathbb{N}}$ and $(\beta_n)_{n\in\mathbb{N}}$ are sequences from $\overline{\mathbb{R}}$ that converge to real numbers α and β, respectively, then $(\alpha_n+\beta_n)_{n\in\mathbb{N}}$ converges to $\alpha+\beta$, $(\alpha_n \vee \beta_n)_{n\in\mathbb{N}}$ converges to $\alpha\vee\beta$, $(\alpha_n\wedge\beta_n)_{n\in\mathbb{N}}$ converges to $\alpha\wedge\beta$, $(|\alpha_n|)_{n\in\mathbb{N}}$ converges to $|\alpha|$, and $(\alpha_n\beta_n)_{n\in\mathbb{N}}$ converges to $\alpha\beta$. □

In $\mathbb{R}$ and in $\overline{\mathbb{R}}$, in fact in any conditionally σ-complete lattice, every bounded sequence has a ***limes superior*** and a ***limes inferior,*** which we now define.

Definition 1.1.19. *Let X be a conditionally σ-complete lattice. For every bounded sequence $(x_n)_{n\in\mathbb{N}}$ from X we define*

$$^X\limsup_{n\to\infty} x_n := \bigwedge_{n\in\mathbb{N}}\left(\bigvee_{m\geq n} x_m\right)$$

$$^X\liminf_{n\to\infty} x_n := \bigvee_{n\in\mathbb{N}}\left(\bigwedge_{m\geq n} x_m\right).$$

We write $\limsup$ *for* $^{\mathbb{R}}\limsup$ *and* $^{\overline{\mathbb{R}}}\limsup$, *and we write* $\liminf$ *for* $^{\mathbb{R}}\liminf$ *and* $^{\overline{\mathbb{R}}}\liminf$. □

The relations between lim, lim sup, and lim inf are described in the following proposition.

Proposition 1.1.20. *Let X be a conditionally σ-complete lattice. The following assertions hold for every bounded sequence $(x_n)_{n\in\mathbb{N}}$ from X.*

(a) $\liminf_{n\to\infty} x_n \leq \limsup_{n\to\infty} x_n$

(b) $(x_n)_{n\in\mathbb{N}}$ *is order-convergent iff*

$$\liminf_{n\to\infty} x_n = \limsup_{n\to\infty} x_n \tag{5}$$

and in this case

$$\lim_{n\to\infty} x_n = \liminf_{n\to\infty} x_n = \limsup_{n\to\infty} x_n. \tag{6}$$

Proof. (a) Assertion (a) follows easily from the definitions.

(b) If (5) holds, then the sequences $(x'_n)_{n\in\mathbb{N}}$ and $(x''_n)_{n\in\mathbb{N}}$ defined by

$$x'_n := \bigwedge_{m\geq n} x_m, \qquad x''_n := \bigvee_{m\geq n} x_m$$

show that $(x_n)_{n\in\mathbb{N}}$ is order-convergent, and (6) holds.

Suppose, on the other hand, that $(x_n)_{n\in\mathbb{N}}$ is order-convergent. Let $(x'_n)_{n\in\mathbb{N}}$ and $(x''_n)_{n\in\mathbb{N}}$ be a pair of sequences that satisfy, relative to the given sequence $(x_n)_{n\in\mathbb{N}}$, the conditions specified in the definition of order-convergence (Definition 13). Then for every n we have

$$x'_n = \bigwedge_{m\geq n} x'_m \leq \bigwedge_{m\geq n} x_m \leq \bigvee_{m\geq n} x_m \leq \bigvee_{m\geq n} x''_m = x''_n.$$

Thus

$$\bigvee_{n\in\mathbb{N}} x'_n \le \liminf_{n\to\infty} x_n \le \limsup_{n\to\infty} x_n \le \bigwedge_{n\in\mathbb{N}} x''_n.$$

Since

$$\bigvee_{n\in\mathbb{N}} x'_n = \bigwedge_{n\in\mathbb{N}} x''_n = \lim_{n\to\infty} x_n$$

we conclude that (5) and (6) both hold. □

EXERCISES

1.1.1(E) Prove the assertions made in Theorem 2. This can be done via the following steps.

(a) For arbitrary finite families of real numbers, $(\alpha_k)_{k\in\mathbb{N}_m}$ and $(\beta_k)_{k\in\mathbb{N}_m}$, and for $\alpha \in \mathbb{R}$, establish these three equalities by complete induction:

(α) $\Sigma_{k=1}^m(\alpha_k + \beta_k) = \Sigma_{k=1}^m \alpha_k + \Sigma_{k=1}^m \beta_k$

(β) $\Sigma_{k=1}^m \alpha\alpha_k = \alpha\Sigma_{k=1}^m \alpha_k$

(γ) If $\alpha_k = \beta$ for every $k \in \mathbb{N}_m$, then $\Sigma_{k=1}^m \alpha_k = m\beta$.

(b) For each finite family of real numbers, $(\alpha_k)_{k\in\mathbb{N}_n}$, and for every bijection $\varphi\colon \mathbb{N}_n \to \mathbb{N}_n$

$$\sum_{k=1}^{n} \alpha_k = \sum_{k=1}^{n} \alpha_{\varphi(k)}.$$

(c) Finally, prove Theorem 2.

1.1.2(E) Prove the assertions made in Propositions 8, 9, and 10. (Bear in mind the suggestions made concerning Propositions 9 and 10).

1.1.3(E) Verify the following statements.

(α) Every totally ordered set is a lattice.

(β) $\mathbb{N}$ and $\mathbb{Z}$ are conditionally complete lattices.

(γ) $\mathbb{Q}$ is a lattice that is not even conditionally σ-complete.

1.1.4(E) Consider the relation, $\preccurlyeq$, on $\mathbb{N}$ defined as follows:

For $m, n \in \mathbb{N}$, $m \preccurlyeq n$ iff for some $k \in \mathbb{N}$, $n = km$.

Show that

(α) $\preccurlyeq$ is an order relation on $\mathbb{N}$ and if $m \preccurlyeq n$, then $m \le n$.

(β) $\mathbb{N}$ is a conditionally complete lattice with respect to $\preccurlyeq$. For every nonempty family $(n_\iota)_{\iota\in I}$ from $\mathbb{N}$, $\bigwedge_{\iota\in I} n_\iota$ is the greatest common divisor of all n_ι's. If $(n_\iota)_{\iota \in I}$ is bounded above, $\bigvee_{\iota \in I} n_\iota$ is the least common multiple of all the n_ι's.

(γ) For $m, n, p \in \mathbb{N}$, $mp \preccurlyeq np$ whenever $m \preccurlyeq n$.

(δ) $1 \preccurlyeq 2$ but not $2 \preccurlyeq 3$. Hence it does not follow from $m \preccurlyeq n$ that $m + p \preccurlyeq n + p$. This shows that the order relation $\preccurlyeq$ does not bear the same relationship to the algebraic operations on $\mathbb{N}$ as does $\leq$.

1.1.5(E) In this exercise, we investigate the relation $\preccurlyeq$ on $\mathbb{R}$ given by the following definition:

$$\text{For } \alpha, \beta \in \mathbb{R},\ \alpha \preccurlyeq \beta \text{ iff for some } \gamma \in \mathbb{R},\ \gamma \geq 1,\ \beta = \alpha\gamma.$$

Verify the following properties of this relation:

(α) $\preccurlyeq$ is an order relation.

(β) For $\alpha, \beta > 0$, $\alpha \preccurlyeq \beta$ iff $\alpha \leq \beta$.

(γ) For $\alpha, \beta < 0$, $\alpha \preccurlyeq \beta$ iff $\alpha \geq \beta$.

(δ) If $\alpha \neq 0$, then neither $\alpha \preccurlyeq 0$ nor $0 \preccurlyeq \alpha$ holds.

(ε) $\mathbb{R}$ is not a lattice with respect to $\preccurlyeq$.

(ζ) For $(\alpha_n)_{n \in \mathbb{N}}$ a sequence from $\mathbb{R}$, $(\alpha_n)_{n \in \mathbb{N}}$ converges to $\alpha \neq 0$ with respect to $\preccurlyeq$ iff $(\alpha_n)_{n \in \mathbb{N}}$ converges to α in the usual sense.

(η) $(0)_{n \in \mathbb{N}}$ is the only sequence which converges to 0 with respect to $\preccurlyeq$.

1.1.6(E) Suppose that $(n_m)_{m \in \mathbb{N}}$ is order-convergent to p in either $\mathbb{N}$ or $\mathbb{Z}$. Then, for some $m_0 \in \mathbb{N}$, $n_m = p$ whenever $m \geq m_0$. Conversely, any sequence with this last-mentioned property is order-convergent.

1.1.7(E) Let X be an ordered set. Prove the following statements:

(α) If a sequence $(x_n)_{n \in \mathbb{N}}$ from X order-converges to x, then every subsequence order-converges to x.

(β) Every constant sequence $(x)_{n \in \mathbb{N}}$ $(x \in X)$ order-converges to x.

(γ) If X is doubly directed and $m \in \mathbb{N}$, and if $(x_n)_{n \in \mathbb{N}}$ is a sequence from X for which $(x_{m+n})_{n \in \mathbb{N}}$ converges to an $x \in X$, then $(x_n)_{n \in \mathbb{N}}$ converges to x.

1.1.8(E) Take an ordered set X. In this set, take three sequences $(x_n)_{n \in \mathbb{N}}$, $(y_n)_{n \in \mathbb{N}}$, and $(z_n)_{n \in \mathbb{N}}$. Suppose that for each $n \in \mathbb{N}$, $x_n \leq y_n \leq z_n$. Suppose further that both $(x_n)_{n \in \mathbb{N}}$ and $(z_n)_{n \in \mathbb{N}}$ are order-convergent to the same $x \in X$. Show that $(y_n)_{n \in \mathbb{N}}$ is also order-convergent to this same x.

1.1.9(E) Consider an ordered set, X, and two sequences, $(x_n)_{n \in \mathbb{N}}$ and $(y_n)_{n \in \mathbb{N}}$, from this set. Suppose that $x_n \leq y_n$ for infinitely many $n \in \mathbb{N}$ (but not necessarily all $n \in \mathbb{N}$). Suppose further that $(x_n)_{n \in \mathbb{N}}$ is order-convergent to x and $(y_n)_{n \in \mathbb{N}}$ to y. Show that $x \leq y$. Give an example in which $x_n < y_n$ for every $n \in \mathbb{N}$ but $x = y$.

1.1.10(E) Let X be an ordered set. A subset A of X is called **closed** iff, for each order-convergent sequence $(x_n)_{n \in \mathbb{N}}$ from A, ${}^X\lim_{n \to \infty} x_n \in A$. Set

$$\mathfrak{T} := \{ A \subset X \mid X \setminus A \text{ is closed} \}$$

and verify the following statements:

(α) $\mathfrak{T}$ is a topology on X.

(β) The closed subsets of X are exactly the closed sets with respect to this topology.

(γ) If a sequence $(x_n)_{n \in \mathbb{N}}$ order-converges to $x \in X$, it converges to x with respect to $\mathfrak{T}$.

(δ) Let $A \subset X$. Then $A \in \mathfrak{T}$ iff, for each order-convergent sequence $(x_n)_{n \in \mathbb{N}}$ from X with ${}^X\lim_{n \to \infty} x_n \in A$, there is $m \in \mathbb{N}$ such that $x_n \in A$ for each $n \in \mathbb{N}$, $n \geq m$.

The type of topology discussed here will be of great importance for our investigations. $\mathfrak{T}$ is called the **order topology** on X. Show that

(ε) the natural topology on $\mathbb{R}$ is the order topology with respect to the natural order on $\mathbb{R}$.

1.1.11(C) This exercise concerns some elements of the theory of ordinals, taking it as far as is required in subsequent exercises. In order to do so, we must extend the framework of sets. We introduce (besides sets) classes, and we operate on classes in much the same way as on sets. As in the text we proceed in naive fashion, but it goes without saying that this process could be grounded in axiomatic set theory.

The theory of ordinals provides in several places a perspective on the theory of integration which is very different from the one adopted in the text. We do not wish to abstain from such alternative viewpoints. Nevertheless, they will be developed completely only in the exercises.

(a) As an introduction to the conceptual world of the theory of ordinals, we first consider the set, $\mathbb{N}$, of natural numbers. The reader should be able to prove the following:

(α) Every nonempty subset of $\mathbb{N}$ has a smallest element.

The significance of this property lies in the fact that it implies the **Principle of Complete Induction,** which is usually expressed in the form:

(β) If $M \subset \mathbb{N}$ such that $1 \in M$ and $n + 1 \in M$ whenever $n \in M$, then $M = \mathbb{N}$.

If M were not equal to $\mathbb{N}$ under the hypotheses of the theorem, then $\mathbb{N} \setminus M \neq \emptyset$. Then, because of ($\alpha$), $\mathbb{N} \setminus M$ would have a smallest element, n. Thus either $n = 1$, contradicting $1 \in M$, or $n - 1 \in M$, leading to $n = (n - 1) + 1 \in M$, which is again a contradiction.

This last argument used the algebraic operations in $\mathbb{N}$. This is not necessary with a different formulation of the Principle of Complete Induction. For convenience set $A_n := \{ m \in \mathbb{N} \mid m < n \}$ for each $n \in \mathbb{N}$. (Note that A_1 is empty.) Prove (β').

(β') Given $M \subset \mathbb{N}$ such that $n \in M$ for each $n \in \mathbb{N}$ with $A_n \subset M$, it follows that $M = \mathbb{N}$.

Indeed, let $M \neq \mathbb{N}$. Then (α) implies the existence of a smallest element $n \in \mathbb{N} \setminus M$. But then, $A_n \subset M$ and so $n \in M$, a contradiction.

Property (β') and its proof lay bare the fact that only the order property expressed in (α) is essential for the Principle of Complete Induction. The reader is probably aware of the importance and significance of complete induction in definitions, constructions and proofs. One aspect important to the theory of integration is the application to definitions. We have already witnessed such applications. For example, recall that we work with a mapping from the set of all finite families $(\alpha_k)_{k \in \mathbb{N}_n} (n \in \mathbb{N})$ of real numbers, associating to $(\alpha_k)_{k \in \mathbb{N}_n}$ the sum $\sum_{k=1}^{n} \alpha_k$. This mapping is defined inductively as follows:

$$\sum_{k=1}^{1} \alpha_k := \alpha_1$$

$$\sum_{k=1}^{n+1} \alpha_k := \sum_{k=1}^{n} \alpha_k + \alpha_{n+1}$$

The completion process in the theory of integration can also be described by a succession of stages, each of which is the starting point of the next one. The essential difference from the example cited earlier lies in the fact that the completions are, in general, not constructions with only countably many stages, so that the stages cannot be indexed by $\mathbb{N}$. This consideration necessitates an extension of the notion of "number" and of the Principle of Complete Induction. Thus we are led to the notion of "ordinal" and to the Principle of Transfinite Induction. We now describe the constructions connected to this extension as far as needed for our purposes.

(b) **The introduction of ordinals**. We start from property (α) and introduce the following definition:

> The ordered set, X, is **well ordered** iff every nonempty subset of X has a smallest element.

The following properties of order isomorphy should not be difficult to check:

(α) Every ordered set is order-isomorphic to itself.

(β) X is order-isomorphic to Y iff Y is order-isomorphic to X.

(γ) If X is order-isomorphic to Y and if Y is order-isomorphic to Z, then X is order-isomorphic to Z.

Hence order isomorphy defines an equivalence relationship. But we cannot leap ahead to the usual sort of constructions given when dealing with an equivalence relationship on a set. We cannot simply partition the "set of all ordered sets" into equivalence classes. If we wish to avoid logical paradoxes (one of them we will meet later), then we cannot speak of "the set of ordered sets" or "the set of well-ordered sets." This does not mean that we cannot talk of the collection of all ordered sets but rather that we can no longer call such a collection a set. We are dealing here with collections that are, so to say, superordinate to sets. These collections are called "classes" to make the distinction. We will have more to say later about this problem. At this stage we simply note that order isomorphy partitions the class of all ordered sets into disjoint classes of order-isomorphic sets just the way one has been accustomed to in the case of equivalence relations on a set. These disjoint classes are called **order types**. The reader should prove:

(δ) If an order type contains a well-ordered set, then all the sets of this order type are well ordered.

Such order types are called **well-order-types**. We note that the Axiom of Choice also applies to classes. In our case this means that there is a class, Ω, of well-ordered sets which contains exactly one representative from each well-order-type. The well-ordered sets contained in this class Ω are called **ordinals.**

Thus, in our treatment, ordinals are fixed representatives of well-order-types, chosen once and for all. We denote them by $\alpha, \beta, \gamma, \ldots$. Thus for each well-ordered set, X, there is exactly one ordinal that is order-isomorphic to X. We call it simply the **ordinal** of X.

(c) We now investigate the new objects in several steps. For X an ordered set and $x \in X$, we set $A_x^X := \{y \in X \mid y < x\}$.

STEP 1: Let X and Y be well-ordered sets. Show that if X is finite, then the next two statements are equivalent:

(α) X and Y are order-isomorphic.

(β) Y has the same number of elements as X.

Consequently the finite ordinals are in bijection to the numbers of $\mathbb{N} \cup \{0\}$. In fact this bijection can be realized by a natural order isomorphism. Notice that $\mathbb{N}$ is itself well ordered, so that there are also infinite ordinals, and we have managed to extend our concept beyond the natural numbers.

STEP 2: Prove the following statements:

(α) Every well-ordered set is totally ordered.

(β) Every subset of a well-ordered set is well ordered by the induced order relation.

(γ) Take a well-ordered set X and $x \in X$. Then either

(γ1) x is the largest element of X, or
(γ2) there is a unique element, $x' \in X$, such that $x < x'$ and, if $z \in X$ with $x < z$, then $x' \leq z$. Call x' the **successor** of x.

(δ) If $(x_n)_{n \in \mathbb{N}}$ is a decreasing sequence in a well-ordered set X, then for some $m \in \mathbb{N}$, $x_n = x_m$ whenever $m \leq n$.

STEP 3: Let X be a well-ordered set, and set $Z := \{A_x \mid x \in X\}$. Prove the statements that follow:

(α) Given $x, y \in X$, $A_x \subset A_y$ iff $x \leq y$.

(β) Z is well ordered by $\subset$.

(γ) Z and X are order-isomorphic.

STEP 4: Let X be well ordered, $Y \subset X$, and $\varphi: X \to Y$ an order isomorphism. Show that

(α) $x \leq \varphi(x)$ for each $x \in X$.

Assume that for some $x \in X$, $\varphi(x) < x$. Then Z, the set of all such elements of X, is not empty. Consequently Z contains a smallest element z. Thus $\varphi(z) < z$, and furthermore $\varphi(\varphi(z)) < \varphi(z)$, proving that $\varphi(z) \in Z$, which contradicts the choice of z.

The next two statements are immediate consequences.

(β) Given a well-ordered set X and $x \in X$, X and A_x are not order-isomorphic.

(γ) If X is a well-ordered set, and $x, y \in X$, then A_x and A_y are order-isomorphic iff $x = y$.

STEP 5: Prove the following statements:

(α) Let $\varphi: X \to Y$ be an order isomorphism between well-ordered sets. Take $x \in X$. Then $\varphi|_{A_x^X}: A_x^X \to A_{\varphi(x)}^Y$ is an order isomorphism.

(β) If X and Y are order-isomorphic, well-ordered sets, then there is exactly one mapping $\varphi: X \to Y$ that is an order isomorphism. [Recall (α) and Step 4 (β).]

STEP 6: Prove the following proposition: If X and Y are well-ordered sets, then exactly one of the following statements is true:

(α) X and Y are order-isomorphic.

(β) There is an $x \in X$ such that A_x^X and Y are order-isomorphic.

(γ) There is a $y \in Y$ such that X and A_y^Y are order-isomorphic.

For the proof, set $X' := \{x \in X \mid \exists y \in Y$ with A_x^X and A_y^Y order-isomorphic$\}$, and set $Y' := \{y \in Y \mid \exists x \in X$ with A_y^Y and A_x^X order-isomorphic$\}$. Verify that either $X = X'$ or $X' = A_{x'}^X$ for some suitable $x' \in X$. (The corresponding statement is true for Y.) Show that X' and Y' are order-isomorphic. To do this, consider the mapping from X' to Y' which associates to each $x \in X'$, the element $y \in Y'$ such that A_x^X and A_y^Y are order-isomorphic. This mapping is a bijection. Next show that it is an order isomorphism. Thus four cases are logically possible:

$X' = X$ and $Y' = Y$.

$X' = A_{x'}^X$ and $Y' = Y$ for a suitable $x' \in X$.

$X' = X$ and $Y' = A_{y'}^Y$ for a suitable $y' \in Y$.

$X' = A_{x'}^X$ and $Y' = A_{y'}^Y$ for suitable $x' \in X$ and $y' \in Y$.

A little thought shows that the fourth case cannot occur. Furthermore the three other cases are the same as the preceding statements (α), (β), and (γ).

STEP 7: Let X and Y be well ordered. X is called **smaller** than Y (written $X \leq Y$) iff X and A_y^Y are order-isomorphic for some $y \in Y$ or if X and Y are isomorphic. The relation $\leq$ is an order relation on Ω. Prove the following statements where Δ denotes a subclass of Ω:

(α) If $\Delta \neq \phi$, then Δ contains a smallest element.

(β) ϕ is the smallest element in Ω.

(γ) Ω is well-ordered by $\leq$.

STEP 8: Every ordinal α is order-isomorphic to A_α^Ω.

STEP 9: Consider now Ω, the class of all ordinals. By Step 7, Ω is well-ordered by $\leq$.

We can now clarify why it was not possible to speak simply of the set of all well-ordered sets. Suppose that the collection of all well-ordered sets forms a set. Then the well-order-types would also form a set. Consequently Ω would be a set. But it has just been shown that Ω is well ordered by $\leq$. Consequently Ω has a (unique) ordinal, α. By Step 8, Ω and A_α^Ω are order-isomorphic, contradicting Step 4, (β). This is the Burali-Forti paradox. It shows that the collection of all ordinals (and a fortiori that of all well-ordered sets, or of all ordered sets) cannot form a set. By this is meant much more than a mere semantic trick. In effect the constructions we are accustomed to executing with sets cannot be performed without further ado on such collections. One is used to carrying out a number of procedures with sets without stumbling into contradictions. Unions, intersections, products, and power sets are regularly constructed without the least fear of contradiction. Objects are frequently collected together into sets without problem. This apparent ease tempts one to see no problems in the construction of sets. Our example shows that such a cavalier attitude is misplaced. Care is needed. There are many other paradoxes, for example, that of G. Cantor or B. Russell. Difficulties arise with the unconditional construction of the set of all sets with a given property. The following example is to clarify what is meant here by unconditional.

Consider a well-ordered set X. Then there is no problem in considering the set of all ordinals smaller than X. The last condition means that we are dealing with a domain determined by (the set) X. Most of the collections of mathematical objects in this book are correspondingly limited, so that we shall be dealing almost exclusively with sets.

Care is called for, however, when we collect together objects that are not subsets of a given, well-defined set. It is exactly this situation that arises in the example just considered. Axiomatic set theory avoids the difficulties mentioned by regulating the construction of sets with set-constructing axioms that block the paradoxes. There are various possible paths, and so there are correspondingly different axiomatizations. We do not go into this area any further, nor do we ask of the reader the knowledge of any particular form of axiomatic set theory. We shall continue to adopt the standpoint of naive set theory. If we carefully avoid the indiscriminant (i.e., unconditional) construction of sets, then we shall not lead ourselves into difficulties.

STEP 10: **The Principle of Transfinite Induction.** Let Φ be a class of ordinals. Given $\Psi \subset \Phi$ such that $\alpha \in \Psi$ whenever $\alpha \in \Phi$ with $A_\alpha^\Phi \subset \Psi$, it follows that $\Psi = \Phi$.

The proof is analogous to that of the Principle of Complete Induction [see (β') in part (a)].

$\mathbb{N}$ is order-isomorphic to the set of all finite nonempty ordinals. In this special case the Principle of Transfinite Induction is exactly that of Complete Induction.

STEP 11: We gather the finite ordinals together into the so-called first number class Z_1. Z_1 is distinguished by the property that each number, other than the very first, has exactly one predecessor. In other words, given $\alpha \in Z_1$, $\alpha \neq \varnothing$, $A_\alpha^{Z_1}$ has a largest element.

Of greater interest to us is the so-called second number class, Z_2, which consists of all countably infinite ordinals. The first element of Z_2 is the ordinal of $\mathbb{N}$, which is denoted by ω_0. The set

$$\left\{\frac{1}{2}, \frac{2}{3}, \ldots, \frac{n}{n+1}, \ldots\right\}$$

has the same ordinal. Hence ω_0 has no predecessor. This leads to distinction between ordinals of the **first kind**—those with an immediate predecessor—and ordinals of the **second kind**—those without a predecessor—also called **limit ordinals**. The finite nonempty ordinals are of the first kind; ω_0 is the first ordinal of the second kind that is not empty.

After ω_0 come $\omega_0 + 1$, $\omega_0 + 2, \ldots, \omega_0 + n, \ldots,$ where $\omega_0 + j$ is the ordinal of

$$\left\{\frac{1}{2}, \frac{2}{3}, \ldots, \frac{n}{n+1}, \ldots, 1, 2, \ldots, j\right\}$$

(The notation $\omega_0 + n$ can be given a deeper meaning. The operations of addition and multiplication in $\mathbb{N}$ can be extended to Ω. As we have no need for it, we do not go into this generalization.)

All ordinals of the form $\omega_0 + n$ are of the first kind. They are followed by $\omega_0 2$, a limit ordinal, which is represented, for example, by

$$\left\{\frac{1}{2}, \frac{2}{3}, \ldots, \frac{n}{n+1}, \ldots, 1, 2, \ldots, n, \ldots\right\}.$$

We can continue, to obtain

$$\omega_0 2 + 1, \ldots, \omega_0 2 + n, \ldots, \omega_0 3, \omega_0 3 + 1, \ldots, \omega_0 3 + m, \ldots, \omega_0 4, \ldots.$$

The ordinal that follows all such numbers is denoted by ω_0^2, which, for example, is the ordinal of $\mathbb{N}^2$ lexicographically ordered:

$$(m, n) \leq (m', n')$$

iff

$$m < m' \quad \text{or} \quad m = m' \quad \text{and} \quad n \leq n'$$

Continuing this way yields

$$\omega_0^2 + 1, \ldots, \omega_0^2 + \omega_0, \ldots, \omega_0^2 + \omega_0 + k, \ldots,$$

$$\omega_0^2 + \omega_0 2, \ldots, \omega_0^2 2, \ldots, \omega_0^2 m + \omega_0 n + k, \ldots .$$

There follows ω_0^3, the ordinal of $\mathbb{N}^3$ lexicographically ordered. Then come $\omega_0^3 + 1, \ldots, \omega_0^4, \ldots, \omega_0^n, \ldots$. This leads naturally to $\omega_0^{\omega_0}$ which is the ordinal of the set of all polynomials in one variable with coefficients in $\mathbb{N}$, ordered by

$$n_0 + n_1 x + \cdots + n_p x^p \leq m_0 + m_1 x + \cdots + m_q x^q$$

iff

(i) $p < q$ or
(ii) $p = q$ and $(n_0, \ldots, n_p) \leq (m_0, \ldots, m_q)$ lexicographically.

Counting further yields

$$\omega_0^{\omega_0} + 1, \ldots, \omega_0^{\omega_0} + \omega_0, \ldots, \omega_0^{\omega_0} + \omega_0 n, \ldots, \omega_0^{\omega_0} + \omega_0^2,$$
$$\omega_0^{\omega_0} + \omega_0^n, \ldots, \omega_0^{\omega_0} 2, \ldots, \omega_0^{\omega_0} n, \ldots, \omega_0^{(\omega_0 + 1)}, \ldots,$$
$$\omega_0^{(\omega_0 + n)}, \ldots, \omega_0^{(\omega_0 2)}, \ldots, \omega_0^{(\omega_0^2)}, \ldots, \omega_0^{(\omega_0^n)}, \ldots,$$
$$\omega_0^{\omega_0^{\omega_0}}, \ldots, \omega_0^{\omega_0^{\cdot^{\cdot^{\cdot^{\omega_0}}}}}$$

The ordinal that comes after all ordinals of the form $\omega_0^{\omega_0^{\cdot^{\cdot^{\cdot^{\omega_0}}}}}$ is denoted by ε. Then come $\varepsilon + 1, \ldots$. This should suffice. All ordinals obtained in this manner are in Z_2. The reader should prove the (quite general) statements

(α) If $\gamma \in Z_2$, then $\gamma + 1 \in Z_2$.
(β) If $Z \subset Z_2$ and Z is countable, and if γ is the first ordinal after all ordinals in Z, then $\gamma \in Z_2$.

It follows from (β) that Z_2 cannot be countable. Prove that if $\alpha \in Z_2$, then A_α^Ω is countable. We shall see that the processes of completion in the theory of integration which we describe by means of ordinals, produce the desired objects by the time Z_2 is exhausted.

(d) As the last part of this exercise, we wish to deduce two equivalents of the Axiom of Choice which play an important role in many mathematical arguments. Prove that the following statements are equivalent:

(α) **Axiom of Choice**. Let $(A_\iota)_{\iota \in I}$ be a family of nonempty sets. Then, there is a map $\varphi: I \to \bigcup_{\iota \in I} A_\iota$ such that $\varphi(\iota) \in A_\iota$ for each $\iota \in I$.
(β) **Zorn Lemma**. Let X be an ordered set. If every well-ordered subset of X is bounded above in X, then for each $x \in X$ there is a maximal element x' in X with $x \leq x'$.

(γ) **Well-Ordering Principle.** Every set can be well ordered.

$(\alpha) \Rightarrow (\beta)$. It suffices to show that if X satisfies (α), then X has a maximal element. Then apply this to $\{y \in X \mid x \leq y\}$.

Set $\mathfrak{A} := \{A \subset X \mid A$ has an upper bound $x \in X \setminus A\}$. Every subset of an element of $\mathfrak{A}$ is itself in $\mathfrak{A}$. For $A \in \mathfrak{A}$, set $\hat{A} := \{x \in X \setminus A \mid z \leq x$ whenever $z \in A\}$. By construction, $\hat{A} \neq \varnothing$. The Axiom of Choice guarantees the existence of a map $\varphi\colon \mathfrak{A} \to \bigcup_{A \in \mathfrak{A}} \hat{A}$ with $\varphi(A) \in \hat{A}$ for each $A \in \mathfrak{A}$.

Now set $\mathfrak{B} := \{Y \in \mathfrak{A} \mid Y$ is well ordered and $\varphi(A_x^Y) = x$ whenever $x \in Y\}$ and take $Z := \bigcup_{Y \in \mathfrak{B}} Y$. We introduce a concept which allows us to divide the rest of the proof into smaller steps. Let W be an ordered set. $V \subset W$ is an **initial segment** of W iff it follows from $v \in V$, $w \in W$ and $w \leq v$ that $w \in V$.

STEP 1: Given $Y, Y' \in \mathfrak{B}$, either Y is an initial segment of Y' or Y' is an initial segment of Y.

To see this, set $A := \{x \in Y \cap Y' \mid \{z \in Y \mid z \leq x\} = \{z' \in Y' \mid z' \leq x\}\}$. A is an initial segment of both Y and Y' and either $A = Y$ or $A = Y'$.

STEP 2: Every $Y \in \mathfrak{B}$ is an initial segment of Z.

Given $x \in Y$ and $z \in Z$ with $z < x$, take $Y' \in \mathfrak{B}$ with $z \in Y'$. Then, by step 1, $z \in Y$.

STEP 3: Z is well ordered.

Let Z' be a nonempty subset of Z. Then there is a $Y \in \mathfrak{B}$ with $Z' \cap Y \neq \varnothing$. Let x be the smallest element in $Z' \cap Y$. Then x is also the smallest element in Z', because given $z \in Z'$, either $z \in Y$ or $z \notin Y$. If $z \in Y$ then $x \leq z$. If $z \notin Y$, then apply step 1 to Y and $Y' \in \mathfrak{B}$ with $z \in Y'$.

STEP 4: If $Z \in \mathfrak{A}$ then $Z \in \mathfrak{B}$.

Given $z \in Z$, there is a $Y \in \mathfrak{B}$ with $z \in Y$. By step 2 $\{y \in Z \mid y < z\} = \{y \in Y \mid y < z\}$. Thus $\varphi(\{y \in Z \mid y < z\}) = \varphi(\{y \in Y \mid y < z\}) = z$.

STEP 5: If $Z \cup \{\varphi(Z)\} \in \mathfrak{A}$, then $Z \cup \{\varphi(Z)\} \in \mathfrak{B}$.

Take $z \in Z \cup \{\varphi(Z)\}$. If $z \in Z$, then $\{y \in Z \cup \{\varphi(Z)\} \mid y < z\} = \{y \in Z \mid y < z\}$. Hence, by step 4 $\varphi(\{y \in Z \cup \{\varphi(Z)\} \mid y < z\}) = z$. If $z \notin Z$, then $z = \varphi(Z)$, and thus $\{y \in Z \cup \{\varphi(Z)\} \mid y < z\} = Z$. So again $\varphi(\{y \in Z \cup \{\varphi(Z)\} \mid y < z\}) = z$.

STEP 6: $Z \cup \{\varphi(Z)\} \notin \mathfrak{A}$.

Otherwise (by step 5) $Z \cup \{\varphi(Z)\} \in \mathfrak{B}$, which would mean that $Z \cup \{\varphi(Z)\} \subset Z$, and thus $\varphi(Z) \in Z$ would follow—a contradiction.

STEP 7: If there is a well-ordered $Y \subset X$ with $Y \notin \mathfrak{A}$, then X has a maximal element.

Since $Y \notin \mathfrak{A}$, Y contains every upper bound of Y in X. Each such upper bound is a maximal element in X.

STEP 8: X has a maximal element.

By step 3, Z is well ordered. If $Z \notin \mathfrak{A}$, then the statement follows from step 6. If $Z \in \mathfrak{A}$, then $Z \cup \{\varphi(Z)\}$ is wll ordered. Our statement then follows from the steps 6 and 7.

$(\beta) \Rightarrow (\gamma)$. Set $\Phi := \{(Y, \leq_Y) \mid Y \subset X$ and Y is well ordered by $\leq_Y\}$. Define $(Y, \leq_Y) \preccurlyeq (Z, \leq_Z)$ iff $Y \subset Z$, $\leq_Z|_Y = \leq_Y$ and Y is an initial segment of Z. $\preccurlyeq$ is then an order relation. Show that each well-ordered subset of Φ is bounded above in Φ. By (β), Φ has a maximal element, say $(\overline{Y}, \leq_{\overline{Y}})$. Furthermore $\overline{Y} = X$.

$(\gamma) \Rightarrow (\alpha)$. Let $(A_\iota)_{\iota \in I}$ be a family of nonempty sets. By (γ) every A_ι can be well ordered and so contains a smallest element x_ι. The map $\varphi\colon I \to \bigcup_{\iota \in I} A_\iota$, $\iota \mapsto x_\iota$ satisfies our requirements.

The reader should look at (γ) in the context of ordinals. (γ) asserts that there are ordinals of arbitrary cardinality.

1.1.12(C) In this Colloquium exercise we investigate some topological properties of the real line $\mathbb{R}$.

(a) **Perfect sets.** Besides the compact sets, there is another important class of closed subsets of $\mathbb{R}$, namely the perfect sets: $A \subset \mathbb{R}$ is **perfect** iff the set of all points of accumulation of A coincides with A itself. Establish the following claims:

(α) Every nonempty compact subset A of $\mathbb{R}$ is either a closed interval or obtained from a closed interval by removal of at most countably many pairwise disjoint open intervals.

(β) Every perfect set is closed.

(γ) Every nonempty bounded perfect set is either a closed interval or obtained from such an interval, I, through the removal of, at most, countably many pairwise disjoint open intervals, none of which have any boundary points in common with any others or with I. Conversely, every set so constructed is perfect.

Observe that every bounded perfect set is compact. The example $[0,1] \cup \{2\}$ shows that not every compact set is perfect. In part (b), we construct a particularly important example of a perfect set. We mention next another example.

(δ) The set of all points in $[0,1]$, in whose decimal expansion the digit $n \in \mathbb{N}_8$ does not occur, is perfect.

(b) **Generalized Cantor sets.** Let $(\alpha_n)_{n \in \mathbb{N}}$ be a sequence of numbers in $]0,1[$. Denote by $\lambda(I)$ the length of the interval $I \subset \mathbb{R}$. Take $a, b \in \mathbb{R}$, $a < b$, and set $\gamma := (a+b)/2$.

From the interval $[a,b]$ remove an open interval, I_{11}, of length $\alpha_1|b-a|$ centered on γ. This leaves two disjoint closed intervals, J_{11} and J_{12}. Now remove from J_{1i} the open interval I_{2i} of length $\alpha_2\lambda(J_{1i})$ centered on the midpoint of J_{1i} ($i = 1,2$). This yields four pairwise disjoint closed intervals, J_{21}, J_{22}, J_{23}, J_{24}. Next remove the open intervals I_{3i} of length $\alpha_3\lambda(J_{2i})$ centered on the midpoint of J_{2i}. This leaves $8(=2^3)$ pairwise disjoint closed intervals $J_{3i}(i \in \{1,2,\dots,8\})$. Continuing in this way, construct, for each

$n \in \mathbb{N}$, 2^n pairwise disjoint closed intervals $J_{ni} (i \in \mathbb{N}_{2^n})$. The set

$$C^{[a,b]}[(\alpha_n)_{n \in \mathbb{N}}] := \bigcap_{n \in \mathbb{N}} \left(\bigcup_{k=1}^{2^n} J_{nk} \right)$$

is called the **generalized Cantor set** determined by $(\alpha_n)_{n \in \mathbb{N}}$. The set $C^{[0,1]}[(1/3)_{n \in \mathbb{N}}]$ is the classical **Cantor set**.

Denote by C_l (resp. C_r) the set of left (resp. right) endpoints of the intervals $J_{ni} (n \in \mathbb{N}, i \in \mathbb{N}_{2^n})$. Prove the statements that follow:

(α) C_l and C_r are countably infinite disjoint subsets of C.

(β) Given $x \in C$ and $\varepsilon > 0$, $]x - \varepsilon, x] \cap C_r \neq \varnothing$ and $[x, x + \varepsilon[\cap C_l \neq \varnothing$.

(γ) Every point of C is a point of accumulation both of C_l and C_r.

(δ) C is a perfect set.

(ε) C has no interior points.

(ζ) C is nowhere dense in $\mathbb{R}$.

(c) Denote by $\mathfrak{c}$ the cardinality of $\mathbb{R}$. The set X has the cardinality $\mathfrak{c}$ iff there is a bijection from X onto $\mathbb{R}$. Prove that every nonempty perfect set has the cardinality $\mathfrak{c}$.

The proof can be executed using the following idea. Let A be a nonempty perfect set. Choose a point, $x \in A$, and an open interval, I, containing x. Then $I \cap A$ contains distinct points, x_0 and x_1. There exist open intervals I_0 and I_1 such that $x_0 \in I_0$, $x_1 \in I_1$ and $\bar{I}_0 \cap \bar{I}_1 = \varnothing$, $\lambda(I_0) < 1$, $\lambda(I_1) < 1$. Next find a point x_{ik} in $I_i \cap A$ and an interval I_{ik} for each $i, k \in \{0,1\}$ such that $I_{ik} \subset I_i$, $x_{ik} \in I_{ik}$, $x_{i0} \neq x_{i1}$, $\bar{I}_{i0} \cap \bar{I}_{i1} = \varnothing$ and $\lambda(I_{ik}) < 1/2$. Continue inductively to obtain for each $n \in \mathbb{N}$ a set of points

$$\left\{ x_{i_1 \dots i_n} \mid i_k \in \{0,1\} \text{ for each } k \in \mathbb{N}_n \right\}$$

and a set of corresponding open intervals $\{ I_{i_0 \dots i_n} \}$ satisfying the following conditions:

(i) $x_{i_1 \dots i_n} \in I_{i_1 \dots i_n} \cap A$.

(ii) $I_{i_1 \dots i_{n-1} i_n} \subset I_{i_1 \dots i_{n-1}}$.

(iii) $\bar{I}_{i_1 \dots i_n} \cap \bar{I}_{i'_1 \dots i'_n} = \varnothing$ whenever $(i_1, \dots, i_n) \neq (i'_1, \dots, i'_n)$.

(iv) $\lambda(I_{i_1 \dots i_n}) < 1/n$.

Each sequence $(i_k)_{k \in \mathbb{N}}$ from $\{0,1\}$ provides a singleton set $\bigcap_{k \in \mathbb{N}} \bar{I}_{i_1 \dots i_k}$. Furthermore different sequences yield different singletons, which proves the claim, since the set of such sequences has the cardinality $\mathfrak{c}$.

(d) **Points of Condensation**. Take $A \subset \mathbb{R}$. The point $x \in \mathbb{R}$ is a **condensation point** of A iff every open interval which contains x also contains uncountably many points of A. Denote by $C(A)$ the set of all condensation points of A. Prove these statements:

(α) If A contains no condensation points, then A is countable.
(For the proof, consider the set, $\mathfrak{R}$, of all open intervals whose intersection with A is countable, and whose boundary points are rational. Show that every point of A is contained in some such interval, that is, $A = \bigcup_{I \in \mathfrak{R}}(A \cap I)$, which is countable since $\mathfrak{R}$ is.)

(β) $A \setminus C(A)$ is countable.

(γ) If A is uncountable, then so is $A \cap C(A)$.

(δ) $C(A)$ is perfect.

(ε) If A is closed, then $A = P \cup B$, with P perfect and B countable.

(This last statement is known as the **Cantor-Bendixsohn Theorem**).

(e) **The Structure of Perfect Sets.** Let A be a bounded nonempty perfect subset of $\mathbb{R}$. Set $a := \inf_{x \in A} x$ and $b := \sup_{x \in A} x$. Then $[a, b] \setminus A$ is an open subset of $\mathbb{R}$, and hence the union of a countable set of pairwise disjoint open intervals $(A_i)_{i \in I}$. Without loss of generality, no A_i is empty. Call the left endpoints of the intervals A_i left endpoints for A. Define an order on I by $i \leq i'$ iff there are $x \in A_i$ and $y \in A_{i'}$ with $x \leq y$. The relation $\leq$ is a total order on I, and the next two statements can be verified if A is nowhere dense:

(α) I has neither a smallest nor a largest element with respect to $\leq$.

(β) Given $i, i' \in I$ with $i < i'$, there is an $i'' \in I$ with $i < i'' < i'$.

Now let D denote the set of all dyadic rational numbers in the interval $]0, 1[$ (i.e., the set of all numbers of the form $m/2^n$, $n \in \mathbb{N}$, $m \in \{1, 3, 5, \ldots, 2^n - 1\}$). We shall construct a bijection φ: $D \to I$ such that $\alpha \leq \beta$ iff $\varphi(\alpha) \leq \varphi(\beta)$.

So choose a bijection ψ: $\mathbb{N} \to I$, $n \mapsto i_n$. Set $\varphi(1/2) := i_1$. Let $\varphi(1/4)$ denote the element $i_n < i_1$ of smallest index n, and $\varphi(3/4)$ the element $i_n > i_1$ with the smallest index n. Next set $\varphi(1/8)$ equal to the element $i_n < \varphi(1/4)$ of smallest index, $\varphi(3/8)$ to the element $i_n \in]\varphi(1/4), \varphi(1/2)[$ of smallest index, $\varphi(5/8)$ to the element $i_n \in]\varphi(1/2), \varphi(3/4)[$ of smallest index and $\varphi(7/8)$ to the element $i_n > \varphi(3/4)$ with the smallest index. Continuing this procedure inductively associates to each $m/2^n \in D$ an index $\varphi(m/2^n) \in I$. This correspondence is the desired bijection.

Now let A and A' be two arbitrary bounded nonempty nowhere dense perfect subsets of $\mathbb{R}$. The preceding constructions lead to the conclusion that the set of left endpoints for A and the set of left endpoints for A' are each order-isomorphic to D. Thus they are order-isomorphic to each other. It is now easy to construct an order isomorphism of A onto A'. We leave the details to the reader. So we have proved (γ), which follows.

(γ) All bounded nonempty nowhere dense perfect subsets are order-isomorphic and therefore homeomorphic with respect to their natural topologies. In particular, they are order-isomorphic and homeomorphic to the Cantor set C.

1.1.13(E) Filters and Ultrafilters. Let X be a lattice and denote by A a subset of X. Define the following notions:

(i) A is a $\vee$**-ideal** iff $A \neq \emptyset$, $x \vee y \in A$ for any $x, y \in A$, and $z \in A$ whenever $z \in X$ and $z \leq w$ for some $w \in A$.

(ii) A is called a $\wedge$**-ideal** iff $A \neq \emptyset$, $x \wedge y \in A$ for any $x, y \in A$, and $z \in A$ whenever $z \in X$ and $z \geq w$ for some $w \in A$.

Since $\vee$- and $\wedge$-ideals are dual objects, it is sufficient to formulate the properties of only one of these two types of ideals. Denote by $\mathfrak{J}(X)$ the set of all $\wedge$-ideals of X. Prove the following propositions.

(α) If X has a largest element, 1, then $\mathfrak{J}(X)$ is a complete lattice with respect to $\subset$. $\mathfrak{J}(X)$ has the following additional properties:

(α1) $\{1\}$ is the smallest element of $\mathfrak{J}(X)$ and X the largest.

(α2) If $(A_\iota)_{\iota \in I}$ is a nonempty family from $\mathfrak{J}(X)$, then $\bigwedge_{\iota \in I}^{\mathfrak{J}(X)} A_\iota = \bigcap_{\iota \in I} A_\iota$.

(α3) If $(A_\iota)_{\iota \in I}$ is a nonempty family from $\mathfrak{J}(X)$, then $\bigvee_{\iota \in I}^{\mathfrak{J}(X)} A_\iota = \{x \in X \mid \bigwedge_{\lambda \in L} x_\lambda \leq x$ for some finite family $(x_\lambda)_{\lambda \in L}$ from $\bigcup_{\iota \in I} A_\iota\}$.

(α4) There is a smallest $\wedge$-ideal of X that contains A. Denote it by $[A]$. It is called the $\wedge$-ideal **generated** by A. We have $[\emptyset] = \{1\}$ and, if $A \neq \emptyset$, $[A] = \{x \in X \mid \bigwedge_{\lambda \in L} x_\lambda \leq x$ for some finite family $(x_\lambda)_{\lambda \in L}$ from $A\}$.

Let X be a set. The $\wedge$-ideals of $\mathfrak{P}(X)$ are called **filters** on X. $\mathfrak{P}(X)$ itself is called the **null filter**. $\mathfrak{F} \subset \mathfrak{P}(X)$ is called a **filter base** iff, for any A, $B \in \mathfrak{F}$, there is $C \in \mathfrak{F}$ with $C \subset A \cap B$, that is, iff $\mathfrak{F}$ is directed downward in $\mathfrak{P}(X)$. Prove assertion (β).

(β) For every nonempty filterbase $\mathfrak{F}$, $\{B \subset X \mid B \supset A$ for some $A \in \mathfrak{F}\}$ is the filter on X generated by $\mathfrak{F}$.

A filter $\mathfrak{F}$ on X is called an **ultrafilter** on X iff $\mathfrak{G} = \mathfrak{P}(X)$ whenever $\mathfrak{G}$ is a filter on X such that $\mathfrak{G} \neq \mathfrak{F}$ and $\mathfrak{F} \subset \mathfrak{G}$.

(γ) The following statements are equivalent for a filter $\mathfrak{F}$ on X.

(γ1) $\mathfrak{F}$ is an ultrafilter.
(γ2) Given $A \subset X$, either $A \in \mathfrak{F}$ or $X \setminus A \in \mathfrak{F}$.
(γ3) Given $A \subset X$ with $A \cap F \neq \emptyset$ for each $F \in \mathfrak{F}$, then $A \in \mathfrak{F}$.

It is easy to show that ultrafilters exist. If $x \in X$, then $\{F \subset X \mid x \in F\}$ is an ultrafilter on X. The ultrafilters of this type are called **trivial**. An ultrafilter is called **free** iff it is not trivial. The existence of free ultrafilters requires the Zorn Lemma (see (β) of Ex. 11(d)). Use this to prove the following:

(δ) For every filterbase $\mathfrak{F} \subset \mathfrak{P}(X)$, there is an ultrafilter $\mathfrak{G}$ on X with $\mathfrak{F} \subset \mathfrak{G}$ iff $\emptyset \notin \mathfrak{F}$.

(ε) For every filter $\mathfrak{F}$ on X, there is an ultrafilter $\mathfrak{G}$ on X with $\mathfrak{F} \subset \mathfrak{G}$ iff $\mathfrak{F} \neq \mathfrak{P}(X)$.

An ultrafilter $\mathfrak{F}$ is called **δ-stable** iff $\bigcap_{G \in \mathfrak{G}} G \in \mathfrak{F}$ for every nonempty countable subset $\mathfrak{G}$ of $\mathfrak{F}$. The existence of δ-stable ultrafilters that are not trivial cannot be proved using the standard axioms of set theory. Of course every trivial ultrafilter is δ-stable. But it is necessary to adjoin an independent axiom that guarantees the existence of free δ-stable ultrafilters. Of course it must be shown that such an additional axiom would be consistent with set theory. This problem is still open. We return to this question in Ex. 2.2.19.

1.2. REAL AND EXTENDED-REAL FUNCTIONS, RIESZ LATTICES, AND EXCEPTIONAL SETS AND EXCEPTIONAL FUNCTIONS

NOTATION FOR SECTION 1.2:

X denotes a set.

This section begins with the definitions and elementary properties of real and extended-real functions. The underlying theme is that algebraic operations and the order relation on such functions are defined pointwise, so the various rules of computation for $\mathbb{R}$ and $\overline{\mathbb{R}}$ yield analogous rules for computing with real and extended-real functions.

After summarizing the elementary properties of real and extended-real functions, we define and briefly discuss Riesz lattices. The function spaces in this book are almost always Riesz lattices.

The most important topic of the section then appears. Associated with every Riesz lattice is a collection of exceptional sets and exceptional functions and a corresponding notion of "almost everywhere" (a.e.). The properties of these exceptional sets and functions, and the results they imply about a.e., are thoroughly investigated, since they will be used again and again throughout the book.

Definition 1.2.1. *An* ***extended-real-valued function*** *on X, or simply an* ***extended-real function*** *on X, is a mapping $f\colon X \to \overline{\mathbb{R}}$. A* ***real-valued function*** *or a* **real function** *on X is a mapping $f\colon X \to \mathbb{R}$. Thus $\overline{\mathbb{R}}^X$ and $\mathbb{R}^X$ denote, respectively, the set of all extended-real functions on X and the set of all real functions on X. Algebraic operations and an order relation on $\overline{\mathbb{R}}^X$ are defined as follows. Let $f, g \in \overline{\mathbb{R}}^X$, and let $\alpha \in \overline{\mathbb{R}}$.*

(*a*) *The sum $f + g$ is defined iff $f(x) + g(x)$ is defined for every x in X, and in that case $f + g\colon X \to \overline{\mathbb{R}}$, $x \mapsto f(x) + g(x)$.*

(*b*) *$fg\colon X \to \overline{\mathbb{R}}$, $x \mapsto f(x)g(x)$.*

(*c*) *$\alpha f\colon X \to \overline{\mathbb{R}}$, $x \mapsto \alpha f(x)$. We set $-f := (-1)f$.*

(*d*) *$f \le g$: $\Leftrightarrow f(x) \le g(x)$ for every x in X.* □

To repeat, operating with functions is just operating pointwise with extended-real numbers. With few exceptions, the rules and properties of $\overline{\mathbb{R}}$ carry over to $\overline{\mathbb{R}}^X$. We do not write out the proofs as they merely invoke the corresponding rules for $\overline{\mathbb{R}}$. Before we can write out the rules, we need a few additional definitions, and we also want to adopt the $\infty - \infty$ convention.

Convention 1.2.2 (***The*** $\infty - \infty$ ***convention for*** $\overline{\mathbb{R}}^X$). *If P is an assertion about extended-real-valued functions, then "P holds" is understood to mean:*

> "*P is true provided that every sum appearing in P, whether of extended-real numbers or of extended-real-valued functions, is defined.*" □

Definition 1.2.3. *For every finite family* $(f_\iota)_{\iota \in I}$ *from* $\mathbb{R}^X$, *define the sum*

$$\sum_{\iota \in I} f_\iota: X \to \mathbb{R}, \qquad x \mapsto \sum_{\iota \in I} f_\iota(x). \qquad \square$$

Although we could define the sum of finitely many *extended-real* functions provided that $\infty - \infty$ does not appear, such generality introduces additional complications and is not needed for doing integration theory. In fact a fully general definition is more easily obtained as a consequence of integration theory.

Definition 1.2.4. *For every set* $A \subset X$, *define*

$$e_A^X: X \to \mathbb{R}, \qquad x \mapsto \begin{cases} 1 & \text{if } x \in A \\ 0 & \text{if } x \in X \setminus A. \end{cases}$$

The function e_A^X *is called the* ***characteristic function*** *of A relative to X. The functions* αe_X $(\alpha \in \overline{\mathbb{R}})$ *are called* ***constant functions*** *on X. For these functions we often follow the general practice and simply write* α. □

No difficulties can arise from this constant-function notation, since $\alpha e_X f = \alpha f$ for every α in $\overline{\mathbb{R}}$ and every f in $\overline{\mathbb{R}}^X$.

Definition 1.2.5. *For each function f in* $\overline{\mathbb{R}}^X$ *we define*

$$f^+: X \to \overline{\mathbb{R}}, \qquad x \mapsto \begin{cases} f(x) & \text{if } f(x) \geq 0 \\ 0 & \text{if } f(x) \leq 0 \end{cases}$$

$$f^-: X \to \overline{\mathbb{R}}, \qquad x \mapsto \begin{cases} 0 & \text{if } f(x) \geq 0 \\ -f(x) & \text{if } f(x) \leq 0 \end{cases}$$

$$|f|: X \to \overline{\mathbb{R}}, \qquad x \mapsto |f(x)|.$$

The functions f^+, f^-, *and* $|f|$ *are called, respectively, the* ***positive part*** *of f, the* ***negative part*** *of f, and the* ***absolute value*** *of f.* □

Definition 1.2.6. *A function $f \in \overline{\mathbb{R}}^X$ is said to be* ***positive (negative)*** *iff $f \geq 0$ ($f \leq 0$). For every set $\mathscr{F} \subset \overline{\mathbb{R}}^X$, we denote by $\mathscr{F}_+$ the set of all positive functions belonging to $\mathscr{F}$.* □

Finally let us state in advance that we follow the same practice used for numbers and write $f - g$ for $f + (-g)$.

We can now formulate the important facts about $\overline{\mathbb{R}}^X$, bearing the $\infty - \infty$ convention in mind. Theorem 7 gives the order properties and the elementary rules of computation.

Theorem 1.2.7.

(*a*) *$\overline{\mathbb{R}}^X$ is a complete lattice with order relation $\leq$. The constant functions ∞ and $-\infty$ are, respectively, the largest and smallest elements of $\overline{\mathbb{R}}^X$. The supremum (infimum) of a family from $\overline{\mathbb{R}}^X$ is the pointwise supremum (infimum). That is, for every set $\mathscr{F} \subset \overline{\mathbb{R}}^X$ and for every $x \in X$,*

$$\left(\bigvee_{f \in \mathscr{F}} f\right)(x) = \sup_{f \in \mathscr{F}} (f(x)) \quad \text{and} \quad \left(\bigwedge_{f \in \mathscr{F}} f\right)(x) = \inf_{f \in \mathscr{F}} (f(x)).$$

(*b*) *For every $f \in \overline{\mathbb{R}}^X$, $f + 0$ is defined and equals f. Moreover 0 is the only element of $\overline{\mathbb{R}}^X$ with this property.*

(*c*) *For each real function f on X, the function $-f$ is uniquely characterized by the equation $f + (-f) = 0$.*

The following rules hold for all $f, g, h \in \overline{\mathbb{R}}^X$ and for all real numbers α, β.

(*d*) $(f + g) + h = f + (g + h)$.

(*e*) $f + g = g + f$.

(*f*) $1f = f$.

(*g*) $(\alpha\beta)f = \alpha(\beta f)$.

(*h*) $(\alpha + \beta)f = \alpha f + \beta f$.

(*i*) $\alpha(f + g) = \alpha f + \alpha g$.

(*j*) $(fg)h = f(gh)$.

(*k*) $fg = gf$.

(*l*) $\alpha(fg) = f(\alpha g) = (\alpha f)g$.

(*m*) $(f + g)h = fh + gh$.

(*n*) $f \leq g \Rightarrow f + h \leq g + h$.

(*o*) $0 \leq f, 0 \leq g \Rightarrow 0 \leq fg$.

(*p*) $f^+ = f \vee 0$, *and* $f^- = (-f) \vee 0$.

(*q*) *The sum $f^+ - f^-$ is defined, and $f = f^+ - f^-$.*

(*r*) *The sum $f^+ + f^-$ is defined, and $|f| = f^+ + f^- = f \vee (-f)$.*

(*s*) $f + g = (f \vee g) + (f \wedge g)$.

(*t*) $|f - g| = (f \vee g) - (f \wedge g)$.
(*u*) $|f + g| \leq |f| + |g|$.
(*v*) $||f| - |g|| \leq |f - g|$.
(*w*) $f \vee g = \frac{1}{2}(f + g + |f - g|)$.
(*x*) $f \wedge g = \frac{1}{2}(f + g - |f - g|)$.
(*y*) *The sum* $|f| + |g|$ *is defined;* $|f \vee g| \leq |f| \vee |g| \leq |f| + |g|$, *and* $|f \wedge g| \leq |f| \vee |g| \leq |f| + |g|$.
(*z*) $|\alpha f| = |\alpha|\,|f|$ *and* $|\infty f| = |-\infty f| = \infty |f|$.
(*a'*) $|fg| = |f|\,|g|$. □

In parallel with Proposition 1.1.9, we have the rules governing sup and inf for families from $\overline{\mathbb{R}}^X$.

Proposition 1.2.8. *Let* $(f_\iota)_{\iota \in I}$ *and* $(g_\kappa)_{\kappa \in K}$ *be nonempty families from* $\overline{\mathbb{R}}^X$, *let* $f \in \overline{\mathbb{R}}^X$, *and let* α *be a real number. Then the following assertions hold.*

(*a*) $(\bigvee_{\iota \in I} f_\iota) \vee (\bigvee_{\kappa \in K} g_\kappa) = \bigvee_{(\iota,\kappa) \in I \times K}(f_\iota \vee g_\kappa)$.
(*b*) $(\bigwedge_{\iota \in I} f_\iota) \wedge (\bigwedge_{\kappa \in K} g_\kappa) = \bigwedge_{(\iota,\kappa) \in I \times K}(f_\iota \wedge g_\kappa)$.
(*c*) $f \wedge (\bigvee_{\iota \in I} f_\iota) = \bigvee_{\iota \in I}(f \wedge f_\iota)$.
(*d*) $f \vee (\bigwedge_{\iota \in I} f_\iota) = \bigwedge_{\iota \in I}(f \vee f_\iota)$.
(*e*) $\bigwedge_{\iota \in I} f_\iota = -\bigvee_{\iota \in I}(-f_\iota)$.
(*f*) $\bigvee_{\iota \in I} f_\iota = -\bigwedge_{\iota \in I}(-f_\iota)$.
(*g*) $f + \bigvee_{\iota \in I} f_\iota = \bigvee_{\iota \in I}(f + f_\iota)$.
(*h*) $f + \bigwedge_{\iota \in I} f_\iota = \bigwedge_{\iota \in I}(f + f_\iota)$.
(*i*) $f(\bigvee_{\iota \in I} f_\iota) = \begin{cases} \bigvee_{\iota \in I}(ff_\iota) & \text{if } f \in \mathbb{R}^X \text{ and } f \geq 0, \\ \bigwedge_{\iota \in I}(ff_\iota) & \text{if } f \in \mathbb{R}^X \text{ and } f \leq 0. \end{cases}$
(*j*) $f(\bigwedge_{\iota \in I} f_\iota) = \begin{cases} \bigwedge_{\iota \in I}(ff_\iota) & \text{if } f \in \mathbb{R}^X \text{ and } f \geq 0, \\ \bigvee_{\iota \in I}(ff_\iota) & \text{if } f \in \mathbb{R}^X \text{ and } f \leq 0. \end{cases}$
(*k*) $\alpha(\bigvee_{\iota \in I} f_\iota) = \begin{cases} \bigvee_{\iota \in I}(\alpha f_\iota) & \text{if } \alpha \geq 0, \\ \bigwedge_{\iota \in I}(\alpha f_\iota) & \text{if } \alpha \leq 0. \end{cases}$
(*l*) $\alpha(\bigwedge_{\iota \in I} f_\iota) = \begin{cases} \bigwedge_{\iota \in I}(\alpha f_\iota) & \text{if } \alpha \geq 0, \\ \bigvee_{\iota \in I}(\alpha f_\iota) & \text{if } \alpha \leq 0. \end{cases}$
(*m*) *If* $(I_\rho)_{\rho \in P}$ *is a family of sets with*

$$\bigcup_{\rho \in P} I_\rho = I$$

then

$$\bigvee_{\iota \in I} f_\iota = \bigvee_{\rho \in P}\left(\bigvee_{\iota \in I_\rho} f_\iota\right)$$

and

$$\bigwedge_{\iota \in I} f_\iota = \bigwedge_{\rho \in P}\left(\bigwedge_{\iota \in I_\rho} f_\iota\right).$$

Also the following assertions hold, for all families $(f_\iota)_{\iota \in I}$ *and* $(g_\iota)_{\iota \in I}$ *from* $\overline{\mathbb{R}}^X$.

(*n*) $\bigvee_{\iota \in I}(f_\iota \vee g_\iota) = (\bigvee_{\iota \in I} f_\iota) \vee (\bigvee_{\iota \in I} g_\iota)$.

(*o*) $\bigwedge_{\iota \in I}(f_\iota \wedge g_\iota) = (\bigwedge_{\iota \in I} f_\iota) \wedge (\bigwedge_{\iota \in I} g_\iota)$. □

In parallel with Proposition 1.1.10, there are additional rules governing sup and inf for monotone sequences.

Proposition 1.2.9. *If* $(f_n)_{n \in \mathbb{N}}$ *and* $(g_n)_{n \in \mathbb{N}}$ *are increasing sequences from* $\overline{\mathbb{R}}^X$, *then*:

(*a*) $\bigvee_{n \in \mathbb{N}}(f_n + g_n) = (\bigvee_{n \in \mathbb{N}} f_n) + (\bigvee_{n \in \mathbb{N}} g_n)$,

(*b*) $\bigvee_{n \in \mathbb{N}}(f_n \wedge g_n) = (\bigvee_{n \in \mathbb{N}} f_n) \wedge (\bigvee_{n \in \mathbb{N}} g_n)$.

Similarly, if $(f_n)_{n \in \mathbb{N}}$ *and* $(g_n)_{n \in \mathbb{N}}$ *are decreasing sequences from* $\overline{\mathbb{R}}^X$, *then*:

(*c*) $\bigwedge_{n \in \mathbb{N}}(f_n + g_n) = (\bigwedge_{n \in \mathbb{N}} f_n) + (\bigwedge_{n \in \mathbb{N}} g_n)$,

(*d*) $\bigwedge_{n \in \mathbb{N}}(f_n \vee g_n) = (\bigwedge_{n \in \mathbb{N}} f_n) \vee (\bigwedge_{n \in \mathbb{N}} g_n)$. □

$\overline{\mathbb{R}}^X$ is an ordered set, so Definition 1.1.13 defines order convergence for sequences from $\overline{\mathbb{R}}^X$. Of course the general rules formulated in Propositions 1.1.14, 1.1.16, and 1.1.17 hold. As a complete lattice, $\overline{\mathbb{R}}^X$ is also σ-complete, and every sequence from $\overline{\mathbb{R}}^X$ has a limes superior and a limes inferior. Since operations in $\overline{\mathbb{R}}^X$ are defined pointwise, order convergence in $\overline{\mathbb{R}}^X$ is just pointwise convergence.

Proposition 1.2.10. *If* $(f_n)_{n \in \mathbb{N}}$ *and* $(g_n)_{n \in \mathbb{N}}$ *are sequences from* $\overline{\mathbb{R}}^X$, *then the following assertions hold.*

(*a*) *For every x in X*,

$$\left(\limsup_{n \to \infty} f_n\right)(x) = \limsup_{n \to \infty} f_n(x),$$

$$\left(\liminf_{n \to \infty} f_n\right)(x) = \liminf_{n \to \infty} f_n(x).$$

(*b*) *The sequence* $(f_n)_{n \in \mathbb{N}}$ *order-converges to* $f \in \overline{\mathbb{R}}^X$ *iff, for every x in X, the sequence* $(f_n(x))_{n \in \mathbb{N}}$ *converges to* $f(x)$. *In this case*

$$\left(\lim_{n \to \infty} f_n\right)(x) = \lim_{n \to \infty} f_n(x) = f(x)$$

for every $x \in X$.

Let $(f_n)_{n\in\mathbb{N}}$ and $(g_n)_{n\in\mathbb{N}}$ be order-convergent sequences from $\overline{\mathbb{R}}^X$, and let α be a real number. Then

(c) $(f_n \vee g_n)_{n\in\mathbb{N}}$ *is order-convergent, and*

$$\lim_{n\to\infty} (f_n \vee g_n) = \Big(\lim_{n\to\infty} f_n\Big) \vee \Big(\lim_{n\to\infty} g_n\Big)$$

(d) $(f_n \wedge g_n)_{n\in\mathbb{N}}$ *is order-convergent, and*

$$\lim_{n\to\infty} (f_n \wedge g_n) = \Big(\lim_{n\to\infty} f_n\Big) \wedge \Big(\lim_{n\to\infty} g_n\Big)$$

(e) $(\alpha f_n)_{n\in\mathbb{N}}$ *is order-convergent, and*

$$\lim_{n\to\infty} (\alpha f_n) = \alpha \lim_{n\to\infty} f_n$$

(f) $(f_n + g_n)_{n\in\mathbb{N}}$ *is order-convergent, and*

$$\lim_{n\to\infty} (f_n + g_n) = \Big(\lim_{n\to\infty} f_n\Big) + \Big(\lim_{n\to\infty} g_n\Big)$$

(*Remember Convention* 1.2.2.). □

Finally, we introduce the notion of uniform convergence, which plays an important role in several investigations.

Definition 1.2.11. *Let A be a subset of X, $(f_n)_{n\in\mathbb{N}}$ a sequence from $\overline{\mathbb{R}}^X$, and f a function belonging to $\overline{\mathbb{R}}^X$. The sequence $(f_n)_{n\in\mathbb{N}}$ is said to* ***converge uniformly*** *on the set A to the function f iff the following condition holds: for every $\varepsilon > 0$ there exists m in $\mathbb{N}$ such that for every $n \geq m$ and for every x in A both $f_n(x)$ and $f(x)$ are real and*

$$|f_n(x) - f(x)| < \varepsilon.$$ □

Evidently, if $(f_n)_{n\in\mathbb{N}}$ converges uniformly on A to f, then

$$\lim_{n\to\infty} f_n(x) = f(x)$$

for every x in A.

Having summarized the properties of real and extended-real functions, we turn to the definition of Riesz lattice. Essentially, the spaces we want to describe are sublattices of $\overline{\mathbb{R}}^X$ that are closed under addition and real scalar multiplication. However, the fact that addition of two functions on X is only defined when their sum is defined pointwise at every point of X creates a minor nuisance. To manage this nuisance, we introduce some useful notation.

Definition 1.2.12. *Let f, $g \in \overline{\mathbb{R}}^X$, and let $\mathscr{F}, \mathscr{G} \subset \overline{\mathbb{R}}^X$. Then*

$$\langle f \dotplus g \rangle := \left\{ h \in \overline{\mathbb{R}}^X \,\middle|\, \begin{array}{l} \text{If } x \in X \text{ and } f(x) + g(x) \text{ is defined,} \\ \text{then } h(x) = f(x) + g(x). \end{array} \right\}$$

$$\langle f \dot{-} g \rangle := \langle f \dotplus (-g) \rangle$$

$$|f \dotplus g| := \{ |h| \mid h \in \langle f \dotplus g \rangle \}$$

$$|f \dot{-} g| := \{ |h| \mid h \in \langle f \dot{-} g \rangle \}$$

$$\mathscr{F} \dotplus \mathscr{G} := \bigcup_{f \in \mathscr{F}, g \in \mathscr{G}} \langle f \dotplus g \rangle$$

$$\mathscr{F} \dot{-} \mathscr{G} := \bigcup_{f \in \mathscr{F}, g \in \mathscr{G}} \langle f \dot{-} g \rangle$$

□

Proposition 1.2.13. *Let f, g, and h be extended-real functions on X. Then*

(*a*) $\langle f \dotplus g \rangle = \langle g \dotplus f \rangle$

(*b*) $h \in \langle f \dotplus g \rangle$ *iff* $g \in \langle h \dot{-} f \rangle$.

Proof. The truth of (a) is evident. To prove (b), suppose that $h \in \langle f \dotplus g \rangle$ and fix x in X. We distinguish three cases.

CASE 1: If $f(x)$ is real, then both $f(x) + g(x)$ and $h(x) - f(x)$ are defined, $h(x) = f(x) + g(x)$, and $g(x) = h(x) - f(x)$.

CASE 2: Suppose that $f(x) = \infty$. If $h(x) = \infty$ also, then $h(x) - f(x)$ is undefined. If, on the other hand, $h(x) < \infty$, then because $h \in \langle f \dotplus g \rangle$ it must be that $f(x) + g(x)$ is not defined. In other words, $g(x) = -\infty$, and we have $g(x) = h(x) - f(x)$.

CASE 3: Suppose finally that $f(x) = -\infty$. If $h(x) = -\infty$, then $h(x) - f(x)$ is not defined. If $h(x) > -\infty$, then $h \in \langle f \dotplus g \rangle$ implies that $f(x) + g(x)$ is undefined, $g(x) = \infty$, and $g(x) = h(x) - f(x)$.

Thus in every case either $h(x) - f(x)$ is undefined or else $g(x) = h(x) - f(x)$. Since x was arbitrary, it follows that $g \in \langle h \dot{-} f \rangle$.

The fact that $g \in \langle h \dot{-} f \rangle$ implies $h \in \langle f \dotplus g \rangle$ can be proved by a similar argument. □

Definition 1.2.14. *A nonempty set $\mathscr{L} \subset \overline{\mathbb{R}}^X$ is called a* **Riesz lattice** *(in $\overline{\mathbb{R}}^X$ or on X) iff it satisfies the following three conditions.*

(*RL*1) *If f, $g \in \mathscr{L}$, then $\langle f \dotplus g \rangle \subset \mathscr{L}$.*

(*RL*2) *If $f \in \mathscr{L}$ and $\alpha \in \mathbb{R}$, then $\alpha f \in \mathscr{L}$.*

(*RL*3) *If f, $g \in \mathscr{L}$, then $f \wedge g \in \mathscr{L}$ and $f \vee g \in \mathscr{L}$.*

A Riesz lattice contained in $\mathbb{R}^X$ is called a ***real Riesz lattice.*** □

To forestall possible confusion, we want to mention that a Riesz lattice is not the same as a vector lattice or (as vector lattices are often called) a Riesz space.

$\overline{\mathbb{R}}^X$ is trivially a Riesz lattice. $\mathbb{R}^X$ is an obvious example of a Riesz lattice that is real. The space of continuous, real-valued functions on an interval of $\overline{\mathbb{R}}$ is a Riesz lattice, as is the set of all piecewise constant real functions on $\mathbb{R}$ or on an interval of $\mathbb{R}$. Additional examples and counterexamples can be found in the exercises.

Why do we use this particular definition for the function classes? Any construction of an integral starts with some convenient class of functions whose integrals are known. The linearity properties of integrals only make sense if this class is closed under addition and scalar multiplication. The class of functions whose integrals have been defined is then expanded, possibly more than once. In this book one of the expansions is achieved via monotone approximation, and axiom $(RL3)$ is used in the monotone-approximation process.

Proposition 1.2.15. *The following assertions hold, for every Riesz lattice $\mathscr{L}$ in $\overline{\mathbb{R}}^X$.*

(*a*) *The zero function, 0, belongs to $\mathscr{L}$.*

(*b*) *For every f in $\mathscr{L}$, $-f$ belongs to $\mathscr{L}$.*

(*c*) *For every f in $\mathscr{L}$, $|f|$, f^+ and f^- belong to $\mathscr{L}_+$.*

(*d*) *If f and g belong to $\mathscr{L}$, then $\langle f \dot{-} g \rangle \subset \mathscr{L}$.*

(*e*) *For every finite family $(f_\iota)_{\iota \in I}$ from $\mathscr{L} \cap \mathbb{R}^X$, the sum $\sum_{\iota \in I} f_\iota$ belongs to $\mathscr{L}$.*

(*f*) *For every nonempty finite family $(f_\iota)_{\iota \in I}$ from $\mathscr{L}$, both $\bigvee_{\iota \in I} f_\iota$ and $\bigwedge_{\iota \in I} f_\iota$ belong to $\mathscr{L}$.*

Proof. The first two assertions follow from $(RL2)$ (remember that $\mathscr{L}$ is nonempty), (c) follows from (a), (b), Theorem 7(p), (r), and $(RL3)$, and (d) follows from (b) and $(RL1)$. The two remaining assertions can be proved by complete induction. □

Proposition 1.2.16. *For every Riesz iattice $\mathscr{L}$ in $\overline{\mathbb{R}}^X$, the set $\mathscr{L} \cap \mathbb{R}^X$ is a real Riesz lattice. A set $\mathscr{L} \subset \mathbb{R}^X$ is a Riesz lattice iff $f + g$, αf, and $|f|$ belong to $\mathscr{L}$ for all $f, g \in \mathscr{L}$ and for every real number α.*

Proof. The first statement is trivially true. To prove the second write $f \wedge g$ and $f \vee g$ in terms of f, g, and $|f - g|$ [use Theorem 7(w), (x)]. □

Many Riesz lattices have special order properties relative to the full space $\overline{\mathbb{R}}^X$. We want to discuss these properties briefly.

Definition 1.2.17. *A set* $\mathscr{F} \subset \overline{\mathbb{R}}^X$ *is said to be:*

(*a*) ***conditionally σ-completely embedded*** *in* $\overline{\mathbb{R}}^X$ *iff, for every nonempty countable family* $(f_\iota)_{\iota \in I}$ *from* $\mathscr{F}$ *that is bounded in* $\mathscr{F}$, *both* $\bigvee_{\iota \in I} f_\iota$ *and* $\bigwedge_{\iota \in I} f_\iota$ *belong to* $\mathscr{F}$;

(*b*) ***σ-completely embedded*** *in* $\overline{\mathbb{R}}^X$ *iff, for every nonempty countable family* $(f_\iota)_{\iota \in I}$ *from* $\mathscr{F}$, *both* $\bigvee_{\iota \in I} f_\iota$ *and* $\bigwedge_{\iota \in I} f_\iota$ *belong to* $\mathscr{F}$. □

Every (conditionally) σ-completely embedded lattice in $\overline{\mathbb{R}}^X$ is itself a (conditionally) σ-complete lattice, with order relation induced by the order relation on $\overline{\mathbb{R}}^X$, but the converse is not true. Obviously every set that is σ-completely embedded in $\overline{\mathbb{R}}^X$ is also conditionally σ-completely embedded.

When the set in question is a Riesz lattice, the test for σ-complete embedding or conditionally σ-complete embedding is easier to perform. One need not test all countable families. It is enough, for instance, to test positive increasing sequences. More precisely, we have the following two propositions.

Proposition 1.2.18. *The following conditions on a Riesz lattice* $\mathscr{L}$ *in* $\overline{\mathbb{R}}^X$ *are equivalent.*

(*a*) $\mathscr{L}$ *is conditionally σ-completely embedded in* $\overline{\mathbb{R}}^X$.

(*b*) *For every sequence* $(f_n)_{n \in \mathbb{N}}$ *from* $\mathscr{L}$ *that is bounded in* $\mathscr{L}$, *both* $\bigvee_{n \in \mathbb{N}} f_n$ *and* $\bigwedge_{n \in \mathbb{N}} f_n$ *belong to* $\mathscr{L}$.

(*c*) *For every sequence* $(f_n)_{n \in \mathbb{N}}$ *from* $\mathscr{L}$ *that decreases and is bounded below in* $\mathscr{L}$, *the function* $\bigwedge_{n \in \mathbb{N}} f_n$ *belongs to* $\mathscr{L}$.

(*d*) *For every sequence* $(f_n)_{n \in \mathbb{N}}$ *from* $\mathscr{L}_+$ *that increases and is bounded above in* $\mathscr{L}$, *the function* $\bigvee_{n \in \mathbb{N}} f_n$ *belongs to* $\mathscr{L}$.

Proof. (a) ⇔ (b) and (b) ⇒ (c) are trivial (and hold for arbitrary subsets $\mathscr{L}$ of $\overline{\mathbb{R}}^X$).

(c) ⇒ (d). If $(f_n)_{n \in \mathbb{N}}$ is a sequence from $\mathscr{L}_+$ that increases and is bounded above in $\mathscr{L}$, then $(-f_n)_{n \in \mathbb{N}}$ is a sequence from $\mathscr{L}$ that decreases and is bounded below in $\mathscr{L}$. Since

$$\bigwedge_{n \in \mathbb{N}} (-f_n) = -\bigvee_{n \in \mathbb{N}} f_n$$

by Proposition 8(e), it follows that $\bigvee_{n \in \mathbb{N}} f_n$ belongs to $\mathscr{L}$.

(d) ⇒ (b). Let $(f_n)_{n \in \mathbb{N}}$ be a sequence from $\mathscr{L}$ bounded in $\mathscr{L}$. We want to construct an increasing sequence of positive functions from $\mathscr{L}$ that is bounded in $\mathscr{L}$ and whose supremum coincides with the supremum of the given sequence or possibly differs from it by a function in $\mathscr{L}$. The sequence

$$\left(\bigvee_{m \leq n} f_m\right)_{n \in \mathbb{N}}$$

increases, but a correction is needed to get a positive sequence. For each n in $\mathbb{N}$, let

$$h_n\colon X \to \overline{\mathbb{R}},$$

$$x \mapsto \begin{cases} \left(\sup_{m \le n} f_m(x)\right) - f_1(x) & \text{if defined} \\ \infty & \text{otherwise.} \end{cases}$$

The sequence $(h_n)_{n \in \mathbb{N}}$ increases, and each h_n is positive. Since

$$h_n \in \left\langle \bigvee_{m \le n} f_m \dot{-} f_1 \right\rangle$$

we see that every h_n is in $\mathscr{L}$ and therefore in $\mathscr{L}_+$. To verify that the sequence $(h_n)_{n \in \mathbb{N}}$ is bounded above in $\mathscr{L}$, note that if g' and g'' are elements of $\mathscr{L}$ that bound the original sequence $(f_n)_{n \in \mathbb{N}}$ from above and below, respectively, then the function $|g'| + |g''|$ is defined, belongs to $\mathscr{L}$, and bounds $(h_n)_{n \in \mathbb{N}}$ from above. Hypothesis (d) implies that $\bigvee_{n \in \mathbb{N}} h_n$ belongs to $\mathscr{L}$. Now

$$\bigvee_{n \in \mathbb{N}} h_n \in \left\langle \bigvee_{n \in \mathbb{N}} f_n \dot{-} f_1 \right\rangle$$

so

$$\bigvee_{n \in \mathbb{N}} f_n \in \left\langle \bigvee_{n \in \mathbb{N}} h_n \dot{+} f_1 \right\rangle$$

by Proposition 13(b), and we conclude that $\bigvee_{n \in \mathbb{N}} f_n$ must belong to $\mathscr{L}$.

To conclude that $\bigwedge_{n \in \mathbb{N}} f_n$ belongs to $\mathscr{L}$, we argue as follows. Since $(f_n)_{n \in \mathbb{N}}$ is bounded in $\mathscr{L}$, so is $(-f_n)_{n \in \mathbb{N}}$. The argument just completed shows that $\bigvee_{n \in \mathbb{N}}(-f_n)$ belongs to $\mathscr{L}$. Now recall that $\bigwedge_{n \in \mathbb{N}} f_n = -\bigvee_{n \in \mathbb{N}}(-f_n)$. □

Proposition 1.2.19. *The following conditions on a Riesz lattice $\mathscr{L}$ in $\overline{\mathbb{R}}^X$ are equivalent.*

- (a) *$\mathscr{L}$ is σ-completely embedded in $\overline{\mathbb{R}}^X$.*
- (b) *For every sequence $(f_n)_{n \in \mathbb{N}}$ from $\mathscr{L}$, both $\bigvee_{n \in \mathbb{N}} f_n$ and $\bigwedge_{n \in \mathbb{N}} f_n$ belong to $\mathscr{L}$.*
- (c) *For every decreasing sequence $(f_n)_{n \in \mathbb{N}}$ from $\mathscr{L}$, the function $\bigwedge_{n \in \mathbb{N}} f_n$ belongs to $\mathscr{L}$.*
- (d) *For every increasing sequence $(f_n)_{n \in \mathbb{N}}$ from $\mathscr{L}_+$, the function $\bigvee_{n \in \mathbb{N}} f_n$ belongs to $\mathscr{L}$.*

Proof. (a) ⇔ (b) and (b) ⇒ (c) are trivial.

(c) ⇒ (d). For every increasing sequence $(f_n)_{n \in \mathbb{N}}$ from $\mathscr{L}_+$, the sequence $(-f_n)_{n \in \mathbb{N}}$ is a decreasing sequence from $\mathscr{L}$ and $\bigvee_{n \in \mathbb{N}} f_n = -\bigwedge_{n \in \mathbb{N}}(-f_n)$.

(d) ⇒ (b). From (d) and the preceding proposition it follows that $\mathscr{L}$ is conditionally σ-completely embedded in $\overline{\mathbb{R}}^X$. It therefore suffices to show that every sequence from $\mathscr{L}$ is bounded in $\mathscr{L}$. Let $(f_n)_{n \in \mathbb{N}}$ be a sequence from $\mathscr{L}$. Then

$$\left(\bigvee_{m \leq n} |f_m| \right)_{n \in \mathbb{N}}$$

is an increasing sequence from $\mathscr{L}_+$, and (d) ensures that its supremum, $\bigvee_{n \in \mathbb{N}} |f_n|$, belongs to $\mathscr{L}$. Since

$$-\bigvee_{n \in \mathbb{N}} |f_n| \leq f_m \leq \bigvee_{n \in \mathbb{N}} |f_n|$$

for every m in $\mathbb{N}$ the proof is complete. □

It is useful to single out those subsets of X on which some element of the Riesz lattice takes infinite values. Indeed, it is imperative that we view these sets as somehow exceptional. Before giving a formal definition, we introduce some notation, notation that is also useful in other connections and will be used extensively.

Definition 1.2.20. *Let $\diamondsuit$ be one of the relations $=, \neq, <, \leq, >, \geq$, and let f and g be arbitrary functions in $\overline{\mathbb{R}}^X$. Then*

$$\{f \diamondsuit g\} := \{x \in X \mid f(x) \diamondsuit g(x)\}.$$ □

Definition 1.2.21. *Let $\mathscr{L}$ be a Riesz lattice in $\overline{\mathbb{R}}^X$. A set $A \subset X$ is called* ***$\mathscr{L}$-exceptional*** *iff there is a function f belonging to $\mathscr{L}$ such that*

$$f(x) = \infty \quad \text{for every } x \text{ in } A.$$

A function $f \in \overline{\mathbb{R}}^X$ is called ***$\mathscr{L}$-exceptional*** *iff $\{f \neq 0\}$ is an $\mathscr{L}$-exceptional set.*

$\mathfrak{N}(\mathscr{L})$ shall denote the collection of all $\mathscr{L}$-exceptional subsets of X, and $\mathscr{N}(\mathscr{L})$ the collection of all $\mathscr{L}$-exceptional functions on X. □

Thus

$$\mathfrak{N}(\mathscr{L}) = \{A \subset X \mid A \subset \{g = \infty\} \text{ for some } g \in \mathscr{L}\}$$

and

$$\mathscr{N}(\mathscr{L}) = \{f \in \overline{\mathbb{R}}^X \mid \{f \neq 0\} \subset \{g = \infty\} \text{ for some } g \in \mathscr{L}\}.$$

If f belongs to $\mathscr{L}$, then each of the sets $\{f = \infty\}$, $\{f = -\infty\}$, and $\{|f| = \infty\}$ is $\mathscr{L}$-exceptional.

There are many ways of characterizing $\mathscr{L}$-exceptional sets and functions, several of which are presented in the next two propositions.

Proposition 1.2.22. *For every Riesz lattice $\mathscr{L}$ in $\overline{\mathbb{R}}^X$, the following conditions on a subset A of X are equivalent.*

(*a*) $A \in \mathfrak{N}(\mathscr{L})$.
(*b*) $\infty e_A \in \mathscr{L}$.
(*c*) $A \subset B$ *for some* $B \in \mathfrak{N}(\mathscr{L})$.
(*d*) $A \subset \{f \neq 0\}$ *for some* $f \in \mathscr{N}(\mathscr{L})$.

Proof. (a) ⇒ (b). By hypothesis, there exists f in $\mathscr{L}$ such that $f(x) = \infty$ for every x in A. For this f,

$$\infty e_A \in \langle f \dot{-} f \rangle$$

so ∞e_A must belong to $\mathscr{L}$.

(b) ⇒ (a). This implication is obvious.

(a) ⇒ (c). Let $B := A$.

(c) ⇒ (d). Let $f := e_B$.

(d) ⇒ (a). By hypothesis, $A \subset \{f \neq 0\} \subset \{g = \infty\}$ for some f in $\overline{\mathbb{R}}^X$ and some g in $\mathscr{L}$. Thus A belongs to $\mathfrak{N}(\mathscr{L})$. □

Proposition 1.2.23. *For every Riesz lattice $\mathscr{L}$ in $\overline{\mathbb{R}}^X$, the following conditions on a function f in $\overline{\mathbb{R}}^X$ are equivalent.*

(*a*) $f \in \mathscr{N}(\mathscr{L})$.
(*b*) $|f| \in \mathscr{N}(\mathscr{L})$.
(*c*) $\infty f \in \mathscr{L}$.
(*d*) $\{f \neq 0\} \subset \{g \neq 0\}$ *for some* $g \in \mathscr{N}(\mathscr{L})$.
(*e*) $\{f \neq 0\} \subset A$ *for some* $A \in \mathfrak{N}(\mathscr{L})$.

Proof. (a) ⇒ (b). After all, $\{|f| \neq 0\} = \{f \neq 0\}$.

(b) ⇒ (c). Write $A := \{|f| \neq 0\}$. By the previous proposition the function ∞e_A belongs to $\mathscr{L}$. Since

$$\infty f \in \langle \infty e_A \dot{-} \infty e_A \rangle$$

∞f must also belong to $\mathscr{L}$.

(c) ⇒ (a). If $\infty f \in \mathscr{L}$, then $|\infty f| \in \mathscr{L}$ and so $f \in \mathscr{N}(\mathscr{L})$.

(a) ⇒ (d). Let $g := f$.

(d) ⇒ (e). Let $A := \{g \neq 0\}$ and use Proposition 22(d) ⇒ (a).

(e) ⇒ (a). By hypothesis, $\{f \neq 0\} \subset A \subset \{g = \infty\}$ for some g in $\mathscr{L}$. Thus $\{f \neq 0\}$ is in $\mathfrak{N}(\mathscr{L})$; that is, f belongs to $\mathscr{N}(\mathscr{L})$. □

The characterizations

$$\mathcal{N}(\mathcal{L}) = \{ f \in \overline{\mathbb{R}}^X \mid \infty f \in \mathcal{L} \} \quad \text{and} \quad \mathfrak{N}(\mathcal{L}) = \{ A \subset X \mid \infty e_A \in \mathcal{L} \}$$

are especially useful.

Proposition 1.2.24. *Let $\mathcal{L}$ be a Riesz lattice in $\overline{\mathbb{R}}^X$. Then $\mathfrak{P}(A) \subset \mathfrak{N}(\mathcal{L})$ for each $A \in \mathfrak{N}(\mathcal{L})$, and the union of every finite family from $\mathfrak{N}(\mathcal{L})$ belongs to $\mathfrak{N}(\mathcal{L})$. If $\mathcal{L}$ is real, then $\mathfrak{N}(\mathcal{L}) = \{\varnothing\}$.*

Proof. The first claim merely restates Proposition 22 (c) ⇒ (a). Let $(A_\iota)_{\iota \in I}$ be a finite family from $\mathfrak{N}(\mathcal{L})$. Since

$$\infty e_{\bigcup_{\iota \in I} A_\iota} = \bigvee_{\iota \in I} \infty e_{A_\iota}$$

we conclude that $\infty e_{\bigcup_{\iota \in I} A_\iota}$ belongs to $\mathcal{L}$ and $\bigcup_{\iota \in I} A_\iota$ belongs to $\mathfrak{N}(\mathcal{L})$. The last claim in the proposition is a trivial consequence of the characterization

$$\mathfrak{N}(\mathcal{L}) = \{ A \subset X \mid \infty e_A \in \mathcal{L} \}. \qquad \square$$

Corollary 1.2.25. *If $\mathcal{L}$ is a Riesz lattice in $\overline{\mathbb{R}}^X$, then the intersection of every nonempty family of $\mathcal{L}$-exceptional subsets of X is $\mathcal{L}$-exceptional, as is $A \setminus B$ for all $A, B \in \mathfrak{N}(\mathcal{L})$.* □

Corollary 1.2.26. *If f and g are arbitrary elements of a Riesz lattice $\mathcal{L}$ in $\overline{\mathbb{R}}^X$, then the set*

$$\{ x \in X \mid f(x) + g(x) \text{ is not defined} \}$$

is $\mathcal{L}$-exceptional.

Proof. The set in question is a subset of the set

$$\{ f = \infty \} \cup \{ g = \infty \}. \qquad \square$$

Corollary 1.2.27. *For every Riesz lattice $\mathcal{L}$ in $\overline{\mathbb{R}}^X$, the set $\mathcal{N}(\mathcal{L})$ is also a Riesz lattice in $\overline{\mathbb{R}}^X$. If $\mathcal{L}$ is real, then $\mathcal{N}(\mathcal{L}) = \{0\}$.*

Proof. The zero function is $\mathcal{L}$-exceptional, so $\mathcal{N}(\mathcal{L})$ is not empty. Suppose that f and g are in $\mathcal{N}(\mathcal{L})$, α is a real number, and $h \in \langle f \dotplus g \rangle$. Notice that

$$\{ h \neq 0 \} \subset \{ f \neq 0 \} \cup \{ g \neq 0 \}$$

$$\{ \alpha f \neq 0 \} \subset \{ f \neq 0 \}$$

$$\{ f \vee g \neq 0 \} \subset \{ f \neq 0 \} \cup \{ g \neq 0 \}$$

$$\{ f \wedge g \neq 0 \} \subset \{ f \neq 0 \} \cup \{ g \neq 0 \}.$$

Using the fact that subsets and finite unions of $\mathscr{L}$-exceptional sets are $\mathscr{L}$-exceptional, we conclude that $\mathscr{N}(\mathscr{L})$ is a Riesz lattice. The second assertion in the corollary merely restates the fact that $\mathfrak{N}(\mathscr{L}) = \{\varnothing\}$ when $\mathscr{L}$ is real. □

The significance for integration theory of the exceptional sets and exceptional functions associated with a Riesz lattice is multifold. These objects appear repeatedly, in various perspectives. Whenever they appear, the notion of almost everywhere also appears.

Definition 1.2.28. *Let $\mathscr{L}$ be a Riesz lattice in $\overline{\mathbb{R}}^X$. A property P that refers to elements of X is said to hold* ***$\mathscr{L}$-almost everywhere*** *(or simply* ***$\mathscr{L}$-a.e.****) iff the set*

$$\{x \in X \mid P(x) \text{ is false or } P(x) \text{ is not defined}\}$$

is an $\mathscr{L}$-exceptional set.

If P holds $\mathscr{L}$-almost everywhere, then we write

$$P \ \mathscr{L}\text{-a.e.}$$

provided that $P(x)$ is defined for every x in X, and we write

$$P(x) \ \mathscr{L}\text{-a.e.}$$

in case $P(x)$ is not necessarily defined for every x in X. □

This definition may seem a bit premature, perhaps even ill-chosen. To say that some property holds almost everywhere usually means that it holds except on a set of "measure" zero (equivalently, except on a set whose characteristic function has "integral" zero). We have no measures yet, and we have defined "almost everywhere" in terms of Riesz lattices. Nevertheless, this definition is appropriate. We shall see in the next section (1.3.3) that every positive linear functional on a Riesz lattice $\mathscr{L}$ assigns to every $\mathscr{L}$-exceptional function the value zero. Thus no $\mathscr{L}$-exceptional set can ever have nonzero "measure." Moreover in those situations where $\mathscr{L}$-exceptional sets and "sets of measure zero" fail to coincide, it is the $\mathscr{L}$-exceptional sets, not the others, that have the characteristics appropriate to exceptional sets in a sense of "almost everywhere."

Example 1.2.29. Operations and relations on the set $\overline{\mathbb{R}}^X$ were defined pointwise. Thus properties of functions in $\overline{\mathbb{R}}^X$ are properties that refer to points of X, and it is meaningful to assert that such a property holds $\mathscr{L}$-almost everywhere. Let f and g be extended-real functions on the set X, let $(f_n)_{n \in \mathbb{N}}$

be a sequence from $\overline{\mathbb{R}}^X$, and let $\mathscr{L}$ be a Riesz lattice in $\overline{\mathbb{R}}^X$. Then

$$f = g \ \mathscr{L}\text{-a.e.} \Leftrightarrow \{f \neq g\} \in \mathfrak{N}(\mathscr{L})$$

$$f \leq g \ \mathscr{L}\text{-a.e.} \Leftrightarrow \{f > g\} \in \mathfrak{N}(\mathscr{L})$$

$$f(x) = \lim_{n\to\infty} f_n(x) \ \mathscr{L}\text{-a.e.} \Leftrightarrow$$

$$\left\{ x \in X \,\middle|\, \begin{array}{l} (f_n(x))_{n\in\mathbb{N}} \text{ does not converge} \\ \text{or } \lim_{n\to\infty} f_n(x) \neq f(x) \end{array} \right\}$$

$$\in \mathfrak{N}(\mathscr{L}).$$

Moreover every function in $\mathscr{L}$ is $\mathscr{L}$-a.e. finite, and $f(x) + g(x)$ is defined $\mathscr{L}$-a.e. for every pair f and g of functions belonging to $\mathscr{L}$. Every $\mathscr{L}$-exceptional function equals zero $\mathscr{L}$-a.e., and conversely. If h is a function in $\langle f \dot{+} g \rangle$, where f and g belong to $\mathscr{L}$, then $h(x) = f(x) + g(x)$ $\mathscr{L}$-a.e. □

Proposition 1.2.30. *The following assertions hold, for every Riesz lattice $\mathscr{L}$ in $\overline{\mathbb{R}}^X$.*

(*a*) *Every $\mathscr{L}$-exceptional function belongs to $\mathscr{L}$; that is, $\mathscr{N}(\mathscr{L}) \subset \mathscr{L}$.*
(*b*) *If A is an $\mathscr{L}$-exceptional subset of X, then fe_A belongs to $\mathscr{N}(\mathscr{L})$ for every extended-real function f on X, and $fe_{X\setminus A}$ belongs to $\mathscr{L}$ if f belongs to $\mathscr{L}$.*
(*c*) *Every function that is $\mathscr{L}$-almost everywhere equal to a function belonging to $\mathscr{L}$ must itself belong to $\mathscr{L}$.*
(*d*) *If f and g belong to $\mathscr{L}$, and if*

$$h(x) = f(x) + g(x) \ \mathscr{L}\text{-a.e.}$$

then h belongs to $\mathscr{L}$.

Proof. (a) If f belongs to $\mathscr{N}(\mathscr{L})$, then ∞f belongs to $\mathscr{L}$ by Proposition 23 (a) $\Rightarrow$ (c). Since

$$f \in \langle \infty f \dot{-} \infty f \rangle$$

f must also belong to $\mathscr{L}$.

(b) Let $A \in \mathfrak{N}(\mathscr{L})$, and let $f \in \overline{\mathbb{R}}^X$. Since

$$\{fe_A \neq 0\} \subset A$$

fe_A belongs to $\mathscr{N}(\mathscr{L})$ by Proposition 23. By (a), fe_A belongs to $\mathscr{L}$. The last claim now follows, since

$$fe_{X\setminus A} \in \langle f \dot{-} fe_A \rangle.$$

(c) Suppose that $f = g$ $\mathscr{L}$-a.e., for some $f \in \mathscr{L}$, $g \in \overline{\mathbb{R}}^X$. Set

$$A := \{f \neq g\}.$$

By (a) and (b), the functions $fe_{X\setminus A}$ and ge_A belong to $\mathscr{L}$. Moreover their sum is defined and equals g. Hence g belongs to $\mathscr{L}$.

(*d*) Let

$$A := \{x \in X \mid f(x) + g(x) \text{ is not defined}\}$$

and let

$$h' := fe_{X\setminus A} + ge_{X\setminus A}.$$

Then A is $\mathscr{L}$-exceptional by Corollary 26, so $h' = h$ $\mathscr{L}$-a.e. In view of (*b*), h' belongs to $\mathscr{L}$. By (*c*), h belongs to $\mathscr{L}$. □

Proposition 1.2.31. *If $\mathscr{L}$ is a Riesz lattice in $\overline{\mathbb{R}}^X$, then $\mathscr{L}$-a.e. equality of functions is an equivalence relation on the set $\overline{\mathbb{R}}^X$.*

Proof. We must verify, for all functions f, g, and h in $\overline{\mathbb{R}}^X$, that the following implications hold:

(*a*) If $f = g$, then $f = g$ $\mathscr{L}$-a.e.

(*b*) If $f = g$ $\mathscr{L}$-a.e., then $g = f$ $\mathscr{L}$-a.e.

(*c*) If $f = g$ $\mathscr{L}$-a.e. and $g = h$ $\mathscr{L}$-a.e., then $f = h$ $\mathscr{L}$-a.e.

For (a), if $f = g$, then the set $\{f \neq g\}$ is empty, hence $\mathscr{L}$-exceptional. That (b) holds is trivial, and (c) follows from the inclusion

$$\{f \neq h\} \subset \{f \neq g\} \cup \{g \neq h\}$$

since the union of two $\mathscr{L}$-exceptional sets is $\mathscr{L}$-exceptional, as are subsets of $\mathscr{L}$-exceptional sets. □

This simple observation will have rather far-reaching consequences in the second part of the book.

Proposition 1.2.32. *Let $\mathscr{L}$ be a Riesz lattice in $\overline{\mathbb{R}}^X$. Then the following implications hold, for all functions f, g, and h in $\overline{\mathbb{R}}^X$.*

(*a*) *If $f \leq g$, then $f \leq g$ $\mathscr{L}$-a.e.*

(*b*) *$f = g$ $\mathscr{L}$-a.e. iff $f \leq g$ $\mathscr{L}$-a.e. and $g \leq f$ $\mathscr{L}$-a.e.*

(*c*) *If $f \leq g$ $\mathscr{L}$-a.e. and $g \leq h$ $\mathscr{L}$-a.e., then $f \leq h$ $\mathscr{L}$-a.e.*

In other words, $\leq$ $\mathscr{L}$-a.e. is a preorder on $\overline{\mathbb{R}}^X$.

Proof. To verify (a), simply note that $\{f > g\}$ is empty if $f \le g$. The equality

$$\{f \neq g\} = \{f > g\} \cup \{f < g\}$$

implies (b), and (c) follows from the inclusion

$$\{f > h\} \subset \{f > g\} \cup \{g > h\}.$$

Of course we have used the fact that subsets and finite unions of $\mathscr{L}$-exceptional sets are $\mathscr{L}$-exceptional. □

Proposition 1.2.33. *Let $\mathscr{L}$ be a Riesz lattice in $\overline{\mathbb{R}}^X$. Suppose that f_1, f_2 and g_1, g_2 are functions in $\overline{\mathbb{R}}^X$ such that*

$$f_i = g_i \ \mathscr{L}\text{-a.e.}$$

for $i = 1, 2$, and let h be an extended-real function on X. Then the following assertions hold.

(*a*) $\alpha f_1 = \alpha g_1$ *$\mathscr{L}$-a.e. for every $\alpha \in \overline{\mathbb{R}}$.*
(*b*) $f_1 f_2 = g_1 g_2$ *$\mathscr{L}$-a.e.*
(*c*) *If $h(x) = f_1(x) + f_2(x)$ $\mathscr{L}$-a.e., then*

$$h(x) = g_1(x) + g_2(x) \ \mathscr{L}\text{-a.e.}$$

Now suppose that $(f_\iota)_{\iota \in I}$ and $(g_\iota)_{\iota \in I}$ are finite families from $\overline{\mathbb{R}}^X$ such that

$$f_\iota = g_\iota \ \mathscr{L}\text{-a.e.}$$

for every ι in I. Then
(*d*) $\bigvee_{\iota \in I} f_\iota = \bigvee_{\iota \in I} g_\iota$ *$\mathscr{L}$-a.e.*
(*e*) $\bigwedge_{\iota \in I} f_\iota = \bigwedge_{\iota \in I} g_\iota$ *$\mathscr{L}$-a.e.*

Proof. The key here is once again the closure of $\mathfrak{N}(\mathscr{L})$ under the taking of subsets and finite unions. Thus the proposition is a consequence of the following five inclusions, one inclusion for each assertion in the proposition:
(a) $\{\alpha f_1 \neq \alpha g_1\} \subset \{f_1 \neq g_1\}$
(b) $\{f_1 f_2 \neq g_1 g_2\} \subset \{f_1 \neq g_1\} \cup \{f_2 \neq g_2\}$
(c) $\{x \in X \mid g_1(x) + g_2(x)$ is undefined or $\neq h(x)\}$
$\subset \{x \in X \mid f_1(x) + f_2(x)$ is undefined or $\neq h(x)\} \cup \{f_1 \neq g_1\} \cup \{f_2 \neq g_2\}$
(d) $\{\bigvee_{\iota \in I} f_\iota \neq \bigvee_{\iota \in I} g_\iota\} \subset \bigcup_{\iota \in I} \{f_\iota \neq g_\iota\}$
(e) $\{\bigwedge_{\iota \in I} f_\iota \neq \bigwedge_{\iota \in I} g_\iota\} \subset \bigcup_{\iota \in I} \{f_\iota \neq g_\iota\}$. □

EXERCISES

1.2.1(E) Let X be a set. Show that the following sets are real Riesz lattices:

(α) $\ell^{\infty}(X) := \{f \in \mathbb{R}^X \mid f \text{ is bounded}\}$.
(β) $\mathscr{k}(X) := \{f \in \mathbb{R}^X \mid \{f \neq 0\} \text{ is finite}\}$.
(γ) $c_0(X) := \{f \in \mathbb{R}^X \mid \{|f| > \varepsilon\} \text{ is finite for each } \varepsilon > 0\}$.
(δ) $c(X) := \{f \in \mathbb{R}^X \mid \exists \alpha \in \mathbb{R}, \{|f - \alpha| > \varepsilon\} \text{ is finite for each } \varepsilon > 0\}$.
(ε) $c_f(X) := \{f \in \mathbb{R}^X \mid \exists \alpha \in \mathbb{R}, \{f \neq \alpha\} \text{ is finite}\}$.

Some of these examples will reappear later in a more general setting. The following sets are also Riesz lattices, but not real ones:

(β') $\bar{\mathscr{k}}(X) := \{f \in \overline{\mathbb{R}}^X \mid \{f \neq 0\} \text{ is finite}\}$.
(γ') $\bar{c}_0(X) := \{f \in \overline{\mathbb{R}}^X \mid \{|f| > \varepsilon\} \text{ is finite for each } \varepsilon > 0\}$.

The following inclusions hold:

$$\mathscr{k}(X) \subset c_0(X) \subset c(X) \subset \ell^{\infty}(X)$$

$$\mathscr{k}(X) \subset c_f(X) \subset c(X)$$

$$\mathscr{k}(X) \subset \bar{\mathscr{k}}(X)$$

$$c_0(X) \subset \bar{c}_0(X)$$

1.2.2(E) Let $\mathfrak{J}$ denote the set of all open intervals in $\mathbb{R}$. Let X be an arbitrary set and $\mathfrak{R}$ a set of subsets of X satisfying

(i) $X \in \mathfrak{R}$.
(ii) If $A, B \in \mathfrak{R}$, then $A \cap B \in \mathfrak{R}$.
(iii) If $(A_\iota)_{\iota \in I}$ is a countable family from $\mathfrak{R}$, then $\bigcup_{\iota \in I} A_\iota \in \mathfrak{R}$.

(α) Prove that

$$\mathscr{L} := \{f \in \mathbb{R}^X \mid f^{-1}(I) \in \mathfrak{R} \text{ for each } I \in \mathfrak{J}\} \tag{1}$$

is a real Riesz lattice.

The proof can be executed with the help of the following steps:

STEP 1: For arbitrary $f, g \in \mathbb{R}^X$,

$$\{f > g\} = \bigcup_{\alpha \in \mathbb{Q}} \bigcup_{n \in \mathbb{N}} \left(f^{-1}(]\alpha, \alpha + n[) \cap g^{-1}(]\alpha - n, \alpha[)\right).$$

Thus, if $f, g \in \mathscr{L}$, then $\{f > g\} \in \mathfrak{R}$.

STEP 2: Given $f, g \in \mathbb{R}^X$ and $A \subset \mathbb{R}$,

$$(f \vee g)^{-1}(A) = \left(f^{-1}(A) \cap g^{-1}(A)\right) \cup \left(f^{-1}(A) \cap \{f > g\}\right) \cup \left(g^{-1}(A) \cap \{f < g\}\right)$$

$$(f \wedge g)^{-1}(A) = \left(f^{-1}(A) \cap g^{-1}(A)\right) \cup \left(f^{-1}(A) \cap \{f < g\}\right) \cup \left(g^{-1}(A) \cap \{f > g\}\right).$$

Thus, if $f, g \in \mathscr{L}$ then $f \vee g,\ f \wedge g \in \mathscr{L}$.

STEP 3: Given $f \in \mathbb{R}^X$, $\alpha \in \mathbb{R}$, and $A \subset \mathbb{R}$,

$$(\alpha f)^{-1}(A) = \begin{cases} X & \text{if } \alpha = 0, 0 \in A \\ \varnothing & \text{if } \alpha = 0, 0 \notin A \\ f^{-1}(\{\gamma/\alpha \mid \gamma \in A\}) & \text{if } \alpha \neq 0 \end{cases}$$

Thus $\alpha f \in \mathscr{L}$ whenever $f \in \mathscr{L}$ and $\alpha \in \mathbb{R}$.

STEP 4: Given $f \in \mathbb{R}^X$, $\alpha \in \mathbb{R}$, and $A \subset \mathbb{R}$,

$$(f + \alpha)^{-1}(A) = f^{-1}(\{\beta - \alpha \mid \beta \in A\})$$

Thus $f + \alpha \in \mathscr{L}$ whenever $f \in \mathscr{L}$ and $\alpha \in \mathbb{R}$.

STEP 5: Given $f, g \in \mathbb{R}^X$ and $\alpha, \beta \in \mathbb{R}$ with $\alpha < \beta$,

$$(f + g)^{-1}(]\alpha, \beta[) = \{\alpha - f < g\} \cap \{f < -g + \beta\}$$

Thus $f + g \in \mathscr{L}$ whenever $f, g \in \mathscr{L}$.

Let $(X, \mathfrak{T})$ be a topological space. Then, as is well known, a real function f on X is continuous iff $f^{-1}(I) \in \mathfrak{T}$ for each $I \in \mathfrak{J}$. Hence we have shown that the set of all continuous real functions on X, which we denote by $\mathscr{C}(X, \mathfrak{T})$, is a real Riesz lattice.

Another example of the situation discussed in this exercise will arise in Chapter 5 where we investigate the "measurable functions" which are also defined in the manner of equation (1).

Let $\mathfrak{R}$ be a set of subsets of a set X that satisfies (i) and (iii) and, instead of (ii),

(ii′) If $(A_\iota)_{\iota \in I}$ is a countable nonempty family from $\mathfrak{R}$, then $\bigcap_{\iota \in I} A_\iota \in \mathfrak{R}$.

(β) Prove, under this assumption, that the space $\mathscr{L}$ defined as in equation (1) is conditionally σ-completely embedded in $\overline{\mathbb{R}}^X$.

For the proof, use Proposition 18 (d) ⇒ (a), and observe that, given an increasing sequence $(f_n)_{n \in \mathbb{N}}$ in $\mathbb{R}^X$ and arbitrary $\alpha, \beta \in \mathbb{R}$ with $\alpha < \beta$,

$$\left(\bigvee_{n \in \mathbb{N}} f_n\right)^{-1}(]\alpha, \beta[) = \bigcup_{k \in \mathbb{N}} \bigcup_{m \in \mathbb{N}} \bigcap_{n \geq m} f_n^{-1}(]\alpha, \beta - 1/k[)$$

Each of the Riesz lattices considered in this example contains all of the real constant functions on X. These are special cases of the spaces we will later designate as Stone lattices, which are of particular importance in measure theory (see Chapters 2 and 5).

1.2.3(E) We continue to consider the situation portrayed in Ex. 1.2.2. Let $\mathfrak{R}$ satisfy (i), (ii), and (iii), and let $\mathscr{L}$ be defined by equation (1).

Let $\mathfrak{K}$ be a set of subsets of X with the property that $A \cup B \in \mathfrak{K}$ whenever $A, B \in \mathfrak{K}$. Define

$$\mathscr{L}_{\mathfrak{K}} := \{f \in \mathscr{L} \mid \exists A \in \mathfrak{K} \text{ with } \{f \neq 0\} \subset A\}$$

and prove

(α) $\mathscr{L}_{\mathfrak{K}}$ is a real Riesz lattice.

(β) If $X \in \mathfrak{K}$, then $\mathscr{L}_{\mathfrak{K}} = \mathscr{L}$

(γ) Let $\mathfrak{K}$ have the property that $\bigcup_{n \in \mathbb{N}} A_n \in \mathfrak{K}$ for each sequence $(A_n)_{n \in \mathbb{N}}$ in $\mathfrak{K}$. Let $\mathfrak{R}$ have the property (ii′) from Ex. 1.2.2. Then $\mathscr{L}_{\mathfrak{K}}$ is conditionally σ-completely embedded in $\overline{\mathbb{R}}^X$.

Letting $\mathfrak{R}$ be the natural topology on $\mathbb{R}$, and $\mathfrak{K}$ the set of all closed and bounded subsets of $\mathbb{R}$, we have an important example. In this case $\mathscr{L}_{\mathfrak{K}}$ is the set of all continuous functions with compact support which we shall study in more detail in Section 2.5.

Let $(X, \mathfrak{T})$ be a Hausdorff space. Denote by $\mathfrak{K}$ the set of all compact subsets of X. We find, as a generalization of the preceding example, that the set of all continuous functions with compact support forms a real Riesz lattice, which is usually denoted by $\mathscr{K}(X, \mathfrak{T})$. If $\mathfrak{T}$ is the discrete topology on X, then $\mathscr{K}(X, \mathfrak{T}) = \mathscr{k}(X)$ (Ex. 1.2.1.) [Recall that for each topological space $(X, \mathfrak{T})$ and each function f on X, the closure of $\{f \neq 0\}$ is called the **support** of f.]

1.2.4(E) Let $\mathscr{L}$ be a real Riesz lattice on X and take $\mathfrak{K} \subset \mathfrak{P}(X)$ such that $A \cup B \in \mathfrak{K}$ whenever $A, B \in \mathfrak{K}$. Set

$$\mathscr{L}^0_{\mathfrak{K}} := \{ f \in \mathscr{L} \mid \text{for each } \varepsilon > 0 \text{ there is an } A \in \mathfrak{K} \text{ with } X \setminus A \subset \{ |f| < \varepsilon \} \}.$$

Prove the following statements:

(α) $\mathscr{L}^0_{\mathfrak{K}}$ is a real Riesz lattice.

(β) If $\mathscr{L}$ is conditionally σ-completely embedded in $\overline{\mathbb{R}}^X$ then $\mathscr{L}^0_{\mathfrak{K}}$ is also conditionally σ-completely embedded in $\overline{\mathbb{R}}^X$.

(γ) If $\mathscr{L}$ is σ-completely embedded in $\overline{\mathbb{R}}^X$ and if $\bigcup_{n \in \mathbb{N}} A_n \in \mathfrak{K}$ for each sequence $(A_n)_{n \in \mathbb{N}}$ from $\mathfrak{K}$, then $\mathscr{L}^0_{\mathfrak{K}}$ is also σ-completely embedded in $\overline{\mathbb{R}}^X$.

Once again we can let $\mathscr{L}$ be the continuous functions on a Hausdorff topological space $(X, \mathfrak{T})$ and $\mathfrak{K}$ the compact subsets of X. This again yields important examples usually denoted by $\mathscr{C}_0(X, \mathfrak{T})$. We have $\mathscr{C}_0(X, \mathfrak{T}) = c_0(X)$ if $\mathfrak{T}$ is the discrete topology on X (Ex. 1.2.1).

1.2.5(E) Let $\mathscr{L}$ be a Riesz lattice on X. Suppose $\mathfrak{R} \subset \mathfrak{P}(X)$, $\mathfrak{R} \neq \varnothing$, satisfies

(i) If $A \in \mathfrak{R}$ and $B \subset A$, then $B \in \mathfrak{R}$.
(ii) If $A, B \in \mathfrak{R}$, then $A \cup B \in \mathfrak{R}$.

Set

$$\mathscr{L}^{\mathfrak{R}} := \{ f \in \overline{\mathbb{R}}^X \mid \{ f \neq g \} \in \mathfrak{R} \text{ for some } g \in \mathscr{L} \}$$

Prove the following statements:

(α) $\mathscr{L}^{\mathfrak{R}}$ is a Riesz lattice.

(β) $\mathfrak{R} \subset \mathfrak{N}(\mathscr{L}^{\mathfrak{R}})$.

We have here an example of the extension of a Riesz lattice by the inclusion of further exceptional sets. This idea will be significant in Chapter 4. Prove the following:

(γ) If $\mathscr{F} := \mathscr{L} \cap \mathbb{R}^X$, then $\mathscr{L} = \mathscr{F}^{\mathfrak{N}(\mathscr{L})}$.

1.2.6(E) The Riemann integral should be familiar to the reader from the differential and integral calculus. Take $a, b \in \mathbb{R}$, $a \leq b$. Take $f \in \mathbb{R}^{[a,b]}$. A **partition** of $[a, b]$ is a family $(x_k)_{k \in \mathbb{N}_n}$ of points from $[a, b]$ such that $x_1 = a$, $x_k \leq x_{k+1}$ for each $k \in \mathbb{N}_{n-1}$ and $x_n = b$. For each such partition, π, define

$$\lambda(\pi) := \sup\{ x_{k+1} - x_k \mid k \in \mathbb{N}_{n-1} \}.$$

f is called **Riemann integrable** on $[a, b]$, and $\alpha \in \mathbb{R}$ the **Riemann integral** of f on $[a, b]$ iff f is bounded, and given $\varepsilon > 0$, there is a $\delta > 0$ so that for any partition $\pi = (x_k)_{k \in \mathbb{N}_n}$ of $[a, b]$ with $\lambda(\pi) < \delta$ and for arbitrary $z_k \in [x_k, x_{k+1}]$,

$$\left| \sum_{k=1}^{n-1} f(z_k)(x_{k+1} - x_k) - \alpha \right| < \varepsilon$$

This notion of integration can be generalized as follows: Let $g: [a, b] \to \mathbb{R}$ be an increasing function. Then f is called **Riemann-Stieltjes integrable** with respect to g on $[a, b]$ and $\alpha \in \mathbb{R}$ is the **Riemann-Stieltjes integral** of f with respect to g iff f is bounded and for each $\varepsilon > 0$ there is $\delta > 0$ such that for every partition $\pi = (x_k)_{k \in \mathbb{N}_n}$ with $\lambda(\pi) < \delta$ and for arbitrary $z_k \in [x_k, x_{k+1}]$,

$$\left| \sum_{k=1}^{n-1} f(z_k)(g(x_{k+1}) - g(x_k)) - \alpha \right| < \varepsilon.$$

The Riemann integral is then simply the Riemann-Stieltjes integral with respect to the function $g: [a, b] \to \mathbb{R}$, $x \mapsto x$.

Prove that if $g: [a, b] \to \mathbb{R}$ is an increasing function, then the set of all functions that are Riemann-Stieltjes integrable with respect to g forms a real Riesz lattice containing $\mathscr{C}([a, b])$.

1.2.7(E) $\mathscr{F} \subset \overline{\mathbb{R}}^X$ is called a **linear function space** iff the following properties hold:

(i) $\langle f \dot{+} g \rangle \subset \mathscr{F}$ whenever $f, g \in \mathscr{F}$.

(ii) $\alpha f \in \mathscr{F}$ whenever $\alpha \in \mathbb{R}$ and $f \in \mathscr{F}$.

Prove:

(α) Every Riesz lattice is a linear function space.

(β) For $n \in \mathbb{N} \cup \{0\}$, denote by $\mathscr{P}^n$ the set of all polynomials $p: \mathbb{R} \to \mathbb{R}$ of degree $\leq n$ (i.e., $p \in \mathscr{P}^n$ iff there is an $m \in \mathbb{N} \cup \{0\}$, $m \leq n$, and

$\exists\ \alpha_0, \alpha_1, \dots, \alpha_m \in \mathbb{R}$ so that for each $x \in \mathbb{R}$, $p(x) = \alpha_0 + \alpha_1 x + \cdots + \alpha_m x^m$). The spaces $\mathscr{P}^n$ are linear function spaces.

(γ) Take $a, b \in \mathbb{R}$, $a < b$, and $n \in \mathbb{N}$. Denote by $\mathscr{C}^n([a,b])$ the set of all functions $f \in \mathbb{R}^{[a,b]}$ that are n-times differentiable. $\mathscr{C}^n([a,b])$ is a linear function space.

Observe that neither $\mathscr{P}^n$ nor $\mathscr{C}^n([a,b])$ are Riesz lattices if $n > 0$.

Now consider a linear function space $\mathscr{L} \subset \overline{\mathbb{R}}^X$. Set

$$\tilde{\mathfrak{R}}(\mathscr{L}) := \{A \mid A = \{f > 0\} \text{ for some } f \in \mathscr{L}\}.$$

Prove:

(δ) If $f \in \mathscr{L}$ and $A \subset \{|f| = \infty\}$, then $A \in \tilde{\mathfrak{R}}(\mathscr{L})$ and $\infty e_A \in \mathscr{L}$.

(ε) If $A \in \tilde{\mathfrak{R}}(\mathscr{L})$, $f \in \mathscr{L}$, and $B \subset \{|f| = \infty\}$, then $A \Delta B \in \tilde{\mathfrak{R}}(\mathscr{L})$.

(ζ) Given $f, g \in \overline{\mathbb{R}}^X$ and $A \subset X$ then

$$fe_A \vee ge_A = fe_{A \setminus \{f = -\infty\}} + (g - f)e_{(A \cap \{g > f\}) \setminus \{f = -\infty\}} + ge_{\{f = -\infty\} \cap A}$$

$$fe_A \wedge ge_A = fe_{A \setminus \{f = \infty\}} - (f - g)e_{(A \cap \{f > g\}) \setminus \{f = \infty\}} + ge_{\{f = \infty\} \cap A}$$

Take now $\mathfrak{S} \subset \mathfrak{P}(X)$ such that these properties hold:

(iii) $\tilde{\mathfrak{R}}(\mathscr{L}) \subset \mathfrak{S}$.

(iv) $A \cap B \in \mathfrak{S}$, $A \setminus B \in \mathfrak{S}$ whenever $A, B \in \mathfrak{S}$.

Denote by $\mathscr{L}(\mathfrak{S})$ the set of all $f \in \overline{\mathbb{R}}^X$ for which there are a finite disjoint family $(A_\iota)_{\iota \in I}$ from $\mathfrak{S}$ and a family $(f_\iota)_{\iota \in I}$ from $\mathscr{L}$ such that $f = \sum_{\iota \in I} f_\iota e_{A_\iota}$. Prove:

(η) $\mathscr{L}(\mathfrak{S})$ is a Riesz lattice.

(ϑ) If $\mathscr{L}$ is real, then so is $\mathscr{L}(\mathfrak{S})$.

The functions in $\mathscr{L}(\mathfrak{S})$ are said to be **piecewise in** $\mathscr{L}$ with respect to $\mathfrak{S}$. We mention, as an example, the Riesz lattice of piecewise linear functions on an interval A obtained by letting $\mathscr{L}$ be the set of all linear functions on A, and $\mathfrak{S}$ the set of all finite unions of subintervals of A.

Now let $\mathscr{L}$ be a Riesz lattice. Suppose that $\mathfrak{S} \subset \mathfrak{P}(X)$ has only property (iv), and define $\mathscr{L}(\mathfrak{S})$ as above. Prove:

(ι) $\mathscr{L}(\mathfrak{S})$ is a Riesz lattice.

If we take the Riesz lattice of all continuous real functions on an interval of $\mathbb{R}$ and the set of all finite unions of subintervals, we get the Riesz lattice of all piecewise continuous functions on this interval.

Hypothesis (iv) can be replaced by a weaker condition:

(iv′) For any $A, B \in \mathfrak{S}$ there are finite disjoint families $(C_\iota)_{\iota \in I}$ and $(D_\lambda)_{\lambda \in L}$ from $\mathfrak{S}$ such that $A \cap B = \bigcup_{\iota \in I} C_\iota$ and $A \setminus B = \bigcup_{\lambda \in L} D_\lambda$.

(κ) Prove that (η)–(ι) remain true if (iv) is replaced by (iv′).

1.2.8(E) Show that the set Δ of all Riesz lattices on a set X is a complete lattice with respect to $\subset$, and prove the following:

(α) $\overline{\mathbb{R}}^X$ is the largest element of Δ, and $\{0\}$ the smallest.

(β) If $(\mathscr{L}_\iota)_{\iota \in I}$ is a nonempty family from Δ, then $\bigwedge_{\iota \in I}\mathscr{L}_\iota = \bigcap_{\iota \in I}\mathscr{L}_\iota$.

As a consequence conclude the following:

(γ) If $\mathscr{F} \subset \overline{\mathbb{R}}^X$, then there is a smallest Riesz lattice $\mathscr{L}$ on X with $\mathscr{F} \subset \mathscr{L}$.

This $\mathscr{L}$ is called the **Riesz lattice generated** by $\mathscr{F}$.

Now prove the analogous results for linear function spaces. The smallest linear function space containing a subset $\mathscr{F}$ of $\overline{\mathbb{R}}^X$ is called the **linear function space generated** by $\mathscr{F}$.

1.2.9(E) Take $\mathscr{F} \subset]-\infty, \infty]^X$, $\mathscr{F} \neq \varnothing$, such that the following properties hold:

(i) $f + g,\ f \vee g \in \mathscr{F}$ whenever $f, g \in \mathscr{F}$.
(ii) $\alpha f \in \mathscr{F}$ whenever $\alpha \in \mathbb{R}_+$ and $f \in \mathscr{F}$.

Prove:

(α) $\mathscr{L} := \bigcup\{\langle f \dot{-} g\rangle \mid f, g \in \mathscr{F}\}$ is the Riesz lattice generated by $\mathscr{F}$.

A proof could proceed along the following lines:

STEP 1: $\langle f \dot{+} g\rangle \subset \mathscr{L}$ whenever $f, g \in \mathscr{L}$.

Let $f_1, f_2, g_1, g_2 \in \mathscr{F}$. Given $f \in \langle f_1 \dot{-} f_2\rangle$ and $g \in \langle g_1 \dot{-} g_2\rangle$, it follows that

$$\{x \mid (f_1 + g_1)(x) - (f_2 + g_2)(x) \text{ is not defined}\}$$

$$\supset \{x \mid f_1(x) - f_2(x) \text{ not defined}\}$$

$$\cup \{x \mid g_1(x) - g_2(x) \text{ not defined}\}$$

$$\cup \{x \mid f_1(x) - g_2(x) \text{ not defined}\}$$

$$\cup \{x \mid g_1(x) - f_2(x) \text{ not defined}\}.$$

Thus

$$\langle f \dot{+} g\rangle \subset \langle (f_1 + g_1) \dot{-} (f_2 + g_2)\rangle.$$

STEP 2: $f \vee g \in \mathscr{L}$ whenever $f, g \in \mathscr{L}$.

Let $f_1, f_2, g_1, g_2 \in \mathscr{F}$. Given $f \in \langle f_1 \dot{-} f_2\rangle$ and $g \in \langle g_1 \dot{-} g_2\rangle$

$$(f_1 + g_2) \vee (g_1 + f_2) \in \langle (f \vee g) \dot{+} (f_2 + g_2)\rangle.$$

STEP 3: $f \wedge g \in \mathscr{L}$ whenever $f, g \in \mathscr{L}$.

STEP 4: Given $f \in \mathscr{L}$ and $\alpha \in \mathbb{R}$ then $\alpha f \in \mathscr{L}$.

Steps 1–4 provide the proof.

Now take $\mathscr{L} \subset \overline{\mathbb{R}}^X$ to be a linear function space (as defined in Ex. 1.2.7) with the property that $\mathscr{L} = \{\langle f \dot{-} g \rangle \mid f, g \in \mathscr{L}_+\}$. Define

$$\mathscr{L}' := \left\{ \bigvee_{\iota \in I} f_\iota \mid (f_\iota)_{\iota \in I} \text{ a nonempty finite family from } \mathscr{L}_+ \right\}$$

and

$$\mathscr{L}'' := \mathscr{L}' \dot{-} \mathscr{L}'.$$

Prove:

(β) $\mathscr{L}''$ is the Riesz lattice generated by $\mathscr{L}$.

First show that $\mathscr{L}'$ fulfills (i) and (ii), observing that

$$\bigvee_{\iota \in I} f_\iota + \bigvee_{\lambda \in L} g_\lambda = \bigvee_{(\iota, \lambda) \in I \times L} (f_\iota + g_\lambda)$$

We could consider, instead of (i), the assumption:

(i′) $f + g, f \wedge g \in \mathscr{F}$ whenever $f, g \in \mathscr{F}$.

Prove:

(γ) If $\mathscr{F} \subset]-\infty, \infty]^X$ satisfies (i′) and (ii), then $\bigcup \{\langle f \dot{-} g \rangle \mid f, g \in \mathscr{F}\}$ is the Riesz lattice generated by $\mathscr{F}$.

1.2.10(E) Let $\mathscr{L}$ be a real Riesz lattice on X. Prove the following statements.

(α) Let $\mathscr{L}'$ equal the set of all real functions on X that are limits of order-convergent sequences from $\mathscr{L}$. Then $\mathscr{L}'$ is a real Riesz lattice on X.

(β) Set $\mathscr{L}''$ equal to the set of all real functions on X that are limits of uniformly convergent sequences from $\mathscr{L}$. Then $\mathscr{L}''$ is a real Riesz lattice on X.

$\mathscr{L}''$ is called the **uniform hull** of $\mathscr{L}$.

(γ) Show that both $\mathscr{L}'$ and $\mathscr{L}''$ are in general not conditionally σ-completely embedded in $\overline{\mathbb{R}}^X$.

1.2.11(E) In this exercise we consider continuous functions on closed intervals of $\mathbb{R}$. The reader should be familiar with this material. Generalizations are stated in Ex. 2.5.1.

Let A be a closed interval of $\mathbb{R}$, and denote by $\mathscr{C}$ the set of all continuous real functions on A. Prove the following statements for $f \in \mathscr{C}$:

(α) There are elements $x, y \in A$ such that

$$f(x) = \sup_{z \in A} f(z) \quad \text{and} \quad f(y) = \inf_{z \in A} f(z).$$

(β) Take $x, y \in A$ such that $x \leq y$. Then for each $w \in f([x, y])$, there is a $z \in [x, y]$ such that $f(z) = w$.

(γ) For all $x, y \in A$ with $x \leq y$, $f([x, y])$ is a closed interval of $\mathbb{R}$.

(δ) Every $f \in \mathscr{C}$ is **uniformly continuous**; that is, for each $\varepsilon \in \mathbb{R}$, $\varepsilon > 0$, there is a $\delta \in \mathbb{R}$, $\delta > 0$ such that $|f(x) - f(y)| < \varepsilon$ whenever $x, y \in A$ with $|x - y| < \delta$.

(ε) **Theorem of Dini**. Let $(f_n)_{n \in \mathbb{N}}$ be a monotonely convergent sequence from $\mathscr{C}$ such that $\lim_{n \to \infty} f_n \in \mathscr{C}$. Then $(f_n)_{n \in \mathbb{N}}$ converges uniformly on A.

An important tool for the proof of these statements is the following theorem of Heine-Borel-Lebsgue:

(ζ) The compact subsets of $\mathbb{R}$ are exactly the closed bounded subsets of $\mathbb{R}$.

The reader should try to prove (α)–(ε) using (ζ).

1.3. FUNCTIONALS AND DANIELL SPACES

NOTATION FOR SECTION 1.3:
X denotes a set.

Real integration theory is in a sense the theory of positive linear functionals. This section discusses the elementary properties of positive linear functionals defined on Riesz lattices, as well as the behavior of such functionals relative to exceptional sets, exceptional functions, and "almost everywhere." The section concludes with a brief discussion of nullcontinuous functionals, including a definition of Daniell space, which is our starting point for constructing an integral.

Definition 1.3.1. *Let $\mathscr{F} \subset \overline{\mathbb{R}}^X$. A **functional** on $\mathscr{F}$ is a real-valued function with domain $\mathscr{F}$. A functional ℓ on $\mathscr{F}$ is said to be:*

(a) ***additive**, if the conditions $f, g, h \in \mathscr{F}$, and $h \in \langle f \dot{+} g \rangle$ always imply that*

$$\ell(h) = \ell(f) + \ell(g)$$

(b) ***homogeneous**, if the conditions $f \in \mathscr{F}$, $\alpha \in \mathbb{R}$, and $\alpha f \in \mathscr{F}$ always imply that*

$$\ell(\alpha f) = \alpha \ell(f)$$

(*c*) ***linear***, *if it is both additive and homogeneous*
(*d*) ***positive***, *if* $\ell(f) \geq 0$ *for every* f *in* $\mathscr{F}_+$
(*e*) ***increasing***, *if it increases relative to the order relations on* $\mathscr{F}$ *and* $\mathbb{R}$. □

Although no occasion for confusion should arise, a word of caution is probably in order. Let $\mathscr{F} \subset \overline{\mathbb{R}}^X$. If ℓ: $\mathscr{F} \to \mathbb{R}$ is viewed as an element of $\mathbb{R}^{\mathscr{F}}$, that is, as a real-valued *function* on $\mathscr{F}$, then ℓ is positive iff $\ell(f) \geq 0$ for every $f \in \mathscr{F}$, according to Definitions 1.2.6 and 1.2.1. However, with ℓ: $\mathscr{F} \to \mathbb{R}$ viewed as a real *functional* on $\mathscr{F}$, ℓ is positive iff $\ell(f) \geq 0$ for every $f \in \mathscr{F}_+$, by Definition 1(d).

Proposition 1.3.2. *The following assertions hold for every functional* ℓ *on a Riesz lattice* $\mathscr{L}$ *in* $\overline{\mathbb{R}}^X$.

(*a*) *If* ℓ *is additive, then*

$$\ell(0) = 0$$

and

$$\ell(-f) = -\ell(f)$$

for every f *in* $\mathscr{L}$. *If* ℓ *is additive,* $f \in \mathscr{L}$, $g \in \mathscr{L}$, *and* $h \in \langle f \dot{-} g \rangle$, *then*

$$\ell(h) = \ell(f) - \ell(g).$$

(*b*) *If* ℓ *is additive and* $(f_\iota)_{\iota \in I}$ *is a finite family from* $\mathscr{L} \cap \mathbb{R}^X$, *then*

$$\ell\left(\sum_{\iota \in I} f_\iota \right) = \sum_{\iota \in I} \ell(f_\iota).$$

If ℓ *is also homogeneous and* $(\alpha_\iota)_{\iota \in I} \in \mathbb{R}^I$, *then*

$$\ell\left(\sum_{\iota \in I} \alpha_\iota f_\iota \right) = \sum_{\iota \in I} \alpha_\iota \ell(f_\iota).$$

(*c*) *If* ℓ *is additive, then* ℓ *is positive iff* ℓ *is increasing.*
(*d*) *If* ℓ *is increasing, then for every nonempty finite family* $(f_\iota)_{\iota \in I}$ *from* $\mathscr{L}$,

$$\ell\left(\bigwedge_{\iota \in I} f_\iota \right) \leq \inf_{\iota \in I} \ell(f_\iota)$$

and

$$\ell\left(\bigvee_{\iota \in I} f_\iota \right) \geq \sup_{\iota \in I} \ell(f_\iota).$$

(*e*) *If ℓ is homogeneous and increasing, then*

$$|\ell(f)| \leq \ell(|f|)$$

for every f in $\mathscr{L}$.

Proof. Since $\mathscr{L}$ is a Riesz lattice, the functional values $\ell(0)$, $\ell(-f)$, and $\ell(h)$ in assertion (a), $\ell(\sum_{\iota \in I} f_\iota)$ in assertion (b), and so on, are all defined.

(a) This assertion is easily verified.

(b) This claim can be proved by complete induction on the number of elements in I.

(c) In view of (a), ℓ additive and ℓ increasing certainly imply ℓ positive. Assume, conversely, that ℓ is additive and positive. Given f and g in $\mathscr{L}$ with $f \leq g$, define

$$h: X \to \overline{\mathbb{R}}, \qquad x \mapsto \begin{cases} g(x) - f(x) & \text{if defined} \\ 0 & \text{otherwise.} \end{cases}$$

Evidently h is a positive function belonging to $\langle g \dot{-} f \rangle$. It follows, by use of (a), that

$$0 \leq \ell(h) = \ell(g) - \ell(f).$$

Thus $\ell(f) \leq \ell(g)$ and ℓ increases.

(d) For every κ in I,

$$\bigwedge_{\iota \in I} f_\iota \leq f_\kappa \leq \bigvee_{\iota \in I} f_\iota$$

and so

$$\ell\Big(\bigwedge_{\iota \in I} f_\iota\Big) \leq \ell(f_\kappa) \leq \ell\Big(\bigvee_{\iota \in I} f_\iota\Big).$$

The required inequalities follow.

(e) From $-|f| \leq f \leq |f|$ we conclude that

$$-\ell(|f|) \leq \ell(f) \leq \ell(|f|). \qquad \square$$

Proposition 1.3.3. *Let ℓ be a positive linear functional on a Riesz lattice $\mathscr{L}$ in $\overline{\mathbb{R}}^X$. Then the following assertions hold, for all $f, g \in \mathscr{L}$, for every $h \in \overline{\mathbb{R}}^X$, and for every $\mathscr{L}$-exceptional set A.*

(*a*) *If f belongs to $\mathscr{N}(\mathscr{L})$, then $\ell(f) = \ell(|f|) = 0$.*

(*b*) $\ell(fe_A) = 0$, $\ell(fe_{X \setminus A}) = \ell(f)$.

(*c*) *If $f \leq g$ $\mathscr{L}$-a.e., then $\ell(f) \leq \ell(g)$.*

(*d*) *If $h = f$ $\mathscr{L}$-a.e., then h belongs to $\mathscr{L}$ and $\ell(h) = \ell(f)$.*

(*e*) *If $h(x) = f(x) + g(x)$ $\mathscr{L}$-a.e., then h belongs to $\mathscr{L}$, and $\ell(h) = \ell(f) + \ell(g)$.*

Proof. (a) If f belongs to $\mathcal{N}(\mathcal{L})$, the functions f, $|f|$, ∞f, and $|\infty f|$ all belong to $\mathcal{L}$ (Propositions 1.2.23, 1.2.30). Now

$$|\infty f| + |\infty f| = |\infty f|$$

so

$$\ell(|\infty f|) + \ell(|\infty f|) = \ell(|\infty f|).$$

Since ℓ takes only real values, we conclude that $\ell(|\infty f|) = 0$. The inequality

$$0 \leq |f| \leq |\infty f|$$

yields

$$0 \leq \ell(|f|) \leq \ell(|\infty f|) = 0$$

[Proposition 2(c)] and therefore $\ell(|f|) = 0$. By Proposition 2(e), $\ell(f) = 0$.

(b) It was shown in Proposition 1.2.30(b) that fe_A belongs to $\mathcal{N}(\mathcal{L})$ and $fe_{X \setminus A}$ belongs to $\mathcal{L}$. Since ℓ is additive and $f = fe_A + fe_{X \setminus A}$, (b) follows from (a).

(c) Let $B := \{f > g\}$. By hypothesis, B is $\mathcal{L}$-exceptional, and

$$fe_{X \setminus B} \leq ge_{X \setminus B}.$$

By Proposition 2(c), ℓ increases. Using (b), we have

$$\ell(f) = \ell(fe_{X \setminus B}) \leq \ell(ge_{X \setminus B}) = \ell(g).$$

(d) We already know that h belongs to $\mathcal{L}$ (Proposition 1.2.30). Since $h = f$ $\mathcal{L}$-a.e. iff $h \leq f$ $\mathcal{L}$-a.e. and $f \leq h$ $\mathcal{L}$-a.e. (Proposition 1.2.32), (d) follows from (c).

(e) That h belongs to $\mathcal{L}$ was established in Proposition 1.2.30. Set

$$B := \{x \in X \mid f(x) + g(x) \text{ is not defined}\}$$

and note that B is $\mathcal{L}$-exceptional. Using (b) and (d), we have

$$\begin{aligned}\ell(h) &= \ell(fe_{X \setminus B} + ge_{X \setminus B}) \\ &= \ell(fe_{X \setminus B}) + \ell(ge_{X \setminus B}) \\ &= \ell(f) + \ell(g).\end{aligned}$$ □

It is now clear why the function values $\pm\infty$ don't disturb the real-valuedness of positive linear functionals on Riesz lattices. For on each Riesz lattice $\mathcal{L}$

there exist only those positive linear functionals that ignore the sets where functions from $\mathscr{L}$ take infinite values.

The significance and the usefulness of the sets $\langle f \dot{+} g \rangle$ and $\langle f \dot{-} g \rangle$ have also become clear. By working with the set $\langle f \dot{+} g \rangle$, we ignore, in effect, those function values that preclude defining the sum $f + g$. Where positive linear functionals on Riesz lattices are concerned, this attitude causes no harm. The class $\langle f \dot{+} g \rangle$ can play the role of the possibly undefined sum $f + g$ because positive linear functionals treat all functions in the class alike. Thus the expression $\ell(f \dot{+} g)$ has a natural meaning, even though $\ell(f + g)$ might have no meaning. The following (very general!) definition makes this precise.

Definition 1.3.4. *Let $(\mathscr{L}_\iota)_{\iota \in I}$ be a family of Riesz lattices, and let Z be a set. Suppose that*

$$\varphi\colon \prod_{\iota \in I} \mathscr{L}_\iota \to Z$$

is a mapping with the following property:

If $(f_\iota)_{\iota \in I}, (g_\iota)_{\iota \in I} \in \prod_{\iota \in I} \mathscr{L}_\iota$, and if $f_\iota = g_\iota$ $\mathscr{L}_\iota$-a.e. for every ι in I, then $\varphi((f_\iota)_{\iota \in I}) = \varphi((g_\iota)_{\iota \in I})$.

Let $(f_\iota)_{\iota \in I}, (g_\iota)_{\iota \in I}$, and $(h_\iota)_{\iota \in I}$ be elements of $\prod_{\iota \in I} \mathscr{L}_\iota$ with $h_\iota \in \langle f_\iota \dot{\pm} g_\iota \rangle$ for every ι in I. Then

$$\varphi\big((f_\iota \dot{\pm} g_\iota)_{\iota \in I}\big) := \varphi\big((h_\iota)_{\iota \in I}\big)$$

$$\varphi\big((|f_\iota \dot{\pm} g_\iota|)_{\iota \in I}\big) := \varphi\big((|h_\iota|)_{\iota \in I}\big). \qquad \square$$

The hypothesized condition on φ is precisely what is needed to ensure that the definition of $\varphi((f_\iota \dot{\pm} g_\iota)_{\iota \in I})$ does not depend on the representative family $(h_\iota)_{\iota \in I}$: every representative family gives the same result. Definition 4 contains, in particular, the definitions of such expressions as $\ell(f \dot{+} g)$ and $\ell(|f \dot{+} g|)$, where ℓ is a positive linear functional on a Riesz lattice $\mathscr{L}$.

Our construction of integrals starts with positive linear functionals that satisfy a rather weak continuity or convergence condition. The condition in question is described in the next definition.

Definition 1.3.5. *A functional ℓ on a set $\mathscr{F} \subset \overline{\mathbb{R}}^X$ is said to be* ***nullcontinuous*** *iff*

$$\lim_{n \in \mathbb{N}} \ell(f_n) = 0$$

for every sequence $(f_n)_{n \in \mathbb{N}}$ from $\mathscr{F}$ that decreases and satisfies

$$\bigwedge_{n \in \mathbb{N}} f_n = 0.$$

A **Daniell space** *is a triple* $(X, \mathscr{L}, \ell)$ *where* $\mathscr{L}$ *is a Riesz lattice in* $\overline{\mathbb{R}}^X$ *and* ℓ *is a positive, linear, nullcontinuous functional on* $\mathscr{L}$. □

It is important in Definition 5 that $\wedge$ is taken $\overline{\mathbb{R}}^X$ and not in $\mathscr{F}$. The name "Daniell space" is chosen to recognize the work of P. J. Daniell, who in 1918 constructed an integral starting with a positive linear nullcontinuous functional on a real Riesz lattice.

Proposition 1.3.6. *The following assertions are equivalent, for every positive linear functional* ℓ *on a Riesz lattice* $\mathscr{L}$ *in* $\overline{\mathbb{R}}^X$.

(*a*) ℓ *is nullcontinuous.*

(*b*) $\inf_{n \in \mathbb{N}} \ell(f_n) = 0$ *for every sequence* $(f_n)_{n \in \mathbb{N}}$ *from* $\mathscr{L}$ *that decreases and satisfies* $\bigwedge_{n \in \mathbb{N}} f_n = 0$.

(*c*) *For every increasing sequence* $(f_n)_{n \in \mathbb{N}}$ *from* $\mathscr{L}$, *if* $\bigvee_{n \in \mathbb{N}} f_n$ *belongs to* $\mathscr{L}$, *then*

$$\ell\Big(\bigvee_{n \in \mathbb{N}} f_n\Big) = \sup_{n \in \mathbb{N}} \ell(f_n).$$

(*d*) *For every decreasing sequence* $(g_n)_{n \in \mathbb{N}}$ *from* $\mathscr{L}$, *if* $\bigwedge_{n \in \mathbb{N}} g_n$ *belongs to* $\mathscr{L}$, *then*

$$\ell\Big(\bigwedge_{n \in \mathbb{N}} g_n\Big) = \inf_{n \in \mathbb{N}} \ell(g_n).$$

Proof. (a) ⇔ (b) follows easily from the fact that ℓ increases [Proposition 2(c)].

(b) ⇒ (c). Let $(f_n)_{n \in \mathbb{N}}$ be an increasing sequence from $\mathscr{L}$ for which

$$f := \bigvee_{n \in \mathbb{N}} f_n$$

belongs to $\mathscr{L}$. The idea readily occurs to apply (b) to the sequence $(f - f_n)_{n \in \mathbb{N}}$. We must be cautious, however, since $f - f_n$ need not be defined and also because $(f(x) - f_n(x))_{n \in \mathbb{N}}$, where defined, will have infimum ∞, not 0, when the sequence $(f_n(x))_{n \in \mathbb{N}}$ lies in $\mathbb{R}$ but increases to ∞. For every n in $\mathbb{N}$, then, define

$$h_n \colon X \to \overline{\mathbb{R}},$$

$$x \mapsto \begin{cases} f(x) - f_n(x) & \text{if defined} \\ 0 & \text{otherwise} \end{cases}$$

and

$$h'_n \colon X \to \mathbb{R},$$

$$x \mapsto \begin{cases} h_n(x) & \text{if } h_n(x) < \infty \\ 0 & \text{if } h_n(x) = \infty. \end{cases}$$

Then $h_n \in \langle f \dot{-} f_n \rangle$, so every h_n belongs to $\mathscr{L}$. Thus for every n,

$$h'_n = h_n \ \mathscr{L}\text{-a.e.},$$

h'_n belongs to $\mathscr{L}$, and $\ell(h'_n) = \ell(h_n)$. The sequences $(h_n)_{n \in \mathbb{N}}$ and $(h'_n)_{n \in \mathbb{N}}$ both decrease, and

$$\bigwedge_{n \in \mathbb{N}} h'_n = 0.$$

Using (b), we have

$$\begin{aligned} 0 &= \inf_{n \in \mathbb{N}} \ell(h'_n) \\ &= \inf_{n \in \mathbb{N}} \ell(h_n) \\ &= \inf_{n \in \mathbb{N}} (\ell(f) - \ell(f_n)) \\ &= \ell(f) + \inf_{n \in \mathbb{N}} (-\ell(f_n)) \\ &= \ell(f) - \sup_{n \in \mathbb{N}} \ell(f_n) \end{aligned}$$

and so

$$\sup_{n \in \mathbb{N}} \ell(f_n) = \ell(f)$$

as required.

(c) $\Rightarrow$ (d). If $(g_n)_{n \in \mathbb{N}}$ is a decreasing sequence from $\mathscr{L}$ whose infimum belongs to $\mathscr{L}$, then $(-g_n)_{n \in \mathbb{N}}$ is an increasing sequence from $\mathscr{L}$ whose supremum belongs to $\mathscr{L}$. In view of (c),

$$\begin{aligned} \ell\left(\bigwedge_{n \in \mathbb{N}} g_n \right) &= \ell\left(- \bigvee_{n \in \mathbb{N}} (-g_n) \right) \\ &= -\ell\left(\bigvee_{n \in \mathbb{N}} (-g_n) \right) \\ &= - \sup_{n \in \mathbb{N}} \ell(-g_n) \\ &= \inf_{n \in \mathbb{N}} \ell(g_n). \end{aligned}$$

(d) $\Rightarrow$ (b) is trivially true. □

EXERCISES

1.3.1(E) Let X be a set and

$$\mathscr{k}(X) := \{ f \in \mathbb{R}^X \mid \{f \neq 0\} \text{ finite} \}.$$

$\mathscr{k}(X)$ is a real Riesz lattice (cf., Ex. 1.2.1(β)). Define

$$\ell : \mathscr{k}(X) \to \mathbb{R}, \qquad f \mapsto \sum_{x \in \{f \neq 0\}} f(x).$$

Show that $(X, \mathscr{k}(X), \ell)$ is a Daniell space. Set

$$\ell^1(X) := \{ f \in \overline{\mathbb{R}}^X \mid \sup\left\{ \sum_{x \in A} |f(x)| \,\middle|\, A \subset X, A \text{ finite} \right\} < \infty \}.$$

Prove:

(α) $\ell^1(X)$ is a real Riesz lattice.
(β) If $f \in \ell^1(X)$, $g \in \mathbb{R}^X$ and $|g| \leq |f|$, then $g \in \ell^1(X)$.
(γ) $\ell^1(X)$ is conditionally σ-completely embedded in $\overline{\mathbb{R}}^X$.

Now define

$$\bar{\ell} : \ell^1(X)_+ \to \mathbb{R}, \qquad f \mapsto \sup\left\{ \sum_{x \in A} f(x) \mid A \subset X, A \text{ finite} \right\}.$$

Verify the next four propositions:

(δ) If $f \in \ell^1(X)_+$, then $\bar{\ell}(f) \geq 0$. Furthermore $\bar{\ell}(f) = 0$ iff $f = 0$.
(ε) If $f, g \in \ell^1(X)_+$, and $\alpha \in \mathbb{R}_+$, then $\bar{\ell}(f+g) = \bar{\ell}(f) + \bar{\ell}(g)$ and $\bar{\ell}(\alpha f) = \alpha \bar{\ell}(f)$.
(ζ) If $f, g, f', g' \in \ell^1(X)_+$ with $f - g = f' - g'$, then $\bar{\ell}(f) - \bar{\ell}(g) = \bar{\ell}(f') - \bar{\ell}(g')$.
(η) If $(f_n)_{n \in \mathbb{N}}$ is a decreasing sequence in $\ell^1(X)_+$ with $\bigwedge_{n \in \mathbb{N}} f_n = 0$, then $\inf_{n \in \mathbb{N}} \bar{\ell}(f_n) = 0$.

For a proof of (η), let $(f_n)_{n \in \mathbb{N}}$ satisfy the hypotheses. Take $\varepsilon > 0$. Then there is a finite $A \subset X$ with $\bar{\ell}(f_1) < \sum_{x \in A} f_1(x) + \varepsilon$. Thus

$$0 \leq \bar{\ell}(f_n) < \sum_{x \in A} f_n(x) + \varepsilon.$$

Now $(f_n e_A)_{n \in \mathbb{N}}$ is a decreasing sequence in $\mathscr{k}(X)$ with $\bigwedge_{n \in \mathbb{N}} f_n e_A = 0$. Consequently

$$\inf_{n \in \mathbb{N}} \left(\sum_{x \in A} f_n(x) \right) = \inf_{n \in \mathbb{N}} \ell(f_n e_A) = 0.$$

Hence $0 \le \inf_{n \in \mathbb{N}} \bar{\ell}(f_n) \le \varepsilon$, establishing the desired conclusion.
Finally, define

$$\bar{\ell}: \ell^1(X) \to \mathbb{R}, \qquad f \mapsto \bar{\ell}(f^+) - \bar{\ell}(f^-).$$

Prove:

(ϑ) $(X, \ell^1(X), \bar{\ell})$ is a Daniell space.

(ι) $(X, k(X), \ell) \preccurlyeq (X, \ell^1(X), \bar{\ell})$.

We have reached $(X, \ell^1(X), \bar{\ell})$ by a direct path in this exercise. In Section 3.4 we shall be examining this space in detail from a general standpoint.

We remind the reader here of the equivalence of the concepts of function and family. They express only different ways of representing the same object. It is then apparent what we mean by the definition

$$\sum_{x \in X} \alpha_x := \bar{\ell}((\alpha_x)_{x \in X})$$

for a family $(\alpha_x)_{x \in X} \in \ell^1(X)$. These families are called **summable**. They will be discussed in Section 3.4.

1.3.2(E) The results of Ex. 1.3.1 lend themselves to generalization. Take $g \in \mathbb{R}_+^X$, and define

$$\ell_g^1(X) := \{ f \in \overline{\mathbb{R}}^X \mid fg \in \ell^1(X) \}$$

$$\ell_g: k(X) \to \mathbb{R}, \qquad f \mapsto \sum_{x \in \{f \neq 0\}} f(x) g(x)$$

and

$$\bar{\ell}_g: \ell_g^1(X) \to \mathbb{R}, \qquad f \mapsto \sum_{x \in X} f(x) g(x).$$

Prove that the following statements are true for an arbitrary $g \in \mathbb{R}_+^X$.

(α) $(X, \ell_g^1(X), \bar{\ell}_g)$ is a Daniell space.

(β) $(X, k(X), \ell_g) \preccurlyeq (X, \ell_g^1(X), \bar{\ell}_g)$.

(γ) If $f \in \ell_g^1(X)$, $h \in \overline{\mathbb{R}}^X$, and $|h| \le |f|$, then $h \in \ell_g^1(X)$.

(δ) $\ell_g^1(X)$ is conditionally σ-completely embedded in $\overline{\mathbb{R}}^X$.

(ε) $\ell_g^1(X)$ is real iff $\{g > 0\} = X$.

(ζ) $\mathfrak{N}(\ell_g^1(X)) = \mathfrak{P}(\{g = 0\})$.

(η) $\mathcal{N}(\ell_g^1(X)) = \{ f \in \overline{\mathbb{R}}^X \mid \{f \neq 0\} \subset \{g = 0\}\}$.

(ϑ) If $g(x) = 1$ for each $x \in X$ then $(X, k(X), \ell_g) = (X, k(X), \ell)$ and $(X, \ell_g^1(X), \bar{\ell}_g) = (X, \ell^1(X), \bar{\ell})$.

The significance of the spaces $(X, \ell_g^1(X), \bar{\ell}_g)$ derived here from $(X, \ell^1(X), \bar{\ell})$ will become apparent in Chapter 9. That chapter deals with the concept of absolute continuity. We shall show that, in the special case considered here, the functionals, ℓ_g, are precisely those that are absolutely continuous with respect to ℓ.

1.3.3(E) The Riemann-Stieltjes Integrals. We continue the line of investigation of Ex. 1.2.6. As in that exercise, take $a, b \in \mathbb{R}$, $a < b$, and an increasing function $g: [a, b] \to \mathbb{R}$. Define

$$\mathscr{L}^g := \{ f \in \mathbb{R}^{[a,b]} \mid f \text{ is Riemann-Stieltjes integrable with respect to } g \}.$$

For $f \in \mathscr{L}^g$ denote by $\ell^g(f)$ the Riemann-Stieltjes integral of f with respect to g. Then $([a, b], \mathscr{L}^g, \ell^g)$ is a Daniell space.

For the proof take the propositions proved about the Riemann integral in the elementary calculus, and generalize these to the Riemann-Stieltjes integral. Null continuity could be a source of difficulty, so we give a few hints. If $(f_n)_{n \in \mathbb{N}}$ is a decreasing sequence of continuous functions on $[a, b]$ with $\bigwedge_{n \in \mathbb{N}} f_n = 0$, then $(f_n)_{n \in \mathbb{N}}$ converges uniformly to 0 by the theorem of Dini (cf., Ex. 1.2.11). Applying the theorem concerning the commutativity of limits with integrals in the case of uniformly convergent sequences of functions yields $\inf_{n \in \mathbb{N}} \ell^g(f_n) = 0$. Now more generally, let $(f_n)_{n \in \mathbb{N}}$ be a decreasing sequence from $\mathscr{L}^g_+$ with $\bigwedge_{n \in \mathbb{N}} f_n = 0$. Take $\varepsilon > 0$. There is, for each $n \in \mathbb{N}$, a continuous function g_n on $[a, b]$ with $0 \le g_n \le f_n$ and with $\ell^g(f_n) - \ell^g(g_n) \le \varepsilon/2^n$. Set $h_n := \bigwedge_{k \le n} g_k$. Then $(h_n)_{n \in \mathbb{N}}$ is a decreasing sequence of continuous functions on $[a, b]$ with $\bigwedge_{n \in \mathbb{N}} h_n = 0$. For $n \in \mathbb{N}$,

$$f_n - h_n \le \sum_{k \le n} (f_k - g_k) \quad \text{and so} \quad \ell^g(f_n) - \ell^g(h_n) < \varepsilon.$$

Thus $0 \le \inf_{n \in \mathbb{N}} \ell^g(f_n) < \varepsilon$, and the null continuity follows since ε was arbitrarily chosen.

1.3.4(E) All the Daniell spaces $(X, \mathscr{L}, \ell)$ discussed up to this point have shared a common property, namely $f \wedge e_X \in \mathscr{L}$ whenever $f \in \mathscr{L}_+$. This is the **Stone property** (M. H. Stone 1948). Such spaces are closely connected to Daniell spaces derived from a measure. We shall go into this question thoroughly in Chapters 2 and 5. It is easy to find examples of Daniell spaces without the Stone property.

Take $f \in \mathbb{R}^X_+$, $f \neq 0$, and define

$$\mathscr{L} := \{ \alpha f \mid \alpha \in \mathbb{R} \}, \qquad \ell : \mathscr{L} \to \mathbb{R}, \qquad \alpha f \mapsto \alpha.$$

Prove:

(α) $\omega : \mathbb{R} \to \mathscr{L}$, $\alpha \mapsto \alpha f$ defines an order isomorphism.

(β) $(X, \mathscr{L}, \ell)$ is a Daniell space.

(γ) $(X, \mathscr{L}, \ell)$ has the Stone property iff for some $\gamma \in \mathbb{R}$ and for some $A \subset X$, $f = \gamma e_A$.

After these introductory examples, we wish to examine more thoroughly the phenomena that arise in connection with positive linear functionals.

1.3.5(E) In this exercise we turn our attention to linear functionals on some of the Riesz lattices discussed in Ex. 1.2.1. As we shall see, such functionals often have a continuity property beyond the null continuity.

An additive functional ℓ on a Riesz lattice $\mathscr{L}$ is called **π-continuous** iff for every nonempty downward directed family $(f_\iota)_{\iota \in I}$ in $\mathscr{L}$ with $\bigwedge_{\iota \in I} f_\iota = 0$, $\inf_{\iota \in I} \ell(f_\iota) = 0$. (The nomenclature π-continuous has been introduced with an eye to the notation to be introduced in Section 6.7, where this notion of continuity will be thoroughly investigated.) π-continuity implies null continuity.

We now examine the situation in the aforementioned function spaces.

(a) **The space $\mathscr{k}(X)$**. Prove:

(α) Every positive linear functional, ℓ, on $\mathscr{k}(X)$ is π-continuous.

(β) For each positive linear functional, ℓ, on $\mathscr{k}(X)$, there is a uniquely determined $g \in \mathbb{R}_+^X$ so that $\ell = \ell_g$ (in the sense of Ex. 1.3.2). g is specified by $g(x) = \ell(e_{\{x\}})$ for each $x \in X$.

We provide some suggestions for (α). Let $(f_\iota)_{\iota \in I}$ be a nonempty downward directed family in $\mathscr{k}(X)$, with $\bigwedge_{\iota \in I} f_\iota = 0$. Take $\iota_0 \in I$. Then $A := \{f_{\iota_0} \neq 0\}$ is finite. Put $I_0 := \{\iota \in I \,|\, f_\iota \leq f_{\iota_0}\}$. Then for $\iota \in I_0$ and $x \in X \setminus A$, $f_\iota(x) = 0$. Thus for arbitrary $\varepsilon > 0$, there is an $\iota_\varepsilon \in I_0$ such that, given $\iota \in I$ with $f_\iota \leq f_{\iota_\varepsilon}$, $f_\iota \leq \varepsilon e_A$. For each such ι it follows that $\ell(f_\iota) \leq \varepsilon \ell(e_A)$. But ε is arbitrary, so that $\inf_{\iota \in I} \ell(f_\iota) = 0$.

The situation for $\mathscr{k}(X)$ is simple. In fact (β) provides a representation for all positive linear functionals on $\mathscr{k}(X)$. We take this opportunity to point out an important aspect. Let $\tilde{\mathscr{k}}(X)$ denote the set of all positive linear functionals on $\mathscr{k}(X)$. Consider the mapping $\varphi : \mathbb{R}_+^X \to \tilde{\mathscr{k}}(X)$, $g \mapsto \ell_g$. A little thought shows that φ is bijective and $\ell_{f+g} = \ell_f + \ell_g$ for $f, g \in \mathbb{R}_+^X$. Furthermore we can transfer the entire order structure of $\mathbb{R}_+^X$ onto $\tilde{\mathscr{k}}(X)$. For $f, g \in \mathbb{R}_+^X$, $f \leq g \Leftrightarrow$ for each $h \in \mathscr{k}(X)_+$, $\ell_f(h) \leq \ell_g(h)$. This last property allows us to define an order relation on $\tilde{\mathscr{k}}(X)$. This makes $\tilde{\mathscr{k}}(X)$ a conditionally complete lattice, isomorphic to $\mathbb{R}_+^X$. $\tilde{\mathscr{k}}(X)$ is not exceptional in this respect. We shall later show that the set of all positive linear functionals on a Riesz lattice, $\mathscr{L}$, can be endowed with a lattice structure in a natural way. This is intimately connected with deep problems in measure theory. In general, this structure is not so easy to describe explicitly as in the case of $\mathscr{k}(X)$. We deal with this in Chapter 8.

(b) **The space $\mathscr{c}_f(X)$**. Prove the following statements:

(α) $\mathscr{c}_f(X) = \{\alpha e_x + g \,|\, \alpha \in \mathbb{R},\ g \in \mathscr{k}(X)\}$.

(β) $\mathscr{c}_f(X) = \mathscr{k}(X)$ whenever X is finite.

Hence the theory of positive linear functionals on a finite set X is completely described by the discussion in (a). Prove the following propositions for a countably infinite set X:

(γ) A positive linear functional on $\mathscr{c}_f(X)$ is π-continuous iff it is null-continuous.

(δ) $\ell : \mathscr{c}_f(X) \to \mathbb{R}$, $\alpha e_X + g \mapsto \alpha$ is a positive linear functional that is not nullcontinuous.

The situation is somewhat different when the set X is uncountable. In this case prove the following statements:

(ε) Every positive linear functional on $c_f(X)$ is nullcontinuous.

(ζ) Let ℓ be a positive linear functional on $c_f(X)$. Then ℓ is π-continuous iff $\ell(e_X) = \sup_{\iota \in I} \ell(f_\iota)$ for every nonempty upward directed family $(f_\iota)_{\iota \in I}$ from $k(X)$ with $\bigvee_{\iota \in I} f_\iota = e_X$.

(η) $\ell : c_f(X) \to \mathbb{R}$, $\alpha e_X + g \mapsto \alpha$ is a positive linear functional on $c_f(X)$ that is not π-continuous.

Thus it can be seen that the situation is already somewhat more complicated. Finally, prove the following representation theorem which holds for any set X:

(ϑ) Let ℓ be a positive linear functional on $c_f(X)$. ℓ is π-continuous iff there is a function $g \in \ell^1(X)$ such that for every $f \in c_f(X)$, $\ell(f) = \sum_{x \in X} f(x) g(x)$. g is uniquely determined by ℓ, through $g(x) = \ell(e_{\{x\}})$ for each $x \in X$.

Assertions analogous to (γ), (δ), (ζ), (η) hold when $c_f(X)$ is replaced by $c(X)$.

(c) **The space $\ell^\infty(X)$**. For X finite $\ell^\infty(X) = k(X)$. Prove as well:

(α) If X is countable, then a positive linear functional on $\ell^\infty(X)$ is π-continuous iff it is nullcontinuous.

(β) The positive linear functional ℓ on $\ell^\infty(X)$ is π-continuous iff there is a function $g \in \ell^1(X)$ such that for every $f \in \ell^\infty(X)$, $\ell(f) = \sum_{x \in X} f(x) g(x)$. Furthermore $g(x) = \ell(e_{\{x\}})$ for each $x \in X$, so that ℓ determines g uniquely.

The question arises: Are there positive linear functionals on $\ell^\infty(X)$ that are not nullcontinuous? Certainly not if X is finite. So assume that X is not finite. Then there is a decreasing sequence of nonempty subsets of X, $(A_n)_{n \in \mathbb{N}}$, with $\bigcap_{n \in \mathbb{N}} A_n = \varnothing$. By Ex. 1.1.13($\delta$) there is an ultrafilter $\mathfrak{F}$ on X such that $A_n \in \mathfrak{F}$ for each $n \in \mathbb{N}$. Prove the following:

(γ) Let $\mathfrak{G}$ be an ultrafilter on X. Define

$$\ell_{\mathfrak{G}} : \ell^\infty(X) \to \mathbb{R}, \qquad f \mapsto \lim_{\mathfrak{G}} f$$

Then ℓ is a positive linear functional on $\ell^\infty(X)$.

(δ) If $\mathfrak{F}$ denotes the ultrafilter just described, $\ell_{\mathfrak{F}}$ is not nullcontinuous.

In fact, $\bigwedge_{n \in \mathbb{N}} e_{A_n} = 0$, but $\ell_{\mathfrak{F}}(e_{A_n}) = 1$ for every $n \in \mathbb{N}$.

Thus on $\ell^\infty(X)$ the functionals that are not nullcontinuous are not so trivial as those found on $c_f(X)$. For a representation theorem we need such theorems as the existence of ultrafilters containing a given filterbase. We need to delve even more deeply if we wish to answer questions concerning the existence of nullcontinuous functionals that are not π-continuous.

We assume that there exists a nontrivial δ-stable ultrafilter $\mathfrak{F}$ on X. (See Ex. 1.1.13 for the definition of this notion.) Prove:

(ε) $\ell_{\mathfrak{F}}$ is nullcontinuous but not π-continuous.

In fact $(e_F)_{F \in \mathfrak{F}}$ is a downward directed family in $\ell^\infty(X)$ with $\bigwedge_{F \in \mathfrak{F}} e_F = 0$, but $\ell_{\mathfrak{F}}(e_F) = 1$ for each $F \in \mathfrak{F}$.

In Ex. 2.2.19 we give a characterization of the sets X with the property that there are positive linear nullcontinuous functionals on $\ell^\infty(X)$ that are not π-continuous.

(d) **The space $c_0(X)$**. Prove the following statements:

(α) $\{f \neq 0\}$ is countable for every $f \in c_0(X)$.

(β) Every positive linear functional on $c_0(X)$ is π-continuous.

(γ) $\ell: c_0(X) \to \mathbb{R}$ is a positive linear functional iff there is a $g \in \ell^1(X)$ such that $\ell(f) = \Sigma_{x \in X} f(x) g(x)$ for each $f \in c_0(X)$. In this case $g(x) = \ell(e_{\{x\}})$ for every $x \in X$, and thus ℓ determines g uniquely.

We provide some suggestions for (γ). Given a positive linear functional, ℓ, on $c_0(X)$, define $g: X \to \mathbb{R}$ by $g(x) := \ell(e_{\{x\}})$. Then, because ℓ is π-continuous, $\ell(f) = \Sigma_{x \in X} f(x) g(x)$ for each $f \in c_0(X)$. A problem arises in checking that g is summable. So assume that g is not summable. Then there is an injection $\omega: \mathbb{N} \to X$, with $g(\omega(k)) > 0$ for each $k \in \mathbb{N}$ and $\sup_{n \in \mathbb{N}} \Sigma_{k=1}^n g(\omega(k)) = \infty$. Now define $f: X \to \mathbb{R}$ by

$$f(x) := \begin{cases} 0 & \text{for } x \in X \setminus \omega(\mathbb{N}) \\ \dfrac{1}{\sum\limits_{k=1}^{n} g(\omega(k))} & \text{for } x = \omega(n) \end{cases}$$

Then $f \in c_0(X)$. We show that fg is not summable, which is our desired contradiction. In fact for all $m, n \in \mathbb{N}$

$$\sum_{k=n+1}^{n+m} fg(\omega(k)) > \frac{\sum\limits_{k=n+1}^{n+m} g(\omega(k))}{\sum\limits_{k=1}^{n+m} g(\omega(k))} = 1 - \frac{\sum\limits_{k=1}^{n} g(\omega(k))}{\sum\limits_{k=1}^{n+m} g(\omega(k))}.$$

For each $n \in \mathbb{N}$ there is an $m \in \mathbb{N}$ such that

$$1 - \frac{\sum\limits_{k=1}^{n} g(\omega(k))}{\sum\limits_{k=1}^{n+m} g(\omega(k))} > \frac{1}{2}$$

which yields the conclusion that

$$\sup\left\{ \sum_{x \in A} fg(x) \mid A \subset X, A \text{ finite} \right\} = \infty.$$

Many interesting aspects of the theory of positive linear functionals have come to light in the current exercise. We summarize some of them:

(i) We have seen that positive linear functionals are not unified objects as it might seem on first appearance. For example, notice the significant differences in the continuity properties.

(ii) We have seen that the space of all positive linear functionals on a Riesz lattice has strong order properties. The reader should take note that the comments added to (a) apply equally to (b)–(d).

(iii) We found representation theorems for the π-continuous functionals in the context of functions in $\ell^1(X)$.

(iv) Several connections between the theory of positive linear functionals and problems of the foundations of mathematics became apparent in (c).

We shall continue to keep these aspects in consideration, and diverse new ones will also make their appearance. The reader now should have a sufficient number of concrete examples at hand, and so we return to problems of a general character.

1.3.6(E) Let $(X, \mathcal{L}, \ell)$ be a Daniell space. Prove the following statements:

(α) Given $f \in \mathcal{L}$, $(f_n)_{n \in \mathbb{N}}$ an increasing sequence (resp. a decreasing sequence) from $\mathcal{L}$ with $f = \bigvee_{n \in \mathbb{N}} f_n$ $\mathcal{L}$-a.e. (resp. $f = \bigwedge_{n \in \mathbb{N}} f_n$ $\mathcal{L}$-a.e.), it follows that $\ell(f) = \sup_{n \in \mathbb{N}} \ell(f_n)$ (resp. $\ell(f) = \inf_{n \in \mathbb{N}} \ell(f_n)$).

(β) For two increasing sequences from $\mathcal{L}$, $(f_n)_{n \in \mathbb{N}}$ and $(g_n)_{n \in \mathbb{N}}$, with $\bigvee_{n \in \mathbb{N}} f_n \leq \bigvee_{n \in \mathbb{N}} g_n$ $\mathcal{L}$-a.e., $\sup_{n \in \mathbb{N}} \ell(f_n) \leq \sup_{n \in \mathbb{N}} \ell(g_n)$. An analogous result holds for decreasing sequences.

(γ) Suppose $(f_n)_{n \in \mathbb{N}}$, $(g_n)_{n \in \mathbb{N}}$ are two increasing sequences from $\mathcal{L}$ with $\bigvee_{n \in \mathbb{N}} f_n = \bigvee_{n \in \mathbb{N}} g_n$ $\mathcal{L}$-a.e. Then $\sup_{n \in \mathbb{N}} \ell(f_n) = \sup_{n \in \mathbb{N}} \ell(g_n)$. An analogous result holds for decreasing sequences.

(δ) If $(f_n)_{n \in \mathbb{N}}$ is an increasing and $(g_n)_{n \in \mathbb{N}}$ a decreasing sequence from $\mathcal{L}$ with $\bigvee_{n \in \mathbb{N}} f_n \geq \bigwedge_{n \in \mathbb{N}} g_n$ $\mathcal{L}$-a.e., then $\sup_{n \in \mathbb{N}} \ell(f_n) \geq \inf_{n \in \mathbb{N}} \ell(g_n)$.

Suggestions:

For (α): construct an increasing sequence $(f_n')_{n \in \mathbb{N}}$ such that $f_n' = f_n$ $\mathcal{L}$-a.e. and $f = \bigvee_{n \in \mathbb{N}} f_n'$ (and similarly for the decreasing case).

For (β): observe that for each $n \in \mathbb{N}$, $f_n = \bigvee_{m \in \mathbb{N}} (f_n \wedge g_m)$ $\mathcal{L}$-a.e.

For (δ): construct an increasing sequence $(h_n)_{n \in \mathbb{N}}$ from $\mathcal{L}$ such that $h_n \in \langle f_n \dot{-} g_n \rangle$ for each $n \in \mathbb{N}$ and $\bigvee_{n \in \mathbb{N}} h_n \geq 0$ $\mathcal{L}$-a.e.; hence $\bigvee_{n \in \mathbb{N}} (h_n \wedge 0) = 0$ $\mathcal{L}$-a.e.

1.3.7(E) We return to the considerations in Ex. 1.2.2. Let X be a set, and let $\mathfrak{R} \subset \mathfrak{P}(X)$ possess the properties (i), (ii), and (iii) formulated in Ex. 1.2.2. Finally let $\mathcal{L}$ be defined as in that exercise. Assume further that $\mathfrak{R}$ has the additional property:

(iv) For each countable family $(A_\iota)_{\iota \in I}$ from $\mathfrak{R}$ with $\bigcup_{\iota \in I} A_\iota = X$ there is a finite subset $J \subset I$ with $\bigcup_{\iota \in J} A_\iota = X$.

Prove:

(α) Every positive additive functional, ℓ, on $\mathscr{L}$ is linear and nullcontinuous.

We offer some suggestions for the proof:

STEP 1: First show by complete induction that $\ell(nf) = n\ell(f)$ for each $f \in \mathscr{L}$ and $n \in \mathbb{N}$.

STEP 2: Now establish the null continuity of ℓ. Take a decreasing sequence $(f_n)_{n \in \mathbb{N}}$ from $\mathscr{L}$ with $\bigwedge_{n \in \mathbb{N}} f_n = 0$. Clearly $e_X \in \mathscr{L}$. Take $m \in \mathbb{N}$ and set $A_{mn} := \{mf_n < e_X\}$. We have $A_{mn} \in \mathfrak{R}$ for all $n \in \mathbb{N}$ and $\bigcup_{n \in \mathbb{N}} A_{mn} = X$. Hence there is a finite set $M_m \subset \mathbb{N}$ such that $\bigcup_{n \in M_m} A_{mn} = X$. We get $mf_n < e_X$ for $n \geq \sup_{p \in M_m} p$ and Step 1 implies $\ell(f_n) \leq (1/m)\ell(e_X)$. This yields the desired conclusion.

STEP 3: The linearity of ℓ now follows easily.

We sharpen (iv) somewhat:

(iv′) Given a family $(A_\iota)_{\iota \in I}$ from $\mathfrak{R}$ with $\bigcup_{\iota \in I} A_\iota = X$, there is a finite subset J of I with $\bigcup_{\iota \in J} A_\iota = X$.

Next show that

(β) every positive additive functional ℓ on $\mathscr{L}$ is π-continuous, in the sense of Ex. 1.3.5.

The foregoing consideration can be generalized further. Let $\mathfrak{R}$, $\mathfrak{K}$, $\mathscr{L}_{\mathfrak{K}}$ be defined as in Ex. 1.2.3. We formulate three stipulations:

(v) For each $A \in \mathfrak{K}$ there is an $f \in \mathscr{L}_{\mathfrak{K}}$ such that $A \subset \{f \geq 1\}$.

(vi) For each $A \in \mathfrak{K}$ and for each countable family $(B_\iota)_{\iota \in I}$ from $\mathfrak{R}$ with $A \subset \bigcup_{\iota \in I} B_\iota$, there is a finite subset J of I with $A \subset \bigcup_{\iota \in J} B_\iota$.

(vi′) For each $A \in \mathfrak{K}$ and for each family $(B_\iota)_{\iota \in I}$ from $\mathfrak{R}$ with $A \subset \bigcup_{\iota \in I} B_\iota$, there is a finite subset J of I with $A \subset \bigcup_{\iota \in J} B_\iota$.

Prove the assertions that follow:

(γ) Given (v) and (vi), every positive additive functional on $\mathscr{L}_{\mathfrak{K}}$ is linear and nullcontinuous.

(δ) Given (v) and (vi′), every positive additive functional on $\mathscr{L}_{\mathfrak{K}}$ is linear and π-continuous.

An important example of (δ) occurs with locally compact spaces. Let $(X, \mathfrak{T})$ be a locally compact space. Set $\mathfrak{R} := \mathfrak{T}$ and $\mathfrak{K} := \{A \subset X \mid A \text{ compact}\}$. In this case $\mathscr{L}_{\mathfrak{K}} = \mathscr{K}(X)$ which yields the following assertion:

(ε) Every positive additive functional on $\mathscr{K}(X)$ is linear and π-continuous.

We met a special case in Ex. 1.3.5(a), where $\mathfrak{T} = \mathfrak{P}(X)$ for a set X. Prove the following statements:

(ζ) The set $A \subset X$ is compact with respect to $\mathfrak{T} = \mathfrak{P}(X)$ iff A is finite.

(η) $(X, \mathfrak{P}(X))$ is a locally compact space.

(ϑ) $\mathscr{k}(X) = \mathscr{K}(X)$.

1.3.8(E) Let $\mathscr{L}$ be a Riesz lattice in $\overline{\mathbb{R}}^X$. Take $\mathscr{F} \subset \mathscr{L}_+$ with the following properties:

(i) Given $f, g \in \mathscr{F}$ and $\alpha \in \mathbb{R}_+$, $f + g \in \mathscr{F}$, and $\alpha f \in \mathscr{F}$.
(ii) For each $f \in \mathscr{L}_+$ there is a $g \in \mathscr{F} \dot{-} \mathscr{F}$ with $f = g$ $\mathscr{L}$-a.e.

Let $\ell_0 : \mathscr{F} \to \mathbb{R}_+$ be a functional satisfying the following properties:

(iii) Given $f, g \in \mathscr{F}$ and $\alpha \in \mathbb{R}_+$, $\ell_0(f + g) = \ell_0(f) + \ell_0(g)$ and $\ell_0(\alpha f) = \alpha \ell_0(f)$.
(iv) Given $f, g \in \mathscr{F}$ with $f = g$ $\mathscr{L}$-a.e., $\ell_0(f) = \ell_0(g)$.

Prove:

(α) There is exactly one positive linear functional, ℓ, on $\mathscr{L}$ such that $\ell|_{\mathscr{F}} = \ell_0$.

We turn to the special case of a real Riesz lattice $\mathscr{L}$. Prove:

(β) Given $\ell_0 : \mathscr{L}_+ \to \mathbb{R}_+$ with $\ell_0(f + g) = \ell_0(f) + \ell_0(g)$ and $\ell_0(\alpha f) = \alpha \ell_0(f)$ for $f, g \in \mathscr{L}_+$ and $\alpha \in \mathbb{R}_+$, there is exactly one positive linear functional, ℓ, on $\mathscr{L}$ with $\ell|_{\mathscr{L}_+} = \ell_0$.
If furthermore, for every decreasing sequence $(f_n)_{n \in \mathbb{N}}$ from $\mathscr{L}_+$ with $\bigwedge_{n \in \mathbb{N}} f_n = 0$, $\inf_{n \in \mathbb{N}} \ell_0(f_n) = 0$, then $(X, \mathscr{L}, \ell)$ is a Daniell space.

1.3.9(E) Let $(X, \mathscr{L}, \ell)$ be a Daniell space. Let $(f_n)_{n \in \mathbb{N}}$ be an increasing sequence from $\mathscr{L}_+$ with $\sup_{n \in \mathbb{N}} \ell(f_n) < \infty$. Define $f := \bigvee_{n \in \mathbb{N}} f_n$ and $A := \{f = \infty\}$, and denote by $\mathscr{L}'$ the Riesz lattice generated by $\mathscr{L} \cup \{f\}$. Set

$$\mathscr{F} := \left\{ g \in \mathscr{L}'_+ \,|\, g = \bigvee_{n \in \mathbb{N}} g_n \text{ for some increasing sequence } (g_n)_{n \in \mathbb{N}} \text{ from } \mathscr{L}_+ \right\}$$

and

$$\ell_0 : \mathscr{F} \to \mathbb{R}, \qquad g = \bigvee_{n \in \mathbb{N}} g_n \mapsto \sup_{n \in \mathbb{N}} \ell(g_n).$$

ℓ_0 is well defined (cf., Ex. 1.3.6(γ)). Prove:

(α) $g \vee g' \in \mathscr{F}$, $g \wedge g' \in \mathscr{F}$, $g + g' \in \mathscr{F}$, and $\alpha g \in \mathscr{F}$ whenever $g, g' \in \mathscr{F}$ and $\alpha \in \mathbb{R}_+$.
(β) For every $g \in \mathscr{L}'$ there is a $g' \in \mathscr{F} \dot{-} \mathscr{F}$ with $g = g'$ $\mathscr{L}'$-a.e.
(γ) $\ell_0(g + g') = \ell_0(g) + \ell_0(g')$ and $\ell_0(\alpha g) = \alpha \ell_0(g)$ whenever $g, g' \in \mathscr{F}$ and $\alpha \in \mathbb{R}_+$.
(δ) $\mathfrak{N}(\mathscr{L}') = \{B \cup C \,|\, B \in \mathfrak{N}(\mathscr{L}), C \subset A\}$.
(ε) $\ell_0(g) = \ell_0(g')$ whenever $g, g' \in \mathscr{F}$ such that $g = g'$ $\mathscr{L}'$-a.e.
(ζ) If $g \in \mathscr{F}$ and $g' \in \overline{\mathbb{R}}_+^X$ such that $\{g \neq g'\} \in \mathfrak{N}(\mathscr{L})$, then $g' \in \mathscr{F}$.
(η) For each $g \in \mathscr{L}'_+$ there is a decreasing sequence $(g_n)_{n \in \mathbb{N}}$ in $\mathscr{F}$ such that $\{g \neq \bigwedge_{n \in \mathbb{N}} g_n\} \subset A$.
(ϑ) There is exactly one positive linear functional ℓ' on $\mathscr{L}'$ with $\ell'|_{\mathscr{L}} = \ell$ and $\ell'(f) = \sup_{n \in \mathbb{N}} \ell(f_n)$.
(ι) $(X, \mathscr{L}', \ell')$ is a Daniell space.

We provide some suggestions: (β) follows from Ex. 1.2.9 and (α).

(ε). Take $g, g' \in \mathscr{F}$ such that $g = g'$ $\mathscr{L}'$-a.e. Take $(g_n)_{n\in\mathbb{N}}$ and $(g'_n)_{n\in\mathbb{N}}$ increasing sequences from $\mathscr{L}_+$ such that $g = \bigvee_{n\in\mathbb{N}} g_n$ and $g' = \bigvee_{n\in\mathbb{N}} g'_n$. Then $g + f = g' + f$ $\mathscr{L}'$-a.e. Now $g + f = \bigvee_{n\in\mathbb{N}}(g_n + f_n)$ and $g' + f = \bigvee_{n\in\mathbb{N}}(g'_n + f_n)$ and each of the sequences $(g_n + f_n)_{n\in\mathbb{N}}$ and $(g'_n + f_n)_{n\in\mathbb{N}}$ is increasing. By Ex. 1.3.6(γ)

$$\begin{aligned}\sup_{n\in\mathbb{N}} \ell(g_n) + \sup_{n\in\mathbb{N}} \ell(f_n) &= \sup_{n\in\mathbb{N}} \ell(g_n + f_n)\\ &= \sup_{n\in\mathbb{N}} \ell(g'_n + f_n)\\ &= \sup_{n\in\mathbb{N}} \ell(g'_n) + \sup_{n\in\mathbb{N}} \ell(f_n)\end{aligned}$$

and hence

$$\ell_0(g) = \sup_{n\in\mathbb{N}} \ell(g_n) = \sup_{n\in\mathbb{N}} \ell(g'_n) = \ell_0(g')$$

proving (ε).

(ζ). Take $g \in \mathscr{F}$ and $g' \in \overline{\mathbb{R}}_+^X$ such that $\{g \neq g'\} \in \mathfrak{N}(\mathscr{L})$. There is an increasing sequence $(g_n)_{n\in\mathbb{N}}$ from $\mathscr{L}_+$ with $g = \bigvee_{n\in\mathbb{N}} g_n$. For each $n \in \mathbb{N}$ define

$$g'_n \colon X \to \overline{\mathbb{R}}$$

$$x \mapsto \begin{cases} g_n(x) & \text{if } g(x) = g'(x)\\ g'(x) & \text{if } g(x) \neq g'(x)\end{cases}$$

Then $(g'_n)_{n\in\mathbb{N}}$ is an increasing sequence from $\mathscr{L}$ too, and $g' = \bigvee_{n\in\mathbb{N}} g'_n$.

(η). Take $g \in \mathscr{L}'_+$. Then there is a $g' \in \mathscr{F} \dot{-} \mathscr{F}$ such that $g = g'$ $\mathscr{L}'$-a.e., and we can choose g' such that $\{g \neq g'\} \subset A$ [because of (ζ)]. There exist increasing sequences $(h_n)_{n\in\mathbb{N}}$ and $(h'_n)_{n\in\mathbb{N}}$ from $\mathscr{L}$ with

$$g' \in \left\langle \bigvee_{n\in\mathbb{N}} h_n \dot{-} \bigvee_{n\in\mathbb{N}} h'_n \right\rangle$$

These sequences can be chosen such that $\sup_{n\in\mathbb{N}} h_n(x) - \sup_{n\in\mathbb{N}} h'_n(x)$ is well defined for every $x \in X\setminus A$, and we can choose $h'_n \in \mathbb{R}^X$ for each $n \in \mathbb{N}$. Then for each $n \in \mathbb{N}$,

$$g_n := \bigvee_{m\in\mathbb{N}} ((h_m - h'_n) \vee 0) \in \mathscr{F}.$$

$(g_n)_{n\in\mathbb{N}}$ has the desired properties.

(ϑ) follows from (α)–(γ), (ε), and Ex. 1.3.8(β).

(ι). We have only to verify the null continuity of ℓ'. Let $(g_n)_{n\in\mathbb{N}}$ be a decreasing sequence from $\mathscr{L}'$ such that $\bigwedge_{n\in\mathbb{N}} g_n = 0$. Using ($\eta$), we get for each $n \in \mathbb{N}$ a decreasing sequence $(g_{np})_{p\in\mathbb{N}}$ from $\mathscr{F}$ such that

$$\left\{ g_n \neq \bigwedge_{p\in\mathbb{N}} g_{np} \right\} \subset A.$$

For each $n \in \mathbb{N}$ define $h_n := \bigwedge_{m \le n} g_{mn}$. $(h_n)_{n \in \mathbb{N}}$ is a decreasing sequence from $\mathscr{F}$ such that

$$\left\{ \bigwedge_{n \in \mathbb{N}} h_n \neq 0 \right\} \subset A$$

and $\{h_n < g_n\} \subset A$ for each $n \in \mathbb{N}$. Our claim is that $\inf_{n \in \mathbb{N}} \ell'(h_n) = 0$.

Take $\varepsilon \in \mathbb{R}$, $\varepsilon > 0$. For each $n \in \mathbb{N}$ there is a $h'_n \in \mathscr{L}_+$ with $h'_n \le h_n$ and $\ell'(h_n) - \ell(h'_n) < \varepsilon/2^n$. For each $n \in \mathbb{N}$ define $k_n := \bigwedge_{m \le n} h'_m$. Then $(k_n)_{n \in \mathbb{N}}$ is a decreasing sequence from $\mathscr{L}$ with $\{\bigwedge_{n \in \mathbb{N}} k_n \neq 0\} \subset A$. Furthermore for each $n \in \mathbb{N}$

$$\ell'(h_n) - \ell(k_n) < \sum_{m=1}^{n} \left(\ell'(h_m) - \ell(h'_m) \right) < \varepsilon$$

and hence

$$\inf_{n \in \mathbb{N}} \ell'(h_n) \le \inf_{n \in \mathbb{N}} \ell(k_n) + \varepsilon.$$

Now for each $\alpha \in \mathbb{R}_+ \setminus \{0\}$

$$\bigwedge_{n \in \mathbb{N}} k_n \le \bigvee_{n \in \mathbb{N}} \alpha f_n$$

and therefore $\inf_{n \in \mathbb{N}} \ell(k_n) \le \alpha \sup_{n \in \mathbb{N}} \ell(f_n)$. We conclude that $\inf_{n \in \mathbb{N}} \ell(k_n) = 0$, and because ε was chosen arbitrarily, $\inf_{n \in \mathbb{N}} \ell'(h_n) = 0$.

1.3.10(E) Let $\mathscr{L}$ be a Riesz lattice on a set X. Let ℓ be a positive linear functional on $\mathscr{L}$, and define

$$d_\ell : \mathscr{L} \times \mathscr{L} \to \mathbb{R}, \qquad (f, g) \mapsto \ell(|f \dot{-} g|).$$

Verify the following properties of d_ℓ, given $f, g, h \in \mathscr{L}$, $\alpha \in \mathbb{R}$:

(α) $d_\ell(f, g) \ge 0$.
(β) $d_\ell(f, f) = 0$.
(γ) $d_\ell(f, g) = d_\ell(g, f)$.
(δ) $d_\ell(f, h) \le d_\ell(f, g) + d_\ell(g, h)$.
(ε) $d_\ell(f, g) = 0$ iff $f = g$ $\mathscr{L}$-a.e.
(ζ) $d_\ell(f, h) = d_\ell(f, g) + d_\ell(g, h)$ if $f \le g \le h$.
(η) $d_\ell(f, g) \le d_\ell(f, h)$ if $f \le g \le h$.
(ϑ) $d_\ell(f \dot{+} g, f \dot{+} h) = d_\ell(g, h)$.
(ι) $d_\ell(\alpha f, \alpha g) = |\alpha| d_\ell(f, g)$.
(κ) $|\ell(f) - \ell(g)| \le d_\ell(f, g)$.

Properties (α)–(δ) imply that d_ℓ is a pseudometric on $\mathscr{L}$. Properties (ζ)–(ι) express important compatibility conditions for d_ℓ with the structure of $\mathscr{L}$. Property (κ) implies the uniform continuity of ℓ.

1.3.11(E) In Ex. 1.3.5 we drew attention to the order properties of the set of all positive linear functionals on some particular Riesz lattices. We wish to pay further attention to this in the current exercise.

Let Φ denote the set of all positive linear functionals on a given Riesz lattice $\mathscr{L}$. For $\ell, \ell' \in \Phi$ write $\ell \leq \ell'$ iff $\ell(f) \leq \ell'(f)$ whenever $f \in \mathscr{L}_+$. Prove the following statements:

(α) $\leq$ is an order relation on Φ.

(β) Φ is a conditionally complete lattice with respect to $\leq$.

(γ) Given $\ell, \ell' \in \Phi$ and $f \in \mathscr{L}_+$,

$$(\ell \vee \ell')(f) = \sup\{\ell(g) + \ell'(h) \mid g, h \in \mathscr{L}_+, g + h = f\}$$

$$(\ell \wedge \ell')(f) = \inf\{\ell(g) + \ell'(h) \mid g, h \in \mathscr{L}_+, g + h = f\}.$$

(δ) Given a family, $(\ell_\iota)_{\iota \in I}$, directed upward in Φ with $\sup_{\iota \in I} \ell_\iota(f) < \infty$ whenever $f \in \mathscr{L}_+$, $\bigvee_{\iota \in I} \ell_\iota$ exists and $(\bigvee_{\iota \in I} \ell_\iota)(f) = \sup_{\iota \in I} \ell_\iota(f)$ for every $f \in \mathscr{L}_+$. An analogous result holds for downward directed families.

Thus Φ has, in all generality, the properties determined for special cases in Ex. 1.3.5. Beyond this, Φ has natural algebraic structures. Prove for $\ell, \ell', \ell'' \in \Phi$ and $\alpha \in \mathbb{R}_+$:

(ε) $\ell + \ell' \in \Phi$ (where $+$ denotes the addition of ℓ and ℓ' as real functions).

(ζ) $\alpha\ell \in \Phi$ (with $\alpha\ell$ denoting the product of the real number α with the real function ℓ).

(η) $\ell + \ell'' \leq \ell' + \ell''$ whenever $\ell \leq \ell'$.

(ϑ) $\alpha\ell \leq \alpha\ell'$ whenever $\ell \leq \ell'$.

Properties (η) and (ϑ) describe the compatibility of the algebraic operations with the order relation $\leq$. The properties of Φ just determined have deep consequences, which we will discuss in Chapter 8. Finally, prove the next two statements:

(ι) Let Φ^σ denote the set of all nullcontinuous positive linear functionals on $\mathscr{L}$. Then Φ^σ is a conditionally complete lattice with respect to $\leq$. If $\ell \in \Phi^\sigma$ and $\ell' \in \Phi$ with $\ell' \leq \ell$, then $\ell' \in \Phi^\sigma$. If $\ell, \ell' \in \Phi^\sigma$ and $\alpha \in \mathbb{R}_+$, then $\alpha\ell$, $\ell + \ell' \in \Phi^\sigma$. Property ($\gamma$) also holds for Φ^σ.

(κ) Let Φ^π denote the set of all π-continuous positive linear functionals on $\mathscr{L}$. Then Φ^π is a conditionally complete lattice with respect to $\leq$. If $\ell \in \Phi^\pi$ and $\ell' \in \Phi$ with $\ell' \leq \ell$, then $\ell' \in \Phi^\pi$. If $\ell, \ell' \in \Phi^\pi$ and $\alpha \in \mathbb{R}_+$, then $\alpha\ell, \ell + \ell' \in \Phi^\pi$. Property ($\gamma$) also applies to Φ^π.

We note that (δ) also applies to both Φ^σ and Φ^π. The interested reader may wish to verify this, even though somewhat greater investment of effort is required than in the proofs of (ι) and (κ).

1.4. CLOSED DANIELL SPACES

NOTATION FOR SECTION 1.4:

X denotes a set.

Linear functionals are of great significance, both within mathematics and in the applications of mathematics. Numerous problems lead naturally to operations with functionals. Such functionals are especially useful when they are compatible with other fundamental structures, in particular with topological structures. Daniell spaces are quite weak in this respect. For example, the Riemann-integrable functions on an interval of $\mathbb{R}$, viewed with the Riemann integral as functional, form a Daniell space. The reader will already have seen how cumbersome the Riemann integral is for working with convergent sequences of functions.

In this section we investigate a stronger functional convergence, one that is suitable when topological compatibility is desired.

Definition 1.4.1. *Let ℓ be an increasing functional on a subset $\mathscr{F}$ of $\overline{\mathbb{R}}^X$. A sequence $(f_n)_{n\in\mathbb{N}}$ from $\mathscr{F}$ is called an **ℓ-sequence** iff $(f_n)_{n\in\mathbb{N}}$ is monotone and $(\ell(f_n))_{n\in\mathbb{N}}$ is bounded in $\mathbb{R}$. The triple $(X, \mathscr{F}, \ell)$ is said to be **closed** iff, for every ℓ-sequence $(f_n)_{n\in\mathbb{N}}$ from $\mathscr{F}$, the function $\lim_{n\to\infty} f_n$ belongs to $\mathscr{F}$ and*

$$\ell\Big(\lim_{n\to\infty} f_n\Big) = \lim_{n\to\infty} \ell(f_n). \qquad \square$$

Notice that for every monotone sequence $(f_n)_{n\in\mathbb{N}}$ from $\overline{\mathbb{R}}^X$, the function $\lim_{n\to\infty} f_n$ exists. Also, if ℓ is an increasing functional on some subset $\mathscr{F}$ of $\overline{\mathbb{R}}^X$, then for every ℓ-sequence $(f_n)_{n\in\mathbb{N}}$ from $\mathscr{F}$, the sequence $(\ell(f_n))_{n\in\mathbb{N}}$ converges in $\mathbb{R}$. For then $(\ell(f_n))_{n\in\mathbb{N}}$ is a monotone sequence of real numbers that is bounded in $\mathbb{R}$.

The closure property just defined is an axiomatization of the theorem from classical integration theory known as the Beppo Levi Theorem (1906). Daniell spaces that are also closed are the primary objects, from an abstract point of view, to be studied in this first part of the book.

Given a Riesz lattice $\mathscr{L}$ in $\overline{\mathbb{R}}^X$ and a positive linear functional ℓ on $\mathscr{L}$, if the triple $(X, \mathscr{L}, \ell)$ happens to be closed, then it is immediate from the definitions that the functional ℓ is nullcontinuous. Thus to show that a given triple is a closed Daniell space, it is redundant to verify the null continuity. In other words, we have the following equivalence.

Proposition 1.4.2. *For every subset $\mathscr{L}$ of $\overline{\mathbb{R}}^X$ and every functional ℓ on $\mathscr{L}$, $(X, \mathscr{L}, \ell)$ is a closed Daniell space iff $\mathscr{L}$ is a Riesz lattice in $\overline{\mathbb{R}}^X$, ℓ is a positive linear functional on $\mathscr{L}$, and the triple $(X, \mathscr{L}, \ell)$ is closed.* $\square$

It is convenient to have even weaker characterizations of closed Daniell spaces. For ℓ a positive linear functional on a Riesz lattice $\mathscr{L}$, it is not necessary to check all ℓ-sequences in order to decide whether $(X, \mathscr{L}, \ell)$ is closed. It suffices, for instance, to consider only increasing ℓ-sequences or even only positive increasing ℓ-sequences.

Proposition 1.4.3. *The following assertions are equivalent, for every positive linear functional ℓ on a Riesz lattice $\mathscr{L}$ in $\overline{\mathbb{R}}^X$.*

(*a*) *The triple $(X, \mathscr{L}, \ell)$ is closed; that is, $(X, \mathscr{L}, \ell)$ is a closed Daniell space.*

(*b*) *For every increasing ℓ-sequence $(f_n)_{n \in \mathbb{N}}$ from $\mathscr{L}$ the function $\bigvee_{n \in \mathbb{N}} f_n$ belongs to $\mathscr{L}$ and*

$$\ell\Big(\bigvee_{n \in \mathbb{N}} f_n\Big) = \sup_{n \in \mathbb{N}} \ell(f_n).$$

(*c*) *For every increasing ℓ-sequence $(f_n)_{n \in \mathbb{N}}$ of positive functions from $\mathscr{L}$, the function $\bigvee_{n \in \mathbb{N}} f_n$ belongs to $\mathscr{L}$ and*

$$\ell\Big(\bigvee_{n \in \mathbb{N}} f_n\Big) = \sup_{n \in \mathbb{N}} \ell(f_n).$$

(*d*) *For every decreasing ℓ-sequence $(f_n)_{n \in \mathbb{N}}$ from $\mathscr{L}$, the function $\bigwedge_{n \in \mathbb{N}} f_n$ belongs to $\mathscr{L}$ and*

$$\ell\Big(\bigwedge_{n \in \mathbb{N}} f_n\Big) = \inf_{n \in \mathbb{N}} \ell(f_n).$$

Proof. (a) $\Rightarrow$ (b) $\Rightarrow$ (c) is evident.

(c) $\Rightarrow$ (d). Let $(f_n)_{n \in \mathbb{N}}$ be a decreasing ℓ-sequence from $\mathscr{L}$. Thus the sequences $(f_n)_{n \in \mathbb{N}}$ and $(\ell(f_n))_{n \in \mathbb{N}}$ both decrease, and $(\ell(f_n))_{n \in \mathbb{N}}$ is bounded in $\mathbb{R}$. For each n in $\mathbb{N}$, define

$$h_n \colon X \to \overline{\mathbb{R}},$$

$$x \mapsto \begin{cases} f_1(x) - f_n(x) & \text{if defined} \\ 0 & \text{otherwise.} \end{cases}$$

Then $h_n \in \langle f_1 \dot{-} f_n \rangle$, so h_n belongs to $\mathscr{L}$ and

$$\ell(h_n) = \ell(f_1) - \ell(f_n).$$

In fact $(h_n)_{n \in \mathbb{N}}$ is an increasing sequence of positive functions from $\mathscr{L}$, and the sequence $(\ell(h_n))_{n \in \mathbb{N}}$ is bounded in $\mathbb{R}$. According to (c), the function

$\bigvee_{n\in\mathbb{N}} h_n$ belongs to $\mathscr{L}$ and

$$\ell\Big(\bigvee_{n\in\mathbb{N}} h_n\Big) = \sup_{n\in\mathbb{N}} \ell(h_n).$$

But

$$\bigwedge_{n\in\mathbb{N}} f_n \in \Big\langle f_1 \dot{-} \bigvee_{n\in\mathbb{N}} h_n \Big\rangle$$

so $\bigwedge_{n\in\mathbb{N}} f_n$ belongs to $\mathscr{L}$. Moreover

$$\begin{aligned}\ell\Big(\bigwedge_{n\in\mathbb{N}} f_n\Big) &= \ell(f_1) - \ell\Big(\bigvee_{n\in\mathbb{N}} h_n\Big)\\ &= \ell(f_1) - \sup_{n\in\mathbb{N}} \ell(h_n)\\ &= \ell(f_1) - \sup_{n\in\mathbb{N}} \big(\ell(f_1) - \ell(f_n)\big)\\ &= \inf_{n\in\mathbb{N}} \ell(f_n).\end{aligned}$$

(d) $\Rightarrow$ (a). Let $(f_n)_{n\in\mathbb{N}}$ be an increasing ℓ-sequence from $\mathscr{L}$. Then the sequence $(-f_n)_{n\in\mathbb{N}}$ decreases, and since

$$\ell(-f_n) = -\ell(f_n)$$

the sequence $(\ell(-f_n))_{n\in\mathbb{N}}$ is bounded in $\mathbb{R}$. In other words, $(-f_n)_{n\in\mathbb{N}}$ is a decreasing ℓ-sequence from $\mathscr{L}$. From (d) we conclude that $\bigwedge_{n\in\mathbb{N}}(-f_n)$ belongs to $\mathscr{L}$ and

$$\ell\Big(\bigwedge_{n\in\mathbb{N}} (-f_n)\Big) = \inf_{n\in\mathbb{N}} \big(-\ell(f_n)\big).$$

But then $\lim_{n\to\infty} f_n$, which can be written as $-\bigwedge_{n\in\mathbb{N}}(-f_n)$, must also belong to $\mathscr{L}$, and

$$\begin{aligned}\ell\Big(\lim_{n\to\infty} f_n\Big) &= -\ell\Big(\bigwedge_{n\in\mathbb{N}} (-f_n)\Big)\\ &= -\inf_{n\in\mathbb{N}} \big(-\ell(f_n)\big)\\ &= \sup_{n\in\mathbb{N}} \ell(f_n)\\ &= \lim_{n\to\infty} \ell(f_n).\end{aligned}$$

What we have just shown, when combined with (d) itself and the fact that ℓ increases, yields (a). $\square$

For closed Daniell spaces the exceptional sets and functions have special properties, as does the behavior of the functional relative to a.e. conditions. The next several propositions describe these special properties.

Proposition 1.4.4. *If $(X, \mathscr{L}, \ell)$ is a closed Daniell space, then the following conditions on a function f in $\overline{\mathbb{R}}^X$ are equivalent.*

(*a*) $f \in \mathscr{N}(\mathscr{L})$.

(*b*) $|f| \in \mathscr{L}$ *and* $\ell(|f|) = 0$.

Proof. (a) ⇒ (b) is true for any positive linear functional ℓ on any Riesz lattice $\mathscr{L}$. (1.3.3 (a)).

(b) ⇒ (a). It suffices, by Proposition 1.2.23, to show that $\infty|f|$ belongs to $\mathscr{L}$. The facts that

$$\infty|f| = \bigvee_{n \in \mathbb{N}} n|f|$$

and

$$\ell(n|f|) = n\ell(|f|) = 0$$

provide what we need. Indeed, the sequence $(n|f|)_{n \in \mathbb{N}}$ is an increasing ℓ-sequence from $\mathscr{L}$. Since the triple $(X, \mathscr{L}, \ell)$ is closed, we conclude that $\infty|f|$ belongs to $\mathscr{L}$. □

Corollary 1.4.5. *If $(X, \mathscr{L}, \ell)$ is a closed Daniell space, then the following conditions on a subset A of X are equivalent.*

(*a*) $A \in \mathfrak{N}(\mathscr{L})$.

(*b*) $e_A \in \mathscr{L}$ *and* $\ell(e_A) = 0$. □

Thus for a closed Daniell space $(X, \mathscr{L}, \ell)$ the exceptional sets associated with the Riesz lattice $\mathscr{L}$ are exactly those subsets of X whose characteristic functions have "integral" zero. The $\mathscr{L}$-exceptional functions are exactly those functions in ℓ whose absolute values have "integral" zero. In this situation "$\mathscr{L}$-a.e." means the same as "except on a set whose characteristic function has 'integral' zero."

Proposition 1.4.6. *Let $(X, \mathscr{L}, \ell)$ be a closed Daniell space, and let f and g be functions in $\mathscr{L}$ such that*

$$f \leq g \text{ and } \ell(f) = \ell(g).$$

Then

$$f = g\ \mathscr{L}\text{-a.e.}$$

Proof. Consider the positive function

$$f': X \to \overline{\mathbb{R}},$$

$$x \mapsto \begin{cases} g(x) - f(x) & \text{if defined} \\ 0 & \text{otherwise.} \end{cases}$$

Evidently $f' \in \langle g \dot{-} f \rangle$. Hence f' is in $\mathscr{L}_+$, and

$$\ell(f') = \ell(g) - \ell(f) = 0.$$

It follows (by Proposition 4) that f' is an $\mathscr{L}$-exceptional function, and so

$$\{f \neq g\} = \{f' \neq 0\} \in \mathfrak{N}(\mathscr{L}).$$

In other words, $f = g$ $\mathscr{L}$-a.e. □

As a consequence of Proposition 6, every function h that is sandwiched between two functions f, g from $\mathscr{L}$ for which $\ell(f) = \ell(g)$ must also belong to $\mathscr{L}$ and satisfy $\ell(h) = \ell(f) = \ell(g)$. It is useful to have a name for this property, so we introduce a definition first.

Definition 1.4.7. *Let ℓ be a functional on a subset $\mathscr{F}$ of $\overline{\mathbb{R}}^X$. The triple $(X, \mathscr{F}, \ell)$ is said to be **complete** iff the following axiom holds: for all f, $g \in \mathscr{F}$ and for every $h \in \overline{\mathbb{R}}^X$, if*

$$f \leq h \leq g \text{ and } \ell(f) = \ell(g)$$

then h belongs to $\mathscr{F}$ and

$$\ell(h) = \ell(f) = \ell(g).$$ □

Corollary 1.4.8. *Every closed Daniell space is complete.*

Proof. Let $(X, \mathscr{L}, \ell)$ be a closed Daniell space, let f and g belong to $\mathscr{L}$, and let h be an extended-real function on X for which

$$f \leq h \leq g \quad \text{and} \quad \ell(f) = \ell(g).$$

By Proposition 6, the functions f and g are $\mathscr{L}$-a.e. equal. The inclusion

$$\{f \neq h\} \subset \{f \neq g\}$$

shows that

$$f = h \ \mathscr{L}\text{-a.e.}$$

Thus h is $\mathscr{L}$-a.e. equal to a function in $\mathscr{L}$. We conclude that h is in $\mathscr{L}$ and

$$\ell(h) = \ell(f) = \ell(g).$$

(Proposition 1.3.3). □

Finite unions of $\mathscr{L}$-exceptional sets are still $\mathscr{L}$-exceptional. When $\mathscr{L}$ appears in some closed Daniell space, "finite" can be replaced by "countable."

Proposition 1.4.9. *If $(X, \mathscr{L}, \ell)$ is a closed Daniell space and $(A_\iota)_{\iota \in I}$ is a countable family from $\mathfrak{N}(\mathscr{L})$, then $\bigcup_{\iota \in I} A_\iota$ also belongs to $\mathfrak{N}(\mathscr{L})$.*

Proof. It suffices to treat the case $I = \mathbb{N}$. So let $(A_n)_{n \in \mathbb{N}}$ be a sequence from $\mathfrak{N}(\mathscr{L})$. The sequence $(f_n)_{n \in \mathbb{N}}$, where

$$f_n := e_{\bigcup_{m \le n} A_m}$$

for all $n \in \mathbb{N}$, is an ℓ-sequence from $\mathscr{L}$: it increases and $\ell(f_n) = 0$ for every n (Proposition 1.2.24, and Corollary 5). Because the triple $(X, \mathscr{L}, \ell)$ is closed, we conclude that the function $\bigvee_{n \in \mathbb{N}} f_n$ belongs to $\mathscr{L}$ and

$$\ell\left(\bigvee_{n \in \mathbb{N}} f_n\right) = \sup_{n \in \mathbb{N}} \ell(f_n) = 0.$$

But $\bigvee_{n \in \mathbb{N}} f_n$ is just the characteristic function of the set $\bigcup_{n \in \mathbb{N}} A_n$. In view of Corollary 5, $\bigcup_{n \in \mathbb{N}} A_n$ belongs to $\mathfrak{N}(\mathscr{L})$. □

Corollary 1.4.10. *If $(X, \mathscr{L}, \ell)$ is a closed Daniell space, then $\mathscr{N}(\mathscr{L})$ is σ-completely embedded in $\overline{\mathbb{R}}^X$.*

Proof. It suffices to show that the supremum of every increasing sequence of positive $\mathscr{L}$-exceptional functions is itself $\mathscr{L}$-exceptional (Proposition 1.2.19). If $(f_n)_{n \in \mathbb{N}}$ is such a sequence, then

$$\left\{\bigvee_{n \in \mathbb{N}} f_n \neq 0\right\} = \bigcup_{n \in \mathbb{N}} \{f_n \neq 0\}$$

and $\bigvee_{n \in \mathbb{N}} f_n$, in view of Proposition 9, must belong to $\mathscr{N}(\mathscr{L})$. □

For an arbitrary Riesz lattice $\mathscr{L}$, function-value differences that occur only on $\mathscr{L}$-exceptional sets can safely be ignored when taking the supremum or infimum of a finite family from $\mathscr{L}$ [Proposition 1.2.33(d), (e)]. As another corollary to 1.4.9, "finite" can be replaced by "countable," provided the context is that of a closed Daniell space.

Corollary 1.4.11. *Suppose that $(X, \mathscr{L}, \ell)$ is a closed Daniell space and that $(f_\iota)_{\iota \in I}$ and $(g_\iota)_{\iota \in I}$ are countable families from $\overline{\mathbb{R}}^X$ such that*

$$f_\iota \leq g_\iota \ \mathscr{L}\text{-a.e.}$$

for every ι in I. Then

$$\bigwedge_{\iota \in I} f_\iota \leq \bigwedge_{\iota \in I} g_\iota \ \mathscr{L}\text{-a.e.}$$

and

$$\bigvee_{\iota \in I} f_\iota \leq \bigvee_{\iota \in I} g_\iota \ \mathscr{L}\text{-a.e.} \qquad \square$$

Having discussed the properties of the exceptional sets and exceptional functions associated with closed Daniell spaces, we turn to the more significant question of the compatibility of closed Daniell spaces and their functionals with the topology of $\overline{\mathbb{R}}^X$. The remainder of this section describes both how the Riesz lattice $\mathscr{L}$ is embedded in $\overline{\mathbb{R}}^X$ and the accompanying convergence behavior of the functional ℓ.

Theorem 1.4.12. *Let $(X, \mathscr{L}, \ell)$ be a closed Daniell space, and let $(f_\iota)_{\iota \in I}$ be a nonempty, countable family from $\mathscr{L}$.*

(*a*) *If the family $(f_\iota)_{\iota \in I}$ is directed upward relative to the relation $\leq \mathscr{L}$-a.e., then the two conditions*

$$\sup_{\iota \in I} \ell(f_\iota) < \infty$$

and

$$\bigvee_{\iota \in I} f_\iota \in \mathscr{L}$$

are equivalent, and each implies that

$$\ell\left(\bigvee_{\iota \in I} f_\iota\right) = \sup_{\iota \in I} \ell(f_\iota).$$

(*b*) *If the family $(f_\iota)_{\iota \in I}$ is directed downward relative to the relation $\leq \mathscr{L}$-a.e., then the two conditions*

$$\inf_{\iota \in I} \ell(f_\iota) > -\infty$$

and

$$\bigwedge_{\iota \in I} f_\iota \in \mathscr{L}$$

are equivalent, and each implies that

$$\ell\left(\bigwedge_{\iota \in I} f_\iota\right) = \inf_{\iota \in I} \ell(f_\iota).$$

Proof. (a) Let $\varphi: \mathbb{N} \to I$ be surjective. Construct an increasing sequence $(g_n)_{n \in \mathbb{N}}$ in $\mathscr{L}$ by setting

$$g_n := \bigvee_{m \leq n} f_{\varphi(m)}.$$

Clearly

$$\bigvee_{n \in \mathbb{N}} g_n = \bigvee_{\iota \in I} f_\iota.$$

Moreover, as we see by complete induction, there exists for each n in $\mathbb{N}$ an index ι_n in I such that

$$g_n \leq f_{\iota_n} \ \mathscr{L}\text{-a.e.}$$

and thus, by Proposition 1.3.3(c), such that

$$\ell(g_n) \leq \ell(f_{\iota_n}).$$

If

$$\sup_{\iota \in I} \ell(f_\iota) < \infty$$

then

$$\sup_{n \in \mathbb{N}} \ell(g_n) < \infty.$$

In this case $(g_n)_{n \in \mathbb{N}}$ is an increasing ℓ-sequence from $\mathscr{L}$. Since the triple $(X, \mathscr{L}, \ell)$ is closed, we conclude in this case that $\bigvee_{n \in \mathbb{N}} g_n$ belongs to $\mathscr{L}$ and thus $\bigvee_{\iota \in I} f_\iota$ belongs to $\mathscr{L}$. Conversely, if we assume that $\bigvee_{\iota \in I} f_\iota$ belongs to $\mathscr{L}$, then the monotonicity of ℓ yields

$$\sup_{\iota \in I} \ell(f_\iota) \leq \ell\left(\bigvee_{\iota \in I} f_\iota\right) < \infty.$$

Finally, if either of the two conditions holds, then so does the other, and

$$\begin{aligned} \ell\left(\bigvee_{\iota \in I} f_\iota\right) &= \ell\left(\bigvee_{n \in \mathbb{N}} g_n\right) \\ &= \sup_{n \in \mathbb{N}} \ell(g_n) \\ &\leq \sup_{n \in \mathbb{N}} \ell(f_{\iota_n}) \\ &\leq \sup_{\iota \in I} \ell(f_\iota) \\ &\leq \ell\left(\bigvee_{\iota \in I} f_\iota\right) \end{aligned}$$

from which it follows that

$$\ell\left(\bigvee_{\iota\in I} f_\iota\right) = \sup_{\iota\in I} \ell(f_\iota).$$

(b) If $(f_\iota)_{\iota\in I}$ is directed downward relative to $\leq \mathscr{L}$-a.e., then the family $(-f_\iota)_{\iota\in I}$ is directed upward relative to $\leq \mathscr{L}$-a.e. Also, as is always the case,

$$\bigwedge_{\iota\in I} f_\iota = -\bigvee_{\iota\in I}(-f_\iota)$$

$$\inf_{\iota\in I} \ell(f_\iota) = -\sup_{\iota\in I} \ell(-f_\iota).$$

Thus (b) can be deduced from (a). □

Proposition 1.4.13. *Let $(X, \mathscr{L}, \ell)$ be a closed Daniell space, and let $(f_\iota)_{\iota\in I}$ be a nonempty, countable family from $\mathscr{L}$.*

(a) If there exists a function g in $\mathscr{L}$ such that $f_\iota \leq g$ $\mathscr{L}$-a.e. for every ι in I, then $\bigvee_{\iota\in I} f_\iota$ belongs to $\mathscr{L}$ and

$$\ell\left(\bigvee_{\iota\in I} f_\iota\right) \geq \sup_{\iota\in I} \ell(f_\iota).$$

(b) If there exists a function g in $\mathscr{L}$ such that $f_\iota \geq g$ $\mathscr{L}$-a.e. for every ι in I, then $\bigwedge_{\iota\in I} f_\iota$ belongs to $\mathscr{L}$ and

$$\ell\left(\bigwedge_{\iota\in I} f_\iota\right) \leq \inf_{\iota\in I} \ell(f_\iota).$$

Proof. (a) Let $\varphi: \mathbb{N} \to I$ be surjective. As in the proof of the preceding theorem we construct an increasing sequence $(g_n)_{n\in\mathbb{N}}$ in $\mathscr{L}$ by putting

$$g_n := \bigvee_{m\leq n} f_{\varphi(m)}.$$

The hypothesis ensures, for every n, that

$$g_n \leq g \ \mathscr{L}\text{-a.e.}$$

hence that

$$\ell(g_n) \leq \ell(g).$$

Once again, $(g_n)_{n\in\mathbb{N}}$ is an increasing ℓ-sequence from $\mathscr{L}$. We conclude that

$$\bigvee_{\iota\in I} f_\iota = \bigvee_{n\in\mathbb{N}} g_n \in \mathscr{L}.$$

The relation

$$\ell\Big(\bigvee_{\iota\in I} f_\iota\Big) \geq \sup_{\iota\in I} \ell(f_\iota)$$

follows from $\bigvee_{\iota\in I} f_\iota \geq f_\kappa$ which holds for every $\kappa \in I$.

(b) One can prove (b) analogously or instead by applying (a) to the family $(-f_\iota)_{\iota\in I}$. □

Corollary 1.4.14. *For every closed Daniell space $(X, \mathscr{L}, \ell)$, the Riesz lattice $\mathscr{L}$ is conditionally σ-completely embedded in $\overline{\mathbb{R}}^X$.* □

Corollary 1.4.15. *Let $(X, \mathscr{L}, \ell)$ be a closed Daniell space, and let $(f_n)_{n\in\mathbb{N}}$ be a sequence from $\mathscr{L}$. If there exists a function g in $\mathscr{L}$ such that*

$$|f_n| \leq g \ \mathscr{L}\text{-a.e.}$$

for every n, then the following assertions hold.

(*a*) $\limsup_{n\to\infty} f_n$ *belongs to* $\mathscr{L}$, *and* $\ell(\limsup_{n\to\infty} f_n) \geq \limsup_{n\to\infty} \ell(f_n)$.

(*b*) $\liminf_{n\to\infty} f_n$ *belongs to* $\mathscr{L}$, *and* $\ell(\liminf_{n\to\infty} f_n) \leq \liminf_{n\to\infty} \ell(f_n)$.

Proof. (a) For every n in $\mathbb{N}$, we set

$$g_n := \bigvee_{m\geq n} f_m.$$

Proposition 13(a) ensures that each g_n belongs to $\mathscr{L}$ and

$$\ell(g_n) \geq \sup_{m\geq n} \ell(f_m).$$

Since

$$\limsup_{n\to\infty} f_n = \bigwedge_{n\in\mathbb{N}} g_n$$

and

$$\limsup_{n\to\infty} \ell(f_n) = \inf_{n\in\mathbb{N}} \sup_{m\geq n} \ell(f_m) \leq \inf_{n\in\mathbb{N}} \ell(g_n)$$

the proof will be finished if we can show that $\bigwedge_{n\in\mathbb{N}} g_n$ belongs to $\mathscr{L}$ and that

$$\ell\Big(\bigwedge_{n\in\mathbb{N}} g_n\Big) \geq \inf_{n\in\mathbb{N}} \ell(g_n).$$

Actually

$$\ell\Big(\bigwedge_{n\in\mathbb{N}} g_n\Big) = \inf_{n\in\mathbb{N}} \ell(g_n) \tag{1}$$

and both of the desired results follow directly from the fact that the triple $(X, \mathcal{L}, \ell)$ is closed. The sequence $(g_n)_{n \in \mathbb{N}}$, which lies in $\mathcal{L}$, decreases. Moreover the sequence $(\ell(g_n))_{n \in \mathbb{N}}$ is bounded, since $g_n \geq -g$ $\mathcal{L}$-a.e. and therefore

$$\ell(g_n) \geq -\ell(g).$$

Thus $(g_n)_{n \in \mathbb{N}}$ is a decreasing ℓ-sequence from $\mathcal{L}$, and it follows, as claimed, that $\bigwedge_{n \in \mathbb{N}} g_n$ belongs to $\mathcal{L}$ and (1) holds.

(b) One proves (b) analogously. □

From this embedding theorem we obtain the following fundamental theorem describing the behavior of the functional ℓ relative to order convergence in $\overline{\mathbb{R}}^X$. Named after H. Lebesgue, the creator of modern integration theory, Theorem 16 is of great significance not only for the theory of integration but also for the applications of the theory.

Theorem 1.4.16 (Lebesgue Dominated Convergence Theorem). *Let $(X, \mathcal{L}, \ell)$ be a closed Daniell space, and let $(f_n)_{n \in \mathbb{N}}$ be a sequence from $\mathcal{L}$. Suppose that f is a function in $\overline{\mathbb{R}}^X$ such that*

$$\lim_{n \to \infty} f_n(x) = f(x) \ \mathcal{L}\text{-}a.e.$$

Suppose also that

$$|f_n| \leq g \ \mathcal{L}\text{-}a.e.$$

for some function g in $\mathcal{L}$ and for every n in $\mathbb{N}$. Then the limit function f belongs to $\mathcal{L}$ and

$$\ell(f) = \lim_{n \to \infty} \ell(f_n).$$

Proof. At every point x in X where $\lim_{n \to \infty} f_n(x)$ exists,

$$\lim_{n \to \infty} f_n(x) = \limsup_{n \to \infty} f_n(x) = \liminf_{n \to \infty} f_n(x).$$

Thus the hypothesis on f says that

$$f = \limsup_{n \to \infty} f_n = \liminf_{n \to \infty} f_n \ \mathcal{L}\text{-a.e.}$$

Corollary 15 implies, for one thing, that both $\limsup_{n \to \infty} f_n$ and $\liminf_{n \to \infty} f_n$ belong to $\mathcal{L}$. Any function that is $\mathcal{L}$-a.e. equal to a function in $\mathcal{L}$ is itself in $\mathcal{L}$ and must be assigned the same value by the functional ℓ. Thus f is in $\mathcal{L}$, and we conclude, using the inequalities resulting from Corollary 15, that

$$\begin{aligned}\limsup_{n\to\infty} \ell(f_n) &\le \ell\Big(\limsup_{n\to\infty} f_n\Big)\\ &= \ell(f)\\ &= \ell\Big(\liminf_{n\to\infty} f_n\Big)\\ &\le \liminf_{n\to\infty} \ell(f_n).\end{aligned}$$

Since the opposite inequality

$$\liminf_{n\to\infty} \ell(f_n) \le \limsup_{n\to\infty} \ell(f_n)$$

always holds, it follows that

$$\ell(f) = \limsup_{n\to\infty} \ell(f_n) = \liminf_{n\to\infty} \ell(f_n) = \lim_{n\to\infty} \ell(f_n). \qquad \square$$

Closed Daniell spaces are the abstract objects studied in the first part of this book. What does this study include?

First it should be stressed that the functionals arising from concrete problems usually do not yield closed triples. However, they are usually nullcontinuous and thus yield Daniell spaces. A fundamental assertion of integration theory is that every Daniell space can be extended to a closed Daniell space. In effect, every nullcontinuous functional can be extended to a functional with the nice convergence properties described in this section.

The extension of nullcontinuous functionals forms the primary task for the first part of the book. Our method relies very heavily on the ideas developed in 1918 by Daniell. Before beginning this construction we familiarize ourselves, in the next chapter, with some nontrivial very important examples of Daniell spaces.

EXERCISES

1.4.1(E) We consider the example from Ex. 1.3.1 once again. We use all the same notation. Let $\Psi(X)$ be the set of all Daniell spaces $(X, \mathcal{L}', \ell')$ with $(X, \mathfrak{k}(X), \ell) \preccurlyeq (X, \mathcal{L}', \ell')$. Prove the following statements:

(α) $(X, \ell^1(X), \bar{\ell})$ is closed.

(β) $(X, \ell^1(X), \bar{\ell})$ is the only closed space in $\Psi(X)$.

(γ) $(X, \ell^1(X), \bar{\ell})$ is the largest element in $\Psi(X)$ with respect to $\preccurlyeq$.

1.4.2(E) We now investigate the generalizations in Ex. 1.3.2 of the examples discussed earlier. We adopt the notation of Ex 1.3.2. Take $g \in \mathbb{R}_+^X$. Denote by $\Psi(X, g)$ the set of all Daniell spaces $(X, \mathscr{L}', \ell')$ with $(X, \mathscr{k}(X), \ell_g) \preccurlyeq (X, \mathscr{L}', \ell')$. Prove the following statements:

(α) $(X, \ell_g^1(X), \bar{\ell}_g)$ is closed.
(β) $(X, \ell_g^1(X), \bar{\ell}_g)$ is a maximal element in $\Psi(X, g)$.

Now set

$$\hat{\mathscr{k}}_g(X) := \left\{ f \in \ell_g^1(X) \mid \{ f \neq 0\} \text{ is countable} \right\}$$

and

$$\hat{\ell}_g := \bar{\ell}_g \mid_{\hat{\mathscr{k}}_g(X)}.$$

Prove the following statements:

(γ) $(X, \hat{\mathscr{k}}_g(X), \hat{\ell}_g) \in \Psi(X, g)$.
(δ) $(X, \hat{\mathscr{k}}_g(X), \hat{\ell}_g)$ is closed.
(ε) The next three propositions are equivalent.

(ε1) $(X, \hat{\mathscr{k}}_g(X), \hat{\ell}_g) = (X, \ell_g^1(X), \bar{\ell}_g)$.
(ε2) $\{g = 0\}$ is countable.
(ε3) $(X, \ell_g^1(X), \bar{\ell}_g)$ is the largest element in $\Psi(X, g)$.

We deduce from this fact that $(X, \ell_g^1(X), \bar{\ell}_g)$ is, in general, not the only closed extension of $(X, \mathscr{k}(X), \ell_g)$, in contrast to the situation with $(X, \ell^1(X), \bar{\ell})$. Thus Daniell spaces can have several distinct closed extensions. Prove the following statements:

(ζ) $(X, \mathscr{k}(X), \ell_g)$ is closed iff X is finite and $g(x) > 0$ for each $x \in X$.
(η) $(X, \ell_g^1(X) \cap \mathbb{R}^X, \bar{\ell}_g)$ is closed iff $g(x) > 0$ for each $x \in X$. (In this case $\ell_g^1(X) \subset \mathbb{R}^X$.)
(ϑ) Setting $\mathscr{F} := \{f \in \mathbb{R}^X \mid \{f \neq 0\}$ countable$\}$, $(X, \ell_g^1(X) \cap \mathscr{F}, \bar{\ell}_g)$ is closed iff $g(x) > 0$ for each $x \in X$. (In this case $\ell_g^1(X) \subset \mathscr{F}$.)
(ι) $(X, \ell_g^1(X) \cap \ell^\infty(X), \bar{\ell}_g)$ is closed iff $\inf_{x \in X} g(x) > 0$.
(κ) $(X, \ell_g^1(X) \cap c_0(X), \bar{\ell}_g)$ is closed iff $\inf_{x \in X} g(x) > 0$.

For the definitions of $\ell^\infty(X)$ and $c_0(X)$, see Ex. 1.2.1.

1.4.3(E) Let $(X, \mathscr{L}, \ell)$ be a Daniell space.

(α) The following propositions are equivalent:

(α1) $\mathscr{L} = \overline{\mathbb{R}}^X$.
(α2) $\mathfrak{P}(X) = \mathfrak{N}(\mathscr{L})$.

In the case of either (α1) or (α2) being fulfilled, $\ell = 0$ and $(X, \mathscr{L}, \ell)$ is closed.

(β) The following propositions are equivalent:

(β1) $\mathscr{L} = \{f \in \overline{\mathbb{R}}^X \mid \{f \neq 0\}$ countable$\}$.
(β2) $\mathfrak{N}(\mathscr{L}) = \{A \subset X \mid A$ countable$\}$.

If either (β1) or (β2) holds, then $\ell = 0$ and $(X, \mathscr{L}, \ell)$ is closed.

1.4.4(E) We return to the Riemann-Stieltjes integral ℓ^g, with respect to the increasing function g defined on the closed interval $[a, b]$, continuing from Ex. 1.3.3. Prove the following statements:

(α) If g is continuous at $x \in [a, b]$, then $e_{\{x\}} \in \mathscr{L}^g$ and $\ell^g(e_{\{x\}}) = 0$.
(β) If g is continuous on $[a, b]$ and $g(a) < g(b)$, then $e_{[a,b] \cap \mathbb{Q}} \notin \mathscr{L}^g$.
(γ) If g is not continuous at $x \in [a, b]$, and $f \in \mathscr{L}^g$, then f is continuous at x.

These properties are immediate from the definitions. Use them to prove the next statement:

(δ) For each g, $([a, b], \mathscr{L}^g, \ell^g)$ is not closed.

This situation highlights the necessity of investigating the existence of closed extensions. We take this question up in the next exercise.

1.4.5(E) **The existence of closed extensions of Daniell spaces**. In Exs. 1.4.1 and 1.4.2 the existence of closed extensions of certain Daniell spaces was proved by explicit construction. Chapter 3 presents a general procedure for the construction of closed extensions of arbitrary Daniell spaces, settling the question of their existence. However, we can already give a proof of the existence here. The elementary extension examined in Ex. 1.3.9 provides the idea. What would happen if we were to continue the process there so long that there were no increasing sequences left to find? We should say immediately that this could not occur, in general, in countably many repetitions of the elementary extension procedure—we need to turn to the ordinals (cf., Ex. 1.1.11).

Let $(X, \mathscr{L}, \ell)$ be a Daniell space. For each ordinal α we can construct a Daniell space $(X, \mathscr{L}_\alpha, \ell_\alpha)$ satisfying

(i) $(X, \mathscr{L}_0, \ell_0) := (X, \mathscr{L}, \ell)$ and $(X, \mathscr{L}_\alpha, \ell_\alpha) \preccurlyeq (X, \mathscr{L}_\beta, \ell_\beta)$ whenever $\alpha < \beta$.

(ii) Given $\alpha \geq 1$, define $\tilde{\mathscr{L}}_\alpha := \bigcup_{\beta < \alpha} \mathscr{L}_\beta$ and $\tilde{\ell}_\alpha$ by $\tilde{\ell}_\alpha \mid_{\mathscr{L}_\beta} = \ell_\beta$ whenever $\beta < \alpha$. If $(X, \tilde{\mathscr{L}}_\alpha, \tilde{\ell}_\alpha)$ is closed, then $(X, \mathscr{L}_\alpha, \ell_\alpha) = (X, \tilde{\mathscr{L}}_\alpha, \tilde{\ell}_\alpha)$. Otherwise $(X, \mathscr{L}_\alpha, \ell_\alpha)$ is an elementary extension of $(X, \tilde{\mathscr{L}}_\alpha, \tilde{\ell}_\alpha)$ in the sense of Ex. 1.3.9.

(iii) Given α, for every $f \in \mathscr{L}_\alpha$ and every $\varepsilon \in \mathbb{R}$, $\varepsilon > 0$, there is a decreasing sequence $(f_n)_{n \in \mathbb{N}}$ from $\mathscr{L} \cap \mathbb{R}^X$ with $\bigwedge_{n \in \mathbb{N}} f_n \leq f$ and $\ell_\alpha(f) - \inf_{n \in \mathbb{N}} \ell(f_n) < \varepsilon$.

(iv) If γ is the first uncountable ordinal, then $(X, \mathscr{L}_\gamma, \ell_\gamma)$ is closed. [Thus $(X, \mathscr{L}_\delta, \ell_\delta) = (X, \mathscr{L}_\gamma, \ell_\gamma)$ for all $\delta > \gamma$; i.e., the procedure ends after γ steps.]

We provide some suggestions. We assume that $\alpha \geq 1$ and that $(X, \mathscr{L}_\beta, \ell_\beta)$ has already been constructed for each $\beta < \alpha$. We wish to show that $(X, \tilde{\mathscr{L}}_\alpha, \tilde{\ell}_\alpha)$ is a Daniell space. The only difficulty arises with the null continuity of $\tilde{\ell}_\alpha$.

Let $(f_n)_{n \in \mathbb{N}}$ be a decreasing sequence from $\tilde{\mathscr{L}}_\alpha$ with $\bigwedge_{n \in \mathbb{N}} f_n = 0$. Take $\varepsilon > 0$. By the inductive hypothesis (iii), there is a decreasing sequence $(g_{nm})_{m \in \mathbb{N}}$ from $\mathscr{L}_+ \cap \mathbb{R}^X$ for each $n \in \mathbb{N}$ satisfying $\bigwedge_{m \in \mathbb{N}} g_{nm} \leq f_n$ and $\tilde{\ell}_\alpha(f_n) - \inf_{m \in \mathbb{N}} \ell(g_{nm}) < \varepsilon/2^n$. For each $m \in \mathbb{N}$ define $h_m := \bigwedge_{k \leq m} g_{km}$. Then $(h_m)_{m \in \mathbb{N}}$ is a decreasing sequence from $\mathscr{L}_+$ and $0 \leq \bigwedge_{m \in \mathbb{N}} h_m \leq \bigwedge_{n \in \mathbb{N}} f_n = 0$ so that $\bigwedge_{m \in \mathbb{N}} h_m = 0$, and thus $\inf_{m \in \mathbb{N}} \ell(h_m) = 0$. On the other hand, for $m \in \mathbb{N}$,

$$f_m - h_m = f_m - \bigwedge_{k \leq m} g_{km} \leq f_m - \bigwedge_{n \in \mathbb{N}} \bigwedge_{k \leq m} g_{kn}$$

$$\leq \sum_{k \leq m} \left| f_k - \bigwedge_{n \in \mathbb{N}} g_{kn} \right|.$$

Thus

$$\tilde{\ell}_\alpha(f_m) - \ell(h_m) = \tilde{\ell}_\alpha(f_m - h_m) \leq \sum_{k \leq m} \frac{\varepsilon}{2^k} < \varepsilon.$$

Hence

$$0 \leq \inf_{m \in \mathbb{N}} \tilde{\ell}_\alpha(f_m) \leq \inf_{m \in \mathbb{N}} \ell(h_m) + \varepsilon,$$

so that $\inf_{m \in \mathbb{N}} \tilde{\ell}_\alpha(f_m) = 0$. Thus $(X, \tilde{\mathscr{L}}_\alpha, \tilde{\ell}_\alpha)$ is a Daniell space.

There are two cases to be distinguished:

CASE 1: $\bigvee_{n \in \mathbb{N}} f_n \in \tilde{\mathscr{L}}_\alpha$ for each increasing sequence $(f_n)_{n \in \mathbb{N}}$ from $\tilde{\mathscr{L}}_\alpha$ with $\sup_{n \in \mathbb{N}} \tilde{\ell}_\alpha(f_n) < \infty$. In this case $(X, \tilde{\mathscr{L}}_\alpha, \tilde{\ell}_\alpha)$ is closed, so set $(X, \mathscr{L}_\alpha, \ell_\alpha) := (X, \tilde{\mathscr{L}}_\alpha, \tilde{\ell}_\alpha)$.

CASE 2: There is an increasing sequence $(f_n)_{n \in \mathbb{N}}$ from $\tilde{\mathscr{L}}_\alpha$ with $\sup_{n \in \mathbb{N}} \tilde{\ell}_\alpha(f_n) < \infty$ but $\bigvee_{n \in \mathbb{N}} f_n \notin \tilde{\mathscr{L}}_\alpha$. In this case take as $(X, \mathscr{L}_\alpha, \ell_\alpha)$ the elementary extension of $(X, \tilde{\mathscr{L}}_\alpha, \tilde{\ell}_\alpha)$, using such a sequence as discussed in Ex. 1.3.9.

So $(X, \mathscr{L}_\alpha, \ell_\alpha)$ has now been constructed. The reader should be able to verify readily that conditions (i)–(iv) are satisfied. We have thus shown that closed extensions do exist. But we only know at this stage that such an extension is available from a transfinite procedure. This highlights the significance of the method originating from Daniell, which allows the construction of the same object in only two steps. It is only with this latter method that the object becomes really manageable and accessible to thorough investigation. We shall deal with Daniell's method in Chapter 3. Nevertheless, one should note that the path to closed extensions followed in this exercise is a very natural one.

1.4.6(E) Let $(X, \mathscr{L}, \ell)$ be a closed Daniell space. Prove the following statements.

(α) Take $f \in \overline{\mathbb{R}}^X$. Let $(f_\iota)_{\iota \in I}$ be a net in $\mathscr{L}$ with $f(x) = \lim_{\iota \in I} f_\iota(x)$ $\mathscr{L}$-a.e. Let $\mathscr{F}$ be an upward directed and $\mathscr{G}$ a downward directed subset of $\mathscr{L}$ with the following properties:

(i) $h \leq g$ whenever $h \in \mathscr{F}$ and $g \in \mathscr{G}$.

(ii) Given $\varepsilon \in \mathbb{R}$, $\varepsilon > 0$, and $\iota \in I$, there are functions $h \in \mathscr{F}$ and $g \in \mathscr{G}$ with $h \leq f_{\iota'} \leq g$ $\mathscr{L}$-a.e. whenever $\iota' \in I$, $\iota' \geq \iota$, and with $\ell(g) - \ell(h) < \varepsilon$.

Then $f \in \mathscr{L}$ and $\ell(f) = \lim_{\iota \in I} \ell(f_\iota)$.

(β) Take $g \in \mathscr{L}_+$. Let $\mathfrak{F}$ be a filter on $\mathscr{L}$ with a countable base such that $\{f \in \overline{\mathbb{R}}^X \mid |f| \leq g \ \mathscr{L}\text{-a.e.}\} \in \mathfrak{F}$. Prove the following statements for $f_0 \in \overline{\mathbb{R}}^X$ with $f_0(x) = \lim_{f \in \mathfrak{F}} f(x)$ $\mathscr{L}$-a.e.:

($\beta 1$) $f_0 \in \mathscr{L}$.

($\beta 2$) $\ell(f_0) = \lim_{f \in \mathfrak{F}} \ell(f)$.

(γ) Take $g \in \mathbb{R}_+^{\mathbb{N}}$. We write $(\mathbb{N}, \ell_g^1(\mathbb{N}), \bar{\ell}_g)$ in the sense of Ex. 1.3.2. For the sequence $(\alpha_n)_{n \in \mathbb{N}} \in \ell_g^1(\mathbb{N})$ define $\Sigma_{n \in \mathbb{N}}^g \alpha_n := \bar{\ell}_g((\alpha_n)_{n \in \mathbb{N}})$. Call a sequence $(f_n)_{n \in \mathbb{N}}$ from $\mathscr{L}$ summable with respect to g iff $(f_n(x))_{n \in \mathbb{N}} \in \ell_g^1(\mathbb{N})$ for each $x \in X$. Define

$$\sum_{n \in \mathbb{N}}^{g} (f_n) : X \to \overline{\mathbb{R}}, \qquad x \mapsto \sum_{n \in \mathbb{N}}^{g} f_n(x).$$

Show that if $(f_n)_{n \in \mathbb{N}}$ is a sequence from $\mathscr{L}$ that is summable with respect to g, then $\Sigma_{n \in \mathbb{N}}^g |f_n| \in \mathscr{L}$ iff $(\ell(|f_n|))_{n \in \mathbb{N}}$ is summable with respect to g. In this case $\Sigma_{n \in \mathbb{N}}^g f_n \in \mathscr{L}$ and $\ell(\Sigma_{n \in \mathbb{N}}^g f_n) = \Sigma_{n \in \mathbb{N}}^g \ell(f_n)$.

1.4.7(E) Let $(X, \mathscr{L}, \ell)$ be a closed Daniell space. In this exercise we investigate the order structure of $\mathscr{L}$, independently of the particularities of the embedding in $\overline{\mathbb{R}}^X$. For the sake of simplicity, we denote by $\leq$ the ordering, $\leq \restriction_{\mathscr{L}}$, on $\mathscr{L}$ inherited from $\overline{\mathbb{R}}^X$. Prove the following statements:

(α) $\mathscr{L}$ is a lattice with respect to $\leq$. Given $f, g \in \mathscr{L}$,

$$f \overset{\mathscr{L}}{\bigvee} g = f \overset{\overline{\mathbb{R}}^X}{\bigvee} g \text{ and } f \overset{\mathscr{L}}{\bigwedge} g = f \overset{\overline{\mathbb{R}}^X}{\bigwedge} g.$$

(This is true for any Riesz lattice.)

(β) Given a nonempty bounded countable family $(f_\iota)_{\iota \in I}$ from $\mathscr{L}$, $\bigvee_{\iota \in I}^{\mathscr{L}} f_\iota$ and $\bigwedge_{\iota \in I}^{\mathscr{L}} f_\iota$ exist, and $\bigvee_{\iota \in I}^{\mathscr{L}} f_\iota = \bigvee_{\iota \in I}^{\overline{\mathbb{R}}^X} f_\iota$ as $\bigwedge_{\iota \in I}^{\mathscr{L}} f_\iota = \bigwedge_{\iota \in I}^{\overline{\mathbb{R}}^X} f_\iota$. ($\mathscr{L}$ is thus a conditionally σ-complete lattice.)

(γ) Given an increasing sequence, $(f_n)_{n \in \mathbb{N}}$, from $\mathscr{L}$ with $\sup_{n \in \mathbb{N}} \ell(f_n) < \infty$, then $\bigvee_{n \in \mathbb{N}}^{\mathscr{L}} f_n$ exists, and $\bigvee_{n \in \mathbb{N}}^{\mathscr{L}} f_n = \bigvee_{n \in \mathbb{N}}^{\overline{\mathbb{R}}^X} f_n$. (The analogous result holds for decreasing sequences.)

(δ) If $(f_n)_{n \in \mathbb{N}}$ is a bounded sequence from $\mathscr{L}$, then ${}^{\mathscr{L}}\limsup_{n \to \infty} f_n$ and ${}^{\mathscr{L}}\liminf_{n \to \infty} f_n$ exist, ${}^{\mathscr{L}}\limsup_{n \to \infty} f_n = {}^{\overline{\mathbb{R}}^X}\limsup_{n \to \infty} f_n$, and ${}^{\mathscr{L}}\liminf_{n \to \infty} f_n = {}^{\overline{\mathbb{R}}^X}\liminf_{n \to \infty} f_n$.

(ε) The sequence $(f_n)_{n\in\mathbb{N}}$ from $\mathscr{L}$ order-converges in $\mathscr{L}$ to $f \in \mathscr{L}$ iff $(f_n)_{n\in\mathbb{N}}$ is bounded in $\mathscr{L}$ and $f(x) = \lim_{n\to\infty} f_n(x)$ for each $x \in X$.

(ζ) ℓ is a continuous function on $\mathscr{L}$ with respect to the order topology.

1.4.8(E) We wish to show in this exercise that the notion of a closed Daniell space can be described topologically.

Let X be a set. For each $(f, \alpha) \in \overline{\mathbb{R}}^X \times \mathbb{R}$ let $\tau(f, \alpha)$ be the set of all sequences $((f_n, \alpha_n))_{n\in\mathbb{N}}$ from $\overline{\mathbb{R}}^X \times \mathbb{R}$ with the following properties:

(i) $(f_n)_{n\in\mathbb{N}}$ converges monotonely to f.

(ii) $(\alpha_n)_{n\in\mathbb{N}}$ is order convergent to α.

A subset A of $\overline{\mathbb{R}}^X \times \mathbb{R}$ is called closed iff $(f, \alpha) \in A$ whenever there is a sequence $((f_n, \alpha_n))_{n\in\mathbb{N}} \in \tau(f, \alpha)$ such that $(f_n, \alpha_n) \in A$ for every $n \in \mathbb{N}$. Define

$$\mathfrak{T} := \left\{ (\overline{\mathbb{R}}^X \times \mathbb{R}) \setminus A \,|\, A \subset \overline{\mathbb{R}} \times \mathbb{R}, A \text{ closed} \right\}.$$

Prove the following statements:

(α) $\mathfrak{T}$ is a topology on $\overline{\mathbb{R}}^X \times \mathbb{R}$.

(β) The Daniell space $(X, \mathscr{L}, \ell)$ is closed iff $\{(f, \ell(f)) \,|\, f \in \mathscr{L}\}$ is closed with respect to $\mathfrak{T}$.

1.4.9(E) This exercise is devoted to revealing the essential nature of null continuity. Let $(X, \mathscr{L}, \ell)$ be a Daniell space. There are several topologies available on $\mathscr{L}$. For example, we could take $\mathscr{L}$ as an ordered set, and consider the order topology. On the other hand, we could take the order topology on $\overline{\mathbb{R}}^X$ and consider its restriction to $\mathscr{L}$. We wish to investigate how ℓ behaves with respect to these topologies.

(a) Let $\mathfrak{T}(\mathscr{L})$ be the order topology on $\mathscr{L}$. Show that ℓ need not be continuous with respect to $\mathfrak{T}(\mathscr{L})$.

We investigate this in the case of the Riemann integral on the interval $[a, b]$ $(a \neq b)$. We denote the Riemann integral by ℓ. Consider a Cantor set $C^{[a,b]}[(\alpha_n)_{n\in\mathbb{N}}]$. Denote by I_{ni} $(n \in \mathbb{N}, i \in \mathbb{N}_{2^{n-1}})$ the open intervals removed from $[a, b]$ and by $\lambda(I_{ni})$ the corresponding lengths—as in Ex. 1.1.12(b). Choose the sequence $(\alpha_n)_{n\in\mathbb{N}}$ so that $\lambda(I_{ni}) = (b-a)/4^n$, and denote the corresponding Cantor set simply by C. The sum of the lengths of these intervals is

$$(b-a)\left(\frac{1}{4} + \frac{2}{4^2} + \frac{2^2}{4^3} + \cdots\right) = \frac{(b-a)}{4}\left(1 + \frac{1}{2} + \frac{1}{2^2} + \frac{1}{2^3} + \cdots\right)$$

$$= \frac{(b-a)}{2} \neq 0.$$

Now construct a decreasing sequence $(f_n)_{n\in\mathbb{N}}$ of continuous real functions on $[a, b]$ such that $e_C = \bigwedge_{n\in\mathbb{N}} f_n$. Take $\beta \in]0, 1[$. Take $n \in \mathbb{N}$. Set $B :=$

$\{f_n > \beta\}$. B is open in the natural topology on $[a, b]$. Thus there is a countable family $(C_\iota)_{\iota \in I}$ of pairwise disjoint subintervals of $[a, b]$ with $B = \bigcup_{\iota \in I} C_\iota$. Observe that all the intervals containing neither a nor b are open. Since C is compact and $C \subset \bigcup_{\iota \in I} C_\iota$, there is a finite subset $J \subset I$ such that $C \subset \bigcup_{\iota \in J} C_\iota$. Furthermore $[a, b] \setminus \bigcup_{\iota \in J} C_\iota \subset [a, b] \setminus C$, from which it follows that $\sum_{\iota \in J} \lambda(C_\iota) \geq (b - a)/2$. From the definition of the Riemann integral, we get $\ell(f_n) \geq \beta(b - a)/2$, where n is arbitrary.

Now consider the Daniell space $([a, b], \mathscr{K}([a, b]), \ell)$ as the restriction of the Riemann integral to the continuous functions on $[a, b]$. Then

$$\bigwedge_{n \in \mathbb{N}}^{\mathscr{K}([a,b])} f_n = 0 \quad \text{but} \quad \inf_{n \in \mathbb{N}} \ell(f_n) \geq \tfrac{1}{2}\beta(b - a) > 0$$

so ℓ is not order continuous on $\mathscr{K}([a, b])$, even though ℓ is nullcontinuous.

(b) Let $\mathfrak{T}$ denote the restriction to $\mathscr{L}$ of the order topology of $\overline{\mathbb{R}}^X$. Show that ℓ need not be continuous with respect to $\mathfrak{T}$.

We again turn to consideration of the example of the Riemann integral on $[a, b]$. Let $(\beta_n)_{n \in \mathbb{N}}$ be a strictly increasing sequence from $[a, b[$ converging to b. For $n \in \mathbb{N}$ define

$$f_n : [a, b] \to \mathbb{R},$$

$$x \mapsto \begin{cases} 0 & \text{for } a \leq x \leq \beta_n \\ (x - \beta_n)\dfrac{4}{(b - \beta_n)^2} & \text{for } \beta_n \leq x \leq \dfrac{\beta_n + b}{2} \\ -(x - b)\dfrac{4}{(b - \beta_n)^2} & \text{for } \dfrac{\beta_n + b}{2} \leq x \leq b \end{cases}$$

Then $(f_n)_{n \in \mathbb{N}}$ is a sequence of Riemann-integrable functions that converges to 0 with respect to the order topology on $\overline{\mathbb{R}}^X$, and so also with respect to the restriction of this topology to the Riemann-integrable functions. But $\ell(f_n) = 1$ for each $n \in \mathbb{N}$.

Thus in neither case does the null continuity imply the continuity with respect to the given topology. In Ex. 1.3.10 we considered a topology with respect to which ℓ is always continuous, namely the one generated by the pseudometric d_ℓ. But the null continuity of ℓ played no part in the proof of this. So what in fact are the nature and meaning of null continuity? Well, they become clearer with the following formulation:

> ℓ is nullcontinuous iff every decreasing sequence that converges to 0 in the order topology on $\overline{\mathbb{R}}^X$ also converges to 0 with respect to the pseudometric d_ℓ.

Viewed in this light, null continuity can be seen to be more a compatibility condition between the order topology of $\overline{\mathbb{R}}^X$ and d_ℓ than a form of continuity.

(c) Finally, we note that if $(X, \mathscr{L}, \ell)$ is a closed Daniell space, then, by Ex. 1.4.7 (ζ), ℓ is continuous in the order topology on $\mathscr{L}$. This is an important reason for the great significance of closed Daniell spaces.

1.4.10(E) We wish to discuss here the problem of the definition of the integral. The consideration in Ex. 1.4.5 shows that each Daniell space has a closed extension. Ex. 1.4.2 shows that, in general, there is more than one such extension. Thus we must make a choice. It is at this point that our problem commences: the particular extension designated as the integral depends critically on the choice made. We wish to illustrate the difficulty using an example.

Take $X := \mathbb{R}$, $\mathscr{L} := \{\alpha e_{\mathbb{R}} + g \mid \alpha \in \mathbb{R}, g \in \mathbb{R}^{\mathbb{R}}, \{g \neq 0\}$ finite$\}$ and $\ell: \mathscr{L} \to \mathbb{R}$, $\alpha e_{\mathbb{R}} + g \mapsto \alpha$. By Ex. 1.3.5, $(X, \mathscr{L}, \ell)$ is a Daniell space. Now define different extensions as follows:

$$\vec{\mathscr{L}} := \left\{ \alpha e_{\mathbb{R}} + g \mid \alpha \in \mathbb{R}, g \in \overline{\mathbb{R}}^{\mathbb{R}}, \{g \neq 0\} \setminus [0, \infty[\text{ countable} \right\}$$

with $\vec{\ell}: \vec{\mathscr{L}} \to \mathbb{R}$, $\quad \alpha e_{\mathbb{R}} + g \mapsto \alpha$.

$$\overleftarrow{\mathscr{L}} := \left\{ \alpha e_{\mathbb{R}} + g \mid \alpha \in \mathbb{R}, g \in \overline{\mathbb{R}}^{\mathbb{R}}, \{g \neq 0\} \setminus]-\infty, 0] \text{ countable} \right\}$$

with $\overleftarrow{\ell}: \overleftarrow{\mathscr{L}} \to \mathbb{R}$, $\quad \alpha e_{\mathbb{R}} + g \mapsto \alpha$. And finally, for each $n \in \mathbb{N}$

$$\mathscr{L}_n := \left\{ \alpha e_{\mathbb{R}} + g \mid \alpha \in \mathbb{R}, g \in \overline{\mathbb{R}}^{\mathbb{R}}, \{g \neq 0\} \setminus]-\infty, n] \text{ countable} \right\}$$

with $\ell_n: \mathscr{L}_n \to \mathbb{R}$, $\alpha e_{\mathbb{R}} + g \mapsto \alpha$.

Show that $(\mathbb{R}, \vec{\mathscr{L}}, \vec{\ell})$, $(\mathbb{R}, \overleftarrow{\mathscr{L}}, \overleftarrow{\ell})$ and each $(\mathbb{R}, \mathscr{L}_n, \ell_n)$ are closed extensions of $(\mathbb{R}, \mathscr{L}, \ell)$. Consider now Φ, the set of all such extensions, ordered by $\preccurlyeq$. Prove the following propositions:

(α) $(\mathbb{R}, \vec{\mathscr{L}}, \vec{\ell})$ and $(\mathbb{R}, \overleftarrow{\mathscr{L}}, \overleftarrow{\ell})$ have no common upper bound in Φ.

(β) $((\mathbb{R}, \mathscr{L}_n, \ell_n))_{n \in \mathbb{N}}$ is an increasing sequence from Φ that is not bounded above.

Observe that as a consequence of (α) it is possible that the extensions of a Daniell space will branch incompatibly. Thus the arbitrariness of the choice at such a branch-point can wreak havoc on the attempt to construct an integral. We would like the integral to depend on the nature of the initial space rather than the roll of dice. So we shall look for the integral among those extensions that are compatible with all others. That problems can arise even within a branch is shown by (β).

These simple examples show that the difficulty in the construction of integrals is based on the poor order properties of the set of all closed extensions. Thus a reasonable choice must begin with the examination of these properties. We shall see in Chapter 4 that this leads to a reasonable choice.

2

MEASURE SPACES, ASSOCIATED DANIELL SPACES, AND STIELTJES FUNCTIONALS

This chapter focuses on the two most important starting points for applying the methods of integration theory—positive measures (Sections 2.1–2.4) and the Stieltjes functionals (Section 2.5).

The concept of measure arose in connection with one of the oldest problems in mathematics, the n-dimensional volume problem in Euclidean space. The object is to assign a volume to each element of a sufficiently large class of subsets of a Euclidean space in such a way that the following conditions are satisfied:

(a) The volume is always positive.

(b) The volume of the unit cube is 1.

(c) The volume of the union of two disjoint sets is the sum of their volumes.

(d) Congruent sets have the same volume.

The "sufficiently large" requirement is important even for practical applications. It is easy to assign a volume (i.e., an area) to figures in the plane such as triangles, rectangles, or polygons, or similarly to simple figures in higher-dimensional Euclidean spaces. Yet such a definition of volume is inadequate even for elementary practical purposes. A reasonable extension is definitely required.

In the volume example the measure-theoretic problem is tied to various special structures of Euclidean space. For instance, (d) makes a connection with the congruence group that operates in Euclidean space, whereas (b) only makes sense when distance has been defined. There are numerous problems similar to the volume problem in Euclidean space but connected with other structures. Thus it is important to find a concept that has the important characteristics of volume yet is not tied to special structures. Such a notion is the abstract concept of measure introduced in mathematics by J. Radon (for Euclidean space, 1913) and M. Fréchet (1915).

2.1. SYSTEMS OF SETS

This section studies the domains on which measures will be defined.

Definition 2.1.1. *A set $\mathfrak{R}$ of sets is called a* **ring of sets** *or a* **set-ring** *iff the following two conditions hold.*

(*a*) *The empty set belongs to $\mathfrak{R}$.*

(*b*) *Both $A \cup B$ and $A \setminus B$ belong to $\mathfrak{R}$ if A and B do.*

A set-ring $\mathfrak{R}$ is called a **δ-ring** *iff*

(*c*) *$\bigcap_{n \in \mathbb{N}} A_n$ belongs to $\mathfrak{R}$ for every sequence $(A_n)_{n \in \mathbb{N}}$ from $\mathfrak{R}$*

and it is called a **σ-ring** *iff*

(*d*) *$\bigcup_{n \in \mathbb{N}} A_n$ belongs to $\mathfrak{R}$ for every sequence $(A_n)_{n \in \mathbb{N}}$ from $\mathfrak{R}$.*

For $\mathfrak{S}$ an arbitrary set of sets, define

$$X(\mathfrak{S}) := \bigcup_{A \in \mathfrak{S}} A.$$

If $\mathfrak{R}$ is a set-ring (δ-ring, σ-ring) and X is any set containing $X(\mathfrak{R})$, then we say that $\mathfrak{R}$ is a **set-ring (δ-ring, σ-ring) on** *X.*

A set-ring on X is said to be **σ-finite** *iff X can be written as the union of countably many elements of the set-ring. If $\mathfrak{R}$ is a σ-finite set-ring on X, we shall also say that the pair $(X, \mathfrak{R})$ is* **σ-finite**. □

Observe that if $\mathfrak{R}$ is a σ-finite set-ring on X, then $X = X(\mathfrak{R})$.

Proposition 2.1.2. *The following assertions hold, for every set-ring $\mathfrak{R}$.*

(*a*) *The union of every finite family from $\mathfrak{R}$ belongs to $\mathfrak{R}$.*

(*b*) *The intersection of every nonempty finite family from $\mathfrak{R}$ belongs to $\mathfrak{R}$.*

(*c*) *If A and B belong to $\mathfrak{R}$, then so does their symmetric difference $A \triangle B$.*

(*d*) *If $\mathfrak{R}$ is a δ-ring, then the intersection of every nonempty countable family from $\mathfrak{R}$ belongs to $\mathfrak{R}$.*

(*e*) *If* $\mathfrak{R}$ *is a* σ*-ring, then the union of every countable family from* $\mathfrak{R}$ *belongs to* $\mathfrak{R}$.

(*f*) *If* $\mathfrak{R}$ *is a* σ*-ring, then* $\mathfrak{R}$ *is also a* δ*-ring.*

Proof. (a) This assertion is proved by complete induction on the number of elements in the family. Note that

$$\bigcup_{\iota \in \varnothing} A_\iota = \varnothing$$

and $\varnothing$ is in $\mathfrak{R}$.

(b) For arbitrary sets A and B,

$$A \cap B = A \setminus (A \setminus B).$$

Thus $A \cap B$ belongs to $\mathfrak{R}$ if A and B do. The full claim now follows by complete induction.

(c), (d), (e) These claims are immediate from the definitions.

(f) Let $(A_n)_{n \in \mathbb{N}}$ be a sequence from $\mathfrak{R}$. Set

$$A := \bigcup_{n \in \mathbb{N}} A_n$$

and observe that

$$\bigcap_{n \in \mathbb{N}} A_n = A \setminus \bigcup_{n \in \mathbb{N}} (A \setminus A_n).$$

Using in turn the definitions of σ-ring and set-ring, we conclude in succession that A belongs to $\mathfrak{R}$, each $A \setminus A_n$ belongs to $\mathfrak{R}$, $\bigcup_{n \in \mathbb{N}}(A \setminus A_n)$ belongs to $\mathfrak{R}$, and, finally, $\bigcap_{n \in \mathbb{N}} A_n$ belongs to $\mathfrak{R}$. □

Thus the implications

$$\sigma\text{-ring} \Rightarrow \delta\text{-ring} \Rightarrow \text{ring of sets}$$

all hold. That the implications in the opposite direction are false is shown by the following simple examples.

Example 2.1.3. (Finite subsets of a set). The set consisting of all finite subsets of a fixed set X is always a δ-ring. It is a σ-ring iff X is finite. We denote this δ-ring by $\mathfrak{F}(X)$. □

Example 2.1.4. (Finite or cofinite subsets of an infinite set). If X is an infinite set, then the set consisting of all subsets A of X for which either A or $X \setminus A$ is finite is a ring of sets but not a δ-ring. □

Rings of sets are very important in elementary constructions and usually arise in a practical problem setting. By contrast, δ- and σ-rings first appear in attempts to construct closed Daniell spaces.

A set-ring, when viewed with set inclusion as an order relation, forms a lattice. Every σ-ring is a σ-complete lattice. The next proposition shows that δ-rings are conditionally σ-complete lattices.

Proposition 2.1.5. *Let $\mathfrak{R}$ be a δ-ring and let $(A_\iota)_{\iota \in I}$ be a countable family from $\mathfrak{R}$. If there exists a set A belonging to $\mathfrak{R}$ such that*

$$\bigcup_{\iota \in I} A_\iota \subset A$$

then $\bigcup_{\iota \in I} A_\iota$ belongs to $\mathfrak{R}$.

Proof. If $I = \varnothing$, there is nothing to prove. If not, use the identity

$$\bigcup_{\iota \in I} A_\iota = A \setminus \bigcap_{\iota \in I} (A \setminus A_\iota). \qquad \square$$

Example 3 shows that Proposition 5 cannot be strengthened.

Every ring of sets has a useful decomposition property. Finite unions decompose into disjoint finite unions, and the decomposition can be achieved one set at a time.

Proposition 2.1.6 (Disjoint decomposition for set-rings). *Suppose that $\mathfrak{R}$ is a ring of sets and that $(A_\iota)_{\iota \in I}$ is a finite family from $\mathfrak{R}$. Then there exists in $\mathfrak{R}$ a finite disjoint family $(B_\kappa)_{\kappa \in K}$ of nonempty sets such that for each ι in I there is a subset K_ι of K with*

$$\bigcup_{\kappa \in K_\iota} B_\kappa = A_\iota.$$

Proof. The reader may wish to construct a proof by induction on the number of elements in I. We prefer to present the following proof, since it makes obvious what the decomposition is. This proof is cumbersome to write out, but it is not difficult to discover. The reader is therefore encouraged, before reading the proof, to construct the required decomposition for two- and three-set-families. To start off, notice, for example, that $A \cup B$ is the same as the union of the three disjoint sets $A \setminus B$, $B \setminus A$, and $A \cap B$, each of which belongs to $\mathfrak{R}$ if A and B do, and $A = (A \cap B) \cup (A \setminus B)$, $B = (A \cap B) \cup (B \setminus A)$.

We begin the proof, then, by setting

$$\mathfrak{K}' := \mathfrak{P}(I) \setminus \{\varnothing\}$$

and by defining, for every $K \in \mathfrak{K}'$,

$$B_K := \Big(\bigcap_{\iota \in K} A_\iota \Big) \setminus \Big(\bigcup_{\iota \in I \setminus K} A_\iota \Big). \tag{1}$$

We shall exhibit a subset $\mathfrak{K}$ of $\mathfrak{K}'$ such that the family $(B_K)_{K \in \mathfrak{K}}$ has the required properties. Each element of $\mathfrak{K}'$ is finite, since I is finite. Accordingly, B_K belongs to $\mathfrak{R}$ for every K in $\mathfrak{K}'$ (by Proposition 2). Moreover the sets B_K are finite in number (I is finite) and pairwise disjoint. To verify the disjointness, suppose that K and $\tilde{K}$ both belong to $\mathfrak{K}'$ with $K \neq \tilde{K}$. At least one of these two sets, K say, contains an element ι that does not belong to the other set $\tilde{K}$. It follows from (1) that B_K is a subset of A_ι while $B_{\tilde{K}}$ is disjoint from A_ι. Thus B_K and $B_{\tilde{K}}$ are disjoint.

Now for each ι in I, we put

$$\mathfrak{K}'_\iota := \{ K \in \mathfrak{K}' \mid \iota \in K \}$$

and we claim that

$$A_\iota = \bigcup_{K \in \mathfrak{K}'_\iota} B_K. \tag{2}$$

Fix an arbitrary index $\iota_0 \in I$. It follows from (1) that $B_K \subset A_{\iota_0}$ for every K in $\mathfrak{K}'_{\iota_0}$, so

$$\bigcup_{K \in \mathfrak{K}'_{\iota_0}} B_K \subset A_{\iota_0}.$$

For the inclusion in the other direction, suppose that $x \in A_{\iota_0}$, and set

$$K(x) := \{ \iota \in I \mid x \in A_\iota \}. \tag{3}$$

Now x belongs to A_{ι_0}, so $\iota_0 \in K(x)$, and so $K(x)$ belongs to $\mathfrak{K}'_{\iota_0}$. From (1) and (3) we see that x belongs to $B_{K(x)}$. Therefore

$$x \in \bigcup_{K \in \mathfrak{K}'_{\iota_0}} B_K.$$

The choice of x in A_{ι_0} was arbitrary, so

$$A_{\iota_0} \subset \bigcup_{K \in \mathfrak{K}'_{\iota_0}} B_K$$

and (2) follows for $\iota = \iota_0$ and hence for every ι in I.

The proof is essentially finished: the remaining catch is that some B_K might be empty. Set

$$\mathfrak{K} := \{ K \in \mathfrak{K}' \mid B_K \neq \varnothing \}$$

and

$$\mathfrak{K}_\iota := \{ K \in \mathfrak{K}'_\iota \mid B_K \neq \varnothing \}$$

for each $\iota \in I$. The proposition now follows.

Of course some index set $\mathfrak{K}_\iota$ might be empty. This occurs, however, only if A_ι is empty, and then it is still true that $A_\iota = \bigcup_{K \in \mathfrak{K}_\iota} B_K$. If every A_ι is empty (or if $I = \varnothing$), then the proposition is true but scarcely interesting, and the proof just given is valid but not necessary. □

The set-systems that arise in concrete examples, for instance the volume problem in $\mathbb{R}^n$, do not always satisfy the conditions required in Definition 1. Of course every set-system $\mathfrak{S}$ can be embedded in a set-ring (or a δ-ring, or a σ-ring). One need only take $\mathfrak{P}(X(\mathfrak{S}))$. Usually, however, $\mathfrak{P}(X(\mathfrak{S}))$ is too large to be useful. The next two results, which will be familiar because of their many analogs in abstract algebra, establish the existence of a smallest set-ring (δ-ring, σ-ring) containing $\mathfrak{S}$.

Proposition 2.1.7. *Let $(\mathfrak{R}_\iota)_{\iota \in I}$ be an arbitrary nonempty family of set-rings (of δ-rings, or σ-rings, respectively). Then*

$$\bigcap_{\iota \in I} \mathfrak{R}_\iota$$

is also a set-ring (a δ-ring, or a σ-ring, respectively).

Proof. The proofs for the three cases are so similar that we carry out the proof only for set-rings, leaving the other two cases to the reader.

Since $\varnothing$ belongs to each $\mathfrak{R}_\iota$, it also belongs to $\bigcap_{\iota \in I} \mathfrak{R}_\iota$. Suppose that A and B belong to $\bigcap_{\iota \in I} \mathfrak{R}_\iota$. Then A and B belong to each $\mathfrak{R}_\iota$, and the definition of set-ring implies that both $A \cup B$ and $A \setminus B$ belong to each $\mathfrak{R}_\iota$. Hence $A \cup B$ and $A \setminus B$ belong to $\bigcap_{\iota \in I} \mathfrak{R}_\iota$, and $\bigcap_{\iota \in I} \mathfrak{R}_\iota$ is a ring of sets. □

Corollary 2.1.8. *Let $\mathfrak{S}$ be an arbitrary set of sets. Then there is exactly one set-ring that contains $\mathfrak{S}$ and is contained in every set-ring containing $\mathfrak{S}$. The set-ring in question is*

$$\bigcap \{ \mathfrak{R} \subset \mathfrak{P}(X(\mathfrak{S})) \mid \mathfrak{R} \text{ is a set-ring containing } \mathfrak{S} \}.$$

These assertions remain true if "set-ring" is replaced in all occurrences by "δ-ring" (or by "σ-ring").

Proof. Here also we prove only the assertions about set-rings. Set

$$\mathfrak{R}' := \bigcap\{\mathfrak{R} \subset \mathfrak{P}(X(\mathfrak{S})) \mid \mathfrak{R} \text{ is a set-ring containing } \mathfrak{S}\}. \tag{4}$$

Since $\mathfrak{P}(X(\mathfrak{S}))$ is a set-ring containing $\mathfrak{S}$ we conclude from the previous proposition that $\mathfrak{R}'$ is a set-ring and that

$$\mathfrak{R}' = \bigcap\{\mathfrak{R} \cap \mathfrak{P}(X(\mathfrak{S})) \mid \mathfrak{R} \text{ is a set-ring and } \mathfrak{S} \subset \mathfrak{R}\}. \tag{5}$$

From (4) it is evident that $\mathfrak{R}'$ contains $\mathfrak{S}$, whereas from (5) we see that every set-ring containing $\mathfrak{S}$ also contains $\mathfrak{R}'$. Finally, the uniqueness is evident, since if $\tilde{\mathfrak{R}}$ is any set-ring with the stipulated properties, then both $\mathfrak{R}' \subset \tilde{\mathfrak{R}}$ and $\tilde{\mathfrak{R}} \subset \mathfrak{R}'$ must hold, so $\tilde{\mathfrak{R}} = \mathfrak{R}'$. □

Definition 2.1.9. *Let $\mathfrak{S}$ be a set of sets. The set-ring characterized in Corollary 8 is denoted by $\mathfrak{S}_r$ and is called the **set-ring generated** by $\mathfrak{S}$. The **δ-ring generated** by $\mathfrak{S}$, denoted $\mathfrak{S}_\delta$, and the **σ-ring generated** by $\mathfrak{S}$, denoted $\mathfrak{S}_\sigma$, are defined analogously.* □

Proposition 2.1.10. *For every ring of sets $\mathfrak{R}$,*

$$\mathfrak{R}_\delta \subset \{A \subset X(\mathfrak{R}) \mid A \subset B \textit{ for some } B \in \mathfrak{R}\}.$$

Proof. Let

$$\mathfrak{S} := \{A \subset X(\mathfrak{R}) \mid A \subset B \text{ for some } B \in \mathfrak{R}\}.$$

Then $\mathfrak{S}$ is a δ-ring, and $\mathfrak{S}$ contains $\mathfrak{R}$. Therefore $\mathfrak{R}_\delta \subset \mathfrak{S}$. □

The set-rings $\mathfrak{S}_r$, $\mathfrak{S}_\delta$, and $\mathfrak{S}_\sigma$ generated by $\mathfrak{S}$ are uniquely determined by $\mathfrak{S}$. One conjectures that their elements must be expressible in terms of the elements of $\mathfrak{S}$. This conjecture is in fact true, but its proof is not so simple as one might think. It should not be assumed, for instance, concerning the δ-ring $\mathfrak{S}_\delta$ and the σ-ring $\mathfrak{S}_\sigma$, that because their defining properties involve countable families, they can be constructed from the elements of $\mathfrak{S}$ in countably many steps. For example, one can form in a first step all intersections of sequences from $\mathfrak{S}$, in a second step all unions of sequences of sets arising from the first step, and continue in this manner, alternating intersection and union. The result after countably infinitely many steps is in general not $\mathfrak{S}_\sigma$, even when the original set $\mathfrak{S}$ was already a set-ring (see the exercises). The situation is somewhat simpler, however, if one begins with a δ-ring. The following characterization will be needed later.

Proposition 2.1.11. *If $\mathfrak{R}$ is a δ-ring, then $\mathfrak{R}_\sigma$ is the set consisting of all unions of increasing sequences from $\mathfrak{R}$.*

Proof. Let $\mathfrak{S}$ be the set of all such unions; that is,

$$\mathfrak{S} := \left\{ A \,\middle|\, \begin{array}{l} \exists \text{ a sequence } (A_n)_{n \in \mathbb{N}} \text{ from } \mathfrak{R} \text{ with } A = \bigcup_{n \in \mathbb{N}} A_n \text{ and } A_n \subset A_{n+1} \\ \text{for each } n \in \mathbb{N} \end{array} \right\}.$$

Certainly $\mathfrak{R} \subset \mathfrak{S} \subset \mathfrak{R}_\sigma$. If we show that $\mathfrak{S}$ is a σ-ring, then the proof will be finished, since $\mathfrak{R}_\sigma$ is contained in every σ-ring that contains $\mathfrak{R}$. Obviously $\emptyset \in \mathfrak{S}$. Let A and B belong to $\mathfrak{S}$. There exist in $\mathfrak{R}$ increasing sequences $(A_n)_{n \in \mathbb{N}}$ and $(B_n)_{n \in \mathbb{N}}$ such that $A = \bigcup_{n \in \mathbb{N}} A_n$ and $B = \bigcup_{n \in \mathbb{N}} B_n$. Now $(A_n \cup B_n)_{n \in \mathbb{N}}$ is an increasing sequence from $\mathfrak{R}$, so its union, $A \cup B$, must belong to $\mathfrak{S}$. Next observe that

$$A \setminus B = \left(\bigcup_{n \in \mathbb{N}} A_n \right) \setminus B$$

$$= \bigcup_{n \in \mathbb{N}} (A_n \setminus B).$$

The sequence $(A_n \setminus B)_{n \in \mathbb{N}}$ increases, and its terms all belong to $\mathfrak{R}$, since

$$A_n \setminus B = \bigcap_{m \in \mathbb{N}} (A_n \setminus B_m)$$

and $\mathfrak{R}$ is a δ-ring. It follows that $A \setminus B$ belongs to $\mathfrak{S}$ and that $\mathfrak{S}$ is a ring of sets.

Finally, let $(C_n)_{n \in \mathbb{N}}$ be a sequence from $\mathfrak{S}$. For each n in $\mathbb{N}$ there exists in $\mathfrak{R}$ an increasing sequence $(C_{n,m})_{m \in \mathbb{N}}$ whose union is C_n. For each m in $\mathbb{N}$, set

$$D_m := \bigcup_{n \leq m} C_{n,m}.$$

The sequence $(D_m)_{m \in \mathbb{N}}$ increases, and its terms belong to $\mathfrak{R}$; therefore its union belongs to $\mathfrak{S}$. But

$$\bigcup_{m \in \mathbb{N}} D_m = \bigcup_{n \in \mathbb{N}} C_n$$

so $\bigcup_{n \in \mathbb{N}} C_n$ belongs to $\mathfrak{S}$. Thus $\mathfrak{S}$ is in fact a σ-ring and $\mathfrak{R}_\sigma = \mathfrak{S}$. □

EXERCISES

2.1.1(E) In this exercise we investigate rings of subsets of a given set X containing X itself. Such systems are of particular interest in several applications of measure theory.

(i) A ring of sets $\mathfrak{R} \subset \mathfrak{P}(X)$ is called an **algebra of sets** on X iff $X \in \mathfrak{R}$.
(ii) A σ-ring $\mathfrak{R} \subset \mathfrak{P}(X)$ is called a **σ-algebra** on X iff $X \in \mathfrak{R}$.

Prove the following propositions:

(α) For every set $\mathfrak{R} \subset \mathfrak{P}(X)$, the following statements are equivalent:

(α1) $\mathfrak{R}$ is an algebra of sets on X.
(α2) $\varnothing \in \mathfrak{R}$ and $A \cup (X \setminus B) \in \mathfrak{R}$ whenever $A, B \in \mathfrak{R}$.
(α3) $X \in \mathfrak{R}$ and $A \cap (X \setminus B) \in \mathfrak{R}$ whenever $A, B \in \mathfrak{R}$.
(α4) $\mathfrak{R} \neq \varnothing$, and $X \setminus A \in \mathfrak{R}$, $A \cup B \in \mathfrak{R}$ whenever $A, B \in \mathfrak{R}$.
(α5) $\mathfrak{R} \neq \varnothing$, and $X \setminus A \in \mathfrak{R}$, $A \cap B \in \mathfrak{R}$ whenever $A, B \in \mathfrak{R}$.

(β) For every set $\mathfrak{R} \subset \mathfrak{P}(X)$, the following statements are equivalent:

(β1) $\mathfrak{R}$ is a σ-algebra on X.
(β2) $\varnothing \in \mathfrak{R}$ and $\bigcup_{n \in \mathbb{N}} A_n \cup (X \setminus B) \in \mathfrak{R}$ whenever $(A_n)_{n \in \mathbb{N}}$ is a sequence from $\mathfrak{R}$ and $B \in \mathfrak{R}$.
(β3) $X \in \mathfrak{R}$ and $\bigcap_{n \in \mathbb{N}} A_n \cap (X \setminus B) \in \mathfrak{R}$ whenever $(A_n)_{n \in \mathbb{N}}$ is a sequence from $\mathfrak{R}$ and $B \in \mathfrak{R}$.
(β4) $\mathfrak{R} \neq \varnothing$, $\bigcup_{n \in \mathbb{N}} A_n \in \mathfrak{R}$ and $X \setminus B \in \mathfrak{R}$ whenever $(A_n)_{n \in \mathbb{N}}$ is a sequence from $\mathfrak{R}$ and $B \in \mathfrak{R}$.
(β5) $\mathfrak{R} \neq \varnothing$, $\bigcap_{n \in \mathbb{N}} A_n \in \mathfrak{R}$ and $X \setminus B \in \mathfrak{R}$ whenever $(A_n)_{n \in \mathbb{N}}$ is a sequence from $\mathfrak{R}$ and $B \in \mathfrak{R}$.

(γ) The set of all algebras of sets on X is a complete lattice with respect to $\subset$. $\{\varnothing, X\}$ is the smallest element and $\mathfrak{P}(X)$ the largest. For every nonempty family $(\mathfrak{R}_\iota)_{\iota \in I}$ of algebras of sets on X we have

$$\bigwedge_{\iota \in I} \mathfrak{R}_\iota = \bigcap_{\iota \in I} \mathfrak{R}_\iota .$$

(δ) For every set $\mathfrak{S} \subset \mathfrak{P}(X)$ there is a smallest algebra of sets on X containing $\mathfrak{S}$. It is called the **algebra of sets** on X **generated** by $\mathfrak{S}$.

(ε) The set of all σ-algebras on X is a complete lattice with respect to $\subset$. $\{\varnothing, X\}$ is the smallest element and $\mathfrak{P}(X)$ the largest. For every nonempty family $(\mathfrak{R}_\iota)_{\iota \in I}$ of σ-algebras on X,

$$\bigwedge_{\iota \in I} \mathfrak{R}_\iota = \bigcap_{\iota \in I} \mathfrak{R}_\iota .$$

(ζ) For every set $\mathfrak{S} \subset \mathfrak{P}(X)$, there is a smallest σ-algebra on X containing $\mathfrak{S}$. It is called the **σ-algebra** on X **generated** by $\mathfrak{S}$.

(η) An algebra of sets on X is a σ-algebra on X iff it is also a δ-ring.

2.1.2(E) Take an arbitrary set of sets, $\mathfrak{S}$. Define

$$\mathfrak{S}_d := \{A \setminus B \mid A, B \in \mathfrak{S}\}$$

$$\mathfrak{S}_i := \{A \cap B \mid A, B \in \mathfrak{S}\}$$

$$\mathfrak{S}_u := \left\{ \bigcup_{\iota \in I} A_\iota \mid (A_\iota)_{\iota \in I} \text{ a finite family from } \mathfrak{S} \right\}$$

$$\mathfrak{S}_v := \left\{ \bigcup_{\iota \in I} A_\iota \mid (A_\iota)_{\iota \in I} \text{ a finite disjoint family from } \mathfrak{S} \right\}.$$

Prove the following statements:

(α) $\mathfrak{S}_{uu} = \mathfrak{S}_u$ and $\mathfrak{S}_{vv} = \mathfrak{S}_v$.
(β) If $\mathfrak{S}_i \subset \mathfrak{S}_u$, then $\mathfrak{S}_{ui} = \mathfrak{S}_u$.
(γ) If $\mathfrak{S}_i = \mathfrak{S}_u = \mathfrak{S}$, then $\mathfrak{S}_{dd} \subset \mathfrak{S}_{dv}$ and $\mathfrak{S}_{di} = \mathfrak{S}_d$.
(δ) If $\mathfrak{S}_d \subset \mathfrak{S}_u$, then $\mathfrak{S}_{ud} = \mathfrak{S}_u$.
(ε) If $\mathfrak{S}_i \subset \mathfrak{S}_v$, then $\mathfrak{S}_{vi} = \mathfrak{S}_v$.
(ζ) If $\mathfrak{S}_d \subset \mathfrak{S}_v$, then $\mathfrak{S}_{vd} = \mathfrak{S}_v$.
(η) If $\mathfrak{S}_i \subset \mathfrak{S}_u$, then $\mathfrak{S}_{udv}$ is the ring of sets generated by $\mathfrak{S}$.
(ϑ) If $\mathfrak{S}_d \subset \mathfrak{S}_v$ and $\mathfrak{S}_i \subset \mathfrak{S}_v$, then $\mathfrak{S}_v$ is the ring of sets generated by $\mathfrak{S}$.

In particular applications we are often not directly led to rings of sets but to systems of sets with weaker properties. That is why the rings of sets generated by such simpler systems are so significant. These are two particularly important examples of these simpler systems:

(i) $\cap$-semi-lattices. A **$\cap$-semi-lattice** is a nonempty set of sets, $\mathfrak{R}$, such that $A \cap B \in \mathfrak{R}$ whenever $A, B \in \mathfrak{R}$.
(ii) Semi-rings. A **semi-ring** is a nonempty set of sets, $\mathfrak{R}$, such that for $A, B \in \mathfrak{R}$ there are finite disjoint families $(A_\iota)_{\iota \in I}$ and $(B_\lambda)_{\lambda \in L}$ from $\mathfrak{R}$ with $A \cap B = \bigcup_{\iota \in I} A_\iota$ and $A \setminus B = \bigcup_{\lambda \in L} B_\lambda$.

The rings of sets generated by $\cap$-semi-lattices (resp. semi-rings) can be characterized by (η) (resp. (ϑ)). Thus the following assertions hold:

(ι) Given the $\cap$-semi-lattice $\mathfrak{S}$, $A \in \mathfrak{S}_r$ iff there are finite sets I, K, L and families $(A_{\iota\kappa})_{(\iota,\kappa) \in I \times K}$ and $(B_{\iota\lambda})_{(\iota,\lambda) \in I \times L}$ from $\mathfrak{S}$ with $A = \bigcup_{\iota \in I}[(\bigcup_{\kappa \in K} A_{\iota\kappa}) \setminus (\bigcup_{\lambda \in L} B_{\iota\lambda})]$.
(κ) Given the semi-ring, $\mathfrak{S}$, $A \in \mathfrak{S}_r$ iff there is a finite disjoint family $(A_\iota)_{\iota \in I}$ from $\mathfrak{S}$ with $A = \bigcup_{\iota \in I} A_\iota$.

In certain situations we have need for the following concept:

(iii) **A lattice of sets** (or **set-lattice**) is a nonempty set $\mathfrak{R}$ of sets with the property that $A \cap B \in \mathfrak{R}$ and $A \cup B \in \mathfrak{R}$ whenever $A, B \in \mathfrak{R}$.

From (ι) we get the next proposition:

(λ) If $\mathfrak{R}$ is a lattice of sets, then $A \in \mathfrak{R}_r$ iff there are finite families $(A_\iota)_{\iota \in I}$ and $(B_\iota)_{\iota \in I}$ from $\mathfrak{R}$ with $A = \bigcup_{\iota \in I}(A_\iota \setminus B_\iota)$.

2.1.3(E) Let $(X, \leq)$ be a totally ordered set. For an arbitrary I, define the **product order** $\leq$ on X^I by $(x_\iota)_{\iota \in I} \leq (y_\iota)_{\iota \in I} :\Leftrightarrow x_\iota \leq y_\iota$ for each $\iota \in I$.

In this exercise we shall be examining some important systems of sets, $\mathfrak{S} \subset \mathfrak{P}(X^I)$. The reader may find it useful to think of $X \subset \mathbb{R}$ to gain a feel for the point of view adopted here. First, consider systems $\mathfrak{S} \subset \mathfrak{P}(X)$:

(α) The following subsets of $\mathfrak{P}(X)$ are $\cap$-semi-lattices:

(α1) The set, $\mathfrak{R}_{[\,]}(X)$, of all closed intervals in X.
(α2) The set, $\mathfrak{R}_{]\,[}(X)$, of all open intervals in X.

(β) The following statements are true:

(β1) $(\mathfrak{R}_{]\,[}(X))_r \subset (\mathfrak{R}_{[\,]}(X))_r$.
(β2) $(\mathfrak{R}_{]\,[}(X))_r = (\mathfrak{R}_{[\,]}(X))_r$ iff X contains neither a largest or a smallest element.

(γ) Suppose that $X \subset \mathbb{Z}$. Then $\mathfrak{R}_{[\,]}(X)$ and $\mathfrak{R}_{]\,[}(X)$ are equal semi-rings.

(δ) The following subsets of $\mathfrak{P}(X)$ are semi-rings:

(δ1) The set, $\mathfrak{R}_{[\,[}(X)$, of right half-open intervals in X.
(δ2) The set, $\mathfrak{R}_{]\,]}(X)$, of left half-open intervals in X.

The following inclusions hold:

(δ3) $(\mathfrak{R}_{[\,[}(X))_r \subset (\mathfrak{R}_{[\,]}(X))_r$.
(δ4) $(\mathfrak{R}_{]\,]}(X))_r \subset (\mathfrak{R}_{[\,]}(X))_r$.

Now consider systems, $\mathfrak{S} \subset \mathfrak{P}(X^I)$, with $I \neq \varnothing$.

(ε) The following subsets of $\mathfrak{P}(X^I)$ are $\cap$-semi-lattices.

(ε1) $\mathfrak{R}_{[\,]}(X^I) := \{\prod_{\iota \in I} A_\iota \mid A_\iota \in \mathfrak{R}_{[\,]}(X) \text{ for each } \iota \in I\}$.
(ε2) $\mathfrak{R}_{]\,[}(X^I) := \{\prod_{\iota \in I} A_\iota \mid A_\iota \in \mathfrak{R}_{]\,[}(X) \text{ for each } \iota \in I\}$.
(ε3) $\mathfrak{R}_{[\,[}(X^I) := \{\prod_{\iota \in I} A_\iota \mid A_\iota \in \mathfrak{R}_{[\,[}(X) \text{ for each } \iota \in I\}$.
(ε4) $\mathfrak{R}_{]\,]}(X^I) := \{\prod_{\iota \in I} A_\iota \mid A_\iota \in \mathfrak{R}_{]\,]}(X) \text{ for each } \iota \in I\}$.

(ζ) $\mathfrak{R}_{[\,[}(X^I)$ and $\mathfrak{R}_{]\,]}(X^I)$ are semi-rings whenever I is finite.

$X = \mathbb{R}$ is a particularly important example in this context.

2.1.4(E) Let $(X, \mathfrak{T})$ be a topological space. Prove the following statements:

(α) The following systems of sets are set-lattices:

(α1) $\mathfrak{T}$.
(α2) The set of all closed subsets of X.

(β) The set of all both closed and open subsets of X is an algebra of sets on X.

A subset A of X is called **nowhere dense** iff the closure of A has empty interior: $\mathring{\bar{A}} = \varnothing$.

(γ) The set of all nowhere dense subsets of X is a δ-ring. Every subset of a nowhere dense set is itself nowhere dense. For every open set U, $\overline{U} \setminus U$ is nowhere dense.

A subset A of X is called of the **first category** or **meager** iff there is a countable family $(A_\iota)_{\iota \in I}$ of nowhere dense subsets of X such that $A = \bigcup_{\iota \in I} A_\iota$.

(δ) The set of all subsets of X of the first category is a σ-ring. Every subset of a set of the first category is of the first category.

(ε) The set of all subsets of X of the first category is the σ-ring generated by the set of all nowhere dense subsets of X. If $(U_n)_{n \in \mathbb{N}}$ is a sequence of open subsets of X, then $\bigcup_{n \in \mathbb{N}} \overline{U}_n \setminus \bigcup_{n \in \mathbb{N}} U_n$ is of the first category.

A subset A of X is called **approximable** iff there exists an open subset B of X such that $A \triangle B$ is of the first category.

(ζ) The set of all approximable subsets of X is a σ-algebra on X.

(η) Show that the following set-systems are identical:

(η1) $\mathfrak{T}_\delta$.
(η2) $\mathfrak{T}_\sigma$.
(η3) The δ-ring generated by the closed subsets of X.
(η4) The σ-ring generated by the closed subsets of X.

The set-system, described in this way is a σ-algebra on X. Its elements are called **Borel sets** on X (or in X).

Borel sets play a very important role in measure theory on topological spaces. Prove:

(ϑ) Every Borel set on X is approximable.

Assume now that $(X, \mathfrak{T})$ is a Hausdorff space. Prove the following statements:

(ι) The set of all compact subsets of X, denoted by $\mathfrak{K}$, is a lattice of sets.

(κ) $\mathfrak{K}_\delta = \{A \in \mathfrak{T}_\delta \mid \overline{A} \in \mathfrak{K}\}$.

$\mathfrak{K}_\delta$ is the set of all relatively compact Borel sets. (Recall that $A \subset X$ is called **relatively compact** iff $\overline{A} \in \mathfrak{K}$). Prove:

(λ) The set of all relatively compact subsets of X is a δ-ring.

2.1.5(E) Let X be a set, $\mathfrak{R}$ an algebra of sets on X, and Φ the set of all filters $\mathfrak{F} \neq \mathfrak{P}(X)$ on X with the property that $\mathfrak{F}$ is the filter generated by $\mathfrak{F} \cap \mathfrak{R}$. Φ is an ordered set with respect to the inclusion $\subset$. Prove the following propositions:

(α) For every $\mathfrak{F} \in \Phi$ there exists a maximal element $\mathfrak{G}$ in Φ such that $\mathfrak{F} \subset \mathfrak{G}$.

(β) The following statements are equivalent for every $\mathfrak{F} \in \Phi$:

(β1) $\mathfrak{F}$ is a maximal element of Φ.
(β2) $A \in \mathfrak{F}$ whenever $A \in \mathfrak{R}$ with $A \cap B \neq \emptyset$ for each $B \in \mathfrak{F}$.
(β3) If $(A_\iota)_{\iota \in I}$ is a finite family from $\mathfrak{R}$ with $\bigcup_{\iota \in I} A_\iota \in \mathfrak{F}$, then there is a $\iota \in I$ such that $A_\iota \in \mathfrak{F}$.
(β4) For all $A, B \in \mathfrak{R}$ with $A \cup B \in \mathfrak{F}$, either $A \in \mathfrak{F}$ or $B \in \mathfrak{F}$.
(β5) For every $A \in \mathfrak{R}$, either $A \in \mathfrak{F}$ or $X \setminus A \in \mathfrak{F}$.

(γ) If $\mathfrak{R} = \mathfrak{P}(X)$, then the maximal elements of Φ are exactly the ultrafilters on X.

(δ) Let Ψ be a set of maximal elements of Φ. Then $\mathfrak{S} := \mathfrak{R} \setminus \bigcup_{\mathfrak{F} \in \Psi} \mathfrak{F}$ is a ring of sets. If $\mathfrak{R}$ is a σ-algebra on X, then $\mathfrak{S}$ is a δ-ring.

(ε) Let Ψ be a set of maximal elements of Φ. Suppose that every filter in Ψ is δ-stable (i.e., $\bigcap_{n \in \mathbb{N}} A_n \in \mathfrak{F}$ whenever $(A_n)_{n \in \mathbb{N}}$ is a sequence from $\mathfrak{F}$). Suppose further that $\mathfrak{R}$ is a σ-algebra on X. Then $\mathfrak{S}$ [as described in (δ)] is a σ-ring.

The reader should formulate (δ) and (ε) in the special case $\mathfrak{R} = \mathfrak{P}(X)$ [taking notice of (γ)].

2.1.6(E) Let I be a set and for every $\iota \in I$ let $\mathfrak{R}_\iota$ be a set of sets. Define

$$X := \{(\iota, x) \mid \iota \in I,\ x \in X(\mathfrak{R}_\iota)\}.$$

For any $A \subset X$ and for any $\iota \in I$ set

$$A_\iota := \{x \in X(\mathfrak{R}_\iota) \mid (\iota, x) \in A\}.$$

Prove the following propositions:

(α) Define $\mathfrak{R} := \{A \subset X \mid A_\iota \in \mathfrak{R}_\iota \text{ for each } \iota \in I\}$. Then $\mathfrak{R}$ is a ring of sets iff $\mathfrak{R}_\iota$ is a ring of sets for every $\iota \in I$. The analogous correspondence holds for δ-rings and σ-rings.

(β) Define $\mathfrak{S} := \{\{\iota\} \times A \mid \iota \in I,\ A \in \mathfrak{R}_\iota\}$. Prove:

($\beta 1$) $\mathfrak{S}$ is a semi-ring iff every $\mathfrak{R}_\iota$ is a semi-ring.

($\beta 2$) If every $\mathfrak{R}_\iota$ is a semi-ring, then

$$\mathfrak{S}_r = \left\{ \bigcup_{(\iota,\lambda) \in J \times L} (\{\iota\} \times A_{\iota\lambda}) \;\middle|\; \begin{array}{l} J, L \text{ finite sets, } J \subset I, \\ A_{\iota\lambda} \in \mathfrak{R}_\iota \text{ for each } (\iota, \lambda) \in J \times L \end{array} \right\}.$$

($\beta 3$) Suppose that every $\mathfrak{R}_\iota$ is a δ-ring. Then $\mathfrak{S}_r$ is a δ-ring.

($\beta 4$) Suppose that every $\mathfrak{R}_\iota$ is a σ-ring. Then

$$\mathfrak{S}_\sigma = \left\{ \bigcup_{\iota \in J} (\{\iota\} \times A_\iota) \mid J \subset I, J \text{ countable, } A_\iota \in \mathfrak{R}_\iota \text{ for each } \iota \in J \right\}.$$

2.1.7(E) Take a set X and a Riesz lattice $\mathscr{L}$ in $\overline{\mathbb{R}}^X$. We define

$$\mathfrak{R}(\mathscr{L}) := \{A \subset X \mid e_A \in \mathscr{L}\}$$

$$\hat{\mathfrak{R}}(\mathscr{L}) := \{A \subset X \mid A = \{f \neq 0\} \text{ for some } f \in \mathscr{L}\}.$$

Prove the following assertions:

(α) $\mathfrak{R}(\mathscr{L})$ is a ring of sets.

(β) $\mathfrak{R}(\mathscr{L})$ is a δ-ring whenever $\mathscr{L}$ is conditionally σ-completely embedded in $\overline{\mathbb{R}}^X$.

(γ) $\mathfrak{R}(\mathscr{L})$ is a σ-ring whenever $\mathscr{L}$ is σ-completely embedded in $\overline{\mathbb{R}}^X$.

(δ) If $\mathscr{L}$ is conditionally σ-completely embedded in $\overline{\mathbb{R}}^X$, then $\hat{\mathfrak{R}}(\mathscr{L})$ is a δ-ring.

(ε) If $\mathscr{L}$ is σ-completely embedded in $\overline{\mathbb{R}}^X$, then $\hat{\mathfrak{R}}(\mathscr{L})$ is a σ-ring.

(ζ) If $(X, \mathscr{L}, \ell)$ is a closed Daniell space, then $\hat{\mathfrak{R}}(\mathscr{L})$ is a σ-ring.

(η) $\mathfrak{N}(\mathscr{L})$ is a δ-ring.

(ϑ) If $\mathscr{L}$ is σ-completely embedded in $\overline{\mathbb{R}}^X$, then $\mathfrak{N}(\mathscr{L})$ is a σ-ring.

(ι) If $(X, \mathscr{L}, \ell)$ is a closed Daniell space, then $\mathfrak{N}(\mathscr{L})$ is a σ-ring.

(κ) Consider the space $\mathscr{K}$ of all continuous functions on $\mathbb{R}$ with compact support. Define

$$\mathscr{L} := \{ f \in \mathbb{R}^{\mathbb{R}} \mid \exists g \in \mathscr{K}_+, \, |f| \leq g \}.$$

Then $\mathfrak{R}(\mathscr{L})$ is a δ-ring. Show that $\mathfrak{R}(\mathscr{L})$ is not a σ-ring, and the same is true for $\hat{\mathfrak{R}}(\mathscr{L})$.

(λ) Suppose $\mathscr{L}$ is the set of all Riemann integrable functions on an interval $[a, b]$; $(a, b \in \mathbb{R},\ a < b)$. Then $\mathfrak{R}(\mathscr{L})$ is an algebra of sets on $[a, b]$ but not a δ-ring.

2.1.8(E) Let $\mathfrak{S}$ be a set of sets. Prove the following statements:

(α) $\mathfrak{S}_r = \{ A \mid \exists \mathfrak{R} \subset \mathfrak{S},\ \mathfrak{R} \text{ finite},\ A \in \mathfrak{R}_r \}$.

(β) $\mathfrak{S}_\delta = \{ A \mid \exists \mathfrak{R} \subset \mathfrak{S},\ \mathfrak{R} \text{ countable},\ A \in \mathfrak{R}_\delta \}$.

(γ) $\mathfrak{S}_\sigma = \{ A \mid \exists \mathfrak{R} \subset \mathfrak{S},\ \mathfrak{R} \text{ countable},\ A \in \mathfrak{R}_\sigma \}$.

(First show that the set of sets with the given property in (α) [resp. (β), (γ)] is a ring of sets (resp. δ-ring, σ-ring), and then show that it must be identical with $\mathfrak{S}_r$ (resp. $\mathfrak{S}_\delta, \mathfrak{S}_\sigma$).)

2.1.9(E) Let $\mathfrak{S}$ be a set of sets. We define recursively a system of sets, $\mathfrak{R}_\alpha$, for each ordinal α. Set

$$\mathfrak{R}_0 := \mathfrak{S} \cup \{\varnothing\}.$$

Suppose that $\mathfrak{R}_\beta$ has been defined for each $\beta < \alpha$. Then define

$$\mathfrak{S}_\alpha := \bigcup_{\beta < \alpha} \mathfrak{R}_\beta$$

$$\mathfrak{S}'_\alpha := \left\{ \bigcup_{n \in \mathbb{N}} A_n \,\middle|\, (A_n)_{n \in \mathbb{N}} \text{ is a sequence from } \mathfrak{S}_\alpha \right\}$$

$$\mathfrak{R}_\alpha := \{ A \setminus B \mid A, B \in \mathfrak{S}'_\alpha \}.$$

Prove the following propositions:

(α) If ω_1 is the first uncountable ordinal, then $\mathfrak{R}_{\omega_1} = \mathfrak{S}_\sigma$.

(β) Replace $\mathfrak{S}'_\alpha$ in the definition of $\mathfrak{R}_\alpha$ by $\{A \in \mathfrak{S}'_\alpha \mid \exists$ a finite family $(A_\iota)_{\iota \in I}$ from $\mathfrak{S}$ with $A \subset \bigcup_{\iota \in I} A_\iota\}$. Then $\mathfrak{R}_{\omega_1} = \mathfrak{S}_\delta$.

Now suppose that $\mathfrak{S}$ is a ring of sets. We define inductively a system of sets, $\mathfrak{R}^\alpha$, for each ordinal α. Set

$$\mathfrak{R}^0 := \mathfrak{S}.$$

Suppose that $\mathfrak{R}^\beta$ has already been defined for each $\beta < \alpha$. Then define

$$\mathfrak{S}^\alpha := \bigcup_{\beta < \alpha} \mathfrak{R}^\beta$$

$$\tilde{\mathfrak{S}}^\alpha := \left\{ \bigcup_{n \in \mathbb{N}} A_n \mid (A_n)_{n \in \mathbb{N}} \text{ an increasing sequence from } \mathfrak{S}^\alpha \right\}$$

$$\mathfrak{R}^\alpha := \left\{ \bigcap_{n \in \mathbb{N}} A_n \mid (A_n)_{n \in \mathbb{N}} \text{ a decreasing sequence from } \tilde{\mathfrak{S}}^\alpha \right\}.$$

Prove the following:

(γ) $\mathfrak{R}^{\omega_1} = \mathfrak{S}_\sigma$.

(δ) Replace $\tilde{\mathfrak{S}}^\alpha$ in the definition of $\mathfrak{R}^\alpha$ by $\{A \in \tilde{\mathfrak{S}}^\alpha \mid \exists B \in \mathfrak{S},\ A \subset B\}$. Then $\mathfrak{R}^{\omega_1} = \mathfrak{S}_\delta$.

2.1.10(E) In the previous exercise, we constructed δ-rings and σ-rings using monotone operations. Such constructions with monotone sequences are important in measure theory. For this reason we introduce the concept of a monotone set.

A monotone set is a set of sets, $\mathfrak{S}$, with the property that if $(A_n)_{n \in \mathbb{N}}$ is a monotone sequence from $\mathfrak{S}$, then $\lim_{n \to \infty} A_n \in \mathfrak{S}$.

Prove the following statements:

(α) Let $\mathfrak{S}$ be a set of sets. Then there exists a monotone set $\mathfrak{S}_m$ such that $\mathfrak{S} \subset \mathfrak{S}_m$ and $\mathfrak{S}_m \subset \mathfrak{R}$ for every monotone set $\mathfrak{R}$ with $\mathfrak{S} \subset \mathfrak{R}$. $\mathfrak{S}_m$ is uniquely determined by $\mathfrak{S}$. $\mathfrak{S}_m$ is called the **monotone set generated** by $\mathfrak{S}$.

(β) Every σ-ring is a monotone set.

(γ) If $\mathfrak{S}$ is both a ring of sets and a monotone set, then $\mathfrak{S}$ is a σ-ring.

(δ) Given the ring of sets, $\mathfrak{R}$, $\mathfrak{R}_m = \mathfrak{R}_\sigma$.

It is possible to prove (δ) using the characterization of $\mathfrak{R}_\sigma$ given in Ex. 2.1.9, but we wish to indicate a proof that does not rely on any transfinite process. Clearly $\mathfrak{R}_m \subset \mathfrak{R}_\sigma$. Given $A \subset X(\mathfrak{R})$, define

$$\mathfrak{S}_A := \{B \subset X(\mathfrak{R}) \mid A \setminus B \in \mathfrak{R}_m,\ B \setminus A \in \mathfrak{R}_m \text{ and } A \cup B \in \mathfrak{R}_m\}.$$

Then, given $A, B \subset X(\mathfrak{R})$, $A \in \mathfrak{S}_B$ iff $B \in \mathfrak{S}_A$. Given a monotone sequence $(A_n)_{n \in \mathbb{N}}$ from $\mathfrak{S}_A$, $\lim_{n \to \infty} A_n \in \mathfrak{S}_A$. Hence $\mathfrak{S}_A$ is a monotone set. For $A \in \mathfrak{R}$, $\mathfrak{R} \subset \mathfrak{S}_A$, and so $\mathfrak{R}_m \subset \mathfrak{S}_A$. Consequently, if $A \in \mathfrak{R}$ and $B \in \mathfrak{R}_m$, then $B \in \mathfrak{S}_A$, and so $A \in \mathfrak{S}_B$. As a result $\mathfrak{R}_m \subset \mathfrak{S}_B$ whenever $B \in \mathfrak{R}_m$, showing that $\mathfrak{R}_m$ is a ring of sets. Using (γ), $\mathfrak{R}_\sigma \subset \mathfrak{R}_m$.

We extend the observations to δ-rings by introducing the next concept:

A **conditionally monotone set** is a set of sets, $\mathfrak{S}$, with the property that $\bigcap_{n \in \mathbb{N}} A_n \in \mathfrak{S}$ whenever $(A_n)_{n \in \mathbb{N}}$ is a decreasing sequence from $\mathfrak{S}$, and $\bigcup_{n \in \mathbb{N}} A_n \in \mathfrak{S}$ whenever $(A_n)_{n \in \mathbb{N}}$ is an increasing bounded sequence from $\mathfrak{S}$.

By analogy to (α)–(δ), the following statements hold:

(α') Let $\mathfrak{S}$ be a set of sets. Then there exists a conditionally monotone set $\mathfrak{S}_{mb}$ such that $\mathfrak{S} \subset \mathfrak{S}_{mb}$ and $\mathfrak{S}_{mb} \subset \mathfrak{R}$ for every conditionally monotone set $\mathfrak{R}$ with $\mathfrak{S} \subset \mathfrak{R}$. $\mathfrak{S}_{mb}$ is uniquely determined by $\mathfrak{S}$. $\mathfrak{S}_{mb}$ is called the **conditionally monotone set generated** by $\mathfrak{S}$.

(β') Every δ-ring is a conditionally monotone set.

(γ') If $\mathfrak{S}$ is a ring of sets and a conditionally monotone set, then $\mathfrak{S}$ is a δ-ring.

(δ') Given a ring of sets, $\mathfrak{R}, \mathfrak{R}_{mb} = \mathfrak{R}_\delta$.

Let $\mathfrak{K}$, $\mathfrak{R}$ be sets of sets with the following properties:

(i) $\mathfrak{K} \subset \mathfrak{R} \subset \mathfrak{K}$.
(ii) $K \cup L \in \mathfrak{K}$ and $K \setminus L \in \mathfrak{R}$ for all $K, L \in \mathfrak{K}$.
(iii) $\bigcap_{n \in \mathbb{N}} A_n \in \mathfrak{R}$ for every sequence $(A_n)_{n \in \mathbb{N}}$ from $\mathfrak{R}$.
(iv) For every sequence $(A_n)_{n \in \mathbb{N}}$ from $\mathfrak{R}$, if there is $K \in \mathfrak{K}$ such that $\bigcup_{n \in \mathbb{N}} A_n \subset K$, then $\bigcup_{n \in \mathbb{N}} A_n \in \mathfrak{R}$.

Prove

(ε) $\mathfrak{R} = \mathfrak{K}_\delta$.

2.1.11(E) This exercise is devoted to questions concerning the cardinality of generated systems of sets.

(a) **Finitely Generated Systems.** Let $(A_k)_{k \in \mathbb{N}_n}$ be a family of nonempty sets. Set $X := \bigcup_{k=1}^n A_k$. To each $x \in X$ associate an element $\vec{x} := (x_1, \ldots, x_n) \in \{0,1\}^n$ where $x_k := e_{A_k}(x)$. For $x, y \in X$ write $x \sim y$ whenever $\vec{x} = \vec{y}$. Then $\sim$ is an equivalence relation on X. Denote by $\mathfrak{S}$ the set of all equivalence classes, and by m the cardinality of $\mathfrak{S}$. Prove the following propositions:

(α) Given $x \in X$,

$$\bigcap_{\substack{k \in \mathbb{N}_n \\ x_k = 1}} A_k \setminus \bigcup_{\substack{k \in \mathbb{N}_n \\ x_k = 0}} A_k$$

is the equivalence class containing x.

(β) Given $k \in \mathbb{N}_n$, A_k is the union of a finite family from $\mathfrak{S}$.

(γ) $\{A_k \mid k \in \mathbb{N}_n\}$ and $\mathfrak{S}$ generate the same ring of sets, $\mathfrak{R}$.

(δ) $\mathfrak{S}$ is a semi-ring.

(ε) $\mathfrak{R}$ is precisely the set of all unions of disjoint families from $\mathfrak{S}$.

(ζ) $0 < m \leq 2^n - 1$.

(η) $\mathfrak{R}$ has 2^m elements.

(ϑ) Construct an example for which the cardinality of $\mathfrak{R}$ is 2^{2^n-1}.

(ι) Given $A \in \mathfrak{R}$ and $B \in \mathfrak{S}$ with $A \subset B$, then $A = B$. (Because of property (ι), the sets in $\mathfrak{S}$ are sometimes called the **minimal terms** (or **minterms**) generated by $(A_k)_{k \in \mathbb{N}_n}$.)

(κ) Every finite ring of sets is both a δ-ring and a σ-ring.

(λ) $\mathfrak{S}_r = \mathfrak{S}_\delta = \mathfrak{S}_\sigma$.

(b) **Countably Infinitely Generated Systems.** Let $\mathfrak{S}$ now be countably infinite. Prove the following statements:

(α) $\mathfrak{S}_r$ is countably infinite.

(β) $\mathfrak{S}_\delta$ has cardinality not exceeding that of the continuum.

(γ) $\mathfrak{S}_\sigma$ always has the same cardinality as the continuum.

We provide some suggestions for (γ). [Observe that (β) follows from (γ).] If $\mathfrak{S}_r$ is countably infinite, then there is a disjoint sequence of nonempty subsets of $\mathfrak{S}_r$, $(A_n)_{n \in \mathbb{N}}$. $\mathfrak{S}$ contains all unions of subsets of $\{A_n \mid n \in \mathbb{N}\}$. But these subsets are in bijection with the sequences $(\alpha_n)_{n \in \mathbb{N}}$ in $\{0,1\}^{\mathbb{N}}$ that are obtained by setting $\alpha_n := 1$ when A_n is a member of the particular subset, and $\alpha_n := 0$ otherwise. Thus $\mathfrak{S}$ has cardinality at least that of the continuum.

On the other hand, the procedure in Ex. 2.1.9, shows that the cardinality cannot exceed that of the continuum, since at no stage of the procedure is the cardinality of the continuum exceeded.

(δ) The δ-ring generated by $\{\{n\} \mid n \in \mathbb{N}\}$ is precisely the set of all finite subsets of $\mathbb{N}$, and consequently countable.

(ε) The δ-ring generated by $\{[\alpha, \beta[\mid \alpha, \beta \in \mathbb{Q}\}$ contains all right half-open intervals of $\mathbb{R}$ and is consequently uncountable.

(c) **Uncountably Infinitely Generated Systems.** Finally let $\mathfrak{S}$ be an uncountably infinite set. Prove the following propositions:

(α) The cardinality of $\mathfrak{S}_\sigma$ is at least that of the continuum.

(β) If the cardinality of $\mathfrak{S}$ is not strictly smaller than that of the continuum, then $\mathfrak{S}$, $\mathfrak{S}_r$, $\mathfrak{S}_\delta$, and $\mathfrak{S}_\sigma$ have the same cardinality.

2.1.12(E) The general notion of a system of sets. In this exercise we denote by M the class of all sets. For any set $\mathscr{F}$ of mappings of $M^{\mathbb{N}}$ into M we set

$$\mathscr{F}_1 := \left\{ \mathfrak{R} \neq \emptyset \mid f\big((A_n)_{n \in \mathbb{N}}\big) \in \mathfrak{R} \text{ for all } f \in \mathscr{F} \text{ and } (A_n)_{n \in \mathbb{N}} \in \mathfrak{R}^{\mathbb{N}} \right\}$$

$$\mathscr{F}_2 := \left\{ \mathfrak{R} \neq \emptyset \,\middle|\, \begin{array}{l} f\big((A_n)_{n \in \mathbb{N}}\big) \in \mathfrak{R} \text{ for all } f \in \mathscr{F} \text{ and } (A_n)_{n \in \mathbb{N}} \in \mathfrak{R}^{\mathbb{N}} \\ \text{for which there is } A \in \mathfrak{R} \text{ with } \bigcup\limits_{n \in \mathbb{N}} A_n \subset A \end{array} \right\}.$$

We say that $\mathscr{F}$ is of **finite type** iff each mapping of $\mathscr{F}$ depends on a finite number of variables only.
We say that $\mathscr{F}$ is **stable** if for every $f \in \mathscr{F}$, for every $(A_n)_{n \in \mathbb{N}}, (B_n)_{n \in \mathbb{N}} \in M^{\mathbb{N}}$, and for every $X \in M$,

$$f((A_n \cap X)_{n \in \mathbb{N}}) = f((A_n)_{n \in \mathbb{N}}) \cap X$$

$$f((A_n \setminus X)_{n \in \mathbb{N}}) = f((A_n)_{n \in \mathbb{N}}) \setminus X$$

$$f(((A_n \cap X) \cup (B_n \setminus X))_{n \in \mathbb{N}}) = (f((A_n)_{n \in \mathbb{N}}) \cap X)$$

$$\cup (f((B_n)_{n \in \mathbb{N}}) \setminus X).$$

Prove the following statements.

(α) To each of the following classes Γ of systems of sets there is a set of mappings, $\mathscr{F}$, of $M^{\mathbb{N}}$ into M such that $\Gamma = \mathscr{F}_1$. ($|\mathscr{F}|$ denotes the cardinality of $\mathscr{F}$).

(α1) The class of all nonempty $\cap$-semi lattices.
($|\mathscr{F}| = 1$, $\mathscr{F}$ is of finite type and stable).
(α2) The class of all nonempty lattices of sets.
($|\mathscr{F}| = 2$, $\mathscr{F}$ is of finite type and stable).
(α3) The class of all rings of sets.
($|\mathscr{F}| = 2$, $\mathscr{F}$ is of finite type and stable).
(α4) The class of all δ-rings. ($|\mathscr{F}| = 3$, $\mathscr{F}$ is stable).
(α5) The class of all σ-rings. ($|\mathscr{F}| = 2$, $\mathscr{F}$ is stable).
(α6) The class of all nonempty monotone sets. ($|\mathscr{F}| = 2$).
(α7) The class of all nonempty pseudomonotone sets. ($|\mathscr{F}| = 2$).
(A **pseudomonotone set** is a set $\mathfrak{R}$ of sets such that

(i) $\bigcap_{n \in \mathbb{N}} A_n \in \mathfrak{R}$ for every sequence $(A_n)_{n \in \mathbb{N}}$ in $\mathfrak{R}$
(ii) $\bigcup_{n \in \mathbb{N}} A_n \in \mathfrak{R}$ for every disjoint sequence $(A_n)_{n \in \mathbb{N}}$ in $\mathfrak{R}$.)

(α8) The class of all Dynkin systems ($|\mathscr{F}| = 2$).
(A **Dynkin system** is a set $\mathfrak{R}$ of sets such that

(i) $\emptyset \in \mathfrak{R}$
(ii) $B \setminus A \in \mathfrak{R}$ for all $A, B \in \mathfrak{R}$ with $A \subset B$
(iii) $\bigcup_{n \in \mathbb{N}} A_n \in \mathfrak{R}$ for every disjoint sequence $(A_n)_{n \in \mathbb{N}}$ in $\mathfrak{R}$.)

(β) To each of the following classes Γ of systems of sets there is a set of mappings $\mathscr{F}$ of $M^{\mathbb{N}}$ into M such that $\Gamma = \mathscr{F}_2$.

(β1) The class of all nonempty conditionally monotone sets. ($|\mathscr{F}| = 2$).
(β2) The class of all nonempty conditionally pseudomonotone sets. ($|\mathscr{F}| = 2$).
(A **conditionally pseudomonotone** set is a set $\mathfrak{R}$ of sets such that

(i) $\bigcap_{n \in \mathbb{N}} A_n \in \mathfrak{R}$ for every sequence $(A_n)_{n \in \mathbb{N}}$ in $\mathfrak{R}$
(ii) $\bigcup_{n \in \mathbb{N}} A_n \in \mathfrak{R}$ for every disjoint sequence $(A_n)_{n \in \mathbb{N}}$ in $\mathfrak{R}$, for which there exists $A \in \mathfrak{R}$ with $\bigcup_{n \in \mathbb{N}} A_n \subset A$.)

($\beta 3$) The class of all conditionally Dynkin systems. ($|\mathscr{F}| = 2$).
(A **conditionally Dynkin system** is a set $\mathfrak{R}$ of sets satisfying conditions (i) and (ii) of a Dynkin system and, instead of (iii),

(iii′) $\bigcup_{n \in \mathbb{N}} A_n \in \mathfrak{R}$ for every disjoint sequence $(A_n)_{n \in \mathbb{N}}$ in $\mathfrak{R}$ for which there exists $A \in \mathfrak{R}$ with $\bigcup_{n \in \mathbb{N}} A_n \subset A$.)

For the following let $i \in \{1, 2\}$ and let $\mathscr{F}$ be a set of mappings of $M^{\mathbb{N}}$ into M such that

$$f((A_n)_{n \in \mathbb{N}}) \subset \bigcup_{n \in \mathbb{N}} A_n$$

for every $f \in \mathscr{F}$ and every $(A_n)_{n \in \mathbb{N}} \in M^{\mathbb{N}}$. Prove:

(γ) For every set X, $\mathfrak{P}(X) \in \mathscr{F}_i$.

(δ) For each nonempty family $(\mathfrak{R}_\iota)_{\iota \in I}$ from $\mathscr{F}_i$,

$$\bigcap_{\iota \in I} \mathfrak{R}_\iota \in \mathscr{F}_i \cup \{\varnothing\}.$$

(ε) To each nonempty set $\mathfrak{G}$ of sets there is a smallest element $\mathfrak{G}_{\mathscr{F}_i}$ in $\mathscr{F}_i$ containing $\mathfrak{G}$.
$\mathfrak{G}_{\mathscr{F}_i}$ is called the **$\mathscr{F}_i$-system generated** by $\mathfrak{G}$.
For every $A \in \mathfrak{G}_{\mathscr{F}_i}$ there is a countable subset $\mathfrak{G}'$ of $\mathfrak{G}$ such that $A \in \mathfrak{G}'_{\mathscr{F}_i}$. If $\mathscr{F}$ is of finite type, then we may replace countable by finite.

As an example of (α) we sketch the proof of ($\alpha 4$). Set

$$f_1 \colon M^{\mathbb{N}} \to M, \qquad (A_n)_{n \in \mathbb{N}} \mapsto A_1 \cup A_2$$

$$f_2 \colon M^{\mathbb{N}} \to M, \qquad (A_n)_{n \in \mathbb{N}} \mapsto A_1 \setminus A_2$$

$$f_3 \colon M^{\mathbb{N}} \to M, \qquad (A_n)_{n \in \mathbb{N}} \mapsto \bigcap_{n \in \mathbb{N}} A_n.$$

As an example of (β) we sketch the proof of ($\beta 1$). Set

$$f_1 \colon M^{\mathbb{N}} \to M, \qquad (A_n)_{n \in \mathbb{N}} \mapsto \begin{cases} \bigcap_{n \in \mathbb{N}} A_n & \text{if } (A_n)_{n \in \mathbb{N}} \text{ is decreasing} \\ A_1 & \text{if } (A_n)_{n \in \mathbb{N}} \text{ is not decreasing} \end{cases}$$

$$f_2 \colon M^{\mathbb{N}} \to M, \qquad (A_n)_{n \in \mathbb{N}} \mapsto \begin{cases} \bigcup_{n \in \mathbb{N}} A_n & \text{if } (A_n)_{n \in \mathbb{N}} \text{ is increasing} \\ A_1 & \text{if } (A_n)_{n \in \mathbb{N}} \text{ is not increasing} \end{cases}$$

(ζ) Let X be a set. Set

$$\mathscr{F}_X := \{\mathfrak{R} \in \mathscr{F}_1 \mid X \in \mathfrak{R} \subset \mathfrak{P}(X)\}.$$

Then prove

(ζ1) $\mathcal{F}_X = \{\mathfrak{R} \in \mathcal{F}_2 \mid X \in \mathfrak{R} \subset \mathfrak{P}(X)\}$.
(ζ2) For $\mathcal{F}$ of (α3), $\mathcal{F}_X$ is the set of algebras of sets on X.
(ζ3) For $\mathcal{F}$ of (α4) or of (α5), $\mathcal{F}_X$ is the set of σ-algebras of sets on X.
(ζ4) The assertions (γ) and (δ) hold if we replace $\mathcal{F}_i$ by $\mathcal{F}_X$. The same is true for (ε) if $\mathfrak{G} \subset \mathfrak{P}(X)$.
(ζ5) $\mathcal{F}_X$, $\{\mathfrak{R} \in \mathcal{F}_i \mid \mathfrak{R} \subset \mathfrak{P}(X)\} \cup \{\varnothing\}$ and $\{\mathfrak{R} \in \mathcal{F}_i \mid \varnothing \in \mathfrak{R} \subset \mathfrak{P}(X)\}$ are complete lattices with respect to $\subset$.
(ζ6) If $\mathfrak{R} \in \mathcal{F}_i$, then $\{A \in \mathfrak{R} \mid A \subset X\} \in \mathcal{F}_i$.
(ζ7) If $\mathcal{F}$ is stable then

$$\{A \cap X \mid A \in \mathfrak{R}\} \in \mathcal{F}_i$$

$$\{A \setminus X \mid A \in \mathfrak{R}\} \in \mathcal{F}_i$$

$$\{(A \cap X) \cup (B \setminus X) \mid A, B \in \mathfrak{R}\} \in \mathcal{F}_i$$

for every $\mathfrak{R} \in \mathcal{F}_i$.

The assertions (ζ6) and (ζ7) still hold if we replace $\mathcal{F}_i$ in all occurrences by the class of all semirings.

(η) Let X, Y be sets and $\varphi \in Y^X$ such that

$$f\big((\varphi^{-1}(A_n))_{n \in \mathbb{N}}\big) = \varphi^{-1}\big(f((A_n)_{n \in \mathbb{N}})\big)$$

for every $f \in \mathcal{F}$ and for every $(A_n)_{n \in \mathbb{N}} \in \mathfrak{P}(Y)^{\mathbb{N}}$. Prove the following assertions:

(η1) For every $\mathfrak{R} \in \mathcal{F}_i$ with $\mathfrak{R} \subset \mathfrak{P}(Y)$, $\{\varphi^{-1}(A) \mid A \in \mathfrak{R}\} \in \mathcal{F}_i$.
(η2) For every $\mathfrak{R} \in \mathcal{F}_i$ with $\mathfrak{R} \subset \mathfrak{P}(X)$, $\{A \subset Y \mid \varphi^{-1}(A) \in \mathfrak{R}\} \in \mathcal{F}_i$.
(η3) For every $\mathfrak{G} \subset \mathfrak{P}(Y)$, $\{\varphi^{-1}(A) \mid A \in \mathfrak{G}\}_{\mathcal{F}_i} = \{\varphi^{-1}(A) \mid A \in \mathfrak{G}_{\mathcal{F}_i}\}$.
(η4) For every $\mathfrak{G} \subset \mathfrak{P}(X)$,

$$\{A \subset Y \mid \varphi^{-1}(A) \in \mathfrak{G}\}_{\mathcal{F}_i} = \{A \subset Y \mid \varphi^{-1}(A) \in \mathfrak{G}_{\mathcal{F}_i}\}.$$

(η5) The above condition is fulfilled for all $\mathcal{F}$ appearing in (α) and (β).

(ϑ) Let $\mathfrak{G}$ be a set of sets such that

(i) $\varnothing \in \mathfrak{G}$
(ii) $A, B \in \mathfrak{G}$ implies $A \setminus B$, $A \cup B \in \mathfrak{G}_m$ (resp. $\mathfrak{G}_{mb}$).

Prove:

($\vartheta 1$) $\mathfrak{G}_m = \mathfrak{G}_\sigma$ (resp. $\mathfrak{G}_{mb} = \mathfrak{G}_\delta$). This assertion still holds if we replace $\mathfrak{G}_m$ (resp. $\mathfrak{G}_{mb}$) in all occurrences by the pseudomonotone set (resp. the conditionally pseudomonotone set) generated by $\mathfrak{G}$.
($\vartheta 2$) Let $\mathfrak{G}$ be a set of sets such that for every $A, B \in \mathfrak{G}$ there exists a disjoint countable family $(C_\iota)_{\iota \in I}$ in $\mathfrak{G}$ with $A \cap B = \bigcup_{\iota \in I} C_\iota$. Then, $\mathfrak{G}_\sigma$ (resp. $\mathfrak{G}_\delta$) is the Dynkin system (resp. the conditionally Dynkin system) generated by $\mathfrak{G}$.

2.1.13(E) Finite Products. This exercise deals with finite products of systems of sets. We use the notions of Ex. 2.1.12. For each family $(\mathfrak{R}_\iota)_{\iota \in I}$ of systems of sets define

$$\mathop{\nabla}_{\iota \in I} \mathfrak{R}_\iota := \left\{ \prod_{\iota \in I} A_\iota \,|\, (A_\iota)_{\iota \in I} \in \prod_{\iota \in I} \mathfrak{R}_\iota \right\}.$$

Prove the following statements for finite families $(\mathfrak{R}_\iota)_{\iota \in I}$ (we use the notation from Ex. 2.1.2):

(α) $\nabla_{\iota \in I} \mathfrak{R}_{\iota u} \subset (\nabla_{\iota \in I} \mathfrak{R}_\iota)_u$.
(β) $\nabla_{\iota \in I} \mathfrak{R}_{\iota v} \subset (\nabla_{\iota \in I} \mathfrak{R}_\iota)_v$.
(γ) $\nabla_{\iota \in I} \mathfrak{R}_{\iota i} = (\nabla_{\iota \in I} \mathfrak{R}_\iota)_i$.
(δ) $(\nabla_{\iota \in I} \mathfrak{R}_\iota)_d \subset (\nabla_{\iota \in I} \mathfrak{R}_{\iota i})_v$.
(ε) If every $\mathfrak{R}_\iota$ is a $\cap$-semi-lattice, then $\nabla_{\iota \in I} \mathfrak{R}_\iota$ is also a $\cap$-semi-lattice.
(ζ) If every $\mathfrak{R}_\iota$ is a semi-ring, then $\nabla_{\iota \in I} \mathfrak{R}_\iota$ is also a semi-ring.

Take now a set $\mathfrak{F}$ of mappings of $M^{\mathbb{N}}$ into M. Define for $i \in \{1,2\}$

$$\mathfrak{F}_i \bigotimes_{\iota \in I} \mathfrak{R}_\iota := \bigotimes_{\iota \in I} \mathfrak{R}_\iota := \left(\mathop{\nabla}_{\iota \in I} \mathfrak{R}_\iota \right)_{\mathfrak{F}_i}.$$

This system is called the $\mathfrak{F}_i$-**product** of $(\mathfrak{R}_\iota)_{\iota \in I}$ or simply the product of $(\mathfrak{R}_\iota)_{\iota \in I}$ if $\mathfrak{F}_i$ is clear from the context.

The next examples are important for us.

(η) For each family of $\cap$-semi-lattices, $(\mathfrak{R}_\iota)_{\iota \in I}$, the product with respect to the class of $\cap$-semi-lattices is equal to $\nabla_{\iota \in I} \mathfrak{R}_\iota$ [see (ε)].
(ϑ) Let $(\mathfrak{R}_\iota)_{\iota \in I}$ be a finite family of rings of sets. Then the following hold:

($\vartheta 1$) $\nabla_{\iota \in I} \mathfrak{R}_\iota$ is a semi-ring.
($\vartheta 2$) $\otimes_{\iota \in I} \mathfrak{R}_\iota$ with respect to the class of rings of sets is equal to the set of all finite unions of disjoint sets from $\nabla_{\iota \in I} \mathfrak{R}_\iota$. (cf., Ex. 2.1.2).

(ι) If $(\mathfrak{R}_\iota)_{\iota \in I}$ is a finite family of δ-rings or σ-rings, the situation, although far more complicated, is not substantially different from that for rings of sets. Again, $\nabla_{\iota \in I} \mathfrak{R}_\iota$ is only a semi-ring. The product with respect to the class of δ-rings (resp. σ-rings) can be determined by the methods in Ex. 2.1.9. or 2.1.10.

2.2. POSITIVE MEASURES

We begin with the definitions and elementary properties of various kinds of set mappings, in particular of positive measures. A discussion of various boundedness conditions follows.

Definition 2.2.1. *Let $\mathfrak{S}$ be a set of sets. A mapping μ: $\mathfrak{S} \to \mathbb{R}$ is said to be:*

(*a*) ***additive***, *if the conditions $A, B, A \cup B \in \mathfrak{S}$ and $A \cap B = \varnothing$ always imply that*

$$\mu(A \cup B) = \mu(A) + \mu(B)$$

(*b*) ***positive***, *if $\mu(A) \geq 0$ for every $A \in \mathfrak{S}$*

(*c*) ***increasing***, *if it increases as a mapping from the ordered set $\mathfrak{S}$ to the ordered set $\mathbb{R}$*

(*d*) ***nullcontinuous***, *if*

$$\lim_{n \in \mathbb{N}} \mu(A_n) = 0$$

for every sequence $(A_n)_{n \in \mathbb{N}}$ from $\mathfrak{S}$ that decreases and satisfies

$$\bigcap_{n \in \mathbb{N}} A_n = \varnothing .$$

*If $\mathfrak{R}$ is a ring of sets and μ: $\mathfrak{R} \to \mathbb{R}$ is an additive, positive, nullcontinuous mapping, then μ is called a **positive measure** on $\mathfrak{R}$. A **positive measure space** (a **positive δ- or σ-measure space**) is a triple $(X, \mathfrak{R}, \mu)$ where X is a set, $\mathfrak{R}$ is a set-ring (a δ-ring or a σ-ring, respectively) on X, and μ is a positive measure on $\mathfrak{R}$.* □

Proposition 2.2.2. *If μ: $\mathfrak{R} \to \mathbb{R}$ is an additive mapping on a ring of sets $\mathfrak{R}$, then the following assertions hold.*

(*a*) $\mu(\varnothing) = 0$.

(*b*) *If $(A_\iota)_{\iota \in I}$ is a finite disjoint family from $\mathfrak{R}$, then*

$$\mu\Big(\bigcup_{\iota \in I} A_\iota\Big) = \sum_{\iota \in I} \mu(A_\iota).$$

(*c*) *If A, B belong to $\mathfrak{R}$ and $A \subset B$, then*

$$\mu(B \setminus A) = \mu(B) - \mu(A).$$

(*d*) *μ is positive iff it is increasing.*

(*e*) *For all $A, B \in \mathfrak{R}$,*

$$\mu(A \cup B) = \mu(A) + \mu(B) - \mu(A \cap B)$$

and if μ is positive, then

$$\mu(A \cup B) \leq \mu(A) + \mu(B).$$

Proof. (a) follows from the identity

$$\mu(\varnothing) = \mu(\varnothing \cup \varnothing) = \mu(\varnothing) + \mu(\varnothing).$$

(b) can easily be proved by complete induction on the number of elements in I. Observe that $\bigcup_{\iota \in \varnothing} A_\iota = \varnothing$, so that (b) follows from (a) in case $I = \varnothing$.

(c) Note that B is the disjoint union of A and $B \setminus A$. Since μ is additive,

$$\mu(B) = \mu(A) + \mu(B \setminus A)$$

and (c) follows.

(d) If μ is positive and $A \subset B$, then

$$\mu(B \setminus A) \geq 0$$

and the inequality

$$\mu(A) \leq \mu(B)$$

follows from (c). Conversely, if μ is increasing, then it follows from (a) that

$$0 = \mu(\varnothing) \leq \mu(A)$$

for every A in $\mathfrak{R}$.

(e) Since $A \cup B$ is the disjoint union of A and $B \setminus A$, we have

$$\mu(A \cup B) = \mu(A) + \mu(B \setminus A).$$

Writing B as the disjoint union of $B \cap A$ and $B \setminus A$, we have

$$\mu(B) = \mu(B \setminus A) + \mu(B \cap A)$$

and (e) follows. □

A bit more should be said concerning the implications of additivity. Additivity is a compatibility condition tying the operation of addition in $\mathbb{R}$ to the operation of forming set-theoretic union. The latter operation possesses certain special properties that are completely obvious. For instance, the union of a family of sets does not depend on the order in which the union is formed nor on how the sets might happen to be grouped. To speak about the additivity of a set mapping requires that corresponding properties hold for finite sums in $\mathbb{R}$: otherwise the notion of an additive set mapping would be meaningless. Thus finite sums of real numbers must obey precisely the rules formulated in Section 1.1. (Proposition 1.1.2 (f)–(j)).

As long as the domain is a ring of sets, various equivalent characterizations of nullcontinuous mappings are possible.

Proposition 2.2.3. *If $\mathfrak{R}$ is a ring of sets, then the following conditions are equivalent, for every positive, additive mapping μ: $\mathfrak{R} \to \mathbb{R}$.*

(*a*) *μ is nullcontinuous.*

(*b*) *$\inf_{n \in \mathbb{N}} \mu(A_n) = 0$ for every sequence $(A_n)_{n \in \mathbb{N}}$ from $\mathfrak{R}$ that decreases and satisfies $\bigcap_{n \in \mathbb{N}} A_n = \varnothing$.*

(*c*) *For every increasing sequence $(A_n)_{n \in \mathbb{N}}$ from $\mathfrak{R}$, if $\bigcup_{n \in \mathbb{N}} A_n$ belongs to $\mathfrak{R}$, then*

$$\mu\left(\bigcup_{n \in \mathbb{N}} A_n\right) = \sup_{n \in \mathbb{N}} \mu(A_n).$$

(*d*) *For every decreasing sequence $(B_n)_{n \in \mathbb{N}}$ from $\mathfrak{R}$, if $\bigcap_{n \in \mathbb{N}} B_n$ belongs to $\mathfrak{R}$, then*

$$\mu\left(\bigcap_{n \in \mathbb{N}} B_n\right) = \inf_{n \in \mathbb{N}} \mu(B_n).$$

Proof. We mimic the proof of Proposition 1.3.6 concerning nullcontinuous functionals.

(a) $\Leftrightarrow$ (b) follows from the fact that μ increases [Proposition 2(d)].

(b) $\Rightarrow$ (c). Let $(A_n)_{n \in \mathbb{N}}$ be an increasing sequence from $\mathfrak{R}$, and assume that

$$A := \bigcup_{n \in \mathbb{N}} A_n$$

belongs to $\mathfrak{R}$. The sequence $(A \setminus A_n)_{n \in \mathbb{N}}$ lies in $\mathfrak{R}$, decreases, and has empty intersection:

$$\bigcap_{n \in \mathbb{N}} (A \setminus A_n) = A \setminus \bigcup_{n \in \mathbb{N}} A_n = A \setminus A = \varnothing.$$

Hypothesis (b) implies that

$$\inf_{n \in \mathbb{N}} \mu(A \setminus A_n) = 0.$$

Since each A_n is a subset of A, we have

$$\mu(A \setminus A_n) = \mu(A) - \mu(A_n)$$

for every n [Proposition 2(c)]. Therefore

$$\begin{aligned} 0 &= \inf_{n\in\mathbb{N}} \mu(A \setminus A_n) \\ &= \inf_{n\in\mathbb{N}} (\mu(A) - \mu(A_n)) \\ &= \mu(A) + \inf_{n\in\mathbb{N}} (-\mu(A_n)) \\ &= \mu(A) - \sup_{n\in\mathbb{N}} \mu(A_n) \end{aligned}$$

and so

$$\mu\Big(\bigcup_{n\in\mathbb{N}} A_n\Big) = \mu(A) = \sup_{n\in\mathbb{N}} \mu(A_n)$$

as required.

(c) $\Rightarrow$ (d). Let $(B_n)_{n\in\mathbb{N}}$ be a decreasing sequence from $\mathfrak{R}$, and assume that

$$B := \bigcap_{n\in\mathbb{N}} B_n$$

belongs to $\mathfrak{R}$. The sequence $(B_1 \setminus B_n)_{n\in\mathbb{N}}$ increases and lies in $\mathfrak{R}$, and its union belongs to $\mathfrak{R}$:

$$\bigcup_{n\in\mathbb{N}} (B_1 \setminus B_n) = B_1 \setminus \bigcap_{n\in\mathbb{N}} B_n = B_1 \setminus B.$$

From (c) we conclude that

$$\mu(B_1 \setminus B) = \mu\Big(\bigcup_{n\in\mathbb{N}} (B_1 \setminus B_n)\Big) = \sup_{n\in\mathbb{N}} \mu(B_1 \setminus B_n).$$

But B_1 contains B and every B_n, so

$$\mu(B_1 \setminus B) = \mu(B_1) - \mu(B)$$

and

$$\mu(B_1 \setminus B_n) = \mu(B_1) - \mu(B_n).$$

Substituting, we have

$$\begin{aligned} \mu(B_1) - \mu(B) &= \sup_{n\in\mathbb{N}} (\mu(B_1) - \mu(B_n)) \\ &= \mu(B_1) + \sup_{n\in\mathbb{N}} (-\mu(B_n)) \\ &= \mu(B_1) - \inf_{n\in\mathbb{N}} \mu(B_n) \end{aligned}$$

from which it follows that

$$\mu\left(\bigcap_{n\in\mathbb{N}} B_n\right) = \mu(B) = \inf_{n\in\mathbb{N}} \mu(B_n).$$

(d) $\Rightarrow$ (b) is trivial. □

Next we turn to the notion of boundedness.

Definition 2.2.4. *For $\mathfrak{R}$ a ring of sets and μ: $\mathfrak{R} \to \mathbb{R}$ a positive additive mapping, μ is said to be:*

(*a*) ***bounded***, *if there exists a real number α such that $\mu(A) \leq \alpha$ for every A in $\mathfrak{R}$;*

(*b*) ***σ-bounded***, *if there exists a countable family $(A_\iota)_{\iota\in I}$ from $\mathfrak{R}$ with*

$$\inf_{\iota\in I} \mu(A \setminus A_\iota) = 0$$

for every A in $\mathfrak{R}$.

A positive measure space $(X, \mathfrak{R}, \mu)$ is called **bounded** *or* **σ-bounded**, *respectively, iff the measure μ is bounded or σ-bounded.* □

Our first observation is an obvious consequence of the definitions.

Proposition 2.2.5. *Every positive measure space $(X, \mathfrak{R}, \mu)$ for which the pair $(X, \mathfrak{R})$ is σ-finite is σ-bounded.* □

We want to show that bounded positive measures are always σ-bounded and that positive measures on σ-rings are always bounded. The proofs make no use of null continuity, so we formulate somewhat more general results first.

Proposition 2.2.6. *Let $\mathfrak{R}$ be a ring of sets. The following assertions hold, for every positive, additive mapping μ: $\mathfrak{R} \to \mathbb{R}$.*

(*a*) *If μ is bounded, then μ is σ-bounded.*

(*b*) *If $\mathfrak{R}$ is a σ-ring, then μ assumes a maximum on $\mathfrak{R}$; i.e., there exists a set $B \in \mathfrak{R}$ such that*

$$\sup_{A\in\mathfrak{R}} \mu(A) = \mu(B) \tag{1}$$

(*c*) *If $\mathfrak{R}$ is a σ-ring and B is any element of $\mathfrak{R}$ for which* (1) *holds, then $\mu(A \setminus B) = 0$ for every $A \in \mathfrak{R}$.*

Proof. (a) Set

$$\alpha := \sup_{A\in\mathfrak{R}} \mu(A)$$

and let $(A_n)_{n \in \mathbb{N}}$ be a sequence from the set-ring $\mathfrak{R}$ such that

$$\sup_{n \in \mathbb{N}} \mu(A_n) = \alpha.$$

By hypothesis, α is a real number. Now let A be an arbitrary element of $\mathfrak{R}$. Writing $A \setminus A_n$ as $(A \cup A_n) \setminus A_n$ and using Proposition 2(c), we see that

$$\begin{aligned} \mu(A \setminus A_n) &= \mu(A \cup A_n) - \mu(A_n) \\ &\leq \alpha - \mu(A_n) \end{aligned}$$

for every n. Take the infimum over n, and use the positivity of μ:

$$\begin{aligned} 0 \leq \inf_{n \in \mathbb{N}} \mu(A \setminus A_n) \leq \inf_{n \in \mathbb{N}} (\alpha - \mu(A_n)) \\ = \alpha - \sup_{n \in \mathbb{N}} \mu(A_n) \\ = 0. \end{aligned}$$

Equality must hold throughout, so

$$\inf_{n \in \mathbb{N}} \mu(A \setminus A_n) = 0.$$

(b) With α and $(A_n)_{n \in \mathbb{N}}$ as in the proof of (a), set

$$B := \bigcup_{n \in \mathbb{N}} A_n.$$

The set B belongs to the σ-ring $\mathfrak{R}$, and $\mu(B) \leq \alpha$. On the other hand,

$$\mu(A_n) \leq \mu(B)$$

for every n, so

$$\alpha = \sup_{n \in \mathbb{N}} \mu(A_n) \leq \mu(B).$$

It follows that $\alpha = \mu(B)$.

(c) Using Proposition 2(c), and with α as in the proof of (a), we have

$$\alpha = \mu(B) \leq \mu(B) + \mu(A \setminus B) = \mu(A \cup B) \leq \alpha$$

so $\mu(A \setminus B) = 0$. □

Corollary 2.2.7. *Every bounded positive measure is σ-bounded. Every positive measure on a σ-ring is bounded.* □

We make one final remark before proceeding to the exercises. It may surprise some readers that we have not included countable additivity in our definition of measure. The reason for not doing so is that a discussion of countable additivity requires certain presuppositions about summable families of real numbers. (See the remarks following Proposition 2, where it was pointed out that additivity is a compatibility condition connecting the operations of addition in $\mathbb{R}$ and set-theoretic union.) It is really not necessary to include such presuppositions in the axioms. Quite the contrary, the concept of a summable number family and all of the accompanying properties will result from integration theory as a special case. The only assumptions that are necessary in this direction are those about sums of finite families, which are spelled out, to the extent they are needed, in Proposition 1.1.2. In a sense, of course, we do not altogether avoid hypothesizing countable additivity. Once the theory of summable families is available, the countable additivity of positive measures will be seen to be equivalent to their null continuity. (See 3.5.1.)

EXERCISES

2.2.1(E) Dirac Measure. Let $\mathfrak{R}$ be a ring of sets and $x \in X(\mathfrak{R})$. Define

$$\delta_x^{\mathfrak{R}}: \mathfrak{R} \to \mathbb{R}, \qquad A \mapsto e_A(x).$$

Verify the following statement:

(α) $\delta_x^{\mathfrak{R}}$ is a positive measure on $\mathfrak{R}$.
$\delta_x^{\mathfrak{R}}$ is called the **Dirac measure** on $\mathfrak{R}$ concentrated in x.

Define now, for all $x, y \in X$, $x \sim y$ iff, for every $A \in \mathfrak{R}$, $x \in A \Leftrightarrow y \in A$. Prove:

(β) $\sim$ is an equivalence relation on $X(\mathfrak{R})$.

(γ) The following properties are equivalent for all $x, y \in X(\mathfrak{R})$:

(γ1) $x \sim y$.
(γ2) $\delta_x = \delta_y$.

$\mathfrak{R}$ is called **separating** iff, for all $x, y \in X(\mathfrak{R})$, $\delta_x = \delta_y \Leftrightarrow x = y$.

(δ) The following properties of $\mathfrak{R}$ are equivalent:

(δ1) $\mathfrak{R}$ is separating.
(δ2) Every equivalence class consists of one element.

Let X be a set, $\mathfrak{R}$ an algebra of sets on X, and $\mathfrak{F}$ a filter on X with the properties described in Ex. 2.1.5 (β). Set

$$\mu: \mathfrak{R} \to \mathbb{R}, \qquad A \mapsto \begin{cases} 1 & \text{if } A \in \mathfrak{F} \\ 0 & \text{if } A \in \mathfrak{R} \setminus \mathfrak{F}. \end{cases}$$

Prove:

(ε) μ is positive.

(ζ) μ is a positive measure on $\mathfrak{R}$ iff the intersection of any sequence in $\mathfrak{F}$ is nonempty. (If $\mathfrak{R} = \mathfrak{P}(X)$, this is equivalent to the assertion that $\mathfrak{F}$ is a δ-stable ultrafilter on X.)

(η) For every $x \in X$, $\mathfrak{F} := \{A \in \mathfrak{R} \mid x \in A\}$ has the property described in (ζ) and μ is the Dirac measure on $\mathfrak{R}$ concentrated in x.

Thus positive measures generated by filters as described in (ζ) are natural generalizations of Dirac measures.

(ϑ) Let X be infinite. Set

$$\mathfrak{R} := \{A \subset X \mid A \text{ finite or } X \setminus A \text{ finite}\}$$

$$\mathfrak{F} := \{A \subset X \mid X \setminus A \text{ finite}\}.$$

Then $\mathfrak{R}$ is an algebra on X and $\mathfrak{F}$ is a filter on X with the properties described in Ex. 2.1.5 (β). The situation described in (ζ) occurs iff X is uncountable.

2.2.2(E) **Counting Measure.** Let X be a set. Let $\mathfrak{F}$ be the set of all finite subsets of X. Define $\chi\colon \mathfrak{F} \to \mathbb{R}$, $A \mapsto \sum_{x \in A} e_A(x)$. Prove the following propositions:

(α) $(X, \mathfrak{F}, \chi)$ is a positive δ-measure space.

(β) Given $A \in \mathfrak{F}$, $\chi(A)$ is the number of elements of A.

(γ) χ is bounded iff X is finite.

(δ) χ is σ-bounded iff X is countable (iff $(X, \mathfrak{F})$ σ-finite).

(ε) Given an increasing sequence $(A_n)_{n \in \mathbb{N}}$ from $\mathfrak{F}$ with $\sup_{n \in \mathbb{N}} \chi(A_n) < \infty$, then $\bigcup_{n \in \mathbb{N}} A_n \in \mathfrak{F}$ and $\chi(\bigcup_{n \in \mathbb{N}} A_n) = \sup_{n \in \mathbb{N}} \chi(A_n)$.

(ζ) If $(X, \mathfrak{R}, \mu)$ is a positive measure space with $(X, \mathfrak{F}, \chi) \preccurlyeq (X, \mathfrak{R}, \mu)$, then $(X, \mathfrak{F}, \chi) = (X, \mathfrak{R}, \mu)$. This is true if $\mu\colon \mathfrak{R} \to \mathbb{R}$ is only positive and additive.

2.2.3(E) (a) We look again at the preceding example, but in a more general setting. Let X again denote a set, and $\mathfrak{F}$ the set of finite subsets of X. Given a function $g \in \mathbb{R}_+^X$, define

$$\chi_g\colon \mathfrak{F} \to \mathbb{R},\ A \mapsto \sum_{x \in A} g(x)$$

Prove the following propositions:

(α) $(X, \mathfrak{F}, \chi_g)$ is a positive δ-measure space.

(β) χ_g is bounded iff $(g(x))_{x \in X}$ is summable (cf., Ex. 1.3.1).

(γ) χ_g is σ-bounded iff $\{g \neq 0\}$ is countable.

(δ) Suppose $\inf_{x \in X} g(x) > 0$. Then the following hold:

($\delta 1$) Given an increasing sequence $(A_n)_{n \in \mathbb{N}}$ from $\mathfrak{F}$ with $\sup_{n \in \mathbb{N}} \chi_g(A_n) < \infty$, $\bigcup_{n \in \mathbb{N}} A_n \in \mathfrak{F}$ and $\chi_g(\bigcup_{n \in \mathbb{N}} A_n) = \sup_{n \in \mathbb{N}} \chi_g(A_n)$.

($\delta 2$) χ_g is bounded iff X is finite.
($\delta 3$) Given a positive measure space $(X, \mathfrak{R}, \mu)$ with $(X, \mathfrak{F}, \chi_g) \preccurlyeq (X, \mathfrak{R}, \mu)$, then $(X, \mathfrak{F}, \chi_g) = (X, \mathfrak{R}, \mu)$.

(b) We now consider an extension of χ_g. To this end, we define

$$\mathfrak{R}_g := \{ A \subset X \,|\, \sup\{ \chi_g(B) \,|\, B \in \mathfrak{F},\, B \subset A \} < \infty \}$$

and

$$\bar{\chi}_g \colon \mathfrak{R}_g \to \mathbb{R}, \qquad A \mapsto \sup\{ \chi_g(B) \,|\, B \in \mathfrak{F},\, B \subset A \}.$$

Prove the statements that follow:

(α) $(X, \mathfrak{R}_g, \bar{\chi}_g)$ is a positive δ-measure space.
(β) $(X, \mathfrak{F}, \chi_g) \preccurlyeq (X, \mathfrak{R}_g, \bar{\chi}_g)$.
(γ) $(X, \mathfrak{R}_g, \bar{\chi}_g) = (X, \mathfrak{F}, \chi_g)$ iff $\inf_{x \in X} g(x) > 0$.
(δ) If $(A_n)_{n \in \mathbb{N}}$ is an increasing sequence from $\mathfrak{R}_g$ with $\sup_{n \in \mathbb{N}} \bar{\chi}_g(A_n) < \infty$, then $\bigcup_{n \in \mathbb{N}} A_n \in \mathfrak{R}_g$ and $\bar{\chi}_g(\bigcup_{n \in \mathbb{N}} A_n) = \sup_{n \in \mathbb{N}} \bar{\chi}_g(A_n)$.
(ε) Given a disjoint family $(A_\iota)_{\iota \in I}$ from $\mathfrak{R}_g$ for which $(\bar{\chi}_g(A_\iota))_{\iota \in I}$ is summable (cf. Ex. 1.3.1), $\bigcup_{\iota \in I} A_\iota \in \mathfrak{R}_g$, and $\bar{\chi}_g(\bigcup_{\iota \in I} A_\iota) = \sum_{\iota \in I} \bar{\chi}_g(A_\iota)$.
(ζ) The following are equivalent:

($\zeta 1$) $(g(x))_{x \in X}$ is summable.
($\zeta 2$) χ_g is bounded.
($\zeta 3$) $\bar{\chi}_g$ is bounded.
($\zeta 4$) $\mathfrak{R}_g$ is a σ-ring.
($\zeta 5$) $X \in \mathfrak{R}_g$.

Hence, in any case, $\mathfrak{R}_g$ is a σ-algebra on X.

(η) $\bar{\chi}_g$ is σ-bounded iff χ_g is σ-bounded.
(ϑ) $A \in \mathfrak{R}_g$ whenever $\chi_g(A \cap B) = 0$ for each $B \in \mathfrak{F}(X)$, and in this case $\bar{\chi}_g(A) = 0$.
(ι) If $(X, \mathfrak{R}, \mu)$ is a positive measure space with $(X, \mathfrak{R}, \mu) \succcurlyeq (X, \mathfrak{R}_g, \bar{\chi}_g)$ then $(X, \mathfrak{R}, \mu) = (X, \mathfrak{R}_g, \bar{\chi}_g)$.

The considerations in (b) show that, in general, the extension $(X, \mathfrak{R}_g, \bar{\chi}_g)$ of $(X, \mathfrak{F}, \chi_g)$ has substantially "better" properties. This is similar to the situation we encountered with Daniell spaces. As we continue to investigate positive measure spaces, we shall find again that a substantial part of the task will be to obtain extensions with desirable properties by suitable choices.

(c) The measures considered in the preceding exercises are special cases of the situation discussed here:

(α) For $g := e_X$ we have $(X, \mathfrak{F}, \chi_g) = (X, \mathfrak{F}, \chi)$.
(β) Take $x \in X$, and define $g := e_{\{x\}}$. Then $(X, \mathfrak{F}, \chi_g) = (X, \mathfrak{F}, \delta_x^{\mathfrak{F}})$.

2.2.4(E) The Positive Measure Induced by a Positive Linear Functional. Let X be a set, $\mathscr{L} \subset \overline{\mathbb{R}}^X$ a Riesz lattice and ℓ a positive linear functional on $\mathscr{L}$. As in Ex. 2.1.7, define $\mathfrak{R}(\mathscr{L}) := \{A \subset X \mid e_A \in \mathscr{L}\}$. Now define μ^ℓ: $\mathfrak{R}(\mathscr{L}) \to \mathbb{R}$, $A \mapsto \ell(e_A)$. Prove the following statements:

(α) μ^ℓ is a positive additive function on the ring of sets, $\mathfrak{R}(\mathscr{L})$.

(β) $\mathfrak{N}(\mathscr{L}) \subset \mathfrak{R}(\mathscr{L})$ and $\mu^\ell(A) = 0$ for each $A \in \mathfrak{N}(\mathscr{L})$.

(γ) If $(X, \mathscr{L}, \ell)$ is a Daniell space, then $(X, \mathfrak{R}(\mathscr{L}), \mu^\ell)$ is a positive measure space.

(δ) If $(X, \mathscr{L}, \ell)$ is a closed Daniell space, then $(X, \mathfrak{R}(\mathscr{L}), \mu^\ell)$ has the following additional properties:

(δ1) $(X, \mathfrak{R}(\mathscr{L}), \mu^\ell)$ is a positive δ-measure space.

(δ2) If $(A_n)_{n \in \mathbb{N}}$ is an increasing sequence from $\mathfrak{R}(\mathscr{L})$ with $\sup_{n \in \mathbb{N}} \mu^\ell(A_n) < \infty$, then $\bigcup_{n \in \mathbb{N}} A_n \in \mathfrak{R}(\mathscr{L})$ and $\mu^\ell(\bigcup_{n \in \mathbb{N}} A_n) = \sup_{n \in \mathbb{N}} \mu^\ell(A_n)$.

(δ3) If $A \in \mathfrak{N}(\mathscr{L})$ with $\mu^\ell(A) = 0$ and $B \subset A$, then $B \in \mathfrak{R}(\mathscr{L})$ and $\mu^\ell(B) = 0$.

The spaces $(X, \mathfrak{R}(\mathscr{L}), \mu^\ell)$ will be treated thoroughly in Section 2.4 and Chapter 5. We simply wish to point out at this juncture that the measure spaces constructed from closed Daniell spaces have much more powerful properties than the measure spaces constructed from other Daniell spaces. Statement (δ) illustrates this and invites the conjecture that the problem of extending positive measure spaces is intimately connected to the problem of extending positive linear functionals.

We now turn to a number of concrete examples (cf., Ex. 1.3.5). We begin by introducing the notion of π-continuity for functions on systems of sets, in analogy with Ex. 1.3.5. Given a set of sets, $\mathfrak{R}$, the function μ: $\mathfrak{R} \to \mathbb{R}_+$ is called **π-continuous** iff given the downward directed family $(A_\iota)_{\iota \in I}$ from $\mathfrak{R}$ with $\bigcap_{\iota \in I} A_\iota = \varnothing$, $\inf_{\iota \in I} \mu(A_\iota) = 0$. We also introduce the convention here that $\mathfrak{F}$ denotes the set of all finite subsets of X.

(a) **Functionals on $\mathscr{k}(X)$.** Prove the statements that follow:

(α) If $g \in \mathbb{R}_+^X$, then $(X, \mathfrak{R}(\mathscr{k}(X)), \mu^{\ell_g}) = (X, \mathfrak{F}, \chi_g)$.

(β) For each additive positive function, μ, on $\mathfrak{F}$, there is exactly one $g \in \mathbb{R}_+^X$ with $\mu = \chi_g$. For each $x \in X$, g is defined by $g(x) = \mu(\{x\})$.

(γ) The association $\ell_g \mapsto \chi_g$ defines an order isomorphism between the set of all positive linear functionals on $\mathscr{k}(X)$ and the set of all additive positive functions on $\mathfrak{F}$. (We have in mind here the order for the functionals defined by $\ell \le \ell'$ iff $\ell(f) \le \ell'(f)$ whenever $f \in \mathscr{k}(X)_+$, and the order relation on the set of all positive additive functions on $\mathfrak{F}$ induced by the order on $\mathbb{R}^{\mathfrak{F}}$.)

(δ) Every additive positive function on $\mathfrak{F}$ is π-continuous.

Now take $g \in \mathbb{R}_+^X$, and consider the extension $(X, \ell_g^1(X), \bar{\ell}_g)$ of $(X, \mathscr{k}(X), \ell_g)$. Show that (with the notation from Ex. 2.2.3) the following statement holds:

(ε) $(X, \mathfrak{R}(\ell_g^1(X)), \mu^{\bar{\ell}_g}) = (X, \mathfrak{R}_g, \bar{\chi}_g)$.

This shows once more the connection between extensions of $(X, \mathscr{L}, \ell)$ and $(X, \mathfrak{R}(\mathscr{L}), \mu^\ell)$ stated earlier. An extension of $(X, \mathscr{L}, \ell)$ with better properties than $(X, \mathscr{L}, \ell)$ itself leads in general to an extension of $(X, \mathfrak{R}(\mathscr{L}), \mu^\ell)$ with analogous advantages. Compare the properties of $(X, \mathfrak{R}_g, \bar{\chi}_g)$ with those of $(X, \mathfrak{F}, \chi_g)$.

(b) **Functionals on $c_f(X)$.** Prove the statements that follow:

(α) $\mathfrak{R}(c_f(X)) = \{A \subset X \mid A \in \mathfrak{F} \text{ or } X \setminus A \in \mathfrak{F}\}$.
(β) If X is finite, then $c_f(X) = \mathbb{R}^X$ and $\mathfrak{R}(c_f(X)) = \mathfrak{P}(X)$.
(γ) If X is countably infinite, then the statements that follow hold:

(γ1) An additive positive function on $\mathfrak{R}(c_f(X))$ is π-continuous iff it is a positive measure.
(γ2) Define

$$\mu: \mathfrak{R}\left(c_f(X)\right) \to \mathbb{R}_+, \qquad A \mapsto \begin{cases} 0 & \text{if } A \in \mathfrak{F} \\ 1 & \text{otherwise} \end{cases}$$

Then μ is an additive positive function on $\mathfrak{R}(c_f(X))$, but not a positive measure.

(δ) If X is uncountable, then the following propositions apply:

(δ1) Every additive positive function on $\mathfrak{R}(c_f(X))$ is a positive measure.
(δ2) The additive positive function μ on $\mathfrak{R}(c_f(X))$ is π-continuous iff $\mu|_{\mathfrak{F}} = \chi_g$ for some $g \in \ell^1(X)$.
(δ3) Every additive positive π-continuous function μ on $\mathfrak{R}(c_f(X))$ has a unique extension to an additive positive π-continuous function on $\mathfrak{P}(X)$.
(δ4) Define μ as in (γ2). Then μ is a positive measure, but not π-continuous.

(ε) $\mathfrak{R}(c(X)) = \mathfrak{R}(c_f(X))$.

(c) **Functionals on $\ell^\infty(X)$.** Prove the statements that follow:

(α) $\mathfrak{R}(\ell^\infty(X)) = \mathfrak{P}(X)$.
(β) Let ℓ be a positive linear functional on $\ell^\infty(X)$. Then μ^ℓ is bounded.
(γ) For the additive positive function $\mu: \mathfrak{P}(X) \to \mathbb{R}$, the following propositions are equivalent:

(γ1) μ is π-continuous.
(γ2) $\mu|_{\mathfrak{F}} = \chi_g$ for some $g \in \ell^1(X)_+$.
(γ3) $\mu = \mu^\ell$ for some positive linear π-continuous functional ℓ on $\ell^\infty(X)$.

If these conditions are fulfilled, then the following statements apply:

(γ4) g (in (γ2)) is uniquely determined by μ. In fact $g(x) = \mu(\{x\})$ for any $x \in X$. The ℓ [in (γ3)] is determined by μ, $\ell = \bar{\ell}_g$, and $\mu^{\bar{\ell}_g}|_{\mathfrak{F}} = \chi_g$.

(δ) If X is countable, then every positive measure on $\mathfrak{P}(X)$ is π-continuous.

(ε) Suppose X is not finite. Let $(A_n)_{n \in \mathbb{N}}$ be a decreasing sequence of nonempty subsets of X such that $\bigcap_{n \in \mathbb{N}} A_n = \varnothing$. Let $\mathfrak{G}$ be an ultrafilter on X with $A_n \in \mathfrak{G}$ for each $n \in \mathbb{N}$. Define

$$\mu : \mathfrak{P}(X) \to \mathbb{R}, \qquad A \mapsto \begin{cases} 1 & \text{if } A \in \mathfrak{G} \\ 0 & \text{if } A \notin \mathfrak{G} \end{cases}$$

Then μ is an additive positive function on $\mathfrak{P}(X)$, but not a positive measure. Show that $\mu = \mu^{\ell}$, where ℓ is the functional on $\ell^{\infty}(X)$ discussed in Ex. 1.3.5(c) (δ).

(ζ) Let $\mathfrak{G}$ be a free δ-stable ultrafilter on X. Define μ as in (ε). Then μ is a positive measure on $\mathfrak{P}(X)$ but is not π-continuous. Show that $\mu = \mu^{\ell}$ where ℓ is the positive linear functional on $\ell^{\infty}(X)$ discussed in Ex. 1.3.5(c) (ε).

(d) **Functionals on c_0(X).** Prove the following statements:

(α) $\mathfrak{R}(c_0(X)) = \mathfrak{F}$.

(β) Let μ be an additive positive function on $\mathfrak{F}$. Then there is a positive linear functional ℓ on $c_0(X)$ with $\mu = \mu^{\ell}$ iff there is $g \in \ell^1(X)_+$ with $\mu \doteq \chi_g$. In this case $g(x) = \mu(\{x\})$ for each $x \in X$, so that μ determines g uniquely. Furthermore $\mu^{\ell_g} = \chi_g$ and $\ell = \bar{\ell}_g$.

Comparing the results in (a)–(d) with those of Ex. 1.3.5 makes it apparent that there is a great deal in common between positive linear functionals on function spaces and additive positive functions on the associated systems of sets.

2.2.5(E) Let μ be an additive real function on $\mathfrak{P}(\mathbb{N})$. Prove the equivalence of the following statements:

(α) μ is nullcontinuous.

(β) The mapping $\{0,1\}^{\mathbb{N}} \to \mathbb{R}$, $f \mapsto \mu(\{f = 1\})$ is continuous. (Here $\{0,1\}$ is equipped with the discrete topology, and $\{0,1\}^{\mathbb{N}}$ with the resulting product topology.)

(γ) The mapping defined in (β) is continuous at at least one point.

2.2.6(E) Let X be a set and $\mathfrak{R}$ a ring of subsets of X. Denote by $\mathfrak{S}$ the σ-algebra on X generated by $\mathfrak{R}$. Prove the equivalence of the following statements:

(α) There is a sequence $(A_n)_{n \in \mathbb{N}}$ from $\mathfrak{R}$ such that $X = \bigcup_{n \in \mathbb{N}} A_n$.

(β) For any two positive measures μ and ν on $\mathfrak{S}$, $\mu = \nu$ iff $\mu|_{\mathfrak{R}} = \nu|_{\mathfrak{R}}$.

(γ) If μ is a positive measure on $\mathfrak{S}$ such that $\mu|_{\mathfrak{R}} = 0$, then $\mu = 0$.

2.2.7(E) Prove the following statements:

(α) Let $\mathfrak{R}$, $\mathfrak{S}$ be rings of sets and μ a positive measure on $\mathfrak{R}$. Show that if $\mathfrak{S} \subset \mathfrak{R}$, then $\mu|_{\mathfrak{S}}$ is a positive measure on $\mathfrak{S}$.

(β) Let $(X, \mathfrak{R}, \mu)$ be a positive measure space. Take $Y \subset X$, and define

$$\mathfrak{R}|_Y := \{A \in \mathfrak{R} \mid A \subset Y\} \quad \text{and} \quad \mu|_Y := \mu|_{\mathfrak{R}|_Y}.$$

Then $(Y, \mathfrak{R}|_Y, \mu|_Y)$ is a positive measure space and is called the **restriction** of $(X, \mathfrak{R}, \mu)$ to Y.

(γ) Let $(X, \mathfrak{R}, \mu)$ be a positive measure space. Let Y be a set and $\varphi: X \to Y$ a mapping. Define

$$\varphi(\mathfrak{R}) := \{A \subset Y \mid \varphi^{-1}(A) \in \mathfrak{R}\}$$

and

$$\varphi(\mu): \varphi(\mathfrak{R}) \to \mathbb{R}, \qquad A \mapsto \mu(\varphi^{-1}(A)).$$

Then $(Y, \varphi(\mathfrak{R}), \varphi(\mu))$ is a positive measure space and is called the **image** of $(X, \mathfrak{R}, \mu)$ under φ.

As we mentioned in Section 2.1 the systems of sets that arise in practical situations often do not yet have all the properties of rings of sets. Naturally, a corresponding problem occurs with positive measures. Functions on semi-rings or $\cap$-semi-lattices must first be extended to measures. We now turn to precisely such problems.

2.2.8(E) Functions on Semi-Rings. Let $\mathfrak{R}$ be a semi-ring and $\mathfrak{R}_r$ the ring of sets generated by $\mathfrak{R}$ (cf., Ex. 2.1.2). By a **content** on $\mathfrak{R}$ we mean a function, $\mu: \mathfrak{R} \to \mathbb{R}$, such that if $A \in \mathfrak{R}$ and if $(A_\iota)_{\iota \in I}$ is a finite disjoint family from $\mathfrak{R}$ with $A = \bigcup_{\iota \in I} A_\iota$, then $\mu(A) = \sum_{\iota \in I} \mu(A_\iota)$. μ is called **σ-additive** if given any disjoint sequence $(A_n)_{n \in \mathbb{N}}$ from $\mathfrak{R}$ with $\bigcup_{n \in \mathbb{N}} A_n \in \mathfrak{R}$, $\mu(\bigcup_{n \in \mathbb{N}} A_n) = \sum_{n \in \mathbb{N}} \mu(A_n)$. (Ex. 1.3.1)

Prove the following statements:

(α) Given a content, μ, on $\mathfrak{R}$, there is exactly one additive function $\tilde{\mu}$ on $\mathfrak{R}_r$ with $\tilde{\mu}|_{\mathfrak{R}} = \mu$.

(β) If μ is a positive content, then $\tilde{\mu}$ is positive.

(γ) If μ is a positive σ-additive content, then $\tilde{\mu}$ is a positive measure.

(δ) Given two additive functions on $\mathfrak{R}_r$, μ, and ν, such that $\mu|_{\mathfrak{R}} = \nu|_{\mathfrak{R}}$, then $\mu = \nu$.

(ε) If $\phi \in \mathfrak{R}$ then each σ-additive real function on $\mathfrak{R}$ is a content.

For the proof of (α), keep in mind the characterization of $\mathfrak{R}_r$ in Ex. 2.1.2(κ). For the proof of (γ), we wish to show that $\tilde{\mu}$ is nullcontinuous. First show that if $A \in \mathfrak{R}_r$ and if $(A_n)_{n \in \mathbb{N}}$ is a disjoint sequence in $\mathfrak{R}$ with $A = \bigcup_{n \in \mathbb{N}} A_n$, then $\tilde{\mu}(A) = \sum_{n \in \mathbb{N}} \mu(A_n)$. In fact there is a finite disjoint family $(B_\iota)_{\iota \in I}$ from $\mathfrak{R}$ such that $A = \bigcup_{\iota \in I} B_\iota$. For each $\iota \in I$ then, $B_\iota = \bigcup_{n \in \mathbb{N}} (A_n \cap B_\iota)$. Thus for each $n \in \mathbb{N}$ and $\iota \in I$ there is a finite disjoint family from $\mathfrak{R}$ whose union is $A_n \cap B_\iota$. Bearing this in mind, there exist for each $\iota \in I$ a disjoint sequence $(D_{\iota k})_{k \in \mathbb{N}}$ from $\mathfrak{R}$ and a sequence $(k_n)_{n \in \mathbb{N}}$ from $\mathbb{N}$ such that $k_1 = 1$ and $A_n \cap B_\iota = \bigcup_{k=k_n}^{k_{n+1}-1} D_{\iota k}$, so that

$B_\iota = \bigcup_{k \in \mathbb{N}} D_{\iota k}$. Thus $\mu(B_\iota) = \Sigma_{k \in \mathbb{N}} \mu(D_{\iota k}) = \Sigma_{n \in \mathbb{N}} (\Sigma_{k=k_n}^{k_{n+1}-1} \mu(D_{\iota k})) = \Sigma_{n \in \mathbb{N}} \tilde{\mu}(A_n \cap B_\iota)$. Hence $\tilde{\mu}(A) = \Sigma_{\iota \in I} \tilde{\mu}(B_\iota) = \Sigma_{\iota \in I} \Sigma_{n \in \mathbb{N}} \tilde{\mu}(A_n \cap B_\iota) = \Sigma_{n \in \mathbb{N}} \Sigma_{\iota \in I} \tilde{\mu}(A_n \cap B_\iota) = \Sigma_{n \in \mathbb{N}} \mu(A_n)$.

Now suppose that $(A_n)_{n \in \mathbb{N}}$ is a decreasing sequence from $\mathfrak{R}_r$ such that $\bigcap_{n \in \mathbb{N}} A_n = \varnothing$. There exist a disjoint sequence $(D_k)_{k \in \mathbb{N}}$ from $\mathfrak{R}$ and a sequence $(n_k)_{k \in \mathbb{N}}$ from $\mathbb{N}$ with $n_1 = 1$ and $A_n = \bigcup_{k \geq k_n} D_k$ for each $n \in \mathbb{N}$. By the previous argument, $\tilde{\mu}(A_n) = \Sigma_{k \geq k_n} \mu(D_k)$. Bringing all of the foregoing together, $\lim_{n \to \infty} \tilde{\mu}(A_n) = \lim_{n \to \infty} \Sigma_{k \geq k_n} \mu(D_k) = 0$, establishing the result.

2.2.9(E) Let $\mathfrak{R}$ be a nonempty set of disjoint sets. Prove the following statements:

(α) $\mathfrak{R}$ is a semi-ring and each real function on $\mathfrak{R}$ is a σ-additive content.

(β) Every additive positive function on $\mathfrak{R}_r$ is a positive measure.

(γ) Given a function $\nu: \mathfrak{R} \to \mathbb{R}_+$, there is exactly one positive measure μ on $\mathfrak{R}_r$ with $\mu|_{\mathfrak{R}} = \nu$.

(δ) For each finite ring of sets, $\mathfrak{S}$, there is a semi-ring $\mathfrak{R}$ of disjoint sets with $\mathfrak{R} \subset \mathfrak{S}$ and $\mathfrak{R}_r = \mathfrak{S}$.

2.2.10(E) **Stieltjes Measures.** An important example of the extension of contents from a semi-ring to the ring of sets generated is provided by the Stieltjes measures, which are the subject of this exercise.

Let A be an open interval in $\overline{\mathbb{R}}$. (In particular, $A = \mathbb{R}$ is allowed.) Denote by $\mathfrak{J}$ the set of all right half-open intervals in A; that is, $\mathfrak{J} = \{[x, y[\mid x, y \in A,\ x \leq y\}$. (This is a special case of the situation investigated in Ex. 2.1.3.) $\mathfrak{J}$ is a semi-ring. The elements of the ring of sets generated by $\mathfrak{J}$ (i.e., the finite unions of intervals in $\mathfrak{J}$) are called the **interval forms** on A.

Now consider an increasing function $g: A \to \mathbb{R}$ that is left-continuous (i.e., given $x \in A$ and a sequence $(x_n)_{n \in \mathbb{N}}$ from A that order converges to x and for which $x_n \leq x$ for each $n \in \mathbb{N}$, $g(x) = \lim_{n \to \infty} g(x_n)$). Define $\nu_g: \mathfrak{J} \to \mathbb{R}, [x, y[\mapsto g(y) - g(x)$. Prove the following propositions:

(α) If $B \in \mathfrak{J}$ and if $(A_\iota)_{\iota \in I}$ is a finite disjoint family from $\mathfrak{J}$ with $\bigcup_{\iota \in I} A_\iota \subset B$, then $0 \leq \Sigma_{\iota \in I} \nu_g(A_\iota) \leq \nu_g(B)$.

(β) If $B \in \mathfrak{J}$ and if $(A_\iota)_{\iota \in I}$ is a finite disjoint family from $\mathfrak{J}$ with $B = \bigcup_{\iota \in I} A_\iota$, then $\nu_g(B) = \Sigma_{\iota \in I} \nu_g(A_\iota)$.

(γ) If $B \in \mathfrak{J}$ and if $(A_\iota)_{\iota \in I}$ is a finite family from $\mathfrak{J}$ with $B \subset \bigcup_{\iota \in I} A_\iota$, then $\nu_g(B) \leq \Sigma_{\iota \in I} \nu_g(A_\iota)$.

(δ) $\nu_g(B) \geq 0$ for any $B \in \mathfrak{J}$.

(ε) ν_g is a σ-additive positive content on $\mathfrak{J}$.

We provide some suggestions for establishing the σ-additivity. Take $[x, y[\in \mathfrak{J}$, and let $([x_n, y_n[)_{n \in \mathbb{N}}$ be a disjoint sequence from $\mathfrak{J}$ with $[x, y[= \bigcup_{n \in \mathbb{N}} [x_n, y_n[$. Take $\varepsilon \in \mathbb{R}$, $\varepsilon > 0$. By the left-continuity of g, there are a $z \in]x, y[$ and $z_n \in A$, $z_n < x_n$ (for each $n \in \mathbb{N}$) with $g(y) - g(z) < \varepsilon/2$ and $g(x_n) - g(z_n) < \varepsilon/2^{n+1}$ for each $n \in \mathbb{N}$. $[x, z] \subset \bigcup_{n \in \mathbb{N}}]z_n, y_n[$, and so there is a finite $N \subset \mathbb{N}$ with $[x, z[\subset \bigcup_{n \in N}]z_n, y_n[$ (Theorem of Heine-Borel-Lebesgue). By (γ), $g(z) - g(x) \leq \Sigma_{n \in N}(g(y_n) - g(z_n))$. Thus

$g(y) - g(x) \leq \sum_{n \in \mathbb{N}} (g(y_n) - g(x_n)) + \varepsilon$. Thus

$$g(y) - g(x) \leq \sum_{n \in \mathbb{N}} (g(y_n) - g(x_n))$$

since ε was arbitrary. The converse follows from (α), showing the σ-additivity.

From Ex. 2.2.8 the next statement follows:

(ζ) If $g \in \mathbb{R}^A$ is an increasing left-continuous function, then there is exactly one positive measure, μ_g, on the ring of sets of the interval forms of A, $\mathfrak{I}_r$, with $\mu_g([x, y[) = g(y) - g(x)$ for each $[x, y[\in \mathfrak{I}$ with $x \leq y$.

μ_g is the **Stieltjes measure** belonging to g. Observe that A need not be an open interval. We could take other sorts of intervals (closed or half-open) instead of open ones. But in those cases certain technical difficulties arise in connection with the endpoints, and we wish to circumvent such difficulties. A complete treatment of positive Stieltjes measures is contained in Chapter 5.

The Stieltjes measures subsume all positive measures on $\mathfrak{I}_r$. To see this, prove the following statements:

(η) Given a positive measure on $\mathfrak{I}_r$, μ, there is an increasing left-continuous function $g \in \mathbb{R}^A$ with $\mu = \mu_g$.

(ϑ) If g and h are increasing left-continuous functions on A with $\mu_g = \mu_h$, then $g = h + \gamma$ for some $\gamma \in \mathbb{R}$.

Finally, we mention the special case of the identity function, id_A. It generates the measure λ on $\mathfrak{I}_r$ which associates to each nonempty interval $[x, y[$ its usual length $y - x$. λ is characterized by its translation invariance; namely, if $[x, y[\in \mathfrak{I}$ and if $[x + z, y + z[\in \mathfrak{I}$ for $z \in \mathbb{R}$, then $\lambda([x + z, y + z[) = \lambda([x, y[)$. This measure is called the **Lebesgue measure** on A.

2.2.11(C) Functions on $\cap$-Semi-Lattices. The extension of functions on $\cap$-semi-lattices is a substantially more difficult matter than the extension of functions defined on semi-rings. We turn to this more difficult task.

Let $\mathfrak{R}$ be a $\cap$-semi-lattice. For X a set, denote by $|X|$ its cardinality. The function ν: $\mathfrak{R} \to \mathbb{R}$ is called **alternating** iff for each finite $\mathfrak{A} \subset \mathfrak{R}$ with $\bigcup_{A \in \mathfrak{A}} A \in \mathfrak{R}$,

$$\nu\Big(\bigcup_{A \in \mathfrak{A}} A\Big) = \sum_{n \in \mathbb{N}} (-1)^{n+1} \sum_{\substack{\mathfrak{B} \subset \mathfrak{A} \\ |\mathfrak{B}| = n}} \nu\Big(\bigcap_{B \in \mathfrak{B}} B\Big). \tag{1}$$

(Observe that in $\Sigma_{n \in \mathbb{N}}$ only finitely many summands are nonzero, because if $n > |\mathfrak{A}|$ then $\{\mathfrak{B} \subset \mathfrak{A} \mid |\mathfrak{B}| = n\}$ is empty, so that

$$\sum_{\substack{\mathfrak{B} \subset \mathfrak{A} \\ |\mathfrak{B}| = n}} \nu(\textstyle\bigcap_{B \in \mathfrak{B}} B) = 0.)$$

Verify the following propositions for the alternating function $\nu: \mathfrak{R} \to \mathbb{R}$:

(α) If $\varnothing \in \mathfrak{R}$, then $\nu(\varnothing) = 0$.

(β) Given a finite disjoint subset $\mathfrak{A}$ of $\mathfrak{R}$ with $\bigcup_{A \in \mathfrak{A}} A \in \mathfrak{R}$, $\nu(\bigcup_{A \in \mathfrak{A}} A) = \Sigma_{A \in \mathfrak{A}} \nu(A)$.

Prove further that if $\mathfrak{R}$ is a lattice of sets, then the following holds:

(γ) $\nu: \mathfrak{R} \to \mathbb{R}$ is alternating iff $\nu(A \cup B) + \nu(A \cap B) = \nu(A) + \nu(B)$ whenever $A, B \in \mathfrak{R}$ and $\nu(\varnothing) = 0$ in case of $\varnothing \in \mathfrak{R}$. (Use complete induction on the cardinality of $\mathfrak{A} \subset \mathfrak{R}$.)

Prove that if $\mathfrak{R}$ is a ring of sets, then the next statement follows:

(δ) $\nu: \mathfrak{R} \to \mathbb{R}$ is alternating iff ν is additive.

(β)–(δ) show that the notion of an alternating function is a natural generalization of the notion of an additive function. Prove the next proposition:

(ε) For the alternating function, ν, on the $\cap$-semi-lattice $\mathfrak{R}$, there is exactly one additive function μ on the ring of sets $\mathfrak{R}_r$ generated by $\mathfrak{R}$ such that $\nu = \mu|_{\mathfrak{R}}$.
The uniqueness of μ follows from (δ) and Ex. 2.1.2(ι).

Prove the existence, first, under the assumption that $\mathfrak{R} = \{A_1, A_2, \ldots, A_n\}$ ($n \in \mathbb{N}$). Put $I :=$

$$\{\vec{i} := (i_1, i_2, \ldots i_n) \mid i_k \in \{0,1\} \text{ for each } k \in \mathbb{N}_n,\ i_k \neq 0 \text{ for some } k \in \mathbb{N}_n\}.$$

For $\vec{i} \in I$, set

$$B_{\vec{i}} := \bigcap_{\substack{k \in \mathbb{N}_n \\ i_k = 1}} A_k \setminus \bigcup_{\substack{k \in \mathbb{N}_n \\ i_k = 0}} A_k .$$

Then $(B_{\vec{i}})_{\vec{i} \in I}$ is a finite disjoint family of sets, and for each $k \in \mathbb{N}_n$,

$$A_k = \bigcup_{\substack{\vec{i} \in I \\ i_k = 1}} B_{\vec{i}}. \tag{2}$$

Note that $C \in \mathfrak{R}_r$ iff C is the union of a family of sets of the form $B_{\vec{i}}$ [cf., Ex. 2.1.11(a)]. By Ex. 2.2.9 every function, λ, on $\{B_{\vec{i}} \mid \vec{i} \in I\}$ generates in a natural way an additive function on $\mathfrak{R}_r$. Our aim is to define a function λ

so that for each $k \in \mathbb{N}_n$,

$$\nu(A_k) = \sum_{\substack{\vec{i} \in I \\ i_k = 1}} \lambda(B_{\vec{i}}).$$

Define $\vec{i} \leq \vec{i}'$ (for $\vec{i}, \vec{i}' \in I$) iff $i_k \leq i'_k$ for every $k \in \mathbb{N}_n$. Then $\leq$ is an order relation on I, with $\vec{1} := (1, 1, \ldots, 1)$ as largest element. For $\vec{i} \in I$, define $L_{\vec{i}} := \{\vec{i}' \in I \mid \vec{i} < \vec{i}'\}$. Next define inductively (by induction on the number of 1's in $\vec{i}$) the numbers $\alpha_{\vec{i}}$:

$$\alpha_{\vec{i}} := \nu\left(\bigcap_{\substack{k \in \mathbb{N}_n \\ i_k = 1}} A_k\right) - \sum_{\vec{i}' \in L_{\vec{i}}} \alpha_{\vec{i}'}. \tag{3}$$

Now verify that $\alpha_{\vec{i}} = 0$ for $\vec{i} \in I$ with $B_{\vec{i}} = \varnothing$. This follows immediately from (α) for the case $\vec{i} = \vec{1}$. So suppose now that $\vec{i} \neq \vec{1}$ and $B_{\vec{i}} = \varnothing$. Define $K := \{k \in \mathbb{N}_n \mid i_k = 1\}$ and $L := \{k \in \mathbb{N}_n \mid i_k = 0\}$. By assumption, $L \neq \varnothing$. $B_{\vec{i}} = \varnothing$ means that $\bigcap_{k \in K} A_k \subset \bigcup_{k \in L} A_k$, and hence, for $A := \bigcap_{k \in K} A_k$, $A = \bigcup_{k \in L}(A \cap A_k)$. Since ν is alternating,

$$\nu(A) = \sum_{m \in \mathbb{N}} (-1)^{m+1} \sum_{\substack{J \subset L \\ |J| = m}} \nu\left(\bigcap_{k \in J} (A \cap A_k)\right). \tag{4}$$

For $J \subset L$, $J \neq \varnothing$, define

$$i_k^J := \begin{cases} 1 & \text{if } k \in J \cup K \\ 0 & \text{if } k \in \mathbb{N}_n \setminus (J \cup K) \end{cases}$$

and

$$\vec{i}^J := (i_1^J, i_2^J, \ldots, i_n^J).$$

By (3) and (4),

$$\nu(A) = \sum_{m \in \mathbb{N}} (-1)^{m+1} \sum_{\substack{J \subset L \\ |J| = m}} \left(\sum_{\vec{i}' \geq \vec{i}^J} \alpha_{\vec{i}'}\right). \tag{5}$$

If $J \subset L$, $J \neq \varnothing$, then $\vec{i}^J > \vec{i}$. Conversely, if $\vec{i}' > \vec{i}$, then there is a $J \subset L$, $J \neq \varnothing$ with $\vec{i}' = \vec{i}^J$. Take such an $\vec{i}' > \vec{i}$. Let $m' := |\{k \in \mathbb{N}_n \mid i'_k = 1\}|$ and $m := |K|$. Then for $k \in \mathbb{N}_{m'-m}$ there are exactly $\binom{m' - m}{k}$ different subsets $J \subset L$ with $|J| = k$ and $\vec{i}' \geq \vec{i}^J$. So, using (5),

$$\nu(A) = \sum_{\vec{i}' \in L_{\vec{i}}} \left(\sum_{k=1}^{m'-m} (-1)^{k+1} \binom{m'-m}{k} \alpha_{\vec{i}'} \right)$$

$$= \sum_{\vec{i}' \in L_{\vec{i}}} \left(\alpha_{\vec{i}'} \sum_{k=1}^{m'-m} (-1)^{k+1} \binom{m'-m}{k} \right)$$

$$= \sum_{\vec{i}' \in L_{\vec{i}}} \alpha_{\vec{i}'}.$$

Thus, using (3), $\alpha_{\vec{i}} = 0$.

Now take $\vec{i}, \vec{i}' \in I$, $\vec{i} \neq \vec{i}'$, with $B_{\vec{i}} = B_{\vec{i}'}$. Then $B_{\vec{i}} = B_{\vec{i}'} = \varnothing$, since $\{B_{\vec{i}} | \vec{i} \in I\}$ is a pairwise disjoint set. Thus $\alpha_{\vec{i}} = \alpha_{\vec{i}'} = 0$. This allows us to set $\alpha(B_{\vec{i}}) := \alpha_{\vec{i}}$ for each $\vec{i} \in I$. Given $B \in \mathfrak{R}_r$, define $\mathfrak{B} := \{B_{\vec{i}} | \vec{i} \in I, B_{\vec{i}} \subset B\}$ and $\lambda(B) := \sum_{B' \in \mathfrak{B}} \alpha(B')$. λ is additive on $\mathfrak{R}_r$. By (2) and (3), $\nu = \lambda|_{\mathfrak{R}}$, completing our discussion for the case of finite $\mathfrak{R}$.

Now let $\mathfrak{R}$ be an arbitrary $\cap$-semi-lattice. Given $A \in \mathfrak{R}_r$, there is a finite $\cap$-semi-lattice $\mathfrak{S} \subset \mathfrak{R}$ with $A \in \mathfrak{S}_r$ (cf., Ex. 2.1.8). By the preceding considerations, $\nu|_{\mathfrak{S}}$ can be extended to an additive function μ on $\mathfrak{S}_r$. By the uniqueness property of the extension, $\mu(A)$ does not depend on the particular $\mathfrak{S}$ chosen. So μ as a function on $\mathfrak{R}_r$ is the extension sought.

It is natural to ask when the extension of an alternating function is a positive one. We now turn to precisely this problem.

Let $\mathfrak{R}$ be a $\cap$-semi-lattice. The function ν: $\mathfrak{R} \to \mathbb{R}$ is called **completely increasing** iff given finite subsets $\mathfrak{A}$ and $\mathfrak{B}$ of $\mathfrak{R}$ with $\bigcup_{A \in \mathfrak{A}} A \subset \bigcup_{B \in \mathfrak{B}} B$,

$$\sum_{n \in \mathbb{N}} (-1)^{n+1} \sum_{\substack{\mathfrak{C} \subset \mathfrak{A} \\ |\mathfrak{C}| = n}} \nu\left(\bigcap_{A \in \mathfrak{C}} A \right) \leq \sum_{n \in \mathbb{N}} (-1)^{n+1} \sum_{\substack{\mathfrak{C} \subset \mathfrak{B} \\ |\mathfrak{C}| = n}} \nu\left(\bigcap_{A \in \mathfrak{C}} A \right).$$

Prove the following statements:

(ζ) If $\mathfrak{R}$ is a lattice of sets, then ν: $\mathfrak{R} \to \mathbb{R}$ is completely increasing iff ν is increasing in the sense of Definition 2.2.1(c).

(η) If $\mathfrak{R}$ is a ring of sets, then ν: $\mathfrak{R} \to \mathbb{R}$ is completely increasing iff ν is positive.

The next proposition is particularly important:

(ϑ) If ν: $\mathfrak{R} \to \mathbb{R}$ is a completely increasing alternating function, then the uniquely determined additive extension to $\mathfrak{R}_r$, μ, is positive.

The proof can be broken down into several steps:

STEP 1: Set $\mathfrak{S} := \{\bigcup_{A \in \mathfrak{A}} A | \mathfrak{A} \in \mathfrak{F}(\mathfrak{R})\}$. (Recall that $\mathfrak{F}(X)$ denotes the set of finite subsets of the set X.) First, show that $\mu|_{\mathfrak{S}}$ is completely increasing. Observe that, according to Ex. 2.1.2(α), (β), $\mathfrak{S}$ is a lattice of sets, so by (ζ) it only remains to show that $\mu|_{\mathfrak{S}}$ is increasing in the sense of

Definition 2.2.1(c). But this is clear from the definition of being completely increasing.

STEP 2: Set $\mathfrak{S}' := \{A \setminus B \mid A, B \in \mathfrak{S}\}$. We show by complete induction that $\mu((\bigcup_{k \le n} C_k) \cup C) \le \mu(D)$ for each family $(C_k)_{k \in \mathbb{N}_n}$ in $\mathfrak{S}'$ and $C, D \in \mathfrak{S}$ such that $(\bigcup_{k \le n} C_k) \cup C \subset D$.

For $n = 1$, set $C_1 := A_1 \setminus B_1$, $A_1, B_1 \in \mathfrak{S}$. Then

$$A_1 \cup C \subset D \cup (A_1 \cap B_1). \tag{6}$$

From (6) it follows that

$$\mu(A_1) - \mu(A_1 \cap B_1) + \mu(C) - \mu(A_1 \cap C) + \mu(A_1 \cap B_1 \cap C) \le \mu(D).$$

Thus

$$\mu((A_1 \setminus B_1) \cup C) \le \mu(D).$$

For $n \Rightarrow n + 1$, suppose the proposition has been proved for n. Take families $(B_k)_{k \in \mathbb{N}_{n+1}}$ and $(A_k)_{k \in \mathbb{N}_{n+1}}$ from $\mathfrak{S}$ such that $(\bigcup_{k \le n+1} (A_k \setminus B_k)) \cup C \subset D$. By the inductive hypothesis,

$$\mu\left(\bigcup_{k \le n} (A_k \setminus B_k) \cup A_{n+1} \cup C\right) \le \mu(D \cup (B_{n+1} \cap A_{n+1})).$$

Consequently

$$\mu\left(\bigcup_{k \le n} (A_k \setminus B_k) \cup C\right) + \mu(A_{n+1}) - \mu(A_{n+1} \cap B_{n+1})$$
$$- \mu\left(\left(\bigcup_{k \le n} (A_k \setminus B_k) \cup C\right) \cap A_{n+1}\right) \le \mu(D) - \mu(D \cap A_{n+1} \cap B_{n+1}). \tag{7}$$

From (7) it follows that

$$\mu\left(\bigcup_{k \le n+1} (A_k \setminus B_k) \cup C\right) \le \mu(D) - \mu(D \cap B_{n+1} \cap A_{n+1})$$
$$+ \mu\left(\left(\bigcup_{k \le n} (A_k \setminus B_k) \cup C\right) \cap A_{n+1} \cap B_{n+1}\right). \tag{8}$$

But

$$\left(\bigcup_{k\le n}(A_k \setminus B_k)\cup C\right)\cap A_{n+1}\cap B_{n+1}\subset D\cap A_{n+1}\cap B_{n+1}.$$

So, using the inductive hypothesis,

$$\mu\left(\left(\bigcup_{k\le n}(A_k \setminus B_k)\cup C\right)\cap A_{n+1}\cap B_{n+1}\right)\le \mu(D\cap A_{n+1}\cap B_{n+1}). \tag{9}$$

Combining (8) and (9)

$$\mu\left(\bigcup_{k\le n+1}(A_k \setminus B_k)\cup C\right)\le \mu(D).$$

STEP 3: If $A \in \Re_r$ and $B \in \mathfrak{S}'$ with $A \subset B$, then $\mu(A) \le \mu(B)$. Given A, there is a family $(A_k)_{k\in \mathbb{N}_n}$ from $\mathfrak{S}'$ with $A = \bigcup_{k\in\mathbb{N}_n} A_k$ [cf., Ex. 2.1.2(γ), (ϑ)]. Given B, there are sets $C, D \in \mathfrak{S}$ with $C \subset D$ and $B = D \setminus C$. Thus $(\bigcup_{k\le n} A_k)\cup C \subset D$ and $(\bigcup_{k\le n} A_k)\cap C = \emptyset$. Using the result in Step 2,

$$\mu(A)+\mu(C)=\mu\left(\bigcup_{k\le n} A_k\right)+\mu(C)\le\mu(D)$$

that is,

$$\mu(A)\le\mu(D)-\mu(C)=\mu(B).$$

STEP 4: We prove by induction on n that $\mu(\bigcup_{k\le n} A_k)\ge 0$ for each family $(A_k)_{k\in\mathbb{N}_n}$ from $\mathfrak{S}'$. We indicate the inductive step. If $(A_k)_{k\in\mathbb{N}_{n+1}}$ is a family from $\mathfrak{S}'$, then

$$\mu\left(\bigcup_{k\le n+1} A_k\right)=\mu\left(\bigcup_{k\le n} A_k\right)+\mu(A_{n+1})-\mu\left(\bigcup_{k\le n}(A_k\cap A_{n+1})\right). \tag{10}$$

But $\bigcup_{k\le n}(A_k\cap A_{n+1})\subset A_{n+1}$, so by Step 3 $\mu(\bigcup_{k\le n}(A_k\cap A_{n+1}))\le \mu(A_{n+1})$. Using (10) and the inductive hypothesis, $\mu(\bigcup_{k\le n+1}A_k)\ge 0$. Thus we establish that μ is positive; by Ex. 2.1.2(ι), every set in $\Re_r$ is of the form considered in Step 4.

It seems natural to raise the question of the null continuity of the extension of ν to $\Re_r$. But to deal with this requires some preparation.

Let $\Re$ be a set of sets. A **triangular family** from $\Re$ is a family $(A_{mn})_{m,n\in\mathbb{N},\, n\le m}$ from $\Re$ satisfying the following conditions:

(i) For each $n\in\mathbb{N}$, $A_n := \bigcup_{m\ge n} A_{mn}\in\Re$.

(ii) There is a finite subset $\mathfrak{S}$ of $\Re$ such that $A_{mn}\subset\bigcup_{B\in\mathfrak{S}} B$ for each $m,n\in\mathbb{N}$, $n\le m$.

Let $(A_{mn})_{m,n \in \mathbb{N},\, n \leq m}$ be a triangular family from $\mathfrak{R}$. A **convergent subfamily** is a family of the form $(A_{m_j n})_{j,n \in \mathbb{N},\, n \leq m_j}$ with $(m_j)_{j \in \mathbb{N}}$ a strictly increasing sequence from $\mathbb{N}$ satisfying the following condition:

(iii) $\lim_{j \to \infty} \bigcup_{n \leq m_j} (A_n \setminus \bigcup_{n \leq i \leq m_j} A_{in}) = \varnothing.$

For the rest, let $\mathfrak{R}$ denote a $\cap$-semi-lattice. Let ν: $\mathfrak{R} \to \mathbb{R}$ be an alternating function and μ the unique extension of ν to $\mathfrak{R}_r$. ν is called **continuous** on $\mathfrak{R}$ iff, for each triangular family $(A_{mn})_{m,n \in \mathbb{N},\, n \leq m}$ from $\mathfrak{R}$ and for each convergent subfamily $(A_{m_j n})_{j,n \in \mathbb{N},\, n \leq m_j}$

(iv) $\lim_{j \to \infty} \mu(\bigcup_{n \leq m_j} (A_n \setminus \bigcup_{n \leq i \leq m_j} A_{in})) = 0.$

Prove the following statement:

(ι) If $\varnothing \in \mathfrak{R}$ and if ν is alternating, completely increasing, and continuous on $\mathfrak{R}$, then μ is a positive measure on $\mathfrak{R}_r$.

We again divide the proof into several steps:

STEP 1: Set $\mathfrak{S} := \{\bigcup_{A \in \mathfrak{A}} A \mid \mathfrak{A} \in \mathfrak{F}(\mathfrak{R})\}$. Show that μ is continuous on $\mathfrak{S}$.

Let $(A_{mn})_{m,n \in \mathbb{N},\, n \leq m}$ be a triangular family from $\mathfrak{S}$. We *first assume* that $A_{mn} \in \mathfrak{R}$ for each $m, n \in \mathbb{N}$, $n \leq m$. Define $A_n := \bigcup_{m \geq n} A_{mn}$ for each $n \in \mathbb{N}$. Then for each $n \in \mathbb{N}$, there is a finite family $(B'_{nk})_{k \in \mathbb{N}_{k_n}}$ from $\mathfrak{R}$, with $A_n = \bigcup_{k \leq k_n} B'_{nk}$. For every $m, n \in \mathbb{N}$, $n \leq m$ we have

$$A_{mn} = \bigcup_{k \leq k_n} A_{mn} \cap B'_{nk}.$$

Construct now a triangular family $(B_{mn})_{m,n \in \mathbb{N},\, n \leq m}$ from $\mathfrak{R}$ and an increasing injective mapping φ: $\mathbb{N} \cup \{0\} \to \mathbb{N} \cup \{0\}$ with $\varphi(0) = 0$ such that for each $n \in \mathbb{N}$

$$\bigcup_{\varphi(n-1) < i \leq \varphi(n)} B_i = A_n$$

$$\bigcup_{\varphi(n-1) < i \leq \varphi(n)} B_{\varphi(n)+k,\, i} = A_{n+k-1,\, n}$$

and

$$\bigcup_{k \leq \varphi(n)} \left(B_k \setminus \bigcup_{k \leq i \leq \varphi(n)} B_{ik} \right) = \bigcup_{k \leq n} \left(A_k \setminus \bigcup_{k \leq i \leq n} A_{ik} \right)$$

where $B_k := \bigcup_{m \geq k} B_{mk}$ for each $k \in \mathbb{N}$. From this we get the desired property (iv), using the continuity of ν.

Now we consider the *general case*. For all $m, n \in \mathbb{N}$, $n \leq m$ there is a finite family $(A_k^{mn})_{k \in K_{mn}}$ from $\mathfrak{R}$ such that $A_{mn} = \bigcup_{k \in K_{mn}} A_k^{mn}$. By this it is not difficult to construct a triangular family $(B_{mn})_{m,n \in \mathbb{N},\, n \leq m}$ from $\mathfrak{S}$ such that $B_{mn} \in \mathfrak{R}$ for all $m, n \in \mathbb{N}$, $n \leq m$, for which there is an increasing injective mapping $\omega\colon \mathbb{N} \to \mathbb{N}$ such that for each $n \in \mathbb{N}$

$$\bigcup_{k \le \omega(n)} \left(B_k \setminus \bigcup_{k \le i \le \omega(n)} B_{ik} \right) = \bigcup_{k \le n} \left(A_k \setminus \bigcup_{k \le i \le n} A_{ik} \right)$$

where $B_k := \bigcup_{m \ge k} B_{mk}$ for every $k \in \mathbb{N}$. It is now easy to prove the desired statement.

STEP 2: If $(C_n)_{n \in \mathbb{N}}$ is a sequence from $\mathfrak{S}$ such that $C := \bigcup_{n \in \mathbb{N}} C_n \in \mathfrak{S}$, then

$$\mu(C) = \sup_{n \in \mathbb{N}} \mu\left(\bigcup_{m \le n} C_m \right).$$

This follows directly from Step 1.

STEP 3: Set $\mathfrak{S}' := \{ A \setminus B \mid A, B \in \mathfrak{S} \}$. Take $C \in \mathfrak{S}$. Let $(C_n)_{n \in \mathbb{N}}$ be a sequence from $\mathfrak{S}'$ with $C = \bigcup_{n \in \mathbb{N}} C_n$. Then

$$\mu(C) = \sup_{n \in \mathbb{N}} \mu\left(\bigcup_{m \le n} C_m \right).$$

Given $n \in \mathbb{N}$, there are sets $A_n, B_n \in \mathfrak{S}$ with $C_n = A_n \setminus B_n$ and $B_n \subset A_n \subset C$. For each $n \in \mathbb{N}$,

$$C \setminus \bigcup_{m \le n} (A_m \setminus B_m)$$

$$= \bigcup_{\substack{K \subset \mathbb{N}_n \\ K \ne \varnothing}} \left(\bigcap_{k \in K} B_k \setminus \bigcup_{k \in \mathbb{N}_n \setminus K} \left(A_k \cap \left(\bigcap_{k \in K} B_k \right) \right) \right) \cup \left(C \setminus \bigcup_{k \le n} A_k \right).$$

Given $n \in \mathbb{N}$, we define an order $\le_n$ on $\mathfrak{P}(\mathbb{N}) \setminus \{\varnothing\}$ by

$$K \le_n K' \quad \text{iff } |K| < |K'| \text{ or } |K| = |K'| \text{ and } \inf_{m \in K \setminus K'} m \le \inf_{m \in K' \setminus K} m.$$

We use this order to define an order $\le$ on $\mathfrak{F}(\mathbb{N}) \setminus \{\varnothing\}$:

$$\begin{aligned} K \le K' \quad \text{iff} \quad & \inf\{ n \in \mathbb{N} \mid K \subset \mathbb{N}_n \} < \inf\{ n \in \mathbb{N} \mid K' \subset \mathbb{N}_n \} \\ \text{or} \quad & \inf\{ n \in \mathbb{N} \mid K \subset \mathbb{N}_n \} = \inf\{ n \in \mathbb{N} \mid K' \subset \mathbb{N}_n \} \\ & \text{and} \quad K \le_n K'. \end{aligned}$$

$\mathfrak{F}(\mathbb{N}) \setminus \{\varnothing\}$ endowed with this ordering is order-isomorphic to $\mathbb{N}$. Let $\omega\colon \mathbb{N} \to \mathfrak{F}(\mathbb{N}) \setminus \{\varnothing\}$ be the uniquely determined order isomorphism. Verify

$$\omega(2^n - 1) = \mathbb{N}_n \quad \text{for each } n \in \mathbb{N}$$

$$\omega(k) \subset \mathbb{N}_n \quad \text{whenever } k \le 2^n - 1.$$

Now construct a triangular family $(A_{mn})_{m,n\in\mathbb{N},n\le m}$ from $\mathfrak{S}$ such that for each $m \in \mathbb{N}$ and $n \le 2^m - 1$

$$\bigcup_{i=n}^{2^m-1} A_{in} = \bigcup_{k\in\mathbb{N}_n\setminus\omega(n)} \left(A_k \cap \left(\bigcap_{j\in\omega(n)} B_j\right)\right).$$

This can be done by setting $A_{in} := A_k \cap (\bigcap_{j\in\omega(n)} B_j)$ for suitable k. The subsequence $(A_{2^m-1,n})_{m,n\in\mathbb{N},\, n<2^m-1}$ is convergent, and for each $n \in \mathbb{N}$ we have

$$\bigcup_{m\ge n} A_{mn} = \bigcap_{i\in\omega(n)} B_i.$$

Take $\varepsilon \in \mathbb{R}$, $\varepsilon > 0$. Step 1 implies the existence of $\bar{n} \in \mathbb{N}$ such that for all $n \in \mathbb{N}$, $n \ge \bar{n}$

$$\mu\left(\bigcup_{\substack{K\subset\mathbb{N}_n\\K\neq\varnothing}}\left(\bigcap_{k\in K} B_k \setminus \bigcup_{k\in\mathbb{N}_n\setminus K}\left(A_k \cap \left(\bigcap_{k\in K} B_k\right)\right)\right)\right) < \frac{\varepsilon}{2}$$

and, as a consequence of Step 2, $\bar{n}$ can be chosen such that for each $n \ge \bar{n}$,

$$\mu\left(C \setminus \bigcup_{k\le n} A_k\right) < \frac{\varepsilon}{2}.$$

(Note that $\bigcup_{k\in\mathbb{N}} A_k = C$.) So we get

$$\mu\left(C \setminus \bigcup_{m\le n} (A_m \setminus B_m)\right) < \varepsilon$$

for each $n \ge \bar{n}$.

STEP 4: If $C \in \mathfrak{S}'$ and $(C_n)_{n\in\mathbb{N}}$ is a sequence from $\mathfrak{S}'$ with $C = \bigcup_{n\in\mathbb{N}} C_n$, then

$$\mu(C) = \sup_{n\in\mathbb{N}} \mu\left(\bigcup_{m\le n} C_m\right).$$

Take $A, B \in \mathfrak{S}$ with $C = A \setminus B$ and $B \subset A$. Then $A = B \cup (\bigcup_{n\in\mathbb{N}} C_n) = \bigcup_{n\in\mathbb{N}} (B \cup C_n)$ and $B \cap C_n = \varnothing$ for any $n \in \mathbb{N}$. Step 3 yields

$$\mu(A) = \sup_{n\in\mathbb{N}} \mu\left(B \cup \left(\bigcup_{m\le n} C_m\right)\right) = \mu(B) + \sup_{n\in\mathbb{N}} \mu\left(\bigcup_{m\le n} C_m\right).$$

Consequently

$$\mu(C) = \mu(A) - \mu(B) = \sup_{n \in \mathbb{N}} \mu\left(\bigcup_{m \le n} C_m \right).$$

STEP 5: If $C \in \mathfrak{S}'$ and if $(C_n)_{n \in \mathbb{N}}$ is an increasing sequence from $\mathfrak{R}_r$ with $C = \bigcup_{n \in \mathbb{N}} C_n$, then

$$\mu(C) = \sup_{n \in \mathbb{N}} \mu(C_n).$$

STEP 6: If $C \in \mathfrak{R}_r$ and if $(C_n)_{n \in \mathbb{N}}$ is an increasing sequence from $\mathfrak{R}_r$ with $C = \bigcup_{n \in \mathbb{N}} C_n$, then

$$\mu(C) = \sup_{n \in \mathbb{N}} \mu(C_n).$$

These last two steps are left entirely to the reader and complete the proof.

2.2.12(C) Regularity Conditions. We continue Ex. 2.2.11. The hypotheses needed can be substantially simplified if certain regularity conditions hold.

For $\mathfrak{R}$ a $\cap$-semi-lattice and $\mathfrak{G}$ a lattice of sets with $\mathfrak{G} \subset \mathfrak{R}$ and $\varnothing \in \mathfrak{G}$, the function ν: $\mathfrak{R} \to \mathbb{R}_+$ is called **$\mathfrak{G}$-regular** iff the following condition is fulfilled:

(i) $\nu(A) - \nu(A \cap B) = \sup\{\nu(G) | G \in \mathfrak{G},\ G \subset A \setminus B\}$ whenever $A, B \in \mathfrak{R}$.

Prove the following statement:

(α) Suppose $\mathfrak{R}$ is a $\cap$-semi-lattice and $\mathfrak{G}$ is a lattice of sets with $\mathfrak{G} \subset \mathfrak{R}$ and $\varnothing \in \mathfrak{G}$. Let ν: $\mathfrak{R} \to \mathbb{R}_+$ be an alternating completely increasing $\mathfrak{G}$-regular function, and denote by μ the uniquely determined additive extension of ν to $\mathfrak{R}_r$. If $\nu|_{\mathfrak{G}}$ is nullcontinuous, then μ is a positive measure on $\mathfrak{R}_r$. μ is $\mathfrak{G}$-regular.

The proof can be broken up into several steps.

STEP 1: Take $A \in \mathfrak{R}$. Let $(B_k)_{k \in \mathbb{N}_n}$ be a family from $\mathfrak{R}$. Then

$$\mu\left(A \setminus \bigcup_{k \in \mathbb{N}_n} B_k\right) = \sup\left\{\nu(G) | G \in \mathfrak{G}, G \subset A \setminus \bigcup_{k \in \mathbb{N}_n} B_k\right\}.$$

Take $\varepsilon \in \mathbb{R}$, $\varepsilon > 0$. Given $k \in \mathbb{N}_n$, there is a $G_k \in \mathfrak{G}$ with $G_k \subset A \setminus B_k$, such that $\mu(A \setminus B_k) \le \nu(G_k) + \varepsilon/2^k$. Set $G := \bigcap_{k \in \mathbb{N}_n} G_k$. Then $G \in \mathfrak{G}$, $G \subset A \setminus \bigcup_{k \in \mathbb{N}_n} B_k$ and

$$\left(A \setminus \bigcup_{k \in \mathbb{N}_n} B_k\right) \setminus G \subset \bigcup_{k \in \mathbb{N}_n} ((A \setminus B_k) \setminus G_k).$$

Hence $\mu(A \setminus \bigcup_{k \in \mathbb{N}_n} B_k) \le \nu(G) + \varepsilon$. But ε is arbitrary, so

$$\mu\left(A \setminus \bigcup_{k \in \mathbb{N}_n} B_k\right) \le \sup\left\{\nu(G) \mid G \in \mathfrak{G},\, G \subset A \setminus \bigcup_{k \in \mathbb{N}_n} B_k\right\}.$$

The opposite inequality is obvious.

STEP 2: Define $\mathfrak{S} := \{\bigcup_{A \in \mathfrak{A}} A \mid \mathfrak{A} \in \mathfrak{F}(\mathfrak{R})\}$. Show that $\mu|_{\mathfrak{S}}$ is $\mathfrak{G}$-regular.

Take $A, B \in \mathfrak{S}$. If $A = \varnothing$, condition (i) is trivially fulfilled. If $B = \varnothing$, it follows from our hypotheses. Assume that $A \ne \varnothing$ and $B \ne \varnothing$. There is $\mathfrak{C} \in \mathfrak{F}(\mathfrak{R})$ such that $A = \bigcup_{C \in \mathfrak{C}} C$, and so

$$A \setminus B = \bigcup_{C \in \mathfrak{C}} (C \setminus B).$$

Take $\varepsilon \in \mathbb{R}$, $\varepsilon > 0$. By Step 1, for each $C \in \mathfrak{C}$ there is a set $G_C \in \mathfrak{G}$ with $G_C \subset C \setminus B$ and

$$\mu(C \setminus B) - \nu(G_C) < \frac{\varepsilon}{|\mathfrak{C}|}$$

($|\mathfrak{C}|$ denotes the cardinality of $\mathfrak{C}$). Define $G := \bigcup_{C \in \mathfrak{C}} G_C$. Then $G \in \mathfrak{G}$, $G \subset A \setminus B$, and

$$\mu(A \setminus B) - \nu(G) \le \sum_{C \in \mathfrak{C}} (\mu(C \setminus B) - \nu(G_C)) < \varepsilon.$$

From this, our assertion follows.

STEP 3: Set $\mathfrak{S}' := \{A \setminus B \mid A, B \in \mathfrak{S}\}$. Let $(C_n)_{n \in \mathbb{N}}$ be a sequence from $\mathfrak{S}'$ such that $C := \bigcup_{n \in \mathbb{N}} C_n \in \mathfrak{S}$. Then

$$\mu(C) = \sup_{n \in \mathbb{N}} \mu\left(\bigcup_{m \le n} C_m\right).$$

For each $n \in \mathbb{N}$ there are sets $A_n, B_n \in \mathfrak{S}$ with $C_n = A_n \setminus B_n$ and $B_n \subset A_n \subset C$. Then for each $n \in \mathbb{N}$,

$$C \setminus \bigcup_{m \le n} C_m = \bigcap_{m \le n} (B_m \cup (C \setminus A_m)).$$

Take $\varepsilon \in \mathbb{R}$, $\varepsilon > 0$. Then for each $n \in \mathbb{N}$ there is a set $G_n \in \mathfrak{G}$ with $G_n \subset B_n \cup (C \setminus A_n)$, and

$$\mu(B_n \cup (C \setminus A_n)) - \nu(G_n) < \frac{\varepsilon}{2^n}.$$

(Use Step 2 and the general hypotheses.) For every $n \in \mathbb{N}$ define $H_n := \bigcap_{m \le n} G_m$. $(H_n)_{n \in \mathbb{N}}$ is a decreasing sequence from $\mathfrak{G}$ such that $\bigcap_{n \in \mathbb{N}} H_n =$

$\varnothing$, and we conclude that $\inf_{n \in \mathbb{N}} \nu(H_n) = 0$. For every $n \in \mathbb{N}$,

$$\mu\Big(\bigcap_{m \le n} (B_m \cup (C \setminus A_m))\Big) - \mu(H_n) < \sum_{m \le n} \frac{\varepsilon}{2^n} < \varepsilon$$

and this implies

$$\inf_{n \in \mathbb{N}} \mu\Big(\bigcap_{m \le n} (B_m \cup (C \setminus A_m))\Big) < \varepsilon.$$

Hence, since ε is arbitrary,

$$\mu(C) - \sup_{n \in \mathbb{N}} \mu\Big(\bigcup_{m \in \mathbb{N}} C_m\Big) = \inf_{n \in \mathbb{N}} \mu\Big(\bigcap_{m \le n} (B_m \cup (C \setminus A_m))\Big) = 0.$$

The rest of the proof is identical with Steps 4–6 of the proof of Ex. 2.2.11(ι).

Of particular importance in this context are functions on lattices of sets. Prove the following result:

(β) If $\mathfrak{R}$ is a lattice of sets with $\varnothing \in \mathfrak{R}$, and if $\nu: \mathfrak{R} \to \mathbb{R}$ is nullcontinuous and $\mathfrak{R}$-regular then there is a unique positive measure μ on $\mathfrak{R}_r$ with $\mu|_{\mathfrak{R}} = \nu$, and μ is $\mathfrak{R}$-regular.

2.2.13(C) **Positive Measures on Hausdorff Spaces.** We give an important application of the results of Ex. 2.2.12.

Let X be a Hausdorff space. Let $\mathfrak{K}$ be the set of all subsets of X that are compact. $\mathfrak{K}$ is a lattice of sets and $\bigcap_{A \in \mathfrak{A}} A \in \mathfrak{K}$ whenever $\varnothing \neq \mathfrak{A} \subset \mathfrak{K}$. Prove the following statement.

Let $\nu: \mathfrak{K} \to \mathbb{R}$ be a function that satisfies these conditions:

(i) $\nu(A \cup B) \le \nu(A) + \nu(B)$ whenever $A, B \in \mathfrak{K}$.

(ii) $\nu(A \cup B) = \nu(A) + \nu(B)$ whenever $A, B \in \mathfrak{K}$, $A \cap B = \varnothing$.

(iii) $\nu(\bigcap_{A \in \mathfrak{A}} A) = \inf_{A \in \mathfrak{A}} \nu(A)$ whenever $\mathfrak{A}$ is a nonempty downward directed subset of $\mathfrak{K}$.

Then there is a unique positive measure on $\mathfrak{K}_r$, μ, such that $\mu|_{\mathfrak{K}} = \nu$. Moreover μ is $\mathfrak{K}$-regular.

We only need to show that ν is $\mathfrak{K}$-regular. Take $A, B \in \mathfrak{K}$ with $A \subset B$. Then

$$\nu(B) - \nu(A) \ge \sup\{\nu(C) \mid C \in \mathfrak{K}, C \subset B \setminus A\}.$$

For the reverse inequality, set

$$\mathfrak{C} := \{C \in \mathfrak{K} \mid A \subset C \subset B, A \cap (\overline{B \setminus C}) = \varnothing\}.$$

Then C is nonempty and directed downward, since if $C', C'' \in \mathfrak{C}$, then $C' \cap C'' \in \mathfrak{C}$. We have $\bigcap_{C \in \mathfrak{C}} C = A$. Thus applying (iii), $\nu(A) = \inf_{C \in \mathfrak{C}} \nu(C)$. On the other hand, if $C \in \mathfrak{C}$, then

$$\nu(B) \leq \nu(C) + \nu(\overline{B \setminus C}) \leq \nu(C) + \sup\{\nu(D) \mid D \in \mathfrak{K}, D \subset B \setminus A\}.$$

Since C is arbitrary,

$$\nu(B) \leq \nu(A) + \sup\{\nu(D) \mid D \in \mathfrak{K}, D \subset B \setminus A\}$$

as required. Hence ν is $\mathfrak{K}$-regular, and so Ex. 2.2.12 can be applied.

This result is very important in measure theory on Hausdorff spaces. That theory will be presented in the final part of the book.

2.2.14(E) (a) Let $\mathfrak{R}$ be a set of sets and μ a real function on $\mathfrak{R}_m$ such that

$$\mu\Big(\lim_{n \to \infty} A_n\Big) = \lim_{n \to \infty} \mu(A_n)$$

for every monotone convergent sequence $(A_n)_{n \in \mathbb{N}}$ from $\mathfrak{R}_m$. Prove

(α) $\mu(\mathfrak{R}_m) \subset \overline{\mu(\mathfrak{R})}$.

(β) μ is positive if $\mu|_{\mathfrak{R}}$ is positive.

(γ) μ is bounded (resp. upper or lower bounded) if $\mu|_{\mathfrak{R}}$ is bounded (resp. upper or lower bounded).

Show that all assertions still hold if we replace $\mathfrak{R}_m$ in all occurrences by $\mathfrak{R}_{mb}$.

(b) Prove the following statements for $\cap$-semi-lattices $\mathfrak{R}$:

(α) If μ and ν are positive measures on $\mathfrak{R}_\delta$ such that $\mu|_{\mathfrak{R}} - \nu|_{\mathfrak{R}}$ is completely increasing, then $\mu - \nu$ is positive.

(β) If μ and ν are positive measures on $\mathfrak{R}_\delta$ with $\mu|_{\mathfrak{R}} = \nu|_{\mathfrak{R}}$ then $\mu = \nu$.

(c) Let $\mathfrak{R}$ be a set of sets such that for every $A, B \in \mathfrak{R}$ there is a disjoint countable family $(C_\iota)_{\iota \in I}$ from $\mathfrak{R}$ such that

$$A \cap B = \bigcup_{\iota \in I} C_\iota,$$

and let μ, ν be additive nullcontinuous real functions on $\mathfrak{R}_\delta$ such that $\mu|_{\mathfrak{R}} = \nu|_{\mathfrak{R}}$.

(α) Prove that $\mu = \nu$.

(Use Ex. 2.1.12 (c).)

2.2.15(E) Define

$$f\colon \mathfrak{P}(\mathbb{N}) \times \mathbb{N} \to \mathbb{N}, \qquad (A,n) \mapsto 1 + \sum_{k\in A,\, k\le n} 2^k$$

$$g\colon \mathfrak{P}(\mathbb{N}) \to \mathfrak{P}(\mathbb{N}), \qquad A \mapsto \{f(A,n) \mid n \in \mathbb{N}\}.$$

Prove the following statements:

(α) Let $A, B \in \mathfrak{P}(\mathbb{N})$ and $m, n \in \mathbb{N}$. Then, if $f(A,m) = f(B,n)$, $\{k \in A \mid k \le \inf(m,n)\} = \{k \in B \mid k \le \inf(m,n)\}$.

(β) $g(A) \cap g(B)$ is finite for any $A, B \in \mathfrak{P}(\mathbb{N})$ with $A \ne B$.

(γ) There is a set $\mathfrak{M}$ of infinite subsets of $\mathbb{N}$ such that $\mathfrak{M}$ has the cardinality of the continuum and such that $A \cap B$ is finite whenever $A, B \in \mathfrak{M}$, $A \ne B$.

2.2.16(E) Let X be a set and $\mathfrak{R}$ a ring of subsets of X with the property that for any disjoint sequence $(A_n)_{n\in\mathbb{N}}$ from $\mathfrak{R}$, there is an infinite subset M of $\mathbb{N}$ for which $\bigcup_{n\in M} A_n \in \mathfrak{R}$. Prove the following:

(α) Every positive measure on $\mathfrak{R}$ is bounded.

(β) Let Φ be a set of ultrafilters on X. Assume that the cardinality of Φ is strictly smaller than that of $\mathbb{R}$. Then, $\mathfrak{R} := \mathfrak{P}(X) \setminus \bigcup_{\mathfrak{F}\in\Phi} \mathfrak{F}$ has the property stated at the beginning of this exercise.

For the proof use Ex. 2.2.15.

2.2.17(E) Let X be a set of cardinality $\aleph_{\omega_0}$. Define

$$\mathfrak{R} := \{A \subset X \mid |A| < \aleph_{\omega_0}\}.$$

($|A|$ denotes the cardinality of A.) Prove:

(α) $\mathfrak{R}$ is a δ-ring but not a σ-ring.

(β) Every positive measure on $\mathfrak{R}$ is bounded.

2.2.18(E) The order structure of the space of all positive linear functionals on a Riesz lattice was investigated in Ex. 1.3.11. We wish to conduct the corresponding investigation here for positive measures.

Let $\mathfrak{R}$ be a ring of sets. Denote by Ψ the set of all additive positive functions on $\mathfrak{R}$. For $\mu, \nu \in \Psi$, define $\mu \le \nu$ iff $\mu(A) \le \nu(A)$ for every $A \in \mathfrak{R}$. Prove the statements that follow:

(α) $\le$ is an order relation on Ψ.

(β) Ψ is a conditionally complete lattice with respect to $\le$. For all $\mu, \nu \in \Psi$ and $A \in \mathfrak{R}$,

$$(\mu \vee \nu)(A) = \sup\{\mu(B) + \nu(A \setminus B) \mid B \in \mathfrak{R},\, B \subset A\}$$

and

$$(\mu \wedge \nu)(A) = \inf\{\mu(B) + \nu(A \setminus B) \mid B \in \mathfrak{R},\ B \subset A\}.$$

(γ) Given a nonempty upward directed family $(\mu_\iota)_{\iota \in I}$ from Ψ with $\sup_{\iota \in I} \mu_\iota(A) < \infty$ whenever $A \in \mathfrak{R}$, $\bigvee_{\iota \in I} \mu_\iota$ exists and $(\bigvee_{\iota \in I} \mu_\iota)(A) = \sup_{\iota \in I} \mu_\iota(A)$ for any $A \in \mathfrak{R}$. An analogous statement is true for downward directed families.

These properties correspond exactly to those we determined for positive linear functionals. The correspondence even extends to the algebraic structures. Show that if $\mu, \nu \in \Psi$ and $\alpha \in \mathbb{R}_+$, then

(δ) $\mu + \nu \in \Psi$.

(ε) $\alpha\mu \in \Psi$.

(ζ) $\mu + \lambda \leq \nu + \lambda$ whenever $\lambda \in \Psi$, $\mu \leq \nu$.

(η) $\alpha\mu \leq \alpha\nu$ whenever $\mu \leq \nu$.

Prove:

(ϑ) Ψ^σ is a conditionally complete lattice with respect to $\leq$, where Ψ^σ denotes the set of all positive measures on $\mathfrak{R}$. If $\mu \in \Psi^\sigma$ and $\nu \in \Psi$ with $\nu \leq \mu$, then $\nu \in \Psi^\sigma$. Given $\mu, \nu \in \Psi^\sigma$ and $\alpha \in \mathbb{R}_+$, $\mu + \nu \in \Psi^\sigma$ and $\alpha\mu \in \Psi^\sigma$. ($\gamma$) holds for Ψ^σ.

(ι) Defining π-continuity as in Ex. 2.2.4, prove that the set Ψ^π of all π-continuous functions of Ψ is a conditionally complete lattice with respect to $\leq$. If $\mu \in \Psi^\pi$ and $\nu \in \Psi$ with $\nu \leq \mu$, then $\nu \in \Psi^\pi$. If $\mu, \nu \in \Psi^\pi$ and $\alpha \in \mathbb{R}_+$, then $\mu + \nu \in \Psi^\pi$ and $\alpha\mu \in \Psi^\pi$. (γ) holds for Ψ^π.

(δ) also holds for Ψ^σ and Ψ^π. We shall deal with this in a general setting in Chapter 8.

2.2.19(C) Ulam Sets (S. Ulam, 1930). We have had occasion to work with δ-stable ultrafilters. But as yet, we know nothing about the existence of nontrivial filters of this kind. In fact we can draw no conclusion about this from the axioms of set theory we have applied until now. (Contrast this to the situation with nontrivial ultrafilters generally.) The problem of the existence of nontrivial δ-stable ultrafilters is closely connected to questions arising in measure theory. We will now discuss the most important facts.

(a) Let $\mathfrak{R}$ be a ring of sets. A positive measure μ on $\mathfrak{R}$ is called a **two-valued positive measure** on $\mathfrak{R}$ iff $\mu(\mathfrak{R}) = \{0, \alpha\}$, $\alpha \neq 0$. A set X is called a **weak Ulam set** iff there is no positive two-valued measure μ on $\mathfrak{P}(X)$ such that $\mu(\{x\}) = 0$ for every $x \in X$.

Prove the following statement:

(α) The following two properties are equivalent for every set X:

($\alpha 1$) X is a weak Ulam set.

($\alpha 2$) There is no nontrivial δ-stable ultrafilter on X.

For (α1) $\Rightarrow$ (α2), Assume that $\mathfrak{F}$ is a nontrivial δ-stable ultrafilter on X. Then,

$$\mu: \mathfrak{P}(X) \to \mathbb{R}, \qquad A \mapsto \begin{cases} 1 & \text{if } A \in \mathfrak{F} \\ 0 & \text{if } A \notin \mathfrak{F} \end{cases}$$

is a positive two-valued measure on $\mathfrak{P}(X)$ with $\mu(\{x\}) = 0$ for all $x \in X$.

For (α2) $\Rightarrow$ (α1), assume that there is a positive two-valued measure μ on $\mathfrak{P}(X)$. Define $\mathfrak{F} := \{A \subset X \mid \mu(A) \neq 0\}$. Then $\mathfrak{F}$ is a δ-stable ultrafilter on X. It is nontrivial if $\mu(\{x\}) = 0$ for all $x \in X$.

Prove now the following statements:

(β) If X is a weak Ulam set, then every set with the cardinality of X is also a weak Ulam set.

(γ) Every countable set is a weak Ulam set.

(δ) Every subset of a weak Ulam set is a weak Ulam set.

(ε) If I is a weak Ulam set and if $(X_\iota)_{\iota \in I}$ is a family of weak Ulam sets, then $\bigcup_{\iota \in I} X_\iota$ is a weak Ulam set.

(ζ) Let X be a set and μ a positive two-valued measure on $\mathfrak{P}(X)$. Let Y be a weak Ulam set and $(A_y)_{y \in Y}$ a family from $\mathfrak{P}(X)$ such that $\mu(A_y) = 0$ for every $y \in Y$. Then $\mu(\bigcup_{y \in Y} A_y) = 0$.

(η) If $(X_\iota)_{\iota \in I}$ is a nonempty family of weak Ulam sets, then $\bigcap_{\iota \in I} X_\iota$ is a weak Ulam set.

(ϑ) If X is a weak Ulam set, then so is $\mathfrak{P}(X)$.

(ι) If X and Y are weak Ulam sets, then X^Y is a weak Ulam set.

(κ) If I is a weak Ulam set and if $(X_\iota)_{\iota \in I}$ is a family of weak Ulam sets, then $\prod_{\iota \in I} X_\iota$ is a weak Ulam set.

We provide some suggestions:

For (ε), set $Z := \{(\iota, x) \mid \iota \in I,\ x \in X_\iota\}$. $\bigcup_{\iota \in I} X_\iota$ has the same cardinality as some subset of Z. Thus it is sufficient to show that Z is a weak Ulam set and then to apply (β) and (δ). So let μ be a two-valued measure on $\mathfrak{P}(Z)$ with the property that $\mu(\{z\}) = 0$ whenever $z \in Z$. For $\iota \in I$, set $Z_\iota := \{(\iota, x) \mid x \in X_\iota\}$. Thus, for each ι, Z_ι has the same cardinality as X_ι and is, consequently, a weak Ulam set. Thus $\mu|_{Z_\iota} = 0$ for $\iota \in I$.

Given $J \subset I$, define $\nu(J) := \mu(\bigcup_{\iota \in J} Z_\iota)$. Then ν is a positive measure on $\mathfrak{P}(I)$ with $\nu(\{\iota\}) = 0$ for every $\iota \in I$ (as has just been shown). But I is, by hypothesis, a weak Ulam set. Thus $\mu(Z) = \nu(I) = 0$, showing, that Z is a weak Ulam set.

For (ζ), assume that $(A_y)_{y \in Y}$ is a disjoint family, and define

$$\nu: \mathfrak{P}(Y) \to \mathbb{R}, \qquad B \mapsto \mu\Big(\bigcup_{y \in B} A_y\Big).$$

ν is a positive measure on $\mathfrak{P}(Y)$ admitting at most two values. But $\nu(\{y\}) = 0$ for every $y \in Y$, and because Y is a weak Ulam set, we conclude that $\nu(Y) = \mu(\bigcup_{y \in Y} A_y) = 0$. For the proof of the general case construct a disjoint family $(B_y)_{y \in Y}$ of subsets of X such that $\bigcup_{y \in Y} B_y = \bigcup_{y \in Y} A_y$.

For (ϑ), let X be a weak Ulam set. To show that $\mathfrak{P}(X)$ is a weak Ulam set, it is sufficient to show that $\{0,1\}^X$ (the set of all functions $f\colon X \to \{0,1\}$) is a weak Ulam set, because $\mathfrak{P}(X)$ and $\{0,1\}^X$ have the same cardinality.

Let μ be a positive measure on $\mathfrak{P}(\{0,1\}^X)$ which takes only the values 0 and 1 such that $\mu(\{f\}) = 0$ for any $f \in \{0,1\}^X$. Assume $\mu(\{0,1\}^X) = 1$. Given $x \in X$, define

$$\mathscr{F}_x := \left\{ f \in \{0,1\}^X \mid f(x) = 1 \right\} \quad \text{and} \quad \mathscr{F}_x^c := \left\{ f \in \{0,1\}^X \mid f(x) = 0 \right\}.$$

Then for each $x \in X$, $\mathscr{F}_x \cup \mathscr{F}_x^c = \{0,1\}^X$ and $\mathscr{F}_x \cap \mathscr{F}_x^c = \varnothing$. Consequently exactly one of $\mathscr{F}_x$ and $\mathscr{F}_x^c$ has measure 1. Denote this one by $\mathscr{G}_x$. Now $\bigcap_{x \in X} \mathscr{G}_x$ has at most one element, and so measure 0. Define $\mathscr{G}_x^c := \{0,1\}^X \setminus \mathscr{G}_x$ for every $x \in X$. Then

$$\{0,1\}^X = \left(\bigcap_{x \in X} \mathscr{G}_x \right) \cup \left(\bigcup_{x \in X} \mathscr{G}_x^c \right)$$

expresses $\{0,1\}^X$ as a union of sets of measure 0. But then (ζ) implies $\mu(\{0,1\}^X) = 0$.

Hence the class of all weak Ulam sets is closed under the important set operations. So the cardinality of a set that is not a weak Ulam set must be extremely large. The existence of such sets cannot be concluded from the axioms of set theory we have applied until now. It would seem possible to introduce it as an additional axiom.

(b) We look at a stronger version of our problem. A set X is called an **Ulam set** iff there is no strictly positive measure μ on $\mathfrak{P}(X)$ such that $\mu(\{x\}) = 0$ for each $x \in X$. Of course every Ulam set is also a weak Ulam set. Is the converse true?

We start our investigations with some propositions that are fully analogous to results stated earlier for weak Ulam sets.

(α) If X is an Ulam set, then every set with the cardinality of X is an Ulam set.

(β) Every countable set is an Ulam set.

(γ) Every subset of an Ulam set is an Ulam set.

(δ) If I is an Ulam set and if $(X_\iota)_{\iota \in I}$ is a family of Ulam sets, then $\bigcup_{\iota \in I} X_\iota$ is an Ulam set.

(ε) Let X be a set and μ a positive measure on $\mathfrak{P}(X)$. Let Y be an Ulam set and $(A_y)_{y \in Y}$ a family from $\mathfrak{P}(X)$ such that $\mu(A_y) = 0$ for every $y \in Y$. Then $\mu(\bigcup_{y \in Y} A_y) = 0$.

We cannot prove a result like (a) (ϑ) for Ulam sets. But the following proposition holds:

(ζ) If X is an Ulam set with cardinality $\aleph_\alpha$, then every set with cardinality $\aleph_{\alpha+1}$ is an Ulam set.

We give some hints for the proof:

For every ordinal γ define $A_\gamma := \{\beta \mid \beta \text{ an ordinal}, \beta < \gamma\}$. A_γ and γ always have the same cardinality. Denote by γ the first ordinal with the cardinality of $\aleph_{\alpha+1}$. Then for every $\beta < \gamma$,

$$A_\gamma = \{\beta\} \cup A_\beta \cup \{\alpha \in A_\gamma \mid \beta < \alpha\}.$$

Let μ be a positive measure on $\mathfrak{P}(A_\gamma)$ such that $\mu(\{\beta\}) = 0$ for each $\beta \in A_\gamma$. For each $\beta \in A_\gamma$ let φ_β: $A_\beta \to X$ be an injective mapping. For $\beta \in A_\gamma$ and $x \in X$ define $A(\beta, x) := \{\alpha \in A_\gamma \mid \beta < \alpha, \varphi_\alpha(\beta) = x\}$. Then $(A(\beta, x))_{\beta \in A_\gamma}$ is, for each $x \in X$, a disjoint family of subsets of A_γ. Consequently for each $x \in X$, $\{A(\beta, x) \mid \mu(A(\beta, x)) > 0\}$ is countable. Thus

$$\{(\beta, x) \in A_\gamma \times X \mid \mu(A(\beta, x)) > 0\}$$

has a cardinal strictly smaller than $\aleph_{\alpha+1}$. Hence, for some $\tilde{\beta} \in A_\gamma$, $\mu(A(\tilde{\beta}, x)) = 0$ for every $x \in X$. Thus $\mu(\bigcup_{x \in X} A(\tilde{\beta}, x)) = 0$ since X is an Ulam set. But $A_{\tilde{\beta}}$ is also an Ulam set, and we conclude

$$\begin{aligned} \mu(A_\gamma) &= \mu(\{\tilde{\beta}\}) + \mu(A_{\tilde{\beta}}) + \mu(\{\beta \in A_\gamma \mid \tilde{\beta} < \beta\}) \\ &\leq \mu(\{\tilde{\beta}\}) + \mu(A_{\tilde{\beta}}) + \mu\Big(\bigcup_{x \in X} A(\tilde{\beta}, x)\Big) = 0 \end{aligned}$$

showing that $\mu(A_\gamma) = 0$.

Now we discuss the question formulated earlier. Prove:

(η) Let X be a weak Ulam set and μ a strictly positive measure on $\mathfrak{P}(X)$ such that $\mu(\{x\}) = 0$ for every $x \in X$. Then for each $\varepsilon \in \mathbb{R}$, $\varepsilon > 0$, there is a subset A of X with $0 < \mu(A) < \varepsilon$.

(ϑ) If $\mathbb{R}$ is an Ulam set, then so is every weak Ulam set.

For (η), using Zorn's lemma, there is a filter $\mathfrak{F}$ on X, maximal with the property that $\mu(F) > 0$ whenever $F \in \mathfrak{F}$. Prove that for every $C \subset X$ either $C \in \mathfrak{F}$ or $X \setminus C \in \mathfrak{F}$. Hence $\mathfrak{F}$ is an ultrafilter. But then there is a decreasing sequence $(F_n)_{n \in \mathbb{N}}$ from $\mathfrak{F}$ such that $\bigcap_{n \in \mathbb{N}} F_n = \emptyset$.

For (ϑ), let X be a weak Ulam set and μ a positive measure on $\mathfrak{P}(X)$ such that $\mu(\{x\}) = 0$ for every $x \in X$. Using (η), construct recursively a sequence $(f_n)_{n \in \mathbb{N}}$ of mappings f_n: $\mathbb{N}^n \to \mathfrak{P}(X)$ such that for each $n \in \mathbb{N}$

(i) $(f_n(k))_{k \in \mathbb{N}^n}$ is a disjoint family with $\bigcup_{k \in \mathbb{N}^n} f_n(k) = X$.

(ii) $\mu(f_n(k)) < 1/2^n$ whenever $k \in \mathbb{N}^n$.

(iii) $f_{n+1}(k, j) \subset f_n(k)$ whenever $k \in \mathbb{N}^n$ and $j \in \mathbb{N}$.

For every $g \in \mathbb{N}^{\mathbb{N}}$ define

$$A_g := \bigcap_{n \in \mathbb{N}} f_n(g(1), g(2), \ldots, g(n)).$$

Then $\mu(A_g) = 0$ whenever $g \in \mathbb{N}^{\mathbb{N}}$. But $X = \bigcup_{g \in \mathbb{N}^{\mathbb{N}}} A_g$, and ($\varepsilon$) implies our assertion.

From (ϑ) and (ζ) we conclude the following:

(ι) If we accept the Continuum Axiom (i.e., that $2^{\aleph_0} = \aleph_1$), then weak Ulam sets and Ulam sets are the same objects.

We consider an application. Prove for a set X:

(κ) The following assertions are equivalent:

(κ1) There is a positive linear nullcontinuous functional on $\ell^\infty(X)$ that is not π-continuous.

(κ2) X is not an Ulam set.

If we accept the Continuum Axiom, these statements both are equivalent to the next assertion.

(κ3) There is a δ-stable free ultrafilter on X.

We give a hint for (κ1) $\Rightarrow$ (κ2): Let ℓ be a positive linear nullcontinuous functional on $\ell^\infty(X)$ that is not π-continuous. Then there is a downward directed family $(f_\iota)_{\iota \in I}$ in $\ell^\infty(X)$ with $\bigwedge_{\iota \in I} f_\iota = 0$ but $\inf_{\iota \in I} \ell(f_\iota) > 0$. For any $n \in \mathbb{N}$ and $\iota \in I$ define $A_{n\iota} := \{f_\iota \geq 1/n\}$. Then for any $n \in \mathbb{N}$, $(A_{n\iota})_{\iota \in I}$ is a downward directed family of sets with $\bigcap_{\iota \in I} A_{n\iota} = \emptyset$. Prove that there is $n \in \mathbb{N}$ such that $\inf_{\iota \in I} \ell(e_{A_{n\iota}}) > 0$. Indeed, define $\alpha :=$ $\inf_{\iota \in I} \ell(f_\iota)$. Suppose that $\inf_{\iota \in I} \ell(e_{A_{n\iota}}) = 0$ for each $n \in \mathbb{N}$. Take $n \in \mathbb{N}$ such that $\ell((1/n)e_X) < \alpha/3$ and $\lambda \in I$. Take $\iota \in I$ with $\ell(e_{A_{n\iota}}) <$ $\alpha/(3 \sup_{x \in X} f_\lambda(x))$ and $f_\iota \leq f_\lambda$. Then we have

$$\ell(f_\iota) \leq \frac{1}{n}\ell(e_X) + \ell(e_{A_{n\iota}}) \sup_{x \in X} f_\lambda(x) < \alpha.$$

This is a contradiction.

Now let B be the set of all $x \in X$ with $\ell(e_{\{x\}}) \neq 0$, and define

$$\mu: \mathfrak{P}(X) \to \mathbb{R}, \qquad A \mapsto \ell(e_A) - \sum_{x \in A \cap B} \ell(e_{\{x\}}).$$

μ is a positive measure on $\mathfrak{P}(X)$ such that $\mu(\{x\}) = 0$ for any $x \in X$. But by Ex. 2.2.4(c) and by our earlier consideration, $\mu \neq 0$. We conclude that X cannot be an Ulam set.

2.2.20(E) Let X be a set and $\mathfrak{R}$ a set-ring on X. Prove the following assertions:

(α) If $f \in \mathbb{R}^{\mathfrak{R}}$ such that

$$\mathfrak{A} := \{A \in \mathfrak{R} \mid f(A) \neq 0\}$$

is finite and $\sum_{A \in \mathfrak{A}} f(A) e_A = 0$, then there is $\alpha \in \mathbb{R} \setminus \{0\}$ such that $\alpha f(A) \in \mathbb{Z}$ for every $A \in \mathfrak{R}$.

The proof can be given by complete induction on the number of elements of A.

(β) There is $\mathfrak{A} \subset \mathfrak{R}$ such that for each $B \in \mathfrak{R}$ there exists exactly one $f_B \in \mathbb{Z}^{\mathfrak{A}}$ with the following properties:

(i) $\mathfrak{A}(B) := \{A \in \mathfrak{A} \mid f_B(A) \neq 0\}$ is finite

(ii) $e_B = \sum_{A \in \mathfrak{A}(B)} f_B(A) e_A$.

Use the Zorn Lemma for the proof.

Fix $\mathfrak{A}$ with the above property. With the notations of (β) prove

(γ) For every $g \in \mathbb{R}^{\mathfrak{A}}$ the map

$$\mathfrak{R} \to \mathbb{R}, \qquad B \mapsto \sum_{A \in \mathfrak{A}(B)} g(A) f_B(A) \tag{1}$$

is additive, and every additive real function on $\mathfrak{R}$ is of this form.

(δ) If $\{\{x\} \mid x \in X\} \subset \mathfrak{R}$, one can choose $\mathfrak{A}$ such that $\{\{x\} \mid x \in X\} \subset \mathfrak{A}$. In this case, if X is infinite, the map

$$\mathfrak{R} \to \mathbb{R}, \qquad B \mapsto \sum_{A \in \mathfrak{A}(B)} f_B(A)$$

is additive and unbounded.

(ε) Let $x \in X$ such that $\{x\} \in \mathfrak{R}$. One can choose $\mathfrak{A}$ such that $\{x\} \in \mathfrak{A}$ and $x \notin A$ for every $A \in \mathfrak{A} \setminus \{x\}$. In this case, the map

$$\mathfrak{R} \to \mathbb{R}, \qquad B \mapsto f_B(\{x\})$$

is additive and bounded.

(ζ) If $\{\{x\} \mid x \in X\} \subset \mathfrak{R}$ and if X is infinite, then there exists $g \in \mathbb{R}^{\mathfrak{A}}$ (resp. $g \in \mathbb{R}^{\mathfrak{A}} \setminus \{0\}$) such that the map (1) is unbounded (resp. bounded).

2.3. STEP FUNCTIONS AND STONE LATTICES

NOTATION FOR SECTION 2.3:

X denotes a set.

We begin the task of connecting positive measures with appropriate positive linear functionals. The present section establishes the correspondence between the domains of definition. Step functions, formed by taking linear combinations of characteristic functions, provide a natural connection. After a few step-function preliminaries, we show (Theorem 5) that every set-ring yields, via its step functions, a real Riesz lattice. This Riesz lattice, it turns out, has an additional, important property, which we call the Stone property. Theorem 10 shows that the correspondence

set-ring → real Riesz lattice of step functions

is reversible, provided that the Riesz lattice in question has the Stone property. Section 2.4 then continues the story, by describing the full correspondence between positive measure spaces and appropriately special Daniell spaces.

Definition 2.3.1. *A **step function** on X is a real function f, defined on X, such that*

$$f = \sum_{\iota \in I} \alpha_\iota e_{A_\iota} \tag{1}$$

for some finite family $(A_\iota)_{\iota \in I}$ *of subsets of* X *and some family* $(\alpha_\iota)_{\iota \in I}$ *of real numbers.*

Given a subset $\mathfrak{S}$ *of* $\mathfrak{P}(X)$ *and a function* f *in* $\mathbb{R}^X$, *the function* f *is called an* $\mathfrak{S}$-***step function*** *on* X, *iff it has a representation* (1) *for which every* A_ι *belongs to* $\mathfrak{S}$. *In this case* (1) *is called an* $\mathfrak{S}$-***representation*** *of* f. *We denote by* $\mathscr{L}_{\mathfrak{S}}^X$ *the set of all* $\mathfrak{S}$*-step functions on* X.

Given a subset $\mathfrak{S}$ *of* $\mathfrak{P}(X)$ *and an* $\mathfrak{S}$*-step function* f *on* X, *if*

$$f = \sum_{\kappa \in K} \beta_\kappa e_{B_\kappa} \tag{2}$$

and $(B_\kappa)_{\kappa \in K}$ *is a disjoint family of nonempty sets belonging to* $\mathfrak{S}$, *then* (2) *is called a **disjoint*** $\mathfrak{S}$-***representation*** *of the* $\mathfrak{S}$*-step function* f. □

The definition of step function given here reflects our intention not to put any special structure on the carrier set X.

We are especially interested in the properties of $\mathfrak{R}$-step functions, with $\mathfrak{R}$ a ring of sets. The most important fact about such functions is that they always have disjoint representations. Occasionally we need a stronger result: the terms of an arbitrary $\mathfrak{R}$-representation can be replaced, one term at a time, by the terms of a disjoint representation. We establish the existence of disjoint representations by proving the stronger result, so we state the stronger result first.

Proposition 2.3.2 (Disjoint representation for $\mathfrak{R}$-step functions, the termwise version). *Let* $\mathfrak{R}$ *be a set-ring on* X, *and let* f *be an* $\mathfrak{R}$*-step function on* X. *If*

$$f = \sum_{\iota \in I} \alpha_\iota e_{A_\iota}$$

is an arbitrary $\mathfrak{R}$-representation of f, then there exist a disjoint $\mathfrak{R}$-representation

$$f = \sum_{\kappa \in K} \beta_\kappa e_{B_\kappa}$$

and an associated family $(K_\iota)_{\iota \in I}$ of subsets of K such that, if

$$\gamma_{\iota,\kappa} := e^K_{K_\iota}(\kappa)$$

then, for every ι in I and every κ in K,

$$e_{A_\iota} = \sum_{\kappa \in K_\iota} e_{B_\kappa}$$

$$\beta_\kappa = \sum_{\iota \in I} \alpha_\iota \gamma_{\iota,\kappa}.$$

Proof. The proof is a consequence of disjoint decomposition for set-rings. According to that proposition (2.1.6) there exist a finite disjoint family $(B_\kappa)_{\kappa \in K}$ of nonempty sets in $\mathfrak{R}$ and an associated family $(K_\iota)_{\iota \in I}$ of subsets of K such that

$$A_\iota = \bigcup_{\kappa \in K_\iota} B_\kappa$$

for every ι in I. Since the B_κ are disjoint, we have

$$e_{A_\iota} = \sum_{\kappa \in K_\iota} e_{B_\kappa} = \sum_{\kappa \in K} \gamma_{\iota,\kappa} e_{B_\kappa}.$$

Summation yields

$$f = \sum_{\iota \in I} \alpha_\iota \sum_{\kappa \in K} \gamma_{\iota,\kappa} e_{B_\kappa} = \sum_{\kappa \in K} \left(\sum_{\iota \in I} \alpha_\iota \gamma_{\iota,\kappa} \right) e_{B_\kappa}$$

which is the required representation. □

Proposition 2.3.3. *The following assertions hold for every set-ring $\mathfrak{R}$ on X and every $\mathfrak{R}$-step function f on X.*

(*a*) *There exists a disjoint $\mathfrak{R}$-representation for f.*

(*b*) *If $f \geq 0$ and if*

$$f = \sum_{\kappa \in K} \beta_\kappa e_{B_\kappa} \tag{3}$$

is an arbitrary disjoint $\mathfrak{R}$-representation of f, then $\beta_\kappa \geq 0$ for every κ in K.

(*c*) *If* $f = 0$ *and if* (3) *is a disjoint* $\mathfrak{R}$*-representation of* f, *then* $\beta_\kappa = 0$ *for every* κ *in* K.

Proof. (a). This assertion has already been established.

(b) Given $\kappa \in K$, choose an element x of the nonempty set B_κ, and use (3) to compute $f(x)$:

$$f(x) = \sum_{\kappa \in K} \beta_\kappa e_{B_\kappa}(x) = \beta_\kappa.$$

Since $f(x) \geq 0$, $\beta_\kappa \geq 0$.

(c) For $\kappa \in K$ and $x \in B_\kappa$, we have

$$0 = f(x) = \beta_\kappa.$$

□

The fact that $\mathfrak{R}$-step functions always have disjoint representations if $\mathfrak{R}$ is a set-ring, simplifies many considerations, as the proofs of the next two results illustrate. Corollary 4 will be used extensively in Chapter 5.

Corollary 2.3.4. *The following assertions hold for every set-ring* $\mathfrak{R}$ *on* X *and every* $\mathfrak{R}$*-step function* f *on* X.

(*a*) $\{f > \alpha\} \in \mathfrak{R}$ *for every real number* $\alpha \geq 0$.

(*b*) $\{f \geq \alpha\} \in \mathfrak{R}$ *for every real number* $\alpha > 0$.

Proof. (a) Let

$$f = \sum_{\kappa \in K} \beta_\kappa e_{B_\kappa} \tag{4}$$

be a disjoint $\mathfrak{R}$-representation for f. If $\alpha \geq 0$, then the representation

$$\{f > \alpha\} = \cup \{B_\kappa \mid \kappa \in K, \beta_\kappa > \alpha\}$$

shows that the set $\{f > \alpha\}$ belongs to $\mathfrak{R}$.

(b) Let (4) represent f disjointly. If $\alpha > 0$, then

$$\{f \geq \alpha\} = \bigcup \{B_\kappa \mid \kappa \in K, \beta_\kappa \geq \alpha\}$$

which shows that $\{f \geq \alpha\}$ belongs to $\mathfrak{R}$. □

Theorem 2.3.5. *The following assertions hold for every set-ring* $\mathfrak{R}$ *on* X.

(*a*) *The set of all* $\mathfrak{R}$*-step functions on* X *is a real Riesz lattice.*

(*b*) *If* f *is an* $\mathfrak{R}$*-step function on* X, *then so is the function* $f \wedge e_X$.

Proof. (a) We use the characterization of real Riesz lattices given in Proposition 1.2.16. That every $\mathfrak{R}$-step function is real is inherent in the definition. The sum of two $\mathfrak{R}$-step functions on X is obviously an $\mathfrak{R}$-step function, as is every real scalar multiple of an $\mathfrak{R}$-step function. So let f be an $\mathfrak{R}$-step function on X. We must show that the same is true for $|f|$. But if

$$f = \sum_{\kappa \in K} \beta_\kappa e_{B_\kappa}$$

is a disjoint representation for f, then

$$|f| = \sum_{\kappa \in K} |\beta_\kappa| e_{B_\kappa}$$

and $|f|$ is displayed as an $\mathfrak{R}$-step function on X.

(b) follows from (a). For if

$$f = \sum_{\iota \in I} \alpha_\iota e_{A_\iota}$$

is an arbitrary $\mathfrak{R}$-step function on X, then

$$f \wedge e_X = f \wedge e_{\bigcup_{\iota \in I} A_\iota}$$

which displays the function $f \wedge e_X$ as the infimum of two functions from the Riesz lattice of all $\mathfrak{R}$-step functions on X. □

Property (b) is called the Stone property, after M. H. Stone (1948), who discovered its significance.

Definition 2.3.6. *For $\mathscr{L}$ a Riesz lattice in $\overline{\mathbb{R}}^X$, we say that $\mathscr{L}$ has the* ***Stone property*** *iff $f \wedge e_X$ belongs to $\mathscr{L}$ for every $f \in \mathscr{L}$. A Riesz lattice in $\overline{\mathbb{R}}^X$ that has the Stone property is also called a* ***Stone lattice***. *A Daniell space $(X, \mathscr{L}, \ell)$ is said to have the* ***Stone property*** *iff its Riesz lattice $\mathscr{L}$ has the Stone property.* □

Thus for any set-ring $\mathfrak{R}$ on X, the set of $\mathfrak{R}$-step functions on X forms a Stone lattice. This correspondence between set-rings, which we can view as domains of measures, and Stone lattices, which we can view as domains of functionals, is reversible. We shall demonstrate that to every Stone lattice $\mathscr{L}$ whose elements are step functions on X, there corresponds a set-ring $\mathfrak{R}(\mathscr{L}) \subset \mathfrak{P}(X)$ such that $\mathscr{L}$ is exactly the collection of all $\mathfrak{R}(\mathscr{L})$-step functions on X. The set-ring in question consists of those subsets of X whose characteristic functions belong to $\mathscr{L}$.

Definition 2.3.7. *For $\mathscr{F} \subset \overline{\mathbb{R}}^X$*

$$\mathfrak{R}(\mathscr{F}) := \{A \subset X \mid e_A \in F\}.$$ □

Proposition 2.3.8. *The following assertions hold, for every Riesz lattice $\mathscr{L}$ in $\overline{\mathbb{R}}^X$.*

(*a*) *$\mathfrak{R}(\mathscr{L})$ is a ring of sets.*

(*b*) *If $\mathscr{L}$ is conditionally σ-completely embedded in $\overline{\mathbb{R}}^X$, then $\mathfrak{R}(\mathscr{L})$ is a δ-ring.*

(*c*) *If $\mathscr{L}$ is σ-completely embedded in $\overline{\mathbb{R}}^X$, then $\mathfrak{R}(\mathscr{L})$ is a σ-ring.*

Proof. (a) The empty set belongs to $\mathfrak{R}(\mathscr{L})$ since $\mathscr{L}$ contains the zero function. Suppose that A and B belong to $\mathfrak{R}(\mathscr{L})$. Then e_A and e_B both belong to $\mathscr{L}$. Since

$$e_{A \cup B} = e_A \vee e_B$$

$$e_{A \cap B} = e_A \wedge e_B$$

$$e_{A \setminus B} = e_A - e_{A \cap B}$$

it follows that $A \cup B$ and $A \setminus B$ both belong to $\mathfrak{R}(\mathscr{L})$. Thus $\mathfrak{R}(\mathscr{L})$ is in fact a ring of sets.

(b) Let $(A_n)_{n \in \mathbb{N}}$ be a sequence from $\mathfrak{R}(\mathscr{L})$. For $n \in \mathbb{N}$, set $B_n := \bigcap_{m \leq n} A_m$. Then $(e_{B_n})_{n \in \mathbb{N}}$ is a sequence from $\mathscr{L}$ that decreases and is bounded below in $\mathscr{L}$, so $\bigwedge_{n \in \mathbb{N}} e_{B_n}$ belongs to $\mathscr{L}$, by Proposition 1.2.18. Since

$$e_{\bigcap_{n \in \mathbb{N}} A_n} = \bigwedge_{n \in \mathbb{N}} e_{B_n}$$

the set $\bigcap_{n \in \mathbb{N}} A_n$ belongs to $\mathfrak{R}(\mathscr{L})$. In view of (a), it follows that $\mathfrak{R}(\mathscr{L})$ is a δ-ring.

(c) If $(A_n)_{n \in \mathbb{N}}$ is a sequence from $\mathfrak{R}(\mathscr{L})$, then $(e_{A_n})_{n \in \mathbb{N}}$ is a sequence from $\mathscr{L}$. By hypothesis $\bigvee_{n \in \mathbb{N}} e_{A_n}$ belongs to $\mathscr{L}$. In other words, $\bigcup_{n \in \mathbb{N}} A_n$ belongs to $\mathfrak{R}(\mathscr{L})$. In view of (b), $\mathfrak{R}(\mathscr{L})$ is a σ-ring. □

Our goal is to show that when $\mathscr{L}$ is a Stone lattice whose elements are step functions on X, then $\mathscr{L}$ is exactly the set of all $\mathfrak{R}(\mathscr{L})$-step functions on X. We require another representation property: (positive) step functions can be built up in successive layers.

Proposition 2.3.9 (Layered representation for positive step functions). *Let f be a positive step function on X such that $f \neq 0$. Then there exist a positive integer n, a family $(C_k)_{k \in \mathbb{N}_n}$ of subsets of X, and a family $(\gamma_k)_{k \in \mathbb{N}_n}$ of real numbers such that the following assertions hold.*

(*a*) $f = \sum_{k=1}^{n} \gamma_k e_{C_k}$.

(*b*) *C_k properly contains C_{k+1} for every $k \in \mathbb{N}_{n-1}$.*

(*c*) *$\gamma_k > 0$ for every $k \in \mathbb{N}_n$.*

Proof. By definition, there exist a finite family $(A_\iota)_{\iota \in I}$ of subsets of X and an associated family $(\alpha_\iota)_{\iota \in I}$ of real numbers such that

$$f = \sum_{\iota \in I} \alpha_\iota e_{A_\iota}.$$

As a subset of $\{\Sigma_{\iota \in J} \alpha_\iota \mid J \subset I\}$, the set $f(X)$ is finite. By hypothesis, $f(X) \setminus \{0\}$ is nonempty. Let

$$f(X) \setminus \{0\} = \{\delta_k \mid k \in \mathbb{N}_n\}$$

where we can and do assume that

$$\delta_k < \delta_{k+1}$$

for every k in $\mathbb{N}_{n-1}$. Define

$$C_k := \{f \geq \delta_k\}, \qquad k \in \mathbb{N}_n$$

$$\gamma_1 := \delta_1$$

$$\gamma_k := \delta_k - \delta_{k-1}, \qquad k \in \mathbb{N}_n, \quad k \neq 1.$$

The reader can readily verify that the families $(C_k)_{k \in \mathbb{N}_n}$ and $(\gamma_k)_{k \in \mathbb{N}_n}$ meet the required conditions. □

Theorem 2.3.10. *If $\mathscr{L}$ is a Stone lattice whose elements are step functions on X, then $\mathscr{L}$ is the set consisting of all $\Re(\mathscr{L})$-step functions on X.*

Proof. Since $\mathscr{L}$ is a Riesz lattice, it is obvious from the definition of $\Re(\mathscr{L})$ that every $\Re(\mathscr{L})$-step function on X belongs to $\mathscr{L}$. We must show that every element of $\mathscr{L}$ is an $\Re(\mathscr{L})$-step function. It suffices to consider only positive functions from $\mathscr{L}$, since every real function f has the representation $f^+ - f^-$. The zero function is trivially $\Re(\mathscr{L})$-step, so suppose that $f \in \mathscr{L}_+$ and $f \neq 0$. Let

$$f = \sum_{k=1}^{n} \gamma_k e_{C_k} \tag{5}$$

be the "layered" representation whose existence is asserted by the previous proposition. We prove inductively that each C_k belongs to $\Re(\mathscr{L})$. Since

$$e_{C_1} = \left(\frac{1}{\gamma_1} f\right) \wedge e_X$$

and $\mathscr{L}$ has the Stone property, e_{C_1} belongs to $\mathscr{L}$ and C_1 is in $\Re(\mathscr{L})$. Suppose then that C_k belongs to $\Re(\mathscr{L})$ for every k in $\mathbb{N}_m$, $m < n$. The key is that we can write

$$e_{C_{m+1}} = \left(\frac{1}{\gamma_{m+1}} g\right) \wedge e_X, \qquad g := \sum_{k=m+1}^{n} \gamma_k e_{C_k} = f - \sum_{k=1}^{m} \gamma_k e_{C_k}.$$

This representation for $e_{C_{m+1}}$ is possible because of the special properties of the layered representation (5). The induction hypothesis ensures that $\Sigma_{k=1}^{m} \gamma_k e_{C_k}$ belongs to $\mathscr{L}$. We conclude successively that g belongs to $\mathscr{L}$, so $e_{C_{m+1}}$ belongs to $\mathscr{L}$, so C_{m+1} belongs to $\Re(\mathscr{L})$. Thus C_k belongs to $\Re(\mathscr{L})$ for every k in $\mathbb{N}_n$. □

EXERCISES

2.3.1(E) Prove that the function f on the set X is a step function on X iff $f(X)$ is finite.

2.3.2(E) Let X be a set and $\Re \subset \mathfrak{P}(X)$ a ring of sets. Let f be a step function on X. Prove the equivalence of the following three statements:

(α1) f is an $\Re$-step function.
(α2) $\{f > \alpha\} \in \Re$ for any $\alpha > 0$.
(α3) $\{f \geq \alpha\} \in \Re$ for any $\alpha > 0$.

The implications one way were proved in Corollary 4. The exercise shows that (α2) and (α3) really are characteristic for $\Re$-step functions.

2.3.3(E) Prove that $\mathbb{R}^X$ consists solely of step functions iff X is finite.

2.3.4(E) Given $f \in \mathbb{R}^X$, show that the following statements are equivalent [recall that $\mathfrak{F}(X) := \{A \in \mathfrak{P}(X) \mid A \text{ finite}\}$]:

(α1) f is an $\mathfrak{F}(X)$-step function.
(α2) $\{f \neq 0\}$ is finite.
(α3) f is a continuous function with compact support with respect to the discrete topology.

2.3.5(E) Given a set X, define $\Re := \mathfrak{F}(X) \cup \{A \in \mathfrak{P}(X) \mid X \setminus A \in \mathfrak{F}(X)\}$. $\Re$ is an algebra on X. Show that

$$\mathscr{L}_{\Re} = \left\{\alpha e_X + g \mid \alpha \in \mathbb{R}, g \in \mathbb{R}^X, \{g \neq 0\} \in \mathfrak{F}(X)\right\}.$$

2.3.6(E) Let X be a set. Let $\Re \subset \mathfrak{P}(X)$ be a ring of sets. Prove the following statements:

(α) Let $f_1, f_2, \ldots, f_n$ be $\Re$-step functions on X. Then there exist a finite family $(A_\iota)_{\iota \in I}$ from $\Re$ and families $(\alpha_{k\iota})_{\iota \in I}$ such that

$$f_k = \sum_{\iota \in I} \alpha_{k\iota} e_{A_\iota}$$

for each $k \in \mathbb{N}_n$. $(A_\iota)_{\iota \in I}$ can be chosen disjoint.

(β) Take $n \in \mathbb{N}$. Let $f: \mathbb{R}^n \to \mathbb{R}$ be a function such that $f(0,0,\ldots,0) = 0$. Then, for each family, $(f_k)_{k \in \mathbb{N}_n}$, of $\mathfrak{R}$-step functions on X,

$$g: X \to \mathbb{R}, \qquad x \mapsto f(f_1(x), f_2(x), \ldots, f_n(x))$$

is also an $\mathfrak{R}$-step function on X.

As special cases prove the statements that follow:

(γ) The product of two $\mathfrak{R}$-step functions on X is an $\mathfrak{R}$-step function on X.

(δ) For each $n \in \mathbb{N}$, the nth power of an $\mathfrak{R}$-step function on X is an $\mathfrak{R}$-step function on X.

More generally, the next statement can be proved:

(ε) For each polynomial, p, in n variables without a free constant, and for each family $(f_k)_{k \in \mathbb{N}_n}$ of $\mathfrak{R}$-step functions on X,

$$X \to \mathbb{R}, \qquad x \mapsto p(f_1(x), f_2(x), \ldots, f_n(x))$$

is an $\mathfrak{R}$-step function on X.

2.3.7(E) Given a set X and a semi-ring $\mathfrak{S}$ of subsets of X, show that the $\mathfrak{S}$-step functions and the $\mathfrak{S}_r$-step functions on X coincide. [Use the characterization of $\mathfrak{S}_r$ for the semi-ring $\mathfrak{S}$ given in Ex. 2.1.2(ϑ).]

2.3.8(E) Take a set X and $\mathfrak{R} \subset \mathfrak{P}(X)$ such that if $A, B \in \mathfrak{R}$, then there is a finite disjoint set $\mathfrak{C} \subset \mathfrak{R}$ with $A \cap B = \bigcup_{C \in \mathfrak{C}} C$. Show that in this case the set of all $\mathfrak{R}$-step functions on X coincides with the set of all $\mathfrak{R}_r$-step functions on X.

It is trivial that $\mathscr{L}_{\mathfrak{R}} \subset \mathscr{L}_{\mathfrak{R}_r}$. The opposite inclusion can be obtained with the following steps:

STEP 1: If $\mathfrak{A} \subset \mathfrak{F}(\mathfrak{R})$, $\mathfrak{A} \neq \emptyset$, then $e_{\bigcap_{A \in \mathfrak{A}} A} \in \mathscr{L}_{\mathfrak{R}}$.

Use induction on the cardinality to establish the existence of a finite disjoint set $\mathfrak{C} \subset \mathfrak{R}$ with $\bigcap_{A \in \mathfrak{A}} A = \bigcup_{C \in \mathfrak{C}} C$. Then

$$e_{\bigcap_{A \in \mathfrak{A}} A} = \sum_{C \in \mathfrak{C}} e_C \in \mathscr{L}_{\mathfrak{R}}.$$

STEP 2: Let $\mathfrak{S}$ denote the set of all finite unions of elements of $\mathfrak{R}$. If $A \in \mathfrak{S}$, then $e_A \in \mathscr{L}_{\mathfrak{R}}$.

If $A = \emptyset$, the proposition is trivial. If $A \neq \emptyset$, there is a family $(A_k)_{k \in \mathbb{N}_n}$ in $\mathfrak{R}$ with $A = \bigcup_{k \in \mathbb{N}_n} A_k$. Then

$$e_A = \sum_{k=1}^{n} \left((-1)^{k+1} \sum_{\substack{K \subset \mathbb{N}_n \\ |K| = k}} e_{\bigcap_{k \in K} A_k} \right)$$

where $|K|$ denotes the cardinality of K.

STEP 3: Define $\mathfrak{S}' := \{A \setminus B \mid A, B \in \mathfrak{S}\}$. Then $e_C \in \mathscr{L}_{\mathfrak{R}}$ whenever $C \in \mathfrak{S}'$.

If $A, B \in \mathfrak{S}$, then $e_{A \setminus B} = e_A - e_{A \cap B}$.

STEP 4: Let $\mathfrak{S}''$ be the set of all finite disjoint unions of elements of $\mathfrak{S}'$. Then $e_A \in \mathscr{L}_{\mathfrak{R}}$ whenever $A \in \mathfrak{S}''$.

STEP 5: $\mathscr{L}_{\mathfrak{R}_r} \subset \mathscr{L}_{\mathfrak{R}}$.
Observe that by Ex. 2.1.2(η), $\mathfrak{S}'' = \mathfrak{R}_r$.

Notice that the step functions with respect to a $\cap$-semi-lattice provide an example for the preceding discussion.

2.3.9(E) Let X, Y be sets. Let $\mathfrak{R} \subset \mathfrak{P}(Y)$ be a ring of sets. Let $\varphi: X \to Y$ be a mapping, and define $\varphi^{-1}(\mathfrak{R}) := \{\varphi^{-1}(A) \mid A \in \mathfrak{R}\}$ [cf., Ex. 2.1.12]. Prove the following statements:

(α) If f is an $\mathfrak{R}$-step function on Y, then $f \circ \varphi$ is a $\varphi^{-1}(\mathfrak{R})$-step function on X.

Define $\omega: \mathscr{L}_{\mathfrak{R}} \to \mathscr{L}_{\varphi^{-1}(\mathfrak{R})}$, $f \mapsto f \circ \varphi$.

(β) $\omega(f+g) = \omega(f) + \omega(g)$, $\omega(f \vee g) = \omega(f) \vee \omega(g)$ and $\omega(f \wedge g) = \omega(f) \wedge \omega(g)$ whenever $f, g \in \mathscr{L}_{\mathfrak{R}}$.

(γ) $\omega(\alpha f) = \alpha\omega(f)$ whenever $f \in \mathscr{L}_{\mathfrak{R}}$ and $\alpha \in \mathbb{R}$.

(β) and (γ) show that ω is a homomorphism of Riesz lattices.

(δ) ω maps $\mathscr{L}_{\mathfrak{R}}$ onto $\mathscr{L}_{\varphi^{-1}(\mathfrak{R})}$.

(ε) If $A \cap \varphi(X) \neq \varnothing$ for any $A \in \mathfrak{R} \setminus \{\varnothing\}$, then ω is also injective. (In this case ω is an isomorphism of Riesz-lattices.)

2.3.10(E) Given sets X, Y, a ring of sets $\mathfrak{R} \subset \mathfrak{P}(X)$, and a mapping $\varphi: X \to Y$, define $\varphi(\mathfrak{R}) := \{A \subset Y \mid \varphi^{-1}(A) \in \mathfrak{R}\}$ [cf., Ex. 2.1.12] and $\omega: \mathscr{L}_{\varphi(\mathfrak{R})} \to \mathbb{R}^X$, $f \mapsto f \circ \varphi$. Prove the propositions that follow:

(α) If f is a $\varphi(\mathfrak{R})$-step function on Y, then $f \circ \varphi$ is an $\mathfrak{R}$-step function on X.

(β) ω is a homomorphism of Riesz lattices (cf., Ex. 2.3.9).

(γ) Assume that $\varphi(x) = \varphi(y)$ implies $x, y \in A$ or $x, y \notin A$ whenever $x, y \in X$ and $A \in \mathfrak{R}$. Then $\omega: \mathscr{L}_{\varphi(\mathfrak{R})} \to \mathscr{L}_{\mathfrak{R}}$ is surjective.

(δ) If the conditions of (γ) are fulfilled and φ is surjective, then $\omega: \mathscr{L}_{\varphi(\mathfrak{R})} \to \mathscr{L}_{\mathfrak{R}}$ is an isomorphism of Riesz lattices.

2.3.11(E) Let X be a set and $(\mathfrak{R}_\iota)_{\iota \in I}$ a nonempty family of rings of subsets of X. Prove the following propositions:

(α) $\mathscr{L}_{\bigcap_{\iota \in I} \mathfrak{R}_\iota} = \bigcap_{\iota \in I} \mathscr{L}_{\mathfrak{R}_\iota}$.

(β) If $\mathfrak{R}$ is the ring of sets generated by $\bigcup_{\iota \in I} \mathfrak{R}_\iota$ and if $\mathscr{L}$ is the Riesz lattice generated by $\bigcup_{\iota \in I} \mathscr{L}_{\mathfrak{R}_\iota}$ (cf., Ex. 1.2.8) then $\mathscr{L}_{\mathfrak{R}} = \mathscr{L}$.

(γ) If $(\mathfrak{R}_\iota)_{\iota \in I}$ is directed upward, then $\bigcup_{\iota \in I} \mathfrak{R}_\iota$ is a ring of sets, and $\mathscr{L}_{\bigcup_{\iota \in I} \mathfrak{R}_\iota} = \bigcup_{\iota \in I} \mathscr{L}_{\mathfrak{R}_\iota}$.

We wish to reformulate these propositions. To this end, denote by $R(X)$ the set of all rings $\mathfrak{R}$ of sets with $\mathfrak{R} \subset \mathfrak{P}(X)$ and by $T(X)$ the set of all Stone lattices consisting of step functions on X.

Verify the following statements:

(δ) $R(X)$ is a complete lattice with respect to $\subset$.

(ε) $T(X)$ is a complete lattice with respect to $\subset$.

(ζ) Define $\omega: R(X) \to T(X)$, $\mathfrak{R} \mapsto \mathscr{L}_{\mathfrak{R}}$. Then ω is an isomorphism of lattices. In other words, ω is a bijection such that $\omega(\mathfrak{R} \vee \mathfrak{S}) = \omega(\mathfrak{R}) \vee \omega(\mathfrak{S})$ and $\omega(\mathfrak{R} \wedge \mathfrak{S}) = \omega(\mathfrak{R}) \wedge \omega(\mathfrak{S})$ whenever $\mathfrak{R}, \mathfrak{S} \in R(X)$. Show that for each family $(\mathfrak{R}_\iota)_{\iota \in I}$ from $R(X)$,

$$\omega\left(\bigvee_{\iota \in I} \mathfrak{R}_\iota\right) = \bigvee_{\iota \in I} \omega(\mathfrak{R}_\iota)$$

and

$$\omega\left(\bigwedge_{\iota \in I} \mathfrak{R}_\iota\right) = \bigwedge_{\iota \in I} \omega(\mathfrak{R}_\iota).$$

(These equalities hold for each isomorphism between lattices X and Y).

2.3.12(E) Take an open interval in $\mathbb{R}$, A. Denote by $\mathfrak{R}$ the ring of sets consisting of the interval forms on A. Denote by $\mathscr{K}(A)$ the Riesz lattice of all continuous functions on A with compact support. Verify the statements that follow:

(α) Given $f \in \mathscr{K}(A)$ and $\varepsilon \in \mathbb{R}$, $\varepsilon > 0$, there are step functions $g, h \in \mathscr{L}_{\mathfrak{R}}$ with $g \le f \le h$ and $\sup_{x \in A}(h(x) - g(x)) < \varepsilon$.

(β) For each $f \in \mathscr{K}(A)$ there exist an increasing sequence $(g_n)_{n \in \mathbb{N}}$ and a decreasing sequence $(f_n)_{n \in \mathbb{N}}$, both from $\mathscr{L}_{\mathfrak{R}}$ and both converging uniformly to f.

(γ) If $f \in \mathscr{L}_{\mathfrak{R}}$, then there are $x, y \in A$, $\gamma \in \mathbb{R}_+$ and a sequence $(f_n)_{n \in \mathbb{N}}$ from $\mathscr{K}(A)$ such that $f = \lim_{n \to \infty} f_n$ and that $|f_n| \le \gamma e_{[x,y[}$ for each $n \in \mathbb{N}$.

(δ) The convergence in (γ) is, in general, neither monotone nor uniform. (Use, for example, the characteristic function of a right half-open interval in A.)

2.3.13(E) Take a set X and a δ-ring $\mathfrak{R} \subset \mathfrak{P}(X)$. Define

$$\mathscr{L}' := \left\{ f \in \mathbb{R}^X \mid \exists (f_n)_{n \in \mathbb{N}} \text{ from } \mathscr{L}_{\mathfrak{R}},\ f = \lim_{n \to \infty} f_n \right\}.$$

Then $\mathscr{L}'$ is a Riesz lattice. Prove the proposition

(α) $\mathscr{L}' = \{f \in \mathbb{R}^X | \{f \leq -\alpha\}$ and $\{f \geq \alpha\} \in \mathfrak{R}_\sigma$ whenever $\alpha > 0\}$.
For the proof, take $f \in \mathscr{L}'$ and $(f_n)_{n \in \mathbb{N}}$ from $\mathscr{L}_{\mathfrak{R}}$ with $f = \lim_{n \to \infty} f_n$. Then

$$\{f \leq -\alpha\} = \bigcap_{\substack{k \in \mathbb{N} \\ 1 < \alpha k}} \bigcup_{n \in \mathbb{N}} \bigcap_{m \geq n} \{f_m < -\alpha + 1/k\} \in \mathfrak{R}_\sigma$$

$$\{f \geq \alpha\} = \bigcap_{\substack{k \in \mathbb{N} \\ 1 < \alpha k}} \bigcup_{n \in \mathbb{N}} \bigcap_{m \geq n} \{f_m > \alpha - 1/k\} \in \mathfrak{R}_\sigma$$

whenever $\alpha > 0$.

For the reverse containment, take $f \in \mathbb{R}^X$ such that $\{f \leq -\alpha\} \in \mathfrak{R}_\sigma$ and $\{f \geq \alpha\} \in \mathfrak{R}_\sigma$ for each $\alpha > 0$. Then

$$f^+ = \lim_{n \to \infty} \sum_{k=1}^{n2^n} \frac{1}{2^n} e_{\{f \geq k/2^n\}}, \qquad f^- = \lim_{n \to \infty} \sum_{k=1}^{n2^n} \frac{1}{2^n} e_{\{f \leq -k/2^n\}}.$$

By Proposition 2.1.11 each $A \in \mathfrak{R}_\sigma$ is the union of an increasing sequence $(A_n)_{n \in \mathbb{N}}$ from $\mathfrak{R}$. Thus for each $n \in \mathbb{N}$ and each $k \in \mathbb{N}_{n2^n}$ there is an increasing sequence $(A_{nkl})_{l \in \mathbb{N}}$ from $\mathfrak{R}$ with $\{f \geq k/2^n\} = \bigcup_{l \in \mathbb{N}} A_{nkl}$. Then

$$f^+ = \lim_{n \to \infty} \sum_{k=1}^{n2^n} \frac{1}{2^n} e_{A_{nkn}} \in \mathscr{L}'$$

and similarly $f^- \in \mathscr{L}'$.
Now define

$$\mathscr{L}'' := \left\{ f \in \mathbb{R}^X | \exists (f_n)_{n \in \mathbb{N}} \text{ from } \mathscr{L}_{\mathfrak{R}},\ f = \lim_{n \to \infty} f_n \text{ uniformly on } X \right\}.$$

$\mathscr{L}''$ is also a Riesz lattice. Verify the statements that follow:

(β) $\mathscr{L}''$ is equal to the set of all functions $f \in \mathbb{R}^X$ with the property that for each $\varepsilon > 0$ there is an $\mathfrak{A} \in \mathfrak{F}(\mathfrak{R})$ with $|f(x) - f(y)| < \varepsilon$ whenever $x, y \in \bigcup_{A \in \mathfrak{A}} A$ and $|f(z)| < \varepsilon$ whenever $z \in X \setminus \bigcup_{A \in \mathfrak{A}} A$.

(γ) Every $f \in \mathscr{L}''$ is bounded (i.e., for every $f \in \mathscr{L}''$ there is a $\gamma \in \mathbb{R}$ with $|f| \leq \gamma$).

2.4. DANIELL SPACES ASSOCIATED WITH MEASURE SPACES, AND VICE VERSA

> NOTATION FOR SECTION 2.4:
>
> X denotes a set.

With each positive measure space are naturally associated a positive linear functional and a corresponding Daniell space, and the details required to prove this correspondence are now easily checked. These details comprise the first part of the section, which culminates in Theorem 3 and Definition 4. In the second part we verify that the correspondence can be reversed, provided certain obvious restrictions are met. The section concludes with a simple, yet important, example.

Suppose we are given a positive measure space $(X, \mathfrak{R}, \mu)$. With the set-ring $\mathfrak{R}$ we associate the Riesz lattice $\mathscr{L}_{\mathfrak{R}}^X$ consisting of all $\mathfrak{R}$-step functions on X. We want to define on $\mathscr{L}_{\mathfrak{R}}^X$ a functional ℓ_μ that is naturally associated with the positive measure μ. "Naturally associated" means that

$$\ell_\mu(e_A) = \mu(A)$$

for every A in $\mathfrak{R}$. If

$$f = \sum_{\iota \in I} \alpha_\iota e_{A_\iota}$$

is an $\mathfrak{R}$-step function on X, linearity requires that

$$\ell_\mu(f) = \sum_{\iota \in I} \alpha_\iota \mu(A_\iota). \tag{1}$$

But $\mathfrak{R}$-step functions have many representations. If (1) is to be used as a definition for ℓ_μ, we need to know that the value of $\ell_\mu(f)$ does not depend on the particular representation used for f. This independence is a direct consequence of disjoint representation, as the proof of the next proposition shows.

Proposition 2.4.1. *Let $(X, \mathfrak{R}, \mu)$ be a positive measure space, and let $(A_\iota)_{\iota \in I}$ and $(\alpha_\iota)_{\iota \in I}$ be finite families from $\mathfrak{R}$ and $\mathbb{R}$, respectively, such that*

$$\sum_{\iota \in I} \alpha_\iota e_{A_\iota} = 0.$$

Then

$$\sum_{\iota \in I} \alpha_\iota \mu(A_\iota) = 0.$$

Proof. Because of termwise disjoint representation [Propositions 2.3.2 and 2.3.3(c)] there exist a finite disjoint family $(B_\kappa)_{\kappa \in K}$ of nonempty sets in $\mathfrak{R}$ and an associated family $(K_\iota)_{\iota \in I}$ of subsets of K such that, for every ι in I

$$A_\iota = \bigcup_{\kappa \in K_\iota} B_\kappa \quad \text{and} \quad \mu(A_\iota) = \sum_{\kappa \in K_\iota} \mu(B_\kappa)$$

and such that

$$0 = \sum_{\iota \in I} \alpha_\iota e_{A_\iota} = \sum_{\kappa \in K} \beta_\kappa e_{B_\kappa}$$

where

$$\beta_\kappa = \sum_{\iota \in I} \alpha_\iota e_{K_\iota}^K(\kappa) = 0$$

for every κ in K. Thus

$$\begin{aligned}
\sum_{\iota \in I} \alpha_\iota \mu(A_\iota) &= \sum_{\iota \in I} \alpha_\iota \sum_{\kappa \in K_\iota} \mu(B_\kappa) \\
&= \sum_{\iota \in I} \alpha_\iota \Big(\sum_{\kappa \in K} e_{K_\iota}^K(\kappa) \mu(B_\kappa) \Big) \\
&= \sum_{\kappa \in K} \Big(\sum_{\iota \in I} \alpha_\iota e_{K_\iota}^K(\kappa) \Big) \mu(B_\kappa) \\
&= \sum_{\kappa \in K} \beta_\kappa \mu(B_\kappa) \\
&= 0.
\end{aligned}$$

□

Corollary 2.4.2. *Let $(X, \mathfrak{R}, \mu)$ be a positive measure space. Suppose that $(A_\iota)_{\iota \in I}$ and $(B_\kappa)_{\kappa \in K}$ are finite families from $\mathfrak{R}$, and that $(\alpha_\iota)_{\iota \in I}$ and $(\beta_\kappa)_{\kappa \in K}$ are families from $\mathbb{R}$, such that*

$$\sum_{\iota \in I} \alpha_\iota e_{A_\iota} = \sum_{\kappa \in K} \beta_\kappa e_{B_\kappa}. \tag{2}$$

Then

$$\sum_{\iota \in I} \alpha_\iota \mu(A_\iota) = \sum_{\kappa \in K} \beta_\kappa \mu(B_\kappa). \tag{3}$$

Proof. Equations (2) and (3) can be rewritten as

$$\sum_{\rho \in P} \gamma_\rho e_{C_\rho} = 0 \tag{2'}$$

$$\sum_{\rho \in P} \gamma_\rho \mu(C_\rho) = 0 \tag{3'}$$

if we set

$$P := \{(1, \iota) \mid \iota \in I\} \cup \{(2, \kappa) \mid \kappa \in K\}$$

$$\gamma_\rho := \begin{cases} \alpha_\iota & \text{for } \rho = (1, \iota),\ \iota \in I \\ -\beta_\kappa & \text{for } \rho = (2, \kappa),\ \kappa \in K \end{cases}$$

$$C_\rho := \begin{cases} A_\iota & \text{for } \rho = (1, \iota),\ \iota \in I \\ B_\kappa & \text{for } \rho = (2, \kappa),\ \kappa \in K. \end{cases}$$

Proposition 1 shows that (2′) implies (3′). □

Theorem 2.4.3. *Given a positive measure space $(X, \mathfrak{R}, \mu)$, denote by $\mathscr{L}_{\mathfrak{R}}^X$ the Stone lattice consisting of all $\mathfrak{R}$-step functions on X. Then there is exactly one functional*

$$\ell_\mu^X : \mathscr{L}_{\mathfrak{R}}^X \to \mathbb{R}$$

for which $(X, \mathscr{L}_{\mathfrak{R}}^X, \ell_\mu^X)$ is a Daniell space and

$$\ell_\mu^X(e_A) = \mu(A) \tag{4}$$

for every $A \in \mathfrak{R}$.

Proof. Define

$$\ell_\mu^X : \mathscr{L}_{\mathfrak{R}}^X \to \mathbb{R}, \qquad f = \sum_{\iota \in I} \alpha_\iota e_{A_\iota} \mapsto \sum_{\iota \in I} \alpha_\iota \mu(A_\iota). \tag{5}$$

By Corollary 2, $\ell_\mu^X(f)$ does not depend on the representation used for f. The functional ℓ_μ^X is evidently linear and satisfies (4). Moreover the positivity of ℓ_μ^X follows from that of μ. For every positive $\mathfrak{R}$-step function has a disjoint representation in which only positive numbers appear as coefficients (Proposition 2.3.3).

It remains to be verified that ℓ_μ^X is nullcontinuous. Let $(f_n)_{n\in\mathbb{N}}$ be a sequence from $\mathscr{L}_{\mathfrak{R}}^X$ that decreases and satisfies

$$\bigwedge_{n\in\mathbb{N}} f_n = 0.$$

The null continuity will follow if we can show that

$$\inf_{n\in\mathbb{N}} \ell_\mu^X(f_n) = 0 \tag{6}$$

(Proposition 1.3.6(b) ⇒ (a)).

Let $\varepsilon > 0$ be given. For each n in $\mathbb{N}$, define

$$A_n := \{f_n > 0\}$$

$$A_{n,\varepsilon} := \{f_n > \varepsilon\}$$

$$\gamma_n := \sup_{x\in X} f_n(x).$$

For every n, both A_n and $A_{n,\varepsilon}$ belong to $\mathfrak{R}$ (Corollary 2.3.4), and γ_n is a real number. The hypotheses on the sequence $(f_n)_{n\in\mathbb{N}}$ imply that the sequences $(A_n)_{n\in\mathbb{N}}$, $(A_{n,\varepsilon})_{n\in\mathbb{N}}$, and $(\gamma_n)_{n\in\mathbb{N}}$ all decrease, and that

$$\bigcap_{n\in\mathbb{N}} A_{n,\varepsilon} = \varnothing.$$

The null continuity of μ yields

$$\inf_{n\in\mathbb{N}} \mu(A_{n,\varepsilon}) = 0$$

which can be used to obtain an estimate for $\inf_{n\in\mathbb{N}} \ell_\mu^X(f_n)$. We have, for every n,

$$\begin{aligned} f_n &\le \varepsilon e_{A_n} + \gamma_n e_{A_{n,\varepsilon}} \\ &\le \varepsilon e_{A_1} + \gamma_1 e_{A_{n,\varepsilon}} \end{aligned}$$

and so

$$\ell_\mu^X(f_n) \le \varepsilon\mu(A_1) + \gamma_1\mu(A_{n,\varepsilon}).$$

Therefore

$$\begin{aligned} \inf_{n\in\mathbb{N}} \ell_\mu^X(f_n) &\le \inf_{n\in\mathbb{N}} \left[\varepsilon\mu(A_1) + \gamma_1\mu(A_{n,\varepsilon})\right] \\ &= \varepsilon\mu(A_1). \end{aligned}$$

Since ε was arbitrary, (6) follows, and ℓ_μ^X is nullcontinuous.

The uniqueness is trivial. Every linear functional on $\mathscr{L}_{\mathfrak{R}}^X$ satisfying (4) must also satisfy (5). □

Definition 2.4.4. *For $(X, \mathfrak{R}, \mu)$ a positive measure space, the functional ℓ_μ^X and the Daniell space $(X, \mathscr{L}_{\mathfrak{R}}^X, \ell_\mu^X)$ described in Theorem 3 are called, respectively, the* ***functional associated*** *with the measure μ and the* ***Daniell space associated*** *with the measure space $(X, \mathfrak{R}, \mu)$.* □

Now consider the opposite situation. Given a Daniell space $(X, \mathscr{L}, \ell)$, it was noted in Section 2.3 that the set

$$\mathfrak{R}(\mathscr{L}) = \{A \subset X \mid e_A \in \mathscr{L}\}$$

is a ring of sets. The functional ℓ induces a positive measure on $\mathfrak{R}(\mathscr{L})$, as we show next.

Proposition 2.4.5. *Let $(X, \mathscr{L}, \ell)$ be a Daniell space, and define*

$$\mu^\ell : \mathfrak{R}(\mathscr{L}) \to \mathbb{R}, \qquad A \mapsto \ell(e_A).$$

Then $(X, \mathfrak{R}(\mathscr{L}), \mu^\ell)$ is a positive measure space.

Proof. By Proposition 2.3.8, $\mathfrak{R}(\mathscr{L})$ is a ring of sets. The positivity of ℓ ensures that μ^ℓ is a positive mapping. If A and B are disjoint sets belonging to $\mathfrak{R}(\mathscr{L})$, then

$$\begin{aligned} \mu^\ell(A \cup B) &= \ell(e_{A \cup B}) \\ &= \ell(e_A + e_B) \\ &= \ell(e_A) + \ell(e_B) \\ &= \mu^\ell(A) + \mu^\ell(B). \end{aligned}$$

Thus the additivity of μ^ℓ follows from that of ℓ. Finally, null continuity for μ^ℓ is also a consequence of the corresponding property for ℓ. For let $(A_n)_{n \in \mathbb{N}}$ be a sequence from $\mathfrak{R}(\mathscr{L})$ that decreases and satisfies

$$\bigcap_{n \in \mathbb{N}} A_n = \varnothing .$$

Then $(e_{A_n})_{n \in \mathbb{N}}$ is a sequence from $\mathscr{L}$ that decreases and satisfies

$$\bigwedge_{n \in \mathbb{N}} e_{A_n} = 0$$

so

$$\inf_{n \in \mathbb{N}} \mu^{\ell}(A_n) = \inf_{n \in \mathbb{N}} \ell(e_{A_n}) = 0.$$

Thus μ^{ℓ} is a positive, additive, nullcontinuous mapping on the ring of sets $\mathfrak{R}(\mathscr{L})$; that is, $(X, \mathfrak{R}(\mathscr{L}), \mu^{\ell})$ is a positive measure space. □

Definition 2.4.6. *For $\mathscr{F} \subset \overline{\mathbb{R}}^X$ and $\ell : \mathscr{F} \to \mathbb{R}$, the mapping*

$$\mu : \mathfrak{R}(\mathscr{F}) \to \mathbb{R}, \qquad A \mapsto \ell(e_A)$$

is called the ***set mapping induced*** *[on $\mathfrak{R}(\mathscr{F})$] by ℓ. If $(X, \mathscr{L}, \ell)$ is a Daniell space and μ is the set mapping induced on $\mathfrak{R}(\mathscr{L})$ by ℓ then we call $(X, \mathfrak{R}(\mathscr{L}), \mu)$ the* ***positive measure space induced*** *by $(X, \mathscr{L}, \ell)$ and μ the* ***positive measure induced*** *by ℓ.* □

The Daniell spaces associated with positive measure spaces always have the Stone property and their Riesz lattices always consist of step functions. Thus there is no hope of recovering a Daniell space from its induced positive measure space unless these two conditions hold. No other conditions are required, however, as we see from the following theorem. [An example in Chapter 5 shows how badly this reversibility can fail when the Stone property is lacking (Example 5.1.1).]

Theorem 2.4.7. *Suppose that $(X, \mathscr{L}, \ell)$ is a Daniell space and $\mathscr{L}$ is a Stone lattice whose elements are step functions. Then there exists a uniquely determined positive measure space whose associated Daniell space is $(X, \mathscr{L}, \ell)$, namely, the positive measure space induced by $(X, \mathscr{L}, \ell)$.*

Proof. Denote by μ^{ℓ} the set mapping induced on $\mathfrak{R}(\mathscr{L})$ by ℓ, so that $(X, \mathfrak{R}(\mathscr{L}), \mu^{\ell})$ is the positive measure space induced by $(X, \mathscr{L}, \ell)$. According to Theorem 2.3.10, $\mathfrak{R}(\mathscr{L})$ is the uniquely determined set-ring $\mathfrak{R}$ for which $\mathscr{L}$ is the set of $\mathfrak{R}$-step functions. By Proposition 5, $(X, \mathfrak{R}(\mathscr{L}), \mu^{\ell})$ is a positive measure space. The definition $\mu^{\ell}(A) = \ell(e_A)$ shows that $(X, \mathfrak{R}(\mathscr{L}), \mu^{\ell})$ is a positive measure space with which the Daniell space $(X, \mathscr{L}, \ell)$ is associated. □

Now that the connection between measures and nullcontinuous positive linear functionals has been spelled out, it becomes clear how measures can be extended. The extensions are obtained with no extra effort by extending the corresponding functionals. As a result we shall obtain measure spaces that have closure properties completely analogous to those for closed Daniell spaces.

The opposite route would also be possible. One could first construct closed measure spaces and then develop the notion of an integral with respect to such a measure. This path is followed in many books but seems to us always to be more complicated than the alternative chosen here.

We conclude this section with a simple example, namely, counting measures and the Daniell spaces associated with them. Although the example itself is rather trivial, its role in the theory is an important one. The counting measure is the starting point in Section 3.4, where we develop the theory of summable families. A related example is studied near the end of Section 4.2.

Example 2.4.8 (Counting measure). Let

$$\mathfrak{F}(X) = \{A \subset X \mid A \text{ is finite}\}$$

$$\mathscr{F}(X) := \{f \in \mathbb{R}^X \mid f \text{ is an } \mathfrak{F}(X)\text{-step function on } X\}.$$

Define

$$\chi_X : \mathfrak{F}(X) \to \mathbb{R}_+, \qquad A \mapsto \sum_{x \in A} 1.$$

It is easy to check that $\mathfrak{F}(X)$ is a δ-ring and χ_X is a positive measure on $\mathfrak{F}(X)$ which assigns to each finite subset of X the number of elements in that subset. According to Theorem 3, there exists a unique positive linear functional ℓ_{χ_X} defined on $\mathscr{F}(X)$ such that

$$\ell_{\chi_X}(e_A) = \chi_X(A)$$

for every $A \in \mathfrak{F}(X)$. The triple $(X, \mathscr{F}(X), \ell_{\chi_X})$ is the Daniell space associated with the positive measure space $(X, \mathfrak{F}(X), \chi_X)$ (Definition 4). A real function f on X belongs to $\mathscr{F}(X)$ iff $\{f \neq 0\}$ is finite. Such a function has the representation

$$f = \sum_{x \in \{f \neq 0\}} f(x) e_{\{x\}}$$

which can be used to compute $\ell_{\chi_X}(f)$:

$$\begin{aligned} \ell_{\chi_X}(f) &= \sum_{x \in \{f \neq 0\}} f(x) \ell_{\chi_X}(e_{\{x\}}) \\ &= \sum_{x \in \{f \neq 0\}} f(x) \chi_X(\{x\}) \\ &= \sum_{x \in \{f \neq 0\}} f(x). \end{aligned}$$

In particular, when X is finite we have

$$\ell_{\chi_X}(f) = \sum_{x \in X} f(x).$$

The measure χ_X just described is called the **counting measure** on the set X and $(X, \mathscr{F}(X), \ell_{\chi_X})$ the **Daniell space associated with the counting measure** on X. □

EXERCISES

2.4.1(E) Let $\mathfrak{R}$ be a ring of sets and $x \in X(\mathfrak{R})$. Denote by δ_x the Dirac measure on $\mathfrak{R}$ concentrated in x (cf., Ex. 2.2.1). Prove:

(α) For every $f \in \mathscr{L}_{\mathfrak{R}}$, $\ell_{\delta_x}(f) = f(x)$.
(β) Set $\mathscr{L} := \{f \in \overline{\mathbb{R}}^{X(\mathfrak{R})} \mid f(x) \in \mathbb{R}\}$ and $\ell(f) := f(x)$ for every $f \in \mathscr{L}$. Then $(X(\mathfrak{R}), \mathscr{L}, \ell)$ is a closed extension of $(X(\mathfrak{R}), \mathscr{L}_{\mathfrak{R}}, \ell_{\delta_x})$.
(γ) $(X(\mathfrak{R}), \mathscr{L}, \ell)$ is a maximal Daniell space extending $(X(\mathfrak{R}), \mathscr{L}_{\mathfrak{R}}, \ell_{\delta_x})$.

2.4.2(E) Let X be a set. Take $g \in \mathbb{R}_+^X$ and χ_g, the positive measure on $\mathfrak{F}(X)$ induced by g (cf., Ex. 2.2.3). By Ex. 2.3.4, $\mathscr{L}_{\mathfrak{F}(X)} = \{f \in \mathbb{R}^X \mid \{f \neq 0\}$ finite$\}$. Prove the propositions that follow:

(α) $\ell_{\chi_g}(f) = \Sigma_{x \in \{f \neq 0\}} f(x) g(x)$ for any $f \in \mathscr{L}_{\mathfrak{F}(X)}$.
(β) $\ell_{\chi}(f) = \Sigma_{x \in \{f \neq 0\}} f(x)$ for any $f \in \mathscr{L}_{\mathfrak{F}(X)}$, where χ denotes the counting measure as in Ex. 2.2.1.

In other words, $\ell_{\chi_g} = \ell_g$ in the sense of Ex. 1.3.2.

2.4.3(E) We take up again the extension $\bar{\chi}_g$ of χ_g to $\mathfrak{R}_g$ discussed in Ex. 2.2.3. Prove the following propositions:

(α) If $f \in \mathscr{L}_{\mathfrak{R}_g}$ with $f \geq 0$, then

$$\ell_{\bar{\chi}_g}(f) = \sup\left\{ \sum_{x \in B} f(x) g(x) \mid B \in \mathfrak{F}(X) \right\}.$$

(β) $\ell_{\bar{\chi}_g}(f) = \Sigma_{x \in X} f(x) g(x)$ for $f \in \mathscr{L}_{\mathfrak{R}_g}$. (Compare with Ex. 1.3.1 and Ex. 1.3.2.)
(γ) $(X, \mathscr{L}_{\mathfrak{R}_g}, \ell_{\chi_g}) \preccurlyeq (X, \ell_g^1, \bar{\ell}_g)$ in the sense of Ex. 1.3.2.
(δ) Given $f \in \ell_g^1$, there is a sequence $(f_n)_{n \in \mathbb{N}}$ from $\mathscr{L}_{\mathfrak{R}_g}$ with $f = \lim_{n \to \infty} f_n$ and $\lim_{n \to \infty} \ell_g(|f - f_n|) = 0$. (This means that $f = \lim_{n \to \infty} f_n$ with respect to the pseudometric d_ℓ in the sense of Ex. 1.3.10.)

For the proof of (α), let $f = \Sigma_{\kappa \in K} \beta_\kappa e_{B_\kappa}$ be a disjoint representation of $f \in \mathscr{L}_{\mathfrak{R}_g}$. Then if $B \in \mathfrak{F}(X)$,

$$\sum_{x \in B} f(x)g(x) = \sum_{\kappa \in K} \beta_\kappa \Bigl(\sum_{x \in B \cap B_\kappa} g(x) \Bigr) \le \sum_{\kappa \in K} \beta_\kappa \overline{\chi}_g(B_\kappa) = \ell_{\overline{\chi}_g}(f).$$

Thus

$$\sup\Bigl\{ \sum_{x \in B} f(x)g(x) \mid B \in \mathfrak{F}(X) \Bigr\} \le \ell_{\overline{\chi}_g}(f).$$

To show the reverse inequality, choose, for each $\kappa \in K$, $A_\kappa \in \mathfrak{F}(B_\kappa)$ and set $A := \bigcup_{\kappa \in K} A_\kappa$. A is finite, and

$$\sum_{\kappa \in K} \beta_\kappa \Bigl(\sum_{x \in A_\kappa} g(x) \Bigr) \le \sup\Bigl\{ \sum_{x \in B} f(x)g(x) \mid B \in \mathfrak{F}(X) \Bigr\}.$$

Thus

$$\begin{aligned} \ell_{\overline{\chi}_g}(f) &= \sum_{\kappa \in K} \beta_\kappa \sup\Bigl\{ \sum_{x \in A_\kappa} g(x) \mid A_\kappa \in \mathfrak{F}(B_\kappa) \Bigr\} \\ &\le \sup\Bigl\{ \sum_{x \in B} f(x)g(x) \mid B \in \mathfrak{F}(X) \Bigr\}. \end{aligned}$$

2.4.4(E) Let X be an infinite set, and put $\mathfrak{R} := \{X \setminus A \mid A \in \mathfrak{F}(X)\} \cup \mathfrak{F}(X)$. By Ex. 2.3.5,

$$\mathscr{L}_{\mathfrak{R}} = \bigl\{ \alpha e_X + g \mid \alpha \in \mathbb{R}, g \in \mathbb{R}^X \text{ with } \{g \ne 0\} \text{ finite} \bigr\}.$$

Define

$$\mu : \mathfrak{R} \to \mathbb{R}, \qquad A \mapsto \begin{cases} 1 & \text{if } X \setminus A \in \mathfrak{F}(X) \\ 0 & \text{if } A \in \mathfrak{F}(X) \end{cases}.$$

For uncountable X, μ is a positive measure on $\mathfrak{R}$. Show that $\ell_\mu(\alpha e_X + g) = \alpha$ for any $\alpha e_X + g \in \mathscr{L}_{\mathfrak{R}}$.

2.4.5(E) **The Functionals for the Stieltjes Measures.** Take a nonempty open interval $A \subset \overline{\mathbb{R}}$. Let $\mathfrak{R}$ be the ring of sets of all interval forms on A, and g an increasing left-continuous function on A. Let μ_g be the Stieltjes measure on $\mathfrak{R}$ belonging to g (cf., Ex. 2.2.10). Prove the following propositions:

(α) $\mathscr{L}_{\mathfrak{R}}$ is equal to the set of all functions $f \in \mathbb{R}^A$ such that there are finite families $(x_\iota)_{\iota \in I}$, $(y_\iota)_{\iota \in I}$ from A and $(\gamma_\iota)_{\iota \in I}$ from $\mathbb{R}$ with $f = \Sigma_{\iota \in I} \gamma_\iota e_{[x_\iota, y_\iota[}$. $(x_\iota)_{\iota \in I}$ and $(y_\iota)_{\iota \in I}$ can always be chosen so that $([x_\iota, y_\iota[)_{\iota \in I}$ is disjoint.

(β) Suppose $f = \Sigma_{\iota \in I} \gamma_\iota e_{[x_\iota, y_\iota[} \in \mathscr{L}_{\mathfrak{R}}$. Then

$$\ell_{\mu_g}(f) = \sum_{\iota \in I} \gamma_\iota (g(y_\iota) - g(x_\iota)).$$

2.4.6(E) Finite Products of Positive Measures. As a first application of the relationship between positive measures and positive linear functionals on spaces of step functions, we establish the existence of finite products of positive measures. The general theory of products of measures is not dealt with until Volume 3, where infinite products also come under scrutiny.

Let $\mathfrak{R}$ and $\mathfrak{S}$ be rings of sets. Define $\mathfrak{R} \nabla \mathfrak{S} := \{A \times B \mid A \in \mathfrak{R}, B \in \mathfrak{S}\}$, and denote by $\mathfrak{R} \otimes \mathfrak{S}$ the ring of sets generated by $\mathfrak{R} \nabla \mathfrak{S}$ (i.e., the product of $\mathfrak{R}$ and $\mathfrak{S}$ in the class of all rings of sets in the sense of Ex. 2.1.13).

Given functions $\mu : \mathfrak{R} \to \mathbb{R}$ and $\nu : \mathfrak{S} \to \mathbb{R}$, define

$$\mu \nabla \nu : \mathfrak{R} \nabla \mathfrak{S} \to \mathbb{R}, \qquad A \times B \mapsto \mu(A)\nu(B).$$

Prove the following propositions (referring to Ex. 2.2.8 for definitions if necessary):

(α) $\mu \nabla \nu$ is a content on the semi-ring of sets, $\mathfrak{R} \nabla \mathfrak{S}$, whenever μ and ν are additive.

(β) $\mu \nabla \nu$ is positive whenever μ and ν are positive.

(γ) If μ and ν are positive measures, then $\mu \nabla \nu$ is a σ-additive positive content on $\mathfrak{R} \nabla \mathfrak{S}$.

To prove (α), let $(A_\iota)_{\iota \in I}$ be a finite disjoint family from $\mathfrak{R} \nabla \mathfrak{S}$, so that $A := \bigcup_{\iota \in I} A_\iota \in \mathfrak{R} \nabla \mathfrak{S}$. Then there are sets, $B \in \mathfrak{R}$ and $C \in \mathfrak{S}$, and families of sets, $(B_\iota)_{\iota \in I}$ and $(C_\iota)_{\iota \in I}$, from $\mathfrak{R}$ and $\mathfrak{S}$ respectively, with $A = B \times C$, and with $A_\iota = B_\iota \times C_\iota$ for each $\iota \in I$. Hence, given $x \in X(\mathfrak{R})$ and $y \in X(\mathfrak{S})$, $e_B(x)e_C(y) = \sum_{\iota \in I} e_{B_\iota}(x)e_{C_\iota}(y)$, so that $e_B(x)e_C = \sum_{\iota \in I} e_{B_\iota}(x)e_{C_\iota}$ for x fixed. Thus

$$e_B(x)\nu(C) = \ell_\nu(e_B(x)e_C) = \sum_{\iota \in I} \ell_\nu(e_{B_\iota}(x)e_{C_\iota}) = \sum_{\iota \in I} e_{B_\iota}(x)\nu(C_\iota).$$

But $x \in X(\mathfrak{R})$ was arbitrary. Hence $\nu(C)e_B = \sum_{\iota \in I} \nu(C_\iota)e_{B_\iota}$, so that $\mu(B)\nu(C) = \sum_{\iota \in I} \mu(B_\iota)\nu(C_\iota)$, showing that $\mu \nabla \nu$ is indeed a content.

To prove (γ), it is necessary to establish the σ-additivity of $\mu \nabla \nu$. Proceed exactly as in (α). Given $A \in \mathfrak{R}$ and $B \in \mathfrak{S}$, take sequences $(A_n)_{n \in \mathbb{N}}$ and $(B_n)_{n \in \mathbb{N}}$ from $\mathfrak{R}$ and $\mathfrak{S}$, respectively, such that $(A_n \times B_n)_{n \in \mathbb{N}}$ is a disjoint sequence and that $A \times B = \bigcup_{n \in \mathbb{N}} A_n \times B_n$. Then given $x \in X(\mathfrak{R})$ and $y \in X(\mathfrak{S})$, $e_A(x)e_B(y) = \sum_{n \in \mathbb{N}} e_{A_n}(x)e_{B_n}(y)$ once again. Applying successively the null continuity of ℓ_μ and ℓ_ν, $e_A(x)\nu(B) = \sum_{n \in \mathbb{N}} e_{A_n}(x)\nu(B_n)$ for each $x \in X(\mathfrak{R})$ and $\mu(A)\nu(B) = \sum_{n \in \mathbb{N}} \mu(A_n)\nu(B_n)$.

Combining the results with those of Ex. 2.2.8 yields the next proposition:

(δ) Given positive measures, μ on $\mathfrak{R}$ and ν on $\mathfrak{S}$, there is exactly one positive measure $\mu \otimes \nu$ on $\mathfrak{R} \otimes \mathfrak{S}$ such that $\mu \otimes \nu(A \times B) = \mu(A)\nu(B)$ for each $A \in \mathfrak{R}$ and $B \in \mathfrak{S}$. $\mu \otimes \nu$ is called the **product** of μ and ν on $\mathfrak{R} \otimes \mathfrak{S}$.

This result can be extended readily to arbitrary finite products by means of complete induction. New problems arise when considering infinite products, so we restrict our attention here to finite products.

Let $(\mathfrak{R}_\iota)_{\iota \in I}$ be a finite nonempty family of rings of sets. Set

$$\underset{\iota \in I}{\nabla} \mathfrak{R}_\iota := \left\{ \prod_{\iota \in I} A_\iota \mid A_\iota \in \mathfrak{R}_\iota \text{ for all } \iota \in I \right\}.$$

Denote by $\otimes_{\iota \in I} \mathfrak{R}_\iota$ the ring of sets generated by $\nabla_{\iota \in I} \mathfrak{R}_\iota$ (the product of $(\mathfrak{R}_\iota)_{\iota \in I}$ in the class of all rings of sets). For each $\iota \in I$ let $\mu_\iota : \mathfrak{R}_\iota \to \mathbb{R}$ be a function. Prove the following statements (note that if $(\alpha_\iota)_{\iota \in I}$ is a finite family of real numbers, then $\prod_{\iota \in I} \alpha_\iota$ denotes the usual arithmetical product):

(ε) If each μ_ι is additive, then

$$\underset{\iota \in I}{\nabla} \mu_\iota : \underset{\iota \in I}{\nabla} \mathfrak{R}_\iota \to \mathbb{R}, \qquad \prod_{\iota \in I} A_\iota \mapsto \prod_{\iota \in I} \mu_\iota(A_\iota)$$

defines a content on the semi-ring of sets, $\nabla_{\iota \in I} \mathfrak{R}_\iota$.

(ζ) If each μ_ι is positive, then so is $\nabla_{\iota \in I} \mu_\iota$.

(η) If each μ_ι is a positive measure, then $\nabla_{\iota \in I} \mu_\iota$ is a positive σ-additive content on $\nabla_{\iota \in I} \mathfrak{R}_\iota$.

(ϑ) For each family of positive measures $(\mu_\iota)_{\iota \in I}$ on the rings of sets $\mathfrak{R}_\iota$, respectively, there is exactly one positive measure $\otimes_{\iota \in I} \mu_\iota$ on $\otimes_{\iota \in I} \mathfrak{R}_\iota$ such that

$$\underset{\iota \in I}{\otimes} \mu_\iota \left(\prod_{\iota \in I} A_\iota \right) = \prod_{\iota \in I} \mu_\iota(A_\iota)$$

for each family $(A_\iota)_{\iota \in I}$ from $\prod_{\iota \in I} \mathfrak{R}_\iota$. $\otimes_{\iota \in I} \mu_\iota$ is called the **product** of $(\mu_\iota)_{\iota \in I}$ on $\otimes_{\iota \in I} \mathfrak{R}_\iota$.

Products of measures will be studied extensively later. At that point it will become evident that the term "product" is more advantageously applied to a different object. Nevertheless $\otimes_{\iota \in I} \mu_\iota$ is usually referred to in the literature as "product," so we have adopted that usage here. We hope no difficulties arise therefrom for the reader.

2.4.7(E) Let $\mathfrak{R} \subset \mathfrak{P}(X)$ be a ring of sets. Let Ψ denote the set of all additive positive functions on $\mathfrak{R}$ and Φ the set of all positive linear functionals on $\mathscr{L}_{\mathfrak{R}}$. (Compare with Exs. 1.3.11 and 2.2.18.) We study the connection between Ψ and Φ in this exercise. In the exercises just cited, it was shown that both Ψ and Φ are naturally endowed with the structure of a conditionally complete lattice. The same was found to be true of Ψ^σ (resp. Φ^σ), the set of all nullcontinuous objects in Ψ (resp. Φ) as well as of Ψ^π (resp. Φ^π), the set of all π-continuous objects in Ψ (resp. Φ). The orderings referred to were

$$\ell \leq \ell' :\Leftrightarrow \ell(f) \leq \ell'(f) \quad \text{for each } f \in \mathscr{L}_{\mathfrak{R}+} \text{ (in } \Phi)$$

$$\mu \leq \mu' :\Leftrightarrow \mu(A) \leq \mu'(A) \quad \text{for each } A \in \mathfrak{R} \text{ (in } \Psi).$$

Now define $\omega : \Psi \to \Phi$, $\mu \mapsto \ell_\mu$. Prove the following propositions:

(α) ω is bijective.

(β) $\omega(\mu) \le \omega(\nu)$ whenever $\mu, \nu \in \Psi$, $\mu \le \nu$.

(γ) $\omega(\mu \vee \nu) = \omega(\mu) \vee \omega(\nu)$ and $\omega(\mu \wedge \nu) = \omega(\mu) \wedge \omega(\nu)$ for all $\mu, \nu \in \Psi$.

(δ) Given a nonempty family $(\mu_\iota)_{\iota \in I}$ from Ψ,

$$\omega\Big(\bigvee_{\iota \in I} \mu_\iota\Big) = \bigvee_{\iota \in I} \omega(\mu_\iota) \quad \text{and} \quad \omega\Big(\bigwedge_{\iota \in I} \mu_\iota\Big) = \bigwedge_{\iota \in I} \omega(\mu_\iota)$$

where the last two equations are to be interpreted as asserting that one side exists iff the other does, and in that case, equality holds.

Those statements assert that ω is an isomorphism of conditionally complete lattices. Prove the next statements:

(ε) $\omega : \Psi^\sigma \to \Phi^\sigma$, $\mu \mapsto \ell_\mu$ is an isomorphism of conditionally complete lattices.

(ζ) $\omega : \Psi^\pi \to \Phi^\pi$, $\mu \mapsto \ell_\mu$ is an isomorphism of conditionally complete lattices.

We also determined that Ψ and Φ bear natural algebraic structures. Prove this final statement:

(η) $\omega(\mu + \nu) = \omega(\mu) + \omega(\nu)$ and $\omega(\alpha\mu) = \alpha\omega(\mu)$ for $\mu, \nu \in \Psi$ and $\alpha \in \mathbb{R}_+$.

2.5. STIELTJES FUNCTIONALS

NOTATION AND TERMINOLOGY FOR SECTION 2.5:

A denotes an open, half-open, or closed interval from $\overline{\mathbb{R}}$, containing neither ∞ nor $-\infty$, and having a nonempty interior.

a is the left endpoint of A.

b is the right endpoint of A.

A is viewed as an ordered set with order relation the restriction to A of the order relation on $\overline{\mathbb{R}}$. For "order-converges in A," "${}^A\lim$," etc., we write "converges in A," "lim," etc.

Elements of A are real numbers, but their role here is that of points in a basic space, so they are denoted by such letters as x, y, z, instead of the usual α, β, γ.

Note that $A \subset \mathbb{R}$ and $-\infty \le a < b \le \infty$.

We come now to a second important class of functionals, namely, the Stieltjes functionals. In the preceding sections the abstract viewpoint was predominant. Here, in contrast, the topological and order structures of the base space come to the foreground.

Some topological properties of intervals and their subsets, and of continuous functions on intervals must be reviewed. The review will be brief, and proofs are sometimes left to the reader, since we assume the material to be familiar to the reader. We include certain properties that are not needed until later sections (in the further investigation of Stieltjes functionals).

Definition 2.5.1. *Let $x, y \in A$. Then*

$$[x, y| := \begin{cases} [x, y[& \text{if } y \neq b \\ [x, b] & \text{if } y = b \end{cases}$$

$$|x, y] := \begin{cases}]x, y] & \text{if } x \neq a \\ [a, y] & \text{if } x = a \end{cases}$$

$$|x, y| := \begin{cases}]x, y[& \text{if } x \neq a,\ y \neq b \\ [a, y| & \text{if } x = a \\ |x, b] & \text{if } y = b. \end{cases}$$

The sets $]x, y[$, $[x, y]$, $[x, y|$, $|x, y]$, and $|x, y|$ are called ***intervals from A***; *we call $|x, y|$ an* ***open interval from A***, *$[x, y]$ a* ***closed interval from A***, *$[x, y|$ a right half-open interval from A, and $|x, y]$ a left half-open interval from A.* □

Definition 2.5.2. *Let X be an ordered set. For $Y \subset X$, the set Y is said to be:*

(*a*) ***closed in X*** *iff every sequence from Y that order-converges in X order-converges to an element of Y;*

(*b*) ***open in X*** *iff $X \setminus Y$ is closed in X;*

(*c*) ***compact in X*** *iff, for every family of subsets of X that are open in X and whose union contains Y, there is a finite subfamily whose union also contains Y;*

(*d*) ***sequentially compact in X*** *iff every sequence from Y has a subsequence that order-converges in X to an element of Y.* □

Proposition 2.5.3. *A sequence from A converges in A iff it converges in $\overline{\mathbb{R}}$ to an element of A.*

For $B \subset A$, the set B is open in A (closed in A) iff $B = A \cap C$ for some set C that is open in $\overline{\mathbb{R}}$ (closed in $\overline{\mathbb{R}}$).

Open intervals from A are open in A, and closed intervals from A are closed in A. □

Note that Theorem 1.1.12 provides an ε-m characterization of convergence in A.

Proposition 2.5.4. *For $B \subset A$, the set B is open in A iff it can be written as the union of a countable, disjoint family of intervals from A that are open in A.* □

Proposition 2.5.5. *For all $x, y \in A$, the closed interval $[x, y]$ is compact in A.*

Proof. The assertion is obviously true in case $x > y$, so assume $x \leq y$. Let $(B_\iota)_{\iota \in I}$ be a family of sets that are open in A and whose union contains $[x, y]$. Set

$$C := \left\{ z \in [x, y] \mid \exists \text{ a finite set } J \subset I \text{ such that } [x, z] \subset \bigcup_{\iota \in J} B_\iota \right\}.$$

C is evidently bounded. C is also nonempty, since x is in C. It suffices to show that $\sup C$ belongs to C and $\sup C = y$. There exist an index ι' in I and a point z in C such that $[z, \sup C] \subset B_{\iota'}$. There exists a finite set $J \subset I$ such that $[x, z]$ is contained in $\bigcup_{\iota \in J} B_\iota$. Since $(\bigcup_{\iota \in J} B_\iota) \cup B_{\iota'}$ contains $[x, \sup C]$ and $J \cup \{\iota'\}$ is finite, $\sup C$ must belong to C. If $\sup C$ were strictly less than y, then $B_{\iota'}$ would contain the interval $[\sup C, z']$ for some z' satisfying $\sup C < z' \leq y$. The inclusion of $[\sup C, z']$ in $B_{\iota'}$ would force z' to belong to C, implying $z' \leq \sup C$. This contradiction shows that $\sup C = y$. □

Theorem 2.5.6. *The following assertions are equivalent, for every set $C \subset A$.*

(*a*) *C is both closed in A and bounded in A.*
(*b*) *C is compact in A.*
(*c*) *C is sequentially compact in A.*

Proof. (a) $\Rightarrow$ (b). Let $(B_\iota)_{\iota \in I}$ be a family of sets that are open in A and whose union contains C. Let $[x, y]$ be a closed interval from A that contains C (such an interval exists by hypothesis). The set $A \setminus C$ is open in A and its union with $\bigcup_{\iota \in I} B_\iota$ contains $[x, y]$. By Proposition 5, I has a finite subset J such that $(\bigcup_{\iota \in J} B_\iota) \cup (A \setminus C)$ contains $[x, y]$. Hence $\bigcup_{\iota \in J} B_\iota$ contains C, and (b) follows.

(b) $\Rightarrow$ (c). We give an indirect proof, using the ε-m characterization of convergence in A (Proposition 3, Theorem 1.1.12). Suppose that $(x_n)_{n \in \mathbb{N}}$ is a sequence from C, no subsequence of which converges to an element of C. Then the set

$$D := \{ x_n \mid n \in \mathbb{N} \}$$

is an infinite subset of C. On the other hand, to each x in C we can associate a strictly positive real number ε_x such that the interval $]x - \varepsilon_x, x + \varepsilon_x[$ contains only finitely many elements of D. Each of the sets $A \cap]x - \varepsilon_x, x + \varepsilon_x[$ is open in A, and their union contains C. By hypothesis, C has a finite subset B such that

$$C \subset \bigcup_{x \in B} (A \cap]x - \varepsilon_x, x + \varepsilon_x[).$$

D is an infinite subset of C, yet the union on the right contains only finitely many elements of D. We are forced into a contradiction.

(c) $\Rightarrow$ (a). This implication follows easily from the definitions. □

Corollary 2.5.7. *Let $(C_\iota)_{\iota \in I}$ be a nonempty family of sets that are compact in A. If every nonempty finite subfamily of $(C_\iota)_{\iota \in I}$ has nonempty intersection, then the intersection $\bigcap_{\iota \in I} C_\iota$ is also nonempty.*

Proof. Suppose $\cap_{\iota \in I} C_\iota$ is empty, and fix $\iota_0 \in I$. For each ι in I the set $A \setminus C_\iota$ is open in A, and de Morgan's principle shows that $\bigcup_{\iota \in I}(A \setminus C_\iota)$ contains C_{ι_0}. It follows that $C_{\iota_0} \subset \bigcup_{\iota \in J}(A \setminus C_\iota)$ for some finite subset J of I, hence that the intersection $C_{\iota_0} \cap (\bigcap_{\iota \in J} C_\iota)$ is empty. □

Definition 2.5.8. *Let $f \in \overline{\mathbb{R}}^A$, and let $\alpha \in \overline{\mathbb{R}}$.*

(*a*) *For $x \in]a, b]$, α is said to be a **left-hand limit** of f at x if $\lim_{n \to \infty} f(x_n) = \alpha$ for every sequence $(x_n)_{n \in \mathbb{N}}$ from $[a, x[\cap A$ that converges to x.*

(*b*) *For $x \in [a, b[$, α is a **right-hand limit** of f at x if $\lim_{n \to \infty} f(x_n) = \alpha$ for every sequence $(x_n)_{n \in \mathbb{N}}$ from $]x, b] \cap A$ that converges to x.* □

Proposition 2.5.9. *The following assertions hold, for every $f \in \overline{\mathbb{R}}^A$ and for all extended real numbers α, β.*

(*a*) *If α and β are both left-hand limits of f at x, for some x in $]a, b]$, or if α and β are both right-hand limits of f at x, for some x in $[a, b[$, then $\alpha = \beta$.*

(*b*) *For $x \in]a, b]$, α is a left-hand limit of f at x iff $\lim_{n \to \infty} f(x_n) = \alpha$ for every increasing sequence $(x_n)_{n \in \mathbb{N}}$ from $[a, x[\cap A$ that converges to x.*

(*c*) *For $x \in [a, b[$, α is a right-hand limit of f at x iff $\lim_{n \to \infty} f(x_n) = \alpha$ for every decreasing sequence $(x_n)_{n \in \mathbb{N}}$ from $]x, b] \cap A$ that converges to x.*

Proof. (a) Use the fact that limits of order-convergent sequences are unique (Proposition 1.1.14).

(b) In one direction the implication is trivial. For the other, note that every sequence $(x_n)_{n \in \mathbb{N}}$ from $[a, x[\cap A$ for which $(x_n)_{n \in \mathbb{N}}$ converges to x but $(f(x_n))_{n \in \mathbb{N}}$ does not converge to α has an increasing subsequence with the same two properties.

(c) Modify the argument used for (b). □

Definition 2.5.10. *Let $f \in \overline{\mathbb{R}}^A$. For $x \in]a, b]$, $f(x-)$ shall denote the left-hand limit of f at x, if that limit exists. For $x \in [a, b[$, $f(x+)$ shall denote the right-hand limit of f at x, if that limit exists. Moreover $f(a-) := f(a)$ if a belongs to A, and $f(b+) := f(b)$ if b belongs to A.*

Let $f \in \overline{\mathbb{R}}^A$, and let $x \in A$. The function f is said to be **continuous from the left** *at x, or simply* **left-continuous** *at x, iff $f(x-)$ is defined and $f(x-) = f(x)$. Similarly f is* **continuous from the right**, *or* **right-continuous**, *at x iff $f(x+)$ is defined and $f(x+) = f(x)$. Finally, f is* **continuous** *at x iff both $f(x+)$ and $f(x-)$ are defined and $f(x+) = f(x-) = f(x)$.*

A function $f \in \overline{\mathbb{R}}^A$ is **continuous** *on A (***left-continuous** *on A,* **right-continuous** *on A) iff f is continuous at x (left-continuous at x, right-continuous at x) for every x in A.*

A function $f \in \mathbb{R}^A$ is **uniformly continuous** *on A iff for every $\varepsilon \in \mathbb{R}$, $\varepsilon > 0$, there exists $\delta \in \mathbb{R}$, $\delta > 0$, such that*

$$|f(x) - f(y)| < \varepsilon$$

for all x, y in A satisfying $|x - y| < \delta$. □

We note some obvious consequences of the definitions.

Proposition 2.5.11. *The following assertions hold for every $f \in \overline{\mathbb{R}}^A$.*

(*a*) *If the left endpoint a belongs to A, then f is left-continuous at a.*

(*b*) *If the right endpoint b belongs to A, then f is right-continuous at b.*

The following assertions are equivalent, for every $x \in A$ and for every $f \in \overline{\mathbb{R}}^A$.

(*c*) *f is continuous at x.*

(*d*) *f is both left-continuous at x and right-continuous at x.*

(*e*) *$\lim_{n\to\infty} f(x_n) = f(x)$ for every sequence $(x_n)_{n\in\mathbb{N}}$ from A that converges to x.*

(*f*) *$\liminf_{n\to\infty} f(x_n) = \limsup_{n\to\infty} f(x_n) = f(x)$ for every sequence $(x_n)_{n\in\mathbb{N}}$ from A that converges to x.*

The following assertions hold, for every $f \in \mathbb{R}^A$.

(*g*) *If f is uniformly continuous on A, then f is continuous on A.*

(*h*) *For $x \in A$, f is continuous at x iff for every real number $\varepsilon > 0$ there exists a real number $\delta > 0$ such that $|f(x) - f(y)| < \varepsilon$ if y belongs to A and satisfies $|x - y| < \delta$.* □

Definition 2.5.12.

$$\mathscr{C}(A) := \{ f \in \mathbb{R}^A \mid f \text{ is continuous on } A \}$$

$$\mathscr{K}(A) := \left\{ f \in \mathbb{R}^A \;\middle|\; \begin{array}{l} f \text{ is continuous on } A \\ \text{and there exist } x, y \in A \\ \text{with } \{f \neq 0\} \subset [x, y] \end{array} \right\}$$ □

Proposition 2.5.13. *The sets $\mathscr{C}(A)$ and $\mathscr{K}(A)$ are both real Riesz lattices.* □

Because of the properties described in the next theorem, the Riesz lattice $\mathscr{K}(A)$, sometimes referred to as the space of functions continuous on A with compact support in A, is of special interest. The first two properties will be familiar. The third property will be used to establish the null continuity of the Stieltjes functionals defined later in this section.

Theorem 2.5.14.

(*a*) *Every function in $\mathscr{K}(A)$ assumes on A a maximum value and a minimum value.*

(*b*) *Every function in $\mathscr{K}(A)$ is uniformly continuous on A.*

(*c*) *For every nonempty family $(f_\iota)_{\iota\in I}$ from $\mathscr{K}(A)_+$ that is directed downward, if*

$$\bigwedge_{\iota\in I} f_\iota = 0 \tag{1}$$

then

$$\inf_{\iota\in I}\left(\sup_{x\in A} f_\iota(x)\right) = 0. \tag{2}$$

Proof. (a) Given f in $\mathscr{K}(A)$, we must show that there exist x, y in A such that

$$\sup_{z\in A} f(z) = f(x), \qquad \inf_{z\in A} f(z) = f(y).$$

Let D be a nonempty, closed interval from A containing $\{f \neq 0\}$. It is possible to choose from A, in fact it is possible to choose from D, a sequence $(x_n)_{n\in\mathbb{N}}$ such that the sequence $(f(x_n))_{n\in\mathbb{N}}$ increases and

$$\sup_{n\in\mathbb{N}} f(x_n) = \sup_{z\in A} f(z).$$

Since D is sequentially compact [Theorem 6 (a) $\Rightarrow$ (c)], this sequence has a subsequence $(x_{n_k})_{k\in\mathbb{N}}$ that converges to some element x of D. Using the continuity of f, we conclude that

$$f(x) = \lim_{k\to\infty} f(x_{n_k}) = \sup_{z\in A} f(z).$$

The argument for the existence of the required y is similar.

(b) Given f in $\mathscr{K}(A)$, let D again be a nonempty, closed interval from A such that $\{f \neq 0\} \subset D$. Suppose that f fails to be uniformly continuous on A. Then there exist a real number $\varepsilon > 0$ and sequences $(x_n)_{n\in\mathbb{N}}, (y_n)_{n\in\mathbb{N}}$ from

A such that

$$|x_n - y_n| < \frac{1}{n}$$

$$|f(x_n) - f(y_n)| > \varepsilon \tag{3}$$

for every n in $\mathbb{N}$. Here too we can assume that the sequence $(x_n)_{n \in \mathbb{N}}$ comes from D. The sequence $(x_n)_{n \in \mathbb{N}}$ therefore has a subsequence $(x_{n_k})_{k \in \mathbb{N}}$ that converges to an element x of D. The corresponding subsequence $(y_{n_k})_{k \in \mathbb{N}}$ must converge to the same x. Since f is continuous on A,

$$f(x) = \lim_{k \to \infty} f(x_{n_k}) = \lim_{k \to \infty} f(y_{n_k}).$$

In view of (3) this is clearly impossible. The contradiction shows that f must be uniformly continuous on A.

(c) Let $(f_\iota)_{\iota \in I}$ be a nonempty family from $\mathscr{K}(A)_+$ that is directed downward, and suppose that (1) holds. Let $\varepsilon > 0$ be given. Fix $\iota_0 \in I$. Let D be a nonempty, closed interval from A such that $\{f_{\iota_0} \neq 0\} \subset D$. For each z in A, $\inf_{\iota \in I} f_\iota(z) = 0$. It is therefore possible to choose, for each z in A, an index ι_z such that $f_{\iota_z} \leq f_{\iota_0}$ and

$$f_{\iota_z}(z) < \varepsilon.$$

The functions f_{ι_z} are all continuous on A, so for each z in A we can choose a real number $\delta_z > 0$ such that for every $z' \in]z - \delta_z, z + \delta_z[\cap A$ we have

$$f_{\iota_z}(z') < \varepsilon.$$

Certainly

$$D \subset \bigcup_{z \in A}]z - \delta_z, z + \delta_z[$$

so A contains a finite subset B such that

$$D \subset \bigcup_{z \in B}]z - \delta_z, z + \delta_z[.$$

Using the directed-downwardness, choose $\tilde{\iota} \in I$ so that $f_{\tilde{\iota}} \leq f_{\iota_z}$ for every z in B. It follows that

$$\sup_{z \in A} f_{\tilde{\iota}}(z) < \varepsilon$$

hence that

$$\inf_{\iota \in I} \sup_{z \in A} f_\iota(z) < \varepsilon$$

and finally, since ε was arbitrary, that (2) holds. □

Having completed the required review, we are ready to construct the Stieltjes functionals.

Definition 2.5.15. *A **partition** of the interval A is a family $(x_k)_{k \in \mathbb{N}_n}$ such that $n \geq 2$, each x_k belongs to A, $x_k \leq x_{k+1}$ for every k in $\mathbb{N}_{n-1}$, $x_1 = a$ if a belongs to A, and $x_n = b$ if b belongs to A.*

*Given two partitions $(x_k)_{k \in \mathbb{N}_n}$ and $(y_l)_{l \in \mathbb{N}_m}$ of the interval A, the partition $(x_k)_{k \in \mathbb{N}_n}$ is said to be **finer** than the partition $(y_l)_{l \in \mathbb{N}_m}$ iff for every l in $\mathbb{N}_m$ there exists k in $\mathbb{N}_n$ such that $y_l = x_k$.*

The set of all partitions of A is denoted by $P(A)$. Given $f \in \mathscr{K}(A)$, we denote by $P(A; f)$ the set of all partitions $(x_k)_{k \in \mathbb{N}_n}$ of A that satisfy $\{f \neq 0\} \subset [x_1, x_n]$.

Given $f \in \mathscr{K}(A)$ and $g \in \mathbb{R}^A$, and given a partition $(x_k)_{k \in \mathbb{N}_n}$, which we call p, belonging to $P(A; f)$, we define, for each k in $\mathbb{N}_{n-1}$,

$$m_k(f) := \inf\{f(x) \mid x \in [x_k, x_{k+1}]\}$$

$$M_k(f) := \sup\{f(x) \mid x \in [x_k, x_{k+1}]\}$$

$$\Delta_k g := g(x_{k+1}) - g(x_k)$$

and we define

$$\varphi_*(f, g; p) := \sum_{k=1}^{n-1} m_k(f) \Delta_k g$$

$$\varphi^*(f, g; p) := \sum_{k=1}^{n-1} M_k(f) \Delta_k g.$$

*The numbers $\varphi_*(f, g; p)$ and $\varphi^*(f, g; p)$ are called, respectively, the **lower and upper Stieltjes sums** for the function f relative to the function g and corresponding to the partition p.* □

Although upper and lower Stieltjes sums have been defined for arbitrary real-valued functions g, it is only when g increases that these sums are of interest. The next proposition summarizes important properties of upper and lower Stieltjes sums for increasing g. These properties all follow easily from the definitions.

Proposition 2.5.16. *The following assertions hold, for every increasing function $g \in \mathbb{R}^A$.*

(*a*) *For every f in $\mathscr{K}(A)$ and for every partition p in $P(A; f)$,*

$$\varphi_*(f, g; p) \le \varphi^*(f, g; p).$$

(*b*) *For every f in $\mathscr{K}(A)$ and for all partitions p_1, p_2 in $P(A; f)$, if p_1 is finer than p_2, then*

$$\varphi_*(f, g; p_1) \ge \varphi_*(f, g; p_2)$$

$$\varphi^*(f, g; p_1) \le \varphi^*(f, g; p_2).$$

(*c*) *For every f in $\mathscr{K}(A)$ and for all partitions p_1, p_2 in $P(A; f)$,*

$$\varphi_*(f, g; p_1) \le \varphi^*(f, g; p_2).$$

(*d*) *For all f_1, f_2 in $\mathscr{K}(A)$, and for every partition p that belongs to both $P(A; f_1)$ and $P(A; f_2)$,*

$$\varphi_*(f_1 + f_2, g; p) \ge \varphi_*(f_1, g; p) + \varphi_*(f_2, g; p)$$

$$\varphi^*(f_1 + f_2, g; p) \le \varphi^*(f_1, g; p) + \varphi^*(f_2, g; p).$$

(*e*) *For every f in $\mathscr{K}(A)$, every positive real number α, and every partition p in $P(A; f)$,*

$$\varphi_*(\alpha f, g; p) = \alpha\varphi_*(f, g; p)$$

$$\varphi^*(\alpha f, g; p) = \alpha\varphi^*(f, g; p)$$

and

$$\varphi_*(-\alpha f, g; p) = -\alpha\varphi^*(f, g; p)$$

$$\varphi^*(-\alpha f, g; p) = -\alpha\varphi_*(f, g; p). \qquad \square$$

Theorem 2.5.17 (F. Stieltjes, 1894). *The following assertions hold for every increasing function $g \in \mathbb{R}^A$.*

(*a*) *For every f in $\mathscr{K}(A)$,*

$$\inf_{p \in P(A; f)} [\varphi^*(f, g; p) - \varphi_*(f, g; p)] = 0.$$

(*b*) *For each f in $\mathcal{K}(A)$ there is exactly one real number $\ell_g(f)$ such that*

$$\varphi_*(f, g; p) \leq \ell_g(f) \leq \varphi^*(f, g; p)$$

for every partition p in $P(A; f)$.

(*c*) *The mapping*

$$\ell_g: \mathcal{K}(A) \to \mathbb{R}, \qquad f \mapsto \ell_g(f)$$

is a nullcontinuous, positive, linear functional.

(*d*) *The triple $(A, \mathcal{K}(A), \ell_g)$ is a Daniell space.*

Proof. (a) Given f in $\mathcal{K}(A)$, let $\varepsilon > 0$ be given. Fix points c and d in A such that $\{f \neq 0\} \subset [c, d]$. Since f is uniformly continuous on A [Theorem 14 (b)], we can choose $\delta > 0$ so that

$$|f(x) - f(y)| < \frac{\varepsilon}{1 + g(d) - g(c)}$$

for all x, y in A satisfying $|x - y| < \delta$. Now choose from $P(A; f)$ a partition $(x_k)_{k \in \mathbb{N}_n}$ such that $|x_{k+1} - x_k| < \delta$ for each k in $\mathbb{N}_{n-1}$ and such that the numbers c and d are two of the partition points: $c = x_{k_1}$ for some k_1 in $\mathbb{N}_n$, $d = x_{k_2}$ for some k_2 in $\mathbb{N}_n$. Call the chosen partition p. Then

$$\begin{aligned} 0 &\leq \varphi^*(f, g; p) - \varphi_*(f, g; p) \\ &= \sum_{k=1}^{n-1} [M_k(f) - m_k(f)]\Delta_k g \\ &\leq \frac{\varepsilon}{1 + g(d) - g(c)}(g(d) - g(c)) \\ &< \varepsilon. \end{aligned}$$

This completes the proof of (a) since $f \in \mathcal{K}(A)$ and $\varepsilon > 0$ were given arbitrarily.

(b) In view of Proposition 16 (c), existence follows from the completeness of $\mathbb{R}$ (Axiom 07 of Definition 1.1.1). Uniqueness then follows from (a).

(c) To show that ℓ_g is additive, let f_1 and f_2 belong to $\mathcal{K}(A)$. Let p_1 be an arbitrary partition in $P(A; f_1)$ and p_2 an arbitrary partition in $P(A; f_2)$. It is easy to find a partition p that belongs to each of the sets $P(A; f_1)$, $P(A; f_2)$, $P(A; f_1 + f_2)$ and is finer than both p_1 and p_2. For such a partition p, we have (using 2.5.16)

$$\begin{aligned}\varphi_*(f_1, g; p_1) + \varphi_*(f_2, g; p_2) &\le \varphi_*(f_1, g; p) + \varphi_*(f_2, g; p)\\ &\le \varphi_*(f_1 + f_2, g; p)\\ &\le \ell_g(f_1 + f_2)\\ &\le \varphi^*(f_1 + f_2, g; p)\\ &\le \varphi^*(f_1, g; p) + \varphi^*(f_2, g; p)\\ &\le \varphi^*(f_1, g; p_1) + \varphi^*(f_2, g; p_2).\end{aligned}$$

Since

$$\begin{aligned}\varphi_*(f_1, g; p_1) + \varphi_*(f_2, g; p_2) &\le \ell_g(f_1) + \ell_g(f_2)\\ &\le \varphi^*(f_1, g; p_1) + \varphi^*(f_2, g; p_2)\end{aligned}$$

we have

$$\begin{aligned}|\ell_g(f_1 + f_2) - (\ell_g(f_1) + \ell_g(f_2))| &\le [\varphi^*(f_1, g; p_1) - \varphi_*(f_1, g; p_1)]\\ &\quad + [\varphi^*(f_2, g; p_2) - \varphi_*(f_2, g; p_2)].\end{aligned}$$

Since p_1 and p_2 were arbitrary, it follows from (a) that

$$\ell_g(f_1 + f_2) = \ell_g(f_1) + \ell_g(f_2).$$

A similar argument shows that

$$\ell_g(\alpha f) = \alpha \ell_g(f)$$

for every f in $\mathscr{K}(A)$ and every real number α. Thus ℓ_g is linear.

It is easy to check that ℓ_g is positive. Let $f \in \mathscr{K}(A)_+$, and choose any partition p in $P(A; f)$. The monotonicity of g implies that

$$0 \le \varphi_*(f, g; p) \le \ell_g(f).$$

Finally, we must verify that ℓ_g is nullcontinuous. Let $(f_n)_{n \in \mathbb{N}}$ be a sequence from $\mathscr{K}(A)$ that decreases and satisfies $\bigwedge_{n \in \mathbb{N}} f_n = 0$. Each partition in $P(A; f_1)$ also belongs to $P(A; f_n)$ for every n. Let $p := (x_k)_{k \in \mathbb{N}_m}$ be such a partition. Then

$$\begin{aligned} 0 &\le \inf_{n\in\mathbb{N}} \ell_g(f_n) \\ &\le \inf_{n\in\mathbb{N}} \varphi^*(f_n, g; p) \\ &\le \inf_{n\in\mathbb{N}} \sup_{x\in A} f_n(x)(g(x_m) - g(x_1)). \end{aligned}$$

Applying Theorem 14 (c), we conclude that

$$\inf_{n\in\mathbb{N}} \ell_g(f_n) = 0$$

hence that ℓ_g is nullcontinuous.

(d) It was previously noted that $\mathscr{K}(A)$ is a Riesz lattice, so (d) follows from (c). □

Definition 2.5.18. *Let g be an increasing real-valued function on A. The functional ℓ_g: $\mathscr{K}(A) \to \mathbb{R}$ described in the preceding theorem is called the* ***Stieltjes functional*** *on $\mathscr{K}(A)$ associated with g.* □

Incidentally, we have not yet introduced the so-called Riemann-Stieltjes integrals, integrals that are defined on a wider class of functions, in general, than the Stieltjes functionals ℓ_g. Riemann-Stieltjes integrals are considerably more difficult to obtain than Stieltjes functionals, and they lost their theoretical usefulness with the appearance of Lebesgue integrals. The extensions that we develop in the next chapter will yield more extensive integrals, with better properties, than Riemann-Stieltjes integrals. For pedagogical reasons, however, Riemann-Stieltjes integrals have never completely disappeared from the literature. Especially in the introductory treatments they continue to play an important role. The relations between these various integrals are thoroughly investigated in Volume 2, where we show why the Riemann-Stieltjes integrals cannot possibly satisfy certain requirements.

The functional ℓ_g corresponding to the identity function g [$g(x) = x$ for all x] is closely related to the volume problem described at the beginning of this chapter. The extension of this functional will yield a usable solution to that problem.

EXERCISES

2.5.1(E) **Continuous Functions with Compact Support.** Some of the propositions concerning $\mathscr{K}(A)$ lend themselves to substantial generalization. Let X be a nonempty locally compact space and denote by $\mathscr{K}(X)$ the set of all continuous functions $f \in \mathbb{R}^X$ with compact support (cf., Ex. 1.2.3). Verify the statements that follow:

(α) If $f \in \mathscr{K}(X)$, then f is bounded, and there are points $x, y \in X$ such that

$$f(x) = \sup_{z \in X} f(z) \quad \text{and} \quad f(y) = \inf_{z \in X} f(z).$$

(β) If $(f_\iota)_{\iota \in I}$ is a downward directed nonempty family from $\mathscr{K}(X)_+$ with $\bigwedge_{\iota \in I} f_\iota = 0$, then $\inf_{\iota \in I}(\sup_{x \in X} f_\iota(x)) = 0$.

(γ) Let $(f_\iota)_{\iota \in I}$ be an upward directed nonempty family from $\mathscr{K}(X)$ such that $\bigvee_{\iota \in I} f_\iota =: f \in \mathscr{K}(X)$. Then for each $\varepsilon \in \mathbb{R}$, $\varepsilon > 0$, there is a $\iota \in I$ such that $|f(x) - f_{\iota'}(x)| < \varepsilon$ for each $\iota' \in I$ with $f_\iota \leq f_{\iota'}$ and for each $x \in X$. An analogous result holds for downward directed families from $\mathscr{K}(X)$.

(δ) If a sequence, $(f_n)_{n \in \mathbb{N}}$, from $\mathscr{K}(X)$ is monotonely convergent to a function $f \in \mathscr{K}(X)$, then it converges uniformly on X.

(α) follows from the well-known fact that the image of a compact set under a continuous real function is itself compact in $\mathbb{R}$. For (β) take $\iota_0 \in I$ and let K be a compact subset of X with $\{f_{\iota_0} \neq 0\} \subset K$. Then, for $\varepsilon \in \mathbb{R}$, $\varepsilon > 0$, $K \subset \bigcup_{\iota \in I,\, f_\iota \leq f_{\iota_0}} \{f_\iota < \varepsilon\}$, and there is a finite $J \subset I$ with $f_\iota \leq f_{\iota_0}$ whenever $\iota \in J$, and $K \subset \bigcup_{\iota \in J}\{f_\iota < \varepsilon\}$. If $\iota' \in I$ and $f_{\iota'} \leq f_\iota$ for each $\iota \in J$, then $\sup_{x \in X} f_{\iota'}(x) < \varepsilon$, so that $\inf_{\iota \in I}(\sup_{x \in X} f_\iota(x)) < \varepsilon$.

Compare now with the results of Ex. 1.2.11, which are important special cases of those stated here.

2.5.2(C) **The Theorems of Stone and Weierstrass-Stone.** This exercise deals with approximation theorems for continuous functions. The first form of such a theorem is due to K. Weierstrass:

(α) **Theorem of Weierstrass** (1885). Take $a, b \in \mathbb{R}$, $a < b$. Let $f\colon [a, b] \to \mathbb{R}$ be a continuous function. Then, given $\varepsilon \in \mathbb{R}$, $\varepsilon > 0$, there is a polynomial p such that $\sup_{x \in [a,b]} |f(x) - p(x)| < \varepsilon$.

We present the proof due to N. Bernstein, which gives the polynomial p explicitly. The proof is outlined in several steps. The reader should complete the details.

STEP 1: For each $x \in \mathbb{R}$,

$$\sum_{k=0}^{n} \binom{n}{k} x^k = (1 + x)^k.$$

STEP 2: Given $x \in \mathbb{R}$,

$$\sum_{k=0}^{n} \binom{n}{k} (k - nx)^2 x^k (1 - x)^{n-k} < \frac{n}{4}.$$

Use the identity

$$\sum_{k=0}^{n} \binom{n}{k} z^k = (1 + z)^n.$$

Differentiate with respect to z. Multiply by z, and repeat the procedure. Now replace z by $x/(1 - x)$, and multiply the resulting equations by $(1 - x)^n$ to get

$$\sum_{k=0}^{n} \binom{n}{k} x^k (1 - x)^{n-k} = 1 \tag{1}$$

$$\sum_{k=0}^{n} k \binom{n}{k} x^k (1 - x)^{n-k} = nx \tag{2}$$

$$\sum_{k=0}^{n} k^2 \binom{n}{k} x^k (1 - x)^{n-k} = nx(1 - x + nx). \tag{3}$$

Multiply (1) by n^2x^2, (2) by $-2nx$, and add the resultants to (3) to get

$$\sum_{k=0}^{n} (k - nx)^2 \binom{n}{k} x^k (1 - x)^{n-k} = nx(1 - x).$$

The claim follows from the inequality $x(1 - x) \leq 1/4$, valid for all $x \in \mathbb{R}$.

Let f be a continuous function on $[0,1]$. Then B_n: $[0,1] \to \mathbb{R}$ defined by

$$B_n(x) := \sum_{k=0}^{n} f\left(\frac{k}{n}\right) \binom{n}{k} x^k (1 - x)^{n-k}$$

is called the ***n*th Bernstein polynomial** of f.

STEP 3: Given f continuous on $[0,1]$, $(B_n)_{n \in \mathbb{N}}$ converges uniformly on $[0,1]$ to f.

Put $\alpha := \sup_{x \in [0,1]} |f(x)|$. Since f is continuous on $[0,1]$, f is uniformly continuous on $[0,1]$. So for $\varepsilon \in \mathbb{R}$, $\varepsilon > 0$ there is a $\delta \in \mathbb{R}$, $\delta > 0$ with $|f(x') - f(x'')| < \varepsilon$ whenever $x', x'' \in [0,1]$ and $|x' - x''| < \delta$. Now take $x \in [0,1]$. By (1)

$$|B_n(x) - f(x)| \leq \sum_{k=0}^{n} \left| f\left(\frac{n}{k}\right) - f(x) \right| \binom{n}{k} x^k (1 - x)^{n-k}.$$

Set $A := \{k \in \{0,1,\ldots n\} \mid |(k/n) - x| < \delta\}$ and $C := \{0,1,\ldots,n\} \setminus A$. If $k \in A$, then $|f(k/n) - f(x)| < \varepsilon$, and so by (1)

$$\sum_{k \in A} \left| f\left(\frac{k}{n}\right) - f(x) \right| \binom{n}{k} x^k (1 - x)^{n-k} < \varepsilon.$$

If $k \in C$, then $(k - nx)^2/n^2\delta^2 \geq 1$. By Step 2

$$\sum_{k \in C} \left| f\left(\frac{k}{n}\right) - f(x) \right| \binom{n}{k} x^k (1-x)^{n-k}$$

$$\leq \frac{2\alpha}{n^2\delta^2} \sum_{k=0}^{n} (k - nx)^2 \binom{n}{k} x^k (1-x)^{n-k} < \frac{\alpha}{2n\delta^2}.$$

Thus summarizing, $|B_n(x) - f(x)| < \varepsilon + \alpha/2n\delta^2$, which establishes Step 3.

STEP 4: If f is continuous on $[a, b]$, then so is g: $[0,1] \to \mathbb{R}$, $x \mapsto f(a + x(b - a))$. By Step 3 there is a polynomial p with $|p(x) - g(x)| < \varepsilon$ for all $x \in [0,1]$. Hence

$$\left| f(x) - p\left(\frac{x-a}{b-a}\right) \right| < \varepsilon \quad \text{for all } x \in [a, b].$$

But p': $[a, b] \to \mathbb{R}$, $x \mapsto p((x - a)/b - a))$ is a polynomial, which completes the proof.

The Weierstrass Theorem can be rephrased:

(α') Given $a, b \in \mathbb{R}$, $a \leq b$, $\mathscr{C}([a, b])$ is precisely the set of all functions $f \in \mathbb{R}^{[a,b]}$ each of which is the limit of a sequence of polynomials that converges uniformly on $[a, b]$.

Denote by $\mathscr{P}([a, b])$ the set of all restrictions of polynomials to the interval $[a, b]$. Then the theorem can be reformulated as follows:

(α'') $\mathscr{C}([a, b])$ is the uniform hull of $\mathscr{P}([a, b])$.

We now turn to two important generalizations of this theorem due to M. H. Stone (1937):

(β) **Stone Theorem**. Let X be a compact space. Let $\mathscr{F} \subset \mathscr{C}(X)$ satisfy the following conditions:

(i) $f \vee g$, $f \wedge g \in \mathscr{F}$ whenever $f, g \in \mathscr{F}$.
(ii) Given $\varepsilon \in \mathbb{R}$, $\varepsilon > 0$, $f \in \mathscr{C}(X)$ and $x, y \in X$, there is $g \in \mathscr{F}$ with $|f(x) - g(x)| < \varepsilon$ and $|f(y) - g(y)| < \varepsilon$.

Then $\mathscr{C}(X)$ is the uniform hull of $\mathscr{F}$.

Take $f \in \mathscr{C}(X)$ and $\varepsilon \in \mathbb{R}$, $\varepsilon > 0$. For $x, y \in X$ take $g_{xy} \in \mathscr{F}$ with

$$|f(x) - g_{xy}(x)| < \varepsilon \quad \text{and} \quad |f(y) - g_{xy}(y)| < \varepsilon.$$

Set $A_{xy} := \{g_{xy} < f + \varepsilon\}$ and $B_{xy} := \{g_{xy} > f - \varepsilon\}$. Then A_{xy} and B_{xy} are open sets, and in particular, $x \in A_{xy}$ and $y \in B_{xy}$. For $y \in X$, $X = \bigcup_{x \in X} A_{xy}$. By the compactness of X, there is a finite set $X_y \subset X$ so that $X = \bigcup_{x \in X_y} A_{xy}$. Set $g_y := \bigwedge_{x \in X_y} g_{xy}$. Then $g_y(x) < f(x) + \varepsilon$ for any $x \in X$. Furthermore, setting $B_y := \bigcap_{x \in X_y} B_{xy}$, $g_y(x) > f(x) - \varepsilon$ for each $x \in B_y$. But $X = \bigcup_{y \in X} B_y$. So, again using the compactness of X, there is a finite set

$X' \subset X$ with $X = \bigcup_{y \in X'} B_y$. Set $g := \bigvee_{y \in X'} g_y$. Then

$$f(x) - \varepsilon < g(x) < f(x) + \varepsilon$$

for each $x \in X$, completing the proof.

Another form of the theorem follows from (β):

(β') Let X be a compact space. Let $\mathscr{F} \subset \mathscr{C}(X)$ satisfy these conditions:

(i) $f \vee g$, $f \wedge g \in \mathscr{F}$ whenever $f, g \in \mathscr{F}$.
(ii) Given $x, y \in X$ and $\alpha, \beta \in \mathbb{R}$ such that $\alpha = \beta$ if $x = y$, there is an $f \in \mathscr{F}$ with $f(x) = \alpha$ and $f(y) = \beta$.

Then $\mathscr{C}(X)$ is the uniform hull of $\mathscr{F}$.

The second generalization of the Weierstrass Theorem is the following:

(δ) **Theorem of Stone-Weierstrass.** Let X be a compact space. Let $\mathscr{F} \subset \mathscr{C}(X)$ satisfy the following conditions:

(i) $\mathscr{F}$ is a linear function space, containing e_X and multiplicatively closed (i.e., $fg \in \mathscr{F}$ if $f, g \in \mathscr{F}$).
(ii) $\mathscr{F}$ separates the points of X [i.e., if $x, y \in X$, $x \neq y$, then for some $f \in \mathscr{F}$, $f(x) \neq f(y)$].

Then $\mathscr{C}(X)$ is the uniform hull of $\mathscr{F}$.

The hypotheses imply that if $x, y \in X$, $x \neq y$, $\alpha, \beta \in \mathbb{R}$, then there is an $f \in \mathscr{F}$ with $f(x) = \alpha$ and $f(y) = \beta$.

Let $\bar{\mathscr{F}}$ be the uniform hull of $\mathscr{F}$. Take $f \in \bar{\mathscr{F}}$ and $\gamma \in \mathbb{R}_+$ with $\{|f| < \gamma\} = X$. Then, given $x \in X$,

$$
\begin{aligned}
|f(x)| &= \sqrt{\gamma^2 - \left(\gamma^2 - f(x)^2\right)} \\
&= \gamma\sqrt{1 - \left(1 - \left(\frac{f(x)}{\gamma}\right)^2\right)} \\
&= \lim_{n \to \infty}\left(1 - \sum_{k=1}^{n} \frac{1 \cdot 1 \cdot 3 \cdots (2k-3)}{2 \cdot 4 \cdot 6 \cdots 2k}\left(1 - \left(\frac{f(x)}{\gamma}\right)^2\right)^k\right)
\end{aligned}
$$

where the last line is obtained by means of the Taylor expansion of the radical. Each

$$g_n := 1 - \sum_{k=1}^{n} \frac{1 \cdot 1 \cdot 3 \cdots (2k-3)}{2 \cdot 4 \cdot 6 \cdots 2k}\left(1 - \left(\frac{f(x)}{\gamma}\right)^2\right)^k$$

is contained in $\bar{\mathscr{F}}$, and $(g_n)_{n \in \mathbb{N}}$ converges uniformly on X to $|f|$. Thus $|f| \in \bar{\mathscr{F}}$. Conclude that $f \vee g$, $f \wedge g \in \bar{\mathscr{F}}$ whenever $f, g \in \bar{\mathscr{F}}$. Hence $\bar{\mathscr{F}}$ satisfies the conditions of the Stone Theorem [in form (β')]. Since $\bar{\mathscr{F}}$ is its own uniform hull, it follows that $\bar{\mathscr{F}} = \mathscr{C}(X)$.

There is another form of the Stone-Weierstrass Theorem:

(δ') Let X be compact. Let $\mathscr{F} \subset \mathscr{C}(X)$ satisfy these conditions:

(i) $\mathscr{F}$ is a multiplicatively closed linear function space.
(ii) Given $x, y \in X$, $x \neq y$, there are $f, g \in \mathscr{F}$ with $f(x)g(y) \neq f(y)g(x)$.
(iii) If X consists of exactly one point, then $\mathscr{F} \neq \{0\}$.

Then $\mathscr{C}(X)$ is the uniform hull of $\mathscr{F}$.

The Stone-Weierstrass Theorem concludes the suit of approximation theorems. The reader should verify that (δ) implies (α). However, (δ) does not by itself permit the explicit description of the approximating polynomials.

2.5.3(C) Positive Linear Functionals on Locally Compact Spaces. In Ex. 1.3.7 we met a large class of function spaces with the property that every positive additive functional on such a space is both linear and nullcontinuous, in fact, even π-continuous if certain conditions are fulfilled. It was determined in that exercise that the space $\mathscr{K}$ of continuous functions with compact support on the locally compact space X satisfies these conditions. This result is of great importance to measure theory, and we wish to present an account of it here, using the property formulated in Ex. 2.5.1 (β). For this, denote by $\mathfrak{K}$ the set of all compact subsets of X. Prove the following statements:

(α) Every positive additive functional ℓ on $\mathscr{K}$ is linear and π-continuous.

Let $(f_\iota)_{\iota \in I}$ be a nonempty downward directed family from $\mathscr{K}$ with $\bigwedge_{\iota \in I} f_\iota = 0$. Then, by Ex. 2.5.1 ($\beta$), $\inf_{\iota \in I} \sup_{x \in X}(f_\iota(x)) = 0$. Take a fixed $\iota_0 \in I$. For each $\varepsilon \in \mathbb{R}$, $\varepsilon > 0$ there is a $\iota_\varepsilon \in I$ with $f_{\iota_\varepsilon} \leq f_{\iota_0}$ and with $f_\iota < \varepsilon^2$ whenever $f_\iota \leq f_{\iota_\varepsilon}$. For each $\iota \in I$ with $f_\iota \leq f_{\iota_\varepsilon}$, $f_\iota^2 \leq f_\iota f_{\iota_\varepsilon} \leq f_\iota f_{\iota_0} \leq \varepsilon^2 f_{\iota_0}$ so that $f_\iota \leq \varepsilon\sqrt{f_{\iota_0}}$. Hence $\inf_{\iota \in I} \ell(f_\iota) \leq \varepsilon \ell(\sqrt{f_{\iota_0}})$. But ε is arbitrary so that $\inf_{\iota \in I} \ell(f_\iota) = 0$.

We wish to establish a relationship between additive positive functionals on $\mathscr{K}$ and certain positive measures on the ring of sets generated by $\mathfrak{K}$. The particular relationship we have in mind is not the one described in this chapter with the help of the step functions but shows another aspect of the intimate connection between the two concepts.

(β) Let ℓ be a positive additive functional on $\mathscr{K}$. Define $\nu \colon \mathfrak{K} \to \mathbb{R}$, $K \mapsto \inf\{\ell(f) \mid f \in \mathscr{K}, e_K \leq f\}$. Prove the assertions that follow:

($\beta 1$) $\nu(K \cup L) \leq \nu(K) + \nu(L)$ for all $K, L \in \mathfrak{K}$.
($\beta 2$) $\nu(K \cup L) = \nu(K) + \nu(L)$ for all $K, L \in \mathfrak{K}$, $K \cap L = \varnothing$.
($\beta 3$) If $\mathfrak{L}$ is a downward directed nonempty subset of $\mathfrak{K}$, then $\nu(\bigcap_{L \in \mathfrak{L}} L) = \inf_{L \in \mathfrak{L}} \nu(L)$.

For ($\beta 1$), define, for $K \in \mathfrak{K}$, $\mathscr{K}_K := \{f \in \mathscr{K} \mid e_K \leq f\}$. Then

$$\{f + g \mid f \in \mathscr{K}_K, g \in \mathscr{K}_L\} \subset \mathscr{K}_{K \cup L}.$$

For ($\beta 2$), recall that if $K \cap L = \varnothing$, then there are functions, $f \in \mathscr{K}_K$ and $g \in \mathscr{K}_L$ with $f \wedge g = 0$. Given $h \in \mathscr{K}_{K \cup L}$, $h \wedge f \in \mathscr{K}_K$ and $h \wedge g \in$

$\mathscr{K}_L$ and $h \wedge f + h \wedge g \leq h$ so that $\nu(K) + \nu(L) \leq \ell(h)$. Thus $\nu(K) + \nu(L) \leq \nu(K \cup L)$.

For ($\beta 3$), let $\mathfrak{L}$ be a nonempty downward directed subset of $\mathfrak{K}$. Take $f \in \mathscr{K}_{\cap_{L \in \mathfrak{L}} L}$ and $K \in \mathfrak{L}$. For each $\alpha \in]0,1[$, $F_\alpha := K \cap \{f \leq \alpha\}$ is compact, and $\cap_{L \in \mathfrak{L}} L \cap F_\alpha = \varnothing$. Hence for some finite $\mathfrak{L}' \subset \mathfrak{L}$, $\cap_{L \in \mathfrak{L}'} L \cap F_\alpha = \varnothing$. Thus there is an $L \in \mathfrak{L}$ with $e_L \leq \alpha^{-1} f$. Then $\inf_{L' \in \mathfrak{L}} \nu(L') \leq \nu(L) \leq \alpha^{-1} \ell(f)$. But $\alpha \in]0,1[$ was arbitrary, so that $\inf_{L \in \mathfrak{L}} \nu(L) \leq \nu(\bigcap_{L \in \mathfrak{L}} L)$. This essentially completes the proof, as the reverse inequality is evident.

Note that ($\beta 1$)–($\beta 3$) correspond precisely to the conditions formulated in Ex. 2.2.13 for the extension of ν to a positive measure on $\mathscr{K}_r$. These results can be combined:

(γ) For each additive positive functional, ℓ, on $\mathscr{K}$ there is exactly one positive measure, μ_ℓ on $\mathfrak{K}_r$ so that

$$\mu_\ell(K) = \inf\{\ell(f) \mid f \in \mathscr{K}, e_K \leq f\}$$

for every $K \in \mathfrak{K}$. μ_ℓ is $\mathfrak{K}$-regular.

The question arises whether two different functionals can give rise to the same measure. The following considerations show that this cannot occur.

(δ) If $f \in \mathscr{K}_+$ and $\alpha \in \mathbb{R}$, $\alpha > 0$, then $\{f \geq \alpha\} \in \mathfrak{K}$.

(ε) If $f \in \mathscr{K}_+$ then $(\sum_{k=1}^{n2^n} (1/2^n) e_{\{f \geq k/2^n\}})_{n \in \mathbb{N}}$ is an increasing sequence of $\mathfrak{K}_r$-step functions.

(ζ) If $f \in \mathscr{K}_+$, then $f = \bigvee_{n \in \mathbb{N}} \sum_{k=1}^{n2^n} (1/2^n) e_{\{f \geq k/2^n\}}$.

(η) If ℓ is an additive positive functional on $\mathscr{K}$ and if $f \in \mathscr{K}_+$ then

$$\ell(f) = \sup_{n \in \mathbb{N}} \sum_{k=1}^{n2^n} \frac{1}{2^n} \mu_\ell(\{f \geq k/2^n\}).$$

(ϑ) If ℓ and ℓ' are additive functionals on $\mathscr{K}$ with $\mu_\ell = \mu_{\ell'}$ then $\ell = \ell'$.

Thus each of ℓ and μ determines the other uniquely. That is not the last we will have to say about the relationship between ℓ and μ.

2.5.4(C) The Stieltjes Functionals. We return to the Stieltjes functionals. We wish to apply the results of Ex. 2.5.3 to establish the connection between positive Stieltjes functionals and positive Stieltjes measures. Let $A :=]a, b[$ with $a, b \in \overline{\mathbb{R}}$ and $a < b$, and let $\mathscr{K}$ be the set of all continuous functions on A with compact support. Theorem 17 asserts that for each increasing function g on A there is a Stieltjes functional ℓ_g on $\mathscr{K}$. We discuss the question of whether all additive positive functionals on $\mathscr{K}$ are of this form.

Let ℓ be an additive positive functional on $\mathscr{K}$. By Ex. 2.5.3 (γ) there is a unique positive measure μ on $\mathfrak{K}_r$ (where $\mathfrak{K}$ is the set of compact subsets of A) with

$$\mu(K) = \inf\{\ell(f) \mid f \in \mathscr{K}, e_K \leq f\}$$

for all $K \in \mathfrak{K}$. Prove the following statement:

(α) $\mathfrak{J} \subset \mathfrak{K}_r$, where $\mathfrak{J}$ denotes the set of all right half-open intervals in A.

Combining this with Ex. 2.2.10 (η), there is an increasing left-continuous function, $g\colon A \to \mathbb{R}$ with $\mu([x, y[) = g(y) - g(x)$ for $x, y \in A$, $x \leq y$. Prove that the following holds:

(β) $\ell = \ell_g$.

On putting this all together, we have the following representation theorem due to F. Riesz (1909):

(γ) For every additive positive functional ℓ on $\mathscr{K}$, there is an increasing left-continuous function g on A with $\ell = \ell_g$. If g and h are two increasing left-continuous functions on A with $\ell_g = \ell_h$, then $g = h + \gamma$ for some suitable $\gamma \in \mathbb{R}$.

A problem arises. We have defined ℓ_g for each increasing function g on A. We have shown that for each such function g there is an increasing left-continuous function, h, on A with $\ell_g = \ell_h$. When do two functions, g and h, give rise to the same functional? We pursue this question in Chapter 5.

3

THE CLOSURE OF A DANIELL SPACE

In Sections 3.1 and 3.2 a general method is developed by which closed extensions of arbitrary Daniell spaces can be constructed. As has already been mentioned, this method was first communicated by the American mathematician Daniell (1918). Based in an essential way on the order structure of the Daniell space, Daniell's method offers various points of connection that allow for generalization.

The reader should note that the closed Daniell spaces constructed in these two sections are not yet integrals in the sense of this book. To obtain integrals, an additional step is generally required. That step, adjoining new null functions, will be studied in Chapter 4. It should be stressed, however, that the construction described in this chapter is the main step in the construction of integrals. It is this closure construction that yields the topological properties of the integral, giving the integral its power for applications.

3.1. CONSTRUCTING THE CLOSURE

NOTATION FOR SECTION 3.1:

X denotes a set.

For every Daniell space $(X, \mathscr{L}, \ell)$ there is a smallest closed Daniell space extending it. It is appropriate to call this extension the closure of the given Daniell space. The present section describes an explicit method, essentially due to Daniell, by which the Daniell-space closure can be constructed. In the next section it is proved that the closed extension constructed here *is* the smallest closed extension.

The Daniell construction can be split into two stages. First, one forms the collections $\mathscr{L}^{\uparrow}$ and $\mathscr{L}^{\downarrow}$ of what one might call the upper and lower functions for the given Daniell space: functions that are suprema of increasing real sequences from $\mathscr{L}$ and functions that are infima of decreasing real sequences from $\mathscr{L}$. One extends the functional ℓ so that it assigns values (but possibly infinite values!) to the upper functions and lower functions. At the second stage one singles out those functions that can be arbitrarily approximated by either upper or lower functions. In this context two functions are close if their functional values are close. Thus one singles out those functions for which there exist a bigger upper function and a smaller lower function such that the upper function and the lower function are arbitrarily close to each other. It is these functions that will appear in the closure $(X, \bar{\mathscr{L}}(\ell), \bar{\ell})$ of the given Daniell space.

The section itself falls into four parts, the first two being the two stages of the Daniell construction. The second stage begins at Definition 3.1.9, the definition of ε-bracket relative to ℓ, and establishes all the properties of ε-brackets that are needed for showing that $\bar{\mathscr{L}}(\ell)$ is a Riesz lattice. The definition and various characterizations of $(X, \bar{\mathscr{L}}(\ell), \bar{\ell})$ comprise the third part of the section. Finally, it is proved that the triple $(X, \bar{\mathscr{L}}(\ell), \bar{\ell})$ is indeed a closed Daniell space extending the original one.

Definition 3.1.1. *For $\mathscr{L}$ a Riesz lattice in $\overline{\mathbb{R}}^X$,*

$$\mathscr{L}^{\uparrow} := \left\{ \bigvee_{n\in\mathbb{N}} f_n \,|\, (f_n)_{n\in\mathbb{N}} \text{ is an increasing sequence from } \mathscr{L}\cap \mathbb{R}^X \right\}. \qquad \square$$

Proposition 3.1.2. *For every Riesz lattice $\mathscr{L}$ in $\overline{\mathbb{R}}^X$, the following assertions concerning $\mathscr{L}^{\uparrow}$ hold.*

(*a*) *For every $f \in \mathscr{L}^{\uparrow}$ and every $x \in X$, $f(x) > -\infty$.*

(*b*) $\mathscr{L}\cap \mathbb{R}^X \subset \mathscr{L}^{\uparrow}$.

(*c*) *If f and g belong to $\mathscr{L}^{\uparrow}$, then $f+g$ is defined and belongs to $\mathscr{L}^{\uparrow}$.*

(*d*) *If f belongs to $\mathscr{L}^{\uparrow}$, then so does αf for every $\alpha \in \mathbb{R}_+$.*

(*e*) *If f and g belong to $\mathscr{L}^{\uparrow}$, then so do $f \vee g$ and $f \wedge g$.*

(*f*) *If the sequence $(f_n)_{n\in\mathbb{N}}$ lies in $\mathscr{L}^{\uparrow}$, then $\bigvee_{n\in\mathbb{N}} f_n$ belongs to $\mathscr{L}^{\uparrow}$. In fact, if, for each n in $\mathbb{N}$, $(f_{n,m})_{m\in\mathbb{N}}$ is an increasing sequence from $\mathscr{L}\cap \mathbb{R}^X$ with*

$$\bigvee_{m\in\mathbb{N}} f_{n,m} = f_n \tag{1}$$

then

$$\left(\bigvee_{k\le n} f_{k,n} \right)_{n\in\mathbb{N}}$$

is an increasing sequence from $\mathscr{L} \cap \mathbb{R}^X$ *whose supremum is the function* $\bigvee_{n \in \mathbb{N}} f_n$.

(g) *For every nonempty, countable family* $(f_\iota)_{\iota \in I}$ *from* $\mathscr{L}^\uparrow$, $\bigvee_{\iota \in I} f_\iota$ *belongs to* $\mathscr{L}^\uparrow$.

Proof. (a) Note that if f belongs to $\mathscr{L}^\uparrow$, then $f \geq g$ for at least one real-valued function g belonging to $\mathscr{L}$. Assertion (a) follows.

(b) By definition, constant sequences are increasing. Every real function in $\mathscr{L}$ is the supremum of a constant sequence from $\mathscr{L} \cap \mathbb{R}^X$. Therefore $\mathscr{L} \cap \mathbb{R}^X \subset \mathscr{L}^\uparrow$.

(c), (d), (e) Suppose that f and g belong to $\mathscr{L}^\uparrow$, and let α be a positive real number. From (a) it follows that $f + g$ is defined. According to the definition of $\mathscr{L}^\uparrow$, there exist in $\mathscr{L} \cap \mathbb{R}^X$ increasing sequences $(f_n)_{n \in \mathbb{N}}$ and $(g_n)_{n \in \mathbb{N}}$ whose suprema are f and g, respectively. The sequences

$$(f_n + g_n)_{n \in \mathbb{N}}, (\alpha f_n)_{n \in \mathbb{N}}, (f_n \vee g_n)_{n \in \mathbb{N}}, \quad \text{and} \quad (f_n \wedge g_n)_{n \in \mathbb{N}}$$

all increase and lie in $\mathscr{L} \cap \mathbb{R}^X$. We easily conclude that

$$\bigvee_{n \in \mathbb{N}} (f_n \vee g_n) = \Big(\bigvee_{n \in \mathbb{N}} f_n\Big) \vee \Big(\bigvee_{n \in \mathbb{N}} g_n\Big) = f \vee g$$

[Proposition 1.2.8 (n)]. Taking into account the monotonicity of the sequences $(f_n)_{n \in \mathbb{N}}$ and $(g_n)_{n \in \mathbb{N}}$ and the positivity of α, we also have

$$\bigvee_{n \in \mathbb{N}} (f_n + g_n) = \Big(\bigvee_{n \in \mathbb{N}} f_n\Big) + \Big(\bigvee_{n \in \mathbb{N}} g_n\Big) = f + g$$

$$\bigvee_{n \in \mathbb{N}} (f_n \wedge g_n) = \Big(\bigvee_{n \in \mathbb{N}} f_n\Big) \wedge \Big(\bigvee_{n \in \mathbb{N}} g_n\Big) = f \wedge g$$

$$\bigvee_{n \in \mathbb{N}} (\alpha f_n) = \alpha \bigvee_{n \in \mathbb{N}} f_n = \alpha f.$$

[Propositions 1.2.9 (a), (b); 1.2.8 (k)]. Thus $f \vee g$, $f + g$, $f \wedge g$, and αf all belong to $\mathscr{L}^\uparrow$.

(f) Let $(f_n)_{n \in \mathbb{N}}$ be a sequence from $\mathscr{L}^\uparrow$, and for each n in $\mathbb{N}$, let $(f_{n,m})_{m \in \mathbb{N}}$ be an increasing sequence from $\mathscr{L} \cap \mathbb{R}^X$ such that (1) holds. For each n, set

$$g_n := \bigvee_{k \leq n} f_{k,n}. \tag{2}$$

The sequence $(g_n)_{n\in\mathbb{N}}$ evidently increases and lies in $\mathscr{L}\cap\mathbb{R}^X$. We must show that

$$\bigvee_{n\in\mathbb{N}} g_n = \bigvee_{n\in\mathbb{N}} f_n. \tag{3}$$

Consider the matrix whose entry in row k, column m is $f_{k,m}$. Row n of this matrix is an increasing sequence whose supremum is f_n. The function g_n is the supremum of the first n entries in the nth column of the matrix. From these two observations it follows that (3) holds.

(g) From (f) we easily obtain (g). □

Definition 3.1.3. *Let $(X,\mathscr{L},\ell)$ be a Daniell space. For every $f\in\mathscr{L}^\uparrow$,*

$$\ell^\uparrow(f) := \sup\{\ell(g)\,|\,g\in\mathscr{L}\cap\mathbb{R}^X,\ g\le f\}. \qquad \square$$

Proposition 3.1.4. *For every Daniell space $(X,\mathscr{L},\ell)$, the following assertions hold for all $f, g\in\mathscr{L}^\uparrow$, and for every sequence $(f_n)_{n\in\mathbb{N}}$ from $\mathscr{L}^\uparrow$.*

(a) $\ell^\uparrow(f) > -\infty$.

(b) *If $f\in\mathscr{L}\cap\mathbb{R}^X$, then $\ell^\uparrow(f)=\ell(f)$.*

(c) *If $f\le g$, then $\ell^\uparrow(f)\le\ell^\uparrow(g)$.*

(d) *If $(f_n)_{n\in\mathbb{N}}$ increases, then*

$$\ell^\uparrow\Big(\bigvee_{n\in\mathbb{N}} f_n\Big) = \sup_{n\in\mathbb{N}} \ell^\uparrow(f_n). \tag{4}$$

(e) $\ell^\uparrow(f+g)=\ell^\uparrow(f)+\ell^\uparrow(g)$.

(f) $\ell^\uparrow(\alpha f)=\alpha\ell^\uparrow(f)$ *for every* $\alpha\in\mathbb{R}_+$.

Proof. (a) There exists h in $\mathscr{L}\cap\mathbb{R}^X$ with $h\le f$. Since the functional ℓ is real valued, the definition of $\ell^\uparrow$ yields

$$-\infty < \ell(h) \le \ell^\uparrow(f).$$

(b) Let $f\in\mathscr{L}\cap\mathbb{R}^X$. Then f also belongs to $\mathscr{L}^\uparrow$, and $\ell(f)\le\ell^\uparrow(f)$. On the other hand, since ℓ is an increasing functional, we have $\ell(h)\le\ell(f)$ for every h in $\mathscr{L}\cap\mathbb{R}^X$ that satisfies $h\le f$. Hence

$$\ell^\uparrow(f) = \sup\{\ell(h)\,|\,h\in\mathscr{L}\cap\mathbb{R}^X,\ h\le f\} \le \ell(f).$$

It follows that $\ell(f)=\ell^\uparrow(f)$.

(c) From the definition of $\ell^\uparrow$ we can easily obtain (c).

(d) Suppose $(f_n)_{n\in\mathbb{N}}$ increases and set $f:=\bigvee_{n\in\mathbb{N}} f_n$. If the given sequence $(f_n)_{n\in\mathbb{N}}$ lies in $\mathscr{L}\cap\mathbb{R}^X$ and the function f also belongs to $\mathscr{L}\cap\mathbb{R}^X$, (4) is a consequence [in view of (b)] of the null continuity of ℓ (Proposition 1.3.6).

In the general case, Proposition 2(f) ensures that f belongs to $\mathscr{L}^{\uparrow}$, and the monotonicity of $\ell^{\uparrow}$ yields

$$\ell^{\uparrow}(f) \geq \sup_{n \in \mathbb{N}} \ell^{\uparrow}(f_n).$$

To get the inequality in the other direction, we first replace $(f_n)_{n \in \mathbb{N}}$ by a suitably chosen increasing sequence from $\mathscr{L} \cap \mathbb{R}^X$. Thus for each n in $\mathbb{N}$, let $(f_{n,m})_{m \in \mathbb{N}}$ be an increasing sequence from $\mathscr{L} \cap \mathbb{R}^X$ whose supremum is f_n. Define, for each n,

$$g_n := \bigvee_{k \leq n} f_{k,n}.$$

Then the sequence $(g_n)_{n \in \mathbb{N}}$ increases and lies in $\mathscr{L} \cap \mathbb{R}^X$, and its supremum is the function f [Proposition 2 (f)]. Moreover the definition of g_n and the monotonicity of the sequence $(f_n)_{n \in \mathbb{N}}$ show that

$$g_n \leq \bigvee_{k \leq n} f_k = f_n \tag{5}$$

for every n in $\mathbb{N}$. Now let h be an arbitrary function in $\mathscr{L} \cap \mathbb{R}^X$ satisfying $h \leq f$. Then we can write

$$h = h \wedge f = h \wedge \left(\bigvee_{n \in \mathbb{N}} g_n \right) = \bigvee_{n \in \mathbb{N}} (h \wedge g_n)$$

[Proposition 1.2.8 (c)]. The sequence $(h \wedge g_n)_{n \in \mathbb{N}}$ increases and lies in $\mathscr{L} \cap \mathbb{R}^X$, and its supremum, h, also belongs to $\mathscr{L} \cap \mathbb{R}^X$. Consequently we can use the null continuity of ℓ to get information about $\ell(h)$. Using (b) and (c), and taking (5) into account, we have

$$\begin{aligned} \ell(h) &= \sup_{n \in \mathbb{N}} \ell(h \wedge g_n) \\ &\leq \sup_{n \in \mathbb{N}} \ell(g_n) \\ &= \sup_{n \in \mathbb{N}} \ell^{\uparrow}(g_n) \\ &\leq \sup_{n \in \mathbb{N}} \ell^{\uparrow}(f_n). \end{aligned}$$

Hence

$$\ell^{\uparrow}(f) = \sup\{\ell(h) \mid h \in \mathscr{L} \cap \mathbb{R}^X, h \leq f\}$$

$$\leq \sup_{n \in \mathbb{N}} \ell^{\uparrow}(f_n)$$

and (d) is established.

Now that (d) has been established, we can compute various values of $\ell^{\uparrow}$ by making appropriate choices of increasing sequences.

(e) From (a) it follows that $\ell^{\uparrow}(f) + \ell^{\uparrow}(g)$ is defined. From Proposition 2 (c) we know that $f + g$ is defined and belongs to $\mathscr{L}^{\uparrow}$, so $\ell^{\uparrow}(f + g)$ is defined. Let $(f_n)_{n \in \mathbb{N}}$ and $(g_n)_{n \in \mathbb{N}}$ be increasing sequences from $\mathscr{L} \cap \mathbb{R}^X$ whose suprema are, respectively, f and g. Then $(f_n + g_n)_{n \in \mathbb{N}}$ is an increasing sequence from $\mathscr{L} \cap \mathbb{R}^X$ whose supremum is $f + g$. Moreover, the sequences $(\ell(f_n))_{n \in \mathbb{N}}$ and $(\ell(g_n))_{n \in \mathbb{N}}$ both increase. It follows from (d) [we also use (b) and Proposition 1.1.10 (a)] that

$$\begin{aligned}
\ell^{\uparrow}(f + g) &= \sup_{n \in \mathbb{N}} \ell(f_n + g_n) \\
&= \sup_{n \in \mathbb{N}} (\ell(f_n) + \ell(g_n)) \\
&= \sup_{n \in \mathbb{N}} \ell(f_n) + \sup_{n \in \mathbb{N}} \ell(g_n) \\
&= \ell^{\uparrow}(f) + \ell^{\uparrow}(g).
\end{aligned}$$

(f) One can prove (f) by an analogous argument. The details are left to the reader. □

The rule that $\ell^{\uparrow}$ commutes with sup when applied to an increasing sequence [Proposition 4 (d)] is especially important.

We introduce $\mathscr{L}^{\downarrow}$ and $\ell^{\downarrow}$ in complete analogy with $\mathscr{L}^{\uparrow}$ and $\ell^{\uparrow}$.

Definition 3.1.5. *Let $(X, \mathscr{L}, \ell)$ be a Daniell space. Then*

$$\mathscr{L}^{\downarrow} := \left\{ \bigwedge_{n \in \mathbb{N}} f_n \mid (f_n)_{n \in \mathbb{N}} \text{ is a decreasing sequence from } \mathscr{L} \cap \mathbb{R}^X \right\}$$

and, for every $f \in \mathscr{L}^{\downarrow}$,

$$\ell^{\downarrow}(f) := \inf\{\ell(g) \mid g \in \mathscr{L} \cap \mathbb{R}^X, g \geq f\}.$$ □

The properties of $(\mathscr{L}^{\downarrow}, \ell^{\downarrow})$ can easily be derived from those of $(\mathscr{L}^{\uparrow}, \ell^{\uparrow})$ by means of the following proposition.

Proposition 3.1.6. *Let $(X, \mathscr{L}, \ell)$ be a Daniell space, and let $f \in \overline{\mathbb{R}}^X$. Then f belongs to $\mathscr{L}^{\downarrow}$ iff $-f$ belongs to $\mathscr{L}^{\uparrow}$, and in this case*

$$\ell^{\downarrow}(f) = -\ell^{\uparrow}(-f).$$

Proof. A sequence $(f_n)_{n \in \mathbb{N}}$ from $\overline{\mathbb{R}}^X$ decreases, after all, iff the sequence $(-f_n)_{n \in \mathbb{N}}$ increases. Moreover $\bigwedge_{n \in \mathbb{N}} f_n = -\bigvee_{n \in \mathbb{N}}(-f_n)$ [Proposition 1.2.8 (e)]. If f belongs to $\mathscr{L}^{\downarrow}$, then

$$\begin{aligned}\ell^{\downarrow}(f) &= \inf\{\ell(g) \mid g \in \mathscr{L} \cap \mathbb{R}^X, \ g \geq f\} \\ &= -\sup\{-\ell(g) \mid g \in \mathscr{L} \cap \mathbb{R}^X, \ g \geq f\} \\ &= -\sup\{\ell(-g) \mid -g \in \mathscr{L} \cap \mathbb{R}^X, \ -g \leq -f\} \\ &= -\ell^{\uparrow}(-f)\end{aligned}$$

[Proposition 1.1.9 (e)]. □

The following corollary summarizes those properties of $\mathscr{L}^{\downarrow}$ and $\ell^{\downarrow}$ that are required in the sequel. The proofs, which are all easy, are left to the reader.

Corollary 3.1.7. *The following assertions hold for every Daniell space $(X, \mathscr{L}, \ell)$.*

(*a*) *$\mathscr{L} \cap \mathbb{R}^X \subset \mathscr{L}^{\downarrow}$, and $\ell^{\downarrow}(f) = \ell(f)$ for every $f \in \mathscr{L} \cap \mathbb{R}^X$.*

(*b*) *For every $f \in \mathscr{L}^{\downarrow}$, $\ell^{\downarrow}(f) < \infty$.*

(*c*) *The conditions $f, g \in \mathscr{L}^{\downarrow}$ and $f \leq g$ always imply that $\ell^{\downarrow}(f) \leq \ell^{\downarrow}(g)$.*

(*d*) *If f and g belong to $\mathscr{L}^{\downarrow}$, then $f + g$ is defined and belongs to $\mathscr{L}^{\downarrow}$, and*

$$\ell^{\downarrow}(f+g) = \ell^{\downarrow}(f) + \ell^{\downarrow}(g).$$

(*e*) *If f belongs to $\mathscr{L}^{\downarrow}$, then for every $\alpha \in \mathbb{R}_+$, αf belongs to $\mathscr{L}^{\downarrow}$ and*

$$\ell^{\downarrow}(\alpha f) = \alpha \ell^{\downarrow}(f).$$

(*f*) *If f and g belong to $\mathscr{L}^{\downarrow}$, then so do $f \vee g$ and $f \wedge g$.*

(*g*) *If $(f_n)_{n \in \mathbb{N}}$ is a sequence from $\mathscr{L}^{\downarrow}$ that decreases, then $\bigwedge_{n \in \mathbb{N}} f_n$ belongs to $\mathscr{L}^{\downarrow}$ and*

$$\ell^{\downarrow}\left(\bigwedge_{n \in \mathbb{N}} f_n\right) = \inf_{n \in \mathbb{N}} \ell^{\downarrow}(f_n).$$ □

Proposition 3.1.8. *Let $(X, \mathscr{L}, \ell)$ be a Daniell space, and let $f \in \mathscr{L}^{\downarrow}$, $g \in \mathscr{L}^{\uparrow}$. If $f \leq g$, then*

$$\ell^{\downarrow}(f) \leq \ell^{\uparrow}(g).$$

Proof. The preceding results make it easy to verify that

$$\ell^{\uparrow}(g) - \ell^{\downarrow}(f) \geq 0.$$

We have $-f \in \mathscr{L}^{\uparrow}$, $g + (-f) \in \mathscr{L}^{\uparrow}$, and $g + (-f) \geq 0$. Thus

$$\begin{aligned} \ell^{\uparrow}(g) - \ell^{\downarrow}(f) &= \ell^{\uparrow}(g) + \ell^{\uparrow}(-f) \\ &= \ell^{\uparrow}(g + (-f)) \\ &\geq \ell^{\uparrow}(0) \\ &= 0. \end{aligned}$$

□

Definition 3.1.9. *Let $(X, \mathscr{L}, \ell)$ be a Daniell space. Suppose that $f \in \overline{\mathbb{R}}^X$ and $\varepsilon \in \mathbb{R}_+$. Then an **ε-bracket** of f relative to ℓ is a pair $(f', f'') \in \mathscr{L}^{\downarrow} \times \mathscr{L}^{\uparrow}$ such that the following hold:*

(*a*) $f' \leq f \leq f''$.
(*b*) $\ell^{\downarrow}(f'), \ell^{\uparrow}(f'') \in \mathbb{R}$.
(*c*) $\ell^{\uparrow}(f'') - \ell^{\downarrow}(f') \leq \varepsilon$. □

Proposition 3.1.10. *Let $(X, \mathscr{L}, \ell)$ be a Daniell space. Suppose that the pairs (f', f'') and (g', g'') both belong to $\mathscr{L}^{\downarrow} \times \mathscr{L}^{\uparrow}$, that $f' \leq f''$ and $g' \leq g''$, and that the numbers $\ell^{\downarrow}(f')$, $\ell^{\downarrow}(g')$, $\ell^{\uparrow}(f'')$, and $\ell^{\uparrow}(g'')$ are all real. Then the following assertions hold:*

(*a*) $\ell^{\downarrow}(f' \vee g')$, $\ell^{\downarrow}(f' \wedge g')$, $\ell^{\uparrow}(f'' \vee g'')$, *and* $\ell^{\uparrow}(f'' \wedge g'')$ *are all real numbers.*
(*b*) $\ell^{\uparrow}(f'' \vee g'') - \ell^{\downarrow}(f' \vee g') \leq [\ell^{\uparrow}(f'') - \ell^{\downarrow}(f')] + [\ell^{\uparrow}(g'') - \ell^{\downarrow}(g')]$.
(*c*) $\ell^{\uparrow}(f'' \wedge g'') - \ell^{\downarrow}(f' \wedge g') \leq [\ell^{\uparrow}(f'') - \ell^{\downarrow}(f')] + [\ell^{\uparrow}(g'') - \ell^{\downarrow}(g')]$.

Proof. Note that the functional values appearing in (a) are all well defined, since both $\mathscr{L}^{\uparrow}$ and $\mathscr{L}^{\downarrow}$ are closed under sup and inf. The key to the whole proof is the fact that for any two functions f, g in $\overline{\mathbb{R}}^X$ we can write

$$(f \vee g) + (f \wedge g) = f + g \tag{6}$$

provided both sums are defined [Theorem 1.2.7 (s), Convention 1.2.2.]. No function from $\mathscr{L}^{\uparrow}$ assumes the value $-\infty$, while no function from $\mathscr{L}^{\downarrow}$ assumes the value ∞. Therefore both sums appearing in (6) are defined when f and g are replaced by f' and g', respectively, or by f'' and g'', respectively.

Both $\ell^{\uparrow}$ and $\ell^{\downarrow}$ are additive, so

$$\ell^{\downarrow}(f' \vee g') + \ell^{\downarrow}(f' \wedge g') = \ell^{\downarrow}(f') + \ell^{\downarrow}(g')$$

$$\ell^{\uparrow}(f'' \vee g'') + \ell^{\uparrow}(f'' \wedge g'') = \ell^{\uparrow}(f'') + \ell^{\uparrow}(g'').$$

By hypothesis, each of these identities has a real right-hand side. Hence each has a real left-hand side. But $\ell^{\downarrow}$ cannot assign the value ∞, and $\ell^{\uparrow}$ cannot assign the value $-\infty$, so the individual terms on the two left-hand sides must each be real. We have verified (a). Now subtract one equation from the other, and group terms to obtain

$$\left[\ell^{\uparrow}(f'' \vee g'') - \ell^{\downarrow}(f' \vee g')\right] + \left[\ell^{\uparrow}(f'' \wedge g'') - \ell^{\downarrow}(f' \wedge g')\right]$$
$$= \left[\ell^{\uparrow}(f'') - \ell^{\downarrow}(f')\right] + \left[\ell^{\uparrow}(g'') - \ell^{\downarrow}(g')\right].$$

From the hypotheses and Proposition 8 we conclude that each of the bracketed terms in the identity just obtained is positive. Thus (b) and (c) hold. □

Proposition 3.1.11. *Let* $(X, \mathscr{L}, \ell)$ *be a Daniell space, and let* $f, g \in \overline{\mathbb{R}}^X$. *Suppose that*

$$h \in \langle f \dotplus g \rangle,$$

that the pairs (f', f'') *and* (g', g'') *belong to* $\mathscr{L}^{\downarrow} \times \mathscr{L}^{\uparrow}$, *and that*

$$f' \leq f \leq f'', \; g' \leq g \leq g''.$$

Then the pair $(f' + g', f'' + g'')$ *belongs to* $\mathscr{L}^{\downarrow} \times \mathscr{L}^{\uparrow}$ *and*

$$f' + g' \leq h \leq f'' + g''.$$

Proof. The first claim has already been established (3.1.2., 3.1.7.). We must show, for every x in X, that

$$f'(x) + g'(x) \leq h(x) \leq f''(x) + g''(x). \tag{7}$$

Let $x \in X$. If $f(x) + g(x)$ is defined, then (7) is an obvious consequence of the hypotheses. So suppose that $f(x) + g(x)$ is undefined. We know that $f'(x) \leq f(x) \leq f''(x)$ and $g'(x) \leq g(x) \leq g''(x)$. We also know that neither $f''(x)$ nor $g''(x)$ equals $-\infty$, and neither $f'(x)$ nor $g'(x)$ equals ∞. Thus $f(x) + g(x)$ undefined implies that

$$f'(x) + g'(x) = -\infty, \; f''(x) + g''(x) = \infty.$$

Hence (7) holds in this case as well. □

Proposition 3.1.12. *Let $(X, \mathscr{L}, \ell)$ be a Daniell space, let $f, g \in \overline{\mathbb{R}}^X$, and let $\alpha \in \mathbb{R}$. Suppose that $h \in \langle f \dot{+} g \rangle$, that $\varepsilon_1, \varepsilon_2 \in \mathbb{R}_+$, and that (f', f'') and (g', g'') are, respectively, an ε_1-bracket of f and an ε_2-bracket of g, both relative to ℓ. Then, relative to ℓ,*

(*a*) $(f' + g', f'' + g'')$ *is an* $(\varepsilon_1 + \varepsilon_2)$*-bracket of* h
(*b*) $(f' \vee g', f'' \vee g'')$ *is an* $(\varepsilon_1 + \varepsilon_2)$*-bracket of* $f \vee g$
(*c*) $(f' \wedge g', f'' \wedge g'')$ *is an* $(\varepsilon_1 + \varepsilon_2)$*-bracket of* $f \wedge g$
(*d*) $(\alpha f', \alpha f'')$ *is an* $(\alpha\varepsilon_1)$*-bracket of* αf *if* $\alpha \geq 0$
(*e*) $(\alpha f'', \alpha f')$ *is an* $(|\alpha|\varepsilon_1)$*-bracket of* αf *if* $\alpha \leq 0$.

Proof. Note first that the pairs named in (a)–(e) all belong to $\mathscr{L}^{\downarrow} \times \mathscr{L}^{\uparrow}$.

(a) The additivity of $\ell^{\downarrow}$ and $\ell^{\uparrow}$, together with the hypotheses, ensure that both $\ell^{\downarrow}(f' + g')$ and $\ell^{\uparrow}(f'' + g'')$ are real. The inequality

$$f' + g' \leq h \leq f'' + g''$$

was established in the previous proposition. Finally,

$$\begin{aligned}\ell^{\uparrow}(f'' + g'') - \ell^{\downarrow}(f' + g') &= \left[\ell^{\uparrow}(f'') - \ell^{\downarrow}(f')\right] + \left[\ell^{\uparrow}(g'') - \ell^{\downarrow}(g')\right] \\ &\leq \varepsilon_1 + \varepsilon_2.\end{aligned}$$

(b), (c) Obviously

$$f' \vee g' \leq f \vee g \leq f'' \vee g''$$

and

$$f' \wedge g' \leq f \wedge g \leq f'' \wedge g''.$$

Assertions (b) and (c) follow from Proposition 10.

(d) Suppose $\alpha \geq 0$. Then $(\alpha f', \alpha f'')$ belongs to $\mathscr{L}^{\downarrow} \times \mathscr{L}^{\uparrow}$, and

$$\alpha f' \leq \alpha f \leq \alpha f''.$$

Since $\ell^{\downarrow}$ and $\ell^{\uparrow}$ are homogeneous with respect to positive scalars, both $\ell^{\downarrow}(\alpha f')$ and $\ell^{\uparrow}(\alpha f'')$ must be real numbers. Finally,

$$\ell^{\uparrow}(\alpha f'') - \ell^{\downarrow}(\alpha f') = \alpha\left[\ell^{\uparrow}(f'') - \ell^{\downarrow}(f')\right] \leq \alpha\varepsilon_1.$$

(e) Suppose $\alpha \leq 0$. Then

$$\alpha f'' \leq \alpha f \leq \alpha f'.$$

Note that

$$(\alpha f'', \alpha f') = (-|\alpha| f'', -|\alpha| f').$$

Using Proposition 6, we conclude that the pair $(\alpha f'', \alpha f')$ belongs to $\mathscr{L}^{\downarrow} \times \mathscr{L}^{\uparrow}$, and both of the numbers $\ell^{\downarrow}(\alpha f''), \ell^{\uparrow}(\alpha f')$ are real. Moreover

$$\ell^{\uparrow}(\alpha f') - \ell^{\downarrow}(\alpha f'') = |\alpha|\left[\ell^{\uparrow}(f'') - \ell^{\downarrow}(f')\right] \leq |\alpha|\varepsilon_1. \qquad \square$$

Assertions (a)–(c) of the preceding proposition can be extended by induction. We require only the extension of (b), which we state here.

Proposition 3.1.13. *Let $(X, \mathscr{L}, \ell)$ be a Daniell space, let $(f_\iota)_{\iota \in I}$ be a finite nonempty family from $\overline{\mathbb{R}}^X$, and let $(\varepsilon_\iota)_{\iota \in I}$ be a family from $\mathbb{R}_+$. Suppose, for every ι in I, that the pair (f_ι', f_ι'') is an ε_ι-bracket of f_ι. Then $(\bigvee_{\iota \in I} f_\iota', \bigvee_{\iota \in I} f_\iota'')$ is a $(\sum_{\iota \in I} \varepsilon_\iota)$-bracket of $\bigvee_{\iota \in I} f_\iota$.* $\square$

The resources are now at hand for extending ℓ to a functional whose properties essentially exceed those hypothesized for ℓ.

Proposition 3.1.14. *Let $(X, \mathscr{L}, \ell)$ be a Daniell space, and let $f \in \overline{\mathbb{R}}^X$. Suppose that for each real number $\varepsilon > 0$ there exists an ε-bracket of f relative to ℓ. Then there exists a unique number $\bar{\ell}(f)$ such that*

$$\ell^{\downarrow}(f') \leq \bar{\ell}(f) \leq \ell^{\uparrow}(f'')$$

for every pair $(f', f'') \in \mathscr{L}^{\downarrow} \times \mathscr{L}^{\uparrow}$ satisfying $f' \leq f \leq f''$. Moreover

$$\bar{\ell}(f) = \inf\{\ell^{\uparrow}(g) \mid g \in \mathscr{L}^{\uparrow}, g \geq f\}$$

$$= \sup\{\ell^{\downarrow}(g) \mid g \in \mathscr{L}^{\downarrow}, g \leq f\}.$$

Proof. Let $\varepsilon > 0$ be given. There exists an ε-bracket (f', f'') for f relative to ℓ. Using Proposition 8, we have

$$-\infty < \ell^{\downarrow}(f') \leq \sup\{\ell^{\downarrow}(g) \mid g \in \mathscr{L}^{\downarrow}, g \leq f\}$$

$$\leq \inf\{\ell^{\uparrow}(g) \mid g \in \mathscr{L}^{\uparrow}, g \geq f\}$$

$$\leq \ell^{\uparrow}(f'') < \infty.$$

But $\ell^{\uparrow}(f'') - \ell^{\downarrow}(f') \leq \varepsilon$ and $\varepsilon > 0$ was given arbitrarily, so

$$\sup\{\ell^{\downarrow}(g) \mid g \in \mathscr{L}^{\downarrow}, g \leq f\} = \inf\{\ell^{\uparrow}(g) \mid g \in \mathscr{L}^{\uparrow}, g \geq f\}.$$

Now if (f', f'') is any pair in $\mathscr{L}^{\downarrow} \times \mathscr{L}^{\uparrow}$ for which $f' \leq f \leq f''$ holds, then

$$\begin{aligned} \ell^{\downarrow}(f') &\leq \sup\{\ell^{\downarrow}(g) \mid g \in \mathscr{L}^{\downarrow},\ g \leq f\} \\ &= \inf\{\ell^{\uparrow}(g) \mid g \in \mathscr{L}^{\uparrow},\ g \geq f\} \\ &\leq \ell^{\uparrow}(f'') \end{aligned}$$

and the proposition is established. □

Definition 3.1.15. *For $(X, \mathscr{F}, \ell)$ a Daniell space,*

$$\bar{\mathscr{L}}(\ell) := \left\{ f \in \overline{\mathbb{R}}^X \,\middle|\, \begin{array}{l} \text{For each } \varepsilon > 0 \text{ there exists} \\ \text{an } \varepsilon\text{-bracket of } f \text{ relative to } \ell. \end{array} \right\}$$

$$\begin{aligned} \bar{\ell} \colon \bar{\mathscr{L}}(\ell) &\to \mathbb{R}, \\ f &\mapsto \sup\{\ell^{\downarrow}(g) \mid g \in \mathscr{L}^{\downarrow},\ g \leq f\} \\ &= \inf\{\ell^{\uparrow}(g) \mid g \in \mathscr{L}^{\uparrow},\ g \geq f\}. \end{aligned}$$ □

There are several useful ways to characterize $(X, \bar{\mathscr{L}}(\ell), \bar{\ell})$. The equivalences in the following proposition are all simple consequences of how the space $\bar{\mathscr{L}}(\ell)$ and the functional $\bar{\ell}$ were defined. Verifications are left to the reader.

Proposition 3.1.16. *Let $(X, \mathscr{L}, \ell)$ be a Daniell space. Then the following assertions are equivalent, for every $f \in \overline{\mathbb{R}}^X$ and for every real number α.*

(*a*) *$f \in \bar{\mathscr{L}}(\ell)$ and $\bar{\ell}(f) = \alpha$.*

(*b*) *For every strictly positive real number ε, there exists relative to ℓ an ε-bracket (f', f'') of f such that*

$$\ell^{\downarrow}(f') \leq \alpha \leq \ell^{\uparrow}(f'');$$

if f is positive then such a bracket can be found in $\mathscr{L}_+^{\downarrow} \times \mathscr{L}_+^{\uparrow}$.

(*c*) *There exist an increasing sequence $(f_n')_{n \in \mathbb{N}}$ from $\mathscr{L}^{\downarrow}$ and a decreasing sequence $(f_n'')_{n \in \mathbb{N}}$ from $\mathscr{L}^{\uparrow}$ such that the sequences $(\ell^{\downarrow}(f_n'))_{n \in \mathbb{N}}$ and $(\ell^{\uparrow}(f_n''))_{n \in \mathbb{N}}$ both lie in $\mathbb{R}$,*

$$\bigvee_{n \in \mathbb{N}} f_n' \leq f \leq \bigwedge_{n \in \mathbb{N}} f_n''$$

and

$$\sup_{n\in\mathbb{N}} \ell^{\downarrow}(f_n') = \inf_{n\in\mathbb{N}} \ell^{\uparrow}(f_n'') = \alpha;$$

if f is positive such sequences can be found in $\mathscr{L}_+^{\downarrow}$ and $\mathscr{L}_+^{\uparrow}$, respectively.

(*d*) $\alpha = \sup\{\ell^{\downarrow}(g) \mid g \in \mathscr{L}^{\downarrow}, g \le f\} = \inf\{\ell^{\uparrow}(g) \mid g \in \mathscr{L}^{\uparrow}, g \ge f\}$.

(*e*) *For every strictly positive real number ε, there exist in $\mathscr{L} \cap \mathbb{R}^X$ an increasing sequence $(f_n')_{n\in\mathbb{N}}$ and a decreasing sequence $(f_n'')_{n\in\mathbb{N}}$ such that*

$$f = \bigvee_{n\in\mathbb{N}} (f_n' \wedge f) = \bigwedge_{n\in\mathbb{N}} (f_n'' \vee f)$$

$$-\infty < \inf_{n\in\mathbb{N}} \ell(f_n'') \le \alpha \le \sup_{n\in\mathbb{N}} \ell(f_n') < \infty$$

$$\varepsilon \ge \sup_{n\in\mathbb{N}} \ell(f_n') - \inf_{n\in\mathbb{N}} \ell(f_n'');$$

if f is positive then such sequences can be found in $(\mathscr{L} \cap \mathbb{R}^X)_+$. □

Note that (e) characterizes the triple $(X, \bar{\mathscr{L}}(\ell), \bar{\ell})$ in terms of $(X, \mathscr{L}, \ell)$ alone, making no use of either the spaces $\mathscr{L}^{\uparrow}$ and $\mathscr{L}^{\downarrow}$ or the mappings $\ell^{\uparrow}$ and $\ell^{\downarrow}$. In fact characterization (e) uses only X, $\mathscr{L} \cap \mathbb{R}^X$, and the restriction of ℓ to $\mathscr{L} \cap \mathbb{R}^X$. Extended-real functions do not enter in at all.

The preliminaries have now been completed. We arrive at our goal for this section.

Theorem 3.1.17. *For every Daniell space $(X, \mathscr{L}, \ell)$ the triple $(X, \bar{\mathscr{L}}(\ell), \bar{\ell})$ is a closed Daniell space extending $(X, \mathscr{L}, \ell)$.*

Proof. Proposition 12 shows that $\bar{\mathscr{L}}(\ell)$ is a Riesz lattice. To establish the properties required of $\bar{\ell}$, we exploit the various characterizations provided by Proposition 16.

We want to verify that $\bar{\ell}$ is a positive, linear functional on $\bar{\mathscr{L}}(\ell)$. Suppose that f and g belong to $\bar{\mathscr{L}}(\ell)$ and that $h \in \langle f \dotplus g \rangle$; then h is also in $\bar{\mathscr{L}}(\ell)$. Using the implication (b) $\Rightarrow$ (a) from Proposition 16, it is easy to show that

$$\bar{\ell}(h) = \bar{\ell}(f) + \bar{\ell}(g).$$

Indeed, let $\varepsilon > 0$ be given. According to the definition of $\bar{\mathscr{L}}(\ell)$ there exist, relative to ℓ, $(\varepsilon/2)$-brackets (f', f'') and (g', g'') of f and g, respectively. The pair $(f' + g', f'' + g'')$ is an ε-bracket of h relative to ℓ [Proposition 12 (a)]. We need only show that

$$\ell^{\downarrow}(f' + g') \le \bar{\ell}(f) + \bar{\ell}(g) \le \ell^{\uparrow}(f'' + g''). \tag{8}$$

The definition of $\bar{\ell}$ yields the inequalities

$$\ell^{\downarrow}(f') \leq \bar{\ell}(f) \leq \ell^{\uparrow}(f'')$$

$$\ell^{\downarrow}(g') \leq \bar{\ell}(g) \leq \ell^{\uparrow}(g'').$$

Inequality (8) follows by summation, in view of the additivity of $\ell^{\downarrow}$ and $\ell^{\uparrow}$. We have established that $\bar{\ell}$ is additive. An analogous argument, which we leave to the reader, shows that $\bar{\ell}$ is homogeneous and therefore linear. The positivity of $\bar{\ell}$ follows from the relation

$$\bar{\ell}(f) = \inf\{\ell^{\uparrow}(g) \mid g \in \mathscr{L}^{\uparrow},\ g \geq f\} \geq 0$$

which holds for every $f \in \bar{\mathscr{L}}(\ell)_{+}$.

A crucial part of the proof is to show that the triple $(X, \bar{\mathscr{L}}(\ell), \bar{\ell})$ is closed. We use the characterization provided by Proposition 1.4.3 (a) $\Leftrightarrow$ (b). To that end, let $(f_n)_{n \in \mathbb{N}}$ be an increasing $\bar{\ell}$-sequence from $\bar{\mathscr{L}}(\ell)$: $(f_n)_{n \in \mathbb{N}}$ increases, and $(\bar{\ell}(f_n))_{n \in \mathbb{N}}$ is bounded above in $\mathbb{R}$. Set

$$f := \bigvee_{n \in \mathbb{N}} f_n.$$

If we can show that f belongs to $\bar{\mathscr{L}}(\ell)$ and

$$\bar{\ell}(f) = \sup_{n \in \mathbb{N}} \bar{\ell}(f_n) \tag{9}$$

then it will follow that the triple $(X, \bar{\mathscr{L}}(\ell), \bar{\ell})$ is closed. We use Proposition 16 (b) $\Rightarrow$ (a): we show that for every $\varepsilon > 0$ there exists, relative to ℓ, an ε-bracket (f', f'') of f for which

$$\ell^{\downarrow}(f') \leq \sup_{n \in \mathbb{N}} \bar{\ell}(f_n) \leq \ell^{\uparrow}(f''). \tag{10}$$

So let $\varepsilon > 0$ be given. For each n in $\mathbb{N}$, there exists an $(\varepsilon/2^{n+1})$-bracket (f_n', f_n'') of f_n, relative to ℓ. For each n the pair

$$\left(\bigvee_{k \leq n} f_k', \bigvee_{k \leq n} f_k''\right)$$

is a $[\sum_{k=1}^{n}(\varepsilon/2^{k+1})]$-bracket, relative to ℓ, of the function $\bigvee_{k \leq n} f_k$, in other words, of the function f_n (Proposition 13). In particular, for every n in $\mathbb{N}$ we have

$$\bar{\ell}(f_n) \leq \ell^{\uparrow}\left(\bigvee_{k \leq n} f_k''\right) \leq \bar{\ell}(f_n) + \sum_{k=1}^{n} \frac{\varepsilon}{2^{k+1}}. \tag{11}$$

Set

$$f'' := \bigvee_{n \in \mathbb{N}} f_n''.$$

Taking the supremum over all n in inequality (11), using Proposition 4 (d) on the middle term, we obtain

$$\sup_{n \in \mathbb{N}} \bar{\ell}(f_n) \leq \ell^{\uparrow}(f'') \leq \frac{\varepsilon}{2} + \sup_{n \in \mathbb{N}} \bar{\ell}(f_n). \tag{12}$$

It follows, since the sequence $(\bar{\ell}(f_n))_{n \in \mathbb{N}}$ is bounded in $\mathbb{R}$, that $\ell^{\uparrow}(f'')$ is real. The function f'' will serve as one of the two functions in the desired ε-bracket of f. To obtain the other function, first choose m in $\mathbb{N}$ so that

$$\sup_{n \in \mathbb{N}} \bar{\ell}(f_n) - \frac{\varepsilon}{4} \leq \bar{\ell}(f_m) \leq \sup_{n \in \mathbb{N}} \bar{\ell}(f_n). \tag{13}$$

Then choose f' from $\mathscr{L}^{\downarrow}$ so that $f' \leq f_m$, $\ell^{\downarrow}(f')$ is real, and

$$\bar{\ell}(f_m) - \frac{\varepsilon}{4} \leq \ell^{\downarrow}(f') \leq \bar{\ell}(f_m). \tag{14}$$

This function f' will do. From (13) and (14) it follows that

$$\sup_{n \in \mathbb{N}} \bar{\ell}(f_n) - \frac{\varepsilon}{2} \leq \ell^{\downarrow}(f') \leq \sup_{n \in \mathbb{N}} \bar{\ell}(f_n). \tag{15}$$

Now (12) and (15) imply that

$$\ell^{\downarrow}(f') \leq \sup_{n \in \mathbb{N}} \bar{\ell}(f_n) \leq \ell^{\uparrow}(f'')$$

and

$$\ell^{\uparrow}(f'') - \ell^{\downarrow}(f') \leq \varepsilon.$$

Clearly $f' \leq f \leq f''$, so (f', f'') is in fact an ε-bracket of f, relative to ℓ, that satisfies (10). Since ε was arbitrary, we conclude that f belongs to $\bar{\mathscr{L}}(\ell)$ and (9) holds. Hence the triple $(X, \bar{\mathscr{L}}(\ell), \bar{\ell})$ is closed. We conclude that $(X, \bar{\mathscr{L}}(\ell), \bar{\ell})$ is a closed Daniell space (Proposition 1.4.2).

It remains to show that $(X, \bar{\mathscr{L}}(\ell), \bar{\ell})$ extends the original Daniell space $(X, \mathscr{L}, \ell)$. So let $f \in \mathscr{L}$. The set

$$A := \{|f| = \infty\}$$

is $\mathscr{L}$-exceptional, so its characteristic function e_A is $\mathscr{L}$-exceptional, and $\ell(e_A) = 0$ [Definition 1.2.21, Proposition 1.3.3 (a)]. But e_A belongs to $\mathscr{L} \cap \mathbb{R}^X$. Hence e_A belongs to both $\mathscr{L}^{\uparrow}$ and $\mathscr{L}^{\downarrow}$, and $\ell^{\uparrow}(e_A) = \ell^{\downarrow}(e_A) = \ell(e_A) = 0$.

We conclude that e_A belongs to $\bar{\mathscr{L}}(\ell)$ and that

$$\bar{\ell}(e_A) = \ell(e_A) = 0.$$

In the closed Daniell space $(X, \bar{\mathscr{L}}(\ell), \bar{\ell})$, the condition $\bar{\ell}(e_A) = 0$ implies that the set A is $\bar{\mathscr{L}}(\ell)$-exceptional (Corollary 1.4.5). Now we use Propositions 1.2.30 (a), (b), and 1.3.3 (b). The function fe_A belongs to $\mathscr{N}(\bar{\mathscr{L}}(\ell))$ and therefore to $\bar{\mathscr{L}}(\ell)$. The function $fe_{X \setminus A}$ belongs to $\mathscr{L} \cap \mathbb{R}^X$ and to $\bar{\mathscr{L}}(\ell)$. Moreover

$$\bar{\ell}(fe_{X \setminus A}) = \ell(fe_{X \setminus A}) = \ell(f).$$

Since f is the sum of $fe_{X \setminus A}$ and fe_A, f belongs to $\bar{\mathscr{L}}(\ell)$, and

$$\bar{\ell}(f) = \bar{\ell}(fe_{X \setminus A}) + \bar{\ell}(fe_A) = \ell(f).$$

We have shown that $\mathscr{L} \subset \bar{\mathscr{L}}(\ell)$ and $\bar{\ell}|_{\mathscr{L}} = \ell$. Thus $(X, \bar{\mathscr{L}}(\ell), \bar{\ell})$ extends $(X, \mathscr{L}, \ell)$, and the proof is complete. □

EXERCISES

3.1.1.(E) Take $g \in \mathbb{R}_+^X$, where X is an arbitrary set. We investigate the extension of $(X, \mathscr{k}(X), \ell_g)$. Such a Daniell space was introduced and discussed in Ex. 1.3.2. Prove the statements that follow:

(α) $f \in \mathscr{k}(X)^{\uparrow}$ iff $\{f \neq -\infty\} = X$, $\{f \neq 0\}$ is countable and $\{f < 0\}$ is finite.

(β) $f \in \mathscr{k}(X)^{\downarrow}$ iff $\{f \neq \infty\} = X$, $\{f \neq 0\}$ is countable and $\{f > 0\}$ is finite.

(γ) If $f \in \mathscr{k}(X)^{\uparrow}$, then

$$\ell_g^{\uparrow}(f) = \sup\left\{\sum_{x \in A} f(x)g(x) \mid A \in \mathfrak{F}(X)\right\} + \sum_{x \in \{f<0\}} f(x)g(x).$$

(δ) If $f \in \mathscr{k}(X)^{\downarrow}$, then

$$\ell_g^{\downarrow}(f) = \inf\left\{\sum_{x \in A} f(x)g(x) \mid A \in \mathfrak{F}(X)\right\} + \sum_{x \in \{f>0\}} f(x)g(x).$$

(ε)
$$\bar{\mathscr{L}}(\ell_g) = \left\{ f \in \bar{\mathbb{R}}^X \;\middle|\; \begin{array}{l} \{f \neq 0\} \text{ countable}, \\ \sup\left\{\sum_{x \in A} |f(x)g(x)| \mid A \in \mathfrak{F}(X)\right\} < \infty \end{array} \right\}$$

(ζ) If $f \in \bar{\mathscr{L}}(\ell_g)$, then fg is summable, and

$$\overline{\ell_g}(f) = \sum_{x \in X} f(x)g(x) = \sup\left\{\sum_{x \in A} f(x)g(x) \mid A \in \mathfrak{F}(X)\right\}$$
$$+ \inf\left\{\sum_{x \in A} f(x)g(x) \mid A \in \mathfrak{F}(X)\right\}.$$

Complete proofs for the special case $g = 1$ are given in Section 3.4.

We next compare $(X, \bar{\mathscr{L}}(\ell_g), \bar{\ell}_g)$ with $(X, \ell_g^1(X), \bar{\ell}_g)$, the extension of $(X, \mathscr{k}(X), \ell_g)$ defined in Ex. 1.3.2. Show that

(η) $(\bar{\mathscr{L}}(\ell_g), \bar{\ell}_g) \preccurlyeq (\ell_g^1(X), \bar{\ell}_g)$ and $\bar{\mathscr{L}}(\ell_g) = \ell_g^1(X)$ iff $\{g = 0\}$ is countable.

Thus $(X, \bar{\mathscr{L}}(\ell_g), \bar{\ell}_g)$ can be extended whenever $\{g = 0\}$ is uncountable, and, in fact, in a very natural way. We shall seal this hole with the definition of the integral in Chapter 4.

3.1.2(E) Let X be a set. Let ℓ be a positive linear nullcontinuous functional on $\ell^\infty(X)$ (cf., Ex. 1.3.5). Prove the following statements:

(α) $\ell^\infty(X)^\uparrow = \{f \in \overline{\mathbb{R}}^X \mid f \geq \alpha$ for some $\alpha \in \mathbb{R}\}$.
(β) $\ell^\infty(X)^\downarrow = \{f \in \overline{\mathbb{R}}^X \mid f \leq \alpha$ for some $\alpha \in \mathbb{R}\}$.
(γ) If $f \in \ell^\infty(X)^\uparrow$ and $\gamma \in \mathbb{R}$, then $f \wedge \gamma \in \ell^\infty(X)$ and $\ell^\uparrow(f) = \sup_{\gamma \in \mathbb{R}} \ell(f \wedge \gamma)$.
(δ) If $f \in \ell^\infty(X)^\downarrow$ and $\gamma \in \mathbb{R}$, then $f \vee \gamma \in \ell^\infty(X)$ and $\ell^\downarrow(f) = \inf_{\gamma \in \mathbb{R}} \ell(f \vee \gamma)$.
(ε) $\bar{\mathscr{L}}(\ell) = \{f \in \overline{\mathbb{R}}^X \mid \sup_{\gamma \in \mathbb{R}} \ell(|f| \wedge \gamma) < \infty\}$.
(ζ) If $f \in \bar{\mathscr{L}}(\ell)$, then $\bar{\ell}(f) = \sup_{\gamma \in \mathbb{R}} \ell(f^+ \wedge \gamma) - \sup_{\gamma \in \mathbb{R}} \ell(f^- \wedge \gamma)$.
(η) If $f \in \bar{\mathscr{L}}(\ell)$ and if $g \in \overline{\mathbb{R}}^X$ with $|g| \leq |f|$ then $g \in \bar{\mathscr{L}}(\ell)$.

3.1.3(E) Let X be a set and ℓ a positive linear nullcontinuous functional on $c_0(X)$ (cf., Ex. 1.3.5). Verify the propositions that follow:

(α) $c_0(X)^\uparrow = \mathscr{k}(X)^\uparrow$ and $c_0(X)^\downarrow = \mathscr{k}(X)^\downarrow$.
(β) $\ell^\uparrow(f) = (\ell|_{\mathscr{k}(X)})^\uparrow(f)$ for $f \in c_0(X)^\uparrow$.
(γ) $\ell^\downarrow(f) = (\ell|_{\mathscr{k}(X)})^\downarrow(f)$ for $f \in c_0(X)^\downarrow$.
(δ) $(X, \bar{\mathscr{L}}(\ell), \bar{\ell}) = (X, \bar{\mathscr{L}}(\ell|_{\mathscr{k}(X)}), \overline{\ell|_{\mathscr{k}(X)}})$.
(ε) $c_0(X) \subset \bigcap\{\bar{\mathscr{L}}(\ell_g) \mid g \in \ell^1(X)_+\} = \ell^\infty(X)$.

3.1.4(E) Let X be a set and ℓ a positive linear nullcontinuous functional on $c(X)$ (cf., Ex. 1.3.5(b)). Verify the propositions that follow:

(α) $c(X)^\uparrow = \{f \in \overline{\mathbb{R}}^X \mid f = \alpha e_X + g$ for $\alpha \in \overline{\mathbb{R}} \setminus \{-\infty\}$ and $g \in \mathscr{k}(X)^\uparrow\}$.
(β) $c(X)^\downarrow = \{f \in \overline{\mathbb{R}}^X \mid f = \alpha e_X + g$ for $\alpha \in \overline{\mathbb{R}} \setminus \{\infty\}$ and $g \in \mathscr{k}(X)^\downarrow\}$.
(γ) If $f = \alpha e_X + g \in c(X)^\uparrow$ for some $g \in \mathscr{k}(X)^\uparrow$, then $\ell^\uparrow(f) = \alpha\ell(e_X) + (\ell|_{\mathscr{k}(X)})^\uparrow(g)$.
(δ) If $f = \alpha e_X + g \in c(X)^\downarrow$ for some $g \in \mathscr{k}(X)^\downarrow$, then $\ell^\downarrow(f) = \alpha\ell(e_X) + (\ell|_{\mathscr{k}(X)})^\downarrow(g)$.
(ε) $\bar{\mathscr{L}}(\ell) = \{\alpha e_X + g \mid \alpha \in \mathbb{R},\ g \in \bar{\mathscr{L}}(\ell|_{\mathscr{k}(X)})\}$.
(ζ) If $f = \alpha e_X + g \in \bar{\mathscr{L}}(\ell)$ for some $g \in \bar{\mathscr{L}}(\ell|_{\mathscr{k}(X)})$, then

$$\bar{\ell}(f) = \alpha\ell(e_X) + \overline{\ell|_{\mathscr{k}(X)}}(g).$$

Now consider the case where X is uncountable. Take a fixed $\gamma \in \mathbb{R}_+ \setminus \{0\}$, and define the functional, ℓ, on $c(X)$ by

$$\ell\colon c(X) \to \mathbb{R}, \qquad \alpha e_X + g \mapsto \alpha\gamma.$$

Verify the following properties of ℓ:

(η) If $f = \alpha e_X + g \in c(X)^{\uparrow}$, then $\ell^{\uparrow}(f) = \alpha\gamma$.

(ϑ) If $f = \alpha e_X + g \in c(X)^{\downarrow}$, then $\ell^{\downarrow}(f) = \alpha\gamma$.

(ι) $\bar{\mathscr{L}}(\ell) = \{\alpha e_X + g \mid \alpha \in \mathbb{R}, \{g \neq 0\}$ countable$\}$.

(κ) If $f = \alpha e_X + g \in \bar{\mathscr{L}}(\ell)$, then $\bar{\ell}(f) = \alpha\gamma$.

3.1.5(E) For the set X define $\mathscr{L} := \{f \in \mathbb{R}^X \mid f(X)$ finite$\}$. Let $\mathfrak{U}$ be an ultrafilter on X which is δ-stable. Show that the following statement holds:

(α) For each $f \in \overline{\mathbb{R}}^X$ there is a unique $\alpha_f \in \overline{\mathbb{R}}$ with $\{f = \alpha_f\} \in \mathfrak{U}$.

For each $n \in \mathbb{N}$ and $m \in \mathbb{Z}$, define $A_{mn} :=]m/2^n, (m+1)/2^n]$. Then

$$\overline{\mathbb{R}} = \bigcup_{m \in \mathbb{Z}} A_{mn} \cup \{-\infty\} \cup \{\infty\}$$

for each $n \in \mathbb{N}$, which expresses $\overline{\mathbb{R}}$ as a disjoint union. Thus

$$X = \bigcup_{m \in \mathbb{Z}} f^{-1}(A_{mn}) \cup f^{-1}(\{-\infty\}) \cup f^{-1}(\{\infty\})$$

which is again a disjoint union. Hence exactly one of the sets to the right of the equality sign belongs to $\mathfrak{U}$. If this set is $f^{-1}(\{\infty\})$ or $f^{-1}(\{-\infty\})$, then (α) is proved. If not, construct a decreasing sequence $(A_{m_n n})_{n \in \mathbb{N}}$ of intervals whose lengths tend to 0, such that $f^{-1}(A_{m_n n}) \in \mathfrak{U}$. Thus $\bigcap_{n \in \mathbb{N}} \overline{A}_{m_n n}$ contains exactly one element, say α_f. Then $\{f = \alpha_f\} = \bigcap_{n \in \mathbb{N}} f^{-1}(A_{m_n n})$, so that $\{f = \alpha_f\} \in \mathfrak{U}$. It is clear that α_f is unique.

Now define $\ell\colon \mathscr{L} \to \mathbb{R}$, $f \mapsto \alpha_f$. Show the following:

(β) $\mathscr{L}^{\uparrow} = \{f \in \overline{\mathbb{R}}^X \mid f \geq \gamma$ for some $\gamma \in \mathbb{R}\}$.

(γ) $\mathscr{L}^{\downarrow} = \{f \in \overline{\mathbb{R}}^X \mid f \leq \gamma$ for some $\gamma \in \mathbb{R}\}$.

(δ) $\bar{\mathscr{L}}(\ell) = \{f \in \overline{\mathbb{R}}^X \mid \alpha_f \in \mathbb{R}\}$.

(ε) $\bar{\ell}(f) = \alpha_f$ for each $f \in \bar{\mathscr{L}}(\ell)$.

3.1.6(E) Suppose that $\mathfrak{R} \subset \mathfrak{P}(X)$ is a ring of sets, where X is a set. Let $\mathscr{L}_{\mathfrak{R}}$ denote, as always, the Riesz lattice of all $\mathfrak{R}$-step functions on X. Define two systems of sets:

$$\check{\mathfrak{R}} := \left\{ \bigcup_{n \in \mathbb{N}} A_n \,\middle|\, (A_n)_{n \in \mathbb{N}} \text{ a sequence from } \mathfrak{R} \right\}$$

$$\hat{\mathfrak{R}} := \left\{ \bigcap_{n \in \mathbb{N}} A_n \,\middle|\, (A_n)_{n \in \mathbb{N}} \text{ a sequence from } \mathfrak{R} \right\}$$

Prove the statements that follow:

(α) $f \in \mathscr{L}_{\mathfrak{R}}^{\uparrow}$ iff $f \in \overline{\mathbb{R}}^X$ and f satisfies the following:

(α1) $\{f > \gamma\} \in \check{\mathfrak{R}}$ whenever $\gamma \in \mathbb{R}$, $\gamma \geq 0$.
(α2) $\{f \leq \gamma\} \in \hat{\mathfrak{R}}$ whenever $\gamma \in \mathbb{R}$, $\gamma < 0$, and $\{f < 0\} \subset A$ for some $A \in \mathfrak{R}$.
(α3) $f \geq \gamma$ for some $\gamma \in \mathbb{R}$.

(β) $f \in \mathscr{L}_{\mathfrak{R}}^{\downarrow}$ iff $f \in \overline{\mathbb{R}}^X$ and f satisfies the following:

(β1) $\{f \geq \gamma\} \in \hat{\mathfrak{R}}$ whenever $\gamma \in \mathbb{R}$, $\gamma > 0$, and $\{f > 0\} \subset A$ for some $A \in \mathfrak{R}$.
(β2) $\{f < \gamma\} \in \check{\mathfrak{R}}$ whenever $\gamma \in \mathbb{R}$, $\gamma \leq 0$.
(β3) $f \leq \gamma$ for some $\gamma \in \mathbb{R}$.

To prove (α), take $f \in \mathscr{L}_{\mathfrak{R}}^{\uparrow}$. Then there is an increasing sequence $(f_n)_{n \in \mathbb{N}}$ from $\mathscr{L}_{\mathfrak{R}}$ such that $f = \bigvee_{n \in \mathbb{N}} f_n$. If $\gamma \in \mathbb{R}_+$ then $\{f_n > \gamma\} \in \mathfrak{R}$ for each $n \in \mathbb{N}$, and therefore $\{f > \gamma\} = \bigcup_{n \in \mathbb{N}} \{f_n > \gamma\} \in \check{\mathfrak{R}}$. If $\gamma \in \mathbb{R}$, $\gamma < 0$, then $\{f_n \leq \gamma\} \in \mathfrak{R}$ for each $n \in \mathbb{N}$, and therefore $\{f \leq \gamma\} = \bigcap_{n \in \mathbb{N}} \{f_n \leq \gamma\} \in \hat{\mathfrak{R}}$.

Conversely take $f \in \overline{\mathbb{R}}^X$ satisfying the conditions (α1)–(α3). First, suppose $f \geq 0$. For $n \in \mathbb{N}$ define

$$g_n := \sum_{k=1}^{n2^n} \frac{1}{2^n} e_{\{f > k/2^n\}}.$$

Then $(g_n)_{n \in \mathbb{N}}$ is an increasing sequence and $f = \bigvee_{n \in \mathbb{N}} g_n$. For each $n \in \mathbb{N}$ there is an increasing sequence $(g_{nm})_{m \in \mathbb{N}}$ from $\mathscr{L}_{\mathfrak{R}}$ such that $g_n = \bigvee_{m \in \mathbb{N}} g_{nm}$. By Proposition 2 (g), $f \in \mathscr{L}_{\mathfrak{R}}^{\uparrow}$.

Now suppose that $f \leq 0$. Take $\gamma \in \mathbb{R}$ such that $f \geq \gamma$ [as in (α3)]. Then, for some $m \in \mathbb{N}$, $-2^m \leq \gamma$. For each $n \in \mathbb{N}$ set $A_{n,1} := A$, where A is the set of (α2). For each $n \in \mathbb{N}$ and for each $k \in \mathbb{N}_{n2^n} \setminus \{1\}$ define $A_{n,k} := \{f \leq -(k-1)/2^n\}$. Next, define

$$h_n := -\sum_{k=1}^{2^{(n+m)}} \frac{1}{2^n} e_{A_{n,k}}.$$

Then $(h_n)_{n \in \mathbb{N}}$ is an increasing sequence with $f = \bigvee_{n \in \mathbb{N}} h_n$. But $h_n \in \mathscr{L}_{\mathfrak{R}}^{\uparrow}$ for each $n \in \mathbb{N}$, so that $f \in \mathscr{L}_{\mathfrak{R}}^{\uparrow}$.

Finally, for an arbitrary f, use the decomposition $f = f^+ - f^-$ and the preceding discussion to complete the proof.

Now turn to the special case $\mathfrak{R} = \mathfrak{P}(X)$. Prove the next two statements:

(γ) $\mathscr{L}_{\mathfrak{R}}^{\uparrow} = \{f \in \overline{\mathbb{R}}^X \mid f \geq \gamma$ for some $\gamma \in \mathbb{R}\}$.
(δ) $\mathscr{L}_{\mathfrak{R}}^{\downarrow} = \{f \in \overline{\mathbb{R}}^X \mid f \leq \gamma$ for some $\gamma \in \mathbb{R}\}$.

Compare with Ex. 3.1.5 (β) and (γ). Observe that $\mathscr{L}_{\mathfrak{P}(X)} = \{f \in \mathbb{R}^X \mid f(X)$ finite$\}$.

3.1.7(E) **The Stieltjes Functionals.** $\mathscr{K}(A)^{\uparrow}$, $\mathscr{K}(A)^{\downarrow}$, $\ell_g^{\uparrow}$, and $\ell_g^{\downarrow}$ for the Stieltjes functional ℓ_g on $\mathscr{K}(A)$ are discussed in Section 3.6. The interested reader is invited to work through that section here.

3.1.8(E) The statements that follow are equivalent for the Daniell space $(X, \mathscr{L}, \ell)$:

(α) $(X, \mathscr{L}, \ell)$ is closed.
(β) $\mathscr{L} = \{f \in \mathscr{L}^{\uparrow} \mid \ell^{\uparrow}(f) < \infty\}$.
(γ) $\mathscr{L} = \{f \in \mathscr{L}^{\downarrow} \mid \ell^{\downarrow}(f) > -\infty\}$.
(δ) $(X, \bar{\mathscr{L}}(\ell), \bar{\ell}) = (X, \mathscr{L}, \ell)$.

3.1.9(E) Let $(X, \mathscr{L}, \ell)$ be a Daniell space. Take $f \in \bar{\mathscr{L}}(\ell)$, and let α, β be real numbers such that $\alpha \le f \le \beta$ and $\alpha \le 0 \le \beta$. Prove that for any $\varepsilon > 0$ there are functions $f' \in \mathscr{L}^{\downarrow} \cap \bar{\mathscr{L}}(\ell)$ and $f'' \in \mathscr{L}^{\uparrow} \cap \bar{\mathscr{L}}(\ell)$ such that the following conditions hold:

(i) $\alpha \le f' \le f \le f'' \le \beta$.
(ii) $\bar{\ell}(f') \ge \bar{\ell}(f) - \varepsilon$.
(iii) $\bar{\ell}(f'') \le \bar{\ell}(f) + \varepsilon$.

3.1.10(E) Let $(X, \mathscr{L}, \ell)$ be a Daniell space. Define

$$\mathscr{L}^{\uparrow}(\ell) := \{f \in \mathscr{L}^{\uparrow} \mid \ell^{\uparrow}(f) < \infty\}.$$

Prove that $\bar{\mathscr{L}}(\ell) = \mathscr{L}^{\uparrow}(\ell) \dot{-} \mathscr{L}^{\uparrow}(\ell)$ and $\bar{\ell}(f \dot{-} g) = \{\ell^{\uparrow}(f) - \ell^{\uparrow}(g)\}$ for $f, g \in \mathscr{L}^{\uparrow}(\ell)$.

We offer some suggestions for a proof.

STEP 1: If $f \in \mathscr{L}^{\uparrow}$ and $g \in \mathscr{L}^{\downarrow}$, then $f - g \in \mathscr{L}^{\uparrow}$ and $\ell^{\uparrow}(f - g) = \ell^{\uparrow}(f) - \ell^{\downarrow}(g)$.

STEP 2: If $f \in \bar{\mathscr{L}}(\ell)$, then there is an increasing sequence $(f_n')_{n \in \mathbb{N}}$ from $\mathscr{L}^{\downarrow}$ with $f \ge f_n'$ for all $n \in \mathbb{N}$ and $\bar{\ell}(f) = \sup_{n \in \mathbb{N}} \ell^{\downarrow}(f_n)$. Define $A := \{f \ne \bigvee_{n \in \mathbb{N}} f_n'\}$. Now define recursively an increasing sequence $(f_n'')_{n \in \mathbb{N}}$ from $\mathscr{L}^{\uparrow}$ with the following properties:

(i) $f_n' \le f_n''$ and $A \subset \{f_n'' = \infty\}$ for every $n \in \mathbb{N}$.
(ii) $(f_n'' - f_n')_{n \in \mathbb{N}}$ is an increasing sequence.
(iii) $\ell^{\uparrow}(f_n'' - f_n') < \Sigma_{k \le n} 1/2^k$ for each $n \in \mathbb{N}$.

Assume that f_n'' has been defined for each $n \le m$. Take $g \in \mathscr{L}$ with $f_{m+1}' \le g$ and $\ell(g) - \ell^{\downarrow}(f_{m+1}') < 1/2^{m+1}$. Define

$$f_{m+1}'' := g + \sum_{k \le m} (f_k'' - f_k').$$

Then $f_{m+1}'' \ge f_m'' \vee f_{m+1}'$, $f_{m+1}'' - f_{m+1}' \ge f_m'' - f_m'$, and $\ell^{\uparrow}(f_{m+1}'' - f_{m+1}') \le \Sigma_{k \le m+1} 1/2^k$. Define $f'' := \bigvee_{n \in \mathbb{N}} f_n''$ and $g'' := \bigvee_{n \in \mathbb{N}} (f_n'' - f_n')$. Then $f'', g'' \in \mathscr{L}^{\uparrow}(\ell)$ and $f \in \langle f'' \dot{-} g'' \rangle$.

The reader has perhaps noticed that the result of this exercise describes a quick way to characterize $(X, \bar{\mathscr{L}}(\ell), \bar{\ell})$. We offer two comments. First, if one were to introduce $\bar{\mathscr{L}}(\ell)$ and $\bar{\ell}$ in this way, attending to all details, then the method would show itself not quite so quick after all. Second, the procedure presented here relies more heavily on the special properties of Daniell spaces than Daniell's classical method. We have always taken pain to use constructions and procedures that suggest ways to generalizations. The reader will appreciate this after completing Ex. 3.1.13.

3.1.11(E) Let $(X, \mathscr{L}, \ell)$ be a Daniell space. Once again put

$$\mathscr{L}^{\uparrow}(\ell) := \{ f \in \mathscr{L}^{\uparrow} \mid \ell^{\uparrow}(f) < \infty \}$$

$$\mathscr{L}^{\downarrow}(\ell) := \{ f \in \mathscr{L}^{\downarrow} \mid \ell^{\downarrow}(f) > -\infty \}.$$

Let $\mathscr{L}_0$ be the set of all $f \in \overline{\mathbb{R}}^X$ for which there exist an increasing sequence $(f_n')_{n \in \mathbb{N}}$ from $\mathscr{L}^{\downarrow}(\ell)$ and a decreasing sequence $(f_n'')_{n \in \mathbb{N}}$ from $\mathscr{L}^{\uparrow}(\ell)$ such that $f = \bigvee_{n \in \mathbb{N}} f_n' = \bigwedge_{n \in \mathbb{N}} f_n''$. Verify the following statements:

(α) $\mathscr{L}_0 \subset \bar{\mathscr{L}}(\ell)$.

(β) If $\ell_0 := \bar{\ell}|_{\mathscr{L}_0}$, then $(X, \mathscr{L}^{\uparrow}(\ell), \ell^{\uparrow}) \preccurlyeq (X, \mathscr{L}_0, \ell_0)$, $(X, \mathscr{L}^{\downarrow}(\ell), \ell^{\downarrow}) \preccurlyeq (X, \mathscr{L}_0, \ell_0)$, and $(X, \mathscr{L}_0, \ell_0) \preccurlyeq (X, \bar{\mathscr{L}}(\ell), \bar{\ell})$.

Observe that $(X, \mathscr{L}_0, \ell_0)$ is, in general, not a Daniell space, since it could happen that $f, g \in \mathscr{L}_0$ but $\langle f \dotplus g \rangle \not\subset \mathscr{L}_0$.

We define, for each ordinal α, $(X, \mathscr{L}_\alpha, \ell_\alpha) \preccurlyeq (X, \bar{\mathscr{L}}(\ell), \bar{\ell})$ as follows: Let $(X, \mathscr{L}_0, \ell_0)$ be the space defined previously. Suppose that $(X, \mathscr{L}_\beta, \ell_\beta)$ has been defined for each $\beta < \alpha, \alpha > 0$. Define

$$\mathscr{L}_\alpha' := \bigcup_{\beta < \alpha} \mathscr{L}_\beta \text{ and } \ell_\alpha'(f) := \ell_\beta(f) \text{ if } f \in \mathscr{L}_\beta, \beta < \alpha$$

$$\mathscr{L}_\alpha^{\uparrow} := \left\{ \bigvee_{n \in \mathbb{N}} f_n \;\middle|\; \begin{array}{l} (f_n)_{n \in \mathbb{N}} \text{ an increasing sequence} \\ \text{from } \mathscr{L}_\alpha' \text{ with } \sup_{n \in \mathbb{N}} \ell_\alpha'(f_n) < \infty \end{array} \right\}$$

$$\mathscr{L}_\alpha^{\downarrow} := \left\{ \bigwedge_{n \in \mathbb{N}} f_n \;\middle|\; \begin{array}{l} (f_n)_{n \in \mathbb{N}} \text{ a decreasing sequence} \\ \text{from } \mathscr{L}_\alpha' \text{ with } \inf_{n \in \mathbb{N}} \ell_\alpha'(f_n) > -\infty \end{array} \right\}.$$

Finally, let $\mathscr{L}_\alpha$ be the set of all $f \in \overline{\mathbb{R}}^X$ for which there exist an increasing sequence $(f_n')_{n \in \mathbb{N}}$ from $\mathscr{L}_\alpha^{\downarrow}$ and a decreasing sequence $(f_n'')_{n \in \mathbb{N}}$ from $\mathscr{L}_\alpha^{\uparrow}$ such that $f = \bigvee_{n \in \mathbb{N}} f_n' = \bigwedge_{n \in \mathbb{N}} f_n''$, and set $\ell_\alpha := \bar{\ell}|_{\mathscr{L}_\alpha}$. Prove the following statements, where ω_1 denotes the first uncountable ordinal:

(γ) $\mathscr{L}_{\omega_1}' = \mathscr{L}_{\omega_1}^{\uparrow} = \mathscr{L}_{\omega_1}^{\downarrow} = \mathscr{L}_{\omega_1}$.

(δ) $\mathscr{L}_{\omega_1 + 1} = \mathscr{L}_{\omega_1}$.

(ε) $(X, \mathscr{L}_{\omega_1}, \ell_{\omega_1})$ is closed but not generally a Daniell space. It is the smallest closed triple extending $(X, \mathscr{L}, \ell)$.

(ζ) $(X, \mathscr{L}, \ell) \preccurlyeq (X, \mathscr{L}_{\omega_1}, \ell_{\omega_1}) \preccurlyeq (X, \bar{\mathscr{L}}(\ell), \bar{\ell})$.

(η) If $f \in \bar{\mathscr{L}}(\ell)$, then there is a $g \in \mathscr{L}_1$ with $f = g$ $\bar{\ell}$-a.e.

$(X, \mathscr{L}_{\omega_1}, \ell_{\omega_1})$ is called the **Borel integral** of $(X, \mathscr{L}, \ell)$. The functions of $\mathscr{L}_{\omega_1}$ are called **Borel integrable**.

The following example shows that, in general, $\mathscr{L}_{\omega_1} \neq \bar{\mathscr{L}}(\ell)$. Let X be a set consisting of two elements. Let $\mathscr{L}$ be the set of all constant real functions on X, and set $\ell := 0$. Then $\bar{\mathscr{L}}(\ell) = \overline{\mathbb{R}}^X$, but $\mathscr{L}_{\omega_1}$ is the set of all constant extended-real functions on X.

3.1.12(E) Prove that if $(X, \mathscr{L}, \ell)$ is a Daniell space, then the space $(X, \mathscr{L}_\gamma, \ell_\gamma)$ defined in the transfinite construction in Ex. 1.4.5 is identical with $(X, \bar{\mathscr{L}}(\ell), \bar{\ell})$.

3.1.13(C) **Generalization of the Construction**. Several aspects of the procedure outlined in Section 3.1 are best discussed in a more general setting. The purpose of this exercise is to start the work toward this goal. In the process we shall see the crystallization of the truly general parts of the procedure and come to a deeper understanding of the peculiarities of linear functionals on Riesz lattices. The concepts introduced here will be of great importance in the sequel—later exercises will contain diverse special cases of them.

We start with the introduction of some notions that are also important in other connections:

(a) **The σ-distributivity**. This property is trivially fulfilled in the case of $\mathbb{R}$ and $\overline{\mathbb{R}}^X$. So it has not appeared as yet.

A lattice X is called **σ-distributive** iff, for every countable family $(x_\iota)_{\iota \in I}$ from X and every $x \in X$ the following conditions are met:

(i) $\bigvee_{\iota \in I}(x_\iota \wedge x)$ exists whenever $\bigvee_{\iota \in I} x_\iota$ exists, and then $\bigvee_{\iota \in I}(x_\iota \wedge x) = (\bigvee_{\iota \in I} x_\iota) \wedge x$.

(ii) $\bigwedge_{\iota \in I}(x_\iota \vee x)$ exists whenever $\bigwedge_{\iota \in I} x_\iota$ exists, and then $\bigwedge_{\iota \in I}(x_\iota \vee x) = (\bigwedge_{\iota \in I} x_\iota) \vee x$.

Prove the following propositions for a σ-distributive lattice X.

(α) If $(x_n)_{n \in \mathbb{N}}$ and $(y_n)_{n \in \mathbb{N}}$ are increasing sequences from X such that $\bigvee_{n \in \mathbb{N}} x_n$ and $\bigvee_{n \in \mathbb{N}} y_n$ exist, then $\bigvee_{n \in \mathbb{N}}(x_n \wedge y_n)$ exists, and $\bigvee_{n \in \mathbb{N}}(x_n \wedge y_n) = (\bigvee_{n \in \mathbb{N}} x_n) \wedge (\bigvee_{n \in \mathbb{N}} y_n)$.

(β) If $(x_n)_{n \in \mathbb{N}}$ and $(y_n)_{n \in \mathbb{N}}$ are decreasing sequences from X such that $\bigwedge_{n \in \mathbb{N}} x_n$ and $\bigwedge_{n \in \mathbb{N}} y_n$ exist, then $\bigwedge_{n \in \mathbb{N}}(x_n \vee y_n)$ exists, and $\bigwedge_{n \in \mathbb{N}}(x_n \vee y_n) = (\bigwedge_{n \in \mathbb{N}} x_n) \vee (\bigwedge_{n \in \mathbb{N}} y_n)$.

(γ) If $(x_n)_{n \in \mathbb{N}}$ and $(y_n)_{n \in \mathbb{N}}$ are order-convergent sequences from X, then $(x_n \vee y_n)_{n \in \mathbb{N}}$ and $(x_n \wedge y_n)_{n \in \mathbb{N}}$ are also order convergent, and

$$\lim_{n \to \infty} (x_n \vee y_n) = \left(\lim_{n \to \infty} x_n\right) \vee \left(\lim_{n \to \infty} y_n\right)$$

$$\lim_{n \to \infty} (x_n \wedge y_n) = \left(\lim_{n \to \infty} x_n\right) \wedge \left(\lim_{n \to \infty} y_n\right).$$

(b) **Metric lattices**. A **metric lattice** is a pair (X, d) with the following properties.

(i) X is a lattice and d a metric on X.

(ii) For any $x, y, z \in X$, $x \leq y \leq z$ implies $d(x, y) \leq d(x, z)$.

(iii) The topology on X generated by d is the order topology. $\leq$ is a closed relation with respect to the product topology on $X \times X$.

Prove: If d denotes the natural metric on $\mathbb{R}$, then $(\mathbb{R}, d)$ is a metric lattice.

Let X be an ordered set. The pair $((x_n)_{n\in\mathbb{N}}, (y_n)_{n\in\mathbb{N}})$ of sequences from X is called **transversal** iff the following conditions are fulfilled:

(iv) $(x_n)_{n\in\mathbb{N}}$ is increasing and $(y_n)_{n\in\mathbb{N}}$ decreasing.
(v) $x = \bigwedge_{n\in\mathbb{N}}(x \vee y_n)$ whenever x is an upper bound of $(x_n)_{n\in\mathbb{N}}$.
(vi) $y = \bigvee_{n\in\mathbb{N}}(x_n \wedge y)$ whenever y is a lower bound of $(y_n)_{n\in\mathbb{N}}$.

If X is a σ-complete lattice, then a pair of sequences $((x_n)_{n\in\mathbb{N}}, (y_n)_{n\in\mathbb{N}})$ is transversal iff it satisfies (iv) and

$$\bigvee_{n\in\mathbb{N}} x_n \geq \bigwedge_{n\in\mathbb{N}} y_n.$$

Now take a lattice Y, a metric lattice (Z, d) and a mapping $\varphi: Y \to Z$. φ is called **weakly increasing** iff the following conditions are fulfilled:

(vii) φ is increasing.
(viii) For any $\varepsilon > 0$ there is a $\delta > 0$ with the following property: If $((x_n)_{n\in\mathbb{N}}, (y_n)_{n\in\mathbb{N}})$ and $((x'_n)_{n\in\mathbb{N}}, (y'_n)_{n\in\mathbb{N}})$ are transversal in Y such that there is an $n \in \mathbb{N}$ with

$$d(\varphi(x_m), \varphi(y_m)) < \delta \quad\text{and}\quad d(\varphi(x'_m), \varphi(y'_m)) < \delta$$

whenever $m \geq n$, then there is an $\bar{n} \in \mathbb{N}$ with

$$d(\varphi(x_m \vee x'_m), \varphi(y_m \vee y'_m)) < \varepsilon$$

and

$$d(\varphi(x_m \wedge x'_m), \varphi(y_m \wedge y'_m)) < \varepsilon$$

whenever $m \geq \bar{n}$.

A pair (δ, ε) with the property (viii) is called a **corresponding pair** for φ. A **φ-sequence** is a sequence $(x_n)_{n\in\mathbb{N}}$ from Y that is monotone and for which $\lim_{n\to\infty}\varphi(x_n)$ exists. (φ-sequences are analogous to ℓ-sequences).

(c) **Lattice Integrals**. A **lattice integral** is a system (X, Z, d, Y, φ) with the following properties:

(i) X is a σ-complete and σ-distributive lattice.
(ii) (Z, d) is a metrically complete, conditionally σ-complete, metric lattice.
(iii) Y is a sublattice of X and $\varphi: Y \to Z$ a weakly increasing mapping with the property that if $((x_n)_{n\in\mathbb{N}}, (y_n)_{n\in\mathbb{N}})$ is a transversal pair of φ-sequences, then

$$\bigvee_{n\in\mathbb{N}} \varphi(x_n) \geq \bigwedge_{n\in\mathbb{N}} \varphi(y_n).$$

A lattice integral (X, Z, d, Y, φ) is called **closed** iff it satisfies the following conditions:

(iv) For every φ-sequence $(x_n)_{n\in\mathbb{N}}$ from Y, $\lim_{n\to\infty} x_n \in Y$ and $\varphi(\lim_{n\to\infty} x_n) = \lim_{n\to\infty}\varphi(x_n)$.

(v) If $x, y \in Y$ with $x \leq y$ and $\varphi(x) = \varphi(y)$, then $z \in Y$ whenever $z \in X$ with $x \leq z \leq y$.

The main theorem of this exercise is the following:

(α) Each lattice integral has a closed extension.

We divide the proof into several steps. Let (X, Z, d, Y, φ) be a lattice integral.

STEP 1: If $(x_n)_{n\in\mathbb{N}}$ is a monotone sequence from Y with $\lim_{n\to\infty} x_n \in Y$, then $\varphi(\lim_{n\to\infty} x_n) = \lim_{n\to\infty}\varphi(x_n)$.

Take an increasing sequence $(x_n)_{n\in\mathbb{N}}$ from Y with $x := \lim_{n\to\infty} x_n \in Y$. Then $((x_n)_{n\in\mathbb{N}}, (x)_{n\in\mathbb{N}})$ is transversal and thus

$$\bigvee_{n\in\mathbb{N}} \varphi(x_n) \geq \bigwedge_{n\in\mathbb{N}} \varphi(x) = \varphi(x) \geq \bigvee_{n\in\mathbb{N}} \varphi(x_n)$$

so that $\varphi(x) = \lim_{n\to\infty}\varphi(x_n)$. The case of decreasing sequences can be treated analogously.

STEP 2: Define

$$Y^{\uparrow} := \left\{ \bigvee_{n\in\mathbb{N}} x_n \,|\, (x_n)_{n\in\mathbb{N}} \text{ is an increasing } \varphi\text{-sequence from } Y \right\}$$

$$Y^{\downarrow} := \left\{ \bigwedge_{n\in\mathbb{N}} x_n \,|\, (x_n)_{n\in\mathbb{N}} \text{ is a decreasing } \varphi\text{-sequence from } Y \right\}.$$

Prove that $Y^{\uparrow}$ and $Y^{\downarrow}$ are sublattices of X.

Take $x, y \in Y^{\uparrow}$. Let $(x_n)_{n\in\mathbb{N}}$ and $(y_n)_{n\in\mathbb{N}}$ be increasing φ-sequences such that $x = \bigvee_{n\in\mathbb{N}} x_n$ and $y = \bigvee_{n\in\mathbb{N}} y_n$. Take a corresponding pair (δ, ε) for φ. There is an $n \in \mathbb{N}$ such that if $m \in \mathbb{N}$, then

$$d(\varphi(x_n), \varphi(x_{n+m})) < \delta \quad \text{and} \quad d(\varphi(y_n), \varphi(y_{n+m})) < \delta.$$

Consequently, there is $\overline{m} \in \mathbb{N}$ such that

$$d(\varphi(x_n \vee y_n), \varphi(x_{n+m} \vee y_{n+m})) < \varepsilon$$

and

$$d(\varphi(x_n \wedge y_n), \varphi(x_{n+m} \wedge y_{n+m})) < \varepsilon$$

for any $m \geq \overline{m}$. Because d is complete, both $\bigvee_{n\in\mathbb{N}}\varphi(x_n \vee y_n)$ and $\bigvee_{n\in\mathbb{N}}\varphi(x_n \wedge y_n)$ exist. Thus, $(x_n \vee y_n)_{n\in\mathbb{N}}$ and $(x_n \wedge y_n)_{n\in\mathbb{N}}$ are φ-sequences. We conclude that $x \vee y = \bigvee_{n\in\mathbb{N}}(x_n \vee y_n) \in Y^{\uparrow}$ and, applying the σ-distributivity of X, $x \wedge y = \bigvee_{n\in\mathbb{N}}(x_n \wedge y_n) \in Y^{\uparrow}$.

STEP 3: Take $x \in Y^{\uparrow}$. Let $(x_n)_{n \in \mathbb{N}}$ and $(y_n)_{n \in \mathbb{N}}$ be increasing φ-sequences with $x = \bigvee_{n \in \mathbb{N}} x_n = \bigvee_{n \in \mathbb{N}} y_n$. Then $\bigvee_{n \in \mathbb{N}} \varphi(x_n) = \bigvee_{n \in \mathbb{N}} \varphi(y_n)$. The analogous proposition holds for $Y^{\downarrow}$.

Using the σ-distributivity, $x_n = \bigvee_{m \in \mathbb{N}} x_n \wedge y_m$ and $y_n = \bigvee_{m \in \mathbb{N}} x_m \wedge y_n$ for each $n \in \mathbb{N}$. By step 1,

$$\varphi(x_n) = \bigvee_{m \in \mathbb{N}} \varphi(x_n \wedge y_m) \leq \bigvee_{m \in \mathbb{N}} \varphi(y_m)$$

and

$$\varphi(y_n) = \bigvee_{m \in \mathbb{N}} \varphi(x_m \wedge y_n) \leq \bigvee_{m \in \mathbb{N}} \varphi(x_m)$$

for each $n \in \mathbb{N}$. Thus

$$\bigvee_{n \in \mathbb{N}} \varphi(x_n) \leq \bigvee_{n \in \mathbb{N}} \varphi(y_n) \leq \bigvee_{n \in \mathbb{N}} \varphi(x_n).$$

From this, the assertion follows.

This allows us to define

$$\varphi^{\uparrow} : Y^{\uparrow} \to Z, \qquad \bigvee_{n \in \mathbb{N}} x_n \mapsto \bigvee_{n \in \mathbb{N}} \varphi(x_n)$$

$$((x_n)_{n \in \mathbb{N}} \text{ an increasing } \varphi\text{-sequence})$$

and

$$\varphi^{\downarrow} : Y^{\downarrow} \to Z, \qquad \bigwedge_{n \in \mathbb{N}} x_n \mapsto \bigwedge_{n \in \mathbb{N}} \varphi(x_n)$$

$$((x_n)_{n \in \mathbb{N}} \text{ a decreasing } \varphi\text{-sequence}).$$

STEP 4: If $(x_n)_{n \in \mathbb{N}}$ is an increasing $\varphi^{\uparrow}$-sequence, then $\bigvee_{n \in \mathbb{N}} x_n \in Y^{\uparrow}$ and $\varphi^{\uparrow}(\bigvee_{n \in \mathbb{N}} x_n) = \bigvee_{n \in \mathbb{N}} \varphi^{\uparrow}(x_n)$. If $(x_n)_{n \in \mathbb{N}}$ is a decreasing $\varphi^{\downarrow}$-sequence, then $\bigwedge_{n \in \mathbb{N}} x_n \in Y^{\downarrow}$ and $\varphi^{\downarrow}(\bigwedge_{n \in \mathbb{N}} x_n) = \bigwedge_{n \in \mathbb{N}} \varphi^{\downarrow}(x_n)$.

The proof is analogous to that in the case of Daniell spaces.

STEP 5: Take $x \in Y^{\downarrow}$ and $y \in Y^{\uparrow}$ with $x \leq y$. Then $\varphi^{\downarrow}(x) \leq \varphi^{\uparrow}(y)$.

Let $(x_n)_{n \in \mathbb{N}}$ be a decreasing φ-sequence and $(y_n)_{n \in \mathbb{N}}$ an increasing φ-sequence such that $x = \bigwedge_{n \in \mathbb{N}} x_n$ and $y = \bigvee_{n \in \mathbb{N}} y_n$. Then $((y_n)_{n \in \mathbb{N}}, (x_n)_{n \in \mathbb{N}})$ is a transversal pair and so

$$\varphi^{\downarrow}(x) = \bigwedge_{n \in \mathbb{N}} \varphi(x_n) \leq \bigvee_{n \in \mathbb{N}} \varphi(y_n) = \varphi^{\uparrow}(y).$$

STEP 6: Take an increasing $\varphi^{\downarrow}$-sequence $(x_n)_{n \in \mathbb{N}}$. Then, for each $\varepsilon > 0$ there is an increasing φ-sequence $(x'_n)_{n \in \mathbb{N}}$ with $x_n \leq x'_n$ for each $n \in \mathbb{N}$

and

$$d\Big(\bigvee_{n\in\mathbb{N}}\varphi^{\downarrow}(x_n),\ \bigvee_{n\in\mathbb{N}}\varphi(x'_n)\Big)\le\varepsilon.$$

An analogous proposition holds for increasing $\varphi^{\uparrow}$-sequences $(x_n)_{n\in\mathbb{N}}$.

Consider the case of an increasing $\varphi^{\downarrow}$-sequence $(x_n)_{n\in\mathbb{N}}$. Set $z:=\bigvee_{n\in\mathbb{N}}\varphi^{\downarrow}(x_n)$ and take $\varepsilon>0$. Construct recursively an increasing sequence $(\delta_n)_{n\in\mathbb{N}}$ of strictly positive real numbers such that (δ_1,ε) is a corresponding pair for φ and (δ_{n+1},δ_n) is a corresponding pair for φ for every $n\in\mathbb{N}$.

For each $n\in\mathbb{N}$ there is a decreasing φ-sequence $(x_{nm})_{m\in\mathbb{N}}$ with $x_n=\bigwedge_{m\in\mathbb{N}}x_{nm}$ and $\varphi^{\downarrow}(x_n)=\bigwedge_{m\in\mathbb{N}}\varphi(x_{nm})$. Given $n\in\mathbb{N}$ there is $m_n\in\mathbb{N}$ with $d(\varphi(x_{nm}),\varphi(x_{nm_n}))<\delta_n$ whenever $m\ge m_n$. By the construction of $(\delta_n)_{n\in\mathbb{N}}$, given $n\in\mathbb{N}$, there is $\overline{m}\in\mathbb{N}$ such that

$$d\Big(\varphi\Big(\bigvee_{k\le n}x_{km}\Big),\varphi\Big(\bigvee_{k\le n}x_{km_k}\Big)\Big)<\varepsilon$$

for any $m\ge\overline{m}$.

Then, for $m\to\infty$,

$$d\Big(\varphi^{\downarrow}(x_n),\varphi\Big(\bigvee_{k\le n}x_{km_k}\Big)\Big)=d\Big(\varphi^{\downarrow}\Big(\bigvee_{k\le n}x_k\Big),\varphi\Big(\bigvee_{k\le n}x_{km_k}\Big)\Big)\le\varepsilon$$

for each $n\in\mathbb{N}$. $(\varphi^{\downarrow}(x_n))_{n\in\mathbb{N}}$ is a Cauchy sequence. Hence, $(\varphi(\bigvee_{k\le n}x_{km_k}))_{n\in\mathbb{N}}$ is also a Cauchy sequence, and since d is complete, $z_\varepsilon:=\bigvee_{n\in\mathbb{N}}\varphi(\bigvee_{k\le n}x_{km_k})$ exists and $d(z,z_\varepsilon)\le\varepsilon$.

Now define $x'_n:=\bigvee_{k\le n}x_{km_k}$ for each $n\in\mathbb{N}$. Then $(x'_n)_{n\in\mathbb{N}}$ is an increasing φ-sequence with $x_n\le x'_n$ for each $n\in\mathbb{N}$ and

$$d\Big(\bigvee_{n\in\mathbb{N}}\varphi^{\downarrow}(x_n),\ \bigvee_{n\in\mathbb{N}}\varphi(x'_n)\Big)=d(z,z_\varepsilon)\le\varepsilon.$$

The other case is treated analogously.

Step 7: Let $(x_n)_{n\in\mathbb{N}}$ be an increasing $\varphi^{\downarrow}$-sequence and $(y_n)_{n\in\mathbb{N}}$ a decreasing $\varphi^{\uparrow}$-sequence. If $((x_n)_{n\in\mathbb{N}},(y_n)_{n\in\mathbb{N}})$ is transversal, then $\bigvee_{n\in\mathbb{N}}\varphi^{\downarrow}(x_n)\ge\bigwedge_{n\in\mathbb{N}}\varphi^{\uparrow}(y_n)$.

Given $\varepsilon>0$, use Step 6 to find elements $x_\varepsilon\in Y^{\uparrow}$ and $y_\varepsilon\in Y^{\downarrow}$ with $x_\varepsilon\ge\bigvee_{n\in\mathbb{N}}x_n$, $y_\varepsilon\le\bigwedge_{n\in\mathbb{N}}y_n$ and

$$d\Big(\varphi^{\uparrow}(x_\varepsilon),\ \bigvee_{n\in\mathbb{N}}\varphi^{\downarrow}(x_n)\Big)<\varepsilon,\qquad d\Big(\varphi^{\downarrow}(y_\varepsilon),\ \bigwedge_{n\in\mathbb{N}}\varphi^{\uparrow}(y_n)\Big)<\varepsilon.$$

Consequently, $x_\varepsilon\ge y_\varepsilon$ and so, using Step 5, $\varphi^{\uparrow}(x_\varepsilon)\ge\varphi^{\downarrow}(y_\varepsilon)$ for all $\varepsilon>0$.

Hence

$$\bigvee_{n\in\mathbb{N}} \varphi^{\downarrow}(x_n) = \lim_{\varepsilon\to 0} \varphi^{\uparrow}(x_\varepsilon) \geq \lim_{\varepsilon\to 0} \varphi^{\downarrow}(y_\varepsilon) = \bigwedge_{n\in\mathbb{N}} \varphi^{\uparrow}(y_n).$$

(Here we use the fact that $\leq$ is closed.)

STEP 8: Let $(x_n)_{n\in\mathbb{N}}$ be an increasing sequence from $Y^{\downarrow}$ and $(y_n)_{n\in\mathbb{N}}$ a decreasing sequence from $Y^{\uparrow}$ such that $\bigvee_{n\in\mathbb{N}} x_n = \bigwedge_{n\in\mathbb{N}} y_n$. Then $\bigvee_{n\in\mathbb{N}}\varphi^{\downarrow}(x_n) = \bigwedge_{n\in\mathbb{N}}\varphi^{\uparrow}(y_n)$.

This follows from the conditional σ-completeness of Z, using Steps 5 and 7.

Now define

$$Y_0 := \left\{ x \in X \,\middle|\, \begin{array}{l} x = \bigvee_{n\in\mathbb{N}} x'_n = \bigwedge_{n\in\mathbb{N}} x''_n, \text{ where } (x'_n)_{n\in\mathbb{N}} \text{ is an} \\ \text{increasing sequence from } Y^{\downarrow} \text{ and } (x''_n)_{n\in\mathbb{N}} \\ \text{a decreasing one from } Y^{\uparrow} \end{array} \right\}.$$

For $x \in Y_0$ define $\varphi_0(x) := \bigvee_{n\in\mathbb{N}}\varphi^{\downarrow}(x'_n) = \bigwedge_{n\in\mathbb{N}}\varphi^{\uparrow}(x''_n)$, where $(x'_n)_{n\in\mathbb{N}}$ and $(x''_n)_{n\in\mathbb{N}}$ are sequences as in the definition of Y_0. Step 8 shows that $\varphi_0(x)$ is well-defined (that is, independent of the choice of $(x'_n)_{n\in\mathbb{N}}$ and $(x''_n)_{n\in\mathbb{N}}$.

STEP 9: $(X, Z, d, Y_0, \varphi_0)$ is a lattice integral.

Take $x, y \in Y_0$. Let $(x'_n)_{n\in\mathbb{N}}$ and $(y'_n)_{n\in\mathbb{N}}$ be increasing sequences from $Y^{\downarrow}$ and $(x''_n)_{n\in\mathbb{N}}$ and $(y''_n)_{n\in\mathbb{N}}$ decreasing sequences from $Y^{\uparrow}$ with $x = \bigvee_{n\in\mathbb{N}} x'_n = \bigwedge_{n\in\mathbb{N}} x''_n$ and $y = \bigvee_{n\in\mathbb{N}} y'_n = \bigwedge_{n\in\mathbb{N}} y''_n$. By the σ-distributivity of X, $x \vee y = \bigvee_{n\in\mathbb{N}}(x'_n \vee y'_n) = \bigwedge_{n\in\mathbb{N}}(x'_n \vee y'_n) \in Y_0$, and $x \wedge y = \bigvee_{n\in\mathbb{N}}(x'_n \wedge y'_n) = \bigwedge_{n\in\mathbb{N}}(x''_n \wedge y''_n) \in Y_0$ so that Y_0 is a sublattice of X.

By Step 5 φ_0 is an increasing mapping. Take a transversal pair $((x_n)_{n\in\mathbb{N}}, (y_n)_{n\in\mathbb{N}})$ of φ_0-sequences and define $z' := \bigvee_{n\in\mathbb{N}}\varphi_0(x_n)$ and $z'' := \bigwedge_{n\in\mathbb{N}}\varphi_0(y_n)$. Choose an increasing sequence $(x'_n)_{n\in\mathbb{N}}$ from $Y^{\downarrow}$ and a decreasing sequence $(y'_n)_{n\in\mathbb{N}}$ from $Y^{\uparrow}$ so that $\bigvee_{n\in\mathbb{N}} x'_n = \bigvee_{n\in\mathbb{N}} x_n$, $\bigwedge_{n\in\mathbb{N}} y'_n = \bigwedge_{n\in\mathbb{N}} y_n$, $z' = \bigvee_{n\in\mathbb{N}}\varphi^{\downarrow}(x'_n)$ and $z'' = \bigwedge_{n\in\mathbb{N}}\varphi^{\uparrow}(y'_n)$. Then $((x'_n)_{n\in\mathbb{N}}, (y'_n)_{n\in\mathbb{N}})$ is a transversal pair of sequences. Thus $z' \geq z''$ by Step 7. Now take a corresponding pair (δ, ε) for φ. Let $((x_n)_{n\in\mathbb{N}}, (y_n)_{n\in\mathbb{N}})$ and $((u_n)_{n\in\mathbb{N}}, (v_n)_{n\in\mathbb{N}})$ be transversal pairs of φ_0-sequences such that for every $n \in \mathbb{N}$

$$d(\varphi_0(x_n), \varphi_0(y_n)) < \delta/3$$

and

$$d(\varphi_0(u_n), \varphi_0(v_n)) < \delta/3.$$

By Step 6 there are $x, u \in Y^{\uparrow}$ and $y, v \in Y^{\downarrow}$ such that

$$y \leq \bigwedge_{n \in \mathbb{N}} y_n \leq \bigvee_{n \in \mathbb{N}} x_n \leq x,$$

$$v \leq \bigwedge_{n \in \mathbb{N}} v_n \leq \bigvee_{n \in \mathbb{N}} u_n \leq u$$

and

$$d\Big(\varphi^{\downarrow}(y), \bigwedge_{n \in \mathbb{N}} \varphi_0(y_n)\Big) < \delta/3, \qquad d\Big(\varphi^{\uparrow}(x), \bigvee_{n \in \mathbb{N}} \varphi_0(x_n)\Big) < \delta/3$$

$$d\Big(\varphi^{\downarrow}(v), \bigwedge_{n \in \mathbb{N}} \varphi_0(v_n)\Big) < \delta/3, \qquad d\Big(\varphi^{\uparrow}(u), \bigvee_{n \in \mathbb{N}} \varphi_0(u_n)\Big) < \delta/3.$$

It follows that

$$d(\varphi^{\uparrow}(x), \varphi^{\downarrow}(y)) < \delta \quad \text{and} \quad d(\varphi^{\uparrow}(u), \varphi^{\downarrow}(v)) < \delta.$$

We conclude that

$$d(\varphi^{\uparrow}(x \vee u), \varphi^{\downarrow}(y \vee v)) < \varepsilon \quad \text{and} \quad d(\varphi^{\uparrow}(x \wedge u), \varphi^{\downarrow}(y \wedge v)) < \varepsilon.$$

But then

$$d\Big(\bigvee_{n \in \mathbb{N}} \varphi_0(x_n \vee u_n), \bigwedge_{n \in \mathbb{N}} \varphi_0(y_n \vee v_n)\Big) \leq \varepsilon$$

and

$$d\Big(\bigvee_{n \in \mathbb{N}} \varphi_0(x_n \wedge u_n), \bigwedge_{n \in \mathbb{N}} \varphi_0(y_n \wedge v_n)\Big) \leq \varepsilon$$

showing that φ_0 is weakly increasing.

STEP 10: Construct for each ordinal α a lattice integral $(X, Z, d, Y_\alpha, \varphi_\alpha)$ with the following properties:

(i) $(X, Z, d, Y_0, \varphi_0)$ is the space defined above.

(ii) If $\alpha \leq \beta$ then $Y_\alpha \subset Y_\beta$ and $\varphi_\beta|_{Y_\alpha} = \varphi_\alpha$.

(iii) If $\alpha > 0$, define $\tilde{Y}_\alpha := \bigcup_{\gamma < \alpha} Y_\gamma$ and $\tilde{\varphi}_\alpha\colon \tilde{Y}_\alpha \to Z$ such that $\tilde{\varphi}_\alpha|_{Y_\gamma} = \varphi_\gamma$ for each $\gamma < \alpha$. Then $(X, Z, d, \tilde{Y}_\alpha, \tilde{\varphi}_\alpha)$ is a lattice integral, and

$$(X, Z, d, Y_\alpha, \varphi_\alpha) = \big(X, Z, d, (\tilde{Y}_\alpha)_0, (\tilde{\varphi}_\alpha)_0\big)$$

(iv) If α is an ordinal, $x \in Y_\alpha$, and $\varepsilon > 0$, then there are $x' \in Y^{\downarrow}$ and $x'' \in Y^{\uparrow}$ such that $x' \leq x \leq x''$ and $d(\varphi^{\downarrow}(x'), \varphi^{\uparrow}(x'')) < \varepsilon$.

The construction can be left to the reader.

STEP 11: Let ω_1 be the first uncountable ordinal. Define

$$\overline{Y}(\varphi) := \left\{ x \in X \;\middle|\; \begin{array}{l} \exists x', x'' \in Y_{\omega_1} \\ \text{with } x' \leq x \leq x'' \text{ and } \varphi_{\omega_1}(x') = \varphi_{\omega_1}(x'') \end{array} \right\}$$

and

$$\overline{\varphi}\colon \overline{Y}(\varphi) \to Z, \qquad x \mapsto \varphi_{\omega_1}(x') = \varphi_{\omega_1}(x'').$$

Prove that $(X, Z, d, \overline{Y}(\varphi), \overline{\varphi})$ is a closed lattice integral and

$$(X, Z, d, Y, \varphi) \preccurlyeq (X, Z, d, \overline{Y}(\varphi), \overline{\varphi}).$$

We leave the details to the reader, who may further prove:

(β) $(X, Z, \overline{Y}(\varphi), \overline{\varphi})$ is the smallest closed extension of (X, Z, d, Y, φ).

Before leaving this exercise, we compare what has just been achieved with the analogous constructions for Daniell spaces. So let $(X, \mathscr{L}, \ell)$ be a Daniell space. Denote by d the natural metric on $\mathbb{R}$. Prove the following statements:

(γ) $(\overline{\mathbb{R}}^X, \mathbb{R}, d, \mathscr{L}, \ell)$ is a lattice integral.

(δ) The object $(\overline{\mathbb{R}}^X, \mathbb{R}, d, \overline{\mathscr{L}}(\ell), \overline{\ell})$ in the sense of this exercise is identical with this 5-tuple viewed in the sense of Section 3.1.

We now turn to a step-by-step comparison of the constructions.

Steps 1–5. It is a prominent feature of our procedure that we worked neither with null continuity nor with monotone continuity, but with a stronger hypothesis for transversal pairs of sequences. The fact that monotone continuity is sufficient in the case of Daniell spaces is connected to the linearity of ℓ. If one wishes to forego the linearity, then the monotone continuity no longer suffices to guarantee the existence of an extension. This confirms our previous observation that the issue is not a continuity property but rather an embeddability condition (cf. Ex. 1.4.10). The stronger condition for the general situation found use in Step 5. The reader should give thought to the way in which the linearity entered into the argument at this stage in the case of Daniell spaces.

Steps 6–9. These steps brought the uniform structure on Z to the fore. The corresponding stages in the case of Daniell spaces make use of the natural uniform structure on $\mathbb{R}$. The necessity of the comparison between neighborhoods of different points is most prominent in Step 6. It is exactly the uniformity that enables such a comparison.

Step 10. The outstanding feature here is the use of transfinite construction. Of course we could have defined objects that correspond in the general case to the ε-brackets: Call $(x', x'') \in Y^{\downarrow} \times Y^{\uparrow}$ an ε-bracket of $x \in X$ iff $x' \leq x \leq x''$ and $d(\varphi^{\downarrow}(x'), \varphi^{\uparrow}(x'')) < \varepsilon (\varepsilon > 0)$. In the case of a Daniell space this leads to a downward directed family of closed intervals of $\mathbb{R}$ whose lengths tend to 0. Using the special topological properties of $\mathbb{R}$ it follows that the intersection of these intervals contains exactly one real number, which then serves to define the extension of the functional. This method could be adopted here, but the ε-brackets are not an easy instrument in the general case. We are obliged to find another way. A comparison with Ex. 3.1.11 shows that the transfinite construction we introduced corresponds exactly to the one discussed there for Daniell spaces.

The general construction will find use at several points in the sequel. We wish to present just one example here. Let $(X, \mathfrak{R}, \mu)$ be a positive measure space. Prove the following propositions:

(ε) $(\mathfrak{P}(X), \mathbb{R}, d, \mathbb{R}, \mu)$ is a lattice integral.

(η) $\overline{\mathfrak{R}}(\mu) = \{A \subset X \mid e_A \in \overline{\mathscr{L}}(\ell_\mu)\}$ and $\bar{\mu}(A) = \overline{\ell}_\mu(e_A)$ for every $A \in \mathfrak{R}(\mu)$. $(X, \overline{\mathfrak{R}}(\mu), \bar{\mu})$ is a positive measure space.

Thus every positive measure space $(X, \mathfrak{R}, \mu)$ can be extended in complete analogy with Daniell spaces. However we shall not follow the path of this analogy in Chapter 5. Instead we shall make use directly of the extension of Daniell spaces.

3.2. THE INDUCTION PRINCIPLE

NOTATION FOR SECTION 3.2:

X denotes a set.

In the preceding section we described a construction method for extending a given Daniell space to a closed Daniell space. Here we examine some of the properties of that extension. One important property has already been alluded to: the method of Section 3.1 constructs the smallest closed Daniell space extending the given Daniell space. The most important result is a key theorem characterizing the properties of the given Daniell space that are preserved in this construction. This result, the Induction Principle, is used extensively in the rest of Chapter 3 and in Chapters 4 and 5.

Proposition 3.2.1. *Let $(X, \mathscr{L}, \ell)$ be a Daniell space and $(X, \mathscr{L}', \ell')$ an arbitrary closed triple extending $(X, \mathscr{L}, \ell)$. Then the following implications hold.*

(a) *If $f \in \mathscr{L}^\uparrow$ and $\ell^\uparrow(f) < \infty$, then $f \in \mathscr{L}'$ and $\ell'(f) = \ell^\uparrow(f)$.*

(b) *If $f \in \mathscr{L}^\downarrow$ and $\ell^\downarrow(f) > -\infty$, then $f \in \mathscr{L}'$ and $\ell'(f) = \ell^\downarrow(f)$.*

Proof. In view of how $(\mathscr{L}^\uparrow, \ell^\uparrow)$ and $(\mathscr{L}^\downarrow, \ell^\downarrow)$ are related (Proposition 3.1.6), it suffices to prove (a). Let $f \in \mathscr{L}^\uparrow$. By hypothesis, f is the supremum of an increasing sequence $(f_n)_{n \in \mathbb{N}}$ from $\mathscr{L} \cap \mathbb{R}^X$. The functional $\ell^{\uparrow}$ agrees on $\mathscr{L} \cap \mathbb{R}^X$ with ℓ, and $\ell^{\uparrow}$ commutes with sup when applied to an increasing sequence [Proposition 3.1.4. (b), (d)]. Therefore

$$\ell^{\uparrow}(f) = \sup_{n \in \mathbb{N}} \ell(f_n).$$

But ℓ' also agrees on $\mathscr{L} \cap \mathbb{R}^X$ with ℓ, so the hypotheses imply that $(f_n)_{n \in \mathbb{N}}$ is an increasing ℓ'-sequence. Since the triple $(X, \mathscr{L}', \ell')$ is closed, we conclude that f belongs to $\mathscr{L}'$ and

$$\begin{aligned} \ell'(f) &= \sup_{n \in \mathbb{N}} \ell'(f_n) \\ &= \sup_{n \in \mathbb{N}} \ell(f_n) \\ &= \ell^{\uparrow}(f). \end{aligned}$$

□

Note that in Proposition 1 the triple $(X, \mathscr{L}', \ell')$ need not be a Daniell space. On the contrary, it is important to allow arbitrary closed extensions.

Corollary 3.2.2. *Let $(X, \mathscr{L}, \ell)$ be a Daniell space. For every $f \in \bar{\mathscr{L}}(\ell)$ and for every real number $\varepsilon > 0$, there exists $g \in \bar{\mathscr{L}}(\ell) \cap \mathscr{L}^{\uparrow}$ such that $g \geq f$ and*

$$\bar{\ell}(g) - \bar{\ell}(f) = \ell^{\uparrow}(g) - \bar{\ell}(f) < \varepsilon.$$

Similarly there exists $h \in \bar{\mathscr{L}}(\ell) \cap \mathscr{L}^{\downarrow}$ with $h \leq f$ and $\bar{\ell}(f) - \bar{\ell}(h) < \varepsilon$. □

Corollary 3.2.3. *Let $(X, \mathscr{L}, \ell)$ be a Daniell space and $(X, \mathscr{L}', \ell')$ an arbitrary closed triple extending $(X, \mathscr{L}, \ell)$. Then for every function f belonging to $\bar{\mathscr{L}}(\ell)$, there exist functions f' and f'', both belonging to $\mathscr{L}'$, such that*

$$f' \leq f \leq f''$$

and

$$\ell'(f') = \ell'(f'') = \bar{\ell}(f).$$

Proof. Let $f \in \bar{\mathscr{L}}(\ell)$. According to Proposition 3.1.16 (a) ⇒ (c), there exist an increasing sequence $(f_n')_{n \in \mathbb{N}}$ from $\mathscr{L}^{\downarrow}$ and a decreasing sequence $(f_n'')_{n \in \mathbb{N}}$ from $\mathscr{L}^{\uparrow}$ such that the sequences $(\ell^{\downarrow}(f_n'))_{n \in \mathbb{N}}$ and $(\ell^{\uparrow}(f_n''))_{n \in \mathbb{N}}$ both lie in $\mathbb{R}$,

$$\bigvee_{n \in \mathbb{N}} f_n' \leq f \leq \bigwedge_{n \in \mathbb{N}} f_n''$$

and

$$\sup_{n\in\mathbb{N}} \ell^{\downarrow}(f_n') = \inf_{n\in\mathbb{N}} \ell^{\uparrow}(f_n'') = \bar{\ell}(f).$$

Using Proposition 1, we conclude that the sequences $(f_n')_{n\in\mathbb{N}}$ and $(f_n'')_{n\in\mathbb{N}}$ are ℓ'-sequences. If we set

$$f' := \bigvee_{n\in\mathbb{N}} f_n', \qquad f'' := \bigwedge_{n\in\mathbb{N}} f_n'',$$

then $f' \le f \le f''$, and the functions f', f'' both belong to $\mathscr{L}'$, since $(X, \mathscr{L}', \ell')$ is closed. Moreover, if we use Proposition 1 with the fact that $(X, \mathscr{L}', \ell')$ is closed, we have

$$\begin{aligned}
\ell'(f') &= \sup_{n\in\mathbb{N}} \ell'(f_n') \\
&= \sup_{n\in\mathbb{N}} \ell^{\downarrow}(f_n') \\
&= \bar{\ell}(f) \\
&= \inf_{n\in\mathbb{N}} \ell^{\uparrow}(f_n'') \\
&= \inf_{n\in\mathbb{N}} \ell'(f_n'') \\
&= \ell'(f'').
\end{aligned}$$

□

The characterizations of the closed Daniell space $(X, \bar{\mathscr{L}}(\ell), \bar{\ell})$ given in the next two theorems now follow easily.

Theorem 3.2.4. *For every Daniell space $(X, \mathscr{L}, \ell)$, $(X, \bar{\mathscr{L}}(\ell), \bar{\ell})$ is the smallest complete, closed triple extending $(X, \mathscr{L}, \ell)$.*

Proof. The triple $(X, \bar{\mathscr{L}}(\ell), \bar{\ell})$ is a closed Daniell space extending $(X, \mathscr{L}, \ell)$ (Theorem 3.1.17). By Corollary 1.4.8 it is also complete. Let $(X, \mathscr{L}', \ell')$ be an arbitrary complete, closed triple such that $(X, \mathscr{L}, \ell) \preccurlyeq (X, \mathscr{L}', \ell')$, and let $f \in \bar{\mathscr{L}}(\ell)$. Since $(X, \mathscr{L}', \ell')$ is closed, Corollary 3 implies the existence of functions f', f'' belonging to $\mathscr{L}'$ such that $f' \le f \le f''$ and

$$\ell'(f') = \ell'(f'') = \bar{\ell}(f).$$

Since $(X, \mathscr{L}', \ell')$ is complete, f must belong to $\mathscr{L}'$ and

$$\ell'(f) = \bar{\ell}(f)$$

(1.4.7). It follows that $(X, \bar{\mathscr{L}}(\ell), \bar{\ell}) \preccurlyeq (X, \mathscr{L}', \ell')$. □

Theorem 3.2.5. *For every Daniell space* $(X, \mathscr{L}, \ell)$, $(X, \bar{\mathscr{L}}(\ell), \bar{\ell})$ *is the smallest closed Daniell space extending* $(X, \mathscr{L}, \ell)$.

Proof. Every closed Daniell space is a closed, complete triple (Corollary 1.4.8). Apply Theorem 4. □

With Theorem 5 available, the following definition is finally appropriate.

Definition 3.2.6. (Daniell-space closure). *For* $(X, \mathscr{L}, \ell)$ *a Daniell space, the closed Daniell space* $(X, \bar{\mathscr{L}}(\ell), \bar{\ell})$ *is called the* **closure** *of* $(X, \mathscr{L}, \ell)$. □

Theorem 4 is more general than its corollary, Theorem 5, since it applies to all complete, closed extensions whether or not they are Daniell spaces. The general theorem yields a very useful characterization of the properties that carry over when the construction method of Section 3.1 is applied. We give two formulations of the principle in question. The first version says any subset of $\bar{\mathscr{L}}(\ell)$ that contains $\mathscr{L}$, that contains all limits of its own $\bar{\ell}$-sequences, and that satisfies a certain completeness condition must actually be all of $\bar{\mathscr{L}}(\ell)$.

Theorem 3.2.7. (Induction Principle, first form). *Let* $(X, \mathscr{L}, \ell)$ *be a Daniell space. Suppose that* $\mathscr{F}$ *is a subset of* $\bar{\mathscr{L}}(\ell)$ *satisfying the following three conditions.*

(*a*) $\mathscr{L} \subset \mathscr{F}$.

(*b*) *If* $(f_n)_{n \in \mathbb{N}}$ *is an* $\bar{\ell}$*-sequence from* $\mathscr{F}$, *then the function* $\lim_{n \to \infty} f_n$ *belongs to* $\mathscr{F}$.

(*c*) *The conditions* $f, g \in \mathscr{F}$, $h \in \bar{\mathscr{L}}(\ell)$, $f \leq h \leq g$, *and* $\bar{\ell}(f) = \bar{\ell}(g)$ *always imply that h belongs to* $\mathscr{F}$.

Then $\mathscr{F} = \bar{\mathscr{L}}(\ell)$.

Proof. Define

$$\ell' : \mathscr{F} \to \mathbb{R}, \qquad f \mapsto \bar{\ell}(f).$$

Note that $(X, \mathscr{L}, \ell) \preccurlyeq (X, \mathscr{F}, \ell')$. To verify that the triple $(X, \mathscr{F}, \ell')$ is closed, let $(f_n)_{n \in \mathbb{N}}$ be an ℓ'-sequence from $\mathscr{F}$. Then $(f_n)_{n \in \mathbb{N}}$ is both an $\bar{\ell}$-sequence from $\mathscr{F}$ and an $\bar{\ell}$-sequence from $\bar{\mathscr{L}}(\ell)$. Using hypothesis (b) and the fact that the triple $(X, \bar{\mathscr{L}}(\ell), \bar{\ell})$ is closed, we conclude that $\lim_{n \to \infty} f_n$ belongs to $\mathscr{F}$ and

$$\ell'\left(\lim_{n \to \infty} f_n\right) = \lim_{n \to \infty} \ell'(f_n).$$

Thus $(X, \mathscr{F}, \ell')$ is a closed triple.

In view of the completeness of $(X, \bar{\mathscr{L}}(\ell), \bar{\ell})$ (Theorem 4), it follows from hypothesis (c) that the triple $(X, \mathscr{F}, \ell')$ is also complete. Thus $(X, \mathscr{F}, \ell')$ is a complete, closed extension of $(X, \mathscr{L}, \ell)$, and Theorem 4 implies, in particular, that $\mathscr{F} = \bar{\mathscr{L}}(\ell)$. □

The completeness condition can be removed from the hypotheses if we compensate by requiring $\bar{\mathscr{L}}(\ell)$-a.e. limits of $\bar{\ell}$-sequences to belong to the subset. More precisely, the second version of the Induction Principle reads as follows.

Theorem 3.2.8. (Induction Principle, second form). *Let $(X, \mathscr{L}, \ell)$ be a Daniell space. Suppose that $\mathscr{F}$ is a subset of $\bar{\mathscr{L}}(\ell)$ satisfying the following two conditions.*

(*a*) $\mathscr{L} \subset \mathscr{F}$.

(*b*) *If $(f_n)_{n \in \mathbb{N}}$ is an $\bar{\ell}$-sequence from $\mathscr{F}$, and if f belongs to $\bar{\mathscr{L}}(\ell)$ and satisfies*

$$f(x) = \lim_{n \to \infty} f_n(x) \quad \bar{\mathscr{L}}(\ell)\text{-}a.e.$$

then f belongs to $\mathscr{F}$.

Then $\mathscr{F} = \bar{\mathscr{L}}(\ell)$.

Proof. Suppose that $f, g \in \mathscr{F}$, $h \in \bar{\mathscr{L}}(\ell)$, $f \le h \le g$, and $\bar{\ell}(f) = \bar{\ell}(g)$. Notice that $h = f$ $\bar{\mathscr{L}}(\ell)$-a.e. (Proposition 1.4.6). For every n in $\mathbb{N}$, set $f_n := f$. Then $(f_n)_{n \in \mathbb{N}}$ is an $\bar{\ell}$-sequence from $\mathscr{F}$ and

$$h = \lim_{n \to \infty} f_n \quad \bar{\mathscr{L}}(\ell)\text{-a.e.}$$

so h belongs to $\mathscr{F}$. The hypotheses of Theorem 7 are satisfied, and it follows that $\mathscr{F} = \bar{\mathscr{L}}(\ell)$. □

We conclude this section by mentioning some straightforward consequences of Theorem 5 and Proposition 1.

Proposition 3.2.9. *Let $(X, \mathscr{L}_1, \ell_1)$ and $(X, \mathscr{L}_2, \ell_2)$ be Daniell spaces such that*

$$(X, \mathscr{L}_1, \ell_1) \preccurlyeq \left(X, \bar{\mathscr{L}}(\ell_2), \bar{\ell}_2\right)$$

and

$$(X, \mathscr{L}_2, \ell_2) \preccurlyeq \left(X, \bar{\mathscr{L}}(\ell_1), \bar{\ell}_1\right).$$

Then

$$\left(X, \bar{\mathscr{L}}(\ell_1), \bar{\ell}_1\right) = \left(X, \bar{\mathscr{L}}(\ell_2), \bar{\ell}_2\right).$$

Proof. The closure of a given Daniell space is the smallest closed Daniell space extending the given Daniell space. The hypotheses therefore imply the inequality

$$\left(X, \bar{\mathscr{L}}(\ell_1), \overline{\ell_1}\right) \preccurlyeq \left(X, \bar{\mathscr{L}}(\ell_2), \overline{\ell_2}\right)$$

and its reverse. □

The construction method of Section 3.1, more precisely the mapping that sends a given Daniell space $(X, \mathscr{L}, \ell)$ to its closure $(X, \bar{\mathscr{L}}(\ell), \bar{\ell})$, has obvious properties of monotonicity and idempotence, which are spelled out in the next proposition.

Proposition 3.2.10.

(*a*) **Monotonicity**. *If* $(X, \mathscr{L}_1, \ell_1)$ *and* $(X, \mathscr{L}_2, \ell_2)$ *are Daniell spaces with*

$$(X, \mathscr{L}_1, \ell_1) \preccurlyeq (X, \mathscr{L}_2, \ell_2)$$

then

$$\left(X, \bar{\mathscr{L}}(\ell_1), \overline{\ell_1}\right) \preccurlyeq \left(X, \bar{\mathscr{L}}(\ell_2), \overline{\ell_2}\right).$$

(*b*) **Idempotence**. *For every Daniell space* $(X, \mathscr{L}, \ell)$,

$$(X, \bar{\mathscr{L}}(\bar{\ell}), \bar{\bar{\ell}}) = (X, \bar{\mathscr{L}}(\ell), \bar{\ell}).$$ □

Part (b) of the preceding proposition shows, in particular, that repeated applications of the construction method of Section 3.1 will not yield new closed Daniell spaces. Additional extensions of a given Daniell space must be obtained in some other way. We shall see in Chapter 4 that they are obtained by introducing new exceptional functions in conjunction with the construction of Section 3.1.

The next theorem shows that the construction method of Section 3.1 introduces no arbitrariness: every closed triple that extends $(X, \mathscr{L}, \ell)$ assigns to functions in $\bar{\mathscr{L}}(\ell)$ the same value $\bar{\ell}$ assigns (if it assigns any value at all).

Theorem 3.2.11. *Let* $(X, \mathscr{L}, \ell)$ *be a Daniell space and* $(X, \mathscr{L}', \ell')$ *an arbitrary closed triple extending* $(X, \mathscr{L}, \ell)$. *Then the two functionals* ℓ' *and* $\bar{\ell}$ *agree on their common domain*:

$$\ell'(f) = \bar{\ell}(f)$$

for every $f \in \mathscr{L}' \cap \bar{\mathscr{L}}(\ell)$.

Proof. Let $f \in \mathscr{L}' \cap \bar{\mathscr{L}}(\ell)$. By Corollary 3, there exist functions f' and f'', both belonging to $\mathscr{L}'$, such that

$$f' \le f \le f'' \quad \text{and} \quad \ell'(f') = \ell'(f'') = \bar{\ell}(f).$$

The monotonicity of ℓ' yields

$$\ell'(f') \le \ell'(f) \le \ell'(f'')$$

and consequently

$$\bar{\ell}(f) = \ell'(f). \qquad \square$$

EXERCISES

3.2.1(E) Before turning to examples illustrating the Principle of Induction, we wish to make a few preliminary observations which are often of great service:

(α) If $(f_n)_{n \in \mathbb{N}}$ is an increasing sequence from $\overline{\mathbb{R}}^X$, then for any $n \in \mathbb{N}$, $-f_1^- \le f_n \le f_n^+ \le f_n \vee f_1^-$.

(β) If $(f_n)_{n \in \mathbb{N}}$ is a decreasing sequence from $\overline{\mathbb{R}}^X$, then for any $n \in \mathbb{N}$, $f_n \wedge (-f_1^+) \le -f_n^- \le f_n \le f_1^+$.

If $\mathscr{L} \subset \overline{\mathbb{R}}^X$ is a Riesz lattice and ℓ a positive linear functional on $\mathscr{L}$, then the next three observations can be made:

(γ) $(f_n \vee 0)_{n \in \mathbb{N}}$ and $(f_n \wedge 0)_{n \in \mathbb{N}}$ are both increasing ℓ-sequences whenever $(f_n)_{n \in \mathbb{N}}$ is.

(δ) $(f_n \vee 0)_{n \in \mathbb{N}}$ and $(f_n \wedge 0)_{n \in \mathbb{N}}$ are both decreasing ℓ-sequences whenever $(f_n)_{n \in \mathbb{N}}$ is.

(ε) If $\mathscr{G}$ is a Riesz lattice such that $f^+ \in \mathscr{G}$ whenever $f \in \mathscr{L}$, then $\mathscr{L} \subset \mathscr{G}$.

3.2.2(E) Take a Daniell space $(X, \mathscr{L}, \ell)$. Suppose $f \in \mathbb{R}^X$ has the property that $fg \in \bar{\mathscr{L}}(\ell)$ for each $g \in \mathscr{L}$. Take $g \in \bar{\mathscr{L}}(\ell)$. Prove that if $\sup_{x \in X} |f(x)| < \infty$, then $fg \in \bar{\mathscr{L}}(\ell)$ and

$$|\bar{\ell}(fg)| \le \bar{\ell}(|fg|) \le \bar{\ell}(g) \sup_{x \in X} |f(x)|.$$

Set $\tilde{\mathscr{L}} := \{g \in \bar{\mathscr{L}}(\ell) \mid fg \in \bar{\mathscr{L}}(\ell)\}$. Let $(g_n)_{n \in \mathbb{N}}$ be an $\bar{\ell}$-sequence from $\tilde{\mathscr{L}}$. Take $g \in \bar{\mathscr{L}}(\ell)$ with $g(x) = \lim_{n \to \infty} g_n(x)$ $\bar{\mathscr{L}}(\ell)$-a.e. Then $fg(x) = \lim_{n \to \infty} fg_n(x)$ $\bar{\mathscr{L}}(\ell)$-a.e., and for each $n \in \mathbb{N}$,

$$|fg_n| \le \left(|\lim_{n \to \infty} g_n| + |g_1|\right) \sup_{x \in X} |f(x)|.$$

We get $fg \in \bar{\mathscr{L}}(\ell)$ by the Lebesgue Dominated Convergence Theorem. The second form of the Induction Principle implies then that $\tilde{\mathscr{L}} = \bar{\mathscr{L}}(\ell)$. The inequality is clear.

3.2.3(E) Let $(X, \mathscr{L}, \ell)$ be a Daniell space. Suppose that, for $f \in \overline{\mathbb{R}}_+^X$, $f \wedge g \in \bar{\mathscr{L}}(\ell)$ for every $g \in \mathscr{L}_+$. Prove that

$$f \wedge g \in \bar{\mathscr{L}}(\ell) \text{ for every } g \in \bar{\mathscr{L}}(\ell).$$

Set $\tilde{\mathscr{L}} := \{g \in \bar{\mathscr{L}}(\ell) \mid f \wedge g \in \bar{\mathscr{L}}(\ell)\}$. Since $f \wedge g = f \wedge g^+ - g^- \in \bar{\mathscr{L}}(\ell)$ for each $g \in \mathscr{L}$, it follows that $\mathscr{L} \subset \tilde{\mathscr{L}}$. So take an $\bar{\ell}$-sequence $(g_n)_{n \in \mathbb{N}}$ from $\tilde{\mathscr{L}}$ and $g \in \bar{\mathscr{L}}(\ell)$ with $g(x) = \lim_{n \to \infty} g_n(x)$ $\bar{\mathscr{L}}(\ell)$-a.e. Then $(f \wedge g_n)_{n \in \mathbb{N}}$ is also an $\bar{\ell}$-sequence, and $(f \wedge g)(x) = \lim_{n \to \infty} (f \wedge g_n)(x) \bar{\mathscr{L}}(\ell)$-a.e. so that $f \wedge g \in \bar{\mathscr{L}}(\ell)$. Thus $g \in \tilde{\mathscr{L}}$, and so $\tilde{\mathscr{L}} = \bar{\mathscr{L}}(\ell)$ according to the second form of the Induction Principle.

3.2.4(E) The Riesz lattice $\mathscr{F}$ on the set X is called **solid** in $\overline{\mathbb{R}}^X$ iff given $f \in \mathscr{F}$ and $g \in \overline{\mathbb{R}}^X$ with $|g| \leq |f|$, $g \in \mathscr{F}$. Prove that if $\mathscr{L} \subset \overline{\mathbb{R}}^X$ is a solid Riesz lattice and if ℓ is a nullcontinuous positive linear functional on $\mathscr{L}$, then $\bar{\mathscr{L}}(\ell)$ is also solid in $\overline{\mathbb{R}}^X$.

Set $\tilde{\mathscr{L}} := \{f \in \bar{\mathscr{L}}(\ell) \mid g \in \bar{\mathscr{L}}(\ell)$ whenever $g \in \overline{\mathbb{R}}_+^X$ and $g \leq f^+\}$. Then $\mathscr{L} \subset \tilde{\mathscr{L}}$. Take $(f_n)_{n \in \mathbb{N}}$ an $\bar{\ell}$-sequence from $\tilde{\mathscr{L}}$. Set $f := \lim_{n \to \infty} f_n$. Then $f^+ = \lim_{n \to \infty} f_n^+$. If $g \in \overline{\mathbb{R}}_+^X$ and $g \leq f^+$, then $g = \lim_{n \to \infty} (f_n^+ \wedge g)$. But by hypothesis, $f_n^+ \wedge g \in \bar{\mathscr{L}}(\ell)$ for each $n \in \mathbb{N}$, and $(f_n^+ \wedge g)_{n \in \mathbb{N}}$ is an $\bar{\ell}$-sequence, so that $g \in \tilde{\mathscr{L}}$. Take $f, g \in \tilde{\mathscr{L}}$, and take h with $f \leq h \leq g$. Then clearly $h \in \tilde{\mathscr{L}}$. So, by the first form of the Induction Principle, $\tilde{\mathscr{L}} = \bar{\mathscr{L}}(\ell)$.

An example of the situation depicted in this exercise is given by the space $\mathscr{k}(X)$ (or equally, $\ell^\infty(X)$ or $c_0(X)$).

3.2.5(E) Take Daniell spaces $(X, \mathscr{L}, \ell)$ and $(X', \mathscr{L}', \ell')$, a number $\alpha \in \mathbb{R}_+$, and a mapping φ: $\overline{\mathbb{R}}^X \to \overline{\mathbb{R}}^{X'}$ with the following properties:

(i) φ is increasing, and $\varphi(\lim_{n \to \infty} f_n) = \lim_{n \to \infty} \varphi(f_n)$ for every monotonely convergent sequence $(f_n)_{n \in \mathbb{N}}$ from $\overline{\mathbb{R}}^X$.
(ii) $\varphi(\mathscr{L}) \subset \bar{\mathscr{L}}(\ell')$.
(iii) For each $f \in \mathscr{L}$, $|\bar{\ell}'(\varphi(f))| \leq \alpha |\ell(f)|$.
(iv) $\varphi(\mathbb{R}^X) \subset \varphi(\mathbb{R}^{X'})$ and $\varphi|_{\mathbb{R}^X}$ is linear.

Prove the following propositions:

(α) $\varphi(\bar{\mathscr{L}}(\ell)) \subset \bar{\mathscr{L}}(\ell')$.
(β) $|\bar{\ell}'(\varphi(f))| \leq \alpha |\bar{\ell}(f)|$ for each $f \in \bar{\mathscr{L}}(\ell)$.
(γ) If $\ell'(\varphi(f)) = \alpha \ell(f)$ for each $f \in \mathscr{L}$, then $\bar{\ell}'(\varphi(f)) = \alpha \bar{\ell}(f)$ for each $f \in \bar{\mathscr{L}}(\ell)$.
(δ) Let ψ: $X' \to X$ be a mapping such that $f \circ \psi \in \bar{\mathscr{L}}(\ell')$ and $|\bar{\ell}'(f \circ \psi))| \leq \alpha |\ell(f)|$ for each $f \in \mathscr{L}$. Then φ: $\overline{\mathbb{R}}^X \to \overline{\mathbb{R}}^{X'}$ defined by $\varphi(f) := f \circ \psi$ satisfies (i)–(iii).

Consider the following important special case. Take an interval A of $\overline{\mathbb{R}}$ having nonempy interior and contained in $\mathbb{R}$. Denote by $\mathscr{K}(A)$ the space of all continuous functions on A with compact support, and by ℓ_A the Lebesgue functional on $\mathscr{K}(A)$. Take $\alpha, \beta \in \mathbb{R}$ and the mapping ω: $\mathbb{R} \to \mathbb{R}$, $x \mapsto \alpha x + \beta$. Such a mapping is called an **affine transformation** on $\mathbb{R}$. Set

$B := \omega(A)$. Prove the following propositions:

(ε) If $\alpha \neq 0$, then ω is a bijection.
(ζ) $f \circ \omega \in \mathscr{K}(A)$ whenever $f \in \mathscr{K}(B)$, and $\ell_A(f \circ \omega) = |\alpha| \ell_B(f)$.
(η) $f \circ \omega \in \bar{\mathscr{L}}(\ell_A)$ whenever $f \in \bar{\mathscr{L}}(\ell_B)$, and $\bar{\ell}_A(f \circ \omega) = |\alpha| \bar{\ell}_B(f)$.

For $A = \mathbb{R}$ and $\alpha = 1$, this property is called the **translation invariance** of the Lebesgue integral. (The mappings $\mathbb{R} \to \mathbb{R}$, $x \mapsto x + \beta$ are **translations** of $\mathbb{R}$.)

3.2.6(E) We consider, once again, restrictions and images of positive measures. The concepts and notations come from Ex. 2.1.12. Prove the following propositions.

(α) **Restrictions**. Given a positive measure space, $(X, \mathfrak{R}, \mu)$ and $Y \subset X$,

$$\bar{\mathscr{L}}(\ell_{\mu|Y}) \subset \{f \in \bar{\mathscr{L}}(\ell_\mu) \mid \{f \neq 0\} \subset Y\}.$$

If $f \in \bar{\mathscr{L}}(\ell_{\mu|Y})$, then $\overline{\ell_{\mu|Y}}(f) = \bar{\ell}_\mu(f)$.

(β) **Images**. Let $(X, \mathfrak{R}, \mu)$ be a positive measure space, Y a set and φ: $X \to Y$ a mapping. Then

$$\bar{\mathscr{L}}(\ell_{\varphi(\mu)}) \subset \{f \in \overline{\mathbb{R}}^Y \mid f \circ \varphi \in \bar{\mathscr{L}}(\ell_\mu)\}$$

and if $f \in \bar{\mathscr{L}}(\ell_{\varphi(\mu)})$, then $\overline{\ell_{\varphi(\mu)}}(f) = \bar{\ell}_\mu(f \circ \varphi)$.

3.2.7(E) Let X be a set, and $\mathscr{L} \subset \overline{\mathbb{R}}^X$ a Riesz lattice. Denote by Φ the set of all nullcontinuous positive linear functionals on $\mathscr{L}$. We know from Ex. 1.3.11 that Φ is a conditionally complete lattice with respect to the order defined by $\ell_1 \leq \ell_2$ iff $\ell_1(f) \leq \ell_2(f)$ for each $f \in \mathscr{L}_+$. We know also that if $\ell_1, \ell_2 \in \Phi$ and $\alpha \in \mathbb{R}_+$, $\alpha\ell_1 \in \Phi$, and $\ell_1 + \ell_2 \in \Phi$. Prove the following propositions:

(α) If $\ell_1, \ell_2 \in \Phi$, $\ell_1 \leq \ell_2$, then $\bar{\mathscr{L}}(\ell_2) \subset \bar{\mathscr{L}}(\ell_1)$ and $\bar{\ell}_1(f) \leq \bar{\ell}_2(f)$ whenever $f \in \bar{\mathscr{L}}(\ell_2)_+$.
(β) If $\ell \in \Phi$ and $\alpha \in \mathbb{R}$, $\alpha > 0$, then $\bar{\mathscr{L}}(\alpha\ell) = \bar{\mathscr{L}}(\ell)$ and $\overline{\alpha\ell} = \alpha\bar{\ell}$.
(γ) If $\ell_1, \ell_2 \in \Phi$, then $\bar{\mathscr{L}}(\ell_1 + \ell_2) = \bar{\mathscr{L}}(\ell_1) \cap \bar{\mathscr{L}}(\ell_2)$ and $\overline{\ell_1 + \ell_2}(f) = \bar{\ell}_1(f) + \bar{\ell}_2(f)$ if $f \in \bar{\mathscr{L}}(\ell_1 + \ell_2)$.
(δ) If $\ell_1, \ell_2 \in \Phi$, then $\bar{\mathscr{L}}(\ell_1 \vee \ell_2) = \bar{\mathscr{L}}(\ell_1) \cap \bar{\mathscr{L}}(\ell_2)$.
(ε) If $\ell_1, \ell_2 \in \Phi$, then $\bar{\mathscr{L}}(\ell_1) \cup \bar{\mathscr{L}}(\ell_2) \subset \bar{\mathscr{L}}(\ell_1 \wedge \ell_2)$.

For (α), define $\tilde{\mathscr{L}} := \{f \in \bar{\mathscr{L}}(\ell_2) \mid f^+ \in \bar{\mathscr{L}}(\ell_1), \bar{\ell}_1(f^+) \leq \bar{\ell}_2(f^+)\}$. Then $\mathscr{L} \subset \tilde{\mathscr{L}}$. If $(f_n)_{n \in \mathbb{N}}$ is an $\bar{\ell}_2$-sequence from $\tilde{\mathscr{L}}$, then $(f_n^+)_{n \in \mathbb{N}}$ is an $\bar{\ell}_1$-sequence. Set $f := \lim_{n \to \infty} f_n$. Then $f^+ = \lim_{n \to \infty} f_n^+ \in \bar{\mathscr{L}}(\ell_1)$ and

$$\bar{\ell}_1(f^+) = \lim_{n \to \infty} \bar{\ell}_1(f_n^+) \leq \lim_{n \to \infty} \bar{\ell}_2(f_n^+) = \bar{\ell}_2(f^+).$$

Thus $f \in \tilde{\mathscr{L}}$. If $f, g \in \tilde{\mathscr{L}}$ and $h \in \overline{\mathbb{R}}^X$ with $f \leq h \leq g$, then $f^+ \leq h^+ \leq g^+$. If $\bar{\ell}_2(f) = \bar{\ell}_2(g)$, then $\bar{\ell}_2(f^+) = \bar{\ell}_2(g^+)$, and so $\bar{\ell}_2(g^+ \dot{-} f^+) = 0$. Consequently $\bar{\ell}_1(g^+ \dot{-} f^+) = 0$. In other words, $\bar{\ell}_1(f^+) = \bar{\ell}_1(g^+)$, so that $h^+ \in \bar{\mathscr{L}}(\ell_1)$ and

$$\bar{\ell}_1(h^+) = \bar{\ell}_1(f^+) \leq \bar{\ell}_2(f^+) = \bar{\ell}_2(h^+).$$

With this, $h \in \bar{\mathscr{L}}$ and so, by the first form of the Induction Principle, $\tilde{\mathscr{L}} = \bar{\mathscr{L}}(\ell)$.

To prove (γ), note that $\bar{\mathscr{L}}(\ell_1 + \ell_2) \subset \bar{\mathscr{L}}(\ell_1) \cap \bar{\mathscr{L}}(\ell_2)$ follows from (α). For the reverse inclusion first show that $(\ell_1 + \ell_2)^{\uparrow} = \ell_1^{\uparrow} + \ell_2^{\uparrow}$ and $(\ell_1 + \ell_2)^{\downarrow} = \ell_1^{\downarrow} + \ell_2^{\downarrow}$. Now take $f \in \bar{\mathscr{L}}(\ell_1) \cap \bar{\mathscr{L}}(\ell_2)$ and $\varepsilon > 0$. Then there exist $\varepsilon/2$-brackets (g_1, h_1) for f with respect to ℓ_1 and (g_2, h_2) for f with respect to ℓ_2. Put $g := g_1 \vee g_2$ and $h := h_1 \wedge h_2$. Then $g \le f \le h$, $g \in \mathscr{L}^{\downarrow}$, $h \in \mathscr{L}^{\uparrow}$ and

$$\begin{aligned}
&(\ell_1 + \ell_2)^{\uparrow}(h) - (\ell_1 + \ell_2)^{\downarrow}(g) \\
&\quad = \left(\ell_1^{\uparrow}(h) - \ell_1^{\downarrow}(g)\right) + \left(\ell_2^{\uparrow}(h) - \ell_2^{\downarrow}(g)\right) \\
&\quad \le \left(\ell_1^{\uparrow}(h_1) - \ell_1^{\downarrow}(g_1)\right) + \left(\ell_2^{\uparrow}(h_2) - \ell_2^{\downarrow}(g_2)\right) \\
&\quad < \varepsilon.
\end{aligned}$$

Thus (g, h) is an ε-bracket for f with respect to $\ell_1 + \ell_2$.

To prove (δ), use (α), (γ), and the fact that $\ell_i \le \ell_1 \vee \ell_2 \le \ell_1 + \ell_2$ $(i = 1, 2)$.

3.2.8(E) We return to the Borel integral of the Daniell space $(X, \mathscr{L}, \ell)$ that was investigated in Ex. 3.1.11. Denote by $\mathscr{B}$ the set of all Borel integrable functions with respect to this Daniell space. Prove the following Induction Principle:

$\mathscr{F} = \mathscr{B}$ if $\mathscr{F} \subset \mathscr{B}$ satisfies the following two conditions.

(i) $\mathscr{L} \subset \mathscr{F}$.

(ii) $\lim_{n \to \infty} f_n \in \mathscr{F}$ for each $\bar{\ell}$-sequence $(f_n)_{n \in \mathbb{N}}$ from $\mathscr{F}$.

We note that behind this proposition lies the Principle of Transfinite Induction, which is applied to the method of construction used in Ex. 3.1.11. Observe that in this case, in contrast to the situation for the Induction Principle for $(X, \bar{\mathscr{L}}(\ell), \bar{\ell})$, the exceptional functions play no role.

3.2.9(C) **Induction Principle for Lattice Integrals.** This exercise is intended for the reader who has worked through Ex. 3.1.13. Let (X, Z, d, Y, φ) be a lattice integral. Suppose $Y' \subset \bar{Y}(\varphi)$ has the following three properties:

(i) $Y \subset Y'$.

(ii) $\lim_{n \to \infty} y_n \in Y'$ whenever $(y_n)_{n \in \mathbb{N}}$ is a φ-sequence from Y'.

(iii) If $x, y \in Y'$ and $x \le y$ with $\bar{\varphi}(x) = \bar{\varphi}(y)$, then $z \in Y'$ for every $z \in X$ with $x \le z \le y$.

Then $Y' = \bar{Y}(\varphi)$.

Let X be a σ-distributive lattice. Take $n \in \mathbb{N}$ and a mapping $\omega: X^n \to X$ with the following property:

(iv) If $k \in \mathbb{N}_n$ and if $(x_{kn})_{n \in \mathbb{N}}$ is a monotonely convergent sequence from X, then $(\omega(x_1, \dots, x_{kn}, \dots, x_n))_{n \in \mathbb{N}}$ is monotonely convergent too and

$$\lim_{n \to \infty} \omega(x_1, \dots, x_{kn}, \dots, x_n) = \omega\Big(x_1, \dots, \lim_{n \to \infty} x_{kn}, \dots x_n\Big)$$

whenever $x_1, \dots, x_{k-1}, x_{k+1}, \dots x_n$ are elements from X.

Such a mapping is called an **admissible lattice operation** on X.

Take now a lattice integral (X, Z, d, Y, φ), and assume that $\omega\colon X^n \to X$ and $\omega'\colon Z^n \to Z$ ($n \in \mathbb{N}$) are admissible lattice operations. Prove the following assertion: If $\omega(Y^n) \subset Y$ and if

$$\varphi(\omega(x_1, \dots, x_n)) = \omega'(\varphi(x_1), \dots, \varphi(x_n))$$

whenever $x_1, \dots, x_n \in Y$, then $\omega(\overline{Y}(\varphi)^n) \subset \overline{Y}(\varphi)$ and

$$\overline{\varphi}(\omega(x_1, \dots, x_n)) = \omega'(\overline{\varphi}(x_1), \dots, \overline{\varphi}(x_n))$$

whenever $x_1, \dots, x_n \in \overline{Y}(\varphi)$.

3.3. UPPER AND LOWER CLOSURE

For $(X, \mathscr{L}, \ell)$ an arbitrary Daniell space, the numbers

$$\ell_*(f) := \sup\{\ell^{\downarrow}(g) \mid g \in \mathscr{L}^{\downarrow},\ g \le f\}$$

and

$$\ell^*(f) := \inf\{\ell^{\uparrow}(g) \mid g \in \mathscr{L}^{\uparrow},\ g \ge f\}$$

are defined for every extended-real function f on X, and functions f in $\overline{\mathscr{L}}(\ell)$ satisfy $\ell_*(f) = \ell^*(f) = \bar{\ell}(f)$. The mappings ℓ_* and ℓ^*, which we might refer to as **upper** and **lower closure**, are important in their own right. In fact many authors use upper and lower closure to define the integral. Since we choose a different path to the definition, upper and lower closure appear here in a derivative role, but they are still very useful. Upper closure, especially, will make numerous appearances in later sections. The present section consists of a straightforward examination of the basic properties of upper and lower closure.

Definition 3.3.1. *Let $(X, \mathscr{L}, \ell)$ be a Daniell space. Then, for $f \in \overline{\mathbb{R}}^X$,*

$$\ell^*(f) := \inf\{\ell^{\uparrow}(g) \mid g \in \mathscr{L}^{\uparrow},\ g \ge f\}. \qquad \square$$

Proposition 3.3.2. *The following assertions hold, for every closed Daniell space $(X, \mathscr{L}, \ell)$.*

- (*a*) *If $f \in \mathscr{L}$, then $\ell^*(f) = \ell(f)$.*
- (*b*) *If $f \in \mathscr{L}^{\uparrow}$, then $\ell^*(f) = \ell^{\uparrow}(f) = \sup\{\ell(g) \mid g \in \mathscr{L} \cap \mathbb{R}^X,\ g \le f\}$.*
- (*c*) *If $f \in \mathscr{L}^{\uparrow}$ and $\ell^{\uparrow}(f) < \infty$, then f belongs to $\mathscr{L}$ and $\ell^*(f) = \ell^{\uparrow}(f) = \ell(f)$.*
- (*d*) *If $f \in \overline{\mathbb{R}}^X$ and $\ell^*(f)$ is real, then $\ell^*(f) = \ell(g)$ for some $g \in \mathscr{L}$ with $g \ge f$.*
- (*e*) *For every $f \in \overline{\mathbb{R}}^X$, $\ell^*(f) = \inf\{\ell^{\uparrow}(g) \mid g \in \mathscr{L}^{\uparrow},\ g \ge f$ $\mathscr{L}$-a.e.$\}$.*

Proof. (a) Since $(X, \mathscr{L}, \ell)$ is its own closure, (a) follows from Proposition 3.1.16 (a) $\Rightarrow$ (d).

(b) That (b) holds follows from the definitions.

(c) According to (b), $\ell^*(f) = \ell^\uparrow(f)$. By Proposition 3.2.1 (a), f belongs to $\mathscr{L}$ and $\ell(f) = \ell^\uparrow(f)$.

(d) By hypothesis, there exists in $\mathscr{L}^\uparrow$ a sequence $(g_n)_{n\in\mathbb{N}}$ such that $g_n \geq f$ for every n and

$$\inf_{n\in\mathbb{N}} \ell^\uparrow(g_n) = \ell^*(f).$$

Moreover the sequence $(g_n)_{n\in\mathbb{N}}$ can be chosen so that $\ell^\uparrow(g_n)$ is real for every n and $(g_n)_{n\in\mathbb{N}}$ decreases. In view of (c), $(g_n)_{n\in\mathbb{N}}$ is an ℓ-sequence from $\mathscr{L}$. Taking $g := \bigwedge_{n\in\mathbb{N}} g_n$, we have $g \in \mathscr{L}$, $g \geq f$, and $\ell(g) = \ell^*(f)$.

(e) Let $f \in \overline{\mathbb{R}}^X$. It suffices to show that $\ell^*(f) \leq \ell^\uparrow(g)$ whenever $g \in \mathscr{L}^\uparrow$ and $g \geq f$ $\mathscr{L}$-a.e. Given such a function g, let $A := \{g < f\}$. By hypothesis, $A \in \mathfrak{N}(\mathscr{L})$. Since $\infty e_A = \bigvee_{n\in\mathbb{N}} n e_A$, the function ∞e_A belongs to $\mathscr{L}^\uparrow$ and $\ell^\uparrow(\infty e_A) = 0$. Thus $g + \infty e_A$ belongs to $\mathscr{L}^\uparrow$ [Proposition 3.1.2 (c)]. Since $f \leq g + \infty e_A$, we have

$$\ell^*(f) \leq \ell^\uparrow(g + \infty e_A) = \ell^\uparrow(g) + \ell^\uparrow(\infty e_A) = \ell^\uparrow(g)$$

[Proposition 3.1.4 (e)]. □

Proposition 3.3.3. *Let $(X, \mathscr{L}, \ell)$ be a closed Daniell space. Then the following assertions hold, for all $f, g, h \in \overline{\mathbb{R}}^X$.*

(*a*) *If $f \leq g$ $\mathscr{L}$-a.e., then $\ell^*(f) \leq \ell^*(g)$.*

(*b*) *If $f = g$ $\mathscr{L}$-a.e., then $\ell^*(f) = \ell^*(g)$.*

(*c*) *If $h(x) = f(x) + g(x)$ $\mathscr{L}$-a.e., and if $\ell^*(f) + \ell^*(g)$ is defined, then*

$$\ell^*(h) \leq \ell^*(f) + \ell^*(g).$$

(*d*) *If $\alpha \in \mathbb{R}_+$, then $\ell^*(\alpha f) = \alpha\ell^*(f)$. If $f \geq 0$, then $\ell^*(\infty f) = \infty\ell^*(f)$.*

Proof. (a) If $\tilde{g}$ is a function in $\mathscr{L}^\uparrow$ with $\tilde{g} \geq g$, then $\tilde{g} \geq f$ $\mathscr{L}$-a.e. and so $\ell^*(f) \leq \ell^\uparrow(\tilde{g})$ by Proposition 2(e). It follows that $\ell^*(f) \leq \ell^*(g)$.

(b) Assertion (b) follows from (a).

(c) If $\tilde{f}, \tilde{g} \in \mathscr{L}^\uparrow$ and $\tilde{f} \geq f$, $\tilde{g} \geq g$, then $\tilde{f} + \tilde{g}$ is defined and belongs to $\mathscr{L}^\uparrow$, $\ell^\uparrow(\tilde{f} + \tilde{g}) = \ell^\uparrow(\tilde{f}) + \ell^\uparrow(\tilde{g})$, and $h \leq \tilde{f} + \tilde{g}$ $\mathscr{L}$-a.e. [Propositions 3.1.2(c), 3.1.4 (e)]. By Proposition 2(e), $\ell^*(h) \leq \ell^\uparrow(\tilde{f}) + \ell^\uparrow(\tilde{g})$. It follows that $\ell^*(h) \leq \ell^*(f) + \ell^*(g)$.

(d) The first assertion is left for the reader to prove. Assume that $f \geq 0$. If $\ell^*(f) > 0$, then

$$\ell^*(\infty f) \geq \ell^*(nf) = n\ell^*(f)$$

for every n in $\mathbb{N}$, and so $\ell^*(\infty f) = \infty = \infty \ell^*(f)$. Suppose that $\ell^*(f) = 0$. According to Proposition 2 (d), there exists a function g in $\mathscr{L}$ such that $g \geq f$ and $\ell^*(f) = \ell(g) = 0$. We conclude successively that g belongs to $\mathscr{N}(\mathscr{L})$, so does f and also ∞f. Hence

$$\ell^*(\infty f) = \ell(\infty f) = 0 = \infty \ell^*(f). \qquad \square$$

Next we describe the important convergence properties of ℓ^*, in the case where $(X, \mathscr{L}, \ell)$ is closed. The Fatou Lemma, assertion (b) of Theorem 4, played a decisive role in earlier treatments of integration theory.

Theorem 3.3.4. *The following assertions hold, for every closed Daniell space* $(X, \mathscr{L}, \ell)$.

(*a*) *For every nonempty, countable, upward-directed family* $(f_\iota)_{\iota \in I}$ *from* $\overline{\mathbb{R}}^X$, *if* $\ell^*(f_{\iota_0}) > -\infty$ *for at least one* $\iota_0 \in I$, *then*

$$\ell^*\Big(\bigvee_{\iota \in I} f_\iota\Big) = \sup_{\iota \in I} \ell^*(f_\iota).$$

(*b*) **Fatou Lemma.** *For every sequence* $(f_n)_{n \in \mathbb{N}}$ *from* $\overline{\mathbb{R}}_+^X$,

$$\ell^*\Big(\liminf_{n \to \infty} f_n\Big) \leq \liminf_{n \to \infty} \ell^*(f_n).$$

Proof. (a) By Proposition 3(a),

$$\ell^*\Big(\bigvee_{\iota \in I} f_\iota\Big) \geq \sup_{\iota \in I} \ell^*(f_\iota).$$

If $\sup_{\iota \in I} \ell^*(f_\iota) = \infty$, equality must hold; so we assume that $\sup_{\iota \in I} \ell^*(f_\iota) < \infty$. Fix $\iota_0 \in I$ with $\ell^*(f_{\iota_0}) > -\infty$, and let $J := \{\iota \in I \mid f_\iota \geq f_{\iota_0}\}$. The family $(f_\iota)_{\iota \in J}$ is directed upward, and $\ell^*(f_\iota)$ is real for every ι in J. According to Proposition 2(d), there exists for each ι in J a function $g_\iota \in \mathscr{L}$ such that

$$g_\iota \geq f_\iota \quad \text{and} \quad \ell(g_\iota) = \ell^*(f_\iota).$$

We claim that the family $(g_\iota)_{\iota \in J}$ is directed upward relative to the preorder $\leq$ $\mathscr{L}$-a.e. Indeed, let ι', ι'' belong to J. There exists ι in I such that $f_{\iota'} \leq f_\iota$, $f_{\iota''} \leq f_\iota$. Now $f_{\iota'} \leq g_{\iota'} \wedge g_\iota$, so

$$\ell^*(f_{\iota'}) \leq \ell(g_{\iota'} \wedge g_\iota) \leq \ell(g_{\iota'}) = \ell^*(f_{\iota'}).$$

Thus $g_{\iota'} = g_{\iota'} \wedge g_\iota$ $\mathscr{L}$-a.e., that is, $g_{\iota'} \le g_\iota$ $\mathscr{L}$-a.e. Similarly $g_{\iota''} \le g_\iota$ $\mathscr{L}$-a.e. Note that

$$\bigvee_{\iota \in I} f_\iota = \bigvee_{\iota \in J} f_\iota$$

by the hypothesis of $(f_\iota)_{\iota \in I}$ being directed upward, and that $\sup_{\iota \in J} \ell(g_\iota) < \infty$. Since $\bigvee_{\iota \in J} f_\iota \le \bigvee_{\iota \in J} g_\iota$, we conclude, using Proposition 3 (a) and Theorem 1.4.12 (a), that $\bigvee_{\iota \in J} g_\iota$ belongs to $\mathscr{L}$ and

$$\ell^*\Big(\bigvee_{\iota \in I} f_\iota\Big) = \ell^*\Big(\bigvee_{\iota \in J} f_\iota\Big) \le \ell\Big(\bigvee_{\iota \in J} g_\iota\Big) = \sup_{\iota \in J} \ell(g_\iota)$$

$$= \sup_{\iota \in J} \ell^*(f_\iota) \le \sup_{\iota \in I} \ell^*(f_\iota).$$

(b) Using (a) and the monotonicity of ℓ^*, we have

$$\ell^*\Big(\liminf_{n \to \infty} f_n\Big) = \sup_{n \in \mathbb{N}} \ell^*\Big(\bigwedge_{m \ge n} f_m\Big)$$

$$\le \sup_{n \in \mathbb{N}} \inf_{m \ge n} \ell^*(f_m)$$

$$= \liminf_{n \to \infty} \ell^*(f_n). \qquad \square$$

We define ℓ_* symmetrically.

Definition 3.3.5. *Let $(X, \mathscr{L}, \ell)$ be a Daniell space. Then for $f \in \overline{\mathbb{R}}^X$,*

$$\ell_*(f) := \sup\{\ell^\downarrow(g) \mid g \in \mathscr{L}^\downarrow, g \le f\}. \qquad \square$$

The properties of ℓ_* can be derived from those of ℓ^* by means of the following proposition. The proposition itself follows readily from the connection between $\ell^\downarrow$ and $\ell^\uparrow$ (Propositions 3.1.6, 3.1.8). Those properties of ℓ_* to be cited later are listed in Proposition 7.

Proposition 3.3.6. *Let $(X, \mathscr{L}, \ell)$ be a Daniell space. Then for every $f \in \overline{\mathbb{R}}^X$,*

$$-\ell^*(-f) = \ell_*(f) \le \ell^*(f). \qquad \square$$

Proposition 3.3.7. *Let $(X, \mathscr{L}, \ell)$ be a closed Daniell space. Then the following assertions hold, for all $f, g, h \in \overline{\mathbb{R}}^X$.*

(a) *If $f \in \mathscr{L}$, then $\ell_*(f) = \ell(f)$.*
(b) *If $f \in \mathscr{L}^\downarrow$, then $\ell_*(f) = \ell^\downarrow(f)$.*
(c) *If $f \in \mathscr{L}^\downarrow$ and $\ell_*(f) > -\infty$, then f belongs to $\mathscr{L}$ and $\ell_*(f) = \ell^\downarrow(f) = \ell(f)$.*
(d) *If $f \le g$ $\mathscr{L}$-a.e., then $\ell_*(f) \le \ell_*(g)$.*
(e) *If $f = g$ $\mathscr{L}$-a.e., then $\ell_*(f) = \ell_*(g)$.*
(f) *If $h(x) = f(x) + g(x)$ $\mathscr{L}$-a.e., and if $\ell_*(f) + \ell_*(g)$ is defined, then $\ell_*(h) \ge \ell_*(f) + \ell_*(g)$.*
(g) *If $\alpha \in \mathbb{R}_+$, then $\ell_*(\alpha f) = \alpha \ell_*(f)$.* □

Closed Daniell spaces $(X, \mathscr{L}, \ell)$ can be characterized in terms of the mappings ℓ_* and ℓ^*.

Proposition 3.3.8. *The following assertions hold, for every closed Daniell space $(X, \mathscr{L}, \ell)$.*

(a) *If $f \in \mathscr{L}$, then $\ell(f) = \ell_*(f) = \ell^*(f)$.*
(b) *$\mathscr{L} = \{f \in \overline{\mathbb{R}}^X \mid \ell^*(f) = \ell_*(f) \in \mathbb{R}\}$.*
(c) *$\mathscr{N}(\mathscr{L}) = \{f \in \overline{\mathbb{R}}^X \mid \ell^*(|f|) = 0\}$.*

Proof. (a) This assertion merely restates Propositions 2(a) and 7(a).

(b) Inclusion $\subset$ follows from (a). Since $(X, \mathscr{L}, \ell)$ is its own closure, $\supset$ follows from Proposition 3.1.16 (d) $\Rightarrow$ (a).

(c) Since $(X, \mathscr{L}, \ell)$ is closed,

$$\mathscr{N}(\mathscr{L}) = \{f \in \overline{\mathbb{R}}^X \mid |f| \in \mathscr{L} \text{ and } \ell(|f|) = 0\}$$

(1.4.4). Moreover

$$0 \le \ell_*(|f|) \le \ell^*(|f|)$$

for every $f \in \overline{\mathbb{R}}^X$. Accordingly, (c) follows from (a) and (b). □

In Chapter 5 we construct Daniell spaces whose functionals cannot be extended, in general, without forfeiting important properties. In these situations ℓ_* and ℓ^* serve as substitutes and Proposition 8 provides the connection to $(X, \mathscr{L}, \ell)$.

So far in this section, except for the definitions, we have assumed that the Daniell space $(X, \mathscr{L}, \ell)$ is closed. The general situation is rather easily described. Indeed, for an arbitrary Daniell space $(X, \mathscr{L}, \ell)$, $\bar{\ell}_* = \ell_*$ and $\bar{\ell}^* = \ell^*$, as we show next.

Proposition 3.3.9. *For every Daniell space* $(X, \mathscr{L}, \ell)$,

$$\ell^* = \bar{\ell}^* \quad \text{and} \quad \ell_* = \bar{\ell}_*.$$

Proof. In view of Proposition 6, it suffices to establish the first relation. That $\ell^* \geq \bar{\ell}^*$ follows immediately from the inclusion

$$\{g \in \mathscr{L}^{\uparrow} \mid g \geq f\} \subset \{g \in \bar{\mathscr{L}}(\ell)^{\uparrow} \mid g \geq f\}$$

which holds for each $f \in \overline{\mathbb{R}}^X$. It follows, for $f \in \overline{\mathbb{R}}^X$, that $\ell^*(f) = \bar{\ell}^*(f)$ if $\bar{\ell}^*(f) = \infty$. Suppose that $\bar{\ell}^*(f)$ is real. By Proposition 2(d), $\bar{\ell}^*(f) = \bar{\ell}(g)$ for some $g \in \bar{\mathscr{L}}(\ell)$ with $g \geq f$. By Proposition 3.1.16 (a) $\Rightarrow$ (d), $\bar{\ell}(g) = \ell^*(g)$, so

$$\ell^*(f) \leq \ell^*(g) = \bar{\ell}(g) = \bar{\ell}^*(f).$$

Finally, suppose that $\bar{\ell}^*(f) = -\infty$. If g belongs to $\bar{\mathscr{L}}(\ell)^{\uparrow}$ and $g \geq f$, then $\bar{\ell}^*(g) = \bar{\ell}^{\uparrow}(g) > -\infty$. Then, using what has already been shown, we have

$$\ell^*(f) \leq \ell^*(g) = \bar{\ell}^*(g) = \bar{\ell}^{\uparrow}(g).$$

Taking the infimum over all such g yields $\ell^*(f) = -\infty$. □

Combining Propositions 8 and 9, we obtain a new characterization of the closure $(X, \bar{\mathscr{L}}(\ell), \bar{\ell})$ of a Daniell space in terms of the mappings ℓ_* and ℓ^*.

Proposition 3.3.10. *The following assertions hold, for every Daniell space* $(X, \mathscr{L}, \ell)$.

(a) *If* $f \in \bar{\mathscr{L}}(\ell)$, *then* $\bar{\ell}(f) = \ell_*(f) = \ell^*(f)$.
(b) $\bar{\mathscr{L}}(\ell) = \{f \in \overline{\mathbb{R}}^X \mid \ell^*(f) = \ell_*(f) \in \mathbb{R}\}$.
(c) $\mathscr{N}(\bar{\mathscr{L}}(\ell)) = \{f \in \overline{\mathbb{R}}^X \mid \ell^*(|f|) = 0\}$. □

Thus ℓ_* and ℓ^* provide an alternate means of constructing the closure. No simplification results from their use, however.

EXERCISE

3.3.1(E) Let $(X, \mathscr{L}, \ell)$ be a Daniell space. Prove the following:

(α) If $(f_\iota)_{\iota \in I}$ is a countable nonempty family from $\bar{\mathscr{L}}(\ell)$ and $f \in \overline{\mathbb{R}}^X$ such that $\ell_*(f) < \infty$ and $f_\iota \leq f$ $\bar{\mathscr{L}}(\ell)$-a.e. for each $\iota \in I$, then $\bigvee_{\iota \in I} f_\iota \in \bar{\mathscr{L}}(\ell)$.

(β) If $(f_\iota)_{\iota \in I}$ is a countable nonempty family from $\bar{\mathscr{L}}(\ell)$ and $f \in \overline{\mathbb{R}}^X$ such that $\ell^*(f) > -\infty$ and $f_\iota \geq f$ $\bar{\mathscr{L}}(\ell)$-a.e. for each $\iota \in I$, then $\bigwedge_{\iota \in I} f_\iota \in \bar{\mathscr{L}}(\ell)$.

(γ) Let $(f_n)_{n \in \mathbb{N}}$ be a sequence from $\bar{\mathscr{L}}(\ell)$ and $f \in \overline{\mathbb{R}}_+^X$ such that $|f_n| \leq f$ for every $n \in \mathbb{N}$. Then, if $\ell_*(f) < \infty$ or equivalently $\ell^*(-f) > -\infty$, the following are true:

($\gamma 1$) $\limsup_{n \to \infty} f_n \in \bar{\mathscr{L}}(\ell)$.
($\gamma 2$) $\liminf_{n \to \infty} f_n \in \bar{\mathscr{L}}(\ell)$.
($\gamma 3$) If $g \in \overline{\mathbb{R}}^X$ such that $g(x) = \lim_{n \to \infty} f_n(x)$ $\bar{\mathscr{L}}(\ell)$-a.e., then $g \in \bar{\mathscr{L}}(\ell)$ and $\bar{\ell}(g) = \lim_{n \to \infty} \bar{\ell}(f_n)$.

3.4. SUMMABLE FAMILIES

NOTATION FOR SECTION 3.4:

X, Y, Z denote sets.

χ denotes the counting measure on the set in question (see Example 2.4.8).

For W a set,

$$\mathfrak{F}(W) := \{A \subset W \mid A \text{ is finite}\}$$

$$\mathscr{F}(W) := \{f \in \mathbb{R}^W \mid f \text{ is an } \mathfrak{F}(W)\text{-step function}\}$$

$$= \{f \in \mathbb{R}^W \mid \{f \neq 0\} \text{ is finite}\}$$

and $(W, \mathscr{F}(W), \ell_\chi)$ denotes the Daniell space associated with the counting measure on W (see Example 2.4.8).

In this section we illustrate the concepts and methods of the preceding sections, using the example of summable families of real numbers. The concepts developed here will also be useful, important tools in many later sections.

Sums of finite families of real numbers are familiar. Important properties of such sums were outlined in Section 1.1. Our task in this section is to extend to infinite real number families the notion of sum. We cannot hope to define a sum for *every* real number family and yet retain all the laws that are valid for sums of finite families: the study of series shows what can go wrong. Consequently there will be a division into summable and nonsummable families. Every family will have an upper sum, but only summable families will have a sum. Which families, then, should be called summable? The example of series cannot be used as an indicator because the index set $\mathbb{N}$ is ordered, a fact that is fully exploited in the definition of convergent series. We want our investigations to be independent of such special properties. The counting measure on the other hand, described in Example 2.4.8, does provide a good starting point.

Before reviewing the counting measure, we remind the reader of the equivalence, described in the preface, between maps and families. A real-valued function and a real number family are just two different presentations of one and the same object, namely, a mapping of some set into $\mathbb{R}$. Every function $f \in \mathbb{R}^X$ is identical with a real family, namely, the family $(f(x))_{x \in X}$. Conversely, every real number family $(\alpha_x)_{x \in X}$ is identical with a function belonging to $\mathbb{R}^X$, namely the function

$$f\colon X \to \mathbb{R}, \qquad x \mapsto \alpha_x.$$

We briefly recapitulate. Given an arbitrary set X, we denote by $\mathfrak{F}(X)$ the set-ring consisting of all finite subsets of X and by χ the counting measure on X; that is, χ is the map from $\mathfrak{F}(X)$ into $\mathbb{R}$ that associates to every finite subset of X the number of elements in that subset. To the positive measure space $(X, \mathfrak{F}(X), \chi)$ is associated a Daniell space, denoted by $(X, \mathscr{F}(X), \ell_\chi)$ and called simply the Daniell space associated with the counting measure on X. Thus $\mathscr{F}(X)$ is the collection of all $\mathfrak{F}(X)$-step functions on X and the functional ℓ_χ satisfies

$$\ell_\chi(e_A) = \chi(A)$$

for every $A \in \mathfrak{F}(X)$. In the special case where X is finite, we have the formula

$$\ell_\chi(f) = \sum_{x \in X} f(x)$$

(Example 2.4.8). Thus the following definition suggests itself.

Definition 3.4.1. *For every family $(\alpha_x)_{x \in X}$ belonging to $\mathscr{F}(X)$,*

$$\sum_{x \in X} \alpha_x := \ell_\chi((\alpha_x)_{x \in X}).$$

The number $\sum_{x \in X} \alpha_x$ is called the **sum** *of the family $(\alpha_x)_{x \in X}$.* □

Of course Definition 1 yields little new as far as defining sums! After all,

$$\mathscr{F}(X) = \{f \in \mathbb{R}^X \mid \{f \neq 0\} \text{ is finite}\}$$

and if $(\alpha_x)_{x \in X}$ belongs to $\mathscr{F}(X)$, then

$$\sum_{x \in X} \alpha_x = \sum_{\substack{x \in X \\ \alpha_x \neq 0}} \alpha_x.$$

What we do gain by this definition is that finite sums are brought into a context that allows a genuine generalization.

We leave it to the reader to prove the following easy generalization of Theorem 1.1.2 (f), (g), (i), and (j).

Proposition 3.4.2. *The following assertions hold.*

(*a*) *If* $\varphi: Y \to X$ *is bijective, then for every family* $(\alpha_x)_{x \in X}$ *belonging to* $\mathscr{F}(X)$, *the family* $(\alpha_{\varphi(y)})_{y \in Y}$ *belongs to* $\mathscr{F}(Y)$ *and*

$$\sum_{y \in Y} \alpha_{\varphi(y)} = \sum_{x \in X} \alpha_x .$$

(*b*) *If* $X = Y \cup Z$ *with* $Y \cap Z = \varnothing$, *then for every family* $(\alpha_x)_{x \in X}$ *belonging to* $\mathscr{F}(X)$, *the family* $(\alpha_y)_{y \in Y}$ *belongs to* $\mathscr{F}(Y)$, *the family* $(\alpha_z)_{z \in Z}$ *belongs to* $\mathscr{F}(Z)$, *and*

$$\sum_{x \in X} \alpha_x = \sum_{y \in Y} \alpha_y + \sum_{z \in Z} \alpha_z .$$

(*c*) *Let* $(\alpha_{x,y})_{(x,y) \in X \times Y}$ *be a family belonging to* $\mathscr{F}(X \times Y)$. *Then* $(\alpha_{x,y})_{x \in X}$ *belongs to* $\mathscr{F}(X)$ *for every* y *in* Y, $(\alpha_{x,y})_{y \in Y}$ *belongs to* $\mathscr{F}(Y)$ *for every* x *in* X, *the family* $(\Sigma_{x \in X} \alpha_{x,y})_{y \in Y}$ *belongs to* $\mathscr{F}(Y)$, *the family* $(\Sigma_{y \in Y} \alpha_{x,y})_{x \in X}$ *belongs to* $\mathscr{F}(X)$, *and*

$$\sum_{(x,y) \in X \times Y} \alpha_{x,y} = \sum_{x \in X} \left(\sum_{y \in Y} \alpha_{x,y} \right) = \sum_{y \in Y} \left(\sum_{x \in X} \alpha_{x,y} \right).$$

(*d*) *If* $(\alpha_x)_{x \in X}$ *belongs to* $\mathscr{F}(X)$ *and* $(\beta_y)_{y \in Y}$ *belongs to* $\mathscr{F}(Y)$, *then the family* $(\alpha_x \beta_y)_{(x,y) \in X \times Y}$ *belongs to* $\mathscr{F}(X \times Y)$ *and*

$$\left(\sum_{x \in X} \alpha_x \right) \left(\sum_{y \in Y} \beta_y \right) = \sum_{(x,y) \in X \times Y} \alpha_x \beta_y . \qquad \square$$

We have taken the first step toward generalizing the notion of sum of a finite real number family. Especially important is the fact that this step moves the summability problem into the effective range of integration theory. The way to proceed is now obvious: we apply to the particular Daniell space $(X, \mathscr{F}(X), \ell_X)$ the method developed in Section 3.1 for constructing its closure. We want to carry out the important steps in that construction in order to obtain a concrete description of the closure $(X, \overline{\mathscr{L}}(\ell_X), \overline{\ell_X})$. Later in the section we discuss upper sums, which result from the upper-closure construction described in Section 3.3. Before proceeding, the reader may want to briefly review Section 3.1, if only to recall the notation and the terminology introduced there.

Proposition 3.4.3. *Let $f \in \overline{\mathbb{R}}^X$. Then the following assertions hold.*

(*a*) *f belongs to $\mathscr{F}(X)^{\uparrow}$ iff $\{f \neq 0\}$ is countable, $\{f < 0\}$ is finite, and $f(x) > -\infty$ for every x in X.*

(*b*) *f belongs to $\mathscr{F}(X)^{\downarrow}$ iff $\{f \neq 0\}$ is countable, $\{f > 0\}$ is finite, and $f(x) < \infty$ for every x in X.*

Proof. (a) Suppose first that f belongs to $\mathscr{F}(X)^{\uparrow}$. By definition there exists an increasing sequence $(f_n)_{n \in \mathbb{N}}$ from $\mathscr{F}(X)$ whose supremum is f (Definition 3.1.1). Evidently

$$\{f \neq 0\} \subset \bigcup_{n \in \mathbb{N}} \{f_n \neq 0\}$$

$$\{f < 0\} \subset \{f_1 < 0\}.$$

Since each set $\{f_n \neq 0\}$ is finite, the countability of $\{f \neq 0\}$ and the finiteness of $\{f < 0\}$ follow. That $f(x) > -\infty$ for every x was already proved [Proposition 3.1.2 (a)].

Conversely, suppose that f has the properties listed. If f were known to be real valued, the argument would be a triviality. Since f can take the value ∞, we must work slightly harder. Choose, as is possible, an increasing sequence $(A_n)_{n \in \mathbb{N}}$ of finite subsets of X such that

$$\{f < 0\} \subset A_1 \subset \bigcup_{n \in \mathbb{N}} A_n = \{f \neq 0\}.$$

The sequence

$$(f \wedge n e_{A_n})_{n \in \mathbb{N}}$$

each term of which belongs to $\mathscr{F}(X)$, increases and has f as its supremum. It follows that f belongs to $\mathscr{F}(X)^{\uparrow}$.

(b) In view of how $\mathscr{F}(X)^{\uparrow}$ and $\mathscr{F}(X)^{\downarrow}$ are related (Proposition 3.1.6), (b) follows from (a). □

Proposition 3.4.4.

(*a*) *If f belongs to $\mathscr{F}(X)^{\uparrow}$ and $\ell_X^{\uparrow}(f) < \infty$, then f is necessarily real valued. If f is a real-valued function belonging to $\mathscr{F}(X)^{\uparrow}$, then $\ell_X^{\uparrow}(f)$ is given by the formula*

$$\ell_X^{\uparrow}(f) = \sum_{x \in \{f < 0\}} f(x) + \sup_{A \in \mathfrak{F}(X)} \sum_{x \in A} f(x). \tag{1}$$

(*b*) *If f belongs to $\mathscr{F}(X)^{\downarrow}$ and $\ell_X^{\downarrow}(f) > -\infty$, then f is real valued. If f is real valued and belongs to $\mathscr{F}(X)^{\downarrow}$, then*

$$\ell_X^{\downarrow}(f) = \sum_{x \in \{f > 0\}} f(x) + \inf_{A \in \mathfrak{F}(X)} \sum_{x \in A} f(x).$$

Proof. (a) Suppose that f belongs to $\mathscr{F}(X)^{\uparrow}$ with $\ell_X^{\uparrow}(f) < \infty$, and let $x \in X$. We know from Proposition 3 that $\{f < 0\}$ is a finite subset of X and $f(x) > -\infty$. Observe that

$$f(x)e_{\{x\}} \leq f - fe_{\{f<0\}}.$$

The functions $fe_{\{f<0\}}$ and $-fe_{\{f<0\}}$ belong to $\mathscr{F}(X)$ and therefore to $\mathscr{F}(X)^{\uparrow}$. The function $f(x)e_{\{x\}}$ belongs to $\mathscr{F}(X)^{\uparrow}$, as does $f - fe_{\{f<0\}}$. Recalling Proposition 3.1.4, we see that

$$f(x) = \ell_X^{\uparrow}\big(f(x)e_{\{x\}}\big) \leq \ell_X^{\uparrow}(f) - \ell_X\big(fe_{\{f<0\}}\big) < \infty. \tag{2}$$

The first claim follows.

Now suppose that f is real valued and belongs to $\mathscr{F}(X)^{\uparrow}$. Thus $\{f < 0\}$ is finite and $\{f \neq 0\}$ is countable. For every finite subset A of X, we have

$$fe_{A \cup \{f<0\}} \leq f$$

and therefore

$$\begin{aligned} \sum_{x \in \{f<0\}} f(x) + \sum_{x \in A} f(x) &\leq \sum_{x \in A \cup \{f<0\}} f(x) \\ &= \ell_X\big(fe_{A \cup \{f<0\}}\big) \\ &\leq \ell_X^{\uparrow}(f). \end{aligned} \tag{3}$$

On the other hand, we can choose an increasing sequence $(A_n)_{n \in \mathbb{N}}$ from $\mathfrak{F}(X)$ such that

$$\{f < 0\} \subset A_1 \subset \bigcup_{n \in \mathbb{N}} A_n = \{f \neq 0\}.$$

Then $(fe_{A_n})_{n \in \mathbb{N}}$ is an increasing sequence whose supremum is f and whose terms belong to $\mathscr{F}(X)$ since f is real valued. Recalling that $\ell_X^{\uparrow}$ agrees on $\mathscr{F}(X)$ with ℓ_X and commutes with sup when applied to an increasing sequence, [Proposition 3.1.4 (b), (d)], we see that

$$\begin{aligned} \ell_X^{\uparrow}(f) &= \sup_{n \in \mathbb{N}} \ell_X\big(fe_{A_n}\big) \\ &= \sum_{x \in \{f<0\}} f(x) + \sup_{n \in \mathbb{N}} \sum_{x \in A_n \setminus \{f<0\}} f(x) \\ &\leq \sum_{x \in \{f<0\}} f(x) + \sup_{A \in \mathfrak{F}(X)} \sum_{x \in A} f(x) \\ &\leq \ell_X^{\uparrow}(f). \end{aligned} \tag{4}$$

In view of (4), the required formula (1) holds.

(b) Again, in view of Proposition 3.1.6, (b) follows from (a). □

The notation $(X, \overline{\mathscr{L}}(\ell_X), \overline{\ell}_X)$ is cumbersome, as is $\overline{\ell}_X{}^*$. Looking ahead, we substitute the notation $(X, \mathscr{L}^1(X), \Sigma)$, Σ^*, and we introduce some related terminology.

Definition 3.4.5. *$(X, \mathscr{L}^1(X), \Sigma)$ shall denote the closure of the Daniell space associated with the counting measure on the set X. That is,*

$$\mathscr{L}^1(X) := \overline{\mathscr{L}}(\ell_X)$$
$$\sum : \mathscr{L}^1(X) \to \mathbb{R}, \qquad f \mapsto \overline{\ell}_X(f).$$

Accordingly,

$$\sum{}^* : \overline{\mathbb{R}}^X \to \overline{\mathbb{R}}, \qquad f \mapsto \sum{}^*(f) = \overline{\ell}_X{}^*(f).$$

A function $f \in \overline{\mathbb{R}}^X$ is said to be ***summable*** *iff it belongs to $\mathscr{L}^1(X)$. For $f \in \mathscr{L}^1(X)$ we write*

$$\sum_{x \in X} f(x) := \sum(f)$$

and for $f \in \overline{\mathbb{R}}^X$ we write

$$\sum_{x \in X}{}^* f(x) := \sum{}^*(f).$$

Equivalently, a family $(\alpha_x)_{x \in X}$ from $\overline{\mathbb{R}}$ is said to be ***summable*** *iff it belongs to $\mathscr{L}^1(X)$. For every summable family $(\alpha_x)_{x \in X}$ from $\overline{\mathbb{R}}$, we write*

$$\sum_{x \in X} \alpha_x := \sum((\alpha_x)_{x \in X})$$

and we call $\Sigma_{x \in X} \alpha_x$ the ***sum*** *of the family $(\alpha_x)_{x \in X}$. In a similar way, for every family $(\alpha_x)_{x \in X}$ from $\overline{\mathbb{R}}$, we write*

$$\sum_{x \in X}{}^* \alpha_x := \sum{}^*((\alpha_x)_{x \in X})$$

and we call $\Sigma^*_{x \in X} \alpha_x$ the ***upper sum*** *of the family.* □

Definition 3.4.6. *A family $(f_\iota)_{\iota \in I}$ from $\overline{\mathbb{R}}^X$ is said to be* ***summable*** *iff, for each $x \in X$, the family $(f_\iota(x))_{\iota \in I}$ is summable. If $(f_\iota)_{\iota \in I}$ is a summable family from $\overline{\mathbb{R}}^X$, then $\Sigma_{\iota \in I} f_\iota$, defined by*

$$\sum_{\iota \in I} f_\iota : X \to \mathbb{R}, \qquad x \mapsto \sum_{\iota \in I} f_\iota(x)$$

is called the ***sum*** *of the family $(f_\iota)_{\iota \in I}$. If $(f_\iota)_{\iota \in I}$ is an arbitrary family from $\overline{\mathbb{R}}^X$,* *then* $\sum^*_{\iota \in I} f_\iota$, *defined by*

$$\sum_{\iota \in I}^* f_\iota \colon X \to \overline{\mathbb{R}}, \qquad x \mapsto \sum_{\iota \in I}{}^* f_\iota(x)$$

is called the **upper sum** *of the family* $(f_\iota)_{\iota \in I}$. □

Corollary 3.4.7. *If* $(\alpha_x)_{x \in X}$ *is a summable family from* $\overline{\mathbb{R}}$, *then*

$$\sum_{x \in X}^* \alpha_x = \sum_{x \in X} \alpha_x.$$

Similarly, if $(f_\iota)_{\iota \in I}$ *is a summable family from* $\overline{\mathbb{R}}^X$, *then*

$$\sum_{\iota \in I}^* f_\iota = \sum_{\iota \in I} f_\iota. \qquad \square$$

Theorem 3.4.8. *The following assertions are equivalent, for every* $f \in \overline{\mathbb{R}}^X$.

(*a*) $f \in \mathscr{L}^1(X)$.
(*b*) $|f| \in \mathscr{L}^1(X)$.
(*c*) f *is real-valued and* $\sup_{A \in \mathfrak{F}(X)} \sum_{x \in A} |f(x)| < \infty$.
(*d*) $|f| \in \mathscr{F}(X)^\uparrow$ *and* $\ell_X^\uparrow(|f|) < \infty$.

Each of the assertions (*a*)–(*d*) *implies both of the following.*

(*e*) $\{|f| \geq 1/n\}$ *is finite for every* $n \in \mathbb{N}$.
(*f*) $\{f \neq 0\}$ *is countable.*

Moreover for every $f \in \mathscr{L}^1(X)$, $\Sigma(f)$ *is given by the formula*

$$\sum_{x \in X} f(x) = \sup_{A \in \mathfrak{F}(X)} \sum_{x \in A} f(x) + \inf_{A \in \mathfrak{F}(X)} \sum_{x \in A} f(x). \tag{5}$$

In particular, if f *belongs to* $\mathscr{L}^1(X)_+$, *then*

$$\sum_{x \in X} f(x) = \sup_{A \in \mathfrak{F}(X)} \sum_{x \in A} f(x). \tag{6}$$

(*Note*: *Another formula for* $\sum_{x \in X} f(x)$ *appears in Corollary* 19.)

Proof. (a) ⇒ (b). Since $\mathscr{L}^1(X)$ is a Riesz lattice, (a) ⇒ (b).

(b) ⇒ (c). There exists a function g belonging to $\mathscr{F}(X)_+^\uparrow$ such that $g \geq |f|$ and $\ell_X^\uparrow(g) < \infty$. By Proposition 4, both g and f are real valued and we have

$$\sup_{A \in \mathfrak{F}(X)} \sum_{x \in A} |f(x)| \leq \sup_{A \in \mathfrak{F}(X)} \sum_{x \in A} g(x)$$

$$= \ell_X^{\uparrow}(g)$$

$$< \infty.$$

(c) $\Rightarrow$ (d), (e), (f). For each n in $\mathbb{N}$, set

$$A_n := \left\{ |f| \geq \frac{1}{n} \right\}.$$

Then

$$\frac{1}{n} e_{A_n} \leq |f|$$

for every n, and

$$\frac{1}{n} \sup_{B \in \mathfrak{F}(X)} \chi(A_n \cap B) = \sup_{B \in \mathfrak{F}(X)} \sum_{x \in B} \frac{1}{n} e_{A_n}(x)$$

$$\leq \sup_{B \in \mathfrak{F}(X)} \sum_{x \in B} |f(x)|$$

$$< \infty,$$

which is only possible if each A_n is a finite set. Since $\{f \neq 0\}$ is a subset of $\bigcup_{n \in \mathbb{N}} A_n$, it follows that $\{f \neq 0\}$ is countable. Proposition 3 now ensures that $|f|$ belongs to $\mathscr{F}(X)^{\uparrow}$. By Proposition 4, $\ell_X^{\uparrow}(|f|) < \infty$.

(d) $\Rightarrow$ (a). First we use Proposition 3 twice. Since $|f| \in \mathscr{F}(X)^{\uparrow}$, $\{|f| \neq 0\}$ is countable. Therefore $\{f^+ \neq 0\}$ and $\{f^- \neq 0\}$ are also countable, so f^+ and f^- belong to $\mathscr{F}(X)^{\uparrow}$. Since $\ell_X^{\uparrow}(|f|) < \infty$, the monotonicity of $\ell_X^{\uparrow}$ shows that $\ell_X^{\uparrow}(f^+) < \infty$ and $\ell_X^{\uparrow}(f^-) < \infty$. Now apply 3.2.1 to conclude that f^+ and f^- belong to $\mathscr{L}^1(X)$. It follows that f belongs to $\mathscr{L}^1(X)$.

Formulas (5) and (6) for $\Sigma(f)$. Now suppose that $f \in \mathscr{L}^1(X)$. By (a) $\Rightarrow$ (c), (f), f is real-valued and $\{f \neq 0\}$ is countable. By 3.4.3, it follows that f^+ and f^- belong to $\mathscr{F}(X)^{\uparrow}$ and $-(f^-)$ belongs to $\mathscr{F}(X)^{\downarrow}$. Moreover, f^+ and f^- are real valued, so we can use the formulas from Proposition 4 to compute $\ell_X^{\uparrow}(f^+)$ and $\ell_X^{\downarrow}(-f^-)$. From (a) $\Rightarrow$ (e), the monotonicity of $\ell_X^{\uparrow}$, and the relation $\ell_X^{\downarrow}(-f^-) = -\ell_X^{\uparrow}(f^-)$, we conclude that $\ell_X^{\uparrow}(f^+) < \infty$ and $\ell_X^{\downarrow}(-f^-) > -\infty$. Consequently $\ell_X^{\uparrow}(f^+) = \overline{\ell}_X(f^+)$ and $\ell_X^{\downarrow}(-f^-) = \overline{\ell}_X(-f^-)$ (3.2.1). In summary,

$$\sum_{x\in X} f(x) = \overline{\ell_X}(f)$$

$$= \overline{\ell_X}(f^+) + \overline{\ell_X}(-f^-)$$

$$= \ell_X^{\uparrow}(f^+) + \ell_X^{\downarrow}(-f^-)$$

$$= \sup_{A\in\mathfrak{F}(X)} \sum_{x\in A} f^+(x) + \inf_{A\in\mathfrak{F}(X)} \sum_{x\in A} -(f^-(x))$$

$$= \sup_{A\in\mathfrak{F}(X)} \sum_{x\in A} f(x) + \inf_{A\in\mathfrak{F}(X)} \sum_{x\in A} f(x). \qquad \square$$

Corollary 3.4.9.

$$\mathfrak{N}(\mathscr{L}^1(X)) = \{\varnothing\}$$

$$\mathscr{N}(\mathscr{L}^1(X)) = \{0\}.$$

Proof. In view of Theorem 8 (c), $\mathscr{L}^1(X)$ is a real Riesz lattice. Apply Proposition 1.2.24 and Corollary 1.2.27. $\square$

The space $(X, \mathscr{L}^1(X), \Sigma)$ is not only the smallest closed Daniell space extending $(X, \mathscr{F}(X), \ell_X)$; it is the only such extension. We now prove this fact, which we already anticipated in introducing sigma notation in Definition 5.

Theorem 3.4.10. *$(X, \mathscr{L}^1(X), \Sigma)$ is the only closed Daniell space that extends the Daniell space $(X, \mathscr{F}(X), \ell_X)$ associated with the counting measure on X.*

Proof. Let $(X, \mathscr{L}, \ell)$ be an arbitrary closed Daniell space extending $(X, \mathscr{F}(X), \ell_X)$. We know that

$$(X, \mathscr{L}^1(X), \Sigma) \preccurlyeq (X, \mathscr{L}, \ell)$$

(Theorem 3.2.5, Definition 5). If we can show that

$$\mathscr{L} \subset \mathscr{L}^1(X) \tag{7}$$

then we can conclude that

$$(X, \mathscr{L}, \ell) = (X, \mathscr{L}^1(X), \Sigma). \tag{8}$$

So let f belong to $\mathscr{L}$. Then $|f|$ also belongs to $\mathscr{L}$. For every x in X, we have

$$|f|(x)e_{\{x\}} \leq |f|$$

and

$$|f|(x) = \ell_X^{\uparrow}\left(|f|(x)e_{\{x\}}\right) \leq \ell(|f|) < \infty$$

so f is real valued. For every finite set $A \subset X$,

$$|f|e_A \leq |f|$$

and

$$\sum_{x \in A} |f(x)| = \ell_X(|f|e_A) \leq \ell(|f|).$$

Thus

$$\sup_{A \in \mathfrak{F}(X)} \sum_{x \in A} |f(x)| \leq \ell(|f|) < \infty.$$

Theorem 8 (c) $\Rightarrow$ (a) shows that f belongs to $\mathscr{L}^1(X)$. Thus (7) and (8) both hold. □

Proposition 3.4.11. *The following assertions hold, for all families* $(\alpha_x)_{x \in X}$ *and* $(\beta_x)_{x \in X}$ *from* $\overline{\mathbb{R}}$.

(*a*) *If* $(\alpha_x)_{x \in X}$ *is summable and if*

$$|\beta_x| \leq |\alpha_x|$$

for every x *in* X, *then* $(\beta_x)_{x \in X}$ *is summable.*

(*b*) *If* $(\alpha_x)_{x \in X}$ *is summable, then so is* $(\beta\alpha_x)_{x \in X}$ *for every real number* β, *and* $\Sigma_{x \in X}\beta\alpha_x = \beta\Sigma_{x \in X}\alpha_x$. *Moreover, if* $(\alpha_x)_{x \in X}$ *is summable and* $(\beta_x)_{x \in X}$ *is bounded in* $\mathbb{R}$, *then the family* $(\alpha_x\beta_x)_{x \in X}$ *is summable. In fact, if* $(\alpha_x)_{x \in X}$ *is summable and if* $|\beta_x| \leq \beta < \infty$ *for every* x *in* X, *then*

$$\sum_{x \in X} |\alpha_x\beta_x| \leq \beta \sum_{x \in X} |\alpha_x|.$$

(*c*) *If* $(\alpha_x)_{x \in X}$ *and* $(\beta_x)_{x \in X}$ *are summable, then* $(\alpha_x + \beta_x)_{x \in X}$ *is defined and summable, and*

$$\sum_{x \in X} (\alpha_x + \beta_x) = \sum_{x \in X} \alpha_x + \sum_{x \in X} \beta_x.$$

Proof. (a) is an easy consequence of the equivalence (a) $\Leftrightarrow$ (c) of Theorem 8.

(b) follows from (a) and Theorem 8.

(c) is evident since Σ is an additive functional on the real Riesz lattice $\mathscr{L}^1(X)$. □

Does the new notion of summability really generalize the previous one? We must check, for one, that the fundamental properties of sums, described in Proposition 2, carry over intact to the newly defined sums. This task can be readily disposed of by using the Induction Principle of Section 3.2. First, however, we want to show that the generalized notion actually yields something new. Recall that $\{x \in X \mid \alpha_x \neq 0\}$ must be finite if the family $(\alpha_x)_{x \in X}$ belongs to $\mathscr{F}(X)$. For $(\alpha_x)_{x \in X}$ to belong to $\mathscr{L}^1(X)$, this requirement no longer holds.

Proposition 3.4.12. *For every countable set $Y \subset X$ there exists a summable family $(\alpha_x)_{x \in X}$ such that $\alpha_x > 0$ for every $x \in Y$.*

Proof. It suffices to consider the case where Y is countably infinite. Then a family of the type required can be obtained by choosing a bijective mapping $\varphi\colon Y \to \mathbb{N}$ and setting

$$\alpha_x := \begin{cases} \dfrac{1}{2^{\varphi(x)}} & x \in Y \\ 0 & x \in X \setminus Y. \end{cases}$$ □

Theorem 3.4.13 [Induction Principle for $\mathscr{L}^1(X)$]. *Let $\mathscr{S}$ be a subset of $\mathscr{L}^1(X)$ that satisfies the following two conditions.*

(*a*) $\{(\alpha_x)_{x \in X} \in \mathbb{R}^X \mid \{x \in X \mid \alpha_x \neq 0\}$ *is finite*$\} \subset \mathscr{S}$; *i.e.* $\mathscr{F}(X) \subset \mathscr{S}$.

(*b*) *For every Σ-sequence $((\alpha_{n,x})_{x \in X})_{n \in \mathbb{N}}$ from $\mathscr{S}$ the family $(\lim_{n \to \infty} \alpha_{n,x})_{x \in X}$ belongs to $\mathscr{S}$.*

Then $\mathscr{S} = \mathscr{L}^1(X)$.

Proof. According to Corollary 9, the Riesz lattice $\mathscr{L}^1(X)$ has no nonempty exceptional sets. Thus Theorem 13 is little more than a restatement, in the present context, of the second form of the Induction Principle (Theorem 3.2.8). □

We show that the sum of the summable family does not depend on the "order" in which the terms are summed.

Corollary 3.4.14. *If* $\varphi: Y \to X$ *is bijective, then for every family* $(\alpha_x)_{x \in X}$ *belonging to* $\mathscr{L}^1(X)$, *the family* $(\alpha_{\varphi(y)})_{y \in Y}$ *belongs to* $\mathscr{L}^1(Y)$ *and*

$$\sum_{y \in Y} \alpha_{\varphi(y)} = \sum_{x \in X} \alpha_x.$$

Proof. Take $\mathscr{S}$ to be the subset of $\mathscr{L}^1(X)$ whose members have the property described. Proposition 2 (a) asserts that $\mathscr{S}$ contains $\mathscr{F}(X)$. Let $((\alpha_{n,x})_{x \in X})_{n \in \mathbb{N}}$ be a Σ-sequence from $\mathscr{S}$, hence from $\mathscr{L}^1(X)$. Then $((\alpha_{n,\varphi(y)})_{y \in Y})_{n \in \mathbb{N}}$ is a Σ-sequence from $\mathscr{L}^1(Y)$. The triples $(X, \mathscr{L}^1(X), \Sigma)$ and $(Y, \mathscr{L}^1(Y), \Sigma)$ are both closed. Therefore the two limit families $(\lim_{n \to \infty} \alpha_{n,x})_{x \in X}$ and $(\lim_{n \to \infty} \alpha_{n,\varphi(y)})_{y \in Y}$ belong to $\mathscr{L}^1(X)$ and $\mathscr{L}^1(Y)$, respectively, and we can interchange lim and Σ to compute:

$$\sum_{y \in Y} \lim_{n \to \infty} \alpha_{n,\varphi(y)} = \lim_{n \to \infty} \sum_{y \in Y} \alpha_{n,\varphi(y)}$$

$$= \lim_{n \to \infty} \sum_{x \in X} \alpha_{n,x}$$

$$= \sum_{x \in X} \lim_{n \to \infty} \alpha_{n,x}.$$

This calculation shows that the family $(\lim_{n \to \infty} \alpha_{n,x})_{x \in X}$ belongs to $\mathscr{S}$. It follows from Theorem 13 that $\mathscr{S} = \mathscr{L}^1(X)$. □

The sum of a summable family is also independent of how the terms are grouped. We formulate this property somewhat more generally than we did in Proposition 2 (b). The reader can carry out the proof, using Theorem 13.

Corollary 3.4.15. *If* $(X_\iota)_{\iota \in I}$ *is a disjoint family of sets with* $X = \bigcup_{\iota \in I} X_\iota$, *then the following assertions hold for every family* $(\alpha_x)_{x \in X}$ *belonging to* $\mathscr{L}^1(X)$.

(*a*) *For every* ι *in* I, *the family* $(\alpha_x)_{x \in X_\iota}$ *belongs to* $\mathscr{L}^1(X_\iota)$.

(*b*) *The family* $(\sum_{x \in X_\iota} \alpha_x)_{\iota \in I}$ *belongs to* $\mathscr{L}^1(I)$.

(*c*) $\sum_{\iota \in I}(\sum_{x \in X_\iota} \alpha_x) = \sum_{x \in X} \alpha_x$. □

Corollary 3.4.16. *Suppose* $X = Y \times Z$. *If* $(\alpha_{y,z})_{(y,z) \in X}$ *belongs to* $\mathscr{L}^1(X)$, *then* $(\alpha_{y,z})_{y \in Y}$ *belongs to* $\mathscr{L}^1(Y)$ *for every* z *in* Z, $(\alpha_{y,z})_{z \in Z}$ *belongs to* $\mathscr{L}^1(Z)$ *for every* y *in* Y, *the family* $(\sum_{y \in Y} \alpha_{y,z})_{z \in Z}$ *belongs to* $\mathscr{L}^1(Z)$, *the family* $(\sum_{z \in Z} \alpha_{y,z})_{y \in Y}$ *belongs to* $\mathscr{L}^1(Y)$, *and*

$$\sum_{(y,z) \in X} \alpha_{y,z} = \sum_{y \in Y} \left(\sum_{z \in Z} \alpha_{y,z} \right) = \sum_{z \in Z} \left(\sum_{y \in Y} \alpha_{y,z} \right).$$

Proof. These assertions follow from the preceding two corollaries. For each z in Z, define

$$X_z := \{(y, z) \mid y \in Y\}$$
$$\varphi_z : Y \to X_z, \qquad y \mapsto (y, z).$$

Then $(X_z)_{z \in Z}$ is a disjoint family of sets whose union is X, and the mappings φ_z are bijective. Now suppose that $(\alpha_{y,z})_{(y,z) \in X}$ is summable. Then for every z in Z the family

$$(\alpha_{y,z})_{(y,z) \in X_z}$$

belongs to $\mathscr{L}^1(X_z)$, by Corollary 15 (a), and the family

$$(\alpha_{y,z})_{y \in Y}$$

belongs to $\mathscr{L}^1(Y)$, by Corollary 14. Moreover

$$\sum_{y \in Y} \alpha_{y,z} = \sum_{(y,z) \in X_z} \alpha_{y,z}$$

and Corollary 15 (b) shows that the family

$$\left(\sum_{y \in Y} \alpha_{y,z} \right)_{z \in Z}$$

belongs to $\mathscr{L}^1(Z)$. Finally, according to Corollary 15 (c),

$$\sum_{z \in Z} \left(\sum_{y \in Y} \alpha_{y,z} \right) = \sum_{z \in Z} \left(\sum_{(y,z) \in X_z} \alpha_{y,z} \right) = \sum_{(y,z) \in X} \alpha_{y,z}.$$

The remaining assertions can be verified in exactly the same way. □

Corollary 3.4.17. *If $(\alpha_x)_{x \in X}$ belongs to $\mathscr{L}^1(X)$ and $(\beta_y)_{y \in Y}$ belongs to $\mathscr{L}^1(Y)$, then the family $(\alpha_x \beta_y)_{(x,y) \in X \times Y}$ belongs to $\mathscr{L}^1(X \times Y)$ and*

$$\sum_{(x,y) \in X \times Y} \alpha_x \beta_y = \left(\sum_{x \in X} \alpha_x \right) \left(\sum_{y \in Y} \beta_y \right). \tag{9}$$

Proof. Note that $\alpha_x \beta_y$ is real for every (x, y) in $X \times Y$ and that

$$\sup_{C \in \mathfrak{F}(X \times Y)} \sum_{(x,y) \in C} |\alpha_x \beta_y| \le \left(\sum_{x \in X} |\alpha_x| \right) \left(\sum_{y \in Y} |\beta_y| \right) < \infty.$$

Thus Theorem 8 shows that the family $(\alpha_x \beta_y)_{(x, y) \in X \times Y}$ is summable, and equation (9) follows from Corollary 16. □

Thus all of the summation properties enunciated in Proposition 2 carry over to the generalized notion of sum. Our use of Σ-notation and the term "summable," as introduced in Definition 5, have thus been fully justified. The notation $\mathscr{L}^1(X)$ will explain itself later.

There are of course other ways to develop the theory of summable families. One possibility is the ε-Y characterization of sum, spelled out in the next theorem.

Theorem 3.4.18. *The following assertions are equivalent, for every family $(\alpha_x)_{x \in X}$ from $\overline{\mathbb{R}}$ and for every real number α.*

(*a*) *$(\alpha_x)_{x \in X}$ is summable and $\Sigma_{x \in X} \alpha_x = \alpha$.*

(*b*) *The family $(\alpha_x)_{x \in X}$ is real, and for every real number $\varepsilon > 0$ there exists a finite set $Y \subset X$ such that the conditions Z finite and $Y \subset Z \subset X$ imply*

$$\left| \alpha - \sum_{x \in Z} \alpha_x \right| < \varepsilon.$$

Proof. *Throughout the proof* we use the following notation. Given a family $(\alpha_x)_{x \in X}$ from $\overline{\mathbb{R}}$, its positive part will be written $(\beta_x)_{x \in X}$, and its negative part will be written $(\gamma_x)_{x \in X}$. For instance, $\beta_x = (\alpha_x)^+ = \alpha_x \vee 0$. (See Definition 1.2.5.)

(a) $\Rightarrow$ (b). By Theorem 8 (a) $\Rightarrow$ (c), $(\alpha_x)_{x \in X}$ is a real family. Let $\varepsilon > 0$ be given. Since the family $(\alpha_x)_{x \in X}$ belongs to the Riesz lattice $\mathscr{L}^1(X)$, its positive part $(\beta_x)_{x \in X}$ and its negative part $(\gamma_x)_{x \in X}$ each belong to $\mathscr{L}^1(X)_+$. According to the formula in Theorem 8, there exist sets B, C belonging to $\mathfrak{F}(X)$ such that

$$\left| \sum_{x \in X} \beta_x - \sum_{x \in B} \beta_x \right| < \frac{\varepsilon}{2} \quad \text{and} \quad \left| \sum_{x \in X} \gamma_x - \sum_{x \in C} \gamma_x \right| < \frac{\varepsilon}{2}.$$

Set $Y := B \cup C$. Y is certainly finite. Moreover, if Z is finite and $Y \subset Z \subset X$, then

$$\begin{aligned} \left| \alpha - \sum_{x \in Z} \alpha_x \right| &\leq \left| \sum_{x \in X} \beta_x - \sum_{x \in Z} \beta_x \right| + \left| \sum_{x \in X} \gamma_x - \sum_{x \in Z} \gamma_x \right| \\ &\leq \left| \sum_{x \in X} \beta_x - \sum_{x \in B} \beta_x \right| + \left| \sum_{x \in X} \gamma_x - \sum_{x \in C} \gamma_x \right| \\ &< \varepsilon. \end{aligned}$$

(b) $\Rightarrow$ (a). First we use Theorem (8) (c) $\Rightarrow$ (a) to show that the family $(\beta_x)_{x \in X}$ is summable. Fix $\varepsilon > 0$, and let Y denote the corresponding set whose existence is asserted by (b). Writing $B := \{x \in X \mid \beta_x > 0\}$, we have for every finite set $A \subset X$:

$$\begin{aligned}
\sum_{x \in A} |\beta_x| &= \sum_{x \in A \cap B} \alpha_x \\
&= \sum_{x \in (A \cap B) \cup Y} \alpha_x - \sum_{x \in Y \setminus (A \cap B)} \alpha_x \\
&\le \left| \sum_{x \in (A \cap B) \cup Y} \alpha_x \right| + \left| \sum_{x \in Y \setminus (A \cap B)} \alpha_x \right| \\
&\le \left| \sum_{x \in Y \cup (A \cap B)} \alpha_x \right| + \sum_{x \in Y} |\alpha_x| \\
&\le |\alpha| + \left| \alpha - \sum_{x \in Y \cup (A \cap B)} \alpha_x \right| + \sum_{x \in Y} |\alpha_x| \\
&< |\alpha| + \varepsilon + \sum_{x \in Y} |\alpha_x|.
\end{aligned}$$

It follows that

$$\sup_{A \in \mathfrak{F}(X)} \sum_{x \in A} |\beta_x| \le |\alpha| + \varepsilon + \sum_{x \in Y} |\alpha_x| < \infty$$

hence that $(\beta_x)_{x \in X}$ belongs to $\mathscr{L}^1(X)$. Similarly $(\gamma_x)_{x \in X}$ belongs to $\mathscr{L}^1(X)$. Hence $(\alpha_x)_{x \in X}$ belongs to $\mathscr{L}^1(X)$.

To show that $\Sigma_{x \in X} \alpha_x = \alpha$, we use the Lebesgue Dominated Convergence Theorem (Theorem 1.4.16). All but countably many α_x must vanish [Theorem 8 (a) $\Rightarrow$ (f)], so we can choose from $\mathfrak{F}(X)$ an increasing sequence $(A_n)_{n \in \mathbb{N}}$ whose union is the set $\{x \in X \mid \alpha_x \neq 0\}$. Using (b), we can choose, for each n in $\mathbb{N}$, a set $Y_n \in \mathfrak{F}(X)$ such that

$$\left| \alpha - \sum_{x \in Z} \alpha_x \right| < \frac{1}{n}$$

for every finite set Z satisfying $Y_n \subset Z \subset X$.

For each n in $\mathbb{N}$ set

$$Z_n := A_n \cup \left(\bigcup_{m \le n} Y_m \right).$$

Then the family $(e_{Z_n}(x)\alpha_x)_{x\in X}$ belongs to $\mathscr{L}^1(X)$, as does the family $(|\alpha_x|)_{x\in X}$. Also

$$\lim_{n\to\infty} e_{Z_n}((\alpha_x)_{x\in X}) = (\alpha_x)_{x\in X}$$

and

$$|e_{Z_n}((\alpha_x)_{x\in X})| \le (|\alpha_x|_{x\in X})$$

for every n in $\mathbb{N}$. By dominated convergence we conclude that

$$\begin{aligned}\sum_{x\in X}\alpha_x &= \lim_{n\to\infty}\sum_{x\in X} e_{Z_n}(x)\alpha_x \\ &= \lim_{n\to\infty}\sum_{x\in Z_n}\alpha_x \\ &= \alpha.\end{aligned}$$

□

Theorem 18 provides additional useful formulas for the sum of a summable real family.

Corollary 3.4.19. *If $(\alpha_x)_{x\in X}$ is a summable family from $\overline{\mathbb{R}}$, then $(\alpha_x)_{x\in X}$ is real and*

$$\begin{aligned}\sum_{x\in X}\alpha_x &= \sup_{Y\in\mathfrak{F}(X)}\ \inf_{\substack{Z\in\mathfrak{F}(X)\\ Y\subset Z}}\ \sum_{x\in Z}\alpha_x \\ &= \inf_{Y\in\mathfrak{F}(X)}\ \sup_{\substack{Z\in\mathfrak{F}(X)\\ Y\subset Z}}\ \sum_{x\in Z}\alpha_x.\end{aligned}$$

□

We now turn to the notion of upper sum. As is true more generally, another approach to summability and summable families would have been via upper and lower sums (i.e., upper and lower closure of the Daniell space associated with the counting measure). We have chosen the closure construction from Section 3.1 rather than the upper and lower closure constructions described in Section 3.3, but we do want to trace the connection. The next theorem does that, as well as yielding a "Cauchy criterion" for summability.

Theorem 3.4.20. *The following assertions hold, for every family $(\alpha_x)_{x\in X}$ from $\overline{\mathbb{R}}$.*

(*a*) *$\sum^*_{x\in X}\alpha_x = \infty$ iff for every n in $\mathbb{N}$ the set X has a finite subset Y with*

$$\sum_{x\in Y}{}^*\alpha_x \ge n.$$

(*b*) $\Sigma^*_{x \in X}\alpha_x = -\infty$ *iff the family* $(\alpha_x^+)_{x \in X}$ *is summable and for every n in* $\mathbb{N}$ *the set X has a finite subset Y with*

$$\sum_{x \in Y}{}^{*}\alpha_x \leq -n.$$

The following assertions are equivalent, for every family $(\alpha_x)_{x \in X}$ *from* $\overline{\mathbb{R}}$.

(*c*) $\Sigma^*_{x \in X}\alpha_x$ *is real.*

(*d*) $(\alpha_x)_{x \in X}$ *is summable.*

(*e*) $\Sigma^*_{x \in X}|\alpha_x| < \infty$.

(*f*) *The family* $(\alpha_x)_{x \in X}$ *is real and for every real number* $\varepsilon > 0$ *the set X has a finite subset Y such that*

$$\left|\sum_{x \in B}\alpha_x\right| < \varepsilon$$

for every finite set $B \subset X \setminus Y$.

Proof. *Throughout the proof* we let

$$\mathscr{A} := \left\{(\beta_x)_{x \in X} \in \mathscr{F}(X)^{\uparrow} \,|\, (\beta_x)_{x \in X} \geq (\alpha_x)_{x \in X}\right\}.$$

An extended-real family $(\beta_x)_{x \in X}$ belongs to $\mathscr{A}$ iff no β_x equals $-\infty$, only finitely many β_x are strictly negative, only countably many β_x are nonzero, and every β_x is at least equal to the corresponding α_x. Note that

$$\sum_{x \in X}{}^{*}\alpha_x = \inf\left\{\ell_X^{\uparrow}\left((\beta_x)_{x \in X}\right) \,|\, (\beta_x)_{x \in X} \in \mathscr{A}\right\} \tag{10}$$

and that for $(\beta_x)_{x \in X} \in \mathscr{A}$,

$$\ell_X^{\uparrow}\left((\beta_x)_{x \in X}\right) = \begin{cases} \infty \text{ if } \beta_x = \infty & \text{for at least one } x \\ \displaystyle\sum_{\substack{x \in X \\ \beta_x < 0}} \beta_x + \sup_{A \in \mathfrak{F}(X)} \sum_{x \in A} \beta_x & \text{otherwise.} \end{cases} \tag{11}$$

Note also that

$$\sum_{x \in \{x_0\}}{}^{*}\alpha_x = \alpha_{x_0} \quad \text{for every } x_0 \in X. \tag{12}$$

(a) $(\Rightarrow)$. Suppose that $\Sigma^*_{x \in X}\alpha_x = \infty$, and let n in $\mathbb{N}$ be given. We distinguish two cases.

CASE 1: $\mathscr{A} = \varnothing$. There are uncountably many strictly positive α_x, so the set

$$A_m := \left\{ x \in X \mid \alpha_x \geq \frac{1}{m} \right\}$$

is uncountable for at least one natural number m. Using that m, take Y to be any finite subset of A_m containing at least mn elements.

CASE 2: $\mathscr{A} \neq \varnothing$. At most countably many α_x are strictly positive, the family $(\alpha_x^+)_{x \in X}$ belongs to $\mathscr{A}$, and

$$\ell_X^{\uparrow}\left((\alpha_x^+)_{x \in X}\right) = \infty.$$

In view of (11) and (12), the existence of the required set Y follows.

($\Leftarrow$). Suppose, conversely, that for each n in $\mathbb{N}$ there is a set $Y_n \in \mathfrak{F}(X)$ with $\Sigma^*_{x \in Y_n} \alpha_x \geq n$. If $\mathscr{A}$ is empty, it follows from (10) that $\Sigma^*_{x \in X} \alpha_x = \infty$. Suppose that $\mathscr{A} \neq \varnothing$, and let $(\beta_x)_{x \in X} \in \mathscr{A}$. If $\beta_x = \infty$ for at least one x then $\ell_X^{\uparrow}((\beta_x)_{x \in X}) = \infty$. If no β_x equals ∞, then we have

$$\sum_{x \in Y_n} \beta_x \geq n$$

for every n in $\mathbb{N}$. By (11), $\ell_X^{\uparrow}((\beta_x)_{x \in X}) = \infty$ in this case as well. It follows from (10) that

$$\sum_{x \in X}{}^* \alpha_x = \infty.$$

(b) ($\Rightarrow$). Suppose that $\Sigma^*_{x \in X} \alpha_x = -\infty$, and let n in $\mathbb{N}$ be given. There exists a family $(\beta_x)_{x \in X}$ belonging to $\mathscr{A}$ for which

$$\ell_X^{\uparrow}\left((\beta_x)_{x \in X}\right) < -n.$$

By Propositions 3.3.2 (c) and 3.3.9, the family $(\beta_x)_{x \in X}$ is summable, and

$$\sum_{x \in X} \beta_x < -n.$$

Thus the family $(\beta_x^+)_{x \in X}$ is summable and, since $0 \leq \alpha_x^+ \leq \beta_x^+$ for every x, so is the family $(\alpha_x^+)_{x \in X}$. The existence of the required set Y is now easy to verify.

($\Leftarrow$). Suppose, conversely, that $(\alpha_x^+)_{x \in X}$ belongs to $\mathscr{L}^1(X)$ and that for each n in $\mathbb{N}$ there is a set $Y_n \in \mathfrak{F}(X)$ with $\Sigma^*_{x \in Y_n} \alpha_x \leq -n$. Then $(\alpha_x^+)_{x \in X}$

certainly belongs to $\mathscr{A}$. Let

$$\alpha := \ell_X^{\uparrow}\left((\alpha_x^+)_{x \in X}\right) = \sum_{x \in X} \alpha_x^+$$

[Propositions 3.3.2 (c), 3.3.9], and note that $\alpha_x \leq \alpha_x^+ \leq \alpha$ for every x in X. For each n in $\mathbb{N}$, define

$$A_n := \{x \in Y_n \mid \alpha_x < 0\}$$

and note that

$$\sum_{x \in A_n}{}^{*} \alpha_x \leq -n - \alpha$$

[Proposition 3.3.3 (a)]. If we set

$$\beta_{n,x} := \begin{cases} \alpha_x \vee (-n-\alpha) & x \in A_n \\ \alpha_x^+ & x \in X \setminus A_n \end{cases}$$

then each family $(\beta_{n,x})_{x \in X}$ belongs to $\mathscr{A}$, and

$$\ell_X^{\uparrow}\left((\beta_{n,x})_{x \in X}\right) \leq -n + \alpha$$

for every n. In view of (10), $\Sigma^*_{x \in X} \alpha_x = -\infty$.

(c) $\Rightarrow$ (d). The idea of the proof is to choose a minimal sequence for the extremal problem (10), note that this sequence is actually a Σ-sequence from $\mathscr{L}^1(X)$, and then verify that its limit, which is summable, is in fact the given family $(\alpha_x)_{x \in X}$.

Let $\alpha := \Sigma^*_{x \in X} \alpha_x$. Since α is real, $\mathscr{A}$ is nonempty, and there exists in $\mathscr{A}$ a sequence

$$\left((\alpha_{n,x})_{x \in X}\right)_{n \in \mathbb{N}} \tag{13}$$

for which

$$\alpha = \inf_{n \in \mathbb{N}} \ell_X^{\uparrow}\left((\alpha_{n,x})_{x \in X}\right).$$

We can and do assume that this sequence decreases and that each of the numbers

$$\ell_X^{\uparrow}\left((\alpha_{n,x})_{x \in X}\right)$$

is real, which implies that the sequence (13) is a Σ-sequence from $\mathscr{L}^1(X)$ and

$$\ell_X^{\uparrow}\left((\alpha_{n,x})_{x\in X}\right) = \sum_{x\in X} \alpha_{n,x}$$

for every n. Writing

$$\alpha_{\infty,x} := \inf_{n\in\mathbb{N}} \alpha_{n,x} \qquad (x \in X)$$

we conclude that the family $(\alpha_{\infty,x})_{x\in X}$ is summable with sum α. We want to show that $\alpha_{\infty,x} = \alpha_x$ for every $x \in X$. Evidently

$$\alpha_{\infty,x} \geq \alpha_x.$$

Suppose that for some $x_0 \in X$ we have

$$\alpha_{\infty,x_0} > \alpha_{x_0}. \tag{14}$$

Set

$$\delta := (\alpha_{\infty,x_0} - \alpha_{x_0}) \wedge 1.$$

Choose $n_0 \in \mathbb{N}$ with

$$\sum_{x\in X} \alpha_{n_0,x} < \alpha + \delta.$$

Now the family

$$\left(\alpha_{n_0,x} - \delta e_{\{x_0\}}(x)\right)_{x\in X}$$

belongs to $\mathscr{A}$; yet the value assigned to this family by the functional $\ell_X^{\uparrow}$ is

$$\sum_{x\in X} \alpha_{n_0,x} - \delta$$

a number strictly smaller than α. This contradiction shows that (14) is impossible; the family $(\alpha_x)_{x\in X}$ is identical with the summable family $(\alpha_{\infty,x})_{x\in X}$.

(d) $\Rightarrow$ (e). By Theorem 8 (a) $\Rightarrow$ (b), $(|\alpha_x|)_{x\in X}$ is summable. By Corollary 7,

$$\sum_{x\in X}{}^{*} |\alpha_x| = \sum_{x\in X} |\alpha_x| < \infty.$$

(e) $\Rightarrow$ (f). Since

$$\sum_{x\in A}{}^{*}|\alpha_x| \le \sum_{x\in X}{}^{*}|\alpha_x|$$

for every finite subset A of X [Proposition 3.3.3 (a)], the hypothesis implies that each α_x is real, as is

$$\sup_{A\in\mathfrak{F}(X)} \sum_{x\in A} |\alpha_x|.$$

It follows that the family $(|\alpha_x|)_{x\in X}$ is summable [Theorem 8 (c) $\Rightarrow$ (b)]. Now let $\varepsilon > 0$ be given. Using the ε-Y characterization of sum (Theorem 18), we can choose a set $Y \in \mathfrak{F}(X)$ such that

$$\sum_{x\in X} |\alpha_x| < \varepsilon + \sum_{x\in Y} |\alpha_x|.$$

If B is any finite subset of $X \setminus Y$, we have

$$\begin{aligned} \left|\sum_{x\in B} \alpha_x\right| &\le \sum_{x\in B} |\alpha_x| \\ &\le \sum_{x\in X\setminus Y} |\alpha_x| \\ &= \sum_{x\in X} |\alpha_x| - \sum_{x\in Y} |\alpha_x| \\ &< \varepsilon \end{aligned}$$

as required.

(f) $\Rightarrow$ (c). We give an indirect proof. Suppose that $\Sigma^*_{x\in X}\alpha_x = \infty$. If some α_x fails to be real, (f) certainly fails, so assume that $(\alpha_x)_{x\in X}$ is real. By (a), given an arbitrary finite subset Y of X, there exists a set $Z \in \mathfrak{F}(X)$ such that

$$\sum_{x\in Z} \alpha_x \ge 1 + \sum_{x\in Y} |\alpha_x|.$$

Setting $B := Z \setminus Y$ we have a finite set $B \subset X \setminus Y$ with

$$\sum_{x\in B} \alpha_x \ge 1.$$

Thus in either case, if $\Sigma^*_{x\in X}\alpha_x = \infty$, then (f) must fail. A similar argument, using (b) in place of (a), shows that if $\Sigma^*_{x\in X}\alpha_x = -\infty$ then (f) is false. □

Besides providing a characterization of summability, the preceding theorem describes the behavior of one particular upper integral. It must be mentioned that this particular upper integral possesses certain special characteristics. For example, the implication (c) $\Rightarrow$ (d), where a finite upper integral implies integrability, does not hold in general.

Several useful corollaries to Theorem 20 should be mentioned.

Corollary 3.4.21 (Sums of finite families). *The following assertions hold for every finite family $(\alpha_x)_{x \in X}$ from $\overline{\mathbb{R}}$.*

(*a*) *$\Sigma^*_{x \in X}\alpha_x = \infty$ iff at least one α_x equals ∞.*
(*b*) *$\Sigma^*_{x \in X}\alpha_x = -\infty$ iff no α_x equals ∞ and at least one α_x equals $-\infty$.*
(*c*) *$\Sigma^*_{x \in X}\alpha_x$ is real and equals $\Sigma_{x \in X}\alpha_x$ iff every α_x is real.*

The proof is left to the reader. □

Corollary 3.4.22. *For every family $(\alpha_x)_{x \in X}$ from $\overline{\mathbb{R}}_+$,*

$$\sum_{x \in X}{}^*\alpha_x = \sup_{A \in \mathfrak{F}(X)} \sum_{x \in A}{}^*\alpha_x. \tag{15}$$

Proof. For $(\alpha_x)_{x \in X}$ summable, (15) restates Theorem 8, formula (6). Otherwise it follows from Theorem 20. □

Of course, if every α_x is real, the Σ^* on the right-hand side of (15) can be replaced by Σ.

Although many nice properties of Σ fail for Σ^*, Σ^* remains independent of order.

Corollary 3.4.23.

(*a*) *If $\varphi: Y \to X$ is bijective, then*

$$\sum_{y \in Y}{}^*\alpha_{\varphi(y)} = \sum_{x \in X}{}^*\alpha_x \tag{16}$$

for every family $(\alpha_x)_{x \in X}$ from $\overline{\mathbb{R}}$.

(*b*) *If $\varphi: K \to I$ is bijective, then*

$$\sum_{\kappa \in K}{}^*f_{\varphi(\kappa)} = \sum_{\iota \in I}{}^*f_\iota$$

for every family $(f_\iota)_{\iota \in I}$ from $\overline{\mathbb{R}}^X$.

Proof. (a) For $(\alpha_x)_{x \in X}$ summable, (16) was established earlier (Corollary 14). Otherwise, (16) follows from Theorem 20.

(b) Assertion (b) follows from (a). □

We consider briefly the special case of convergent series. The reader knows from analysis that there exist number sequences $(\alpha_n)_{n\in\mathbb{N}}$ for which the sum $\sum_{n=1}^{\infty}\alpha_n$ can be defined even though the sequence $(\alpha_n)_{n\in\mathbb{N}}$ is not summable in the sense of Definition 5. This situation occurs precisely because of the order structure of $\mathbb{N}$. A real series $\sum_{n=1}^{\infty}\alpha_n$ is said to be convergent iff its partial-sum sequence $(\sum_{k=1}^{n}\alpha_k)_{n\in\mathbb{N}}$ converges in $\mathbb{R}$, and then one defines

$$\sum_{k=1}^{\infty}\alpha_k := \lim_{n\to\infty}\sum_{k=1}^{n}\alpha_k.$$

A sequence $(\alpha_n)_{n\in\mathbb{N}}$ from $\mathbb{R}$ is summable iff the series $\sum_{k=1}^{\infty}|\alpha_k|$ is convergent, and then, as we verify below,

$$\sum_{k=1}^{\infty}\alpha_k = \sum_{k\in\mathbb{N}}\alpha_k.$$

Thus when the sequence $(\alpha_n)_{n\in\mathbb{N}}$ is summable, the series $\sum_{n=1}^{\infty}\alpha_n$ is usually referred to in the literature as an absolutely convergent series, and the term-sequence $(\alpha_n)_{n\in\mathbb{N}}$ as an absolutely summable sequence.

For the sum of a convergent series we intentionally write $\sum_{k=1}^{\infty}\alpha_k$ and not $\sum_{k\in\mathbb{N}}\alpha_k$. For one thing, these two numbers are defined and coincide only when the series is absolutely convergent. For another, the sum of a series that is convergent but not absolutely convergent depends in an essential way on the order of the terms. Thus Corollary 14 fails for the term-sequence of such a series. In fact one can show for every real series $\sum_{k=1}^{\infty}\alpha_k$ that is convergent but not absolutely convergent and for every real number α, that there exists a bijective mapping $\varphi\colon \mathbb{N}\to\mathbb{N}$ rearranging the given series to make its sum α:

$$\sum_{k=1}^{\infty}\alpha_{\varphi(k)} = \alpha.$$

Theorem 10 can be viewed as asserting that in any attempt to extend further the notion of summable family, some familiar properties will inevitably be lost.

We do not want to delve any further into the details of nonabsolutely convergent series. Since the theory developed here is independent of any special structure on the underlying index set, series that are convergent but not absolutely convergent are of no interest to us.

The next proposition collects some special properties of summable sequences.

Proposition 3.4.24. *The following assertions hold, for every summable sequence $(\alpha_n)_{n\in\mathbb{N}}$ of real numbers.*

(a) $\sum_{n\in\mathbb{N}}\alpha_n = \lim_{m\to\infty}\sum_{k=1}^{m}\alpha_k$.

(b) $\inf_{n\in\mathbb{N}}(\sum_{m>n}|\alpha_m|) = 0$.

(c) $\lim_{n\to\infty}\alpha_n = 0$.

Proof. Let $(\alpha_n)_{n\in\mathbb{N}}$ be a summable sequence, and note that $(|\alpha_n|)_{n\in\mathbb{N}}$ is also summable (Theorem 8).

(a) The idea here is to apply Lebesgue Dominated Convergence to the sequence whose mth term is the truncated sequence $(\alpha_1, \alpha_2, \ldots, \alpha_m, 0, 0, \ldots)$, that is, to the sequence $(e_{\mathbb{N}_m}((\alpha_n)_{n\in\mathbb{N}}))_{m\in\mathbb{N}}$. This sequence converges to the sequence $(\alpha_n)_{n\in\mathbb{N}}$, and its terms certainly belong to $\mathscr{L}^1(\mathbb{N})$. Moreover

$$|e_{\mathbb{N}_m}((\alpha_n)_{n\in\mathbb{N}})| \leq (|\alpha_n|)_{n\in\mathbb{N}}$$

for every m in $\mathbb{N}$. Thus (a) follows by dominated convergence (Theorem 1.4.16).

(b) By Corollary 15

$$\sum_{m>n} |\alpha_m| = \sum_{n\in\mathbb{N}} |\alpha_n| - \sum_{k=1}^{n} |\alpha_k|$$

so

$$\inf_{n\in\mathbb{N}}\left(\sum_{m>n} |\alpha_m|\right) = \sum_{n\in\mathbb{N}} |\alpha_n| - \sup_{n\in\mathbb{N}}\left(\sum_{k=1}^{n} |\alpha_k|\right)$$

(Proposition 1.1.9). Using (a) for the summable sequence $(|\alpha_n|)_{n\in\mathbb{N}}$, we compute

$$\sup_{n\in\mathbb{N}}\left(\sum_{k=1}^{n} |\alpha_k|\right) = \lim_{n\to\infty}\left(\sum_{k=1}^{n} |\alpha_k|\right) = \sum_{n\in\mathbb{N}} |\alpha_n|$$

and (b) follows.

(c) Assertion (c) follows easily from (b). □

Corollary 3.4.25. *A sequence $(\alpha_n)_{n\in\mathbb{N}}$ of real numbers is summable iff*

$$\lim_{n\to\infty} \sum_{k=1}^{n} |\alpha_k| < \infty. \tag{17}$$

In particular, a sequence $(\alpha_n)_{n\in\mathbb{N}}$ from $\mathbb{R}_+$ is summable iff its partial-sum sequence $(\sum_{k=1}^{n}\alpha_k)_{n\in\mathbb{N}}$ is bounded in $\mathbb{R}$.

Proof. The sequence $(\alpha_n)_{n\in\mathbb{N}}$ is summable iff the sequence $(|\alpha_n|)_{n\in\mathbb{N}}$ is summable. In view of part (a) of the preceding proposition, summability of $(\alpha_n)_{n\in\mathbb{N}}$ implies that (17) holds, and the converse follows from Theorem 8 (c) $\Rightarrow$ (a). □

Corollary 3.4.26. *For every sequence from* $\overline{\mathbb{R}}_+$,

$$\sum_{n\in\mathbb{N}}{}^{*}\alpha_n = \sup_{n\in\mathbb{N}} \sum_{k=1}^{n}{}^{*}\alpha_k. \tag{18}$$

Proof. For $(\alpha_n)_{n\in\mathbb{N}}$ summable, (18) restates Proposition 24 (a). Otherwise, it follows from Theorem 20 (a). □

Here, too, if $(\alpha_n)_{n\in\mathbb{N}}$ is real, the Σ^* on the right-hand side of (18) can be replaced by Σ.

EXERCISES

3.4.1(E) Let X be a set and $f \in \mathbb{R}^X$.

(α) Prove that the following assertions are equivalent:

(α1) $f \in \mathscr{L}^1(X)$.
(α2) $\Sigma^*_{x\in\{f>0\}} f(x) < \infty$, and $\Sigma^*_{x\in\{f<0\}}(-f(x)) < \infty$.

(β) Prove that any of the assertions in (α) implies

$$\sum_{x\in X} f(x) = \sum_{x\in\{f>0\}} f(x) - \sum_{x\in\{f<0\}} (-f(x)).$$

3.4.2(E) Let $(\alpha_\iota)_{\iota\in I}$ be a family of real numbers, and let A, B be sets such that

$$\{\iota \in I \,|\, \alpha_\iota < 0\} \subset A \subset \{\iota \in I \,|\, \alpha_\iota \le 0\}$$

and

$$\{\iota \in I \,|\, \alpha_\iota > 0\} \subset B \subset \{\iota \in I \,|\, \alpha_\iota \ge 0\}.$$

(α) Prove that the following assertions are equivalent:

(α1) $(\alpha_\iota)_{\iota\in I}$ is summable.
(α2) $(\alpha_\iota)_{\iota\in A}$ and $(\alpha_\iota)_{\iota\in B}$ are summable.

(β) Prove that any of the assertions in (α) implies

$$\sum_{\iota\in I} \alpha_\iota = \sum_{\iota\in A} \alpha_\iota + \sum_{\iota\in B} \alpha_\iota.$$

3.4.3(E) Let $(\alpha_\iota)_{\iota\in I}$ and $(\beta_\iota)_{\iota\in I}$ be summable families of real numbers. Let $(\gamma_\iota)_{\iota\in I}$ be a family from $\mathbb{R}$ such that $\alpha_\iota \le \gamma_\iota \le \beta_\iota$ whenever $\iota \in I$. Prove the following two statements:

(α) $(\gamma_\iota)_{\iota\in I}$ is summable.
(β) $\Sigma_{\iota\in I}\alpha_\iota \le \Sigma_{\iota\in I}\gamma_\iota \le \Sigma_{\iota\in I}\beta_\iota$.

3.4.4(E) Let $(\alpha_\iota)_{\iota \in I}$ be a family from $\mathbb{R}$.

(α) Prove the equivalence of the following assertions:

$(\alpha 1)$ $(\alpha_\iota)_{\iota \in I}$ is summable.
$(\alpha 2)$ $(\alpha_\iota)_{\iota \in J}$ is summable for every countable subset J of I.
$(\alpha 3)$ $\{\Sigma_{\iota \in J}\alpha_\iota \mid J \subset I, J$ finite$\}$ is bounded.

(β) Prove that every of the assertions in (α) implies that $\{\Sigma_{\iota \in J}\alpha_\iota \mid J \subset I\}$ and $\{\Sigma_{\iota \in J}\alpha_\iota \mid J \subset I, J$ countable$\}$ are compact subsets of $\mathbb{R}$.

3.4.5(E) Let $(\alpha_\iota)_{\iota \in I}$ be a family of real numbers. Let $\mathfrak{F}$ be the set of all finite subsets of I, ordered by $\subset$. Denote by $\mathfrak{G}$ the section filter on $\mathfrak{F}$, that is, the filter on $\mathfrak{F}$ generated by the filter base

$$\{\{J \in \mathfrak{F} \mid K \subset J\} \mid K \in \mathfrak{F}\}.$$

(α) Prove that the following statements are equivalent:

$(\alpha 1)$ $(\alpha_\iota)_{\iota \in I}$ is summable.
$(\alpha 2)$ $\lim_{J, \mathfrak{G}} \Sigma_{\iota \in J}\alpha_\iota$ exists.

(β) Each of the equivalent statements $(\alpha 1)$, $(\alpha 2)$ implies

$$\sum_{\iota \in I} \alpha_\iota = \lim_{J, \mathfrak{G}} \sum_{\iota \in J} \alpha_\iota .$$

3.4.6(C) This colloquium is devoted to the study of a special case of summable families, namely, summable sequences. The theory of summable sequences has a long history. Many notable mathematicians have contributed to its development, so that we now have a treasury of deep results. It is a regrettable fact of current mathematical training that this theory receives less and less attention. We have two principal reasons for devoting an exercise to this theory. On the one hand, we would like to present to the reader at least some of the beautiful results. On the other hand, the theory of summable sequences is particularly well suited to demonstrate that mathematical objects whose construction is very simple, and which one believes to have well in grasp, can become multifold when viewed from a different standpoint.

From this vast store we have chosen a small part containing criteria for the summability of sequences. Necessary and sufficient conditions are formulated in Theorem 8. These conditions are very general, and not particularly well suited to check summability in concrete cases. A usable criterion is formulated in Proposition 11. It is called a comparison criterion of the first kind. Propositions of this kind can be generalized somewhat. We formulate the generalization for sequences.

Comparison Criterion of the First Kind

Let $(\alpha_n)_{n \in \mathbb{N}}$ and $(\beta_n)_{n \in \mathbb{N}}$ be sequences from $\mathbb{R}$. Then

(α) $(\beta_n)_{n \in \mathbb{N}}$ is summable whenever $|\beta_n| \le |\alpha_n|$ for each $n \in \mathbb{N}$ and $(\alpha_n)_{n \in \mathbb{N}}$ is summable.

(β) $(\beta_n)_{n \in \mathbb{N}}$ is not summable whenever $|\alpha_n| \le |\beta_n|$ for each $n \in \mathbb{N}$ and $(\alpha_n)_{n \in \mathbb{N}}$ is not summable.

Many concrete criteria are based on this comparison test. Another important source is provided by the comparison criterion of the second kind.

Comparison Criterion of the Second Kind

Let $(\alpha_n)_{n \in \mathbb{N}}$ and $(\beta_n)_{n \in \mathbb{N}}$ be sequences from $\mathbb{R}_+$ such that for each $n \in \mathbb{N}$ $\alpha_n \neq 0$ and $\beta_n \neq 0$. Then

(α') $(\beta_n)_{n \in \mathbb{N}}$ is summable whenever $(\alpha_n)_{n \in \mathbb{N}}$ is summable and $\beta_{n+1}/\beta_n \le \alpha_{n+1}/\alpha_n$ for each $n \in \mathbb{N}$.

(β') $(\beta_n)_{n \in \mathbb{N}}$ is not summable whenever $(\alpha_n)_{n \in \mathbb{N}}$ is not summable and $\alpha_{n+1}/\alpha_n \le \beta_{n+1}/\beta_n$ for each $n \in \mathbb{N}$.

For the application of this criterion, it is important to have a store of test sequences to serve as comparison. The reader is probably familiar with the application of the summable sequence $(\beta^n)_{n \in \mathbb{N}}$ ($\beta \in]-1,1[$) as a test sequence, which leads to the following criteria:

(γ) **Cauchy Radical Criterion.** Take $(\alpha_n)_{n \in \mathbb{N}}$ a sequence from $\mathbb{R}$ and $\beta \in [0,1[$. If $\sqrt[n]{|\alpha_n|} \le \beta$ for each $n \in \mathbb{N}$, then $(\alpha_n)_{n \in \mathbb{N}}$ is summable. On the other hand, if $\sqrt[n]{|\alpha_n|} \ge 1$ for each $n \in \mathbb{N}$, then $(\alpha_n)_{n \in \mathbb{N}}$ is not summable.

(δ) **Cauchy Quotient Criterion.** Take $(\alpha_n)_{n \in \mathbb{N}}$ a sequence from $\mathbb{R} \setminus \{0\}$, and $\beta \in [0,1[$. If $|\alpha_{n+1}|/|\alpha_n| \le \beta$ for each $n \in \mathbb{N}$, then $(\alpha_n)_{n \in \mathbb{N}}$ is summable. If $|\alpha_{n+1}|/|\alpha_n| \ge 1$ for each $n \in \mathbb{N}$, then $(\alpha_n)_{n \in \mathbb{N}}$ is not summable.

Observe that the requirement "for each $n \in \mathbb{N}$" may be replaced by "for all but finitely many $n \in \mathbb{N}$," but for the sake of simplicity we shall not continually insist on this phrasing.

We shall deduce a sequence of criteria such that each is stronger than its predecessor. To this end, note that both of the above criteria are relatively weak, and are inadequate to decide the summability of the sequence $(1/n)_{n \in \mathbb{N}}$. The criteria we deduce are all stronger than the radical and the quotient criteria, but we shall discuss this later. Our arguments are based on the following **Theorem of Abel and Dini**:

(ε) Take a nonsummable sequence $(\alpha_n)_{n \in \mathbb{N}}$ from $\mathbb{R}_+$. For each $n \in \mathbb{N}$ define $\sigma_n := \sum_{k \in \mathbb{N}_n} \alpha_k$, and suppose that $\sigma_n \neq 0$ for every $n \in \mathbb{N}$. (σ_n is called the nth partial sum of $(\alpha_n)_{n \in \mathbb{N}}$). Take $\alpha \in \mathbb{R}$, $\alpha > 0$. Then $(\alpha_n/\sigma_n^\alpha)_{n \in \mathbb{N}}$ is summable iff $\alpha > 1$.

For the proof, consider first the case that $\alpha = 1$. For $m, n \in \mathbb{N}$,

$$\frac{\alpha_{n+1}}{\sigma_{n+1}} + \cdots + \frac{\alpha_{n+m}}{\sigma_{n+m}} \ge \frac{\alpha_{n+1} + \cdots + \alpha_{n+m}}{\sigma_{n+m}} = 1 - \frac{\sigma_n}{\sigma_{n+m}}.$$

Now, since $(\alpha_n)_{n \in \mathbb{N}}$ is not summable, there is for each $n \in \mathbb{N}$ an $m \in \mathbb{N}$

such that

$$\frac{\alpha_{n+1}}{\sigma_{n+1}} + \cdots + \frac{\alpha_{n+m}}{\sigma_{n+m}} > \frac{1}{2}$$

so that $(\alpha_n/\sigma_n)_{n \in \mathbb{N}}$ is not summable.

The required result for the case $\alpha < 1$ follows from the preceding case by an application of the general comparison criterion of the first kind.

Finally consider $\alpha > 1$. Set $\beta := \alpha - 1$. Consider the function

$$f: [\sigma_n, \sigma_{n+1}] \to \mathbb{R}, \qquad x \mapsto -\frac{1}{\beta}x^{-\beta}.$$

By the mean-value theorem of the differential calculus, there is a $\tau_{n+1} \in [\sigma_n, \sigma_{n+1}]$ with

$$\frac{1}{\beta}\left(\frac{1}{\sigma_n^\beta} - \frac{1}{\sigma_{n+1}^\beta}\right) = \frac{\alpha_{n+1}}{\tau_{n+1}^\alpha} \geq \frac{\alpha_{n+1}}{\sigma_{n+1}^\alpha}.$$

But for each $n \in \mathbb{N}$,

$$\sum_{k=1}^{n} \frac{1}{\beta}\left(\frac{1}{\sigma_k^\beta} - \frac{1}{\sigma_{k+1}^\beta}\right) = \frac{1}{\beta}\left(\frac{1}{\sigma_1^\beta} - \frac{1}{\sigma_{n+1}^\beta}\right).$$

Thus $((1/\beta)(1/\sigma_n^\beta - 1/\sigma_{n+1}^\beta))_{n \in \mathbb{N}}$ is summable, so that by the general comparison criterion of the first kind, $(\alpha_n/\sigma_n^\alpha)_{n \in \mathbb{N}}$ is also summable.

The scope of this proposition can be extended. To assist in this we introduce the concept of asymptotically equal sequences. Two sequences $(\alpha_n)_{n \in \mathbb{N}}$ and $(\beta_n)_{n \in \mathbb{N}}$ from $\mathbb{R}_+ \setminus \{0\}$ are **asymptotically equal** iff $\lim_{n \to \infty} \alpha_n/\beta_n = 1$. The reader should verify that the asymptotic equality is an equivalence relation and then prove the following generalization of (ε).

(ζ) Given $(\alpha_n)_{n \in \mathbb{N}}$, $(\sigma_n)_{n \in \mathbb{N}}$ and α as in (ε), let $(\beta_n)_{n \in \mathbb{N}}$ be a sequence from $\mathbb{R}_+ \setminus \{0\}$ that is asymptotically equal to $(\sigma_n)_{n \in \mathbb{N}}$. Then $(\alpha_n/\beta_n^\alpha)_{n \in \mathbb{N}}$ is summable iff $\alpha > 1$.

A beautiful and particularly important example of asymptotic equivalence of sequences is obtained when one treats the question of the relationship between $(\sigma_n)_{n \in \mathbb{N}}$, the sequence of partial sums of $(\alpha_n)_{n \in \mathbb{N}}$, and $(\alpha_n/\sigma_n)_{n \in \mathbb{N}}$. For the investigation we need the next result (O. Toeplitz):

(η) Let $(\gamma_n)_{n \in \mathbb{N}}$ be a sequence from $\mathbb{R}$ that converges to 0. Let $(\alpha_{nm})_{m, n \in \mathbb{N}, m \leq n}$ be a family from $\mathbb{R}$ with the following properties:

(i) $\lim_{n \to \infty} \alpha_{nm} = 0$ for each $m \in \mathbb{N}$.

(ii) For some $\gamma \in \mathbb{R}_+$, $\Sigma_{m \in \mathbb{N}_n} |\alpha_{nm}| < \gamma$ for each $n \in \mathbb{N}$.

Then $(\Sigma_{m \in \mathbb{N}_n} \alpha_{nm}\gamma_m)_{n \in \mathbb{N}}$ converges to 0.

The proof is simple. Take $\varepsilon \in \mathbb{R}$, $\varepsilon > 0$. Choose $m \in \mathbb{N}$ with $|\gamma_n| < \varepsilon/2\gamma$ whenever $n \geq m$. Thus, given such an n,

$$\left| \sum_{k \in \mathbb{N}_n} \alpha_{nk} \gamma_k \right| \leq \left| \sum_{k \in \mathbb{N}_m} \alpha_{nk} \gamma_k \right| + \frac{\varepsilon}{2}.$$

Now choose $m' \geq m$ such that for each $n \geq m'$,

$$\left| \sum_{k \in \mathbb{N}_m} \alpha_{nk} \gamma_k \right| < \frac{\varepsilon}{2}$$

Then for each $n \geq m'$,

$$\left| \sum_{k \in \mathbb{N}_n} \alpha_{nk} \gamma_k \right| < \epsilon.$$

This lemma has a number of consequences:

(ϑ) If $(\gamma_n)_{n \in \mathbb{N}}$ is a sequence from $\mathbb{R}$ converging to $\gamma \in \mathbb{R}$ and if $(\alpha_{nm})_{n, m \in \mathbb{N},\, n \geq m}$ is a family from $\mathbb{R}$ that satisfies (i) and (ii) and fulfills the additional condition that

(iii) $\lim_{n \to \infty} (\Sigma_{k \in \mathbb{N}_n} \alpha_{nk}) = 1$,

then $(\Sigma_{k \in \mathbb{N}_n} \alpha_{nk} \gamma_k)_{n \in \mathbb{N}}$ also converges to γ.

To see this, observe $\Sigma_{k \in \mathbb{N}_n} \alpha_{nk} \gamma_k = \gamma + \Sigma_{k \in \mathbb{N}_n} \alpha_{nk}(\gamma_k - \gamma)$.

(ι) Let $(\alpha_n)_{n \in \mathbb{N}}$ be a nonsummable sequence from $\mathbb{R}_+ \setminus \{0\}$ and $(\beta_n)_{n \in \mathbb{N}}$ a sequence from $\mathbb{R}$ such that $(\beta_n/\alpha_n)_{n \in \mathbb{N}}$ converges to $\gamma \in \mathbb{R}$. Then

$$\lim_{n \to \infty} \frac{\beta_1 + \beta_2 + \cdots + \beta_n}{\alpha_1 + \alpha_2 + \cdots + \alpha_n} = \gamma.$$

Set $\gamma_n := \beta_n/\alpha_n$ and $\alpha_{nm} := \alpha_m/(\alpha_1 + \alpha_2 + \cdots + \alpha_n)$, and apply ($\vartheta$).

Now prove the following conjecture:

(κ) Let $(\alpha_n)_{n \in \mathbb{N}}$ be a nonsummable sequence from $\mathbb{R}_+$. For each $n \in \mathbb{N}$ set $\sigma_n := \Sigma_{k \in \mathbb{N}_n} \alpha_k$. Assume that $\sigma_n > 0$ for each $n \in \mathbb{N}$ and that $\lim_{n \to \infty} \alpha_n/\sigma_n = 0$. Suppose further that $(\beta_n)_{n \in \mathbb{N}}$ is a sequence from $\mathbb{R}_+ \setminus \{0\}$ that is asymptotically equal to $(\sigma_n)_{n \in \mathbb{N}}$. Then $(\Sigma_{k \in \mathbb{N}_n} \alpha_k/\beta_k)_{n \in \mathbb{N}}$ and $(\ln \beta_n)_{n \in \mathbb{N}}$ are asymptotically equal.

Using de l'Hopitals rule,

$$\lim_{\substack{x \to 0 \\ x < 1}} \frac{x}{\ln\left(\dfrac{1}{1 - x}\right)} = 1$$

so that

$$\lim_{n\to\infty}\left[\frac{(\alpha_{n+1}/\sigma_{n+1})}{\ln(\sigma_{n+1}/\sigma_n)}\right]=1.$$

Thus

$$\lim_{n\to\infty}\left[\frac{(\alpha_{n+1}/\beta_{n+1})}{\ln(\beta_{n+1}/\beta_n)}\right]=1.$$

Now apply (ι):

$$1=\lim_{n\to\infty}\frac{(\alpha_1/\beta_1)+\cdots+(\alpha_{n+1}/\beta_{n+1})}{\ln\beta_1+\ln(\beta_2/\beta_1)+\cdots+\ln(\beta_{n+1}/\beta_n)}$$

$$=\lim_{n\to\infty}\frac{1}{\ln\beta_{n+1}}\left(\frac{\alpha_1}{\beta_1}+\cdots+\frac{\alpha_{n+1}}{\beta_{n+1}}\right).$$

We are now in a position to construct an important sequence of test sequences. We begin with the nonsummable sequence $(1)_{n\in\mathbb{N}}$. By (ε), $(1/n^\alpha)_{n\in\mathbb{N}}$ is summable iff $\alpha>1$. Thus, in particular, $(1/n)_{n\in\mathbb{N}}$ is not summable, and by (κ), $(\Sigma_{k\in\mathbb{N}_n}1/k)_{n\in\mathbb{N}}$ and $(\ln n)_{n\in\mathbb{N}}$ are asymptotically equal. Now take the nonsummable sequence $(1/n)_{n\in\mathbb{N}}$, and use (ζ) and the asymptotic equality just established to get that $(1/n(\ln n)^\alpha)_{n\in\mathbb{N}\setminus\{1\}}$ is summable iff $\alpha>1$. In particular, $(1/n\ln n)_{n\in\mathbb{N}\setminus\{1\}}$ is not summable, and by (κ) $(\Sigma_{k=2}^n 1/k\ln k)_{n\in\mathbb{N}}$ and $(\ln(\ln n))_{n\in\mathbb{N}\setminus\{1\}}$ are asymptotically equal. It is possible to continue in this way. To simplify the notation, we introduce the following notation: $\ln_1:=\ln,\ldots,\ln_{k+1}:=\ln(\ln_k)$ for all $k\in\mathbb{N}$. For each $k\in\mathbb{N}$ denote by m_k the first natural number, m, for which $\ln_k$ is defined.

(λ) For each $k\in\mathbb{N}$

$$\left(\frac{1}{n\ln_1 n\ln_2 n\cdots\ln_{k-1}n(\ln_k n)^\alpha}\right)_{n\in\mathbb{N}\setminus\mathbb{N}_{m_k}}$$

is summable iff $\alpha>1$. In particular,

$$\left(\frac{1}{n\ln_1 n\ln_2 n\cdots\ln_{k-1}n\ln_k n}\right)_{n\in\mathbb{N}\setminus\mathbb{N}_{m_k}}$$

is not summable. The sequences

$$\left(\sum_{m\in\mathbb{N}_n\setminus\mathbb{N}_{m_k}}\frac{1}{m\ln_1 m\cdots\ln_k m}\right)_{n\in\mathbb{N}\setminus\mathbb{N}_{m_{k+1}}}$$

$$\text{and}\quad(\ln_{k+1}n)_{n\in\mathbb{N}\setminus\mathbb{N}_{m_{k+1}}}$$

are asymptotically equal.

This collection of test sequences enables us to give, for each k, one comparison criterion of the first kind and one of the second kind. We take a closer look at the criteria of the second kind.

Suppose that $(\alpha_n)_{n\in\mathbb{N}}$ is a sequence from $\mathbb{R}$ such that for $n \geq n_k$

$$\left|\frac{\alpha_{n+1}}{\alpha_n}\right| \leq \frac{n}{n+1}\frac{\ln_1 n}{\ln_1(n+1)}\cdots\frac{\ln_{k-1} n}{\ln_{k-1}(n+1)}\left(\frac{\ln_k n}{\ln_k(n+1)}\right)^{\alpha}$$

with $\alpha > 1$. Then $(\alpha_n)_{n\in\mathbb{N}}$ is summable. If, on the other hand,

$$\left|\frac{\alpha_{n+1}}{\alpha_n}\right| \geq \frac{n}{n+1}\frac{\ln_1 n}{\ln_1(n+1)}\cdots\left(\frac{\ln_k n}{\ln_k(n+1)}\right)^{\alpha}$$

with $\alpha \leq 1$, then $(\alpha_n)_{n\in\mathbb{N}}$ is not summable.

These criteria can be reformulated. To prepare the reformulation, prove that if $k \in \mathbb{N} \cup \{0\}$ and if $\alpha \in \mathbb{R}$, then there is a bounded sequence $(\gamma_n)_{n\in\mathbb{N}\setminus\mathbb{N}_{m_k}}$ from $\mathbb{R}_+$ such that if $n \geq m_k$, then

$$\left(\frac{\ln_k(n-1)}{\ln_k n}\right)^{\alpha} = 1 - \frac{\alpha}{\ln_0 n \ln_1 n \cdots \ln_k n} - \frac{\gamma_n}{n^2}$$

where $\ln_0 x := x$, $m_0 := 0$, and $\mathbb{N}_{m_0} := \varnothing$. (Use Taylor's formula for the proof.)

By this proposition

$$\frac{\ln_0 n}{\ln_0(n+1)}\frac{\ln_1 n}{\ln_1(n+1)}\cdots\frac{\ln_{k-1} n}{\ln_{k-1}(n+1)}\left(\frac{\ln_k n}{\ln_k(n+1)}\right)^{\alpha}$$

$$= 1 - \frac{1}{n+1} - \frac{1}{(n+1)\ln(n+1)}$$

$$- \cdots - \frac{\alpha}{(n+1)\ln(n+1)\cdots\ln_k(n+1)}$$

$$- \frac{\gamma'_{n+1}}{(n+1)^2}$$

for each $n \geq m_k$, where $(\gamma'_n)_{n\in\mathbb{N}\setminus\mathbb{N}_{m_k}}$ is bounded.

We obtain now the following criteria:

(μ) Let $(\alpha_n)_{n\in\mathbb{N}}$ be a sequence from $\mathbb{R} \setminus \{0\}$. Suppose that

$$\left|\frac{\alpha_{n+1}}{\alpha_n}\right| \leq 1 - \frac{1}{n} - \frac{1}{n\ln n} - \cdots - \frac{\alpha}{n\ln n\cdots\ln_k n}$$

for some $\alpha > 1$ and for each $n \geq m_k$. Then $(\alpha_n)_{n\in\mathbb{N}}$ is summable. If, on the other hand,

$$\left|\frac{\alpha_{n+1}}{\alpha_n}\right| \geq 1 - \frac{1}{n} - \frac{1}{n \ln n} - \cdots - \frac{\alpha}{n \ln n \cdots \ln_k n}$$

for some $\alpha \leq 1$ and for each $n \geq m_k$, then $(\alpha_n)_{n \in \mathbb{N}}$ is not summable.

The sequence of comparison criteria just described is known as the **logarithmic scale**. Several well-known criteria are just special cases.

The Raabe Criterion

Let $(\alpha_n)_{n \in \mathbb{N}}$ be a sequence from $\mathbb{R} \setminus \{0\}$. Suppose that for some m_0 and $\alpha < -1$

$$n\left(\left|\frac{\alpha_{n+1}}{\alpha_n}\right| - 1\right) \leq \alpha$$

for every $n \geq m_0$. Then $(\alpha_n)_{n \in \mathbb{N}}$ is summable. If, on the other hand,

$$n\left(\left|\frac{\alpha_{n+1}}{\alpha_n}\right| - 1\right) \geq -1$$

for every $n \geq m_0$, then $(\alpha_n)_{n \in \mathbb{N}}$ is not summable.

The Bertrand Criterion

Let $(\alpha_n)_{n \in \mathbb{N}}$ be a sequence from $\mathbb{R} \setminus \{0\}$. Suppose that for some m_1 and $\alpha > 1$,

$$\left|\frac{\alpha_{n+1}}{\alpha_n}\right| \leq 1 - \frac{1}{n} - \frac{\alpha}{n \ln n}$$

for every $n \geq m_1$. Then $(\alpha_n)_{n \in \mathbb{N}}$ is summable. If, on the other hand, for some $\alpha \leq 1$

$$\left|\frac{\alpha_{n+1}}{\alpha_n}\right| \geq 1 - \frac{1}{n} - \frac{\alpha}{n \ln n}$$

for every $n \geq m_1$, then $(\alpha_n)_{n \in \mathbb{N}}$ is not summable.

The Gauss Criterion

Let $(\alpha_n)_{n \in \mathbb{N}}$ be a sequence from $\mathbb{R} \setminus \{0\}$ such that there are real numbers $\gamma > 1$ and α, and a bounded sequence $(\beta_n)_{n \in \mathbb{N}}$ from $\mathbb{R}$ with

$$\left|\frac{\alpha_{n+1}}{\alpha_n}\right| = 1 - \frac{\alpha}{n} - \frac{\beta_n}{n^\gamma}$$

for every $n \in \mathbb{N}$. Then $(\alpha_n)_{n \in \mathbb{N}}$ is summable iff $\alpha > 1$.

Every criterion from the logarithmic scale has a greater range of application than any of its predecessors. Thus, if the summability of a sequence can be proved with the kth criterion, then it can be proved with the nth criterion for any $n \geq k$. It is then natural to ask if there is a best possible comparison criterion that can be used to decide the summability of any sequence. There is no such criterion.

3.5. APPLICATIONS OF SUMMABILITY

NOTATION FOR SECTION 3.5:

For W a set, $\mathfrak{F}(W) := \{A \subset W \mid A \text{ is finite}\}$

First we close a gap that was left open in the discussion of positive measures. That countable additivity (for positive set mappings) is equivalent to null continuity plus finite additivity was alluded to earlier (see the remark at the end of Section 2.2). We now have the wherewithal both to formulate and to prove this equivalence.

Theorem 3.5.1. *The following assertions are equivalent, for every ring of sets* $\mathfrak{R}$ *and every positive mapping* $\mu\colon \mathfrak{R} \to \mathbb{R}$.

(*a*) μ *is a positive measure on* $\mathfrak{R}$.

(*b*) **Countable additivity**. *For every countable, disjoint family* $(A_\iota)_{\iota \in I}$ *from* $\mathfrak{R}$, *if* $\bigcup_{\iota \in I} A_\iota$ *belongs to* $\mathfrak{R}$, *then the family* $(\mu(A_\iota))_{\iota \in I}$ *is summable and*

$$\mu\left(\bigcup_{\iota \in I} A_\iota\right) = \sum_{\iota \in I} \mu(A_\iota).$$

(*c*) *For every disjoint sequence* $(A_n)_{n \in \mathbb{N}}$ *from* $\mathfrak{R}$, *if* $\bigcup_{n \in \mathbb{N}} A_n$ *belongs to* $\mathfrak{R}$, *then the sequence* $(\mu(A_n))_{n \in \mathbb{N}}$ *is summable, and*

$$\mu\left(\bigcup_{n \in \mathbb{N}} A_n\right) = \sum_{n \in \mathbb{N}} \mu(A_n).$$

Each of these assertions implies, for every countable family $(A_\iota)_{\iota \in I}$ *from* $\mathfrak{R}$, *that*

$$\mu\left(\bigcup_{\iota \in I} A_\iota\right) \leq \sum_{\iota \in I} \mu(A_\iota)$$

whenever both sides are defined.

Proof. (a) $\Rightarrow$ (b). Let $(A_\iota)_{\iota \in I}$ be a countable, disjoint family from $\mathfrak{R}$ whose union belongs to $\mathfrak{R}$. It suffices to consider the case that I is countably infinite, so let $\varphi\colon \mathbb{N} \to I$ be bijective. Construct an increasing sequence from $\mathfrak{R}$ whose union is the same as $\bigcup_{\iota \in I} A_\iota$ by setting, for each n in $\mathbb{N}$, $B_n := \bigcup_{k \leq n} A_{\varphi(k)}$. Since μ is nullcontinuous, additive, and positive, we have

$$\begin{aligned} \infty > \mu\left(\bigcup_{\iota \in I} A_\iota\right) &= \mu\left(\bigcup_{n \in \mathbb{N}} B_n\right) \\ &= \sup_{n \in \mathbb{N}} \mu(B_n) \\ &= \lim_{n \to \infty} \mu(B_n) \\ &= \lim_{n \to \infty} \sum_{k=1}^{n} \mu(A_{\varphi(k)}). \end{aligned}$$

It follows that the sequence $(\mu(A_{\varphi(k)}))_{k \in \mathbb{N}}$ is summable (Corollary 3.4.25) and that the family $(\mu(A_\iota))_{\iota \in I}$ is summable (Corollary 3.4.14). Moreover

$$\begin{aligned} \mu\left(\bigcup_{\iota \in I} A_\iota\right) &= \lim_{n \to \infty} \sum_{k=1}^{n} \mu(A_{\varphi(k)}) \\ &= \sum_{k \in \mathbb{N}} \mu(A_{\varphi(k)}) \\ &= \sum_{\iota \in I} \mu(A_\iota). \end{aligned}$$

(b) $\Rightarrow$ (c). This implication is evident.

(c) $\Rightarrow$ (a). First notice that $\mu(\varnothing) = 0$. Indeed, hypothesis (c), together with Proposition 3.4.24 (a), implies that $\mu(\varnothing) = \lim_{n \to \infty} n\mu(\varnothing)$, which is only possible if $\mu(\varnothing) = 0$. The additivity of μ then follows from the assumed additivity. We must verify that μ is nullcontinuous, for which we use the characterization in Proposition 2.2.3 (c) $\Rightarrow$ (a). Let $(A_n)_{n \in \mathbb{N}}$ be an increasing sequence from $\mathfrak{R}$ whose union belongs to $\mathfrak{R}$. Construct in $\mathfrak{R}$ a disjoint sequence $(B_n)_{n \in \mathbb{N}}$ with the same union as the given sequence by setting $B_1 := A_1$ and $B_n := A_n \setminus A_{n-1}$ for $n \in \mathbb{N} \setminus \{1\}$. Then

$$\begin{aligned} \mu\left(\bigcup_{n \in \mathbb{N}} A_n\right) &= \sum_{n \in \mathbb{N}} \mu(B_n) \\ &= \lim_{n \to \infty} \sum_{k=1}^{n} \mu(B_k) \\ &= \sup_{n \in \mathbb{N}} \sum_{k=1}^{n} \mu(B_k) \\ &= \sup_{n \in \mathbb{N}} \mu(A_n) \end{aligned}$$

as required.

The reader can readily verify the final claim. □

Theorem 3.5.2. *Let $(X, \mathscr{L}, \ell)$ be a closed Daniell space. Then the following assertions are equivalent, for every countable family $(f_\iota)_{\iota \in I}$ from $\mathscr{L}$.*

(*a*) *The family $(\ell(|f_\iota|))_{\iota \in I}$ is summable.*
(*b*) $\Sigma^*_{\iota \in I} |f_\iota|$ *belongs to* $\mathscr{L}$.

If these assertions hold, then so do all of the following.

(*c*) $\ell(\Sigma^*_{\iota \in I} |f_\iota|) = \Sigma_{\iota \in I} \ell(|f_\iota|)$.
(*d*) *The family $(f_\iota)_{\iota \in I}$ is summable $\mathscr{L}$-a.e.*
(*e*) *The family $(\ell(f_\iota))_{\iota \in I}$ is summable.*
(*f*) *For $f \in \overline{\mathbb{R}}^X$, if $f = \Sigma^*_{\iota \in I} f_\iota$ $\mathscr{L}$-a.e., then f belongs to $\mathscr{L}$ and $\ell(f) = \Sigma_{\iota \in I} \ell(f_\iota)$.*

Proof.

CASE 1: *I finite.* If I is finite, there is little to prove. Indeed, (a) and (e) always hold in this case. Functions that are $\mathscr{L}$-a.e. equal to functions from $\mathscr{L}$ themselves belong to $\mathscr{L}$, and the set

$$A := \bigcup_{\iota \in I} \{ |f_\iota| = \infty \}$$

is $\mathscr{L}$-exceptional. Using these two facts and the Riesz-lattice structure of $\mathscr{L}$, it is easy to check that the remaining assertions all hold as well.

CASE 2: *I infinite.* Let $\varphi: \mathbb{N} \to I$ be bijective. Set

$$A := \bigcup_{\iota \in I} \{ |f_\iota| = \infty \}$$

and note that A is $\mathscr{L}$-exceptional.

(a) $\Rightarrow$ (b), (c). The sequence $(\Sigma^*_{k \in \mathbb{N}_n} |f_{\varphi(k)}|)_{n \in \mathbb{N}}$, whose terms belong to $\mathscr{L}$, increases and

$$\begin{aligned} 0 \le \ell\left(\sum_{k=1}^{n}{}^{*} |f_{\varphi(k)}| \right) &= \sum_{k=1}^{n} \ell(|f_{\varphi(k)}|) \\ &\le \sum_{\iota \in I} \ell(|f_\iota|) \\ &< \infty. \end{aligned}$$

Thus $(\Sigma^*_{k \in \mathbb{N}_n} | f_{\varphi(k)} |)_{n \in \mathbb{N}}$ is an ℓ-sequence from $\mathscr{L}$. Therefore $\Sigma^*_{k \in \mathbb{N}} | f_{\varphi(k)} |$ belongs to $\mathscr{L}$ and

$$\ell\left(\sum_{k \in \mathbb{N}}{}^* | f_{\varphi(k)} | \right) = \sum_{k \in \mathbb{N}} \ell(| f_{\varphi(k)} |)$$

$$= \sum_{\iota \in I} \ell(| f_\iota |)$$

[Corollary 3.4.26, Proposition 1.3.3(d), Corollary 3.4.14]. Since $\Sigma^*_{\iota \in I} | f_\iota |$ is equal to $\Sigma^*_{k \in \mathbb{N}} | f_{\varphi(k)} |$, both (b) and (c) follow [Proposition 1.3.3(d)].

(b) $\Rightarrow$ (a). For every n in $\mathbb{N}$, $\Sigma^*_{k \in \mathbb{N}_n} | f_{\varphi(k)} |$ belongs to $\mathscr{L}$, and

$$\sum_{k=1}^{n} \ell(| f_{\varphi(k)} |) = \ell\left(\sum_{k=1}^{n}{}^* | f_{\varphi(k)} | \right)$$

$$\leq \ell\left(\sum_{\iota \in I}{}^* | f_\iota | \right)$$

$$< \infty.$$

It follows that $(\ell(| f_{\varphi(k)} |))_{k \in \mathbb{N}}$ is summable (Corollary 3.4.25). Therefore so is $(\ell(| f_\iota |))_{\iota \in I}$ (Corollary 3.4.14).

(b) $\Rightarrow$ (d). Since the function $\Sigma^*_{\iota \in I} | f_\iota |$ belongs to $\mathscr{L}$, it is real $\mathscr{L}$-a.e. Thus (d) follows from Theorem 3.4.20 (c) $\Rightarrow$ (d).

(a) $\Rightarrow$ (e). Since $|\ell(f_\iota)| \leq \ell(| f_\iota |)$ for every ι, (e) follows easily from (a) [Proposition 3.4.11(a)].

(a), (b), (d) $\Rightarrow$ (f). We use the Lebesgue Dominated Convergence Theorem (Theorem 1.4.16). In view of Proposition 1.3.3 (d), it suffices to show that $\Sigma^*_{\iota \in I} f_\iota$ belongs to $\mathscr{L}$ and

$$\ell\left(\sum_{\iota \in I}{}^* f_\iota \right) = \sum_{\iota \in I} \ell(f_\iota).$$

Set

$$B := \{ x \in X \,|\, (| f_\iota(x) |)_{\iota \in I} \text{ is not summable} \}$$

and note that B is $\mathscr{L}$-exceptional. We have

$$\sum_{\iota \in I}{}^* f_\iota = \lim_{n \to \infty} \sum_{k=1}^{n} f_{\varphi(k)} e_{X \setminus B} \quad \mathscr{L}\text{-a.e.}$$

For every n in $\mathbb{N}$ we have

$$\left| \sum_{k=1}^{n} f_{\varphi(k)} e_{X \setminus B} \right| \leq \sum_{\iota \in I} |f_\iota e_{X \setminus B}| \leq \sum_{\iota \in I}{}^{*} |f_\iota|.$$

Thus $\Sigma^*_{\iota \in I} f_\iota$ is the $\mathscr{L}$-a.e. limit of a sequence from $\mathscr{L}$ that is bounded in $\mathscr{L}$. Using the Dominated Convergence Theorem, we conclude that $\Sigma^*_{\iota \in I} f_\iota$ belongs to $\mathscr{L}$ and

$$\ell\left(\sum_{\iota \in I}{}^{*} f_\iota \right) = \lim_{n \to \infty} \sum_{k=1}^{n} \ell(f_{\varphi(k)})$$

$$= \sum_{k \in \mathbb{N}} \ell(f_{\varphi(k)})$$

$$= \sum_{\iota \in I} \ell(f_\iota).$$ □

Theorem 2 has two corollaries that will prove useful in Chapter 5.

Corollary 3.5.3. *Let $(X, \mathscr{L}, \ell)$ be a closed Daniell space. Suppose, for some sequence $(f_n)_{n \in \mathbb{N}}$ from $\mathscr{L}$ and for some function f belonging to $\mathscr{L}$, that the sequence $(\ell(|f \dot{-} f_n|))_{n \in \mathbb{N}}$ is summable. Then the sequence $(f_n)_{n \in \mathbb{N}}$ is bounded in $\mathscr{L}$, and*

$$f(x) = \lim_{n \to \infty} f_n(x) \ \mathscr{L}\text{-a.e.} \tag{1}$$

Proof. Define

$$h_n\colon X \to \overline{\mathbb{R}}, \qquad x \mapsto \begin{cases} f(x) - f_n(x) & \text{if defined} \\ \infty & \text{otherwise.} \end{cases}$$

$$h := \sum_{n \in \mathbb{N}}{}^{*} |h_n|.$$

Then $(\ell(|h_n|))_{n \in \mathbb{N}}$ is summable. By Theorem 2, $(h_n)_{n \in \mathbb{N}}$ is summable $\mathscr{L}$-a.e. and h belongs to $\mathscr{L}$. At each x for which $(h_n(x))_{n \in \mathbb{N}}$ is summable we have

$$\lim_{n \to \infty} (f_n(x) - f(x)) = 0$$

so (1) holds. Evidently

$$-(h + |f|) \leq f_n \leq h + |f|$$

for every $n \in \mathbb{N}$. Since $h + |f|$ belongs to $\mathscr{L}$, the sequence $(f_n)_{n \in \mathbb{N}}$ is bounded in $\mathscr{L}$. □

Corollary 3.5.4. *Let $(X, \mathscr{L}, \ell)$ be a Daniell space. Then every function in $\bar{\mathscr{L}}(\ell)$ is the $\bar{\mathscr{L}}(\ell)$-a.e. limit of a sequence from $\mathscr{L}$ that is bounded in $\bar{\mathscr{L}}(\ell)$.*

Proof. Let $f \in \bar{\mathscr{L}}(\ell)$. For each n in $\mathbb{N}$ there exists a function g_n belonging to $\bar{\mathscr{L}}(\ell) \cap \mathscr{L}^{\uparrow}$ such that $f \leq g_n$ and

$$\bar{\ell}(g_n) - \bar{\ell}(f) = \ell^{\uparrow}(g_n) - \bar{\ell}(f)$$
$$< \frac{1}{2 \cdot 2^n}$$

(Corollary 3.2.2), and there exists a function f_n belonging to $\mathscr{L} \cap \mathbb{R}^X$ such that $f_n \leq g_n$ and

$$\bar{\ell}(g_n) - \ell(f_n) < \frac{1}{2 \cdot 2^n}.$$

For every n in $\mathbb{N}$,

$$\bar{\ell}(|f - f_n|) \leq \bar{\ell}(|f \dot{-} g_n|) + \bar{\ell}(|g_n - f_n|)$$
$$< \frac{1}{2^n}.$$

Since the sequence $(1/2^n)_{n \in \mathbb{N}}$ is summable, the sequence $(\bar{\ell}(|f - f_n|))_{n \in \mathbb{N}}$ is summable. Now apply Corollary 3 to conclude that the sequence $(f_n)_{n \in \mathbb{N}}$ is bounded in $\bar{\mathscr{L}}(\ell)$ and

$$f(x) = \lim_{n \to \infty} f_n(x) \; \bar{\mathscr{L}}(\ell)\text{-a.e.}$$ □

Proposition 3.5.5. *Let $(X, \mathscr{L}, \ell)$ be a closed Daniell space. Then*

$$\ell_*\left(\sum_{\iota \in I}{}^{*} f_\iota \right) \geq \sum_{\iota \in I}{}^{*} \ell_*(f_\iota) \tag{2}$$

for every family $(f_\iota)_{\iota \in I}$ from $\overline{\mathbb{R}}_+^X$, and

$$\ell^*\left(\sum_{\iota \in I}{}^{*} f_\iota \right) \leq \sum_{\iota \in I}{}^{*} \ell^*(f_\iota) \tag{3}$$

for every countable family $(f_\iota)_{\iota \in I}$ from $\overline{\mathbb{R}}_+^X$.

Proof. For $(f_\iota)_{\iota \in I}$ a finite family from $\overline{\mathbb{R}}_+^X$, inequalities (2) and (3) follow by complete induction from Propositions 3.3.3(c) and 3.3.7(f).

Let $(f_\iota)_{\iota \in I}$ be a countably infinite family from $\overline{\mathbb{R}}_+^X$, and let $\varphi: \mathbb{N} \to I$ be bijective. Using (3) for finite families, together with Theorem 3.3.4(a), Corollary 3.4.14, and Corollary 3.4.26, we have

$$\begin{aligned}
\ell^*\left(\sum_{\iota \in I}{}^* f_\iota\right) &= \ell^*\left(\sum_{n \in \mathbb{N}}{}^* f_{\varphi(n)}\right) \\
&= \ell^*\left(\sup_{n \in \mathbb{N}} \sum_{k=1}^{n}{}^* f_{\varphi(k)}\right) \\
&= \sup_{n \in \mathbb{N}} \ell^*\left(\sum_{k=1}^{n}{}^* f_{\varphi(k)}\right) \\
&\leq \sup_{n \in \mathbb{N}} \sum_{k=1}^{n}{}^* \ell^*\left(f_{\varphi(k)}\right) \\
&= \sum_{n \in \mathbb{N}}{}^* \ell^*\left(f_{\varphi(n)}\right) \\
&= \sum_{\iota \in I}{}^* \ell^*(f_\iota).
\end{aligned}$$

Finally, let $(f_\iota)_{\iota \in I}$ be an arbitrary family from $\overline{\mathbb{R}}_+^X$. According to Corollary 3.4.22,

$$\sum_{\iota \in I}{}^* \ell_*(f_\iota) = \sup_{J \in \mathfrak{F}(I)} \sum_{\iota \in J}{}^* \ell_*(f_\iota).$$

For every finite subset J of I,

$$\begin{aligned}
\sum_{\iota \in J}{}^* \ell_*(f_\iota) &\leq \ell_*\left(\sum_{\iota \in J}{}^* f_\iota\right) \\
&\leq \ell_*\left(\sum_{\iota \in I}{}^* f_\iota\right).
\end{aligned}$$

Inequality (2) follows. □

EXERCISES

3.5.1(E) Let $\mathfrak{R}$ be a ring of sets and μ: $\mathfrak{R} \to \overline{\mathbb{R}}_+$ a function with the following properties:

(i) $\mu(\varnothing) = 0$.
(ii) $\mu(\bigcup_{n \in \mathbb{N}} A_n) = \sum^*_{n \in \mathbb{N}} \mu(A_n)$ whenever $(A_n)_{n \in \mathbb{N}}$ is a disjoint sequence from $\mathfrak{R}$ with $\bigcup_{n \in \mathbb{N}} A_n \in \mathfrak{R}$.

Define $\mathfrak{S} := \{A \in \mathfrak{R} \mid \mu(A) < \infty\}$. Then

(α) $\mathfrak{S}$ is a ring of sets.
(β) $\mu|_{\mathfrak{S}}$ is a positive measure on $\mathfrak{S}$.

3.5.2(C) **Multipliable Families, Infinite Products.** We arrived at the notion of summable families in Section 3.4 by generalizing the addition of real numbers. It seems natural enough to ask if the second arithmetic operation on $\mathbb{R}$, multiplication, cannot similarly be generalized. In fact it can, and this generalization will be of significance when we turn to the discussion of the product of measures. The generalization of multiplication by analogy with the generalization of addition leads to the concept of multipliable families. We have no intention to develop an independent theory of such families but treat them as an application of summable families and make use of the knowledge acquired there.

(a) The starting point is the extension of the product of two numbers to the product of finitely many factors. Prove the following propositions:

(α) Let I be a finite set. Then there is a unique mapping

$$\mathbb{R}^I \to \mathbb{R}, \qquad (\alpha_\iota)_{\iota \in I} \mapsto \prod_{\iota \in I} \alpha_\iota$$

with the following properties:

(α1) $\prod_{\iota \in I}(\alpha_\iota \beta_\iota) = (\prod_{\iota \in I} \alpha_\iota)(\prod_{\iota \in I} \beta_\iota)$ for any $(\alpha_\iota)_{\iota \in I}$ and $(\beta_\iota)_{\iota \in I}$ from $\mathbb{R}^I$.
(α2) Given $\lambda \in I$ and $\alpha \in \mathbb{R}$, define

$$\delta^{\lambda, \alpha}: I \to \mathbb{R}, \qquad \iota \mapsto \begin{cases} \alpha & \text{if } \iota = \lambda \\ 1 & \text{otherwise} \end{cases}.$$

Then $\prod_{\iota \in I} \delta^{\lambda, \alpha}(\iota) = \alpha$.
(α3) $\prod_{\iota \in \varnothing} \alpha_\iota = 1$.

Verify the following additional properties of this function:

(β) $\prod_{\iota \in I} \alpha_\iota = \alpha^n$ for the family $(\alpha_\iota)_{\iota \in I}$ given by $\alpha_\iota = \alpha$ for each $\iota \in I$, where n denotes the number of elements of I.

(γ) $\prod_{\iota \in I} \alpha_{\varphi(\iota)} = \prod_{\iota \in I} \alpha_\iota$ for any bijection φ: $I \to I$.

(δ) Given a sequence $(\alpha_k)_{k \in \mathbb{N}_n}$ from $\mathbb{R}$, define $\prod_{k=1}^{1}\alpha_k := \alpha_1$ and $\prod_{k=1}^{m+1}\alpha_k := \alpha_{m+1}\prod_{k=1}^{m}\alpha_k$ for $m \in \mathbb{N}_{n-1}$. If $\varphi\colon \mathbb{N}_n \to I$ is a bijection, then $\prod_{\iota \in I}\alpha_\iota = \prod_{k=1}^{n}\alpha_{\varphi(k)}$ for every $(\alpha_\iota)_{\iota \in I} \in \mathbb{R}^I$.

(b) We now give a direct definition of the notion of a multipliable family.

Let X be a set. The family $(\alpha_x)_{x \in X}$ of real numbers is called a **multipliable family** and $\alpha \in \mathbb{R}$ its **product** iff, given any $\varepsilon \in \mathbb{R}$ with $\varepsilon > 0$, there is a finite subset $Y \subset X$ such that $|\alpha - \prod_{z \in Z}\alpha_z| < \varepsilon$ for every finite set Z with $Y \subset Z \subset X$. Write $\prod_{x \in X}\alpha_x$ for the product.

The reader should verify that this notion is compatible with that introduced for finite sets in (a).

Denote by $\mathscr{M}(X)$ the set of all multipliable families $(\alpha_x)_{x \in X}$. Our principle interest is in $\mathscr{M}(X)_+$, so we restrict our attention to $\mathscr{M}(X)_+$ in this exercise.

(c) For the sequel, define $\ln 0 := -\infty$ and $e^{-\infty} := 0$. The connection with summable families then takes the following form:

(α) Let $(\alpha_x)_{x \in X}$ be a family from $\mathbb{R}_+$. Then the following statements are equivalent:

($\alpha 1$) $(\alpha_x)_{x \in X} \in \mathscr{M}(X)_+$.
($\alpha 2$) $\Sigma_{*x \in X}\ln \alpha_x < \infty$.

If these hold, then $\prod_{x \in X}\alpha_x = \exp(\Sigma_{*x \in X}\ln \alpha_x) \in \mathbb{R}_+$.

As a consequence of these properties, show that the following hold:

(β) If $(\alpha_x)_{x \in X} \in \mathscr{M}(X)_+$, then exactly one of the following two statements is true:

($\beta 1$) $\Sigma_{*x \in X}\ln \alpha_x = -\infty$. (This is true exactly when $\prod_{x \in X}\alpha_x = 0$.)
($\beta 2$) $(\ln \alpha_x)_{x \in X} \in \ell^1(X)$. (This is true exactly when $\prod_{x \in X}\alpha_x > 0$.)

(d) The characterizations given in (c) can be used to prove the following propositions, for $(\alpha_x)_{x \in X}$ and $(\beta_x)_{x \in X} \in \mathscr{M}(X)_+$:

(α) $(\alpha_x\beta_x)_{x \in X} \in \mathscr{M}(X)_+$, and $\prod_{x \in X}\alpha_x\beta_x = (\prod_{x \in X}\alpha_x)(\prod_{x \in X}\beta_x)$.
(β) $(\alpha_x/\beta_x)_{x \in X} \in \mathscr{M}(X)_+$ whenever $\prod_{x \in X}\beta_x \neq 0$ and $\prod_{x \in X}(\alpha_x/\beta_x) = \prod_{x \in X}\alpha_x/\prod_{x \in X}\beta_x$.
(γ) $(\alpha_x \vee \beta_x)_{x \in X} \in \mathscr{M}(X)_+$ and $(\alpha_x \wedge \beta_x)_{x \in X} \in \mathscr{M}(X)_+$.
(δ) $(\alpha_x^\gamma)_{x \in X} \in \mathscr{M}(X)_+$ for any $\gamma \in \mathbb{R}$, and $\prod_{x \in X}(\alpha_x^\gamma) = (\prod_{x \in X}\alpha_x)^\gamma$.
(ε) If $(\gamma_x)_{x \in X} \in \mathbb{R}_+^X$ with $\gamma_x \leq \alpha_x$ for each $x \in X$, then $(\gamma_x)_{x \in X} \in \mathscr{M}(X)_+$ and $\prod_{x \in X}\gamma_x \leq \prod_{x \in X}\alpha_x$.
(ζ) If $\prod_{x \in X}\alpha_x \neq 0$, then $\{x \in X \mid \alpha_x \neq 1\}$ is countable.
(η) If $\varphi\colon X \to X$ is a bijection, then $(\alpha_{\varphi(x)})_{x \in X} \in \mathscr{M}(X)_+$ and $\prod_{x \in X}\alpha_{\varphi(x)} = \prod_{x \in X}\alpha_x$.
(ϑ) If $(X_\iota)_{\iota \in I}$ is a disjoint family with $X = \bigcup_{\iota \in I}X_\iota$, then the following hold:

($\vartheta 1$) $(\alpha_x)_{x \in X_\iota} \in \mathscr{M}(X_\iota)_+$ for each $\iota \in I$.

($\vartheta 2$) $(\prod_{x \in X_\iota} \alpha_x)_{\iota \in I} \in \mathcal{M}(I)_+$.
($\vartheta 3$) $\prod_{\iota \in I}(\prod_{x \in X_\iota} \alpha_x) = \prod_{x \in X} \alpha_x$.

(ι) If $X = Y \times Z$, writing $(\alpha_{yz})_{(y,z) \in Y \times Z}$ for $(\alpha_x)_{x \in X}$, we can state the following:

($\iota 1$) $(\alpha_{yz})_{z \in Z} \in \mathcal{M}(Z)_+$ for each $y \in Y$.
($\iota 2$) $(\alpha_{yz})_{y \in Y} \in \mathcal{M}(Y)_+$ for each $z \in Z$.
($\iota 3$) $(\prod_{z \in Z} \alpha_{yz})_{y \in Y} \in \mathcal{M}(Y)_+$
($\iota 4$) $(\prod_{y \in Y} \alpha_{yz})_{z \in Z} \in \mathcal{M}(Z)_+$
($\iota 5$) $\prod_{(y,z) \in Y \times Z} \alpha_{yz} = \prod_{y \in Y}(\prod_{z \in Z} \alpha_{yz}) = \prod_{z \in Z}(\prod_{y \in Y} \alpha_{yz})$.

(κ) If $((\gamma_{nx})_{x \in X})_{n \in \mathbb{N}}$ is an increasing sequence from $\mathcal{M}(X)_+$ with $0 < \sup_{n \in \mathbb{N}}(\prod_{x \in X} \gamma_{nx}) < \infty$, then $(\sup_{n \in \mathbb{N}} \gamma_{nx})_{x \in X} \in \mathcal{M}(X)_+$, and $\prod_{x \in X} \sup_{n \in \mathbb{N}} \gamma_{nx} = \sup_{n \in \mathbb{N}}(\prod_{x \in X} \gamma_{nx})$. (An analogous result can be formulated for decreasing sequences.)

We wish to look at the relationship between $\mathcal{M}(X)_+$ and $\ell^1(X)$ in a different light. It is often more convenient to investigate $(1 + \alpha_x)_{x \in X}$ instead of $(\alpha_x)_{x \in X}$ itself. Prove the following assertion:

(ν) For any family $(\alpha_x)_{x \in X}$ from $\mathbb{R}_+$, the following are equivalent:

($\nu 1$) $(1 + \alpha_x)_{x \in X} \in \mathcal{M}(X)_+$.
($\nu 2$) $(\alpha_x)_{x \in X} \in \ell^1(X)$.

Observe that, on the one hand, $\ln(1 + \alpha_x) \leq \alpha_x$ for each $x \in X$ and that, on the other hand, $\prod_{z \in Z}(1 + \alpha_z) \geq \sum_{z \in Z} \alpha_z$ for any $Z \in \mathfrak{F}(X)$.

If we drop the positiveness of $(\alpha_x)_{x \in X}$, the following can be proved:

(o) Let $(\alpha_x)_{x \in X}$ be a family from $\mathbb{R}$ such that $(1 + \alpha_x)_{x \in X} \in \mathbb{R}_+^X$. Then

($o1$) $(1 + \alpha_x)_{x \in X} \in \mathcal{M}(X)_+$ whenever $(\alpha_x)_{x \in X} \in \ell^1(X)$.
($o2$) $(\alpha_x)_{x \in X} \in \ell^1(X)$ whenever $(1 + \alpha_x)_{x \in X} \in \mathcal{M}(X)_+$ and $\prod_{x \in X}(1 + \alpha_x) \neq 0$.

3.6. THE CLOSURE OF THE STIELTJES FUNCTIONALS

NOTATION AND TERMINOLOGY FOR SECTION 3.6:

The same as for Section 2.5.

We conclude this chapter by applying the construction method of Sections 3.1 and 3.2 to the Stieltjes functionals introduced in Section 2.5. Some additional review of topological notions is required first, so we begin by defining and discussing the notion of semicontinuity. Theorem 6 uses this notion to characterize the spaces $\mathcal{K}(A)^{\uparrow}$ and $\mathcal{K}(A)^{\downarrow}$. Corollary 7 concludes the section with a description of the closure of the Daniell space $(A, \mathcal{K}(A), \ell_g)$.

Definition 3.6.1. *A function $f \in \overline{\mathbb{R}}^A$ is said to be:*

(*a*) ***lower semicontinuous*** *at x, for $x \in A$, iff*

$$f(x) \leq \liminf_{n \to \infty} f(x_n)$$

for every sequence $(x_n)_{n \in \mathbb{N}}$ from A that converges to x;

(*b*) ***upper semicontinuous*** *at x, for $x \in A$, iff*

$$f(x) \geq \limsup_{n \to \infty} f(x_n)$$

for every sequence $(x_n)_{n \in \mathbb{N}}$ from A that converges to x;

(*c*) ***lower semicontinuous*** *on A iff f is lower semicontinuous at x for every x in A;*

(*d*) ***upper semicontinuous*** *on A iff f is upper semicontinuous at x for every x in A.* □

Proposition 3.6.2. *A function $f \in \overline{\mathbb{R}}^A$ is continuous at x, for $x \in A$, iff f is both lower semicontinuous at x and upper semicontinuous at x. A function $f \in \overline{\mathbb{R}}^A$ is continuous on A iff it is both lower semicontinuous on A and upper semicontinuous on A.*

Proof. Let $f \in \overline{\mathbb{R}}^A$ and let $x \in A$. If f is continuous at x, then

$$f(x) = \lim_{n \to \infty} f(x_n) = \liminf_{n \to \infty} f(x_n) = \limsup_{n \to \infty} f(x_n)$$

for every sequence $(x_n)_{n \in \mathbb{N}}$ from A that converges to x, and f is both upper and lower semicontinuous at x. Suppose, conversely that f is both lower semicontinuous at x and upper semicontinuous at x. If $(x_n)_{n \in \mathbb{N}}$ is a sequence from A that converges to x, then

$$f(x) \leq \liminf_{n \to \infty} f(x_n) \leq \limsup_{n \to \infty} f(x_n) \leq f(x)$$

so

$$\liminf_{n \to \infty} f(x_n) = \limsup_{n \to \infty} f(x_n) = f(x)$$

and

$$\lim_{n \to \infty} f(x_n) = f(x).$$

Thus f is continuous at x.

The second assertion is a consequence of the first. □

The properties of upper semicontinuous functions can be derived from those of lower semicontinuous functions, and vice versa, as the following, easily verified proposition shows.

Proposition 3.6.3. *For $x \in A$ and $f \in \overline{\mathbb{R}}^A$, the function f is upper semicontinuous at x iff the function $-f$ is lower semicontinuous at x.* □

For real-valued functions there is the following ε-δ characterization of semicontinuity.

Proposition 3.6.4. *For $x \in A$ and $f \in \mathbb{R}^A$, the function f is:*

(*a*) *lower semicontinuous at x iff for every $\varepsilon \in \mathbb{R}$, $\varepsilon > 0$, there exists $\delta \in \mathbb{R}$, $\delta > 0$, such that*

$$f(y) > f(x) - \varepsilon$$

for every $y \in]x - \delta, x + \delta[\cap A$;

(*b*) *upper semicontinuous at x iff for every $\varepsilon \in \mathbb{R}$, $\varepsilon > 0$, there exists $\delta \in \mathbb{R}$, $\delta > 0$, such that*

$$f(y) < f(x) + \varepsilon$$

for every $y \in]x - \delta, x + \delta[\cap A$.

Proof. (a) Let f be a real-valued function on A that is lower semicontinuous at x. If the characterization in (a) were false, then there would exist $\varepsilon > 0$ and a sequence $(x_n)_{n \in \mathbb{N}}$ from A such that

$$|x - x_n| < \frac{1}{n}, \qquad f(x_n) \leq f(x) - \varepsilon$$

for every n. For this sequence we would have

$$\lim_{n \to \infty} x_n = x$$

but

$$\liminf_{n \to \infty} f(x_n) \leq f(x) - \varepsilon$$

which contradicts the lower semicontinuity. Conversely, the ε-δ condition stated in (a) obviously implies that

$$f(x) \leq \liminf_{n \to \infty} f(x_n)$$

for every sequence $(x_n)_{n \in \mathbb{N}}$ from A converging to x.

(b) In view of Proposition 3, (b) follows from (a). □

Proposition 3.6.5. *The following assertions hold, for every $f \in \overline{\mathbb{R}}^A$ and for every real number α.*

(*a*) *If f is lower semicontinuous on A, then $\{f \le \alpha\}$ is closed in A and $\{f > \alpha\}$ is open in A.*

(*b*) *If f is upper semicontinuous on A, then $\{f \ge \alpha\}$ is closed in A and $\{f < \alpha\}$ is open in A.*

Proof. (a) It suffices to verify that $\{f \le \alpha\}$ is closed in A. Let $x \in A$, and let $\{x_n\}$ be a sequence from $\{f \le \alpha\}$ that converges to x. Using the definition of lower semicontinuity, we have

$$f(x) \le \liminf_{n \to \infty} f(x_n) \le \alpha.$$

(b) Either use the analogous argument, or use the fact that f is upper semicontinuous iff $-f$ is lower semicontinuous. □

We can now present a description of $\mathscr{K}(A)^{\uparrow}$ and $\mathscr{K}(A)^{\downarrow}$.

Theorem 3.6.6. *The following assertions hold, for every function $f \in \overline{\mathbb{R}}^A$.*

(*a*) *$f \in \mathscr{K}(A)^{\uparrow}$ iff f is lower semicontinuous on A and $f \ge h$ for some $h \in \mathscr{K}(A)$.*

(*b*) *$f \in \mathscr{K}(A)^{\downarrow}$ iff f is upper semicontinuous on A and $f \le h$ for some $h \in \mathscr{K}(A)$.*

Proof. (a) Let $f \in \mathscr{K}(A)^{\uparrow}$. By definition, there exists an increasing sequence $(f_n)_{n \in \mathbb{N}}$ from $\mathscr{K}(A)$ whose supremum is the given f. We need only show that f is lower semicontinuous at every x in A. Given x in A, let $(x_n)_{n \in \mathbb{N}}$ be a sequence from A that converges to x. Since

$$f(y) \ge f_k(y)$$

for every y in A and for every k in $\mathbb{N}$, we have for every k,

$$\begin{aligned}\liminf_{n \to \infty} f(x_n) &\ge \liminf_{n \to \infty} f_k(x_n) \\ &= f_k(x).\end{aligned}$$

It follows that

$$\liminf_{n \to \infty} f(x_n) \ge f(x)$$

so f is lower semicontinuous at x.

Conversely, assume that f is lower semicontinuous on A and that $f \ge h$ for some h in $\mathscr{K}(A)$. For now, we assume also that f is real valued. First, we construct an increasing sequence of continuous functions on A whose supremum is the given f. For each k in $\mathbb{N}$ define

$$f_k: A \to \mathbb{R}, \qquad x \mapsto \inf_{z \in A} (f(z) + k|x - z|).$$

Note that the sequence $(f_k)_{k \in \mathbb{N}}$ increases and that $f_k \leq f$ for every k. It is not difficult to verify that each f_k is continuous on A. In fact, to see that each f_k is uniformly continuous on A, we can argue as follows. Fix $k \in \mathbb{N}$, let $\varepsilon > 0$ be given, and suppose that x, y belong to A and satisfy $|x - y| < \varepsilon/2k$. There exists $z \in A$ such that

$$f_k(x) \geq f(z) + k|x - z| - \frac{\varepsilon}{2}.$$

Since

$$\begin{aligned} f_k(y) &\leq f(z) + k|y - z| \\ &\leq f(z) + k(|y - x| + |x - z|) \end{aligned}$$

we have

$$f_k(y) - f_k(x) \leq k|y - x| + \frac{\varepsilon}{2} < \varepsilon.$$

Interchanging x and y, we conclude by symmetry that

$$|f_k(x) - f_k(y)| < \varepsilon.$$

It follows that f_k is uniformly continuous on A.

Evidently $\bigvee_{k \in \mathbb{N}} f_k \leq f$. We want to show that $\bigvee_{k \in \mathbb{N}} f_k \geq f$. Fix $x \in A$, and let $\varepsilon > 0$ be given. Since f is lower semicontinuous at x, there exists $\delta > 0$ such that $f(y) > f(x) - \varepsilon$ for every y in $]x - \delta, x + \delta[\cap A$. For each k, therefore, we have

$$f(y) + k|x - y| > f(x) - \varepsilon, \qquad y \in]x - \delta, x + \delta[\cap A.$$

If y belongs to $A \setminus]x - \delta, x + \delta[$, then for each k we have

$$\begin{aligned} f(y) + k|x - y| &\geq h(y) + k\delta \\ &\geq \inf_{z \in A} h(z) + k\delta. \end{aligned}$$

Hence there exists k_0 in $\mathbb{N}$ such that for $k \geq k_0$,

$$f(y) + k|x - y| > f(x) - \varepsilon, \qquad y \in A \setminus]x - \delta, x + \delta[.$$

It follows for $k \geq k_0$ that

$$f_k(x) \geq f(x) - \varepsilon.$$

Thus $\bigvee_{k \in \mathbb{N}} f_k(x) \geq f(x) - \varepsilon$. Since ε was given arbitrarily, we have

$$\bigvee_{k \in \mathbb{N}} f_k(x) \geq f(x)$$

as required.

We need yet an increasing sequence from $\mathscr{K}(A)$ whose supremum is the given f. Because $f \geq h$, we have, for every k, $\inf_{z \in A} f_k(z) > -\infty$, and $\{f_k < 0\} \subset [c, d]$ for appropriate points c, d belonging to A. Thus for each k in $\mathbb{N}$ it is easy to find an increasing sequence $(f_{k,n})_{n \in \mathbb{N}}$ from $\mathscr{K}(A)$ whose supremum is the continuous function f_k. If we set

$$h_n := \bigvee_{k \leq n} f_{k,n} \qquad n \in \mathbb{N}$$

then $(h_n)_{n \in \mathbb{N}}$ provides the required sequence from $\mathscr{K}(A)$.

We have shown that if f satisfies the conditions stated in (a) and is also real valued, then f must belong to $\mathscr{K}(A)^{\uparrow}$. Suppose then that f is not necessarily real valued but does satisfy the conditions stated in (a), and consider the approximating sequence $(f \wedge m)_{m \in \mathbb{N}}$. Each of the functions $f \wedge m$ is real valued and, as we show next, satisfies the conditions from (a). Evidently $f \wedge m \geq h \wedge m$ if $f \geq h$, and $h \wedge m$ belongs to $\mathscr{K}(A)$ if h does. Let $x \in A$, and let $(x_n)_{n \in \mathbb{N}}$ be a sequence from A that converges to x. From the lower semicontinuity of f at x we have

$$f(x) \leq \liminf_{n \to \infty} f(x_n) = \sup_{n \in \mathbb{N}} \inf_{k \geq n} f(x_k)$$

and so

$$\begin{aligned}(f \wedge m)(x) &= f(x) \wedge m \\ &\leq \left(\sup_{n \in \mathbb{N}} \inf_{k \geq n} f(x_k)\right) \wedge m \\ &= \sup_{n \in \mathbb{N}} \inf_{k \geq n} \left(f(x_k) \wedge m\right) \\ &= \liminf_{n \to \infty} (f \wedge m)(x_n).\end{aligned}$$

Thus the lower semicontinuity of $f \wedge m$ follows from that of f. Applying to the functions $f \wedge m$ what we have already proved, we conclude that each

$f \wedge m$ belongs to $\mathscr{K}(A)^{\uparrow}$. As the supremum of an increasing sequence from $\mathscr{K}(A)^{\uparrow}$, f itself must also belong to $\mathscr{K}(A)^{\uparrow}$ [Proposition 3.1.2. (f)].

(b) In view of Proposition 3 and Proposition 3.1.6, (b) follows from (a).

□

What can be said, then, about the closure $(A, \bar{\mathscr{L}}(\ell_g), \overline{\ell}_g)$ of the Daniell space $(A, \mathscr{K}(A), \ell_g)$ for the Stieltjes functional ℓ_g? A function f belonging to $\bar{\mathscr{L}}(\ell_g)$ need not be semicontinuous, but as far as $\overline{\ell}_g(f)$ is concerned, every f in $\bar{\mathscr{L}}(\ell_g)$ can be approximated arbitrarily closely by either an upper semicontinuous or a lower semicontinuous function. The following corollary makes this description precise.

Corollary 3.6.7. *Let g be an increasing real-valued function on A, and suppose that f belongs to the Riesz lattice $\bar{\mathscr{L}}(\ell_g)$. Then for each $\varepsilon > 0$ there exist in $\bar{\mathscr{L}}(\ell_g)$ a function f', upper semicontinuous on A, and a function f'', lower semicontinuous on A, such that*

$$f' \leq f \leq f'' \quad \text{and} \quad \overline{\ell}_g(f'') - \overline{\ell}_g(f') < \varepsilon.$$

The functions f' and f'' can be chosen so that for every $\alpha \in \mathbb{R}$, $\alpha > 0$ the sets $\{f' \geq \alpha\}$, $\{f'' \leq -\alpha\}$ are bounded in A.

Proof. By Corollary 3.2.2 there exist functions

$$f' \in \bar{\mathscr{L}}(\ell_g) \cap \mathscr{K}(A)^{\downarrow}$$

$$f'' \in \bar{\mathscr{L}}(\ell_g) \cap \mathscr{K}(A)^{\uparrow}$$

such that

$$f' \leq f \leq f'' \quad \text{and} \quad \overline{\ell}_g(f'') - \overline{\ell}_g(f') < \varepsilon.$$

By Theorem 6, f' is upper semicontinuous, f'' is lower semicontinuous, and there exist functions $h', h'' \in \mathscr{K}(A)$ such that

$$f' \leq h', \qquad h'' \leq f''.$$

If $\alpha > 0$, then $\{f' \geq \alpha\} \subset \{h' \neq 0\}$ and $\{f'' \leq -\alpha\} \subset \{h'' \neq 0\}$, so these sets are bounded in A. □

4

THE INTEGRAL FOR A DANIELL SPACE

We have seen how to extend a given Daniell space so that it acquires the stronger convergence properties possessed by closed Daniell spaces. It does sometimes happen that the closure of a given Daniell space is the only closed extension of the original Daniell space, but this is rarely the case. When additional closed extensions exist, how are we to choose among them which extension deserves to be called "the integral for the given Daniell space"—if indeed there is only one extension that can rightfully play that role? What requirements should the integral satisfy?

In the historical development of integration theory, new requirements arose continually. The new requirements were followed by new extensions designed to take them into account. As a result many distinct definitions of the integral are currently in use. One goal, an important goal, for our theory is to bring all of these definitions under one hat.

One important difference between various closed extensions is that of scope: which functions are "integrable" for the one extension or the other? The other difference is in the "integrals" assigned to the various integrable functions. A natural goal, especially if we hope to bring several different definitions under one hat, is that there be as many integrable functions as possible. On the other hand, we want to exclude pure chance from playing a role in the assignment of the values of the integral. Otherwise, the resulting "integral" is only partly determined by the original functional. These two goals conflict to some extent, since one way to acquire a large class of integrable functions is to allow considerable leeway in how the original functional is extended.

Having stated our goals, let us describe how we proceed. Suppose a Daniell space $(X, \mathscr{L}, \ell)$ is given. We consider the collection of all closed extensions of $(X, \mathscr{L}, \ell)$, each of which automatically extends the closure $(X, \bar{\mathscr{L}}(\ell), \bar{\ell})$. By now it is clear that it is not the individual functions in $\bar{\mathscr{L}}(\ell)$ that are important but the equivalence classes relative to the equivalence relation of $\bar{\mathscr{L}}(\ell)$-a.e. equality. If $(X, \tilde{\mathscr{L}}, \tilde{\ell})$ is any closed extension of $(X, \mathscr{L}, \ell)$, there is a natural map from the set of equivalence classes of $\bar{\mathscr{L}}(\ell)$ modulo $\bar{\mathscr{L}}(\ell)$-a.e.

equality to the set of equivalence classes of $\tilde{\mathscr{L}}$ modulo $\tilde{\mathscr{L}}$-a.e. equality. We require this natural map to be bijective. Any other possibility, it is easy to check, would necessarily involve arbitrariness (of one of two kinds). We shall call closed extensions that meet this requirement admissible. Every Daniell space obviously has its closure as one admissible extension.

Again, it may happen that the closure of a given Daniell space is the only admissible extension of the original Daniell space. Generally, however, this is also not the case. One must decide among the various admissible extensions which one, if indeed there is only one, deserves to be called the integral.

The functionals for two admissible extensions will share some common domain, a domain that includes but is generally larger than $\bar{\mathscr{L}}(\ell)$. The two functionals will automatically agree on $\bar{\mathscr{L}}(\ell)$. A natural requirement is that they agree throughout their common domain: otherwise, a single function could be assigned different functional values by different extensions. Let us say that an admissible extension $(X, \mathscr{L}', \ell')$ of $(X, \mathscr{L}, \ell)$ is determined by $(X, \mathscr{L}, \ell)$ if, for every admissible extension $(X, \tilde{\mathscr{L}}, \tilde{\ell})$ of $(X, \mathscr{L}, \ell)$, the functionals ℓ' and $\tilde{\ell}$ agree on their common domain.

We restrict our attention, then, to extensions of $(X, \mathscr{L}, \ell)$ that are determined by $(X, \mathscr{L}, \ell)$. Arbitrariness in assigning functional values has been ruled out, and we want the class of integrable functionals to be as large as possible. Happily, it turns out that for every Daniell space $(X, \mathscr{L}, \ell)$ the class of determined extensions forms a complete lattice: the smallest element is the closure of the given Daniell space; the largest element rightfully deserves to be called the integral for the given Daniell space.

Before proceeding with the theory, we illustrate the notions of admissible extension and determined extension with some very simple examples. Let $X = \{a, b, c\}$, and let

$$\mathscr{L} = \{f \in \overline{\mathbb{R}}^X \mid f(a) = 0, f(b) = f(c) \in \mathbb{R}\}; \quad \ell: \mathscr{L} \to \mathbb{R}, \qquad f \mapsto f(b).$$

The triple $(X, \mathscr{L}, \ell)$ is itself a closed Daniell space, but it has many extensions, most of which are also closed. We look at a few such extensions. Let

$$\mathscr{L}' = \{f \in \mathbb{R}^X \mid f(b) = f(c)\}; \quad \ell': \mathscr{L}' \to \mathbb{R}, \qquad f \mapsto f(b)$$

$$\mathscr{L}'' = \{f \in \mathbb{R}^X \mid f(b) = f(c)\}; \quad \ell'': \mathscr{L}'' \to \mathbb{R}, \qquad f \mapsto f(a) + f(b)$$

$$\mathscr{L}_b = \{f \in \overline{\mathbb{R}}^X \mid f(b) \in \mathbb{R}\}; \quad \ell_b: \mathscr{L}_b \to \mathbb{R}, \qquad f \mapsto f(b)$$

$$\mathscr{L}_c = \{f \in \overline{\mathbb{R}}^X \mid f(c) \in \mathbb{R}\}; \quad \ell_c: \mathscr{L}_c \to \mathbb{R}, \qquad f \mapsto f(c)$$

$$\tilde{\mathscr{L}} = \{f \in \overline{\mathbb{R}}^X \mid f(b) = f(c) \in \mathbb{R}\}; \quad \tilde{\ell}: \tilde{\mathscr{L}} \to \mathbb{R}, \qquad f \mapsto f(b).$$

Each of the triples $(X, \mathscr{L}', \ell')$, $(X, \mathscr{L}'', \ell'')$, and so forth, is a Daniell space extending $(X, \mathscr{L}, \ell)$. The extension $(X, \mathscr{L}', \ell')$ is not closed. The extension

$(X, \mathscr{L}'', \ell'')$ is closed but not admissible. How senseless and arbitrary nonadmissible extensions can be is clear from this example. We have $\ell''(e_{\{a\}}) = 1$. Moreover, it would have been possible, for each positive real number γ, to define ℓ'' in such a way that $\ell''(e_{\{a\}}) = \gamma$. The extension chosen to be the integral should certainly be admissible. Both $(X, \mathscr{L}_b, \ell_b)$ and $(X, \mathscr{L}_c, \ell_c)$ are admissible but not determined. These examples illustrate that admissibility alone does not rule out all arbitrariness, hence the additional restriction to determined extensions. The only two extensions of $(X, \mathscr{L}, \ell)$ that are both admissible and determined are $(X, \tilde{\mathscr{L}}, \tilde{\ell})$ and $(X, \mathscr{L}, \ell)$ itself. Thus $(X, \mathscr{L}, \ell)$ is the smallest admissible determined extension, and $(X, \tilde{\mathscr{L}}, \tilde{\ell})$ the largest, of the Daniell space $(X, \mathscr{L}, \ell)$. According to the definition eventually given in Section 4.2, $(X, \tilde{\mathscr{L}}, \tilde{\ell})$ is the integral for $(X, \mathscr{L}, \ell)$.

4.1. ADMISSIBLE EXTENSIONS OF A DANIELL SPACE

NOTATION FOR SECTION 4.1:

X denotes a set.

This section takes up the task of describing the various admissible extensions of a given Daniell space. In brief, these extensions are obtained by adjoining new exceptional functions to the original Riesz lattice, extending the Daniell-space functional in an obvious way, and then taking Daniell-space closure.

The section begins with some definitions and preliminary observations. With preliminaries out of the way, we prove what might be called the main theorem on adjoining exceptional functions: under appropriate hypotheses there is a unique positive linear functional extending the original functional to the bigger Riesz lattice obtained by adjoining new exceptional functions, and the resulting triple is a Daniell space extending the original Daniell space. Moreover, the new Daniell space is closed if the original was.

A corollary to the Main Theorem 10 is the fact that (again, under appropriate hypotheses) the operation of adjoining new exceptional functions and the formation of Daniell-space closure commute. Theorem 12, also corollary to Theorem 10 but very important in its own right, characterizes admissible extensions as exactly those obtained by adjoining exceptional functions and taking Daniell-space closure.

The section ends with two important properties of admissible extensions. One is the induction principle for admissible extensions. The other is a formula, analogous to the formula established in Theorem 3.4.8 for the sum of a summable family, for computing the values of the extension functional.

To begin, we introduce some notation that will simplify the statements of many later results.

Definition 4.1.1. *Let* $\mathscr{F} \subset \overline{\mathbb{R}}^X$. *Then*

$$\hat{\mathfrak{R}}(\mathscr{F}) := \{\{f \neq 0\} \mid f \in \mathscr{F}\}. \qquad \square$$

Proposition 4.1.2. *The following assertions hold, for every Riesz lattice* $\mathscr{L}$ *in* $\overline{\mathbb{R}}^X$.

(*a*) *If* f *and* g *belong to* $\mathscr{L}$, *then each of the sets*

$$\{f < g\}, \{f > g\}, \{f \neq g\}$$

belongs to $\hat{\mathfrak{R}}(\mathscr{L})$.

(*b*) $\mathfrak{N}(\mathscr{L}) \subset \hat{\mathfrak{R}}(\mathscr{L})$.

(*c*) $\hat{\mathfrak{R}}(\mathscr{L}) = \hat{\mathfrak{R}}(\mathscr{L} \cap \mathbb{R}^X)$.

Proof. (a) Given f and g in $\mathscr{L}$, define

$$h \colon X \to \overline{\mathbb{R}}, \qquad x \mapsto \begin{cases} f(x) - g(x) & \text{if defined} \\ 0 & \text{otherwise.} \end{cases}$$

Then

$$\{f < g\} = \{h \wedge 0 \neq 0\}$$

$$\{f > g\} = \{h \vee 0 \neq 0\}$$

$$\{f \neq g\} = \{h \neq 0\}$$

and the functions h, $h \wedge 0$, $h \vee 0$, all belong to $\mathscr{L}$.

(b) $\mathfrak{N}(\mathscr{L}) = \{A \subset X \mid \infty e_A \in \mathscr{L}\} \subset \mathfrak{R}(\mathscr{L}) \subset \hat{\mathfrak{R}}(\mathscr{L})$.

(c) Given f in $\mathscr{L}$, define

$$g \colon X \to \mathbb{R}, \qquad x \mapsto \begin{cases} f(x) & \text{if } |f(x)| \neq \infty \\ 1 & \text{otherwise.} \end{cases}$$

Then

$$\{f \neq 0\} = \{g \neq 0\}$$

and g belongs to $\mathscr{L}$, since g and f are $\mathscr{L}$-a.e. equal. The required assertion follows. $\square$

As was indicated in the introductory paragraphs of this chapter, we shall call an extension $(X, \tilde{\mathscr{L}}, \tilde{\ell})$ of a Daniell space $(X, \mathscr{L}, \ell)$ admissible iff $(X, \tilde{\mathscr{L}}, \tilde{\ell})$ is a closed Daniell space and the natural map from the set of equivalence classes of $\overline{\mathscr{L}}(\ell)$ modulo $\overline{\mathscr{L}}(\ell)$-a.e. equality to the set of equivalence classes of $\tilde{\mathscr{L}}$ modulo $\tilde{\mathscr{L}}$-a.e. equality is bijective. Actually this map is injective for every closed Daniell space $(X, \tilde{\mathscr{L}}, \tilde{\ell})$ extending $(X, \mathscr{L}, \ell)$, as Proposition 4 (c) asserts. To require that it also be surjective is to require that every function in $\tilde{\mathscr{L}}$ be $\tilde{\mathscr{L}}$-a.e. equal to some function in $\bar{\mathscr{L}}(\ell)$. We take this latter, easier to use formulation as the definition.

Definition 4.1.3. *Let $(X, \mathscr{L}, \ell)$ be a Daniell space. An* ***admissible extension*** *of $(X, \mathscr{L}, \ell)$ is a closed Daniell space $(X, \tilde{\mathscr{L}}, \tilde{\ell})$ extending $(X, \mathscr{L}, \ell)$ such that for every f belonging to $\tilde{\mathscr{L}}$ there is a function g belonging to $\bar{\mathscr{L}}(\ell)$ with $f = g$ $\tilde{\mathscr{L}}$-a.e.* □

Every Daniell space evidently has its closure as one admissible extension.

Proposition 4.1.4. *Let $(X, \mathscr{L}, \ell)$ be a Daniell space. Then the following assertions hold, for every closed Daniell space $(X, \tilde{\mathscr{L}}, \tilde{\ell})$ that extends $(X, \mathscr{L}, \ell)$.*

(*a*) $\mathscr{N}(\tilde{\mathscr{L}}) \cap \bar{\mathscr{L}}(\ell) = \mathscr{N}(\bar{\mathscr{L}}(\ell))$.

(*b*) $\mathfrak{N}(\tilde{\mathscr{L}}) \cap \hat{\mathfrak{R}}(\bar{\mathscr{L}}(\ell)) = \mathfrak{N}(\bar{\mathscr{L}}(\ell))$.

(*c*) *The conditions $f, g \in \bar{\mathscr{L}}(\ell)$, $f = g$ $\tilde{\mathscr{L}}$-a.e. always imply that $f = g$ $\bar{\mathscr{L}}(\ell)$-a.e.*

(*d*) *$(X, \tilde{\mathscr{L}}, \tilde{\ell})$ is an admissible extension of $(X, \mathscr{L}, \ell)$ iff the natural map from the set of equivalence classes of $\bar{\mathscr{L}}(\ell)$ modulo $\bar{\mathscr{L}}(\ell)$-a.e. equality to the set of equivalence classes of $\tilde{\mathscr{L}}$ modulo $\tilde{\mathscr{L}}$-a.e. equality is bijective.*

Proof. (a) Since $(X, \tilde{\mathscr{L}}, \tilde{\ell})$ automatically extends $(X, \bar{\mathscr{L}}(\ell), \bar{\ell})$, by Theorem 3.2.5, and both $(X, \bar{\mathscr{L}}(\ell), \bar{\ell})$ and $(X, \tilde{\mathscr{L}}, \tilde{\ell})$ are closed Daniell spaces, we have

$$\begin{aligned}\mathscr{N}(\tilde{\mathscr{L}}) \cap \bar{\mathscr{L}}(\ell) &= \{f \in \bar{\mathscr{L}}(\ell) \mid \tilde{\ell}(|f|) = 0\} \\ &= \{f \in \bar{\mathscr{L}}(\ell) \mid \bar{\ell}(|f|) = 0\} \\ &= \mathscr{N}(\bar{\mathscr{L}}(\ell))\end{aligned}$$

(Proposition 1.4.4).

(b) That $\mathfrak{N}(\bar{\mathscr{L}}(\ell)) \subset \mathfrak{N}(\tilde{\mathscr{L}}) \cap \hat{\mathfrak{R}}(\bar{\mathscr{L}}(\ell))$ follows immediately from (a). To verify the opposite inclusion, let $A \in \mathfrak{N}(\tilde{\mathscr{L}}) \cap \hat{\mathfrak{R}}(\bar{\mathscr{L}}(\ell))$. Then e_A belongs to $\mathscr{N}(\tilde{\mathscr{L}})$, and there exists f belonging to $\bar{\mathscr{L}}(\ell)$ such that $A = \{f \neq 0\}$ and hence $f = fe_A$. By Proposition 1.2.30 (b) f belongs to $\mathscr{N}(\tilde{\mathscr{L}})$. By (a), f belongs to $\mathscr{N}(\bar{\mathscr{L}}(\ell))$. Thus A belongs to $\mathfrak{N}(\bar{\mathscr{L}}(\ell))$.

(c) By hypothesis,

$$\{f \neq g\} \in \mathfrak{N}(\tilde{\mathscr{L}}) \cap \hat{\mathfrak{R}}(\bar{\mathscr{L}}(\ell))$$

[Proposition 2(a)]. Thus (c) follows from (b).

(d) Assertion (d) is merely a restatement of (c) together with the definition of admissibility. □

As the reader can readily verify, the defining condition for admissible extension can be weakened as follows.

Proposition 4.1.5. *Let* $(X, \mathscr{L}, \ell)$ *be a Daniell space and* $(X, \tilde{\mathscr{L}}, \tilde{\ell})$ *a closed Daniell space with* $(X, \mathscr{L}, \ell) \preccurlyeq (X, \tilde{\mathscr{L}}, \tilde{\ell})$. *Then* $(X, \tilde{\mathscr{L}}, \tilde{\ell})$ *is an admissible extension of* $(X, \mathscr{L}, \ell)$ *iff for every positive function f in* $\tilde{\mathscr{L}}$ *there exists a positive function g in* $\bar{\mathscr{L}}(\ell)$ *such that* $f = g$ $\tilde{\mathscr{L}}$*-a.e.* □

Intuitively, we imagine an admissible extension $(X, \tilde{\mathscr{L}}, \tilde{\ell})$ of a Daniell space $(X, \mathscr{L}, \ell)$ being related to $(X, \bar{\mathscr{L}}(\ell), \bar{\ell})$ as follows. Expand the class of $\bar{\mathscr{L}}(\ell)$-exceptional functions to include all of $\mathscr{N}(\tilde{\mathscr{L}})$. Expand $\bar{\mathscr{L}}(\ell)$ accordingly. Extend the positive linear functional $\bar{\ell}$ in the natural way. The resulting triple should be $(X, \tilde{\mathscr{L}}, \tilde{\ell})$.

This description can be made precise, as we soon show. The process of extending a Daniell space $(X, \mathscr{L}, \ell)$ by adjoining new exceptional functions to the Riesz lattice $\mathscr{L}$ and then taking the natural extension of the positive linear functional ℓ can be carried out more generally. In fact we use the more general construction in obtaining a full description of the integral for the Daniell space $(X, \mathscr{L}, \ell)$. We therefore develop a theory for the general situation. We begin with the following definition.

Definition 4.1.6. *Let* $\mathscr{L}$ *and* $\mathscr{F}$ *be Riesz lattices in* $\overline{\mathbb{R}}^X$. *Then*

$$\mathscr{L} \vdash \mathscr{F} := \{ f \in \overline{\mathbb{R}}^X \mid \exists g \in \mathscr{L} \text{ with } f = g \ \mathscr{F}\text{-a.e.} \}. \qquad \square$$

In the particular situation described previously, where we adjoin to $\bar{\mathscr{L}}(\ell)$ the $\tilde{\mathscr{L}}$-exceptional functions, we know that

$$\mathscr{N}(\tilde{\mathscr{L}}) \cap \bar{\mathscr{L}}(\ell) = \mathscr{N}(\bar{\mathscr{L}}(\ell)).$$

The construction also uses three additional properties possessed by $\mathscr{N}(\tilde{\mathscr{L}})$. For one, $\mathscr{N}(\tilde{\mathscr{L}})$ is a Riesz lattice in $\overline{\mathbb{R}}^X$. Second, $\mathscr{N}(\tilde{\mathscr{L}})$ is σ-completely embedded in $\overline{\mathbb{R}}^X$. Finally, any extended-real function g on X such that

$$\{g \neq 0\} \subset \{f \neq 0\}$$

for some $f \in \mathscr{N}(\tilde{\mathscr{L}})$ must itself belong to $\mathscr{N}(\tilde{\mathscr{L}})$. Since it is useful to have a name for this last property we make another definition.

Definition 4.1.7. *A Riesz lattice* $\mathscr{L}$ *in* $\overline{\mathbb{R}}^X$ *is called a* ***tower*** *(in* $\overline{\mathbb{R}}^X$ *or on X) iff the conditions*

$$f \in \mathscr{L}, \qquad g \in \overline{\mathbb{R}}^X, \qquad \{g \neq 0\} \subset \{f \neq 0\}$$

always imply that g belongs to $\mathscr{L}$. □

Proposition 4.1.8. *For every Riesz lattice $\mathscr{L}$ in $\overline{\mathbb{R}}^X$, assertions (a) through (c) are equivalent and imply (d). For every Riesz lattice $\mathscr{L}$ that is σ-completely embedded in $\overline{\mathbb{R}}^X$, all four of the following assertions are equivalent.*

(*a*) *$\mathscr{L}$ is a tower.*

(*b*) *$\mathscr{N}(\mathscr{L}) = \mathscr{L}$.*

(*c*) *$\mathfrak{N}(\mathscr{L}) = \hat{\mathfrak{R}}(\mathscr{L})$.*

(*d*) *The conditions*

$$f \in \mathscr{L}, \qquad g \in \overline{\mathbb{R}}^X, \qquad |g| \le |f|$$

always imply that g belongs to $\mathscr{L}$.

Proof. (a) ⇒ (b). Always $\mathscr{N}(\mathscr{L}) \subset \mathscr{L}$. If f belongs to $\mathscr{L}$, then (a) implies that ∞f also belongs to $\mathscr{L}$, which implies that f belongs to $\mathscr{N}(\mathscr{L})$.

(b) ⇒ (c). This implication is immediate from the definitions.

(c) ⇒ (a). This implication follows from the fact that subsets of $\mathscr{L}$-exceptional sets are $\mathscr{L}$-exceptional.

(a) ⇒ (d). The inequality $|g| \le |f|$ implies $\{g \ne 0\} \subset \{f \ne 0\}$.

(d) ⇒ (a) when $\mathscr{L}$ is σ-completely embedded. Suppose, first, that $f \in \mathscr{L}$, $g \in \overline{\mathbb{R}}^X_+$, and $\{g \ne 0\} \subset \{f \ne 0\}$. Then $|f|$ belongs to $\mathscr{L}$, and we conclude that $g \wedge n|f|$ belongs to $\mathscr{L}$ for every n in $\mathbb{N}$. Since

$$g = \bigvee_{n \in \mathbb{N}} (g \wedge n|f|)$$

and $\mathscr{L}$ is σ-completely embedded, it follows that g belongs to $\mathscr{L}$. For an arbitrary g, write $g = g^+ - g^-$, and note that the sets $\{g^+ \ne 0\}$, $\{g^- \ne 0\}$ are both subsets of $\{g \ne 0\}$. □

Proposition 4.1.9. *Let $(X, \mathscr{L}, \ell)$ be a Daniell space, and let $\mathscr{F}$ be a tower in $\overline{\mathbb{R}}^X$ that satisfies the compatibility condition*

$$\mathscr{F} \cap \bar{\mathscr{L}}(\ell) = \mathscr{N}(\bar{\mathscr{L}}(\ell)).$$

Then the following assertions hold.

(*a*) *$\mathfrak{N}(\mathscr{L}) \subset \mathfrak{N}(\mathscr{F}) \cap \hat{\mathfrak{R}}(\mathscr{L}) \subset \mathfrak{N}(\mathscr{F})$.*

(*b*) *$\mathscr{L} \cup \mathscr{F} \subset \mathscr{L} \vdash \mathscr{F}$.*

(*c*) *$\mathscr{L} \vdash \mathscr{F} = \{f \in \overline{\mathbb{R}}^X \mid \exists g \in \mathscr{L} \cap \mathbb{R}^X$ with $f = g$ $\mathscr{F}$-a.e.$\}$.*

(*d*) *If f_1 and f_2 belong to $\mathscr{L} \vdash \mathscr{F}$ and $h \in \langle f_1 \dot{+} f_2 \rangle$, then $h(x) = f_1(x) + f_2(x)$ $\mathscr{F}$-a.e.*

(*e*) *$\mathscr{L} \vdash \mathscr{F}$ is a Riesz lattice.*

(*f*) *$\mathfrak{N}(\mathscr{L} \vdash \mathscr{F}) = \mathfrak{N}(\mathscr{F}) = \hat{\mathfrak{R}}(\mathscr{F})$ and $\mathscr{N}(\mathscr{L} \vdash \mathscr{F}) = \mathscr{F}$.*

(*g*) *For every decreasing sequence $(f_n)_{n \in \mathbb{N}}$ from $\mathscr{L} \vdash \mathscr{F}$, there is a decreasing sequence $(g_n)_{n \in \mathbb{N}}$ from $\mathscr{L}$ such that $f_n = g_n$ $\mathscr{F}$-a.e. for every n in $\mathbb{N}$. The analogous assertion holds for increasing sequences from $\mathscr{L} \vdash \mathscr{F}$.*

Proof. (a) The only inclusion that requires checking is $\mathfrak{N}(\mathscr{L}) \subset \mathfrak{N}(\mathscr{F})$. Let $A \in \mathfrak{N}(\mathscr{L})$. We can argue in succession that

$$e_A \in \mathscr{N}(\mathscr{L})$$

$$\ell(e_A) = \bar{\ell}(e_A) = 0$$

$$e_A \in \mathscr{N}(\bar{\mathscr{L}}(\ell)) \subset \mathscr{F} = \mathscr{N}(\mathscr{F}).$$

Hence A belongs to $\mathfrak{N}(\mathscr{F})$.

(b) That $\mathscr{L} \subset \mathscr{L} \vdash \mathscr{F}$ is obvious. Since $\mathscr{N}(\mathscr{F}) = \mathscr{F}$, every function in $\mathscr{F}$ is $\mathscr{F}$-a.e. equal to a function in $\mathscr{L}$, namely, the zero function. Thus $\mathscr{F} \subset \mathscr{L} \vdash \mathscr{F}$.

(c) In one direction the inclusion is trivial. For the other inclusion we use the facts that membership in $\mathscr{L}$ is not affected when function values on an $\mathscr{L}$-exceptional set are altered and every function in $\mathscr{L}$ is $\mathscr{L}$-a.e. real, hence by (a) $\mathscr{F}$-a.e. real. Indeed, given $f \in \mathscr{L} \vdash \mathscr{F}$, let g' be a function in $\mathscr{L}$ for which $f = g'$ $\mathscr{F}$-a.e., and set

$$g := g' e_{\{|g'| \neq \infty\}}.$$

Then g belongs to $\mathscr{L} \cap \mathbb{R}^X$ and $f = g$ $\mathscr{F}$-a.e.

(d) Given f_1 and f_2 in $\mathscr{L} \vdash \mathscr{F}$, there exist, according to (c), functions g_1, g_2 in $\mathscr{L} \cap \mathbb{R}^X$ with $f_i = g_i$ $\mathscr{F}$-a.e. $(i = 1, 2)$. The function $g_1 + g_2$ is defined, so $f_1(x) + f_2(x)$ is defined $\mathscr{F}$-a.e. The assertion follows.

(e) Assertion (e) follows from (d) and familiar properties of $\mathscr{F}$-a.e. equality, in particular Proposition 1.2.33.

(f) From (b) we conclude that $\mathfrak{N}(\mathscr{F}) \subset \mathfrak{N}(\mathscr{L} \vdash \mathscr{F})$. Suppose A belongs to $\mathfrak{N}(\mathscr{L} \vdash \mathscr{F})$. Then ∞e_A belongs to $\mathscr{L} \vdash \mathscr{F}$. In view of (c) there is a real-valued function g with

$$\infty e_A = g \quad \mathscr{F}\text{-a.e.}$$

and the inclusion

$$A \subset \{\infty e_A \neq g\}$$

shows that A is $\mathscr{F}$-exceptional. Thus $\mathfrak{N}(\mathscr{L} \vdash \mathscr{F}) = \mathfrak{N}(\mathscr{F})$. Since $\mathscr{N}(\mathscr{F}) = \mathscr{F}$ and $\mathfrak{N}(\mathscr{F}) = \hat{\mathfrak{R}}(\mathscr{F})$, the remaining assertions follow.

(g) It is not difficult to prove (g). We use Proposition 1.2.33 (e). Let $(f_n)_{n \in \mathbb{N}}$ be a decreasing sequence from $\mathscr{L} \vdash \mathscr{F}$. For each n in $\mathbb{N}$ choose $g'_n \in \mathscr{L}$ such that $f_n = g'_n$ $\mathscr{F}$-a.e., and set

$$g_n := \bigwedge_{m \leq n} g'_m.$$

Then $(g_n)_{n \in \mathbb{N}}$ is a decreasing sequence from $\mathscr{L}$. Since $\bigwedge_{m \leq n} f_m = f_n$, Proposition 1.2.33 (e) ensures that $g_n = f_n\ \mathscr{F}$-a.e. For $(f_n)_{n\in\mathbb{N}}$ increasing, modify the proof just given and use 1.2.33 (d). □

Theorem 4.1.10 **(Main Theorem on adjoining exceptional functions).** *Let $(X, \mathscr{L}, \ell)$ be a Daniell space, and let $\mathscr{F}$ be a tower that is σ-completely embedded in $\overline{\mathbb{R}}^X$. Suppose also that $\mathscr{F}$ satisfies the compatibility condition:*

$$\mathscr{F} \cap \bar{\mathscr{L}}(\ell) = \mathscr{N}(\bar{\mathscr{L}}(\ell)). \tag{1}$$

Then the following assertions hold. [In (b)–(d), $\ell_{\mathscr{F}}$ denotes the functional characterized in (a).]

(a) *There is exactly one positive linear functional $\ell_{\mathscr{F}}$ that is defined on $\mathscr{L} \vdash \mathscr{F}$ and agrees on $\mathscr{L}$ with ℓ.*

(b) *$(X, \mathscr{L} \vdash \mathscr{F}, \ell_{\mathscr{F}})$ is a Daniell space that extends $(X, \mathscr{L}, \ell)$.*

(c) *$\mathfrak{N}(\bar{\mathscr{L}}(\ell_{\mathscr{F}})) = \mathfrak{N}(\mathscr{F}) = \hat{\mathfrak{R}}(\mathscr{F})$, and $\mathscr{N}(\bar{\mathscr{L}}(\ell_{\mathscr{F}})) = \mathscr{F}$.*

(d) *If $(X, \mathscr{L}, \ell)$ is closed, then so is $(X, \mathscr{L} \vdash \mathscr{F}, \ell_{\mathscr{F}})$.*

Proof. We remind the reader that $\mathscr{N}(\mathscr{F}) = \mathscr{F}$ (Proposition 8).

(a) Given f in $\mathscr{F}$, the natural definition for $\ell_{\mathscr{F}}(f)$ is achieved by taking g in $\mathscr{L}$ with $f = g\ \mathscr{F}$-a.e. and setting $\ell_{\mathscr{F}}(f) = \ell(g)$. It is more farsighted to take the function g to be real valued [Proposition 9 (c)]. The compatibility condition (1) will guarantee that $\ell_{\mathscr{F}}(f)$ is well defined. Suppose that g_1 and g_2 are real-valued functions in $\mathscr{L}$ that are both $\mathscr{F}$-a.e. equal to f. Then $g_1 = g_2\ \mathscr{F}$-a.e., so $g_1 - g_2$ (certainly defined) belongs to $\mathscr{N}(\mathscr{F})$ and therefore to $\mathscr{F}$. Since $g_1 - g_2$ also belongs to $\mathscr{L}$, (1) implies that $g_1 - g_2$ is $\bar{\mathscr{L}}(\ell)$-exceptional. Hence

$$0 = \bar{\ell}(g_1 - g_2) = \ell(g_1 - g_2) = \ell(g_1) - \ell(g_2).$$

Thus

$$\ell_{\mathscr{F}}\colon \mathscr{L} \vdash \mathscr{F} \to \mathbb{R}, \qquad f \mapsto \ell(g) \text{ where } g \in \mathscr{L} \cap \mathbb{R}^X \text{ and } f = g\ \mathscr{F}\text{-a.e.}$$

is a well-defined functional on $\mathscr{L} \vdash \mathscr{F}$.

It will be a straightforward task to verify that the functional is positive and linear. The required computations rely on the properties of $\mathscr{F}$-a.e. equality and of course on the fact that ℓ is positive and linear. Concerning $\mathscr{F}$-a.e. equality, we refer to Proposition 1.2.33 and to Proposition 9 (d).

$\ell_{\mathscr{F}}$ is additive. Suppose that f_1, f_2 belong to $\mathscr{L} \vdash \mathscr{F}$ and that $h \in \langle f_1 \dot{+} f_2 \rangle$. Choose functions g_1, g_2 in $\mathscr{L} \cap \mathbb{R}^X$ such that $f_i = g_i\ \mathscr{F}$-a.e. $(i = 1, 2)$. Then $g_1 + g_2$ is defined and belongs to $\mathscr{L} \cap \mathbb{R}^X$, and $h = g_1 + g_2\ \mathscr{F}$-a.e. Thus

$$\begin{aligned} \ell_{\mathscr{F}}(h) &= \ell(g_1 + g_2) \\ &= \ell(g_1) + \ell(g_2) \\ &= \ell_{\mathscr{F}}(f_1) + \ell_{\mathscr{F}}(f_2). \end{aligned}$$

$\ell_{\mathscr{F}}$ *is homogeneous*. Given f in $\mathscr{L} \vdash \mathscr{F}$ and given a real number α, choose g in $\mathscr{L} \cap \mathbb{R}^X$ with $f = g$ $\mathscr{F}$-a.e. Then $\alpha f = \alpha g$ $\mathscr{F}$-a.e. and

$$\ell_{\mathscr{F}}(\alpha f) = \ell(\alpha g) = \alpha \ell(g) = \alpha \ell_{\mathscr{F}}(f).$$

$\ell_{\mathscr{F}}$ *is positive*. Given f in $(\mathscr{L} \vdash \mathscr{F})_+$, choose g in $\mathscr{L} \cap \mathbb{R}^X$ with $f = g$ $\mathscr{F}$-a.e. The function g^+ belongs to $\mathscr{L} \cap \mathbb{R}^X$ and $f = g^+$ $\mathscr{F}$-a.e. Hence

$$\ell_{\mathscr{F}}(f) = \ell(g^+) \geq 0.$$

Thus $\ell_{\mathscr{F}}$ is a positive linear functional on $\mathscr{L} \vdash \mathscr{F}$. To show that $\ell_{\mathscr{F}}$ agrees with ℓ on $\mathscr{L}$, let $f \in \mathscr{L}$. Set

$$A := \{ |f| = \infty \} \qquad g := f e_{X \setminus A}.$$

Then g belongs to $\mathscr{L} \cap \mathbb{R}^X$ [Proposition 1.2.30 (c)] and $f = g$ $\mathscr{L}$-a.e., hence $\mathscr{F}$-a.e. [Proposition 9 (a)]. Therefore

$$\ell_{\mathscr{F}}(f) = \ell(g) = \ell(f)$$

as required.

Finally, let ℓ' be an arbitrary positive linear extension of ℓ to $\mathscr{L} \vdash \mathscr{F}$. Given f in $\mathscr{L} \vdash \mathscr{F}$, choose g in $\mathscr{L} \cap \mathbb{R}^X$ with $g = f$ $\mathscr{F}$-a.e.. In view of Proposition 9 (f), g and f are also $(\mathscr{L} \vdash \mathscr{F})$-a.e. equal, and so

$$\ell'(f) = \ell'(g) = \ell(g) = \ell_{\mathscr{F}}(f)$$

[Proposition 1.3.3 (d)]. Thus $\ell' = \ell_{\mathscr{F}}$ and $\ell_{\mathscr{F}}$ is uniquely determined.

(b) We already know that $\mathscr{L} \vdash \mathscr{F}$ is a Riesz lattice containing $\mathscr{L}$ [Proposition 9 (b), (e)] and $\ell_{\mathscr{F}}$ a positive linear extension of ℓ to $\mathscr{L} \vdash \mathscr{F}$. It remains to show that $\ell_{\mathscr{F}}$ is nullcontinuous. Besides the hypotheses on $\mathscr{F}$, the key tool is the null continuity of $\bar{\ell}$. Let $(f_n)_{n \in \mathbb{N}}$ be a decreasing sequence from $\mathscr{L} \vdash \mathscr{F}$ whose infimum is the zero function. We must show that

$$\inf_{n \in \mathbb{N}} \ell_{\mathscr{F}}(f_n) = 0. \tag{2}$$

By Proposition 9 (g), there exists a decreasing sequence $(g_n)_{n \in \mathbb{N}}$ from $\mathscr{L}$ with $f_n = g_n$ $\mathscr{F}$-a.e. for every n. In view of (1),

$$g_n \geq 0 \ \bar{\mathscr{L}}(\ell)\text{-a.e.}$$

Thus $(g_n)_{n \in \mathbb{N}}$ is an $\bar{\ell}$-sequence from $\bar{\mathscr{L}}(\ell)$. It follows that

$$g := \bigwedge_{n \in \mathbb{N}} g_n$$

belongs to $\bar{\mathscr{L}}(\ell)$. Since

$$\{g \neq 0\} \subset \bigcup_{n \in \mathbb{N}} \{f_n \neq g_n\}$$

and $\mathfrak{N}(\mathscr{F})$ is a σ-ring (Proposition 1.4.9), g must also belong to $\mathscr{N}(\mathscr{F})$ and therefore to $\mathscr{F}$. Using (1), we conclude that $g = 0\ \bar{\mathscr{L}}(\ell)$-a.e. and

$$\begin{aligned} 0 = \bar{\ell}(g) &= \inf_{n \in \mathbb{N}} \bar{\ell}(g_n) \\ &= \inf_{n \in \mathbb{N}} \ell(g_n) \\ &= \inf_{n \in \mathbb{N}} \ell_{\mathscr{F}}(f_n) \end{aligned}$$

establishing (2).

(c) We need only prove that $\mathfrak{N}(\bar{\mathscr{L}}(\ell_{\mathscr{F}})) = \mathfrak{N}(\mathscr{F})$. Both claims then follow, since $\hat{\mathfrak{R}}(\mathscr{F}) = \mathfrak{N}(\mathscr{F})$ and $\mathscr{N}(\mathscr{F}) = \mathscr{F}$ (Proposition 8). The inclusion $\mathfrak{N}(\mathscr{F}) \subset \mathfrak{N}(\bar{\mathscr{L}}(\ell_{\mathscr{F}}))$ follows from the inclusions $\mathscr{F} \subset \mathscr{L} \vdash \mathscr{F} \subset \bar{\mathscr{L}}(\ell_{\mathscr{F}})$. So let $A \in \mathfrak{N}(\bar{\mathscr{L}}(\ell_{\mathscr{F}}))$. The function ∞e_A belongs to $\bar{\mathscr{L}}(\ell_{\mathscr{F}})$, and Proposition 3.1.16 (a) $\Rightarrow$ (e) guarantees the existence of an increasing sequence $(f_n)_{n \in \mathbb{N}}$ from $\mathscr{L} \vdash \mathscr{F}$ such that

$$\bigvee_{n \in \mathbb{N}} f_n \geq \infty e_A \qquad \sup_{n \in \mathbb{N}} \ell_{\mathscr{F}}(f_n) < \infty .$$

By Proposition 9 (g) we can choose an increasing sequence $(g_n)_{n \in \mathbb{N}}$ from $\mathscr{L}$ with

$$g_n = f_n\ \mathscr{F}\text{-a.e.}$$

for every n. Moreover

$$\begin{aligned} \sup_{n \in \mathbb{N}} \bar{\ell}(g_n) &= \sup_{n \in \mathbb{N}} \ell(g_n) \\ &= \sup_{n \in \mathbb{N}} \ell_{\mathscr{F}}(f_n) \\ &< \infty . \end{aligned}$$

Thus $(g_n)_{n \in \mathbb{N}}$ is an $\bar{\ell}$-sequence from $\bar{\mathscr{L}}(\ell)$, and its supremum, which we call g, must belong to $\bar{\mathscr{L}}(\ell)$. Using (1) again, we conclude that the set $\{g = \infty\}$ belongs to $\mathfrak{N}(\mathscr{F})$. Now

$$A \subset \{g = \infty\} \cup \left(\bigcup_{n \in \mathbb{N}} \{f_n \neq g_n\} \right)$$

and $\mathfrak{N}(\mathscr{F})$ is a σ-ring. We conclude that A belongs to $\mathfrak{N}(\mathscr{F})$.

(d) Assume that $(X, \mathscr{L}, \ell)$ is a closed Daniell space. To show that the triple $(X, \mathscr{L} \vdash \mathscr{F}, \ell_{\mathscr{F}})$ is closed, we use the characterization from Proposition 1.4.3 (d) $\Rightarrow$ (a). Let $(f_n)_{n \in \mathbb{N}}$ be a decreasing $\ell_{\mathscr{F}}$-sequence from $\mathscr{L} \vdash \mathscr{F}$, and set $f := \bigwedge_{n \in \mathbb{N}} f_n$. Choose a decreasing sequence $(g_n)_{n \in \mathbb{N}}$ from $\mathscr{L}$ with $f_n = g_n$ $\mathscr{F}$-a.e., and set $g := \bigwedge_{n \in \mathbb{N}} g_n$. Note that

$$\{f \neq g\} \subset \bigcup_{n \in \mathbb{N}} \{f_n \neq g_n\}$$

and $\mathfrak{N}(\mathscr{F})$ is a σ-ring, so $f = g$ $\mathscr{F}$-a.e. Moreover $\ell(g_n) = \ell_{\mathscr{F}}(f_n)$ for every n, so $(g_n)_{n \in \mathbb{N}}$ is a decreasing ℓ-sequence from $\mathscr{L}$. Since the triple $(X, \mathscr{L}, \ell)$ is closed, we conclude that g belongs to $\mathscr{L}$, so f belongs to $\mathscr{L} \vdash \mathscr{F}$, and that

$$\begin{aligned} \ell_{\mathscr{F}}(f) &= \ell(g) \\ &= \inf_{n \in \mathbb{N}} \ell(g_n) \\ &= \inf_{n \in \mathbb{N}} \ell_{\mathscr{F}}(f_n). \end{aligned}$$ □

Prior to Definition 6 we described how to view the relation between a Daniell space $(X, \mathscr{L}, \ell)$ and an admissible extension $(X, \tilde{\mathscr{L}}, \tilde{\ell})$ of it: $\tilde{\mathscr{L}}$ is the Riesz lattice $\bar{\mathscr{L}}(\ell) \vdash \mathscr{N}(\tilde{\mathscr{L}})$, and $\tilde{\ell}$ is the natural extension of $\bar{\ell}$. One consequence of Theorem 10 is that $(X, \tilde{\mathscr{L}}, \tilde{\ell})$ could also be achieved by first forming $\mathscr{L} \vdash \mathscr{N}(\tilde{\mathscr{L}})$, extending ℓ in the natural way, and then forming the closure of the resulting Daniell space; that is, the processes of adjoining new exceptional functions and taking closure commute. More precisely we have the following corollary to Theorem 10.

Corollary 4.1.11. *Let $(X, \mathscr{L}, \ell)$ be a Daniell space, and let $\mathscr{F}$ be a tower that is σ-completely embedded in $\overline{\mathbb{R}}^X$ and satisfies*

$$\mathscr{F} \cap \bar{\mathscr{L}}(\ell) = \mathscr{N}(\bar{\mathscr{L}}(\ell)). \tag{1}$$

Denote by $\ell_{\mathscr{F}}$ and $\bar{\ell}_{\mathscr{F}}$ the uniquely determined positive linear extensions of ℓ and $\bar{\ell}$, respectively, to $\mathscr{L} \vdash \mathscr{F}$ and $\bar{\mathscr{L}}(\ell) \vdash \mathscr{F}$, respectively. Then the Daniell space $(X, \bar{\mathscr{L}}(\ell) \vdash \mathscr{F}, \bar{\ell}_{\mathscr{F}})$ is the closure of the Daniell space $(X, \mathscr{L} \vdash \mathscr{F}, \ell_{\mathscr{F}})$.

Proof. Note that

$$(X, \mathscr{L} \vdash \mathscr{F}, \ell_{\mathscr{F}}) \preccurlyeq (X, \bar{\mathscr{L}}(\ell) \vdash \mathscr{F}, \bar{\ell}_{\mathscr{F}}).$$

Also

$$(X, \bar{\mathscr{L}}(\ell), \bar{\ell}) \preccurlyeq (X, \bar{\mathscr{L}}(\ell_{\mathscr{F}}), \overline{\ell_{\mathscr{F}}}) \tag{3}$$

and

$$(X, \bar{\mathscr{L}}(\ell_{\mathscr{F}}), \overline{\ell_{\mathscr{F}}}) \preccurlyeq (X, \bar{\mathscr{L}}(\ell) \vdash \mathscr{F}, \bar{\ell}_{\mathscr{F}}) \tag{4}$$

(in each case the right-hand side is a closed extension of the Daniell space whose closure appears on the left-hand side). In view of (4) we need only show that

$$\bar{\mathscr{L}}(\ell) \vdash \mathscr{F} \subset \bar{\mathscr{L}}(\ell_{\mathscr{F}}). \tag{5}$$

By (3) we know that $\bar{\mathscr{L}}(\ell) \subset \bar{\mathscr{L}}(\ell_{\mathscr{F}})$. By Theorem 10 (c)

$$\mathfrak{N}(\mathscr{F}) = \mathfrak{N}(\bar{\mathscr{L}}(\ell_{\mathscr{F}})).$$

Using the definitions together with Proposition 1.2.30 (c), we see that (5) holds. □

Given a Daniell space $(X, \mathscr{L}, \ell)$, consider a closed Daniell space $(X, \tilde{\mathscr{L}}, \tilde{\ell}) \succcurlyeq (X, \mathscr{L}, \ell)$. We have already noted that $\mathscr{N}(\tilde{\mathscr{L}})$ satisfies all of the hypotheses placed on $\mathscr{F}$ in the preceding development. Thus we can adjoin the $\tilde{\mathscr{L}}$-exceptional functions to $\mathscr{L}$ via Definition 6, extend the functional ℓ to a positive linear functional on $\mathscr{L} \vdash \mathscr{N}(\tilde{\mathscr{L}})$, and then take the closure of the resulting Daniell space (or vice versa). Little more than checking definitions is now required in order to verify: this closed Daniell space is an admissible extension of $(X, \mathscr{L}, \ell)$, and it coincides with the original extension $(X, \tilde{\mathscr{L}}, \tilde{\ell})$ iff $(X, \tilde{\mathscr{L}}, \tilde{\ell})$ is admissible.

Theorem 4.1.12. *Let $(X, \mathscr{L}, \ell)$ be a Daniell space and $(X, \tilde{\mathscr{L}}, \tilde{\ell})$ a closed Daniell space extending $(X, \mathscr{L}, \ell)$. Denote by $\ell_{\mathscr{N}(\tilde{\mathscr{L}})}$ and by $\bar{\ell}_{\mathscr{N}(\tilde{\mathscr{L}})}$ the uniquely determined positive linear extensions of ℓ to $\mathscr{L} \vdash \mathscr{N}(\tilde{\mathscr{L}})$ and of $\bar{\ell}$ to $\bar{\mathscr{L}}(\ell) \vdash \mathscr{N}(\tilde{\mathscr{L}})$, respectively. Then the following assertions are equivalent.*

(*a*) *$(X, \tilde{\mathscr{L}}, \tilde{\ell})$ is an admissible extension of $(X, \mathscr{L}, \ell)$.*
(*b*) *$(X, \tilde{\mathscr{L}}, \tilde{\ell}) = (X, \bar{\mathscr{L}}(\ell) \vdash \mathscr{N}(\tilde{\mathscr{L}}), \bar{\ell}_{\mathscr{N}(\tilde{\mathscr{L}})})$.*
(*c*) *$(X, \tilde{\mathscr{L}}, \tilde{\ell})$ is the closure of $(X, \mathscr{L} \vdash \mathscr{N}(\tilde{\mathscr{L}}), \ell_{\mathscr{N}(\tilde{\mathscr{L}})})$.*

Proof. That (b) and (c) are equivalent follows from Corollary 11. The key to (a) ⇔ (b) is the equivalence $\tilde{\mathscr{L}}$-a.e. ⇔ $\mathscr{N}(\tilde{\mathscr{L}})$-a.e. One need only check the definitions, taking closure characteristics (in particular, Theorem 3.2.5) into account. □

Corollary 4.1.13. *Let $(X, \mathscr{L}, \ell)$ be a Daniell space, and let $\mathscr{F}$ be a tower that is σ-completely embedded in $\bar{\mathbb{R}}^X$ and satisfies*

$$\mathscr{F} \cap \bar{\mathscr{L}}(\ell) = \mathscr{N}(\bar{\mathscr{L}}(\ell)).$$

Denote by $\ell_{\mathscr{F}}$ the uniquely determined positive linear extension of ℓ to $\mathscr{L} \vdash \mathscr{F}$. Then the closure of the Daniell space $(X, \mathscr{L} \vdash \mathscr{F}, \ell_{\mathscr{F}})$ is an admissible extension of $(X, \mathscr{L}, \ell)$.

Proof. Let $(X, \tilde{\mathscr{L}}, \tilde{\ell})$ denote the closure of $(X, \mathscr{L} \vdash \mathscr{F}, \ell_{\mathscr{F}})$. By Theorem 10 (c), $\mathscr{F} = \mathscr{N}(\tilde{\mathscr{L}})$. Apply Theorem 12 (c) $\Rightarrow$ (a). □

According to Theorem 12, all admissible extensions of a given Daniell space are obtained in the manner Corollary 13 describes.

To prove the functional formula of Proposition 17, and also for other purposes, we require more information about the sets $\hat{\mathfrak{R}}(\mathscr{L})$.

Proposition 4.1.14. *Let $\mathscr{L}$ be a Riesz lattice that is conditionally σ-completely embedded in $\overline{\mathbb{R}}^X$. Then the conditions $f \in \mathscr{L}$ and $A \in \hat{\mathfrak{R}}(\mathscr{L})$ imply that fe_A and $fe_{X \setminus A}$ both belong to $\mathscr{L}$.*

Proof. Since $f = f^+ - f^-$, it suffices to consider positive functions. Given $A \in \hat{\mathfrak{R}}(\mathscr{L})$, there exists g in $\mathscr{L}$ with

$$A = \{g \neq 0\} = \{|g| \neq 0\}.$$

Given $f \in \mathscr{L}_+$, note that

$$fe_A = \bigvee_{n \in \mathbb{N}} (f \wedge n|g|).$$

Since $f \wedge n|g| \leq f$ for every n, and each $f \wedge n|g|$ belongs to $\mathscr{L}$, the σ-complete embedding ensures that fe_A belongs to $\mathscr{L}$. Moreover

$$fe_{X \setminus A} \in \langle f \dot{-} fe_A \rangle$$

so $fe_{X \setminus A}$ belongs to $\mathscr{L}$. □

Proposition 4.1.15.

(*a*) *If $\mathscr{L}$ is a Riesz lattice that is conditionally σ-completely embedded in $\overline{\mathbb{R}}^X$, then $\hat{\mathfrak{R}}(\mathscr{L})$ is a δ-ring.*

(*b*) *If $(X, \mathscr{L}, \ell)$ is a closed Daniell space, then $\hat{\mathfrak{R}}(\mathscr{L})$ is a σ-ring.*

Proof. (a) Since the zero function belongs to $\mathscr{L}$, the empty set belongs to $\hat{\mathfrak{R}}(\mathscr{L})$. Given $A, B \in \hat{\mathfrak{R}}(\mathscr{L})$, let f, g be functions in $\mathscr{L}$ with

$$A = \{f \neq 0\} \qquad B = \{g \neq 0\}.$$

Since

$$A \cup B = \{|f| + |g| \neq 0\}$$

$A \cup B$ certainly belongs to $\hat{\mathfrak{R}}(\mathscr{L})$. In view of the preceding proposition fe_B belongs to $\mathscr{L}$. Since

$$fe_{A \setminus B} \in \langle f \dot{-} fe_B \rangle$$

it follows that $fe_{A \setminus B}$ belongs to $\mathscr{L}$ and $A \setminus B$ belongs to $\hat{\mathfrak{R}}(\mathscr{L})$.

Finally, let $(A_n)_{n\in\mathbb{N}}$ be a sequence from $\hat{\mathfrak{R}}(\mathscr{L})$. Choose $g \in \mathscr{L}_+$ such that $A_1 = \{g \neq 0\}$. By what has already been shown we can also choose a sequence $(g_n)_{n\in\mathbb{N}}$ from $\mathscr{L}_+$ such that $\{g_n \neq 0\} = A_1 \setminus A_n$. The sequence $(g \wedge g_n)_{n\in\mathbb{N}}$ is bounded in $\mathscr{L}$ so its supremum belongs to $\mathscr{L}$. Moreover

$$\left\{\bigvee_{n\in\mathbb{N}} (g \wedge g_n) \neq 0\right\} = \bigcup_{n\in\mathbb{N}} \{g \wedge g_n \neq 0\}$$

$$= A_1 \cap \bigcup_{n\in\mathbb{N}} \{g_n \neq 0\}.$$

It follows that $A_1 \cap \bigcup_{n\in\mathbb{N}}\{g_n \neq 0\}$ belongs to $\hat{\mathfrak{R}}(\mathscr{L})$, hence that $A_1 \setminus \bigcup_{n\in\mathbb{N}}\{g_n \neq 0\}$ belongs to $\hat{\mathfrak{R}}(\mathscr{L})$. But

$$A_1 \setminus \bigcup_{n\in\mathbb{N}} \{g_n \neq 0\} = \bigcap_{n\in\mathbb{N}} A_n.$$

We have shown that $\hat{\mathfrak{R}}(\mathscr{L})$ is a δ-ring.

(b) It follows from (a) that $\hat{\mathfrak{R}}(\mathscr{L})$ is a ring of sets (Corollary 1.4.14). Given a sequence $(A_n)_{n\in\mathbb{N}}$ from $\hat{\mathfrak{R}}(\mathscr{L})$, set

$$B_1 := A_1$$

$$B_n := A_n \setminus \bigcup_{k=1}^{n-1} A_k \qquad (n \in \mathbb{N} \setminus \{1\})$$

to obtain a disjoint sequence from $\hat{\mathfrak{R}}(\mathscr{L})$ with the same union as the original sequence. Let $(f_n)_{n\in\mathbb{N}}$ be a sequence from $\mathscr{L} \cap \mathbb{R}^X$ with

$$B_n = \{f_n \neq 0\}$$

for every n [Proposition 2 (c)]. For each n in $\mathbb{N}$, choose a real number $\alpha_n > 0$ so that

$$\ell(\alpha_n |f_n|) < \frac{1}{2^n}.$$

The function $\sum_{n\in\mathbb{N}} \alpha_n |f_n|$ is certainly defined. According to Theorem 3.5.2, it belongs to $\mathscr{L}$. Since

$$\bigcup_{n\in\mathbb{N}} A_n = \left\{\sum_{n\in\mathbb{N}} \alpha_n |f_n| \neq 0\right\}$$

the set $\bigcup_{n\in\mathbb{N}} A_n$ belongs to $\hat{\mathfrak{R}}(\mathscr{L})$. Thus $\hat{\mathfrak{R}}(\mathscr{L})$ is a σ-ring. □

Proposition 4.1.16. *Let $(X, \tilde{\mathscr{L}}, \tilde{\ell})$ be an admissible extension of a Daniell space $(X, \mathscr{L}, \ell)$. Then the following assertions hold, for every $A \in \hat{\mathfrak{R}}(\tilde{\mathscr{L}})$.*

(*a*) *There exist an increasing sequence $(A_n)_{n \in \mathbb{N}}$ from $\hat{\mathfrak{R}}(\mathscr{L})$ and an $\tilde{\mathscr{L}}$-exceptional set B such that*

$$A \subset \left(\bigcup_{n \in \mathbb{N}} A_n \right) \cup B.$$

(*b*) *If the difference of two sets belonging to $\hat{\mathfrak{R}}(\mathscr{L})$ always belongs to $\hat{\mathfrak{R}}(\mathscr{L})$, in particular, if $\mathscr{L}$ is conditionally σ-completely embedded in $\overline{\mathbb{R}}^X$ or if the Daniell space $(X, \mathscr{L}, \ell)$ is closed, then there exist a disjoint sequence $(B_n)_{n \in \mathbb{N}}$ from $\hat{\mathfrak{R}}(\mathscr{L})$ and an $\tilde{\mathscr{L}}$-exceptional set B such that*

$$A \subset \left(\bigcup_{n \in \mathbb{N}} B_n \right) \cup B.$$

Proof. (a) Given $A \in \hat{\mathfrak{R}}(\tilde{\mathscr{L}})$, let f be a function in $\tilde{\mathscr{L}}$ such that

$$A = \{ f \neq 0 \}.$$

By admissibility, there exists g in $\bar{\mathscr{L}}(\ell)$ such that the set

$$B := \{ f \neq g \}$$

is $\tilde{\mathscr{L}}$-exceptional. Let h be a function in $\mathscr{L}^{\uparrow}$ with $h \geq |g|$, let $(h_n)_{n \in \mathbb{N}}$ be an increasing sequence from $\mathscr{L}_+$ whose supremum is h, and set

$$A_n := \{ h_n \neq 0 \} \qquad (n \in \mathbb{N})$$

Then $(A_n)_{n \in \mathbb{N}}$ is an increasing sequence from $\hat{\mathfrak{R}}(\mathscr{L})$ whose union is the set $\{h \neq 0\}$. Evidently

$$A \subset \left(\bigcup_{n \in \mathbb{N}} A_n \right) \cup B.$$

(b) follows from (a). We need only set

$$B_1 := A_1$$

$$B_n := A_n \setminus A_{n-1} \qquad (n \in \mathbb{N} \setminus \{1\}). \qquad \square$$

We are ready to prove a formula analogous to the formula for the functional Σ (Theorem 3.4.8). The formula for Σ is simpler than its analog for admissible-extension functionals: in Theorem 3.4.8 the analog of what is called fe_A here actually belongs to the original Daniell space, which is not true in general.

Proposition 4.1.17 (Formula for admissible-extension functionals). *Let $(X, \tilde{\mathscr{L}}, \tilde{\ell})$ be an admissible extension of a Daniell space $(X, \mathscr{L}, \ell)$. Then for every $f \in \tilde{\mathscr{L}}$,*

$$\tilde{\ell}(f) = \sup_{A \in \hat{\mathfrak{R}}(\mathscr{L})} \tilde{\ell}(fe_A) + \inf_{A \in \hat{\mathfrak{R}}(\mathscr{L})} \tilde{\ell}(fe_A).$$

In particular, if $f \in \tilde{\mathscr{L}}_+$, then

$$\tilde{\ell}(f) = \sup_{A \in \hat{\mathfrak{R}}(\mathscr{L})} \tilde{\ell}(fe_A) \tag{6}$$

Proof. Suppose first that $f \in \tilde{\mathscr{L}}_+$. If A belongs to $\hat{\mathfrak{R}}(\mathscr{L})$, then fe_A is in $\tilde{\mathscr{L}}_+$, by Proposition 14 and $fe_A \le f$. Thus $\tilde{\ell}(fe_A) \le \tilde{\ell}(f)$ for every A in $\hat{\mathfrak{R}}(\mathscr{L})$. Applying Proposition 16 (a) to the set $\{f \ne 0\}$, we can find an increasing sequence $(A_n)_{n \in \mathbb{N}}$ from $\hat{\mathfrak{R}}(\mathscr{L})$ such that

$$f = \bigvee_{n \in \mathbb{N}} fe_{A_n} \quad \tilde{\mathscr{L}}\text{-a.e.}$$

We have

$$\tilde{\ell}(f) = \sup_{n \in \mathbb{N}} \tilde{\ell}(fe_{A_n}) \le \sup_{A \in \hat{\mathfrak{R}}(\mathscr{L})} \tilde{\ell}(fe_A) \le \tilde{\ell}(f)$$

and (6) follows.

Now let f be an arbitrary function in $\tilde{\mathscr{L}}$, and set $B := \{f > 0\}$. Then, using what we have just proved,

$$\begin{aligned}
\tilde{\ell}(f^+) &= \sup_{A \in \hat{\mathfrak{R}}(\mathscr{L})} \tilde{\ell}(f^+ e_A) \\
&\ge \sup_{A \in \hat{\mathfrak{R}}(\mathscr{L})} \tilde{\ell}(fe_A) \\
&\ge \sup_{A \in \hat{\mathfrak{R}}(\mathscr{L})} \tilde{\ell}(fe_{A \cap B}) \\
&= \sup_{A \in \hat{\mathfrak{R}}(\mathscr{L})} \tilde{\ell}(f^+ e_{A \cap B}) \\
&= \sup_{A \in \hat{\mathfrak{R}}(\mathscr{L})} \tilde{\ell}(f^+ e_A).
\end{aligned}$$

It follows that

$$\tilde{\ell}(f^+) = \sup_{A \in \hat{\mathfrak{R}}(\mathscr{L})} \tilde{\ell}(fe_A).$$

Accordingly,

$$\begin{aligned}\tilde{\ell}(f^-) &= \tilde{\ell}((-f)^+)\\ &= \sup_{A\in\hat{\mathfrak{R}}(\mathscr{L})} \tilde{\ell}(-fe_A)\\ &= -\inf_{A\in\hat{\mathfrak{R}}(\mathscr{L})} \tilde{\ell}(fe_A).\end{aligned}$$

Hence

$$\begin{aligned}\tilde{\ell}(f) &= \tilde{\ell}(f^+) - \tilde{\ell}(f^-)\\ &= \sup_{A\in\hat{R}(\mathscr{L})} \tilde{\ell}(fe_A) + \inf_{A\in\hat{R}(\mathscr{L})} \tilde{\ell}(fe_A).\end{aligned}$$ □

For admissible extensions the Induction Principle takes the following form.

Theorem 4.1.18 (Induction Principle for admissible extensions). *Let $(X, \tilde{\mathscr{L}}, \tilde{\ell})$ be an admissible extension of the Daniell space $(X, \mathscr{L}, \ell)$. Suppose that $\mathscr{F}$ is a subset of $\tilde{\mathscr{L}}$ satisfying the following conditions.*

(a) $\mathscr{L} \subset \mathscr{F}$.

(b) *If $(f_n)_{n\in\mathbb{N}}$ is an $\tilde{\ell}$-sequence from $\mathscr{F}$, and if f belongs to $\tilde{\mathscr{L}}$ and satisfies*

$$f(x) = \lim_{n\to\infty} f_n(x)\ \tilde{\mathscr{L}}\text{-}a.e. \tag{7}$$

then f belongs to $\mathscr{F}$.

Then $\mathscr{F} = \tilde{\mathscr{L}}$.

Proof. Let $\ell_{\mathscr{N}(\tilde{\mathscr{L}})}$ denote the uniquely determined positive linear extension of ℓ to $\mathscr{L} \vdash \mathscr{N}(\tilde{\mathscr{L}})$. According to Theorem 12, $(X, \tilde{\mathscr{L}}, \tilde{\ell})$ is the closure of the Daniell space $(X, \mathscr{L} \vdash \mathscr{N}(\tilde{\mathscr{L}}), \ell_{\mathscr{N}(\tilde{\mathscr{L}})})$. The present theorem will follow from Theorem 3.2.8 (Induction Principle, second form) if we can show that $\mathscr{F}$ contains $\mathscr{L} \vdash \mathscr{N}(\tilde{\mathscr{L}})$. Let $f \in \mathscr{L} \vdash \mathscr{N}(\tilde{\mathscr{L}})$. Then f is $\tilde{\mathscr{L}}$-a.e. equal to some function g in $\mathscr{L}$. If we set $f_n := g$ for every n, then $(f_n)_{n\in\mathbb{N}}$ is an ℓ-sequence satisfying (7). By hypothesis, f belongs to $\mathscr{F}$. □

EXERCISES

4.1.1(E) Let X be a set, and consider the positive linear functionals on $\mathscr{k}(X)$. By Ex. 1.3.5 (a), every such functional ℓ is nullcontinuous (even π-continuous) and is representable by some $g \in \mathbb{R}_+^X$ (in the sense that $\ell = \ell_g$). Let $(X, \mathscr{L}, \ell)$ be a closed extension of $(X, \mathscr{k}(X), \ell_g)$. Prove the equivalence of the following statements:

(α1) $(X, \mathscr{L}, \ell)$ is admissible.
(α2) $\{f \in \mathscr{L} \mid \{f \neq 0\} \subset \{g = 0\}\} \subset \mathscr{N}(\mathscr{L})$.
(α3) $\mathscr{L} = \bar{\mathscr{L}}(\ell_g) \vdash \{f \in \mathscr{N}(\mathscr{L}) \mid \{f \neq 0\} \subset \{g = 0\}\}$.

Also prove the following propositions:

(β) $\mathscr{N} := \{f \in \overline{\mathbb{R}}^X \mid \{f \neq 0\} \subset \{g = 0\}\}$ is a tower completely embedded in $\overline{\mathbb{R}}^X$.

(γ) If $\mathscr{L}' := \bar{\mathscr{L}}(\ell_g) \vdash \mathscr{N}$ and if ℓ' is the (uniquely determined) positive linear extension of ℓ_g to $\mathscr{L}'$, then

$$(X, \mathscr{L}', \ell') = \left(X, \ell_g^1(X), \overline{\ell_g}\right)$$

and this is the largest admissible extension of $(X, \mathscr{k}(X), \ell_g)$.

(δ) $\{g \neq 0\}$ is countable iff $(X, \bar{\mathscr{L}}(\ell_g), \overline{\ell_g})$ is the only admissible extension of $(X, \mathscr{k}(X), \ell_g)$.

The considerations in this exercise also characterize the admissible extensions of $(X, \mathscr{L}_{\mathfrak{F}}, \ell_{\chi_g})$, where $\mathfrak{F}$ denotes the ring of all finite subsets of X and χ_g the positive measure associated with $g \in \mathbb{R}_+^X$ (cf., Ex. 2.2.3).

4.1.2(E) Let ℓ be a positive linear nullcontinuous functional on $\ell^\infty(X)$ [cf., Ex. 1.3.5 (c)]. Prove the following statement: $(X, \bar{\mathscr{L}}(\ell), \bar{\ell})$ is the only closed extension of $(X, \ell^\infty(X), \ell)$. [In particular, $(X, \bar{\mathscr{L}}(\ell), \bar{\ell})$ is also the only admissible extension of $(X, \ell^\infty(X), \ell)$.]

4.1.3(E) Given an uncountable set X, consider $c_f(X)$ and the functional $\ell: c_f(X) \to \mathbb{R}$, $\alpha e_X + g \to \alpha$ [cf., Ex. 1.3.5 (b)]. Prove that if $(X, \mathscr{L}', \ell')$ is a closed extension of $(X, c_f(X), \ell)$, then the following are equivalent:

(α1) $(X, \mathscr{L}', \ell')$ is admissible.
(α2) $\mathscr{L}' = \{\alpha e_X + g \mid \alpha \in \mathbb{R},\ g \in \mathscr{N}(\mathscr{L}')\}$.

If these conditions are fulfilled and $\alpha e_X + g \in \mathscr{L}'$, then the following holds:

(β) $\ell'(\alpha e_X + g) = \alpha$.

4.1.4(E) Set $X := [0,1]$. For each $\alpha \in \mathbb{R}$, define f_α: $X \to \mathbb{R}$, $x \mapsto \alpha x$. Let $\mathscr{L} := \{f_\alpha \mid \alpha \in \mathbb{R}\}$, and let ℓ: $\mathscr{L} \to \mathbb{R}$, $f_\alpha \mapsto \alpha/2$. Then $(X, \mathscr{L}, \ell)$ is a Daniell space. Denote by $\mathscr{K}$ the set of all continuous real functions on $[0,1]$ and by $\mathscr{k}$ the Riemann integral on $\mathscr{K}$. Prove the propositions that follow:

(α) $(X, \mathscr{L}, \ell) \preccurlyeq (X, \mathscr{K}, \mathscr{k})$.
(β) $(X, \bar{\mathscr{L}}(\mathscr{k}), \bar{\mathscr{k}})$ is a closed extension of $(X, \mathscr{L}, \ell)$ that is not admissible.

4.1.5(E) Take a Daniell space $(X, \mathscr{L}, \ell)$. Prove the two statements that follow:

(α) If $(X, \mathscr{L}', \ell') \preccurlyeq (X, \mathscr{L}'', \ell'')$ for two closed extensions of $(X, \mathscr{L}, \ell)$, then $(X, \mathscr{L}', \ell')$ is admissible whenever $(X, \mathscr{L}'', \ell'')$ is.

(β) If $(X, \mathscr{L}', \ell')$ is an admissible extension of $(X, \mathscr{L}, \ell)$ and $(X, \mathscr{L}'', \ell'')$ an admissible extension of $(X, \mathscr{L}', \ell')$, then $(X, \mathscr{L}'', \ell'')$ is an admissible extension of $(X, \mathscr{L}, \ell)$.

4.1.6(E) Let $(X, \mathscr{L}, \ell)$ be a Daniell space. Take a set A with the property that $\mathfrak{P}(A) \cap \hat{\mathfrak{R}}(\bar{\mathscr{L}}(\ell)) \subset \mathfrak{N}(\bar{\mathscr{L}}(\ell))$. Define

$$\mathscr{N} := \{ g \in \overline{\mathbb{R}}^X \mid \{g \neq 0\} \setminus A \in \mathfrak{N}(\bar{\mathscr{L}}(\ell)) \}.$$

Prove the following statements:

(α) $\mathscr{N}$ is a σ-completely embedded tower in $\overline{\mathbb{R}}^X$.

(β) There is a unique admissible extension $(X, \mathscr{L}', \ell')$ of $(X, \mathscr{L}, \ell)$ such that $\mathscr{L}' = \bar{\mathscr{L}}(\ell) \vdash \mathscr{N}$. Moreover, $A \in \mathfrak{N}(\mathscr{L}')$.

4.1.7(E) Let $(X, \mathscr{L}, \ell)$ be a Daniell space. Set $A := \bigcap_{f \in \mathscr{L}} \{f = 0\}$ and let $\mathscr{N} := \{ g \in \overline{\mathbb{R}}^X \mid \{g \neq 0\} \subset A \}$. Prove the following statements:

(α) $\mathscr{N}$ is a tower completely embedded in $\overline{\mathbb{R}}^X$.

(β) Given an admissible extension $(X, \mathscr{L}', \ell')$ of $(X, \mathscr{L}, \ell)$, there is a unique admissible extension $(X, \mathscr{L}'', \ell'')$ of $(X, \mathscr{L}, \ell)$ such that $\mathscr{L}'' = \mathscr{L}' \vdash \mathscr{N}$.

4.1.8(E) Let $(X, \mathscr{L}, \ell)$ be a Daniell space. Take $f \in \overline{\mathbb{R}}_+^X$ with

$$\beta := \inf\{ \bar{\ell}(g) \mid g \in \bar{\mathscr{L}}(\ell), g \geq f \} \geq \sup\{ \bar{\ell}(g) \mid g \in \bar{\mathscr{L}}(\ell), g \leq f \} =: \alpha.$$

Prove the following statements:

(α) If $\beta < \infty$, then there is a $g^* \in \bar{\mathscr{L}}(\ell)$ such that $g^* \geq f$ and $\bar{\ell}(g^*) = \beta$. Furthermore, if $h \in \langle g^* \dot{-} f \rangle$, then $\{ h' \in \bar{\mathscr{L}}(\ell) \mid |h'| \leq |h| \} \subset \mathscr{N}(\bar{\mathscr{L}}(\ell))$.

(β) If g^* and h are as in (α), and if

$$\{ h' \in \bar{\mathscr{L}}(\ell) \mid \{h' \neq 0\} \subset \{h \neq 0\} \} \subset \mathscr{N}(\bar{\mathscr{L}}(\ell))$$

then there is an admissible extension $(X, \mathscr{L}', \ell')$ of $(X, \mathscr{L}, \ell)$ such that $f \in \mathscr{L}'$ and $\ell'(f) = \beta$.

(γ) There is a $g_* \in \bar{\mathscr{L}}(\ell)$ such that $g_* \leq f$ and $\bar{\ell}(g_*) = \alpha$. Furthermore, if $h \in \langle f \dot{-} g_* \rangle$, then $\{ h' \in \bar{\mathscr{L}}(\ell) \mid |h'| \leq |h| \} \subset \mathscr{N}(\bar{\mathscr{L}}(\ell))$.

(δ) If g_* and h are as in (γ), and if

$$\{ h' \in \bar{\mathscr{L}}(\ell) \mid \{h' \neq 0\} \subset \{h \neq 0\} \} \subset \mathscr{N}(\bar{\mathscr{L}}(\ell))$$

then there is an admissible extension $(X, \mathscr{L}', \ell')$ of $(X, \mathscr{L}, \ell)$ such that $f \in \mathscr{L}'$ and $\ell'(f) = \alpha$.

Suppose now that $g \in \bar{\mathscr{L}}(\ell)$ and $A \subset X$. Define $f := ge_A$ and α, β as before. Prove the following assertions:

(ε) There is an admissible extension $(X, \mathscr{L}', \ell')$ of $(X, \mathscr{L}, \ell)$ such that $f \in \mathscr{L}'$ and $\ell'(f) = \beta$.

(ζ) There is an admissible extension $(X, \mathscr{L}', \ell')$ of $(X, \mathscr{L}, \ell)$ such that $f \in \mathscr{L}'$ and $\ell'(f) = \alpha$.

(η) There exists an admissible extension $(X, \mathscr{L}', \ell')$ of $(X, \mathscr{L}, \ell)$ for which $f \in \mathscr{L}'$ and $\ell'(f) \notin \{\alpha, \beta\}$.

4.1.9(E) Let $(X, \mathscr{L}, \ell)$ be a closed Daniell space. Take $A \subset X$ such that $\mathfrak{P}(A) \cap \hat{\mathfrak{R}}(\mathscr{L}) \subset \mathfrak{N}(\mathscr{L})$.

(α) Define $\mathscr{L}_1 := \{fe_{X \setminus A} \mid f \in \mathscr{L}\}$ and $\ell_1\colon \mathscr{L}_1 \to \mathbb{R}$, $fe_{X \setminus A} \mapsto \ell(f)$. Prove that ℓ_1 is well defined; that is, show that if $fe_{X \setminus A} = ge_{X \setminus A}$ for $f, g \in \mathscr{L}$, then $\ell(f) = \ell(g)$. Verify that $(X, \mathscr{L}_1, \ell_1)$ is a closed Daniell space.

Assume now that $\{f \in \mathscr{L} \mid \{f \neq 0\} \supset A\} = \varnothing$:

(β) Define $\mathscr{L}_2 := \{\alpha e_A + g \mid \alpha \in \mathbb{R},\ g \in \overline{\mathbb{R}}^X,\ \exists B \in \hat{\mathfrak{R}}(\mathscr{L}) \text{ with } \{g \neq 0\} \subset A \cap B\}$ and $\ell_2\colon \mathscr{L}_2 \to \mathbb{R}$, $\alpha e_A + g \mapsto \alpha\gamma$ where $\gamma \in \mathbb{R}_+$ is fixed. Show that if $\alpha e_A + g \in \mathscr{L}_2$ and $\alpha' e_A + g' \in \mathscr{L}_2$ with $\alpha e_A + g = \alpha' e_A + g'$, then $\alpha = \alpha'$ so that ℓ_2 is well defined. Verify that $(X, \mathscr{L}_2, \ell_2)$ is a closed Daniell space.

(γ) Show that if $f \in \mathscr{L}_1$ and $g \in \mathscr{L}_2$, then $|f| \wedge |g| = 0$.

(δ) Define $\tilde{\mathscr{L}} := \{f + g \mid f \in \mathscr{L}_1,\ g \in \mathscr{L}_2\}$ and $\tilde{\ell}(f + g) := \ell_1(f) + \ell_2(g)$. Show that $(X, \tilde{\mathscr{L}}, \tilde{\ell})$ is a closed Daniell space and $(X, \mathscr{L}, \ell) \preccurlyeq (X, \tilde{\mathscr{L}}, \tilde{\ell})$.

(ε) Show that $(X, \tilde{\mathscr{L}}, \tilde{\ell})$ is an admissible extension of $(X, \mathscr{L}, \ell)$ iff $\gamma = 0$.

4.1.10(E) Given a Riesz lattice, $\mathscr{L}$, and a positive linear functional, ℓ, we defined in Ex. 1.3.10 the pseudometric d_ℓ. Now take a Daniell space $(X, \mathscr{L}, \ell)$, and a closed extension $(X, \mathscr{L}', \ell')$ of $(X, \mathscr{L}, \ell)$.

(α) Prove that the following statements are equivalent:

(α1) $(X, \mathscr{L}', \ell')$ is an admissible extension of $(X, \mathscr{L}, \ell)$.
(α2) $(\mathscr{L}, d_\ell)$ is dense in $(\mathscr{L}', d_{\ell'})$.
(α3) If $f \in \mathscr{L}'$, then there is a sequence $(f_n)_{n \in \mathbb{N}}$ from $\mathscr{L} \cap \mathbb{R}^X$ with $\lim_{n \to \infty} d_{\ell'}(f - f_n) = 0$.

Every pseudometric d on a set defines an equivalence relation on that set by $x \sim y$ iff $d(x, y) = 0$. For every $f \in \mathscr{L}$, let $\dot{f}^{(\ell)}$ be the equivalence class with respect to d_ℓ containing f. Denote by $\mathscr{L}/_{\sim \ell}$ the set of all equivalence classes with respect to d_ℓ of functions of $\mathscr{L}$.

(β) Prove the equivalence of the following statements:

(β1) $(X, \mathscr{L}', \ell')$ is admissible.
(β2) There is an isometry φ of $\bar{\mathscr{L}}(\ell)/_{\sim \bar{\ell}}$ onto $\mathscr{L}'/\sim \ell'$ such that $\varphi(\dot{f}^{(\ell)}) = \dot{f}^{(\ell')}$ for every $f \in \mathscr{L}$.

4.1.11(E) Set $X := \mathbb{N}_n \times \{0,1\}$ and $\mathfrak{R} := \{A \times \{0,1\} \mid A \subset \mathbb{N}_n\}$. Let μ be a positive measure on $\mathfrak{R}$ such that $\mu(\{m\} \times \{0,1\}) > 0$ for all $m \in \mathbb{N}_n$. Denote by Φ the set of all admissible extensions $(X, \mathscr{L}, \ell)$ of $(X, \mathscr{L}_{\mathfrak{R}}, \ell_\mu)$ with $\mathscr{L} = \mathbb{R}^X$. Prove:

(α) Φ has exactly 2^n elements.

(β) $\{\ell(\mathbb{N}_n \times \{0\}) \mid (X, \mathscr{L}, \ell) \in \Phi\} = \mu(\mathfrak{R})$.

4.1.12(E) Let $(X, \mathfrak{R}, \mu)$ be a δ-measure space. Take $A \in \mathfrak{R}$ with $(\ell_\mu)_*(e_A) = 0$, and $\alpha \in \mathbb{R}_+$. Prove the equivalence of the following propositions:

(α) There is an admissible extension $(X, \mathscr{L}, \ell)$ of $(X, \mathscr{L}_{\mathfrak{R}}, \ell_\mu)$ with $e_A \in \mathscr{L}$ and $\ell(e_A) = \alpha$.

(β) There is an increasing sequence $(A_n)_{n \in \mathbb{N}}$ from $\mathfrak{R}$ with $\sup_{n \in \mathbb{N}} \ell_\mu^*(e_{A \cap A_n}) = \alpha$.

4.2. DETERMINED EXTENSIONS OF A DANIELL SPACE AND THE DANIELL-SPACE INTEGRAL

NOTATION FOR SECTION 4.2:

X denotes a set.

For W a set,

$$\mathfrak{F}(W) := \{A \subset W \mid A \text{ is finite}\}$$

$$\mathscr{F}(W) := \left\{ f \in \mathbb{R}^W \;\middle|\; \begin{array}{l} f \text{ is an } \mathfrak{F}(W)\text{-step} \\ \text{function on } W \end{array} \right\} = \left\{ f \in \mathbb{R}^W \mid \{f \neq 0\} \text{ is finite} \right\}.$$

Already the field of candidates from which to choose the integral for a Daniell space has been narrowed: only admissible extensions are to be considered. Admissibility rules out arbitrariness, however, only insofar as the collection of equivalence classes modulo a.e.-equality is concerned. The danger remains that in different admissible extensions of a given Daniell space, one and the same function might be assigned different functional values. The determined extensions studied in this section are those extensions for which the second kind of arbitrariness has been ruled out.

A brief outline of the section may be useful. The section begins with the definition and elementary properties of determined extensions, culminating in a proof that closed determined extensions are always admissible. Main Theorem 4.2.7 collects properties of determined extensions that are of key importance in the sequel. (Its proof is long and possibly difficult, but the reader will find the going much easier after that.) The task of defining the integral can then be taken up with confidence. Indeed, by that point an idea will already have emerged as to how one might construct a largest determined extension. The collection of new exceptional functions that one should adjoin, the "locally null" functions referred to in the Volume 1 introduction, are presented in Definition 9. Theorem 11 confirms the hunch that adjoining these new exceptional functions and taking Daniell-space closure yields the largest determined extension, which is then defined to be the Daniell-space integral. Several easy consequences of the definition of the integral are then spelled out, including the fact that when σ-finiteness is hypothesized, the Daniell-space integral defined here and the Daniell-space closure constructed in Chapter 3 (which is the integral of classical abstract real integration theory) coincide. The final part of the section also includes a few examples.

Definition 4.2.1. *Let* $(X, \mathscr{L}, \ell)$ *and* $(X, \mathscr{L}', \ell')$ *be Daniell spaces with* $(X, \mathscr{L}', \ell')$ *closed and* $(X, \mathscr{L}, \ell) \preccurlyeq (X, \mathscr{L}', \ell')$. *We say that* $(X, \mathscr{L}', \ell')$ *is* ***determined*** *by* $(X, \mathscr{L}, \ell)$, *or* $(X, \mathscr{L}', \ell')$ *is a* ***determined extension*** *of* $(X, \mathscr{L}, \ell)$, *iff for every admissible extension* $(X, \tilde{\mathscr{L}}, \tilde{\ell})$ *of* $(X, \mathscr{L}, \ell)$ *the functionals* ℓ' *and* $\tilde{\ell}$ *agree on their common domain.* □

Although nonadmissible extensions are no longer of interest to us, Definition 1 does not require $(X, \mathscr{L}', \ell')$ to be admissible. Our first goal is to show that determined extensions of $(X, \mathscr{L}, \ell)$ are automatically admissible extensions.

In order to exploit the definition of determined extension, we will need enough admissible extensions for purposes of comparison. The next two results enable us to construct such extensions. The idea behind this construction is as follows. Suppose we have a Daniell space $(X, \mathscr{L}, \ell)$. If, for some subset A of X, the values assumed at points of A by a given function f do not influence the value of $\bar{\ell}(f)$, then some admissible extension of $(X, \mathscr{L}, \ell)$ should encompass all functions from $\overline{\mathbb{R}}^X$ that vanish outside A. The functional for that extension should continue to assign the value zero to any function that vanishes outside A. For instance, with $X := \{a, b, c\}$, $\mathscr{L} := \{f \in \overline{\mathbb{R}}^X \mid f(a) = 0,\ f(b) = f(c) \in \mathbb{R}\}$, $\ell(f) := f(b) = f(c)$ for every f in $\mathscr{L}$, and $A := \{a\}$, we have exactly this situation. For this example we need only drop the restriction $f(a) = 0$, which amounts to having $\{a\}$ be a new exceptional set, and we have the admissible extension. The general situation is basically the same.

Proposition 4.2.2. *Let* $(X, \mathscr{L}, \ell)$ *be a Daniell space. Then the following assertions are equivalent, for every* $A \subset X$.

(*a*) $\mathfrak{P}(A) \cap \hat{\mathfrak{R}}(\bar{\mathscr{L}}(\ell)) \subset \mathfrak{N}(\bar{\mathscr{L}}(\ell))$.

(*b*) $\bar{\ell}(|f|) = 0$ *for every function* $f \in \bar{\mathscr{L}}(\ell)$ *that vanishes throughout* $X \setminus A$.

If either assertion holds, then the set

$$\mathscr{F}_A := \{f \in \overline{\mathbb{R}}^X \mid \{f \neq 0\} \subset A \cup B \text{ for some } B \in \mathfrak{N}(\bar{\mathscr{L}}(\ell))\}$$

is a tower that is σ-*completely embedded in* $\overline{\mathbb{R}}^X$ *and satisfies the compatibility condition*

$$\mathscr{F}_A \cap \bar{\mathscr{L}}(\ell) = \mathscr{N}(\bar{\mathscr{L}}(\ell)).$$

Proof. The details, although numerous, are easy to check. We leave them to the reader. □

Corollary 4.2.3. *Let $(X, \mathscr{L}, \ell)$ be a Daniell space, and let A be a subset of X such that $\bar{\ell}(|f|) = 0$ for every $f \in \bar{\mathscr{L}}(\ell)$ that vanishes throughout $X \setminus A$. Then $(X, \mathscr{L}, \ell)$ has an admissible extension $(X, \tilde{\mathscr{L}}, \tilde{\ell})$ for which*

$$\{f \in \overline{\mathbb{R}}^X \mid \{f \neq 0\} \subset A\} \subset \mathscr{N}(\tilde{\mathscr{L}}).$$

Proof. Use $\mathscr{F}_A$ from Proposition 2 for $\mathscr{F}$ in Corollary 4.1.13 and Theorem 4.1.10. □

Proposition 4.2.4. *Let $(X, \mathscr{L}', \ell')$ be a determined extension of a Daniell space $(X, \mathscr{L}, \ell)$. Then every set belonging to $\hat{\mathfrak{R}}(\mathscr{L}')$ can be written as the disjoint union of a set belonging to $\hat{\mathfrak{R}}(\bar{\mathscr{L}}(\ell))$ and an $\mathscr{L}'$-exceptional set.*

Proof. Let $C \in \hat{\mathfrak{R}}(\mathscr{L}')$, and choose $f \in \mathscr{L}'_+$ so that

$$C = \{f \neq 0\}.$$

The idea is to find a subset B of C that belongs to $\hat{\mathfrak{R}}(\bar{\mathscr{L}}(\ell))$ and is somehow maximal. Then the difference $C \setminus B$ should be $\mathscr{L}'$-exceptional. Define

$$\alpha := \sup\{\ell'(fe_{B'}) \mid B' \in \hat{\mathfrak{R}}(\bar{\mathscr{L}}(\ell)),\ B' \subset C\}.$$

Note that $\hat{\mathfrak{R}}(\bar{\mathscr{L}}(\ell)) \subset \hat{\mathfrak{R}}(\mathscr{L}')$ (Theorem 3.2.5) so $fe_{B'}$ belongs to $\mathscr{L}'$ for every B' in $\hat{\mathfrak{R}}(\bar{\mathscr{L}}(\ell))$, by 4.1.14. Moreover $fe_{B'} \leq f$ for every B' in $\hat{\mathfrak{R}}(\bar{\mathscr{L}}(\ell))$, so $\alpha < \infty$. Choose a sequence $(B_n)_{n \in \mathbb{N}}$ from $\hat{\mathfrak{R}}(\bar{\mathscr{L}}(\ell))$, with $B_n \subset C$ for every n, such that

$$\sup_{n \in \mathbb{N}} \ell'(fe_{B_n}) = \alpha.$$

Set

$$B := \bigcup_{n \in \mathbb{N}} B_n \qquad A := C \setminus B.$$

Since $\hat{\mathfrak{R}}(\bar{\mathscr{L}}(\ell))$ is a σ-ring (Proposition 4.1.15), B belongs to $\hat{\mathfrak{R}}(\bar{\mathscr{L}}(\ell))$. Moreover A belongs to $\hat{\mathfrak{R}}(\mathscr{L}')$. Note that fe_A therefore belongs to $\mathscr{L}'$ (Proposition 4.1.14). To show that A is $\mathscr{L}'$-exceptional, we argue as follows:

(i) A satisfies the hypothesis of Corollary 3.

(ii) By Corollary 3, $(X, \mathscr{L}, \ell)$ has an admissible extension $(X, \tilde{\mathscr{L}}, \tilde{\ell})$ for which $\tilde{\ell}(fe_A) = 0$.

(iii) Since $(X, \mathscr{L}', \ell')$ is a determined extension of $(X, \mathscr{L}, \ell)$ and f is positive,

$$\tilde{\ell}(fe_A) = \ell'(|fe_A|) = 0$$

and $fe_A \in \mathscr{N}(\mathscr{L}')$.

(iv) Since $A = \{fe_A \neq 0\}$, $A \in \mathfrak{N}(\mathscr{L}')$.

Of these four assertions only (i) needs any proof. We show that A satisfies Proposition 2 (a). Let D be a subset of A that belongs to $\hat{\mathfrak{R}}(\bar{\mathscr{L}}(\ell))$. For every $n \in \mathbb{N}$ we have

$$\begin{aligned} \ell'(fe_{B_n}) &\leq \ell'(fe_B) \\ &\leq \ell'(fe_B) + \ell'(fe_D) \\ &= \ell'(fe_{B \cup D}) \\ &\leq \alpha. \end{aligned}$$

Thus

$$\begin{aligned} \alpha &= \sup_{n \in \mathbb{N}} \ell'(fe_{B_n}) \\ &\leq \ell'(fe_B) + \ell'(fe_D) \\ &\leq \alpha \end{aligned}$$

and

$$\ell'(fe_D) = 0,\ D \in \mathfrak{N}(\mathscr{L}').$$

Choose g in $\bar{\mathscr{L}}(\ell)_+$ such that $D = \{g \neq 0\}$. Then

$$\bar{\ell}(g) = \ell'(g) = 0$$

so g belongs to $\mathscr{N}(\bar{\mathscr{L}}(\ell))$, and D belongs to $\mathfrak{N}(\bar{\mathscr{L}}(\ell))$, as required. □

Theorem 4.2.5. *Let $(X, \mathscr{L}, \ell)$ and $(X, \mathscr{L}', \ell')$ be closed Daniell spaces such that*

$$(X, \mathscr{L}, \ell) \preccurlyeq (X, \mathscr{L}', \ell')$$

and

$$\hat{\mathfrak{R}}(\mathscr{L}') \subset \hat{\mathfrak{R}}(\mathscr{L}).$$

Then $(X, \mathscr{L}, \ell) = (X, \mathscr{L}', \ell')$.

Proof. It suffices to show that $\mathscr{L}'_+ \subset \mathscr{L}$. Let $g \in \mathscr{L}'_+$. First we approximate g from below by a function in $\mathscr{L}$ in such a way that the value assigned to that function by ℓ is as close as possible to $\ell'(g)$. Set

$$\alpha := \sup\{\ell(h) \mid h \in \mathscr{L}_+, h \leq g\}$$

and note that α is real. Choose an increasing sequence $(h_n)_{n \in \mathbb{N}}$ from $\mathscr{L}_+$ with $h_n \leq g$ for every n and

$$\sup_{n \in \mathbb{N}} \ell(h_n) = \alpha.$$

Then $(h_n)_{n \in \mathbb{N}}$ is an increasing ℓ-sequence, so $\bigvee_{n \in \mathbb{N}} h_n$ belongs to both $\mathscr{L}_+$ and $\mathscr{L}'_+$, $\bigvee_{n \in \mathbb{N}} h_n \leq g$, and

$$\ell\left(\bigvee_{n \in \mathbb{N}} h_n\right) = \alpha.$$

Now we look at the difference between the given function g and the approximating function $\bigvee_{n \in \mathbb{N}} h_n$. Construct a positive function g' in $\langle g \dot{-} \bigvee_{n \in \mathbb{N}} h_n \rangle$ by defining

$$g' \colon X \to \overline{\mathbb{R}}, \qquad x \mapsto \begin{cases} g(x) - \bigvee_{n \in \mathbb{N}} h_n(x) & \text{if defined} \\ 0 & \text{otherwise.} \end{cases}$$

Note that g' belongs to $\mathscr{L}'$. We want to show that g' belongs to $\mathscr{N}(\mathscr{L})$ and therefore to $\mathscr{L}$. First, we claim that every f in $\mathscr{L}$ satisfying $0 \leq f \leq g'$ is $\mathscr{L}$-exceptional. Indeed, for each such f, $f + \bigvee_{n \in \mathbb{N}} h_n$ belongs to $\mathscr{L}_+$ and satisfies $f + \bigvee_{n \in \mathbb{N}} h_n \leq g$. Hence

$$\begin{aligned} \alpha &= \ell\left(\bigvee_{n \in \mathbb{N}} h_n\right) \\ &\leq \ell\left(\bigvee_{n \in \mathbb{N}} h_n\right) + \ell(f) \\ &= \ell\left(f + \bigvee_{n \in \mathbb{N}} h_n\right) \\ &\leq \alpha \end{aligned}$$

so $\ell(f) = 0$ and f is $\mathscr{L}$-exceptional. Finally, we bring the condition $\hat{\mathfrak{R}}(\mathscr{L}') \subset \hat{\mathfrak{R}}(\mathscr{L})$ into play. Choose $f \in \mathscr{L}_+$ with

$$\{f \neq 0\} = \{g \neq 0\}.$$

Set

$$A_n := \left\{ \frac{1}{n} f < g' \right\} \qquad (n \in \mathbb{N}).$$

Each A_n belongs to $\hat{\mathfrak{R}}(\mathscr{L}')$, hence to $\hat{\mathfrak{R}}(\mathscr{L})$. Therefore the functions $(1/n)fe_{A_n}$ all belong to $\mathscr{L}_+$ (4.1.14). Moreover

$$\frac{1}{n} fe_{A_n} \leq g'$$

for every n. According to the claim just proved, each A_n belongs to $\mathfrak{N}(\mathscr{L})$. Since

$$\{g' \neq 0\} = \bigcup_{n \in \mathbb{N}} A_n$$

we conclude that g' belongs to $\mathscr{N}(\mathscr{L})$ and therefore to $\mathscr{L}$.

It now follows that since $g \in \langle g' \dotplus \bigvee_{n \in \mathbb{N}} h_n \rangle$, g belongs to $\mathscr{L}$. □

Our first goal can now be achieved.

Corollary 4.2.6. *Let $(X, \mathscr{L}, \ell)$ and $(X, \mathscr{L}', \ell')$ be Daniell spaces with $(X, \mathscr{L}, \ell) \preccurlyeq (X, \mathscr{L}', \ell')$. If $(X, \mathscr{L}', \ell')$ is a determined extension of $(X, \mathscr{L}, \ell)$, then it is also an admissible extension.*

Proof. We use the characterization provided by Theorem 4.1.12 (a) ⇔ (b). The closed Daniell space $(X, \mathscr{L}', \ell')$ automatically extends $(X, \bar{\mathscr{L}}(\ell), \bar{\ell})$, and $(X, \bar{\mathscr{L}}(\ell), \bar{\ell})$ is its own closure, so $\mathscr{N}(\mathscr{L}')$ satisfies, relative to $\bar{\mathscr{L}}(\ell)$, the hypotheses placed on $\mathscr{F}$ in Theorem 4.1.10. Denote by $\bar{\ell}_{\mathscr{N}(\mathscr{L}')}$ the uniquely determined positive linear extension of $\bar{\ell}$ to $\bar{\mathscr{L}}(\ell) \vdash \mathscr{N}(\mathscr{L}')$. The definitions show that

$$\left(X, \bar{\mathscr{L}}(\ell) \vdash \mathscr{N}(\mathscr{L}'), \bar{\ell}_{\mathscr{N}(\mathscr{L}')}\right) \preccurlyeq (X, \mathscr{L}', \ell').$$

Recall that $\mathfrak{N}(\bar{\mathscr{L}}(\ell) \vdash \mathscr{N}(\mathscr{L}')) = \mathfrak{N}(\mathscr{L}')$ (Theorem 4.1.10, Corollary 4.1.11) and $\hat{\mathfrak{R}}(\mathscr{L}')$ is a ring of sets. Hence Proposition 4 shows that

$$\hat{\mathfrak{R}}(\mathscr{L}') \subset \hat{\mathfrak{R}}(\bar{\mathscr{L}}(\ell) \vdash \mathscr{N}(\mathscr{L}')).$$

By Theorem 5,

$$(X, \mathscr{L}', \ell') = \left(X, \bar{\mathscr{L}}(\ell) \vdash \mathscr{N}(\mathscr{L}'), \bar{\ell}_{\mathscr{N}(\mathscr{L}')}\right)$$

and by Theorem 4.1.12, $(X, \mathscr{L}', \ell')$ is admissible. □

We are ready to prove the most important properties of determined extensions. The next theorem is key for much of what follows.

Theorem 4.2.7 (Main Theorem on determined extensions). *Let $(X, \mathscr{L}, \ell)$ be a Daniell space. Then the following assertions are equivalent, for every closed Daniell space $(X, \mathscr{L}', \ell') \succcurlyeq (X, \mathscr{L}, \ell)$.*

(a) *$(X, \mathscr{L}', \ell')$ is determined by $(X, \mathscr{L}, \ell)$.*

(b) *For each $f \in \mathscr{L}'$ there exists a set $A \in \hat{\mathfrak{R}}(\bar{\mathscr{L}}(\ell))$ such that fe_A belongs to $\bar{\mathscr{L}}(\ell)$ and $fe_{X \setminus A}$ is $\mathscr{L}'$-exceptional.*

In (b) the set A can be chosen so that $A \subset \{f \neq 0\}$.

Each of the assertions (a), (b) implies all of the following.

(c) *For every $f \in \mathscr{L}'$ and for every $C \in \hat{\mathfrak{R}}(\bar{\mathscr{L}}(\ell))$ the function fe_C belongs to $\bar{\mathscr{L}}(\ell)$.*

(d) *For every $B \in \hat{\mathfrak{R}}(\mathscr{L}')$ and for every $C \in \hat{\mathfrak{R}}(\bar{\mathscr{L}}(\ell))$, the set $B \cap C$ belongs to $\hat{\mathfrak{R}}(\bar{\mathscr{L}}(\ell))$.*

(e) *For every $B \in \mathfrak{N}(\mathscr{L}')$ and for every $C \in \hat{\mathfrak{R}}(\bar{\mathscr{L}}(\ell))$, the set $B \cap C$ belongs to $\mathfrak{N}(\bar{\mathscr{L}}(\ell))$.*

(f) *For every $f \in \mathscr{N}(\mathscr{L}')$ and for every $C \in \hat{\mathfrak{R}}(\bar{\mathscr{L}}(\ell))$, the function fe_C belongs to $\mathscr{N}(\bar{\mathscr{L}}(\ell))$.*

If $(X, \mathscr{L}', \ell')$ is an admissible extension of $(X, \mathscr{L}, \ell)$, then the assertions (c)–(f) are each equivalent to the assertions (a), (b).

Proof. (a) $\Rightarrow$ (b). Let $f \in \mathscr{L}'_+$. Every determined extension of $(X, \mathscr{L}, \ell)$ is also admissible (Corollary 6), so there exists $g \in \bar{\mathscr{L}}(\ell)_+$ with $f = g$ $\mathscr{L}'$-a.e. Set

$$A := \{g \neq 0\} \cap \{f = g\}$$

$$B := \{g \neq 0\} \cap \{f \neq g\}.$$

Note that

$$fe_A = ge_A$$

$$g = ge_A + ge_B.$$

The major portion of the proof consists of verifying the following

Claim: $ge_B \in \mathscr{N}(\bar{\mathscr{L}}(\ell))$ and $B \in \mathfrak{N}(\bar{\mathscr{L}}(\ell))$.

Approximate ge_B from above by a function in $\bar{\mathscr{L}}(\ell)$ as follows. Set

$$\mathscr{H} := \{h' \mid h' \in \bar{\mathscr{L}}(\ell),\ h' \geq ge_B\}$$

$$\alpha := \inf_{h' \in \mathscr{H}} \bar{\ell}(h').$$

Note that $\mathscr{H}$ is nonempty, since g is in $\mathscr{H}$, so α is real. Let $(h_n)_{n \in \mathbb{N}}$ be a decreasing sequence from $\mathscr{H}$ with

$$\inf_{n \in \mathbb{N}} \bar{\ell}(h_n) = \alpha$$

and set

$$h := g \wedge \bigwedge_{n \in \mathbb{N}} h_n.$$

The function h belongs to $\bar{\mathscr{L}}(\ell)$, therefore also to $\mathscr{L}'$. We have $\bar{\ell}(h) = \alpha$, $0 \leq ge_B \leq h$, and

$$h = he_A + he_B. \tag{1}$$

Since $B \subset \{f \neq g\}$, we have

$$he_B \in \mathscr{N}(\mathscr{L}')$$

from which we conclude that he_A belongs to $\mathscr{L}'_+$. Next we show that the hypotheses of Corollary 3 are satisfied for the set $A \cap \{h \neq 0\}$. Equivalently (Proposition 2) we show

$$\mathfrak{P}(A \cap \{h \neq 0\}) \cap \hat{\mathfrak{R}}(\bar{\mathscr{L}}(\ell)) \subset \mathfrak{N}(\bar{\mathscr{L}}(\ell)). \tag{2}$$

Let $C \in \hat{\mathfrak{R}}(\bar{\mathscr{L}}(\ell))$ with $C \subset A \cap \{h \neq 0\}$. Then he_C belongs to $\bar{\mathscr{L}}(\ell)$ (Proposition 4.1.14), and $he_{X \setminus C}$ belongs to $\mathscr{H}$. Therefore

$$\begin{aligned} \alpha &\leq \bar{\ell}(he_{X \setminus C}) \\ &\leq \bar{\ell}(he_{X \setminus C}) + \bar{\ell}(he_C) \\ &= \bar{\ell}(h) \\ &= \alpha \end{aligned}$$

and $\bar{\ell}(he_C) = 0$. Since $C = \{he_C \neq 0\}$, it follows that C belongs to $\mathfrak{N}(\bar{\mathscr{L}}(\ell))$ and (2) holds. According to Corollary 3, $(X, \mathscr{L}, \ell)$ has an admissible extension $(X, \tilde{\mathscr{L}}, \tilde{\ell})$ for which the positive function he_A is $\tilde{\mathscr{L}}$-exceptional. Since he_A belongs to $\mathscr{L}'_+$, the hypothesis that $(X, \mathscr{L}', \ell')$ is determined implies that he_A is also $\mathscr{L}'$-exceptional. In view of (1), h belongs to $\mathscr{N}(\mathscr{L}')$. But h also belongs to $\bar{\mathscr{L}}(\ell)$, so h must belong to $\mathscr{N}(\bar{\mathscr{L}}(\ell))$. Finally, the inequality $0 \leq ge_B \leq h$ shows that ge_B belongs to $\mathscr{N}(\bar{\mathscr{L}}(\ell))$ and B belongs to $\mathfrak{N}(\bar{\mathscr{L}}(\ell))$, as was claimed.

It now follows that fe_A belongs to $\bar{\mathscr{L}}(\ell)$. Since $A = \{g \neq 0\} \setminus B$, and $\hat{\mathfrak{R}}(\bar{\mathscr{L}}(\ell))$ is a ring of sets (Proposition 4.1.15), the set A belongs to $\hat{\mathfrak{R}}(\bar{\mathscr{L}}(\ell))$.

Finally, the computation

$$\ell'(fe_A) = \bar{\ell}(fe_A) = \bar{\ell}(ge_A) = \bar{\ell}(g) = \ell'(g) = \ell'(f)$$

shows that the positive function $fe_{X \setminus A}$ is $\mathcal{L}'$-exceptional. Thus we have shown that

$$A \in \hat{\mathfrak{R}}(\bar{\mathcal{L}}(\ell)) \qquad fe_A \in \bar{\mathcal{L}}(\ell) \qquad fe_{X \setminus A} \in \mathcal{N}(\mathcal{L}').$$

Note in addition that $A \subset \{f \neq 0\}$.

Now let f be an arbitrary function in $\mathcal{L}'$. By what we have just proved, there exist sets A_1 and A_2 belonging to $\hat{\mathfrak{R}}(\bar{\mathcal{L}}(\ell))$, such that

$$f^+ e_{A_1} \in \bar{\mathcal{L}}(\ell) \qquad f^+ e_{X \setminus A_1} \in \mathcal{N}(\mathcal{L}') \qquad A_1 \subset \{f^+ \neq 0\}$$

$$f^- e_{A_2} \in \bar{\mathcal{L}}(\ell) \qquad f^- e_{X \setminus A_2} \in \mathcal{N}(\mathcal{L}') \qquad A_2 \subset \{f^- \neq 0\}.$$

Note that $A_1 \cap A_2 = \varnothing$, and set $A := A_1 \cup A_2$. Then A belongs to $\hat{\mathfrak{R}}(\bar{\mathcal{L}}(\ell))$, $A \subset \{f \neq 0\}$, and

$$fe_A = f^+ e_{A_1} - f^- e_{A_2}$$

$$fe_{X \setminus A} = f^+ e_{X \setminus A_1} - f^- e_{X \setminus A_2}.$$

Thus fe_A belongs to $\bar{\mathcal{L}}(\ell)$, and $fe_{X \setminus A}$ belongs to $\mathcal{N}(\mathcal{L}')$.

It has now been established that (a) $\Rightarrow$ (b). The proof also shows that the set A can be chosen to satisfy $A \subset \{f \neq 0\}$.

(b) $\Rightarrow$ (a). Let $(X, \tilde{\mathcal{L}}, \tilde{\ell})$ be an arbitrary admissible extension of $(X, \mathcal{L}, \ell)$. We must show that $\tilde{\ell}$ and ℓ' agree on $\mathcal{L}' \cap \tilde{\mathcal{L}}$. As a first step in this direction we show that

$$\mathcal{N}(\mathcal{L}') \cap \tilde{\mathcal{L}} \subset \mathcal{N}(\tilde{\mathcal{L}}) \tag{3}$$

for which it suffices to show that

$$\mathcal{N}(\mathcal{L}') \cap \tilde{\mathcal{L}}_+ \subset \mathcal{N}(\tilde{\mathcal{L}}).$$

Let $f \in \mathcal{N}(\mathcal{L}') \cap \tilde{\mathcal{L}}_+$. By admissibility, there exists $g \in \bar{\mathcal{L}}(\ell)_+$ with

$$f = g \ \tilde{\mathcal{L}}\text{-a.e.} \tag{4}$$

Note that

$$ge_{\{f \neq 0\}} \in \mathcal{N}(\mathcal{L}') \qquad ge_{\{f = 0\}} \in \mathcal{L}'.$$

To the function $ge_{\{f=0\}}$ there exists, by hypothesis, a set A in $\hat{\mathfrak{R}}(\bar{\mathscr{L}}(\ell))$ with

$$ge_{\{f=0\}}e_A \in \bar{\mathscr{L}}(\ell) \qquad ge_{\{f=0\}}e_{X \setminus A} \in \mathscr{N}(\mathscr{L}').$$

In view of (4),

$$ge_{\{f=0\}}e_A \in \mathscr{N}(\tilde{\mathscr{L}})$$

so

$$ge_{\{f=0\}}e_A \in \mathscr{N}(\bar{\mathscr{L}}(\ell)) \subset \mathscr{N}(\mathscr{L}').$$

Thus

$$ge_{\{f=0\}} \in \mathscr{N}(\mathscr{L}') \qquad g \in \mathscr{N}(\mathscr{L}').$$

Hence

$$g \in \mathscr{N}(\bar{\mathscr{L}}(\ell)) \subset \mathscr{N}(\tilde{\mathscr{L}}).$$

In view of (4), f belongs to $\mathscr{N}(\tilde{\mathscr{L}})$, as we wanted to show.

Now that (3) has been verified, let $f \in \mathscr{L}' \cap \tilde{\mathscr{L}}$. By admissibility, there exists g in $\bar{\mathscr{L}}(\ell)$ with

$$f = g \ \tilde{\mathscr{L}}\text{-a.e.} \tag{4'}$$

By hypothesis, there exists a set A in $\hat{\mathfrak{R}}(\bar{\mathscr{L}}(\ell))$ such that

$$fe_A \in \bar{\mathscr{L}}(\ell)$$

$$fe_{X \setminus A} \in \mathscr{N}(\mathscr{L}'). \tag{5}$$

By Proposition 4.1.14,

$$g, ge_A, ge_{X \setminus A} \in \bar{\mathscr{L}}(\ell) \subset \mathscr{L}' \cap \tilde{\mathscr{L}}. \tag{6}$$

In view of (4′),

$$ge_{X \setminus A}e_{\{f=0\}} \in \mathscr{N}(\tilde{\mathscr{L}}) \tag{7}$$

so

$$ge_{X \setminus A}e_{\{f \neq 0\}} \in \tilde{\mathscr{L}}.$$

Because of (5),

$$ge_{X \setminus A}e_{\{f \neq 0\}} \in \mathscr{N}(\mathscr{L}').$$

Now (3) shows that

$$ge_{X \setminus A} e_{\{f \neq 0\}} \in \mathscr{N}(\tilde{\mathscr{L}})$$

which together with (6) and (7) shows that

$$ge_{X \setminus A} \in \mathscr{N}(\bar{\mathscr{L}}(\ell)).$$

The functions fe_A and ge_A both belong to $\bar{\mathscr{L}}(\ell)$. Since they are $\tilde{\mathscr{L}}$-a.e. equal, they must also be $\bar{\mathscr{L}}(\ell)$-a.e. equal. Now we can compute as follows:

$$\tilde{\ell}(f) = \tilde{\ell}(g) = \bar{\ell}(ge_A) = \bar{\ell}(fe_A) = \ell'(f).$$

(b) $\Rightarrow$ (c). Let $f \in \mathscr{L}'$, and let $C \in \hat{\mathfrak{R}}(\bar{\mathscr{L}}(\ell))$. By hypothesis, there exists $A \in \hat{\mathfrak{R}}(\bar{\mathscr{L}}(\ell))$ such that fe_A belongs to $\bar{\mathscr{L}}(\ell)$ and $fe_{X \setminus A}$ is $\mathscr{L}'$-exceptional. The function $fe_A e_C$ belongs to $\bar{\mathscr{L}}(\ell)$ (Proposition 4.1.14). The set $C \setminus A$ belongs to $\hat{\mathfrak{R}}(\bar{\mathscr{L}}(\ell))$ (Proposition 4.1.15) and therefore to $\hat{\mathfrak{R}}(\mathscr{L}')$. The set $\{fe_{C \setminus A} \neq 0\}$ belongs to $\mathfrak{N}(\mathscr{L}')$ and therefore to $\hat{\mathfrak{R}}(\mathscr{L}')$. Thus $(C \setminus A) \setminus \{fe_{C \setminus A} \neq 0\}$ also belongs to $\hat{\mathfrak{R}}(\mathscr{L}')$, and we can choose a function g in $\mathscr{L}'$ with

$$\{g \neq 0\} = (C \setminus A) \setminus \{fe_{C \setminus A} \neq 0\} = \{f = 0\} \cap (C \setminus A).$$

By hypothesis, there exists a set $B \in \hat{\mathfrak{R}}(\bar{\mathscr{L}}(\ell))$ with $ge_B \in \bar{\mathscr{L}}(\ell)$ and $ge_{X \setminus B} \in \mathscr{N}(\mathscr{L}')$. We can and do assume that $B \subset \{g \neq 0\}$. The set $\{C \setminus A) \setminus B$ belongs to $\hat{\mathfrak{R}}(\bar{\mathscr{L}}(\ell))$. Moreover

$$(C \setminus A) \setminus B \subset \{fe_{X \setminus A} \neq 0\} \cup \{ge_{X \setminus B} \neq 0\}$$

so $(C \setminus A) \setminus B$ belongs to $\mathfrak{N}(\mathscr{L}')$. It follows that $(C \setminus A) \setminus B$ belongs to $\mathfrak{N}(\bar{\mathscr{L}}(\ell))$ [Proposition 4.1.4 (b)], so $fe_{(C \setminus A) \setminus B}$ belongs to $\mathscr{N}(\bar{\mathscr{L}}(\ell))$ and therefore to $\bar{\mathscr{L}}(\ell)$. Note that $fe_{B \cap (C \setminus A)} = 0$, so

$$fe_C = fe_C e_A + fe_{(C \setminus A) \setminus B}.$$

We conclude that fe_C belongs to $\bar{\mathscr{L}}(\ell)$.

The implications (c) $\Rightarrow$ (d) $\Rightarrow$ (e) $\Rightarrow$ (f) are easy to check.

(f) plus admissibility $\Rightarrow$ (b). Given $f \in \mathscr{L}'_+$, there exists $g \in \bar{\mathscr{L}}(\ell)$ with $f = g$ $\mathscr{L}'$-a.e. Set

$$A := \{g \neq 0\} \cap \{f = g\}.$$

The inclusion

$$\{fe_{X \setminus A} \neq 0\} \subset \{f \neq g\}$$

shows that $fe_{X \setminus A}$ is $\mathscr{L}'$-exceptional. Note that

$$fe_A = ge_{\{f=g\}}.$$

Since

$$ge_{\{f \neq g\}} = ge_{\{f \neq g\}} e_{\{g \neq 0\}}$$

it follows from (f), using $C := \{g \neq 0\}$, that $ge_{\{f \neq g\}}$ is $\bar{\mathscr{L}}(\ell)$-exceptional. Consequently $ge_{\{f=g\}}$ belongs to $\bar{\mathscr{L}}(\ell)$; in other words fe_A belongs to $\bar{\mathscr{L}}(\ell)$. The definition of A shows that

$$A = \{fe_A \neq 0\}.$$

Consequently A belongs to $\hat{\mathfrak{R}}(\bar{\mathscr{L}}(\ell))$ and $A \subset \{f \neq 0\}$.

For f not necessarily positive, the argument needed to complete the proof is the same as that used earlier in the proof of (a) $\Rightarrow$ (b). □

Corollary 4.2.8. *Let $(X, \mathscr{L}, \ell)$ and $(X, \mathscr{L}', \ell')$ be Daniell spaces with $(X, \mathscr{L}, \ell) \preccurlyeq (X, \mathscr{L}', \ell')$. If $(X, \mathscr{L}', \ell')$ is a determined extension of $(X, \mathscr{L}, \ell)$, then the following assertions hold.*

(*a*) *For every $A \in \hat{\mathfrak{R}}(\bar{\mathscr{L}}(\ell))$ and every $B \in \hat{\mathfrak{R}}(\mathscr{L}')$, if $B \subset A$ then B belongs to $\hat{\mathfrak{R}}(\bar{\mathscr{L}}(\ell))$.*

(*b*) *For every $A \in \hat{\mathfrak{R}}(\bar{\mathscr{L}}(\ell))$ and every $B \in \mathfrak{N}(\mathscr{L}')$, if $B \subset A$ then B belongs to $\mathfrak{N}(\bar{\mathscr{L}}(\ell))$.*

Proof. Apply (d) and (e) from the preceding theorem. □

We can now describe the structure of the set of all determined extensions of a given Daniell space.

Definition 4.2.9. *For $(X, \mathscr{L}, \ell)$ a Daniell space,*

$$\mathscr{N}(\ell) := \{f \in \overline{\mathbb{R}}^X \mid fe_A \in \mathscr{N}(\bar{\mathscr{L}}(\ell)) \text{ for every } A \in \hat{\mathfrak{R}}(\mathscr{L})\}$$

$$\mathfrak{N}(\ell) := \{B \subset X \mid B \cap A \in \mathfrak{N}(\bar{\mathscr{L}}(\ell)) \text{ for every } A \in \hat{\mathfrak{R}}(\mathscr{L})\}. \qquad \square$$

Proposition 4.2.10. *The following assertions hold, for every Daniell space $(X, \mathscr{L}, \ell)$.*

(*a*) *For $f \in \overline{\mathbb{R}}^X$, $f \in \mathscr{N}(\ell)$ iff $\{f \neq 0\} \in \mathfrak{N}(\ell)$.*

(*b*) *$\mathscr{N}(\ell)$ is a tower that is σ-completely embedded in $\overline{\mathbb{R}}^X$.*

(*c*) *$\mathfrak{N}(\ell) = \mathfrak{N}(\mathscr{N}(\ell))$.*

(*d*) *$\mathscr{N}(\ell) = \mathscr{N}(\bar{\mathscr{L}}(\ell) \vdash \mathscr{N}(\ell))$.*

(*e*) *$\mathscr{N}(\ell) \cap \bar{\mathscr{L}}(\ell) = \mathscr{N}(\bar{\mathscr{L}}(\ell))$.*

Proof. Several requirements must be verified, but the verifications are all straightforward and are left to the reader. □

Theorem 4.2.11. *Let $(X, \mathscr{L}, \ell)$ be a Daniell space. Denote by ℓ' the uniquely determined positive linear extension of ℓ to $\mathscr{L} \vdash \mathscr{N}(\ell)$. Let Ω denote the set of all closed Daniell spaces that extend $(X, \mathscr{L}, \ell)$ and are determined by $(X, \mathscr{L}, \ell)$. Then the following assertions hold.*

(*a*) *The closure of $(X, \mathscr{L} \vdash \mathscr{N}(\ell), \ell')$ belongs to Ω.*

(*b*) *Relative to $\preccurlyeq$, $(X, \bar{\mathscr{L}}(\ell), \bar{\ell})$ is the smallest element in Ω.*

(*c*) *Relative to $\preccurlyeq$, the closure of $(X, \mathscr{L} \vdash \mathscr{N}(\ell), \ell')$ is the largest element in Ω.*

(*d*) *Ω is a complete lattice.*

Proof. (a) According to Corollary 4.1.13 and Proposition 10 (b), (e), $(X, \bar{\mathscr{L}}(\ell'), \overline{\ell'})$ is an admissible extension of $(X, \mathscr{L}, \ell)$. Theorem 4.1.10 (c) says that $\mathscr{N}(\bar{\mathscr{L}}(\ell')) = \mathscr{N}(\ell)$. Thus (f) of Theorem 7 holds, relative to the closed Daniell space $(X, \bar{\mathscr{L}}(\ell'), \overline{\ell'})$. It follows that $(X, \bar{\mathscr{L}}(\ell'), \overline{\ell'})$ is determined by $(X, \mathscr{L}, \ell)$.

(b) Assertion (b) follows from the fact that the closure of $(X, \mathscr{L}, \ell)$ is the smallest closed extension of $(X, \mathscr{L}, \ell)$.

(c) Let $(X, \tilde{\mathscr{L}}, \tilde{\ell}) \in \Omega$. Theorem 7 (a) $\Rightarrow$ (f) shows that $\mathscr{N}(\tilde{\mathscr{L}}) \subset \mathscr{N}(\ell)$. Then Theorem 7 (a) $\Rightarrow$ (b) shows that

$$(X, \tilde{\mathscr{L}}, \tilde{\ell}) \preccurlyeq (X, \bar{\mathscr{L}}(\ell'), \overline{\ell'}).$$

(d) Let $((X, \mathscr{L}_\iota, \ell_\iota))_{\iota \in I}$ be a nonempty family from Ω. Fix an arbitrary index ι_0 in I. Then

$$\bigwedge_{\iota \in I}^{\Omega} (X, \mathscr{L}_\iota, \ell_\iota) = \left(X, \bigcap_{\iota \in I} \mathscr{L}_\iota, \ell_{\iota_0} \Big|_{\bigcap_{\iota \in I} \mathscr{L}_\iota} \right).$$

Set

$$\Omega' := \{(X, \tilde{\mathscr{L}}, \tilde{\ell}) \in \Omega \mid (X, \mathscr{L}_\iota, \ell_\iota) \preccurlyeq (X, \tilde{\mathscr{L}}, \tilde{\ell}), \forall \iota \in I\}.$$

Note that Ω' is nonempty since $(X, \bar{\mathscr{L}}(\ell'), \overline{\ell'})$ belongs to Ω'. It follows that

$$\bigvee_{\iota \in I}^{\Omega} (X, \mathscr{L}_\iota, \ell_\iota) = \bigwedge^{\Omega} \{(X, \tilde{\mathscr{L}}, \tilde{\ell}) \mid (X, \tilde{\mathscr{L}}, \tilde{\ell}) \in \Omega'\}.$$ □

We are ready to formulate the definition of "the integral for the Daniell space $(X, \mathscr{L}, \ell)$". The preceding theorem makes clear: the object we want is the closure of $(X, \mathscr{L} \vdash \mathscr{N}(\ell), \ell')$, where ℓ' is the unique positive linear extension of ℓ to $\mathscr{L} \vdash \mathscr{N}(\ell)$.

Definition 4.2.12 (Daniell-space integral). *Let $(X, \mathscr{L}, \ell)$ be a Daniell space. Denote by ℓ' the uniquely determined positive linear extension of ℓ to $\mathscr{L} \vdash \mathscr{N}(\ell)$. Then*

$$\mathscr{L}^1(\ell) := \bar{\mathscr{L}}(\ell')$$

$$\int_\ell : \mathscr{L}^1(\ell) \to \mathbb{R}, \qquad f \mapsto \int_\ell f := \overline{\ell'}(f).$$

*Functions belonging to $\mathscr{L}^1(\ell)$ are said to be **ℓ-integrable**. The number $\int_\ell f$ is called the **ℓ-integral** of f. The triple*

$$\left(X, \mathscr{L}^1(\ell), \int_\ell \right)$$

*is called the **integral** for the Daniell space $(X, \mathscr{L}, \ell)$.*

*Functions belonging to $\mathscr{N}(\ell)$, that is, to $\mathscr{N}(\mathscr{L}^1(\ell))$, are called **$\ell$-exceptional** functions. Sets belonging to $\mathfrak{N}(\ell)$, that is, to $\mathfrak{N}(\mathscr{L}^1(\ell))$, are called **$\ell$-exceptional** sets. In other words, we call the $\mathscr{L}^1(\ell)$-exceptional sets and functions simply ℓ-exceptional.*

*A property P that refers to elements of X is said to hold **ℓ-almost everywhere** (or simply ℓ-a.e.) iff it holds $\mathscr{L}^1(\ell)$-almost everywhere. Rather than write P $\mathscr{L}^1(\ell)$-a.e. and $P(x) \mathscr{L}^1(\ell)$-a.e., we write, respectively, $P\,\ell$-a.e. and $P(x)\,\ell$-a.e.*

□

For the Daniell space $(X, \mathscr{L}, \ell)$ discussed in the paragraph preceding Proposition 4.2.2, the reader can verify that

$$\mathscr{L}^1(\ell) = \tilde{\mathscr{L}} = \{ f \in \overline{\mathbb{R}}^X \mid f(b) = f(c) \in \mathbb{R} \}$$

$$\int_\ell f = \tilde{\ell}(f) = f(b) = f(c) \qquad (f \in \mathscr{L}^1(\ell)).$$

The construction of the integral for a given Daniell space has the same kind of idempotence as the construction of the closure.

Proposition 4.2.13 (Idempotence of integral). *For every Daniell space $(X, \mathscr{L}, \ell)$, the integral of $(X, \mathscr{L}, \ell)$ is its own integral:*

$$\left(X, \mathscr{L}^1\left(\int_\ell \right), \int_{\int_\ell} \right) = \left(X, \mathscr{L}^1(\ell), \int_\ell \right).$$

Proof. That

$$\left(X, \mathscr{L}^1\left(\int_\ell\right), \int_{\int_\ell}\right) \succcurlyeq \left(X, \mathscr{L}^1(\ell), \int_\ell\right)$$

is immediate. It suffices therefore to show that

$$\mathscr{L}^1\left(\int_\ell\right) \subset \mathscr{L}^1(\ell). \tag{8}$$

Note that $(X, \mathscr{L}^1(\ell), \int_\ell)$ is its own closure. Let $f \in \mathscr{N}(\int_\ell)$, and let $A \in \hat{\mathfrak{R}}(\mathscr{L})$. Then A belongs to $\hat{\mathfrak{R}}(\mathscr{L}^1(\ell))$, and the definition of $\mathscr{N}(\int_\ell)$ shows that fe_A belongs to $\mathscr{N}(\ell)$. Since $fe_A = fe_A e_A$, the definition of $\mathscr{N}(\ell)$ then shows that fe_A belongs to $\mathscr{N}(\bar{\mathscr{L}}(\ell))$. Thus fe_A belongs to $\mathscr{N}(\bar{\mathscr{L}}(\ell))$ for every A in $\hat{\mathfrak{R}}(\mathscr{L})$, so f belongs to $\mathscr{N}(\ell)$. We have shown that

$$\mathscr{N}\left(\int_\ell\right) \subset \mathscr{N}(\ell).$$

Now let $f \in \mathscr{L}^1(\int_\ell)$. There exists a set $B \in \hat{\mathfrak{R}}(\mathscr{L}^1(\ell))$ such that

$$fe_B \in \mathscr{L}^1(\ell) \qquad fe_{X \setminus B} \in \mathscr{N}\left(\int_\ell\right)$$

[Theorem 7 (a) $\Rightarrow$ (b)]. According to the preceding argument, $fe_{X \setminus B}$ belongs to $\mathscr{N}(\ell)$. Therefore f belongs to $\mathscr{L}^1(\ell)$, and (8) holds. □

In numerous important cases the integral and the closure of a Daniell space are identical. This occurs, for example, in the case of summable families (Theorem 3.4.10). The following property often proves to be the reason.

Definition 4.2.14. *Let $\mathscr{F} \subset \overline{\mathbb{R}}^X$. Then $\mathscr{F}$ is said to be* **σ-finite** *iff there is a countable family $(f_\iota)_{\iota \in I}$ from $\mathscr{F}$ such that*

$$X = \bigcup_{\iota \in I} \{f_\iota \neq 0\}.$$ □

Proposition 4.2.15. *Let $(X, \mathscr{L}, \ell)$ be a Daniell space. If $\mathscr{L}$ is σ-finite, then the closure of $(X, \mathscr{L}, \ell)$ and the integral of $(X, \mathscr{L}, \ell)$ are identical:*

$$(X, \bar{\mathscr{L}}(\ell), \bar{\ell}) = \left(X, \mathscr{L}^1(\ell), \int_\ell\right).$$

Proof. That $(X, \bar{\mathscr{L}}(\ell), \bar{\ell}) \preccurlyeq (X, \mathscr{L}^1(\ell), \int_\ell)$ is immediate. We need only show that

$$\mathscr{L}^1(\ell) \subset \bar{\mathscr{L}}(\ell)$$

for which, since $\mathscr{L}^1(\ell) = \bar{\mathscr{L}}(\ell) \vdash \mathscr{N}(\ell)$ (Theorem 4.1.12), it suffices to show that

$$\mathfrak{N}(\ell) \subset \mathfrak{N}(\bar{\mathscr{L}}(\ell)). \tag{9}$$

Let $(f_\iota)_{\iota \in I}$ be a countable family from $\mathscr{L}$ such that

$$X = \bigcup_{\iota \in I} \{ f_\iota \neq 0 \}.$$

Given $B \in \mathfrak{N}(\ell)$, use Theorem 7 (a) $\Rightarrow$ (e) to conclude that each $B \cap \{ f_\iota \neq 0\}$ belongs to $\mathfrak{N}(\bar{\mathscr{L}}(\ell))$. Since

$$B = \bigcup_{\iota \in I} (B \cap \{ f_\iota \neq 0\})$$

it follows that B belongs to $\mathfrak{N}(\bar{\mathscr{L}}(\ell))$. Thus (9) holds. □

Example 4.2.16 (Stieltjes functionals). Among the cases to which the preceding proposition applies are the Daniell spaces for the Stieltjes functionals ℓ_g on arbitrary intervals A. (See Sections 2.5 and 3.6.) The Riesz lattice $\mathscr{K}(A)$ is obviously σ-finite, so the integral for the Daniell space $(A, \mathscr{K}(A), \ell_g)$ is simply its closure, which was described in Corollary 3.6.7. No further investigation of the Stieltjes functionals is required. □

The theory of summable families, which took the counting measure as its starting point, has a natural generalization. Weighted counting measures and the theory that flows from them provide an interesting example of closure and integral. This example, which we take up next, will enable us, among other things, both to illustrate and to emphasize one respect in which the Daniell-space integral has significantly better properties than the Daniell-space closure.

Definition 4.2.17. *For each positive real function g on X,*

$$\chi_g \colon \mathfrak{F}(X) \to \mathbb{R}_+, \qquad A \mapsto \sum_{x \in A} g(x). \qquad \square$$

Definition 4.2.18. *The measures χ_g, with g a positive real function on X are called* ***weighted counting measures*** *on X. For each such g we denote by ℓ_{χ_g} the functional associated with $(X, \mathfrak{F}(X), \chi_g)$, and we call $(X, \mathscr{F}(X), \ell_{\chi_g})$ the* ***Daniell space*** *associated with the weighted counting measure χ_g.* □

Proposition 4.2.19 (Weighted counting measure: closure). *Let g be a positive real function on X. Then the following assertions are equivalent, for every*

$f \in \overline{\mathbb{R}}^X$.

(a) $f \in \overline{\mathscr{L}}(\ell_{\chi_g})$.

(b) $fg \in \mathscr{L}^1(X)$ *and* $\{f \neq 0\}$ *is countable.*

Moreover, if $f \in \overline{\mathscr{L}}(\ell_{\chi_g})$, *then* $\overline{\ell_{\chi_g}}(f)$ *is given by*

$$\overline{\ell_{\chi_g}}(f) = \sum_{x \in X} f(x)g(x). \tag{10}$$

Proof. Formula (10) and (a) $\Rightarrow$ (b). We use the second form of the Induction Principle (Theorem 3.2.8). Set

$$\mathscr{G} := \left\{ f \in \overline{\mathscr{L}}\left(\ell_{\chi_g}\right) \mid fg \in \mathscr{L}^1(X), \overline{\ell_{\chi_g}}(f) = \sum_{x \in X} f(x)g(x) \right\}.$$

For $f \in \mathscr{F}(X)$, $\{f \neq 0\}$ is finite, and we have

$$f = \sum_{x \in \{f \neq 0\}} f(x) e_{\{x\}}$$

$$fg = \sum_{x \in \{f \neq 0\}} f(x)g(x)e_{\{x\}} \in \mathscr{L}^1(X)$$

$$\begin{aligned} \ell_{\chi_g}(f) &= \sum_{x \in \{f \neq 0\}} f(x)\ell_{\chi_g}(e_{\{x\}}) \\ &= \sum_{x \in \{f \neq 0\}} f(x)\chi_g(\{x\}) \\ &= \sum_{x \in \{f \neq 0\}} f(x)g(x). \end{aligned}$$

We conclude that $\mathscr{F}(X) \subset \mathscr{G}$. Let $(f_n)_{n \in \mathbb{N}}$ be an $\overline{\ell_{\chi_g}}$-sequence from $\mathscr{G}$. Then the sequence $(f_n g)_{n \in \mathbb{N}}$ is a Σ-sequence from $\mathscr{L}^1(X)$ (see Section 3.4) and $(\lim_{n \to \infty} f_n)g = \lim_{n \to \infty} f_n g$. It follows that $(\lim_{n \to \infty} f_n)g$ belongs to $\mathscr{L}^1(X)$ and

$$\begin{aligned} \overline{\ell_{\chi_g}}\left(\lim_{n \to \infty} f_n\right) &= \lim_{n \to \infty} \overline{\ell_{\chi_g}}(f_n) \\ &= \lim_{n \to \infty} \sum_{x \in X} f_n(x)g(x) \\ &= \sum_{x \in X} \left(\lim_{n \to \infty} f_n\right)(x)g(x) \end{aligned}$$

so $\lim_{n \to \infty} f_n$ belongs to $\mathscr{G}$. Now let f be a function in $\overline{\mathscr{L}}(\ell_{\chi_g})$ such that

$$f = \lim_{n \to \infty} f_n \quad \overline{\mathscr{L}}\left(\ell_{\chi_g}\right)\text{-a.e.}$$

Certainly

$$f(x) = \lim_{n\to\infty} f_n(x) \quad \text{if } g(x) \neq 0$$

so

$$fg = \left(\lim_{n\to\infty} f_n\right) g.$$

Therefore fg belongs to $\mathscr{L}^1(X)$ and

$$\overline{\ell_{\chi_g}}(f) = \overline{\ell_{\chi_g}}\left(\lim_{n\to\infty} f_n\right) = \sum_{x\in X}\left(\lim_{n\to\infty} f_n\right)(x)g(x)$$

$$= \sum_{x\in X} f(x)g(x).$$

It follows that f belongs to $\mathscr{G}$. According to Theorem 3.2.8, $\mathscr{G} = \bar{\mathscr{L}}(\ell_{\chi_g})$. We have established formula (10) and shown that (a) implies fg summable. Since f belongs to $\bar{\mathscr{L}}(\ell_{\chi_g})$, so does $|f|$ and there exists a function h in $\mathscr{F}(X)^{\uparrow}$ such that $h \geq |f|$. For h to belong to $\mathscr{F}(X)^{\uparrow}$ requires that $\{h \neq 0\}$ be countable (Proposition 3.4.3). Thus $\{f \neq 0\}$ is countable.

(b) $\Rightarrow$ (a). First let f be a positive function in $\overline{\mathbb{R}}^X$ such that fg belongs to $\mathscr{L}^1(X)$ and $\{f \neq 0\}$ is countable. From $\mathfrak{F}(X)$ choose an increasing sequence $(A_n)_{n\in\mathbb{N}}$ whose union is the set $\{f \neq 0\}$. Construct in $\mathscr{F}(X)$ an increasing sequence $(f_n)_{n\in\mathbb{N}}$ whose supremum is f by defining

$$f_n\colon X \to \mathbb{R}, \qquad x \mapsto \begin{cases} n & \text{if } x \in A_n,\ f(x) = \infty \\ f(x) & \text{if } x \in A_n,\ f(x) < \infty \\ 0 & \text{if } x \in X \setminus A_n. \end{cases}$$

For every n we have

$$0 \leq \overline{\ell_{\chi_g}}(f_n) = \ell_{\chi_g}(f_n) = \sum_{x\in X} f_n(x)g(x)$$

$$\leq \sum_{x\in X} f(x)g(x) < \infty.$$

Thus $(f_n)_{n\in\mathbb{N}}$ is an $\overline{\ell_{\chi_g}}$-sequence, and we conclude that f belongs to $\bar{\mathscr{L}}(\ell_{\chi_g})$.

If f is not necessarily positive but satisfies (b), then f^+ and f^- also satisfy (b). Therefore f^+, f^-, and f all belong to $\bar{\mathscr{L}}(\ell_{\chi_g})$. □

Let us look more closely at this characterization of $\bar{\mathscr{L}}(\ell_{\chi_g})$. It is not at all clear what the countability of $\{f \neq 0\}$ has to do with anything. To be sure, this condition is necessary in the case of $\mathscr{L}^1(X)$ (i.e., $g = e_X$). How absurd it is in the general case becomes very clear if we take g to be the zero function, so that

$\ell_{X_g}(f) = 0$ for every f in $\mathscr{F}(X)$. One naturally expects the integral of ℓ_{X_g} to be defined on all of $\overline{\mathbb{R}}^X$ and to assign to every function $f \in \overline{\mathbb{R}}^X$ the value zero. That $\{f \neq 0\}$ be countable is a totally meaningless requirement. This weakness exhibited by the Daniell-space closure is fully overcome in the Daniell-space integral, both in the specific example just discussed and in general. Indeed, the reader can readily verify the following results.

Theorem 4.2.20 (Weighted counting measure: integral). *The following assertions hold, for every positive real function g on X.*

(a) $\mathscr{N}(\ell_{X_g}) = \{f \in \overline{\mathbb{R}}^X \mid fg = 0\}$.

(b) $\mathfrak{N}(\ell_{X_g}) = \{A \subset X \mid A \subset \{g = 0\}\}$.

(c) *For every $f \in \overline{\mathbb{R}}^X$, f is ℓ_{X_g}-integrable iff fg is summable.*

(d) *For every $f \in \mathscr{L}^1(\ell_{X_g})$,*

$$\int_{\ell_{X_g}}(f) = \sum_{x \in X} f(x)g(x).$$

(e) *The integral $(X, \mathscr{L}^1(\ell_{X_g}), \int_{\ell_{X_g}})$ and the closure $(X, \overline{\mathscr{L}}(\ell_{X_g}), \overline{\ell_{X_g}})$ coincide iff $\{g = 0\}$ is countable.* □

By means of Theorem 20 (c) and (d), all properties of the integral for ℓ_{X_g} can be derived from those of summable families.

Proposition 4.2.21. *The following assertions are equivalent, for every Daniell space $(X, \mathscr{L}, \ell)$.*

(a) $\ell = 0$.

(b) $\mathscr{N}(\ell) = \overline{\mathbb{R}}^X$.

(c) $\mathfrak{N}(\ell) = \mathfrak{P}(X)$.

(d) $(X, \mathscr{L}^1(\ell), \int_\ell) = (X, \overline{\mathbb{R}}^X, 0)$. □

EXERCISES

4.2.1(E) Set $X := \{1,2\}$ and $\mathscr{L} := \{\alpha e_{\{1\}} \mid \alpha \in \mathbb{R}\}$. Define $\ell: \mathscr{L} \to \mathbb{R}$, $\alpha e_{\{1\}} \mapsto \alpha$. Prove the following:

(α) $(X, \mathscr{L}, \ell)$ is a Daniell space.

(β) $(X, \overline{\mathscr{L}}(\ell), \bar{\ell}) = (X, \mathscr{L}, \ell)$. In particular $\mathscr{N}(\overline{\mathscr{L}}(\ell)) = \{0\}$ and $\mathfrak{N}(\overline{\mathscr{L}}(\ell)) = \{\varnothing\}$.

(γ) $\mathscr{N}(\ell) = \{\alpha e_{\{2\}} \mid \alpha \in \overline{\mathbb{R}}\}$ and $\mathfrak{N}(\ell) = \{\varnothing, \{2\}\}$.

(δ) $\mathscr{L}^1(\ell) = \{f \in \overline{\mathbb{R}}^{\{1,2\}} \mid f(1) \in \mathbb{R}\}$ and $\int_\ell(f) = f(1)$ for any $f \in \mathscr{L}^1(\ell)$.

(ε) $(X, \mathscr{L}^1(\ell), \int_\ell)$ is the only admissible extension of $(X, \mathscr{L}, \ell)$.

(ζ) Define ℓ': $\mathbb{R}^{\{1,2\}} \to \mathbb{R}$, $f \mapsto f(1) + f(2)$. Then $(X, \mathbb{R}^{\{1,2\}}, \ell')$ is not an admissible extension of $(X, \mathscr{L}, \ell)$.

4.2.2(E) Let X be a nonempty set, and $(Y, \mathscr{L}, \ell)$ a Daniell space. Given $x \in X$, define $i_x\colon Y \to X \times Y$, $y \mapsto (x, y)$. Given $\mathscr{F} \subset \overline{\mathbb{R}}^Y$ and $x \in X$, define

$$\mathscr{F}_x := \left\{ f \in \overline{\mathbb{R}}^{X \times Y} \,|\, f \circ i_x \in \mathscr{F} \text{ and } f \circ i_{x'} = 0 \text{ whenever } x' \in X,\ x' \neq x \right\}.$$

Define

$$\mathscr{L}' := \left\{ \sum_{z \in Z}^{*} f_z \,|\, Z \in \mathfrak{F}(X),\ f_z \in \mathscr{L}_z \text{ for } z \in Z \right\}$$

and

$$\ell'\colon \mathscr{L}' \to \mathbb{R},\quad \sum_{z \in Z}^{*} f_z \mapsto \sum_{z \in Z} \ell(f_z \circ i_z).$$

Prove the following:

(α) $(X \times Y, \mathscr{L}', \ell')$ is a Daniell space.

(β) $\bar{\mathscr{L}}(\ell')$ is the set of all $f \in \overline{\mathbb{R}}^{X \times Y}$ of the form $f = \Sigma^*_{z \in Z} f_z$ such that Z is countable, $f_z \in \bar{\mathscr{L}}(\ell)_z$ for each $z \in Z$, and $(\bar{\ell}(|f_z \circ i_z|))_{z \in Z}$ is summable. For each $f \in \bar{\mathscr{L}}(\ell')$, $\bar{\ell}'(f) = \Sigma_{z \in Z} \bar{\ell}(f_z \circ i_z)$.

(γ) $\mathscr{N}(\ell') = \{ f \in \overline{\mathbb{R}}^{X \times Y} \,|\, f = \Sigma^*_{z \in Z} f_z$ where $Z \subset X$ and $f_z \in \mathscr{N}(\ell)_z$ for each $z \in Z \}$.

(δ) $\mathfrak{N}(\ell') = \{ A \subset X \times Y \,|\, i_x^{-1}(A) \in \mathfrak{N}(\ell)$ for each $x \in X \}$.

(ε) $\mathscr{L}^1(\ell')$ is the set of all $f \in \overline{\mathbb{R}}^{X \in Y}$ of the form $f = \Sigma^*_{z \in Z} f_z$ where $Z \subset X$, $f_z \in \mathscr{L}^1(\ell)_z$ for each $z \in Z$ and $(\int_\ell(|f_z \circ i_z|))_{z \in Z}$ is summable. If $f \in \mathscr{L}^1(\ell')$, then $\int_{\ell'}(f) = \Sigma_{z \in Z} \int_\ell(f_z \circ i_z)$ where $f = \Sigma^*_{z \in Z} f_z$ as above.

Ex. 4.2.1 and Ex. 4.2.2 present the types of sets which come into consideration as possible additional sets of measure 0 when constructing the integral.

4.2.3(E) Set $X := \{1, 2\}$, and define

$$\mathscr{L} := \left\{ \alpha e_{\{1,2\}} \,|\, \alpha \in \mathbb{R} \right\} \quad \text{and} \quad \ell\colon \mathscr{L} \to \mathbb{R},\ \alpha e_{\{1,2\}} \mapsto \alpha$$

$$\mathscr{L}_1 := \left\{ \alpha e_{\{1\}} \,|\, \alpha \in \mathbb{R} \right\} \quad \text{and} \quad \ell_1\colon \mathscr{L}_1 \to \mathbb{R},\ \alpha e_{\{1\}} \mapsto \alpha$$

$$\mathscr{L}_2 := \left\{ \alpha e_{\{2\}} \,|\, \alpha \in \mathbb{R} \right\} \quad \text{and} \quad \ell_2\colon \mathscr{L}_2 \to \mathbb{R},\ \alpha e_{\{2\}} \mapsto \alpha.$$

Prove the following statements (cf., Ex. 4.2.1):

(α) $(X, \mathscr{L}, \ell)$, $(X, \mathscr{L}_1, \ell_1)$, and $(X, \mathscr{L}_2, \ell_2)$ are Daniell spaces.

(β) $(X, \mathscr{L}^1(\ell_1), \int_{\ell_1})$ and $(X, \mathscr{L}^1(\ell_2), \int_{\ell_2})$ are admissible extensions of $(X, \mathscr{L}, \ell)$.

(γ) $(X, \mathscr{L}^1(\ell_1), \int_{\ell_1})$ and $(X, \mathscr{L}^1(\ell_2), \int_{\ell_2})$ are not determined by $(X, \mathscr{L}, \ell)$.

(δ) Define $\ell^1\colon \mathbb{R}^{\{1,2\}} \to \mathbb{R}$, $f \mapsto (1/2)(f(1) + f(2))$. Then $(X, \mathbb{R}^{\{1,2\}}, \ell^1)$ is not an admissible extension of $(X, \mathscr{L}, \ell)$.

(ϵ) $(X, \mathscr{L}, \ell)$ is its own integral.

4.2.4(E) We investigate the integrals of the functionals in Ex. 1.3.5.

(a) Functionals on $\mathscr{k}(X)$

Every positive linear functional on $\mathscr{k}(X)$ is nullcontinuous (even π-continuous) and of the form ℓ_g for some $g \in \mathbb{R}_+^X$. Prove the following:

(α) $(X, \mathscr{L}^1(\ell_g), \int_{\ell_g}) = (X, \ell_g^1(X), \overline{\ell_g})$ where $(X, \ell_g^1(X), \overline{\ell_g})$ is the Daniell space considered in Ex. 1.3.2.

(β) $\mathscr{N}(\ell_g) = \{f \in \overline{\mathbb{R}}^X \mid \{f \neq 0\} \subset \{g = 0\}\} = \{f \in \overline{\mathbb{R}}^X \mid fg = 0\}$.

(γ) $\mathfrak{N}(\ell_g) = \mathfrak{P}(\{g = 0\})$.

(δ) $(X, \mathscr{L}^1(\ell_g), \int_{\ell_g}) = (X, \mathscr{L}^1(\ell_{\chi_g}), \int_{\ell_{\chi_g}})$.

(b) Functionals on $c_0(X)$

Every positive linear functional ℓ on $c_0(X)$ is nullcontinuous and satisfies $\ell(f) = \Sigma_{x \in X} f(x)g(x)$ for a $g \in \ell^1(X)_+$. Show that the following hold:

(α) $c_0(X) \subset \overline{\mathscr{L}}(\ell_g) \subset \mathscr{L}^1(\ell_g)$

(β) $(X, \mathscr{L}^1(\ell), \int_\ell) = (X, \mathscr{L}^1(\ell_g), \int_{\ell_g})$

(γ) $\mathscr{N}(\ell) = \mathscr{N}(\ell_g)$

(δ) $\mathfrak{N}(\ell) = \mathfrak{N}(\ell_g)$.

(c) Functionals on $\ell^\infty(X)$

Every π-continuous positive linear functional ℓ on $\ell^\infty(X)$ satisfies $\ell(f) = \Sigma_{x \in X} f(x)g(x)$ for some $g \in \ell^1(X)$. Show that if ℓ is such a functional, then these four assertions hold:

(α) $\ell^\infty(X) \subset \mathscr{L}^1(\ell_g)$

(β) $(X, \mathscr{L}^1(\ell), \int_\ell) = (X, \mathscr{L}^1(\ell_g), \int_{\ell_g})$

(γ) $\mathscr{N}(\ell) = \mathscr{N}(\ell_g)$

(δ) $\mathfrak{N}(\ell) = \mathfrak{N}(\ell_g)$.

We have also seen that given certain assumptions about the underlying set theory, there are nullcontinuous positive linear functionals on $\ell^\infty(X)$ that are not π-continuous (cf., Exs. 1.3.5 and 2.2.19). Such functionals are not of the form ℓ_g. Now take an arbitrary nullcontinuous positive linear functional ℓ on $\ell^\infty(X)$, and prove the next three assertions:

(ε) $\mathscr{L}^1(\ell)_+ = \{f \in \overline{\mathbb{R}}_+^X \mid \sup_{\alpha \in \mathbb{R}_+} \ell(f \wedge \alpha) < \infty\}$.

(ζ) $\mathscr{N}(\ell) = \{f \in \overline{\mathbb{R}}^X \mid \ell(e_{\{f \neq 0\}}) = 0\}$.

(η) $\mathfrak{N}(\ell) = \{A \subset X \mid \ell(e_A) = 0\}$.

(d) Functionals on $c_f(X)$

Let ℓ be a π-continuous positive linear functional on $c_f(X)$. Then $\ell(f) = \Sigma_{x \in X} f(x)g(x)$ for some $g \in \ell^1(X)$. We have $c_f(X) \subset \ell^\infty(X)$, and it is easy to see that statements can be made about the integral obtained from ℓ which correspond precisely to (α)–(δ) for the case $\ell^\infty(X)$.

Now consider the functional $\ell\colon c_f(X) \to \mathbb{R}$, $\alpha e_X + h \mapsto \alpha$, with X uncountable and $h \in \mathcal{k}(X)$. Show the following:

(α) $\mathscr{L}^1(\ell) = \{\alpha e_X + h \mid h \in \bar{\mathscr{L}}(\ell \mid_{\mathcal{k}(X)})\} = \{\alpha e_X + h \mid h \in \overline{\mathbb{R}}^X, \ \{h \neq 0\}$ countable$\}$.

(β) $\mathscr{N}(\ell) = \mathscr{N}(\bar{\mathscr{L}}(\ell \mid_{\mathcal{k}(X)})) = \bar{\mathscr{L}}(\ell \mid_{\mathcal{k}(X)}) = \{f \in \overline{\mathbb{R}}^X \mid \{f \neq 0\}$ countable$\}$.

(γ) $\mathfrak{N}(\ell) = \mathfrak{N}(\bar{\mathscr{L}}(\ell \mid_{\mathcal{k}(X)})) = \{A \subset X \mid A$ countable$\}$.

4.2.5(E) Let $(X, \mathscr{L}, \ell)$ be a Daniell space. Set

$$\Phi := \{(X, \mathscr{L}', \ell') \mid (X, \mathscr{L}', \ell') \text{ is an admissible extension of } (X, \mathscr{L}, \ell)\}.$$

This exercise is devoted to investigating the order properties of Φ with respect to $\preccurlyeq$.

(a) Take a nonempty family $((X, \mathscr{L}_\iota, \ell_\iota))_{\iota \in I}$ from Φ. Given $f \in \bar{\mathscr{L}}(\ell)$, define

$$\mathscr{L}_f := \left\{ g \in \bigcap_{\iota \in I} \mathscr{L}_\iota \mid f = g \ \mathscr{L}_\iota\text{-a.e. for each } \iota \in I \right\}$$

and

$$\tilde{\mathscr{L}} := \bigcup_{f \in \bar{\mathscr{L}}(\ell)} \mathscr{L}_f, \quad \tilde{\ell}\colon \tilde{\mathscr{L}} \to \mathbb{R},\ g \mapsto \bar{\ell}(f)$$

where $g \in \mathscr{L}_f$. Prove that $\tilde{\ell}$ is well defined and verify the assertions that follow:

(α) $(X, \tilde{\mathscr{L}}, \tilde{\ell}) \in \Phi$

(β) $(X, \tilde{\mathscr{L}}, \tilde{\ell}) = \bigwedge^{\Phi}_{\iota \in I} (X, \mathscr{L}_\iota, \ell_\iota)$.

(b) Suppose that for each $(X, \mathscr{L}', \ell') \in \Phi$ there is a maximal element $(X, \mathscr{L}'', \ell'')$ in Φ with $(X, \mathscr{L}', \ell') \preccurlyeq (X, \mathscr{L}'', \ell'')$. Define

$$\Phi_{\max} := \{(X, \mathscr{L}', \ell') \in \Phi \mid (X, \mathscr{L}', \ell') \text{ maximal in } \Phi\}.$$

Prove that

$$\left(X, \mathscr{L}^1(X), \int_\ell\right) = \bigwedge^{\Phi} \{(X, \mathscr{L}', \ell') \mid (X, \mathscr{L}', \ell') \in \Phi_{\max}\}.$$

(c) Prove the equivalence of the following statements:

(α) Every totally ordered subset of Φ is bounded above.

(β) Every well-ordered subset of Φ is bounded above.

(γ) Every increasing sequence from Φ is bounded above.
(δ) If $A \in \hat{\mathfrak{R}}(\bar{\mathscr{L}}(\ell))$ and $B \subset A$, then either $B \in \hat{\mathfrak{R}}(\bar{\mathscr{L}}(\ell))$ or B is finite.

(γ) $\Rightarrow$ (δ) can be proved indirectly. Take $A \in \hat{\mathfrak{R}}(\bar{\mathscr{L}}(\ell))$ and $B \subset A$ such that B is infinite and $B \notin \hat{\mathfrak{R}}(\bar{\mathscr{L}}(\ell))$. Take $f \in \bar{\mathscr{L}}(\ell)_+$ such that $A = \{f \neq 0\}$. Then

$$\alpha := \inf\{\bar{\ell}(g) \mid g \in \bar{\mathscr{L}}(\ell),\, g \geq fe_B\} \geq \sup\{\bar{\ell}(g) \mid g \in \bar{\mathscr{L}}(\ell),\, g \leq fe_B\}.$$

Take $g \in \bar{\mathscr{L}}(\ell)$ with $g \geq fe_B$ and $\bar{\ell}(g) = \alpha$. Define $C := \{g \neq 0\} \setminus B$. Then there is an $(X, \mathscr{L}', \ell') \in \Phi$ such that

$$\mathscr{L}' = \bar{\mathscr{L}}(\ell) \vdash \{h \in \overline{\mathbb{R}}^X \mid \{h \neq 0\} \subset C\}.$$

Then $fe_B \in \mathscr{L}'$ and $\ell'(fe_B) = \alpha$. On the other hand, there is an increasing sequence $(B_n)_{n \in \mathbb{N}}$ of subsets of B such that $B_n \neq B_{n+1}$ for each $n \in \mathbb{N}$ and $B = \bigcup_{n \in \mathbb{N}} B_n$. Define for each $n \in \mathbb{N}$

$$\beta_n := \sup\{\bar{\ell}(g) \mid g \in \bar{\mathscr{L}}(\ell),\, g \leq fe_{B_n}\}.$$

Now construct an increasing sequence $((X, \mathscr{L}_n, \ell_n))_{n \in \mathbb{N}}$ from Φ with the following properties:

(i) $(X, \mathscr{L}', \ell') \preccurlyeq (X, \mathscr{L}_1, \ell_1)$.
(ii) If $n \in \mathbb{N}$ and $k \in \mathbb{N}_n$, then $fe_{B_k} \in \mathscr{L}_n$ and $\ell_n(fe_{B_k}) = \beta_k$.

$((X, \mathscr{L}_n, \ell_n))_{n \in \mathbb{N}}$ cannot be bounded above in Φ.

(δ) $\Rightarrow$ (α) can also be proved indirectly. Let Φ' be a totally ordered subset of Φ that is not bounded above. So set $\tilde{\mathscr{L}} := \bigcup\{\mathscr{L}' \mid (X, \mathscr{L}', \ell') \in \Phi'\}$, and define $\tilde{\ell}\colon \tilde{\mathscr{L}} \to \mathbb{R}$ by $\tilde{\ell}|_{\mathscr{L}'} = \ell'$ for each $(X, \mathscr{L}', \ell') \in \Phi'$. By assumption, $(X, \tilde{\mathscr{L}}, \tilde{\ell})$ is not a Daniell space. Thus for some $f \in \tilde{\mathscr{L}}_+$ there is an increasing sequence $(f_n)_{n \in \mathbb{N}}$ from $\tilde{\mathscr{L}}_+$ with $f = \bigvee_{n \in \mathbb{N}} f_n$ and $\tilde{\ell}(f) > \sup_{n \in \mathbb{N}} \tilde{\ell}(f_n)$. We may assume that $f \in \bar{\mathscr{L}}(\ell)$. Then define for each $n \in \mathbb{N}$

$$\Phi_n := \{(X, \mathscr{L}', \ell') \in \Phi \mid f_k \in \mathscr{L}' \text{ and } \ell'(f_k) = \tilde{\ell}(f_k) \text{ whenever } k \in \mathbb{N}_n\}.$$

Define

$$(X, \mathscr{L}_n', \ell_n') := \bigwedge^{\Phi}\{(X, \mathscr{L}', \ell') \mid (X, \mathscr{L}', \ell') \in \Phi_n\}.$$

$((X, \mathscr{L}_n', \ell_n'))_{n \in \mathbb{N}}$ is an increasing sequence from Φ. A little thought shows that there is a subsequence $((X, \mathscr{L}_{n_k}', \ell_{n_k}'))_{k \in \mathbb{N}}$ with $(X, \mathscr{L}_{n_k}', \ell_{n_k}') \prec (X, \mathscr{L}_{n_{k+1}}', \ell_{n_{k+1}}')$ for each $k \in \mathbb{N}$. Set $A := \{f \neq 0\}$. By Theorem 5 there is a sequence of pairwise different subsets of A, $(D_n)_{n \in \mathbb{N}}$, such that $D_k \notin \hat{\mathfrak{R}}(\bar{\mathscr{L}}(\ell))$ $(k \in \mathbb{N})$. Define $D := \bigcup_{n \in \mathbb{N}} D_n$. Then D is an infinite subset of A.

This leaves two possibilities:

CASE 1: $D \notin \hat{\mathfrak{R}}(\bar{\mathscr{L}}(\ell))$. In this case ($\delta$) does not hold.

CASE 2: $D \in \hat{\mathfrak{R}}(\bar{\mathscr{L}}(\ell))$. Take $k \in \mathbb{N}$. Then either D_k or $D \setminus D_k$ is infinite and not in $\hat{\mathfrak{R}}(\bar{\mathscr{L}}(\ell))$, so that again, ($\delta$) does not hold.

4.2.6(E) Functionals on $\mathbb{R}^X$. Let X be a set and denote by Φ the set of all δ-stable ultrafilters on X. Prove that for any positive linear functional ℓ on $\mathbb{R}^X$ there is $g \in \mathscr{k}(\Phi)_+$ such that

$$\ell(f) = \sum_{\mathfrak{F} \in \Phi} g(\mathfrak{F}) \lim_{\mathfrak{F}} f$$

for each $f \in \mathbb{R}^X$. In particular, ℓ is nullcontinuous.

We give some hints for the proof. Define

$$g: \Phi \to \mathbb{R}, \qquad \mathfrak{F} \mapsto \inf_{F \in \mathfrak{F}} \ell(e_F).$$

Prove that $\{g \neq 0\}$ is finite. Then define

$$\ell': \mathbb{R}^X \to \mathbb{R}, \qquad f \mapsto \ell(f) - \sum_{\mathfrak{F} \in \Phi} g(\mathfrak{F}) \lim_{\mathfrak{F}} f.$$

ℓ' is a positive linear functional on $\mathbb{R}^X$, and for every $\mathfrak{F} \in \Phi$ exists an $A \in \mathfrak{F}$ with $\ell'(e_A) = 0$. Let Ψ be the set of all ultrafilters on X. Prove that for any $f \in \mathbb{R}^X_+$ and for any $\mathfrak{F} \in \Psi$, there is an $A(\mathfrak{F})$ in $\mathfrak{F}$ such that $\ell'(fe_{A(\mathfrak{F})}) = 0$. Indeed, if $\ell'(fe_A) > 0$ for each $A \in \mathfrak{F}$, we can conclude that $\mathfrak{F} \notin \Phi$. Hence there is a decreasing sequence $(A_n)_{n \in \mathbb{N}}$ from $\mathfrak{F}$ such that $\bigcap_{n \in \mathbb{N}} A_n = \varnothing$. Then a function $g \in \mathbb{R}^X_+$ can be constructed with $\ell'(g) = \infty$. But this is not possible since ℓ' is real valued. Next prove that there is a finite subset Ψ_0 of Ψ with

$$\{f \neq 0\} \setminus \bigcup_{\mathfrak{F} \in \Psi_0} A(\mathfrak{F}) = \varnothing. \tag{1}$$

Otherwise, the set

$$\mathfrak{G} := \left\{ \{f \neq 0\} \setminus \bigcup_{\mathfrak{F} \in \Psi'} A(\mathfrak{F}) \mid \Psi' \subset \Psi, \Psi' \text{ finite} \right\}$$

would be a filterbase on X. We could take $\mathfrak{F} \in \Psi$ with $\mathfrak{G} \subset \mathfrak{F}$. But then

$$\{f \neq 0\} \setminus A(\mathfrak{F}) \in \mathfrak{G} \subset \mathfrak{F} \quad \text{and} \quad A(\mathfrak{F}) \in \mathfrak{F}$$

which is a contradiction. From (1) it follows that $\ell'(f) = 0$, and we conclude that $\ell' = 0$.

4.2.7(E) Prove that for any Daniell space $(X, \mathscr{L}, \ell)$ the following assertions are equivalent:

(α) $A \in \mathfrak{N}(\ell) \setminus \mathfrak{N}(\bar{\mathscr{L}}(\ell))$.

(β) $A \cap \{f \neq 0\} \in \mathfrak{N}(\bar{\mathscr{L}}(\ell))$ for every $f \in \mathscr{L}$ and $A \setminus \bigcup_{f \in \mathscr{F}} \{f \neq 0\} \neq \varnothing$ for every countable subset $\mathscr{F}$ of $\mathscr{L}$.

4.2.8(E) Let $(X, \mathscr{L}, \ell)$ be a Daniell space. Prove that the following are equivalent.

(α) Every closed extension of $(X, \mathscr{L}, \ell)$ is admissible.

(β) $\{B \in \mathfrak{P}(X) \mid \mathfrak{P}(B) \cap \hat{\mathfrak{R}}(\bar{\mathscr{L}}(\ell)) \subset \mathfrak{N}(\bar{\mathscr{L}}(\ell))\} \subset \hat{\mathfrak{R}}(\bar{\mathscr{L}}(\ell))$.

(γ) $(X, \bar{\mathscr{L}}(\ell), \bar{\ell})$ is the only closed extension of $(X, \mathscr{L}, \ell)$.

For (α) $\Rightarrow$ (β), see Exs. 4.1.8 and 4.1.9.

For (β) $\Rightarrow$ (γ), take a closed extension $(X, \tilde{\mathscr{L}}, \tilde{\ell})$ of $(X, \mathscr{L}, \ell)$. Given $A \in \hat{\mathfrak{R}}(\tilde{\mathscr{L}})$ there is a $B \in \hat{\mathfrak{R}}(\bar{\mathscr{L}}(\ell))$ with $B \subset A$ and $\mathfrak{P}(A \setminus B) \cap \hat{\mathfrak{R}}(\bar{\mathscr{L}}(\ell)) \subset \mathfrak{N}(\bar{\mathscr{L}}(\ell))$. Hence $A \setminus B \in \hat{\mathfrak{R}}(\bar{\mathscr{L}}(\ell))$, so that $\hat{\mathfrak{R}}(\tilde{\mathscr{L}}) \subset \hat{\mathfrak{R}}(\bar{\mathscr{L}}(\ell))$. Now apply Theorem 5.

4.2.9(E) Let $(X, \mathscr{L}, \ell)$ be a Daniell space, and $\mathscr{L}$ a Stone lattice. $(X, \mathscr{L}, \ell)$ is said to be **bounded** iff $\sup\{\ell(g) \mid g \in \mathscr{L}_+,\ g \leq e_X\} < \infty$.

The **uniform hull** of $\mathscr{L}$ is defined to be the set of all $f \in \overline{\mathbb{R}}^X$ which are the uniform limit of a sequence from $\mathscr{L}$. Prove that the following statements are equivalent:

(α) $(X, \mathscr{L}, \ell)$ is bounded.

(β) $e_X \in \mathscr{L}^1(\ell)$.

(γ) $(X, \mathscr{L}^1(\ell), \int_\ell)$ is bounded.

(δ) The uniform hull of $\mathscr{L}^1(\ell)$ is $\mathscr{L}^1(\ell)$ itself.

(ε) The uniform hull of $\mathscr{L}$ is contained in $\mathscr{L}^1(\ell)$.

4.2.10(C) Let $(X, \mathscr{L}, \ell)$ be a Daniell space. We have already characterized the integral of $(X, \mathscr{L}, \ell)$ as the greatest admissible extension of $(X, \mathscr{L}, \ell)$ determined by $(X, \mathscr{L}, \ell)$. What happens when we ignore admissibility? Denote by $\Phi(X, \mathscr{L}, \ell)$ the set of all Daniell spaces extending $(X, \mathscr{L}, \ell)$. Call an element $(X, \mathscr{L}', \ell')$ of $\Phi(X, \mathscr{L}, \ell)$ **strongly determined** by $(X, \mathscr{L}, \ell)$ iff, for every $(X, \mathscr{L}'', \ell'') \in \Phi(X, \mathscr{L}, \ell)$, $\ell'|_{\mathscr{L}' \cap \mathscr{L}''} = \ell''|_{\mathscr{L}' \cap \mathscr{L}''}$. Is there a greatest stongly determined element in $\Phi(X, \mathscr{L}, \ell)$?

The answer to this question completely clarifies the significance of admissibility for the construction of the integral:

$(X, \bar{\mathscr{L}}(\ell), \bar{\ell})$ is the largest strongly determined element in $\Phi(X, \mathscr{L}, \ell)$.

The proof of this statement is somewhat long. We give it in its entirety. But because it contains no novelties, we advise the reader who has carefully worked through Section 4.2 to attempt it independently.

Let $(X, \mathscr{L}', \ell') \in \Phi(X, \mathscr{L}, \ell)$ be closed. Then $(X, \bar{\mathscr{L}}(\ell), \bar{\ell}) \preccurlyeq (X, \mathscr{L}', \ell')$. Assume that the inequality is strict. By Theorem 5 there is a set $A \in \hat{\mathfrak{R}}(\mathscr{L}') \setminus \hat{\mathfrak{R}}(\bar{\mathscr{L}}(\ell))$. Take $f \in \mathscr{L}'_+$ such that $A = \{f \neq 0\}$. Set

$$\alpha := \sup\{\ell'(fe_B) \mid B \in \hat{\mathfrak{R}}(\bar{\mathscr{L}}(\ell)),\ B \subset A\}.$$

Then $\alpha \in \mathbb{R}$, and there is an increasing sequence $(B_n)_{n \in \mathbb{N}}$ from $\hat{\mathfrak{R}}(\bar{\mathscr{L}}(\ell))$ with $B_n \subset A$ for each $n \in \mathbb{N}$ and $\alpha = \sup_{n \in \mathbb{N}} \ell'(fe_{B_n})$. Set $B := \bigcup_{n \in \mathbb{N}} B_n$. Then $B \in \hat{\mathfrak{R}}(\bar{\mathscr{L}}(\ell))$, $B \subset A$, and $A \setminus B \neq \varnothing$. Take $g \in \bar{\mathscr{L}}(\ell)$ such that $\{g \neq 0\} \subset A \setminus B$. Then $fe_{\{g \neq 0\}} \in \mathscr{N}(\mathscr{L}')$, so that $\{g \neq 0\} \in \mathfrak{N}(\mathscr{L}')$. Consequently $\{g \neq 0\} \in \mathfrak{N}(\bar{\mathscr{L}}(\ell))$ and $g \in \mathscr{N}(\bar{\mathscr{L}}(\ell))$. Setting $\mathscr{F}' := \{h \in \overline{\mathbb{R}}^X \mid \{h \neq 0\} \subset A \setminus B\}$, $\mathscr{F}' \cap \bar{\mathscr{L}}(\ell) \subset \mathscr{N}(\bar{\mathscr{L}}(\ell))$, and $\mathscr{F} \cap \bar{\mathscr{L}}(\ell) = \mathscr{N}(\bar{\mathscr{L}}(\ell))$, where $\mathscr{F} := \mathscr{F}' \vdash \mathscr{N}(\bar{\mathscr{L}}(\ell))$.

It is not too difficult to show that $\mathscr{F}$ is a tower σ-completely embedded in $\overline{\mathbb{R}}^X$. Using Corollary 4.1.11, it follows that if $\tilde{\ell}$ denotes the uniquely determined positive linear extension of $\bar{\ell}$ to $\bar{\mathscr{L}}(\ell) \vdash \mathscr{F}$, then $(X, \bar{\mathscr{L}}(\ell) \vdash \mathscr{F}, \tilde{\ell})$ is a closed Daniell space in $\Phi(X, \mathscr{L}, \ell)$, $fe_{A \setminus B} \in (\bar{\mathscr{L}}(\ell) \vdash \mathscr{F}) \cap \mathscr{L}'$ and $\bar{\tilde{\ell}}(fe_{A \setminus B}) = 0$. Were $(X, \mathscr{L}', \ell')$ determined by $(X, \mathscr{L}, \ell)$, then $fe_{A \setminus B} \in \mathscr{N}(\mathscr{L}')$. We assume that this latter is the case and show that $(X, \mathscr{L}', \ell')$ is not strongly determined by $(X, \mathscr{L}, \ell)$ even then.

There are two cases to distinguish:

Case 1: $A \setminus B \subset \{h \neq 0\}$ for some $h \in \bar{\mathscr{L}}(\ell)_+$.

Take such an h. Set $f' := f \wedge h$ and $\beta := \inf\{\bar{\ell}(h') \mid h' \in \bar{\mathscr{L}}(\ell),\ h' \geq f'e_{A \setminus B}\}$. Then $\beta \in \mathbb{R}$, $\beta > 0$, and there is a decreasing sequence $(h_n)_{n \in \mathbb{N}}$ from $\bar{\mathscr{L}}(\ell)$ such that $h_n \geq f'e_{A \setminus B}$ for each $n \in \mathbb{N}$ and $\inf_{n \in \mathbb{N}} \bar{\ell}(h_n) = \beta$. Set $\tilde{h} := \bigwedge_{n \in \mathbb{N}} h_n$. Then $\tilde{h} \in \bar{\mathscr{L}}(\ell)$, $\tilde{h} \geq f'e_{A \setminus B}$, and $\bar{\ell}(\tilde{h}) = \beta > 0$. Set $D := \{\tilde{h} \neq 0\} \setminus (A \setminus B)$. Then $D \neq \varnothing$. Now define

$$\mathscr{G} := \mathscr{N}(\bar{\mathscr{L}}(\ell)) \vdash \left\{ g \in \overline{\mathbb{R}}^X \mid \{g \neq 0\} \subset D \right\}.$$

Then $\mathscr{G}$ is a tower σ-completely embedded in $\overline{\mathbb{R}}^X$, and $\mathscr{G} \cap \bar{\mathscr{L}}(\ell) = \mathscr{N}(\bar{\mathscr{L}}(\ell))$.

We generate a closed Daniell space $(X, \bar{\mathscr{L}}(\ell) \vdash \mathscr{G}, \tilde{\ell})$ in this way for which D is an exceptional set. $A \setminus B$ cannot be an exceptional set in this space, so that $(X, \mathscr{L}', \ell')$ is not determined by $(X, \mathscr{L}, \ell)$.

Case 2: $\{h \in \bar{\mathscr{L}}(\ell) \mid A \setminus B \subset \{h \neq 0\}\} = \varnothing$.

Take $g \in \bar{\mathscr{L}}(\ell)$ with $ge_{A \setminus B} \in \bar{\mathscr{L}}(\ell)$. Then $ge_{A \setminus B} \in \mathscr{N}(\bar{\mathscr{L}}(\ell))$ so that $\bar{\ell}(ge_{A \setminus B}) = 0$ and $\bar{\ell}(ge_{X \setminus (A \setminus B)}) = \bar{\ell}(g)$. Take $g, h \in \bar{\mathscr{L}}(\ell)$ with $ge_{X \setminus (A \setminus B)} = he_{X \setminus (A \setminus B)}$. Then $\{g \neq h\} \subset A \setminus B$ so that $\{g \neq h\} \in \mathfrak{N}(\bar{\mathscr{L}}(\ell))$. Thus $\bar{\ell}(g) = \bar{\ell}(h)$. Define

$$\mathscr{L}_1 := \left\{ ge_{X \setminus (A \setminus B)} \mid g \in \bar{\mathscr{L}}(\ell) \right\} \text{ and } \ell_1 \colon \mathscr{L}_1 \to \mathbb{R},\ ge_{X \setminus (A \setminus B)} \mapsto \bar{\ell}(g).$$

It is easy to see that $(X, \mathscr{L}_1, \ell_1)$ is a closed Daniell space. Define further

$$\mathscr{L}_2 := \left\{ \alpha fe_{A \setminus B} + g \,\middle|\, \begin{array}{l} \alpha \in \mathbb{R},\ g \in \mathbb{R}^X,\ \exists h \in \bar{\mathscr{L}}(\ell) \\ \text{with } \{g \neq 0\} \subset \{h \neq 0\} \cap (A \setminus B) \end{array} \right\}$$

and

$$\ell_2 \colon \mathscr{L}_2 \to \mathbb{R}, \qquad \alpha fe_{A \setminus B} + g \mapsto \alpha.$$

It is a simple matter to check that $(X, \mathscr{L}_2, \ell_2)$ is a closed Daniell space. Now set $\tilde{\mathscr{L}} := \{g + h \mid g \in \mathscr{L}_1,\ h \in \mathscr{L}_2\}$, and define $\tilde{\ell}(g+h) := \ell_1(g) + \ell_2(h)$ for each $g \in \mathscr{L}_1$ and $h \in \mathscr{L}_2$. Then $(X, \tilde{\mathscr{L}}, \tilde{\ell})$ is a Daniell space in $\Phi(X, \mathscr{L}, \ell)$. Furthermore $fe_{A \setminus B} \in \tilde{\mathscr{L}}$, and $\ell(fe_{A \setminus B}) = 1$. Thus $(X, \mathscr{L}', \ell')$ is not strongly determined by $(X, \mathscr{L}, \ell)$. This completes our proof.

Thus it is precisely the admissibility that distinguishes $(X, \mathscr{L}^1(\ell), \int_\ell)$ from $(X, \bar{\mathscr{L}}(\ell), \bar{\ell})$.

4.2.11(C) The General Notion of Determinedness. The notion of determinedness is not only useful for the definition of integrals. It can be brought in a very general form by the following definition.

Let X be an ordered set. An element x of X is called **determined** in X iff, for each $y \in X$, there is a $z \in X$ such that $x \leq z$ and $y \leq z$. Thus, determinedness is a notion in the context of ordered sets. Prove the following statements for a Daniell space $(X, \mathscr{L}, \ell)$:

(α) Let Ψ be the set of all admissible extensions of $(X, \mathscr{L}, \ell)$, ordered by $\preccurlyeq$. Then $(X, \mathscr{L}^1(\ell), \int_\ell)$ is the greatest determined element in Ψ with respect to this order relation, in the sense of the preceding definition.

(β) Let Φ be the set of all Daniell spaces extending $(X, \mathscr{L}, \ell)$, ordered by $\preccurlyeq$. Then $(X, \bar{\mathscr{L}}(\ell), \bar{\ell})$ is the greatest determined element in Φ with respect to this order relation (see Ex. 4.2.10).

In this form we will use the notion of determinedness later in several investigations.

Prove now the following general results for an ordered set X and the set of all determined elements of X, denoted by X^d:

(γ) If $x \in X^d$ and $y \in X$, $y \leq x$, then $y \in X^d$.

(δ) If x is the smallest element of X, then $x \in X^d$.

(ε) If X is directed upward, then $X = X^d$.

(ζ) Let Y be a finite subset of X^d. Then, for each $x \in X$, there is a $z \in X$ such that $y \leq z$ for each $y \in Y$ and $x \leq z$.

(η) Assume that $X^d \neq \varnothing$. Assume further that every well-ordered subset of X is bounded above in X. Then for every well-ordered subset Y of X^d and for every $x \in X$, there is a $z \in X$ with $x \leq z$ and $y \leq z$ for each $y \in Y$. If under the preceding assumptions $\bigvee^X_{y \in Y} y$ exists, then $\bigvee^X_{y \in Y} y \in X^d$.

(ϑ) Let $X^d \neq \varnothing$. Assume that $\bigvee^X_{y \in Y} y$ exists for every well-ordered nonempty subset Y of X. Then the following assertions are equivalent:

(ϑ1) X^d is directed upward.

(ϑ2) There is a greatest element $\bar{x}$ in X^d.

If either of these statements is fulfilled, then

$$\bar{x} = \bigwedge\{x \in X \mid x \text{ maximal in } X\}.$$

(ι) Assume that for every $x \in X$ there is a maximal element z in X with $x \leq z$. Then the following assertions are equivalent:

($\iota 1$) $\bigvee_{x \in X^d} x$ exists.
($\iota 2$) $\bigwedge\{x \in X \mid x \text{ maximal in } X\}$ exists.

If either of these elements exists, it is the greatest determined element of X.

(κ) Assume that for every $x \in X$ there is a maximal element z in X with $x \leq z$, and assume further that $\bigwedge_{z \in Z} z$ exists for every nonempty subset Z of X. Prove the following statements for $X \neq \varnothing$:

($\kappa 1$) X^d is a complete lattice with respect to the order relation inherited from X.
($\kappa 2$) $\bigwedge\{x \in X \mid x \text{ maximal in } X\}$ is the greatest determined element of X.

We want to describe now our integral using the general results just stated.

Let $(X, \mathscr{L}, \ell)$ be a Daniell space. Denote by Φ the set of all Daniell spaces $(X, \mathscr{L}', \ell')$ with the following properties:

(i) $(X, \mathscr{L}, \ell) \preccurlyeq (X, \mathscr{L}', \ell')$.
(ii) For every $f \in \mathscr{L}'$ there is a Cauchy sequence $(f_n)_{n \in \mathbb{N}}$ from $\mathscr{L}$ with respect to d_ℓ such that $f = \lim_{n \to \infty} f_n$ $\mathscr{L}'$-a.e. and

$$\lim_{n \to \infty} d_{\ell'}(f, f_n) = 0.$$

Prove the next statement:

(λ) $(X, \mathscr{L}^1(\ell), \int_\ell)$ is the greatest determined element of Φ with respect to the order relation $\preccurlyeq$

The existence of a greatest determined element of Φ can be proved by means of (κ). But the general theory does not give the desired characterization. For this purpose use the methods of Section 4.2. Nevertheless, we get a new description of the integral by ($\kappa 2$).

Now prove this statement.

(μ) Let Y and Z be ordered subsets of the ordered set X. Assume that for every $y \in Y$ there is $z \in Z$ such that $y \leq z$ and for each $z \in Z$ there is $y \in Y$ with $z \leq y$. Then the following assertions hold:

($\mu 1$) $Y^d \cap Z = Y \cap Z^d$.
($\mu 2$) Let y be the greatest element in Y^d. Then, if

$$\{z \in Y \mid z \geq y\} = \{z \in Z \mid z \geq y\}$$

y is the greatest element of Z^d.

Conclude using these results:

(ν) Let $(X, \mathscr{L}, \ell)$ and $(X, \mathscr{L}', \ell')$ be Daniell spaces. Suppose that for every admissible extension $(X, \mathscr{L}'', \ell'')$ of $(X, \mathscr{L}, \ell)$ there is an admissible extension $(X, \tilde{\mathscr{L}}, \tilde{\ell})$ of $(X, \mathscr{L}', \ell')$ with $(X, \mathscr{L}'', \ell'') \preccurlyeq (X, \tilde{\mathscr{L}}, \tilde{\ell})$ and that for every admissible extension $(X, \mathscr{L}'', \ell'')$ of $(X, \mathscr{L}', \ell')$ there is an admissible extension $(X, \tilde{\mathscr{L}}, \tilde{\ell})$ of $(X, \mathscr{L}, \ell)$ such that $(X, \mathscr{L}'', \ell'') \preccurlyeq (X, \tilde{\mathscr{L}}, \tilde{\ell})$. Then

$$\left(X, \mathscr{L}^1(\ell), \int_\ell\right) = \left(X, \mathscr{L}^1(\ell'), \int_{\ell'}\right)$$

4.2.12(C) A general notion of an integral of a positive linear functional. We have defined the integral under the hypothesis that functionals are nullcontinuous. What happens if this condition fails? Our constructions then make no sense. Nevertheless, using the method of admissibility and determinedness, we can define a natural notion of integral in any case. Let us develop the basic ideas.

We define a **weak Daniell space** to be a triple $(X, \mathscr{L}, \ell)$ with the properties that $\mathscr{L}$ is a Riesz lattice on X and ℓ a positive linear functional on $\mathscr{L}$. Of course, a weak Daniell space $(X, \mathscr{L}, \ell)$ is a Daniell space iff ℓ is nullcontinuous. For every weak Daniell space $(X, \mathscr{L}, \ell)$ there is a certain degree of continuity of ℓ. Our first aim is to characterize such a degree. For this purpose we consider sets F of sequences from $\mathscr{L} \cap \mathbb{R}^X$ with the following properties:

(i) $(f_n + g_n)_{n \in \mathbb{N}} \in F$, $(f_n \vee g_n)_{n \in \mathbb{N}} \in F$, $(f_n \wedge g_n)_{n \in \mathbb{N}} \in F$ and $(\alpha f_n)_{n \in \mathbb{N}} \in F$ whenever $(f_n)_{n \in \mathbb{N}} \in F$, $(g_n)_{n \in \mathbb{N}} \in F$, and $\alpha \in \mathbb{R}$.

(ii) If $(f_n)_{n \in \mathbb{N}}$ is a sequence from $\mathscr{L} \cap \mathbb{R}^X$ such that there is $f \in \mathscr{L}$ with $f = \lim_{n \to \infty} f_n$ $\mathscr{L}$-a.e. and $\lim_{n \to \infty} \ell(|f - f_n|) = 0$, then $(f_n)_{n \in \mathbb{N}} \in F$.

(iii) There is a weak Daniell space $(X, \mathscr{L}', \ell')$ with $(X, \mathscr{L}, \ell) \preccurlyeq (X, \mathscr{L}', \ell')$ such that there is, for every sequence $(f_n)_{n \in \mathbb{N}} \in F$, a function $f \in \mathscr{L}'$ with $f = \lim_{n \to \infty} f_n$ $\mathscr{L}'$-a.e. and $\lim_{n \to \infty} \ell'(|f - f_n|) = 0$.

Such a set F of sequences is called a **degree of continuity** of $(X, \mathscr{L}, \ell)$. We denote by $C(X, \mathscr{L}, \ell)$ the set of all degrees of continuity of $(X, \mathscr{L}, \ell)$. In the following we denote by $\Phi(X, \mathscr{L}, \ell)$ the set of all weak Daniell spaces extending $(X, \mathscr{L}, \ell)$. $\Phi(X, \mathscr{L}, \ell)$ is ordered by $\preccurlyeq$.

Prove the following statements:

(α) For every degree F of continuity of $(X, \mathscr{L}, \ell)$, there is a smallest element $(X, \mathscr{L}_F, \ell_F)$ in $\Phi(X, \mathscr{L}, \ell)$ for which (iii) holds.

(β) For all $F, F' \in C(X, \mathscr{L}, \ell)$, $(X, \mathscr{L}_F, \ell_F) \preccurlyeq (X, \mathscr{L}_{F'}, \ell_{F'})$ whenever $F \subset F'$.

(γ) Every well-ordered subset of $C(X, \mathscr{L}, \ell)$ is bounded above.

(δ) If $(F_\iota)_{\iota \in I}$ is a nonempty family from $C(X, \mathscr{L}, \ell)$, then $\bigcap_{\iota \in I} F_\iota \in C(X, \mathscr{L}, \ell)$.

Using Ex. 4.2.11 (κ), we deduce the next statement:

(ε) There is a greatest determined element $F(\ell)$ in $C(X,\mathscr{L},\ell)$ with respect to $\subset$.

$F(\ell)$ is called the **natural degree of continuity** of $(X,\mathscr{L},\ell)$. There is a suggestion that we are interested only in such extensions of $(X,\mathscr{L},\ell)$ that have at least the same degree of continuity as $(X,\mathscr{L},\ell)$ itself. So let $\Psi(X,\mathscr{L},\ell)$ be the set of all elements $(X,\mathscr{L}',\ell')$ of $\Phi(X,\mathscr{L},\ell)$ with $F(\ell)\subset F(\ell')$. Prove the following assertions:

(ζ) Let $F_0(\ell)$ be the set of all sequences $(f_n)_{n\in\mathbb{N}}$ from $\mathscr{L}\cap\mathbb{R}^X$ for which there is $f\in\mathscr{L}$ such that $f=\lim_{n\to\infty}f_n$ $\mathscr{L}$-a.e. and $\lim_{n\to\infty}\ell(|f-f_n|)=0$. Then $F_0(\ell)\in C(X,\mathscr{L},\ell)$.

(η) $(X,\mathscr{L}_{F(\ell)},\ell_{F(\ell)})\preccurlyeq(X,\mathscr{L}_{F(\ell')},\ell_{F(\ell')})$ whenever $(X,\mathscr{L}',\ell')\in\Psi(X,\mathscr{L},\ell)$.

(ϑ) Let $(X,\mathscr{L}',\ell')$ be an element of $\Psi(X,\mathscr{L},\ell)$ for which (iii) holds with respect to $F(\ell)$. Then $(X,\mathscr{L}',\ell')\in\Psi(X,\mathscr{L},\ell)$.

(ι) $(X,\mathscr{L}_{F(\ell')},\ell_{F(\ell')})\in\Psi(X,\mathscr{L},\ell)$ whenever $(X,\mathscr{L}',\ell')\in\Psi(X,\mathscr{L},\ell)$.

(κ) Let $(X,\mathscr{L}',\ell')$ be a maximal element of $\Psi(X,\mathscr{L},\ell)$. Then $(X,\mathscr{L}',\ell')=(X,\mathscr{L}_{F(\ell')},\ell_{F(\ell')})$.

Our next goal is to prove the existence of a greatest determined element in $\Psi(X,\mathscr{L},\ell)$ for any weak Daniell space $(X,\mathscr{L},\ell)$. Let $\mathfrak{S}$ be the set of all subsets A of X with the property that for any $f\in\mathbb{R}^X$ with $\{f\neq 0\}\subset A$ there is a sequence $(g_n)_{n\in\mathbb{N}}$ from $\mathscr{L}_+$ such that $|f|\le\bigwedge_{n\in\mathbb{N}}g_n$ and $\inf_{n\in\mathbb{N}}\ell(g_n)=0$. Prove the next statement:

(λ) There is a smallest element $(X,\mathscr{L}_0,\ell_0)\in\Phi(X,\mathscr{L},\ell)$ such that $\mathfrak{S}\subset\mathfrak{N}(\mathscr{L}_0)$. We have $\mathfrak{S}=\mathfrak{N}(\mathscr{L}_0)$ and $(X,\mathscr{L}_0,\ell_0)\in\Psi(X,\mathscr{L},\ell)$.

$(X,\mathscr{L}_0,\ell_0)$ is called the **zero-extension** of $(X,\mathscr{L},\ell)$. Now take a weak Daniell space $(X,\mathscr{L},\ell)$, denote by $(X,\tilde{\mathscr{L}},\tilde{\ell})$ the zero-extension of $(X,\mathscr{L}_{F(\ell)},\ell_{F(\ell)})$, and define

$$\bar{\mathscr{L}}(\ell):=\left\{f\in\overline{\mathbb{R}}^X\;\middle|\;\begin{array}{l}\inf\{\tilde{\ell}(g)\,|\,g\in\tilde{\mathscr{L}},\,g\ge f\}\\ =\sup\{\tilde{\ell}(g)\,|\,g\in\tilde{\mathscr{L}},\,g\le f\}\in\mathbb{R}\end{array}\right\}$$

and, for every $f\in\bar{\mathscr{L}}(\ell)$, $\bar{\ell}(f):=\inf\{\tilde{\ell}(g)\,|\,g\in\tilde{\mathscr{L}},\ g\ge f\}$. Prove the following assertions:

(μ) $(X,\bar{\mathscr{L}}(\ell),\bar{\ell})\preccurlyeq(X,\bar{\mathscr{L}}(\ell'),\overline{\ell'})$ whenever $(X,\mathscr{L}',\ell')\in\Psi(X,\mathscr{L},\ell)$.

(ν) $(X,\bar{\mathscr{L}}(\ell'),\overline{\ell'})\in\Psi(X,\mathscr{L},\ell)$ whenever $(X,\mathscr{L}',\ell')\in\Psi(X,\mathscr{L},\ell)$.

(ξ) Let $(X,\mathscr{L}',\ell')$ be a maximal element of $\Psi(X,\mathscr{L},\ell)$. Then $(X,\bar{\mathscr{L}}(\ell'),\overline{\ell'})=(X,\mathscr{L}',\ell')$.

We can now prove the following result:

(o) $(X,\bar{\mathscr{L}}(\ell),\bar{\ell})$ is the greatest determined element in $\Psi(X,\mathscr{L},\ell)$.

We give some hints for the proof. From (μ) and (ν) it follows that

$(X, \bar{\mathcal{L}}(\ell), \bar{\ell})$ is determined. We have to prove that it is the greatest determined element.

Let $(X, \mathcal{L}', \ell') \in \Psi(X, \mathcal{L}, \ell)$ such that $\mathcal{L}' \setminus \bar{\mathcal{L}}(\ell) \neq \emptyset$. Then there is $f \in (\mathcal{L}' \setminus \bar{\mathcal{L}}(\ell')) \cap \mathbb{R}_+^X$. Denote by $\mathcal{F}$ the Riesz lattice generated by $\bar{\mathcal{L}}(\ell) \cup \{f\}$. From the definition of $(X, \bar{\mathcal{L}}(\ell), \bar{\ell})$ it follows that

$$\alpha^*(f) := \inf\{\bar{\ell}(g) \mid g \in \bar{\mathcal{L}}(\ell),\, g \geq f\}$$

$$> \sup\{\bar{\ell}(g) \mid g \in \bar{\mathcal{L}}(\ell),\, g \leq f\} =: \alpha_*(f)$$

Choose $\gamma \in]\alpha_*(f), \alpha^*(f)[$ such that $\gamma \neq \ell'(f)$ and define

$$\mathcal{H} := \{g + \alpha f \mid g \in \bar{\mathcal{L}}(\ell),\, \alpha \in \mathbb{R}\}$$

and $k(g + \alpha f) := \ell(g) + \alpha\gamma$ whenever $g \in \bar{\mathcal{L}}(\ell)$ and $\alpha \in \mathbb{R}$. k is an increasing linear functional on the linear function space $\mathcal{H}$ (cf., Ex. 1.2.7). We have

$$(X, \bar{\mathcal{L}}(\ell), \bar{\ell}) \preccurlyeq (X, \mathcal{H}, k) \quad \text{and} \quad \mathcal{H} \subset \mathcal{F}.$$

Let Ω be the set of all triples $(X, \mathcal{H}', k')$ such that k' is an increasing linear functional on the linear function space $\mathcal{H}'$ and

$$(X, \mathcal{H}, k) \preccurlyeq (X, \mathcal{H}', k') \qquad \mathcal{H}' \subset \mathcal{F}.$$

By the Zorn Lemma there is a maximal element in Ω. Denote by $(X, \bar{\mathcal{H}}, \bar{k})$ such an element. Assume that there is $h \in (\mathcal{F} \setminus \bar{\mathcal{H}}) \cap \mathbb{R}^X$. Then

$$\beta^*(h) := \inf\{\bar{k}(g) \mid g \in \bar{\mathcal{H}},\, g \geq h\}$$

$$\geq \sup\{\bar{k}(g) \mid g \in \bar{\mathcal{H}},\, g \leq h\} =: \beta_*(h).$$

In the same way we could extend $\bar{k}$ to the linear function space generated by $\bar{\mathcal{H}} \cup \{h\}$ such that the extension is increasing and linear. But this is a contradiction since $(X, \bar{\mathcal{H}}, \bar{k})$ is maximal. We conclude that $(\mathcal{F} \setminus \bar{\mathcal{H}}) \cap \mathbb{R}^X = \emptyset$. But from the construction it follows that $\mathfrak{N}(\bar{\mathcal{H}}) = \mathfrak{N}(\mathcal{F}) = \mathfrak{N}(\bar{\mathcal{L}}(\ell))$, and we find $\bar{\mathcal{H}} = \mathcal{F}$.

Prove now that $(X, \bar{\mathcal{H}}, \bar{k}) \in \Psi(X, \mathcal{L}, \ell)$. But $(X, \bar{\mathcal{H}}, \bar{k})$ and $(X, \mathcal{L}', \ell')$ cannot have any common upper bound in $\Psi(X, \mathcal{L}, \ell)$. From this our assertion follows.

The extension $(X, \bar{\mathcal{L}}(\ell), \bar{\ell})$ of $(X, \mathcal{L}, \ell)$ is called the **closure** of $(X, \mathcal{L}, \ell)$. Prove the following:

(π) If $(X, \mathcal{L}, \ell)$ is a Daniell space then the following is true.

(π1) $F(\ell)$ is the set of all d_ℓ-Cauchy sequences from $\mathcal{L} \cap \mathbb{R}^X$ converging $\bar{\mathcal{L}}(\ell)$-a.e. ($\bar{\mathcal{L}}(\ell)$ is the closure in the sense of Section 3.1).

(π2) The space $(X, \bar{\mathcal{L}}(\ell), \bar{\ell})$ defined in this exercise is identical with the closure in the sense of Section 3.1.

It is now easy to define the integral for weak Daniell spaces.

Let $(X, \mathscr{L}, \ell)$ be a weak Daniell space. An element $(X, \mathscr{L}', \ell')$ of $\Psi(X, \mathscr{L}, \ell)$ is called **admissible** iff, for any $f \in \mathscr{L}'$, there is a $g \in \bar{\mathscr{L}}(\ell)$ such that $f = g$ $\overline{\mathscr{L}'}$-a.e. Denote by $\Psi_a(X, \mathscr{L}, \ell)$ the set of all admissible elements of $\Psi(X, \mathscr{L}, \ell)$. Define

$$\mathfrak{N}^*(\ell) := \{ A \subset X \mid \exists f \in \bar{\mathscr{L}}(\ell), \{ f \neq 0 \} \subset A, \bar{\ell}(f) \neq 0 \}$$

$$\mathfrak{N}(\ell) := \{ A \subset X \mid \forall B \in \mathfrak{N}^*(\ell), \exists C \in \mathfrak{N}^*(\ell), C \subset B \setminus A \}$$

$$\mathscr{L}^1(\ell) := \left\{ f \in \overline{\mathbb{R}}^X \mid \exists g \in \bar{\mathscr{L}}(\ell), \{ f \neq g \} \in \mathfrak{N}(\ell) \right\}.$$

Next show that the definition $\int_\ell(f) := \ell(g)$ for every $f \in \mathscr{L}^1(\ell)$ is independent of the choice of $g \in \bar{\mathscr{L}}(\ell)$ with $\{ f \neq g \} \in \mathfrak{N}(\ell)$. Finally prove this statement:

(ρ) $(X, \mathscr{L}^1(\ell), \int_\ell)$ is the greatest determined element of $\Psi_a(X, \mathscr{L}, \ell)$.

The extension $(X, \mathscr{L}^1(\ell), \int_\ell)$ of $(X, \mathscr{L}, \ell)$ is called the **integral** of $(X, \mathscr{L}, \ell)$. Prove this final assertion:

(σ) The objects $\mathfrak{N}(\ell)$ and $(X, \mathscr{L}^1(\ell), \int_\ell)$ coincide with the analogous objects defined in Section 4.2 for Daniell spaces $(X, \mathscr{L}, \ell)$.

5

INTEGRALS DERIVED FROM MEASURES AND MEASURABILITY

5.1. CLOSED POSITIVE MEASURE SPACES

NOTATION FOR SECTION 5.1:
X denotes a set.

The connection between positive measures and positive linear functionals was described in Chapter 2. In the meanwhile we have studied extensions of positive linear functionals. We want to investigate the implications of this theory for positive measures.

Given a Daniell space $(X, \mathscr{L}, \ell)$, recall how to define its induced measure space $(X, \Re(\mathscr{L}), \mu^{\ell})$:

$$\Re(\mathscr{L}) := \{A \subset X \mid e_A \in \mathscr{L}\}$$

$$\mu^{\ell}: \Re(\mathscr{L}) \to \mathbb{R}, \qquad A \mapsto \ell(e_A)$$

(Definition 2.4.6). In general the positive measure space induced by a Daniell space can be quite arbitrary, as the next example shows.

Example 5.1.1. For each real number α, define

$$f_\alpha: [0,1] \to \mathbb{R}, \qquad x \mapsto \alpha x.$$

Let $\mathscr{L} := \{f_\alpha \mid \alpha \in \mathbb{R}\}$, and define

$$\ell : \mathscr{L} \to \mathbb{R}, \qquad f_\alpha \mapsto \alpha.$$

$\mathscr{L}$ is a real Riesz lattice and $([0,1], \mathscr{L}, \ell)$ is a closed Daniell space. $\mathfrak{R}(\mathscr{L})$ contains only the empty set, and the measure induced on $\mathfrak{R}(\mathscr{L})$ by ℓ is the trivial one: $\mu^\ell(\varnothing) = 0$. The Daniell space $([0,1], \mathscr{L}, \ell)$ cannot be recovered from its induced positive measure space $([0,1], \mathfrak{R}(\mathscr{L}), \mu^\ell)$. □

The reader will notice that the Riesz lattice in the preceding example does not have the Stone property. The Stone property played an important role in Chapter 2, and it will continue to play that role here, in ensuring that the association between Daniell space and positive measure space is reversible. We begin by showing that the Stone property is always preserved when one takes an admissible extension.

Proposition 5.1.2. *The following assertions hold, for every Daniell space* $(X, \mathscr{L}, \ell)$.

(*a*) *If* $\mathscr{L}$ *is a Stone lattice, then so is* $\bar{\mathscr{L}}(\ell)$.

(*b*) *If* $\mathscr{L}$ *is a Stone lattice and* $(X, \tilde{\mathscr{L}}, \tilde{\ell})$ *is an admissible extension of* $(X, \mathscr{L}, \ell)$, *then* $\tilde{\mathscr{L}}$ *is a Stone lattice. In particular, if* $\mathscr{L}$ *is a Stone lattice, then so is* $\mathscr{L}^1(\ell)$.

Proof. (a) We use the Induction Principle 3.2.8. Let

$$\mathscr{F} := \{f \mid f \in \bar{\mathscr{L}}(\ell), \qquad f \wedge e_X \in \bar{\mathscr{L}}(\ell)\}.$$

According to the hypothesis, $\mathscr{L} \subset \mathscr{F}$. Let $(f_n)_{n \in \mathbb{N}}$ be an $\bar{\ell}$-sequence from $\mathscr{F}$ and f an extended-real function on X such that

$$\lim_{n \to \infty} f_n = f \quad \bar{\mathscr{L}}(\ell)\text{-a.e.}$$

We must show that f belongs to $\mathscr{F}$. Note that both $\lim_{n \to \infty} f_n$ and f belong to $\bar{\mathscr{L}}(\ell)$. If $(f_n)_{n \in \mathbb{N}}$ increases, then

$$f_k \wedge e_X \leq f_{k+1} \wedge e_X \leq \lim_{n \to \infty} f_n$$

for every k in $\mathbb{N}$. If $(f_n)_{n \in \mathbb{N}}$ decreases, then

$$f_k \wedge e_X \geq f_{k+1} \wedge e_X \geq 0 \wedge \lim_{n \to \infty} f_n$$

for every k. In either case $(f_n \wedge e_X)_{n \in \mathbb{N}}$ is an $\bar{\ell}$-sequence from $\bar{\mathscr{L}}(\ell)$, so its limit, which is the same as $(\lim_{n\to\infty} f_n) \wedge e_X$, must belong to $\bar{\mathscr{L}}(\ell)$. Since

$$f \wedge e_X = \left(\lim_{n\to\infty} f_n\right) \wedge e_X \quad \bar{\mathscr{L}}(\ell)\text{-a.e.}$$

it follows that $f \wedge e_X$ belongs to $\bar{\mathscr{L}}(\ell)$ [Proposition 1.2.30 (c)] and f belongs to $\mathscr{F}$. According to Theorem 3.2.8,

$$\mathscr{F} = \bar{\mathscr{L}}(\ell).$$

(b) According to Theorem 4.1.12, the admissible extension $(X, \tilde{\mathscr{L}}, \tilde{\ell})$ is the closure of the Daniell space $(X, \mathscr{L} \vdash \mathscr{N}(\tilde{\mathscr{L}}), \ell')$, where ℓ' denotes the positive linear extension of ℓ to $\mathscr{L} \vdash \mathscr{N}(\tilde{\mathscr{L}})$. In view of (a) it suffices to show that $\mathscr{L} \vdash \mathscr{N}(\tilde{\mathscr{L}})$ has the Stone property. Let $f \in \mathscr{L} \vdash \mathscr{N}(\tilde{\mathscr{L}})$. There exists a function g in $\mathscr{L}$ such that

$$f = g \quad \mathscr{N}(\tilde{\mathscr{L}})\text{-a.e.}$$

hence such that

$$f \wedge e_X = g \wedge e_X \quad \mathscr{N}(\tilde{\mathscr{L}})\text{-a.e.}$$

By hypothesis, $g \wedge e_X$ belongs to $\mathscr{L}$. By definition, therefore, $f \wedge e_X$ belongs to $\mathscr{L} \vdash \mathscr{N}(\tilde{\mathscr{L}})$. Thus $\mathscr{L} \vdash \mathscr{N}(\tilde{\mathscr{L}})$ has the Stone property. □

We shall need some facts about approximation by step functions.

Proposition 5.1.3. *Let $f \in \overline{\mathbb{R}}_+^X$. For n in $\mathbb{N}$, define*

$$f_n := \sum_{k=1}^{n2^n} \frac{1}{2^n} e_{\{f \geq k/2^n\}}.$$

Then $(f_n)_{n\in\mathbb{N}}$ is an increasing sequence of positive step functions on X whose supremum is f. If A is a subset of X for which

$$\sup_{x\in A} f(x) < \infty$$

then the sequence $(f_n)_{n\in\mathbb{N}}$ converges uniformly on A to f.

Proof. Each f_n is evidently a positive step function on X. Let $x \in X$. If $f(x) = \infty$, then $f_n(x) = n$ for every n and $(f_n(x))_{n\in\mathbb{N}}$ increases to $f(x)$. So suppose that $f(x) < \infty$. There exists a smallest integer n_0 such that $f(x) < n_0$.

Denoting by $[2^n f(x)]$ the greatest integer not exceeding $2^n f(x)$, we have

$$f_n(x) = \begin{cases} n & \text{if } n < n_0 \\ \dfrac{[2^n f(x)]}{2^n} & \text{if } n \geq n_0. \end{cases}$$

For $n \geq n_0$, we have

$$f_n(x) = \frac{[2^n f(x)]}{2^n} \leq f(x) \leq \frac{1 + [2^n f(x)]}{2^n} \tag{1}$$

which shows that

$$\lim_{n \to \infty} f_n(x) = f(x).$$

Furthermore

$$f_{n-1}(x) = n - 1 < n = f_n(x) \qquad \text{if } 2 \leq n < n_0$$

$$f_{n_0 - 1}(x) = n_0 - 1 \leq \frac{[2^{n_0} f(x)]}{2^{n_0}} = f_{n_0}(x) \qquad \text{if } n_0 > 1$$

$$f_{n-1}(x) = \frac{[2^{n-1} f(x)]}{2^{n-1}} = \frac{2[2^{n-1} f(x)]}{2^n} \leq \frac{[2^n f(x)]}{2^n} = f_n(x) \qquad \text{if } n > n_0.$$

Thus the sequence $(f_n(x))_{n \in \mathbb{N}}$ increases.

Suppose, finally, that

$$\sup_{x \in A} f(x) = \alpha < \infty.$$

Then (1) holds for every $n > \alpha$ and for every x in A. The uniform convergence of $(f_n)_{n \in \mathbb{N}}$ on A follows. □

Proposition 5.1.4. *The following assertions hold, for every Stone lattice $\mathscr{L}$ that is conditionally σ-completely embedded in $\overline{\mathbb{R}}^X$.*

(*a*) *For every $f \in \mathscr{L}$ and every real number $\alpha > 0$, the function $f \wedge \alpha$ belongs to $\mathscr{L}$.*

(*b*) *For every $f \in \mathscr{L}$ and every real number $\alpha > 0$, the set $\{f \geq \alpha\}$ belongs to $\mathfrak{R}(\mathscr{L})$.*

(*c*) *Every positive function belonging to $\mathscr{L}$ is the supremum of an increasing sequence of positive $\mathfrak{R}(\mathscr{L})$-step functions on X.*

(*d*) *Every set belonging to $\hat{\mathfrak{R}}(\mathscr{L})$ can be written as the union of an increasing sequence from $\mathfrak{R}(\mathscr{L})$.*

Proof. (a) This assertion is true for every Stone lattice $\mathscr{L}$. Simply note that (with $\alpha > 0$)

$$f \wedge \alpha = \alpha\left(\left(\frac{1}{\alpha} f\right) \wedge e_X\right).$$

(b) For $\alpha > 0$,

$$e_{\{f \geq \alpha\}} = \bigwedge_{n \in \mathbb{N}} \frac{n+1}{\alpha}\left[(f \wedge \alpha) - \left(f \wedge \frac{n\alpha}{n+1}\right)\right]$$

so (b) follows from (a) and the conditionally σ-complete embedding.

(c) In view of (b), (c) follows from Proposition 3.

(d) Let $A \in \hat{\Re}(\mathscr{L})$, and choose $f \in \mathscr{L}_+$ such that

$$A = \{f \neq 0\}.$$

By (c), there is an increasing sequence $(f_n)_{n \in \mathbb{N}}$ of positive $\Re(\mathscr{L})$-step functions on X such that

$$f = \bigvee_{n \in \mathbb{N}} f_n.$$

Set

$$A_n := \{f_n > 0\}.$$

Note that each A_n belongs to $\Re(\mathscr{L})$ [Corollary 2.3.4 (a)]. Evidently $(A_n)_{n \in \mathbb{N}}$ is an increasing sequence whose union is A. □

Corollary 5.1.5. *Let $(X, \mathscr{L}_1, \ell_1)$ and $(X, \mathscr{L}_2, \ell_2)$ be closed Daniell spaces such that $\mathscr{L}_1$ and $\mathscr{L}_2$ are Stone lattices. Then the following assertions are equivalent*

(a) $(X, \mathscr{L}_1, \ell_1) \preccurlyeq (X, \mathscr{L}_2, \ell_2)$.

(b) $\Re(\mathscr{L}_1) \subset \Re(\mathscr{L}_2)$ *and* $\ell_1(e_A) = \ell_2(e_A)$ *for every* A *in* $\Re(\mathscr{L}_1)$.

(c) The positive measure space induced by $(X, \mathscr{L}_2, \ell_2)$ extends the positive measure space induced by $(X, \mathscr{L}_1, \ell_1)$.

Proof. The implications (a) $\Rightarrow$ (b) $\Leftrightarrow$ (c) are trivial.

(b) $\Rightarrow$ (a). Let $f \in \mathscr{L}_{1+}$. Then f is the supremum of an increasing sequence $(f_n)_{n \in \mathbb{N}}$ of positive $\Re(\mathscr{L}_1)$-step functions on X [Proposition 4 (c)]. This same sequence is both an ℓ_2-sequence from $\mathscr{L}_2$ and an ℓ_1-sequence from $\mathscr{L}_1$. We conclude that f belongs to $\mathscr{L}_2$ and

$$\ell_2(f) = \sup_{n \in \mathbb{N}} \ell_2(f_n) = \sup_{n \in \mathbb{N}} \ell_1(f_n) = \ell_1(f).$$

An arbitrary f in $\mathscr{L}_1$ can be written as $f^+ - f^-$. Thus (a) follows. □

For closed Daniell spaces and their induced measure spaces, the mutual association that is ensured by the Stone property can now be described.

Theorem 5.1.6. *Let $\mathscr{L}$ be a Riesz lattice in $\overline{\mathbb{R}}^X$. Let ℓ be a positive linear functional on $\mathscr{L}$, and denote by μ the positive mapping induced on $\mathfrak{R}(\mathscr{L})$ by ℓ (Definition 2.4.6.). Then the following assertions are equivalent.*

(*a*) $(X, \mathscr{L}, \ell)$ *is a closed Daniell space and $\mathscr{L}$ is a Stone lattice.*

(*b*) $(X, \mathfrak{R}(\mathscr{L}), \mu)$ *is a positive measure space whose associated Daniell space has closure $(X, \mathscr{L}, \ell)$.*

Proof. (a) $\Rightarrow$ (b). It has already been shown that $(X, \mathfrak{R}(\mathscr{L}), \mu)$ is a positive measure space (Proposition 2.4.5). Denote by $(X, \mathscr{L}', \ell')$ the Daniell space associated with $(X, \mathfrak{R}(\mathscr{L}), \mu)$. Thus $\mathscr{L}'$ consists of the $\mathfrak{R}(\mathscr{L})$-step functions on X and

$$\ell'(e_A) = \mu(A) = \ell(e_A)$$

for every A in $\mathfrak{R}(\mathscr{L})$. It is evident from the definitions that

$$(X, \mathscr{L}', \ell') \preccurlyeq (X, \mathscr{L}, \ell).$$

Hence

$$(X, \bar{\mathscr{L}}(\ell'), \overline{\ell'}) \preccurlyeq (X, \mathscr{L}, \ell).$$

It is also evident from the definitions that

$$\mathfrak{R}(\mathscr{L}) \subset \mathfrak{R}(\mathscr{L}') \subset \mathfrak{R}(\bar{\mathscr{L}}(\ell'))$$

and

$$\overline{\ell'}(e_A) = \mu(A) = \ell(e_A)$$

for every A in $\mathfrak{R}(\mathscr{L})$. By Corollary 5 (b) $\Rightarrow$ (a),

$$(X, \mathscr{L}, \ell) \preccurlyeq (X, \bar{\mathscr{L}}(\ell'), \overline{\ell'}).$$

(b) $\Rightarrow$ (a). Again, denote by $(X, \mathscr{L}', \ell')$ the Daniell space associated with $(X, \mathfrak{R}(\mathscr{L}), \mu)$. According to Theorem 2.3.5, $\mathscr{L}'$ is a Stone lattice. Thus (a) follows from Proposition 2 (a). □

Definition 5.1.7. *A positive measure space $(X, \mathfrak{R}, \mu)$ is said to be* ***closed*** *iff it is induced by a closed Daniell space $(X, \mathscr{L}, \ell)$ for which $\mathscr{L}$ is a Stone lattice.* □

In other words, $(X, \mathfrak{R}, \mu)$ is closed iff there exists a closed Daniell space $(X, \mathscr{L}, \ell)$ with the Stone property such that

$$\mathfrak{R} = \{A \subset X \mid e_A \in \mathscr{L}\}$$

and

$$\mu(A) = \ell(e_A)$$

for every A in $\mathfrak{R}$. If there exists such a Daniell space $(X, \mathscr{L}, \ell)$, then by Theorem 6 it is uniquely determined: $(X, \mathscr{L}, \ell)$ is the closure of the Daniell space associated with $(X, \mathfrak{R}, \mu)$.

Closed positive measure spaces have many of the same advantages, relative to arbitrary positive measure spaces, that closed Daniell spaces have, relative to arbitrary Daniell spaces. Indeed, they inherit these advantages from the corresponding Daniell spaces. We conclude this section by summarizing the important properties of closed positive measure spaces.

Theorem 5.1.8 (Main Theorem on closed positive measure spaces). *The following assertions hold, for every closed positive measure space $(X, \mathfrak{R}, \mu)$. In (c) and (d), $\mathscr{L}$ is the Stone lattice described in (a).*

(a) *There exists a unique closed Daniell space $(X, \mathscr{L}, \ell)$, namely, the closure of the Daniell space associated with $(X, \mathfrak{R}, \mu)$, such that $\mathscr{L}$ is a Stone lattice and $(X, \mathfrak{R}, \mu)$ is induced by $(X, \mathscr{L}, \ell)$.*

(b) *$\mathfrak{R}$ is a δ-ring.*

(c) $\hat{\mathfrak{R}}(\mathscr{L}) = \mathfrak{R}_\sigma$.

(d) $\mathfrak{N}(\mathscr{L}) = \{A \subset X \mid A \in \mathfrak{R},\ \mu(A) = 0\}$.

(e) *The conditions $A \in \mathfrak{R}$, $\mu(A) = 0$, $B \subset A$, always imply that B belongs to $\mathfrak{R}$ and $\mu(B) = 0$.*

(f) *For every nonempty countable family $(A_\iota)_{\iota \in I}$ from $\mathfrak{R}$ that is directed upward, the conditions*

$$\sup_{\iota \in I} \mu(A_\iota) < \infty$$

and

$$\bigcup_{\iota \in I} A_\iota \in \mathfrak{R}$$

are equivalent and imply that

$$\mu\Big(\bigcup_{\iota \in I} A_\iota\Big) = \sup_{\iota \in I} \mu(A_\iota).$$

(*g*) *For every nonempty, countable family* $(A_\iota)_{\iota \in I}$ *from* $\mathfrak{R}$ *that is directed downward, the set* $\bigcap_{\iota \in I} A_\iota$ *belongs to* $\mathfrak{R}$ *and*

$$\mu\left(\bigcap_{\iota \in I} A_\iota\right) = \inf_{\iota \in I} \mu(A_\iota).$$

(*h*) *For every countable disjoint family* $(A_\iota)_{\iota \in I}$ *from* $\mathfrak{R}$, *the conditions*

$$(\mu(A_\iota))_{\iota \in I} \quad \textit{summable}$$

and

$$\bigcup_{\iota \in I} A_\iota \in \mathfrak{R}$$

are equivalent and imply that

$$\mu\left(\bigcup_{\iota \in I} A_\iota\right) = \sum_{\iota \in I} \mu(A_\iota).$$

(*i*) *If* $(A_n)_{n \in \mathbb{N}}$ *is a sequence from* $\mathfrak{R}$ *that is bounded above in* $\mathfrak{R}$, *then the sets*

$$^{\mathfrak{P}(X)}\limsup_{n \to \infty} A_n \qquad ^{\mathfrak{P}(X)}\liminf_{n \to \infty} A_n$$

both belong to $\mathfrak{R}$. *If in addition there is a set* A *satisfying*

$$A = {}^{\mathfrak{P}(X)}\lim_{n \to \infty} A_n$$

then A *belongs to* $\mathfrak{R}$ *and*

$$\mu(A) = \lim_{n \to \infty} \mu(A_n).$$

Proof.

(a) Assertion (a) is little more than a restatement of Theorem 6.

(b) Assertion (b) is a consequence of Corollary 1.4.14 and Proposition 2.3.8 (b).

(c) Since $\mathfrak{R}$ is a δ-ring, $\mathfrak{R}_\sigma$ is the set consisting of all unions of increasing sequences from $\mathfrak{R}$ (Proposition 2.1.11). Proposition 4 (d) says that

$$\hat{\mathfrak{R}}(\mathscr{L}) \subset \mathfrak{R}_\sigma.$$

According to 4.1.15, $\hat{\mathfrak{R}}(\mathscr{L})$ is also a σ-ring, and $\hat{\mathfrak{R}}(\mathscr{L})$ certainly contains $\mathfrak{R}$. Therefore (c) holds.

The remaining assertions can be verified by using the map $A \mapsto e_A$ to translate the assertion in question into a familiar fact about closed Daniell spaces. In particular, (d) follows from Corollary 1.4.5 and implies (e), whereas (f) and (g) follow from Theorem 1.4.12. As for (h), the implications in one direction are valid for every positive measure space (Theorem 3.5.1); the opposite implications follow from Theorem 3.5.2. Finally, (i) is a consequence of Corollary 1.4.15 and Theorem 1.4.16. □

EXERCISES

Before proceeding to the exercises of this chapter, the reader is advised to look again at Ex. 2.2.4 which dealt in detail with the generation of positive measures by positive linear functionals.

5.1.1(E) Consider the positive measure space $(X, \mathfrak{F}, \chi_g)$ defined in Ex. 2.2.3 and its extension $(X, \mathfrak{R}_g, \bar{\chi}_g)$. For convenience, write ℓ_g for ℓ_{χ_g} and μ_g for μ^{ℓ_g} (cf., Ex. 2.4.2). Prove the following statements:

(α) $(X, \mathfrak{R}_g, \bar{\chi}_g) = (X, \mathfrak{R}(\mathscr{L}^1(\ell_g)), \mu_g)$.
(β) $(X, \mathfrak{R}_g, \bar{\chi}_g)$ is closed.

Define $\mathfrak{R}'_g := \{A \in \mathfrak{R}_g \mid A \text{ countable}\}$ and $\chi'_g := \bar{\chi}_g|_{\mathfrak{R}'_g}$. Prove the next three statements:

(γ) $(X, \mathfrak{R}'_g, \chi'_g) = (X, \mathfrak{R}(\bar{\mathscr{L}}(\ell_g)), \mu^{\bar{\ell}_g})$.
(δ) $(X, \mathfrak{R}'_g, \chi'_g)$ is closed.
(ε) $(X, \mathfrak{R}'_g, \chi'_g) = (X, \mathfrak{R}_g, \bar{\chi}_g)$ iff $\{g = 0\}$ is countable.

Thus the same phenomena occur with closed extensions of positive measure spaces as do with the closed extensions of Daniell spaces.

5.1.2(E) Let ℓ be a nullcontinuous positive linear functional on $\ell^\infty(X)$. Show that the following statement holds:

(α) $(X, \mathfrak{R}(\ell^\infty(X)), \mu^\ell)$ is a closed positive measure space, $\mathfrak{R}(\ell^\infty(X)) = \mathfrak{P}(X)$.

This shows that the positive measure space induced by a Daniell space can be closed even when the Daniell space itself is not closed! Prove the next assertion:

(β) For the Daniell space $(X, \mathscr{L}, \ell)$ the following are equivalent if $\mathscr{L}$ is a Stone lattice:

($\beta 1$) $(X, \mathfrak{R}(\mathscr{L}), \mu^\ell)$ is closed.
($\beta 2$) $\mathfrak{R}(\mathscr{L}) = \mathfrak{R}(\bar{\mathscr{L}}(\ell))$.

5.1.3(E) Let $(X, \mathfrak{R}, \mu)$ be a positive measure space. Prove the equivalence of the following:

(α) $(X, \mathfrak{R}, \mu)$ is closed.

(β) $\bigcup_{n \in \mathbb{N}} A_n \in \mathfrak{R}$ for every increasing sequence $(A_n)_{n \in \mathbb{N}}$ from $\mathfrak{R}$ with $\sup_{n \in \mathbb{N}} \mu(A_n) < \infty$, and $\mathfrak{P}(A) \subset \mathfrak{R}$ for every $A \in \mathfrak{R}$ with $\mu(A) = 0$.

(γ) Given a disjoint sequence $(A_n)_{n \in \mathbb{N}}$ from $\mathfrak{R}$ for which $(\mu(A_n))_{n \in \mathbb{N}}$ is summable, $\bigcup_{n \in \mathbb{N}} A_n \in \mathfrak{R}$, and $\mathfrak{P}(A) \subset \mathfrak{R}$ for every $A \in \mathfrak{R}$ with $\mu(A) = 0$.

Observe that the statement for decreasing sequences analogous to (β) is not sufficient to ensure that $(X, \mathfrak{R}, \mu)$ be closed. To see this, let X be an infinite set, $\mathfrak{F}$ the set of finite subsets of X, and $g \in \ell^1(X)$. Then $\mathfrak{F}$ is a ring of sets and χ_g a positive measure on $\mathfrak{F}$. Show that the next statements hold:

(δ) $\bigcap_{n \in \mathbb{N}} A_n \in \mathfrak{F}$ for every decreasing sequence $(A_n)_{n \in \mathbb{N}}$ in $\mathfrak{F}$.

(ε) $(X, \mathfrak{F}, \chi_g)$ is not closed.

Finally, prove the following statements:

(ζ) Let $(X, \mathfrak{R}, \mu)$ be a positive measure space. If $\mathfrak{R}$ is a σ-ring, then the following are equivalent:

(ζ1) $(X, \mathfrak{R}, \mu)$ is closed.

(ζ2) If $A \in \mathfrak{R}$ and $\mu(A) = 0$, then $\mathfrak{P}(A) \subset \mathfrak{R}$.

(η) Every positive measure space $(X, \mathfrak{P}(X), \mu)$ is closed.

(ϑ) Let $\mathfrak{R}$ be a σ-ring such that $\mathfrak{P}(A) \subset \mathfrak{R}$ for every $A \in \mathfrak{R}$. Then $(X, \mathfrak{R}, \mu)$ is closed for any positive measure μ on $\mathfrak{R}$.

(ι) Let $\mathfrak{R}$ be the set of all countable subsets of the set X. Then $(X, \mathfrak{R}, \mu)$ is closed for any positive measure μ on $\mathfrak{R}$.

5.1.4(E) Prove the following statement:

Let $(X, \mathfrak{R}, \mu)$ be a positive measure space. Show that if $(X, \mathfrak{R}', \mu')$ is a closed positive measure space with $(X, \mathfrak{R}, \mu) \preccurlyeq (X, \mathfrak{R}', \mu')$, then $(X, \mathfrak{R}(\bar{\mathscr{L}}(\ell_\mu)), \mu^{\bar{\ell}_\mu}) \preccurlyeq (X, \mathfrak{R}', \mu')$. In other words, the positive measure space $(X, \mathfrak{R}(\bar{\mathscr{L}}(\ell_\mu)), \mu^{\bar{\ell}_\mu})$ associated to $(X, \mathfrak{R}, \mu)$ is the smallest closed extension of $(X, \mathfrak{R}, \mu)$.

In the following sections we shall go into other characterizations of this smallest closed extension of a positive measure space.

5.1.5(E) Let X be a set and $\mathfrak{R} \subset \mathfrak{P}(X)$ a ring of subsets of X. Let μ and ν be positive measures on $\mathfrak{R}$. Prove the following statements:

(α) If $\mu \leq \nu$ and $(X, \mathfrak{R}, \mu)$ is closed, then so are $(X, \mathfrak{R}, \nu)$ and $(X, \mathfrak{R}, \alpha\mu)$ for any $\alpha \in \mathbb{R}$, $\alpha > 0$.

(β) If $(X, \mathfrak{R}, \mu)$ and $(X, \mathfrak{R}, \nu)$ are closed positive measure spaces, then so is $(X, \mathfrak{R}, \mu + \nu)$.

(γ) If $(X, \mathfrak{R}, \mu)$ and $(X, \mathfrak{R}, \nu)$ are closed positive measure spaces, then so is $(X, \mathfrak{R}, \mu \vee \nu)$.

We remark that $(X, \mathfrak{R}, \mu \wedge \nu)$ is also closed if $(X, \mathfrak{R}, \mu)$ and $(X, \mathfrak{R}, \nu)$ are, but the proof is more complicated. For the operations $\mu \vee \nu$ and $\mu \wedge \nu$, see Ex. 2.2.18.

5.1.6(E) Let $(X, \mathscr{L}, \ell)$ be a Daniell space, $\mathscr{L}$ a Stone lattice. Prove that the following are equivalent:

(α) $(X, \mathscr{L}, \ell)$ is bounded (in the sense of Ex. 4.2.9).
(β) $(X, \mathfrak{R}(\mathscr{L}), \mu^{\ell})$ is bounded.

5.1.7(E) Let $(X, \mathfrak{R}, \mu)$ be a closed positive measure space. We wish to show that $(X, \mathfrak{R}, \mu)$ has convergence properties similar to those of Daniell spaces as formulated and proved in Section 1.4. $\mathfrak{P}(X)$ is naturally ordered by $\subset$. But our propositions can be expressed more generally if we introduce the following preorder: Let $A \leq B$ in $\mathfrak{P}(X)$ iff $A \setminus B \in \mathfrak{R}$ and $\mu(A \setminus B) = 0$. Prove the following:

(α) $\leq$ is a preorder on $\mathfrak{P}(X)$.
(β) $A \leq B$ and $B \leq A$ iff $A \triangle B \in \mathfrak{R}$ and $\mu(A \triangle B) = 0$.
(γ) If $A, B \in \mathfrak{R}$, then $A \leq B$ whenever $A \subset B$.

Prove now the following propositions for $(X, \mathfrak{R}, \mu)$ with respect to the preorder defined here:

(δ) $\{A \in \mathfrak{R} \mid \mu(A) = 0\} = \{A \in \mathfrak{P}(X) \mid A \leq \varnothing\}$ is a σ-ring that contains every subset of each of its elements.
(ε) If $(A_\iota)_{\iota \in I}$ is a nonempty countable family from $\mathfrak{R}$ directed upward, then the following are equivalent:

(ε1) $\sup_{\iota \in I} \mu(A_\iota) < \infty$.
(ε2) $\bigcup_{\iota \in I} A_\iota \in \mathfrak{R}$.

In this case $\mu(\bigcup_{\iota \in I} A_\iota) = \sup_{\iota \in I} \mu(A_\iota)$.
(ζ) If $(A_\iota)_{\iota \in I}$ is a nonempty countable family from $\mathfrak{R}$ bounded above in $\mathfrak{R}$, then $\bigcup_{\iota \in I} A_\iota \in \mathfrak{R}$ and $\mu(\bigcup_{\iota \in I} A_\iota) \geq \sup_{\iota \in I} \mu(A_\iota)$.
(η) If $(A_\iota)_{\iota \in I}$ is a nonempty countable family from $\mathfrak{R}$ directed downward, then $\bigcap_{\iota \in I} A_\iota \in \mathfrak{R}$ and $\mu(\bigcap_{\iota \in I} A_\iota) = \inf_{\iota \in I} \mu(A_\iota)$.
(ϑ) If $(A_\iota)_{\iota \in I}$ is a nonempty countable family from $\mathfrak{R}$, then $\bigcap_{\iota \in I} A_\iota \in \mathfrak{R}$ and $\mu(\bigcap_{\iota \in I} A_\iota) \leq \inf_{\iota \in I} \mu(A_\iota)$.
(ι) If $(A_n)_{n \in \mathbb{N}}$ is a sequence from $\mathfrak{R}$ bounded above in $\mathfrak{R}$, and if $A =^{\mathfrak{P}(X)} \limsup_{n \to \infty} A_n$, then $A \in \mathfrak{R}$ and $\mu(A) \geq \limsup_{n \to \infty} \mu(A_n)$.
(κ) If $(A_n)_{n \in \mathbb{N}}$ is a sequence from $\mathfrak{R}$ and if $A =^{\mathfrak{P}(X)} \liminf_{n \to \infty} A_n$ then $A \in \mathfrak{R}$ and $\mu(A) \leq \liminf_{n \to \infty} \mu(A_n)$.
(λ) If $(A_n)_{n \in \mathbb{N}}$ is a sequence from $\mathfrak{R}$ bounded above in $\mathfrak{R}$, and if $A =^{\mathfrak{P}(X)} \lim_{n \to \infty} A_n$, then $A \in \mathfrak{R}$ and $\mu(A) = \lim_{n \to \infty} \mu(A_n)$.

5.1.8(E) We investigate restrictions and images of positive measures, in the sense of Ex. 2.2.7.

Let $(X, \mathfrak{R}, \mu)$ be a positive measure space, $Y \subset X$ and φ: $X \to Z$ a mapping into the set Z. Prove the following statements:

(α) $(Y, \mathfrak{R}|_Y, \mu|_Y)$ is a closed positive measure space, whenever $(X, \mathfrak{R}, \mu)$ is closed.

(β) $(Z, \varphi(\mathfrak{R}), \varphi(\mu))$ is a closed positive measure space, whenever $(X, \mathfrak{R}, \mu)$ is closed.

Finally show that neither converse holds, even if φ is surjective.

5.1.9(E) It was shown in Ex. 1.3.10 that every positive linear functional ℓ on a Riesz lattice $\mathscr{L}$ generates a pseudometric, d_ℓ. We wish to present the analogous considerations for the case of additive functions on rings of sets.

Let μ be an additive function on the ring of sets, $\mathfrak{R}$. Define d_μ: $\mathfrak{R} \times \mathfrak{R} \to \mathbb{R}$, $(A, B) \mapsto \mu(A \triangle B)$. Prove the following statements, for $A, B, C \in \mathfrak{R}$:

(α) $(\mathfrak{R}, d_\mu)$ is a pseudometric space.

(β) $d_\mu(A, B) = \mu(A \setminus B)$ if $A \supset B$ and $d_\mu(A, \varnothing) = \mu(A)$.

(γ) If $A \subset B \subset C$, then $d_\mu(A, C) = d_\mu(A, B) + d_\mu(B, C)$.

(δ) If $A \subset B \subset C$, then $d_\mu(A, B) \leq d_\mu(A, C)$.

Take a set X such that $\mathfrak{R} \subset \mathfrak{P}(X)$. Let ℓ_μ denote, as usual, the positive linear functional on $\mathscr{L}_\mathfrak{R}$ (the space of $\mathfrak{R}$-step functions) associated to μ. Define φ: $\mathfrak{R} \to \mathscr{L}_\mathfrak{R}$, $A \mapsto e_A$. Prove the following statements:

(ε) φ is an injective mapping.

(ζ) $d_\mu(A, B) = d_{\ell_\mu}(e_A, e_B)$.

(η) φ is an isometry of $(\mathfrak{R}, d_\mu)$ onto $(\varphi(\mathfrak{R}), d_{\ell_\mu})$.

5.1.10(E) Let $(X, \mathscr{L}, \ell)$ be a Daniell space and $\mathscr{L}$ a Stone lattice. Define sets $\mathfrak{R}^\downarrow(\mathscr{L})$, $\mathfrak{R}^\uparrow(\mathscr{L})$ and mappings $\mu^\downarrow, \mu^\uparrow$ as follows:

(i) $\mathfrak{R}^\downarrow(\mathscr{L}) := \{A \subset X \mid e_A = \bigwedge_{n \in \mathbb{N}} f_n$ for some decreasing sequence $(f_n)_{n \in \mathbb{N}}$ in $\mathscr{L}\}$

(ii) $\mathfrak{R}^\uparrow(\mathscr{L}) := \{A \subset X \mid e_A = \bigvee_{n \in \mathbb{N}} f_n$ for some increasing sequence $(f_n)_{n \in \mathbb{N}}$ in $\mathscr{L}$ with $\sup_{n \in \mathbb{N}} \ell(f_n) < \infty\}$

(iii) $\mu^\downarrow$: $\mathfrak{R}^\downarrow(\mathscr{L}) \to \mathbb{R}$, $\quad A \mapsto \inf\{\ell(f) \mid f \in \mathscr{L},\ e_A \leq f\}$
$\mu^\uparrow$: $\mathfrak{R}^\uparrow(\mathscr{L}) \to \mathbb{R}$, $\quad A \mapsto \sup\{\ell(f) \mid f \in \mathscr{L},\ f \leq e_A\}$

Prove the following:

(α) There is exactly one positive measure, μ, on $\mathfrak{R}^\downarrow(\mathscr{L})_\delta$ with the property that $(X, \mathscr{L}, \ell) \preccurlyeq (X, \bar{\mathscr{L}}(\ell_\mu), \bar{\ell}_\mu)$. For this μ, $\mu|_{\mathfrak{R}^\downarrow(\mathscr{L})} = \mu^\downarrow$ and $\mu(A) = \sup\{\mu^\downarrow(B) \mid B \in \mathfrak{R}^\downarrow(\mathscr{L}),\ B \subset A\}$ for any $A \in \mathfrak{R}^\downarrow(\mathscr{L})_\delta$.

(β) There is exactly one positive measure, ν, on $\mathfrak{R}^\uparrow(\mathscr{L})_\delta$ with the property that $(X, \mathscr{L}, \ell) \preccurlyeq (X, \bar{\mathscr{L}}(\ell_\nu), \bar{\ell}_\nu)$. For this ν, $\nu|_{\mathfrak{R}^\uparrow(\mathscr{L})} = \mu^\uparrow$ and $\nu(A) = \inf\{\mu^\uparrow(B) \mid B \in \mathfrak{R}^\uparrow(\mathscr{L}),\ A \subset B\}$ for any $A \in \mathfrak{R}^\uparrow(\mathscr{L})_\delta$.

(γ) $(X, \mathfrak{R}^\downarrow(\mathscr{L})_\delta, \mu) \preccurlyeq (X, \mathfrak{R}^\uparrow(\mathscr{L})_\delta, \nu) \preccurlyeq (X, \mathfrak{R}(\bar{\mathscr{L}}(\ell)), \mu^{\bar{\ell}})$.

5.2. MEASURE-SPACE CLOSURE AND COMPLETION AND THE MEASURE-SPACE INTEGRAL

NOTATION FOR SECTION 5.2:
X denotes a set.

The preceding section investigated general relationships between positive measure spaces and closed Daniell spaces with the Stone property. We return now to the problem of extending a given positive measure μ from the set-ring $\mathfrak{R}$ on which it is defined to some larger set-ring. If we fix a set $X \supset X(\mathfrak{R})$ and consider only extensions within $\mathfrak{P}(X)$, then this extension problem is completely analogous, in view of Theorem 2.4.3, to the corresponding problem for functionals. The one additional factor, which must not be lost sight of, is the dependence of the extensions on the set X.

Our first goal is to show that there is always a smallest closed extension.

Proposition 5.2.1. *Let μ be a positive measure on a set-ring $\mathfrak{R}$, and let X_1, X_2 be sets containing $X(\mathfrak{R})$. For $k = 1, 2$, denote by $(X_k, \mathscr{L}_k, \ell_k)$ the Daniell space associated with the positive measure space $(X_k, \mathfrak{R}, \mu)$, and denote by $(X_k, \mathfrak{R}_k, \mu_k)$ the closed positive measure space induced by $(X_k, \bar{\mathscr{L}}(\ell_k), \overline{\ell_k})$. Then the following assertions hold.*

(*a*) $\mathfrak{R}_1, \mathfrak{R}_2 \subset \mathfrak{P}(X(\mathfrak{R}))$.

(*b*) $(\mathfrak{R}_1, \mu_1) = (\mathfrak{R}_2, \mu_2)$. (1)

Proof. (a) Let $k \in \{1,2\}$, and let $A \in \mathfrak{R}_k$. By definition $e_A^{X_k}$ belongs to $\bar{\mathscr{L}}(\ell_k)$. Thus there exists in $\mathscr{L}_k$ an increasing sequence $(f_n)_{n \in \mathbb{N}}$ with

$$e_A^{X_k} \leq \bigvee_{n \in \mathbb{N}} f_n.$$

[Proposition 3.1.16 (a) $\Rightarrow$ (*e*)]. For each n, the set $\{f_n > 0\}$ belongs to $\mathfrak{R}$ [Corollary 2.3.4 (a)]. Since

$$A \subset \bigcup_{n \in \mathbb{N}} \{f_n > 0\}$$

A must be a subset of $X(\mathfrak{R})$.

(b) Schematically we have

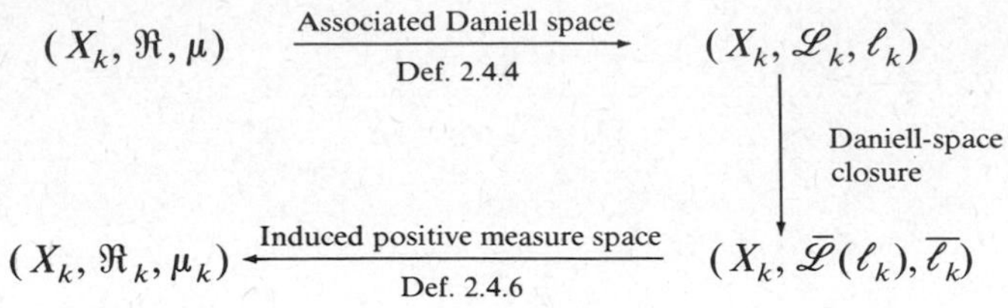

$$\begin{array}{ccc} (X_k, \mathfrak{R}, \mu) & \xrightarrow[\text{Def. 2.4.4}]{\text{Associated Daniell space}} & (X_k, \mathscr{L}_k, \ell_k) \\ & & \Big\downarrow \text{Daniell-space closure} \\ (X_k, \mathfrak{R}_k, \mu_k) & \xleftarrow[\text{Def. 2.4.6}]{\text{Induced positive measure space}} & (X_k, \bar{\mathscr{L}}(\ell_k), \overline{\ell_k}) \end{array}$$

That (1) holds is a consequence of the following observations. There is a natural isomorphism between the Daniell spaces $(X_1, \mathscr{L}_1, \ell_1)$ and $(X_2, \mathscr{L}_2, \ell_2)$ (obtained by associating $e_A^{X_1}$ with $e_A^{X_2}$ for each A in $\mathfrak{R}$). This isomorphism extends to the Daniell-space closures, as can be seen by using the characterization of Daniell-space closure from Proposition 3.1.16 (a) $\Leftrightarrow$ (e). Then the definition of induced positive measure space shows that $(X_1, \bar{\mathscr{L}}(\ell_1), \overline{\ell_1})$ and $(X_2, \bar{\mathscr{L}}(\ell_2), \overline{\ell_2})$ induce the same positive measure. □

Proposition 5.2.2. *Let μ be a positive measure on a set-ring $\mathfrak{R}$. Denote by $(X(\mathfrak{R}), \mathfrak{R}(\mu), \bar{\mu})$ the closed positive measure space induced by the closure of the Daniell space associated with $(X(\mathfrak{R}), \mathfrak{R}, \mu)$. Then, for every $X \supset X(\mathfrak{R})$, $(X, \mathfrak{R}(\mu), \bar{\mu})$ is the smallest closed positive measure space extending $(X, \mathfrak{R}, \mu)$.*

Proof. Let $X \supset X(\mathfrak{R})$, and denote by $(X, \mathscr{L}, \ell)$ the Daniell space associated with the positive measure space $(X, \mathfrak{R}, \mu)$. According to Proposition 1, $(X, \mathfrak{R}(\mu), \bar{\mu})$ is the closed positive measure space induced by $(X, \bar{\mathscr{L}}(\ell), \bar{\ell})$. Evidently $(X, \mathfrak{R}, \mu) \preccurlyeq (X, \mathfrak{R}(\mu), \bar{\mu})$. Now let $(X, \mathfrak{R}', \mu')$ be an arbitrary closed positive measure space for which

$$(\mathfrak{R}, \mu) \preccurlyeq (\mathfrak{R}', \mu'). \tag{2}$$

We must show that

$$(\mathfrak{R}(\mu), \bar{\mu}) \preccurlyeq (\mathfrak{R}', \mu'). \tag{3}$$

According to Theorem 5.1.8 (a), there is a unique closed Daniell space $(X, \mathscr{L}', \ell')$ with the Stone property such that $(X, \mathfrak{R}', \mu')$ is the positive measure space induced by $(X, \mathscr{L}', \ell')$. It follows from (2) that

$$(X, \mathscr{L}, \ell) \preccurlyeq (X, \mathscr{L}', \ell')$$

and therefore

$$(X, \bar{\mathscr{L}}(\ell), \bar{\ell}) \preccurlyeq (X, \mathscr{L}', \ell')$$

(Theorem 3.2.5). Now (3) follows from Proposition 5.1.2 (a) and Corollary 5.1.5 (a) ⇒ (c). □

The following definition is now appropriate.

Definition 5.2.3 (Measure-space closure). *Let μ be a positive measure on a set-ring $\mathfrak{R}$. Then $(X(\mathfrak{R}), \mathfrak{R}(\mu), \bar{\mu})$ denotes the closed positive measure space induced by the closure of the Daniell space associated with $(X(\mathfrak{R}), \mathfrak{R}, \mu)$. We call $\bar{\mu}$ the* ***closure*** *of μ and $(\mathfrak{R}(\mu), \bar{\mu})$ the* ***closure*** *of $(\mathfrak{R}, \mu)$. For each $X \supset X(\mathfrak{R})$, we call $(X, \mathfrak{R}(\mu), \bar{\mu})$ the* ***closure*** *of $(X, \mathfrak{R}, \mu)$.* □

It may be useful to diagram the definition. Denote by $(X, \mathscr{L}, \ell)$ the Daniell space associated with $(X, \mathfrak{R}, \mu)$. Then measure-space closure is achieved by following the arrows in

$$\begin{array}{ccc} (X, \mathfrak{R}, \mu) & \xrightarrow{\text{Associated Daniell space}} & (X, \mathscr{L}, \ell) \\ & & \Big\downarrow \text{Daniell-space closure} \\ (X, \mathfrak{R}(\mu), \bar{\mu}) & \xleftarrow{\text{Induced positive measure space}} & (X, \bar{\mathscr{L}}(\ell), \bar{\ell}) \end{array}$$

Bear in mind that

$$\mathscr{L} = \mathfrak{R}\text{-step functions on } X$$
$$\ell(e_A) = \mu(A) \qquad (A \in \mathfrak{R})$$

$$\mathfrak{R}(\mu) = \mathfrak{R}(\bar{\mathscr{L}}(\ell)) = \{A \subset X \mid e_A \in \bar{\mathscr{L}}(\ell)\}$$
$$\bar{\mu}(A) = \bar{\ell}(e_A) \qquad (A \in \mathfrak{R}(\mu)).$$

Measure-space closures exhibit many characteristics analogous to those of Daniell-space closures. Accordingly, they serve, as Daniell-space closures do, as an important tool in our theory. For a start we have the following analogue to Theorem 3.2.11 showing that measure-space closures are as strongly determined as Daniell-space closures.

Corollary 5.2.4. *For every positive measure space $(X, \mathfrak{R}, \mu)$, if $(X, \mathfrak{S}, \nu)$ is a positive measure space extending $(X, \mathfrak{R}, \mu)$, and if $\mathfrak{S} \subset \mathfrak{R}(\mu)_\sigma$, then $\mathfrak{S} \subset \mathfrak{R}(\mu)$ and*

$$\nu = \bar{\mu}|_{\mathfrak{S}}.$$

In particular, $\bar{\mu}$ and $\bar{\mu}|_{\mathfrak{R}_\delta}$ are the only positive measures on $\mathfrak{R}(\mu)$ and $\mathfrak{R}_\delta$, respectively, that extend μ.

Proof. We have

$$(X, \mathfrak{R}, \mu) \preccurlyeq (X, \mathfrak{S}, \nu) \preccurlyeq (X, \mathfrak{R}(\nu), \bar{\nu})$$

and $(X, \mathfrak{R}(\nu), \bar{\nu})$ is closed. It follows that

$$(X, \mathfrak{R}(\mu), \bar{\mu}) \preccurlyeq (X, \mathfrak{R}(\nu), \bar{\nu}).$$

In particular,

$$\bar{\nu}|_{\mathfrak{R}(\mu)} = \bar{\mu}.$$

Let $A \in \mathfrak{S}$. By hypothesis, A belongs to $\mathfrak{R}(\mu)_\sigma$. According to Proposition 2.1.11 and Theorem 5.1.8(b), we can choose from $\mathfrak{R}(\mu)$ an increasing sequence $(A_n)_{n \in \mathbb{N}}$ whose union is A. Using null continuity, we have

$$\sup_{n \in \mathbb{N}} \bar{\mu}(A_n) = \sup_{n \in \mathbb{N}} \bar{\nu}(A_n) = \bar{\nu}(A) < \infty.$$

By Theorem 5.1.8(f), A belongs to $\mathfrak{R}(\mu)$ and $\nu(A) = \bar{\nu}(A) = \bar{\mu}(A)$. □

Some authors require the domains for positive measures to be δ-rings. In view of Corollary 4, allowing arbitrary set-rings as domains does not expand the collection of closed positive measure spaces. What *is* gained is greater ease in defining a particular measure with which one wants to work.

We now turn our attention to measure extensions induced by integrals. As has already been mentioned, given a positive measure space $(X, \mathfrak{R}, \mu)$, if we consider only extensions of μ to set-rings contained in $\mathfrak{P}(X)$, then the problems of admissibility and determinedness are completely analogous to those for functionals. Thus these questions require no further investigation. In fact the following definition suggests itself.

Definition 5.2.5 (Measure-space integral and completion). *Given a positive measure space $(X, \mathfrak{R}, \mu)$, denote by $(X, \mathscr{L}(X, \mu), \ell_{X,\mu})$ the Daniell space associated with $(X, \mathfrak{R}, \mu)$. Then*

$$\mathscr{L}^1(X, \mu) := \mathscr{L}^1(\ell_{X,\mu})$$

$$\mathfrak{L}(X, \mu) := \mathfrak{R}(\mathscr{L}^1(X, \mu))$$

$$\mathscr{N}(X, \mu) := \mathscr{N}(\mathscr{L}^1(X, \mu)) = \mathscr{N}(\ell_{X,\mu})$$

$$\mathfrak{N}(X, \mu) := \mathfrak{N}(\mathscr{L}^1(X, \mu)) = \mathfrak{N}(\ell_{X,\mu}).$$

For $f \in \mathscr{L}^1(X, \mu)$ we define

$$\int_{X,\mu} f := \int_X f\, d\mu := \int_{\ell_{X,\mu}} f$$

and we call $\int_X f\, d\mu$ the ***μ-integral*** *of f. For $A \in \mathfrak{L}(X, \mu)$ we define*

$$\mu^X(A) := \int_X e_A\, d\mu$$

and we call $\mu^X(A)$ the ***μ-measure*** *of A in X.*

Functions belonging to $\mathscr{L}^1(X, \mu)$ are called ***μ-integrable functions*** *on X, and functions belonging to $\mathscr{N}(X, \mu)$ are called* ***μ-null functions*** *on X. Similarly sets belonging to $\mathfrak{L}(X, \mu)$ are called* ***μ-integrable sets*** *in X or* ***μ-integrable subsets*** *of X, and sets belonging to $\mathfrak{N}(X, \mu)$ are called* ***μ-null sets*** *in X or* ***μ-null subsets*** *of X.*

The integral $(X, \mathscr{L}^1(X, \mu), \int_{X,\mu})$ is called the ***integral*** *for $(X, \mathfrak{R}, \mu)$ or the* ***integral*** *on X associated with μ.*

The closed positive measure space $(X, \mathfrak{L}(X, \mu), \mu^X)$ is called the ***completion*** *of $(X, \mathfrak{R}, \mu)$ and μ^X is called the* ***completion*** *of μ on X or simply the completion of μ. If $(X, \mathfrak{R}, \mu) = (X, \mathfrak{L}(X, \mu), \mu^X)$, then we call the measure space $(X, \mathfrak{R}, \mu)$* ***complete***, *and we say that μ is a* ***complete measure*** *on X.*

We call $\int_{\ell_{X,\mu}}$-sequences ***μ-sequences***; *that is, a monotone sequence $(f_n)_{n\in\mathbb{N}}$ from $\mathscr{L}^1(X, \mu)$ for which $(\int_X f_n\, d\mu)_{n\in\mathbb{N}}$ is bounded in $\mathbb{R}$ is a μ-sequence. Monotone sequences $(A_n)_{n\in\mathbb{N}}$ from $\mathfrak{L}(X, \mu)$ for which $(\mu^X(A_n))_{n\in\mathbb{N}}$ is bounded in $\mathbb{R}$ will also be called* ***μ-sequences***.

Finally, a property P that refers to elements of X is said to hold ***μ-almost everywhere (μ-a.e.)*** *iff it holds $\ell_{X,\mu}$-a.e. We write P μ-a.e. and $P(x)$ μ-a.e., respectively, for P $\ell_{X,\mu}$-a.e. and $P(x)$ $\ell_{X,\mu}$-a.e.* □

Once again, we diagram the definition. The integral on X associated with μ is achieved by following the first two arrows, and the completion of $(X, \mathfrak{R}, \mu)$ is achieved by following all three arrows in

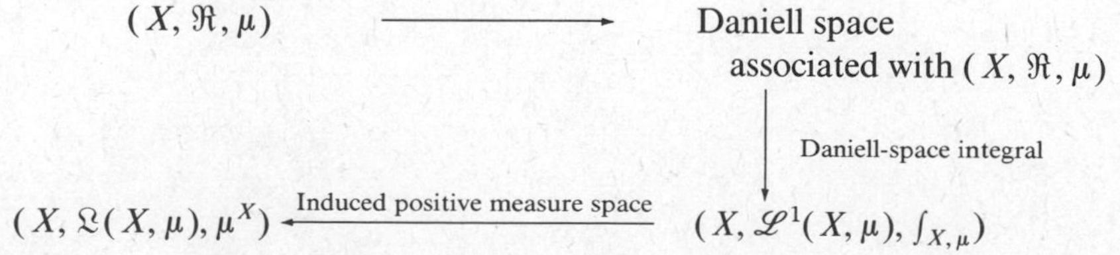

$(X, \mathfrak{R}, \mu)$ ⟶ Daniell space associated with $(X, \mathfrak{R}, \mu)$

↓ Daniell-space integral

$(X, \mathfrak{L}(X, \mu), \mu^X)$ ⟵ (Induced positive measure space) $(X, \mathscr{L}^1(X, \mu), \int_{X,\mu})$

The connections between closure and completion are shown next, where we have used $(X, \mathscr{L}, \ell)$ to denote the Daniell space associated with $(X, \mathfrak{R}, \mu)$:

$(X, \mathfrak{R}, \mu)$ —Associated Daniell space→ $(X, \mathscr{L}, \ell)$

从 ←- - - Smallest closed extension - - -→ 从 (Daniell-space closure)

$(X, \mathfrak{R}(\mu), \bar{\mu})$ ←Induced positive measure space— $(X, \bar{\mathscr{L}}(\ell), \bar{\ell})$

从   从 (Daniell-space integral)

$(X, \mathfrak{L}(X, \mu), \mu^X)$ ←Induced positive measure space— $(X, \mathscr{L}^1(X, \mu), \int_{X,\mu})$

从 : Largest determined extension (from $(X, \mathscr{L}, \ell)$ to $(X, \mathscr{L}^1(X, \mu), \int_{X,\mu})$)

These measure-specific definitions and notation prove to be useful since we work almost exclusively with integrals derived from measures. Moreover fitting the notation to the case of measure-derived integrals is in accord with common usage.

The next theorem summarizes the important properties of measure-space integral and completion, properties that are all more or less immediate consequences of the theory developed in Chapter 4. Since $(X, \mathscr{L}^1(X, \mu), \int_{X,\mu})$ is a closed Daniell space, $\mathscr{L}^1(X, \mu)$ and $\mathscr{L}(X, \mu)$ have many additional properties, which were discussed in earlier chapters.

Theorem 5.2.6 (Main Theorem on measure-space completion and integral). *Let $(X, \mathfrak{R}, \mu)$ be a positive measure space, and denote by $(X, \mathscr{L}, \ell)$ its associated Daniell space. Then the following assertions hold.*

(*a*) *$(X, \mathfrak{L}(X, \mu), \mu^X)$ is a closed positive measure space extending $(X, \mathfrak{R}, \mu)$. As a closed positive measure space it has all of the properties described in Theorem* 5.1.8.

(*b*) *$(X, \mathscr{L}^1(X, \mu), \int_{X,\mu})$ is the largest closed Daniell space extending $(X, \mathscr{L}, \ell)$ that is determined by $(X, \mathscr{L}, \ell)$. The Induction Principle for admissible extensions, Theorem* 4.1.18, *holds.*

(*c*)

$$\mathscr{N}(X, \mu) = \mathscr{N}(\mathscr{L}^1(X, \mu))$$

$$= \left\{ f \in \mathscr{L}^1(X, \mu) \mid \int_X |f| \, d\mu = 0 \right\}$$

$$= \{ f \in \overline{\mathbb{R}}^X \mid f e_A \in \mathscr{N}(\bar{\mathscr{L}}(\ell)), \forall A \in \mathfrak{R} \}$$

$$= \{ f \in \overline{\mathbb{R}}^X \mid f e_A \in \mathscr{N}(X, \mu), \forall A \in \mathfrak{R} \}$$

(d)
$$\begin{aligned}\mathfrak{N}(X,\mu) &= \mathfrak{N}(\mathscr{L}^1(X,\mu))\\ &= \{B \in \mathfrak{L}(X,\mu) \mid \mu^X(B) = 0\}\\ &= \{B \subset X \mid B \cap A \in \mathfrak{N}(\bar{\mathscr{L}}(\ell)), \forall A \in \mathfrak{R}\}\\ &= \{B \subset X \mid B \cap A \in \mathfrak{N}(X,\mu), \forall A \in \mathfrak{R}\}\\ &= \{B \subset X \mid \mu^X(B \cap A) = 0, \forall A \in \mathfrak{R}\}\end{aligned}$$

(e) *If f belongs to $\mathscr{L}^1(X,\mu)$, then there exist disjoint sets B and C such that*

$$f = fe_B + fe_C \qquad B \in \hat{\mathfrak{R}}(\bar{\mathscr{L}}(\ell)) \qquad C \in \mathfrak{N}(X,\mu).$$

Moreover fe_B belongs to $\bar{\mathscr{L}}(\ell)$ for every B in $\hat{\mathfrak{R}}(\bar{\mathscr{L}}(\ell))$.

(f) *A set A belongs to $\mathfrak{L}(X,\mu)$ iff there exist disjoint sets B and C such that*

$$A = B \cup C \qquad B \in \mathfrak{R}(\mu) \qquad C \in \mathfrak{N}(X,\mu)$$

If A belongs to $\mathfrak{L}(X,\mu)$, then $A \cap B$ belongs to $\mathfrak{R}(\mu)$ for every B in $\mathfrak{R}(\mu)$.

(g) *If f belongs to $\mathscr{L}^1(X,\mu)$, then there exist a sequence $(A_n)_{n\in\mathbb{N}}$ from $\mathfrak{R}$ and a set B belonging to $\mathfrak{N}(X,\mu)$ such that*

$$\{f \neq 0\} \subset \left(\bigcup_{n\in\mathbb{N}} A_n\right) \cup B.$$

The sequence $(A_n)_{n\in\mathbb{N}}$ can be chosen increasing or disjoint.

(h) *If A belongs to $\mathfrak{L}(X,\mu)$, then there exist a sequence $(A_n)_{n\in\mathbb{N}}$ from $\mathfrak{R}$ and a set B belonging to $\mathfrak{N}(X,\mu)$ such that*

$$A \subset \left(\bigcup_{n\in\mathbb{N}} A_n\right) \cup B.$$

The sequence $(A_n)_{n\in\mathbb{N}}$ can be chosen increasing or disjoint.

(i) *For every f belonging to $\mathscr{L}^1(X,\mu)$,*

$$\int_X f\,d\mu = \sup_{A\in\mathfrak{R}} \int_X fe_A\,d\mu + \inf_{A\in\mathfrak{R}} \int_X fe_A\,d\mu.$$

For every positive f in $\mathscr{L}^1(X,\mu)$,

$$\int_X f\,d\mu = \sup_{A\in\mathfrak{R}} \int_X fe_A\,d\mu.$$

(j) For every B belonging to $\mathfrak{L}(X, \mu)$,

$$\mu^X(B) = \sup_{A \in \mathfrak{R}} \mu^X(A \cap B).$$

Proof. Note that $\hat{\mathfrak{R}}(\mathscr{L}) = \mathfrak{R}$ in this instance.

(a) is an immediate consequence of the definitions.

(b) follows from the definitions together with Theorem 4.2.11(c) and Corollary 4.2.6.

(c) and (d) follow from the definitions.

(e) and (f) follow from (b) together with Theorem 4.2.7 (a) $\Rightarrow$ (b), (c).

(g) and (h) are consequences of Proposition 4.1.16 and Corollary 4.2.6.

(i) and (j) follow from Proposition 4.1.17. □

Given a positive measure μ on a set-ring $\mathfrak{R}$, each set X containing $X(\mathfrak{R})$ gives rise to a positive measure space $(X, \mathfrak{R}, \mu)$, hence to an integral $(X, \mathscr{L}^1(X, \mu), \int_{X,\mu})$ and a completion $(X, \mathfrak{L}(X, \mu), \mu^X)$. In contrast to the situation for measure-space closures, the dependence on the set X of the measure-space integral and the measure-space completion is not just apparent. On the other hand, it is quite what one would expect: enlarging X merely enlarges null sets.

Proposition 5.2.7. *Let μ be a positive measure on a set-ring $\mathfrak{R}$, and let X, Y be sets such that*

$$X \supset Y \supset X(\mathfrak{R}).$$

Then the following assertions hold.

(a) The set $X \setminus Y$ belongs to $\mathfrak{N}(X, \mu)$.

(b) $\mathfrak{N}(X, \mu) = \{A \subset X \mid A \cap Y \in \mathfrak{N}(Y, \mu)\}$.

(c) $\mathscr{N}(X, \mu) = \{f \in \overline{\mathbb{R}}^X \mid f|_Y \in \mathscr{N}(Y, \mu)\}$.

(d) $\mathfrak{L}(X, \mu) = \{A \subset X \mid A \cap Y \in \mathfrak{L}(Y, \mu)\}$.

(e) $\mathscr{L}^1(X, \mu) = \{f \in \overline{\mathbb{R}}^X \mid f|_Y \in \mathscr{L}^1(Y, \mu)\}$.

(f) For every $A \in \mathfrak{L}(X, \mu)$, $\mu^X(A) = \mu^Y(A \cap Y)$.

(g) For every $f \in \mathscr{L}^1(X, \mu)$, $\int_X f\,d\mu = \int_Y f|_Y\,d\mu$.

Proof. (a) For every A in $\mathfrak{R}$, $(X \setminus Y) \cap A = \varnothing$. According to Theorem 6 (d), $X \setminus Y$ belongs to $\mathfrak{N}(X, \mu)$.

Throughout the remainder of the proof we denote by $(X, \mathscr{L}(X), \ell_X)$ and $(Y, \mathscr{L}(Y), \ell_Y)$ the Daniell spaces associated with the positive measure spaces $(X, \mathfrak{R}, \mu)$ and $(Y, \mathfrak{R}, \mu)$, respectively.

(b) Let $A \in \mathfrak{N}(X, \mu)$. In other words, suppose that

$$A \cap B \in \mathfrak{N}(\overline{\mathscr{L}}(\ell_X))$$

for each B in $\mathfrak{R}$. From Proposition 1 and Theorem 6 (d) we have

$$\mathfrak{N}(\bar{\mathscr{L}}(\ell_X)) = \mathfrak{N}(\bar{\mathscr{L}}(\ell_Y)) \subset \mathfrak{P}(Y).$$

It follows that

$$(A \cap Y) \cap B \in \mathfrak{N}(\bar{\mathscr{L}}(\ell_Y)) \tag{4}$$

for each B in $\mathfrak{R}$, hence that $A \cap Y$ belongs to $\mathfrak{N}(Y, \mu)$. Suppose, conversely, that A is a subset of X whose intersection with Y belongs to $\mathfrak{N}(Y, \mu)$. Then (4) holds for each B in $\mathfrak{R}$, and $A \cap Y$ must belong to $\mathfrak{N}(X, \mu)$. Since $A \setminus Y$ also belongs to $\mathfrak{N}(X, \mu)$, by (a), we conclude that A belongs to $\mathfrak{N}(X, \mu)$.

(c) Assertion (c) follows from (b).

(d), (f) Let $A \in \mathfrak{L}(X, \mu)$. According to Theorem 6 (f), there exist disjoint sets $B \in \mathfrak{R}(\mu)$ and $C \in \mathfrak{N}(X, \mu)$ such that $A = B \cup C$. By (b), $C \cap Y$ belongs to $\mathfrak{N}(Y, \mu)$. Invoking Theorem 6 (f) again, we conclude that $A \cap Y$ belongs to $\mathfrak{L}(Y, \mu)$. Moreover

$$\mu^X(A) = \bar{\mu}(B) = \mu^Y(A \cap Y).$$

The argument in the other direction is similar.

(e), (g) Let $\tilde{\mathscr{L}} := \{ f \in \overline{\mathbb{R}}^X \mid f|_Y \in \mathscr{L}^1(Y, \mu)\}$, and define

$$\tilde{\ell} : \tilde{\mathscr{L}} \to \mathbb{R}, \qquad f \mapsto \int_Y f|_Y \, d\mu.$$

It is easy to see that $(X, \tilde{\mathscr{L}}, \tilde{\ell})$ is a closed Daniell space with the Stone property and that

$$(X, \mathscr{L}(X), \ell_X) \preccurlyeq (X, \tilde{\mathscr{L}}, \tilde{\ell}).$$

It follows that

$$(X, \bar{\mathscr{L}}(\ell_X), \overline{\ell_X}) \preccurlyeq (X, \tilde{\mathscr{L}}, \tilde{\ell})$$

(Theorem 3.2.5). In view of (c) and Theorem 6 (d),

$$\left(X, \mathscr{L}^1(X, \mu), \int_{X, \mu} \right) \preccurlyeq (X, \tilde{\mathscr{L}}, \tilde{\ell}).$$

On the other hand, using (d), we see that

$$\mathfrak{R}(\tilde{\mathscr{L}}) = \mathfrak{L}(X, \mu).$$

Using Theorem 5.1.8 (c), we conclude that

$$\hat{\mathfrak{R}}(\tilde{\mathscr{L}}) = \hat{\mathfrak{R}}(\mathscr{L}^1(X, \mu)).$$

Now Theorem 4.2.5 shows that

$$\left(X, \mathscr{L}^1(X, \mu), \int_{X, \mu}\right) = (X, \tilde{\mathscr{L}}, \tilde{\ell}). \qquad \square$$

Part (a) of the preceding proposition shows that $\mathfrak{L}(X, \mu)$ differs from $\mathfrak{L}(Y, \mu)$, and $\mathscr{L}^1(X, \mu)$ differs from $\mathscr{L}^1(Y, \mu)$, unless $X = Y$. Thus there is good reason for including X in the denotations for the various objects connected with measures.

A new uniqueness question, distinct from the question of determinedness, arises when one works with different measure spaces where neither is an extension of the other. The question concerns whether two such measures have the same integral. The next few results deal with this matter.

Proposition 5.2.8. *Let* $(X, \mathfrak{R}, \mu)$ *and* $(X, \mathfrak{S}, \nu)$ *be positive measure spaces with*

$$(\mathfrak{R}, \mu) \preccurlyeq (\mathfrak{L}(X, \nu), \nu^X).$$

Suppose that for every set A *in* $\mathfrak{S}$ *there exists a countable family* $(A_\iota)_{\iota \in I}$ *from* $\mathfrak{R}$ *such that*

$$A \setminus \bigcup_{\iota \in I} A_\iota \in \mathfrak{N}(X, \nu). \tag{6}$$

Then

$$\left(X, \mathscr{L}^1(X, \mu), \int_{X, \mu}\right) \preccurlyeq \left(X, \mathscr{L}^1(X, \nu), \int_{X, \nu}\right)$$

and

$$(X, \mathfrak{L}(X, \mu), \mu^X) \preccurlyeq (X, \mathfrak{L}(X, \nu), \nu^X).$$

Proof. Denote by $(X, \mathscr{L}, \ell)$ the Daniell space associated with $(X, \mathfrak{R}, \mu)$. The hypotheses imply

$$(X, \mathscr{L}, \ell) \preccurlyeq \left(X, \mathscr{L}^1(X, \nu), \int_{X, \nu}\right)$$

and consequently

$$(X, \bar{\mathscr{L}}(\ell), \bar{\ell}) \preccurlyeq \left(X, \mathscr{L}^1(X,\nu), \int_{X,\nu} \right) \tag{7}$$

(Theorem 3.2.5).

Now we use Theorem 6 (c) to show that

$$\mathscr{N}(X,\mu) \subset \mathscr{N}(X,\nu). \tag{8}$$

Indeed, let $f \in \mathscr{N}(X,\mu)_+$. Given A belonging to $\mathfrak{S}$, let $(A_\iota)_{\iota \in I}$ be a countable family from $\mathfrak{R}$ for which (6) holds. Then

$$fe_A = \bigvee_{\iota \in I} fe_{A \cap A_\iota} \ \nu\text{-a.e.}$$

Each fe_{A_ι} belongs to $\mathscr{N}(\bar{\mathscr{L}}(\ell))$, hence so does each function $fe_{A \cap A_\iota}$. But $fe_{A \cap A_\iota}$ also belongs to $\mathscr{L}^1(X,\nu)$ and therefore to $\mathscr{N}(X,\nu)$. It follows that fe_A belongs to $\mathscr{N}(X,\nu)$. Using Theorem 6 (c), we conclude that f belongs to $\mathscr{N}(X,\nu)$ and (8) follows.

Now let $f \in \mathscr{L}^1(X,\mu)$. According to Theorem 6 (e), we have

$$f = fe_A + fe_B$$

for some $A \in \hat{\mathfrak{R}}(\bar{\mathscr{L}}(\ell))$, $B \in \mathfrak{N}(X,\mu)$, with A and B disjoint. In view of Theorem 6 (e) and (7), fe_A must belong to $\mathscr{L}^1(X,\nu)$. The argument in the preceding paragraph shows that fe_B belongs to $\mathscr{N}(X,\nu)$. It follows that f belongs to $\mathscr{L}^1(X,\nu)$. Moreover

$$\int_X f\,d\nu = \int_X fe_A\,d\nu = \bar{\ell}(fe_A) = \int_X f\,d\mu.$$

We have verified that

$$\left(X, \mathscr{L}^1(X,\mu), \int_{X,\mu} \right) \preccurlyeq \left(X, \mathscr{L}^1(X,\nu), \int_{X,\nu} \right).$$

The second assertion follows from the first. □

Corollary 5.2.9. *Let $(X, \mathfrak{R}, \mu)$ and $(X, \mathfrak{S}, \nu)$ be positive measure spaces. If*

$$(\mathfrak{R},\mu) \preccurlyeq (\mathfrak{L}(X,\nu), \nu^X)$$

and

$$(\mathfrak{S},\nu) \preccurlyeq (\mathfrak{L}(X,\mu), \mu^X)$$

then $(X, \mathfrak{R}, \mu)$ *and* $(X, \mathfrak{S}, \nu)$ *have the same integral and the same completion*:

$$\left(X, \mathscr{L}^1(X,\mu), \int_{X,\mu}\right) = \left(X, \mathscr{L}^1(X,\nu), \int_{X,\nu}\right)$$

$$(X, \mathfrak{L}(X,\mu), \mu^X) = (X, \mathfrak{L}(X,\nu), \nu^X).$$

Proof. Let $A \in \mathfrak{S}$. By hypothesis, A belongs to $\mathfrak{L}(X,\mu)$, and Theorem 6 (h) ensures the existence of a countable family $(A_\iota)_{\iota \in I}$ from $\mathfrak{R}$ such that

$$A \setminus \bigcup_{\iota \in I} A_\iota \in \mathfrak{N}(X,\mu).$$

In order to apply Proposition 8, we want to show that

$$A \setminus \bigcup_{\iota \in I} A_\iota \in \mathfrak{N}(X,\nu). \tag{9}$$

Since $\mathfrak{R} \subset \mathfrak{L}(X,\nu)$ and $\mathfrak{L}(X,\nu)$ is a δ-ring [Theorem 5.1.8 (b)], the set $A \setminus \bigcup_{\iota \in I} A_\iota$, which can be written as $\bigcap_{\iota \in I}(A \setminus A_\iota)$, must belong to $\mathfrak{L}(X,\nu)$. According to Theorem 6 (f), there exist disjoint sets B in $\mathfrak{R}(\nu)$ and C in $\mathfrak{N}(X,\nu)$ such that

$$A \setminus \bigcup_{\iota \in I} A_\iota = B \cup C.$$

Corollary 4 implies that

$$(\mathfrak{R}(\nu), \bar{\nu}) \preccurlyeq (\mathfrak{L}(X,\mu), \mu^X)$$

from which we conclude that B belongs to $\mathfrak{L}(X,\mu)$ and $\mu^X(B) = \bar{\nu}(B)$. Since B is a subset of $A \setminus \bigcup_{\iota \in I} A_\iota$, which is a μ-null subset of X, it must be that B belongs to $\mathfrak{N}(X,\nu)$. Consequently (9) holds.

Applying Proposition 8, we conclude that

$$\left(X, \mathscr{L}^1(X,\mu), \int_{X,\mu}\right) \preccurlyeq \left(X, \mathscr{L}^1(X,\nu), \int_{X,\nu}\right).$$

Reversing the argument, we obtain the opposite inequality and the corollary follows. □

Corollary 5.2.10. *Let* $(X, \mathfrak{R}, \mu)$ *and* $(X, \mathfrak{S}, \nu)$ *be positive measure spaces. If*

$$\mathfrak{R} \subset \mathfrak{S}_\delta \subset \mathfrak{L}(X,\mu)$$

and

$$\mu^X|_{\mathfrak{S}_\delta} = \nu^X|_{\mathfrak{S}_\delta}$$

then $(X, \mathfrak{R}, \mu)$ *and* $(X, \mathfrak{S}, \nu)$ *have the same integral and the same completion.*

Proof. Let $\tilde{\nu} := \nu^X|_{\mathfrak{S}_\delta}$. Then

$$(\mathfrak{R}, \mu) \preccurlyeq (\mathfrak{L}(X, \tilde{\nu}), \tilde{\nu}^X)$$

and

$$(\mathfrak{S}_\delta, \tilde{\nu}) \preccurlyeq (\mathfrak{L}(X, \mu), \mu^X).$$

Using Corollary 9, we conclude that

$$\left(X, \mathscr{L}^1(X, \mu), \int_{X,\mu}\right) = \left(X, \mathscr{L}^1(X, \tilde{\nu}), \int_{X,\tilde{\nu}}\right)$$

$$(X, \mathfrak{L}(X, \mu), \mu^X) = (X, \mathfrak{L}(X, \tilde{\nu}), \tilde{\nu}^X).$$

On the other hand,

$$(\mathfrak{S}, \nu) \preccurlyeq (\mathfrak{L}(X, \tilde{\nu}), \tilde{\nu}^X)$$

$$(\mathfrak{S}_\delta, \tilde{\nu}) \preccurlyeq (\mathfrak{L}(X, \nu), \nu^X).$$

Using Corollary 9 again, we have

$$\left(X, \mathscr{L}^1(X, \nu), \int_{X,\nu}\right) = \left(X, \mathscr{L}^1(X, \tilde{\nu}), \int_{X,\tilde{\nu}}\right)$$

$$(X, \mathfrak{L}(X, \nu), \nu^X) = (X, \mathfrak{L}(X, \tilde{\nu}), \tilde{\nu}^X).$$

The corollary follows. □

A small consequence is the idempotence of measure-space integrals and completions.

Corollary 5.2.11. *For every positive measure space* $(X, \mathfrak{R}, \mu)$

$$\left(X, \mathscr{L}^1(X, \mu^X), \int_{X,\mu^X}\right) = \left(X, \mathscr{L}^1(X, \mu), \int_{X,\mu}\right)$$

$$(X, \mathfrak{L}(X, \mu^X), (\mu^X)^X) = (X, \mathfrak{L}(X, \mu), \mu^X).$$

Proof. Apply Corollary 10 using $(X, \mathfrak{S}, \nu) := (X, \mathfrak{L}(X, \mu), \mu^X)$. According to Theorem 5.1.8 (b), $\mathfrak{L}(X, \mu)$ is a δ-ring, and trivially

$$\mu^X|_{\mathfrak{L}(X,\mu)} = (\mu^X)^X|_{\mathfrak{L}(X,\mu)} = \mu^X. \qquad \square$$

Corollary 5.2.12. *Let $(X, \mathfrak{R}, \mu)$ and $(X, \mathfrak{S}, \nu)$ be positive measure spaces with*

$$(\mathfrak{R}, \mu) \preccurlyeq (\mathfrak{L}(X, \nu), \nu^X).$$

Suppose that for every set A in $\mathfrak{S}$ there exists a countable family $(A_\iota)_{\iota \in I}$ from $\mathfrak{R}$ such that

$$A \triangle \bigcup_{\iota \in I} A_\iota \in \mathfrak{N}(X, \nu).$$

Then $(X, \mathscr{L}^1(X, \nu), \int_{X,\nu})$ is an admissible extension of the Daniell space associated with $(X, \mathfrak{R}, \mu)$.

Proof. Let $(X, \mathscr{L}, \ell)$ denote the Daniell space associated with $(X, \mathfrak{R}, \mu)$. By Proposition 8,

$$\left(X, \mathscr{L}^1(X, \mu), \int_{X,\mu}\right) \preccurlyeq \left(X, \mathscr{L}^1(X, \nu), \int_{X,\nu}\right).$$

Let

$$\mathscr{L}' := \mathscr{L}^1(X, \mu) \vdash \mathscr{N}(X, \nu)$$

and denote by ℓ' the uniquely determined positive linear extension to $\mathscr{L}'$ of $\int_{X,\mu}$ (4.1.10 (a)). Then

$$(X, \mathscr{L}', \ell') \preccurlyeq \left(X, \mathscr{L}^1(X, \nu), \int_{X,\nu}\right). \tag{10}$$

We show that

$$\hat{\mathfrak{R}}(\mathscr{L}^1(X, \nu)) \subset \hat{\mathfrak{R}}(\mathscr{L}'). \tag{11}$$

Let $A \in \hat{\mathfrak{R}}(\mathscr{L}^1(X, \nu))$. By Theorem 6 (g) there exist a sequence $(A_n)_{n \in \mathbb{N}}$ from $\mathfrak{S}$ and a set $B \in \mathfrak{N}(X, \nu)$ such that

$$A \subset \left(\bigcup_{n \in \mathbb{N}} A_n\right) \cup B.$$

In view of the hypotheses, for every n in $\mathbb{N}$ there exists a countable family $(B_{n\iota})_{\iota \in I_n}$ from $\mathfrak{R}$ such that

$$A_n \triangle \left(\bigcup_{\iota \in I_n} B_{n\iota}\right) \in \mathfrak{N}(X, \nu)$$

for every n in $\mathbb{N}$. We conclude that

$$A \triangle \left(\bigcup_{n \in \mathbb{N}} \bigcup_{\iota \in I_n} B_{n\iota} \right) \subset \left(\bigcup_{n \in \mathbb{N}} \left(A_n \triangle \left(\bigcup_{\iota \in I_n} B_{n\iota} \right) \right) \right) \cup B$$

and

$$A \triangle \left(\bigcup_{n \in \mathbb{N}} \bigcup_{\iota \in I_n} B_{n\iota} \right) \in \mathfrak{N}(X, \nu).$$

It follows from the definitions that A belongs to $\hat{\mathfrak{R}}(\mathscr{L}')$. Having established (10) and (11), we note that $(X, \mathscr{L}', \ell')$ is a closed Daniell space [Theorem 4.1.10 (d)] and apply Theorem 4.2.5 to conclude that

$$(X, \mathscr{L}', \ell') = \left(X, \mathscr{L}^1(X, \nu), \int_{X, \nu} \right).$$

Thus for every $f \in \mathscr{L}^1(X, \nu)$ there exists $g \in \mathscr{L}^1(X, \mu)$ with $f = g$ ν-a.e. It follows by Theorem 6 (e) that for every $f \in \mathscr{L}^1(X, \nu)$ there exists a $g \in \bar{\mathscr{L}}(\ell)$ with $f = g$ ν-a.e.; that is, $(X, \mathscr{L}^1(X, \nu), \int_{X,\nu})$ is an admissible extension of $(X, \mathscr{L}, \ell)$. □

Finally, continuing a theme that was begun in Section 2.2 and last touched on in Proposition 4.2.15, we investigate what consequences arise for measure-space integrals when the underlying measure space satisfies various boundedness conditions. For a start, we verify that when $(X, \mathfrak{R})$ is σ-finite, the closure of μ and the completion of μ are identical, and the integral of μ is merely the closure of the Daniell space associated with $(X, \mathfrak{R}, \mu)$.

Proposition 5.2.13. *Let $(X, \mathfrak{R}, \mu)$ be a positive measure space, and denote by $(X, \mathscr{L}, \ell)$ its associated Daniell space. If the pair $(X, \mathfrak{R})$ is σ-finite, then the following assertions hold.*

(*a*) $(X, \mathscr{L}^1(X, \mu), \int_{X,\mu}) = (X, \bar{\mathscr{L}}(\ell), \bar{\ell})$.
(*b*) $(X, \mathfrak{L}(X, \mu), \mu^X) = (X, \mathfrak{R}(\mu), \bar{\mu})$.

Proof. By hypothesis, there exists a countable family $(A_\iota)_{\iota \in I}$ from $\mathfrak{R}$ whose union is X. The functions e_{A_ι} belong to $\mathscr{L}$ and

$$X = \bigcup_{\iota \in I} \{ e_{A_\iota} \neq 0 \}.$$

In other words, the Riesz lattice $\mathscr{L}$ is σ-finite. We conclude (a) from Proposition 4.2.15, and we conclude (b) from (a). □

Proposition 5.2.14. *The following assertions are equivalent, for every positive measure space* $(X, \mathfrak{R}, \mu)$.

(*a*) $(X, \mathfrak{R}, \mu)$ *is* σ*-bounded.*

(*b*) *There exists a countable family* $(A_\iota)_{\iota \in I}$ *from* $\mathfrak{R}$ *such that* $X \setminus \bigcup_{\iota \in I} A_\iota \in \mathfrak{N}(X, \mu)$.

(*c*) X *belongs to* $\hat{\mathfrak{R}}(\mathscr{L}^1(X, \mu))$.

(*d*) *The pair* $(X, \mathfrak{L}(X, \mu))$ *is* σ*-finite.*

Proof. (a) $\Rightarrow$ (b). By definition (2.2.4) there exists a countable family $(A_\iota)_{\iota \in I}$ from $\mathfrak{R}$ with

$$\inf_{\iota \in I} \mu(A \setminus A_\iota) = 0$$

for every A in $\mathfrak{R}$. Now $\mathfrak{L}(X, \mu)$ is a δ-ring [Theorem 5.1.8 (b)] containing $\mathfrak{R}$. Since

$$A \setminus \bigcup_{\iota \in I} A_\iota = \bigcap_{\iota \in I} (A \setminus A_\iota)$$

we conclude, for every A in $\mathfrak{R}$, that

$$A \setminus \bigcup_{\iota \in I} A_\iota \in \mathfrak{L}(X, \mu)$$

$$\mu^X\left(A \setminus \bigcup_{\iota \in I} A_\iota\right) = 0.$$

Since

$$\left(X \setminus \bigcup_{\iota \in I} A_\iota\right) \cap A = A \setminus \bigcup_{\iota \in I} A_\iota$$

Theorem 6 (d) now shows that the set $X \setminus \bigcup_{\iota \in I} A_\iota$ belongs to $\mathfrak{N}(X, \mu)$.

(b) $\Rightarrow$ (c) follows from the relation $\hat{\mathfrak{R}}(\mathscr{L}^1(X, \mu)) = \mathfrak{L}(X, \mu)_\sigma$ [Theorem 5.1.8 (c)].

(c) $\Rightarrow$ (d) is a consequence of Theorem 6 (g).

(d) $\Rightarrow$ (a). According to the hypothesis, there exists a countable family from $\mathfrak{L}(X, \mu)$ whose union is X. Thus X itself belongs to $\hat{\mathfrak{R}}(\mathscr{L}^1(X, \mu))$, and by Theorem 6 (g) there exist an increasing sequence $(A_n)_{n \in \mathbb{N}}$ from $\mathfrak{R}$ and a set B belonging to $\mathfrak{N}(X, \mu)$ such that

$$X = \left(\bigcup_{n \in \mathbb{N}} A_n\right) \cup B.$$

It follows that $A \setminus \bigcup_{n \in \mathbb{N}} A_n$ belongs to $\mathfrak{N}(X, \mu)$ for every A in $\mathfrak{R}$ and that

$$\inf_{n \in \mathbb{N}} \mu(A \setminus A_n) = \mu^X\Big(\bigcap_{n \in \mathbb{N}} (A \setminus A_n)\Big)$$

$$= \mu^X\Big(A \setminus \bigcup_{n \in \mathbb{N}} A_n\Big)$$

$$= 0. \qquad \square$$

Bounded measures play a very special role. The completions of such measures provide a new kind of set system.

Definition 5.2.15. *For $\mathfrak{R} \subset \mathfrak{P}(X)$, $\mathfrak{R}$ is said to be a **σ-algebra on X** iff $\mathfrak{R}$ is a σ-ring and X belongs to $\mathfrak{R}$.* □

Thus σ-algebras are only defined relative to some underlying set X. They can be characterized in various ways.

Proposition 5.2.16. *The following assertions are equivalent, for every set of sets $\mathfrak{R}$ and every set $X \supset X(\mathfrak{R})$.*

(*a*) *$\mathfrak{R}$ is a σ-algebra on X.*
(*b*) *$\mathfrak{R}$ is a δ-ring, and X belongs to $\mathfrak{R}$.*
(*c*) *The empty set belongs to $\mathfrak{R}$, $X \setminus A$ belongs to $\mathfrak{R}$ for every A in $\mathfrak{R}$, and $\bigcup_{n \in \mathbb{N}} A_n$ belongs to $\mathfrak{R}$ for every sequence $(A_n)_{n \in \mathbb{N}}$ from $\mathfrak{R}$.*

Proof. (a) ⇒ (b) is trivial.

(b) ⇒ (c) is a consequence of Proposition 2.1.5.

(c) ⇒ (a). The hypotheses readily imply that X belongs to $\mathfrak{R}$, and that $A \cup B$ belongs to $\mathfrak{R}$ whenever A and B do. To verify that $A \setminus B$ belongs to $\mathfrak{R}$ if A and B do, merely note that

$$A \setminus B = X \setminus [(X \setminus A) \cup B]. \qquad \square$$

Proposition 5.2.17. *The following assertions are equivalent, for every positive measure space $(X, \mathfrak{R}, \mu)$.*

(*a*) *$(X, \mathfrak{R}, \mu)$ is bounded.*
(*b*) *X belongs to $\mathfrak{L}(X, \mu)$.*
(*c*) *$\mathfrak{L}(X, \mu)$ is a σ-algebra on X.*
(*d*) *$(X, \mathfrak{L}(X, \mu), \mu^X)$ is bounded.*

Each of these assertions implies

(*e*) $\sup_{A \in \mathfrak{R}} \mu(A) = \sup_{A \in \mathfrak{L}(X,\mu)} \mu^X(A) = \mu^X(X)$.

Proof. (a) $\Rightarrow$ (b). Since μ is bounded it is also σ-bounded (Corollary 2.2.7), so X belongs to $\hat{\mathfrak{R}}(\mathcal{L}^1(X,\mu))$, by Proposition 14. According to Theorem 6 (g), there exist an increasing sequence $(A_n)_{n\in\mathbb{N}}$ from $\mathfrak{R}$ and a set $B \in \mathfrak{N}(X,\mu)$ such that

$$X = \left(\bigcup_{n\in\mathbb{N}} A_n\right) \cup B.$$

Since

$$\sup_{n\in\mathbb{N}} \mu(A_n) < \infty$$

Theorem 5.1.8 (f) shows that $\bigcup_{n\in\mathbb{N}} A_n$ belongs to $\mathfrak{L}(X,\mu)$. Thus X belongs to $\mathfrak{L}(X,\mu)$.

(b) $\Rightarrow$ (c) is a consequence of Theorem 5.1.8 (b) and Proposition 16 (b) $\Rightarrow$ (a).

(c) $\Rightarrow$ (d). From (c) we conclude that

$$\mu^X(A) \le \mu^X(X)$$

for every A in $\mathfrak{L}(X,\mu)$.

(d) $\Rightarrow$ (a) is trivial.

(e) Theorem 6 (j) shows that (e) follows from (b). □

Proposition 5.2.18. *Let $(X,\mathfrak{R},\mu)$ be a positive measure space for which $\mathfrak{R}$ is a σ-ring. Then assertions (a)–(e) of Proposition 17 all hold. Moreover there exists a set $B \in \mathfrak{R}$ such that $X \setminus B$ is μ-null and*

$$\mu^X(A \setminus B) = 0 \qquad \mu^X(A) = \mu^X(A \cap B)$$

for every $A \in \mathfrak{L}(X,\mu)$.

Proof. That Proposition 17(a) holds has already been noted (Corollary 2.2.7). Hence Proposition 17(b)–(e) also hold. The remaining assertion is an easy consequence of Proposition 2.2.6 (b), (c), and the characterization

$$\mathfrak{N}(X,\mu) = \{A \subset X \mid \mu^X(A \cap C) = 0, \forall C \in \mathfrak{R}\}.$$ □

EXERCISES

5.2.1(E) Let $(X,\mathfrak{R},\mu)$ be a positive measure space with a finite ring of sets, $\mathfrak{R}$. Describe the closure and the completion of $(X,\mathfrak{R},\mu)$.

5.2.2(E) Let $(X,\mathfrak{R},\mu)$ be a bounded positive measure space. Prove that there is a unique extension of μ to $\mathfrak{R}_\sigma$.

5.2.3(E) Let $(X, \mathfrak{R}, \mu)$ be a positive measure space. Prove that the following statements are equivalent:

(α) $\mu = 0$.
(β) $X \in \mathfrak{N}(X, \mu)$.
(γ) $\overline{\mathbb{R}}^X = \mathcal{N}(X, \mu)$.
(δ) $\overline{\mathbb{R}}^X = \mathcal{L}^1(X, \mu)$.
(ε) $\int f \, d\mu = 0$ whenever $f \in \mathcal{L}^1(X, \mu)$.

5.2.4(E) Let $((X_\iota, \mathfrak{R}_\iota, \mu_\iota))_{\iota \in I}$ be a family of positive measure-spaces such that $(X_\iota)_{\iota \in I}$ is disjoint. Define $X := \bigcup_{\iota \in I} X_\iota$ and $\mathfrak{R} := (\bigcup_{\iota \in I} \mathfrak{R}_\iota)_r$. Prove the following assertions:

(α) $\bigcup_{\iota \in I} \mathfrak{R}_\iota$ is a semi-ring. For every $A \in \mathfrak{R}$, there is a unique finite family $(A_\iota)_{\iota \in J}$ such that $J \subset I$, $A_\iota \in \mathfrak{R}_\iota \setminus \{\varnothing\}$ for every $\iota \in J$ and $A = \bigcup_{\iota \in J} A_\iota$.
(β) There is a unique positive measure μ on $\mathfrak{R}$ such that $\mu|_{\mathfrak{R}_\iota} = \mu_\iota$ for every $\iota \in I$.
(γ) The following assertions are equivalent for any $f \in \overline{\mathbb{R}}^X$:

($\gamma 1$) $f \in \mathcal{L}^1(\mu)$.
($\gamma 2$) $f|_{X_\iota} \in \mathcal{L}^1(\mu_\iota)$ for every $\iota \in I$ and $(\int f|_{X_\iota} \, d\mu_\iota)_{\iota \in I}$ is summable.

If one of these statements holds, then the next does too:

($\gamma 3$) $\int f \, d\mu = \sum_{\iota \in I} \int f|_{X_\iota} \, d\mu_\iota$.

5.2.5(E) Let $(X, \mathfrak{R}, \mu)$ be a positive measure space with the property that $\bigcup_{n \in \mathbb{N}} A_n \in \mathfrak{R}$ for each increasing sequence $(A_n)_{n \in \mathbb{N}}$ in $\mathfrak{R}$ with $\sup_{n \in \mathbb{N}} \mu(A_n) < \infty$. Prove that $(X, \mathfrak{R}, \mu)$ is then a positive δ-measure space. If in addition $(X, \mathfrak{R}, \mu)$ is bounded, then $(X, \mathfrak{R}, \mu)$ is a σ-measure space.

5.2.6(E) Let $(X, \mathfrak{R}, \mu)$ be a positive measure space. Prove that the following assertions are equivalent:

(α) $(X, \mathfrak{R}, \mu)$ is complete.
(β) $(X, \mathfrak{R}, \mu)$ is closed and $A \in \mathfrak{R}$ whenever $A \subset X$ such that $A \cap B \in \mathfrak{R}$ and $\mu(A \cap B) = 0$ for any $B \in \mathfrak{R}$.

5.2.7(E) Define $\mathfrak{R} := \{A \subset \mathbb{R} \mid A \text{ is countable or } \mathbb{R} \setminus A \text{ is countable}\}$, $\mathfrak{S} := \{A \subset \mathbb{R} \mid A \text{ is finite or } \mathbb{R} \setminus A \text{ is finite}\}$,

$$\mu: \mathfrak{R} \to \mathbb{R}, \qquad A \mapsto \begin{cases} 0 & \text{A countable} \\ 1 & \text{otherwise} \end{cases}$$

$$\nu := \mu|_{\mathfrak{S}}, \quad \mathfrak{S}_{\mathbb{R} \setminus \mathbb{Q}} := \{A \subset \mathbb{R} \setminus \mathbb{Q} \mid A \in \mathfrak{S}\}, \quad \nu_{\mathbb{R} \setminus \mathbb{Q}} := \nu|_{\mathfrak{S}_{\mathbb{R} \setminus \mathbb{Q}}}.$$

Prove the following statements:

(α) $\mathfrak{R}$ is a σ-algebra on $\mathbb{R}$.

(β) $\mathfrak{S}$ is a ring of sets. $\mathfrak{S}_{\mathbb{R} \setminus \mathbb{Q}}$ is a δ-ring.

(γ) μ, ν, and $\nu_{\mathbb{R} \setminus \mathbb{Q}}$ are positive measures.

(δ) μ and ν generate the same integral on $\mathbb{R}$.

(ε) $e^{\mathbb{R}}_{\mathbb{R} \setminus \mathbb{Q}} \in \mathscr{L}^1(\nu)$, $e^{\mathbb{R} \setminus \mathbb{Q}}_{\mathbb{R} \setminus \mathbb{Q}} \in \mathscr{L}^1(\nu_{\mathbb{R} \setminus \mathbb{Q}})$, and $\int e^{\mathbb{R}}_{\mathbb{R} \setminus \mathbb{Q}} \, d\nu = 1$, $\int e^{\mathbb{R} \setminus \mathbb{Q}}_{\mathbb{R} \setminus \mathbb{Q}} \, d\nu_{\mathbb{R} \setminus \mathbb{Q}} = 0$.

5.2.8(E) We wish to show that both the closure and the completion of a positive measure space can be characterized by means of admissibility and determinedness, in analogy with the corresponding situation with a Daniell space.

Let $(X, \mathfrak{R}, \mu)$ be a positive measure space. Set $\Psi := \{(X, \mathfrak{R}', \mu') \mid (X, \mathfrak{R}', \mu')$ is a closed positive measure space and $(X, \mathfrak{R}, \mu) \preccurlyeq (X, \mathfrak{R}', \mu')\}$. $(X, \mathfrak{R}', \mu') \in \Psi$ is an **admissible extension** of $(X, \mathfrak{R}, \mu)$ iff given $A \in \mathfrak{R}'$, there is a $B \in \mathfrak{R}(\mu)$ such that $\mu'(A \triangle B) = 0$. Set $\Psi_a := \{(X, \mathfrak{R}', \mu') \in \Psi \mid (X, \mathfrak{R}', \mu')$ admissible$\}$. $(X, \mathfrak{R}', \mu') \in \Psi_a$ is **determined** by $(X, \mathfrak{R}, \mu)$ iff given $(X, \mathfrak{R}'', \mu'') \in \Psi_a$, $\mu'|_{\mathfrak{R}' \cap \mathfrak{R}''} = \mu''|_{\mathfrak{R}' \cap \mathfrak{R}''}$. Set $\Psi_d := \{(X', \mathfrak{R}', \mu') \in \Psi_a \mid (X', \mathfrak{R}', \mu')$ is determined by $(X, \mathfrak{R}, \mu)\}$.

To see the correspondence with Daniell spaces, denote by Φ_a the set of all admissible extensions of $(X, \mathscr{L}_{\mathfrak{R}}, \ell_\mu)$ and by Φ_d the set of all spaces in Φ_a determined by $(X, \mathscr{L}_{\mathfrak{R}}, \ell_\mu)$. Prove the following statements:

(α) If $(X, \mathfrak{R}', \mu') \in \Psi$, then the following are equivalent.

(α1) $(X, \mathfrak{R}', \mu') \in \Psi_a$.

(α2) $(X, \overline{\mathscr{L}}(\ell_{\mu'}), \overline{\ell_{\mu'}}) \in \Phi_a$.

(β) If $(X, \mathfrak{R}', \mu') \in \Psi$, then the following are equivalent.

(β1) $(X, \mathfrak{R}', \mu') \in \Psi_d$.

(β2) $(X, \overline{\mathscr{L}}(\ell_{\mu'}), \overline{\ell_{\mu'}}) \in \Phi_d$.

Define $\psi : \Psi_a \to \Phi_a$, $(X, \mathfrak{R}', \mu') \mapsto (X, \overline{\mathscr{L}}(\ell_{\mu'}), \overline{\ell_{\mu'}})$. Prove that the following statements hold:

(γ) ψ is an order isomorphism.

(δ) $\psi|_{\Psi_d}$ is an order isomorphism of Ψ_d onto Φ_d.

Notice that in general, ψ cannot be extended to an order isomorphism of Ψ onto Φ, the set of all closed extensions of $(X, \mathscr{L}_{\mathfrak{R}}, \ell_\mu)$. This is because there may be spaces $(X, \mathscr{L}, \ell)$ in Φ for which $\mathscr{L}$ is not a Stone lattice. But by Proposition 5.1.2 this cannot occur in Φ_a. Now prove the next statement:

(ε) $(X, \mathfrak{L}(\mu), \mu^X)$ is the greatest extension of $(X, \mathfrak{R}, \mu)$ determined by $(X, \mathfrak{R}, \mu)$.

Finally, consider what happens when we forgo the requirement of admissibility and consider determinedness in Ψ. (The analogous problem for Daniell spaces was discussed in Ex. 4.2.10.) Call $(X, \mathfrak{R}', \mu')$ **strongly determined** by

$(X, \mathfrak{R}, \mu)$ iff given any $(X, \mathfrak{R}'', \mu'') \in \Psi$, $\mu'|_{\mathfrak{R}' \cap \mathfrak{R}''} = \mu''|_{\mathfrak{R}' \cap \mathfrak{R}''}$. Prove the final statement:

(ζ) $(X, \mathfrak{R}(\mu), \bar{\mu})$ is the greatest strongly determined element of Ψ.

To prove (ζ), the relationship to Daniell spaces cannot be applied directly because, as we have just commented, ψ does not admit an extension to an order isomorphism of Ψ onto Φ. Thus the reader is advised to try to assemble a proof in analogy with Ex. 4.2.10.

5.2.9 (E) Let $(X, \mathfrak{R}, \mu)$ and $(X, \mathfrak{S}, \nu)$ be positive measure spaces. Use the notions of admissibility and determinedness in the sense of Ex. 5.2.8. Prove the following statement:

Suppose that for every admissible extension $(X, \mathfrak{R}', \mu')$ of $(X, \mathfrak{R}, \mu)$ there is an admissible extension $(X, \mathfrak{S}', \nu')$ of $(X, \mathfrak{S}, \nu)$ such that $(X, \mathfrak{R}', \mu') \preccurlyeq (X, \mathfrak{S}', \nu')$, and that for every admissible extension $(X, \mathfrak{S}', \nu')$ of $(X, \mathfrak{S}, \nu)$ there is an admissible extension $(X, \mathfrak{R}', \mu')$ of $(X, \mathfrak{R}, \mu)$ with $(X, \mathfrak{S}', \nu') \preccurlyeq (X, \mathfrak{R}', \mu')$. Then

$$(X, \mathfrak{L}(\mu), \mu^X) = (X, \mathfrak{L}(\nu), \nu^X)$$

Compare with Ex. 4.2.11 (ν). Of course the reverse is also true.

5.2.10(E) Let μ be a positive measure on the ring of sets $\mathfrak{R}$. The set of sets $\mathfrak{N}$ is called **solid** iff $\mathfrak{P}(A) \subset \mathfrak{N}$ for each $A \in \mathfrak{N}$. Prove the following statements:

(α) If $\mathfrak{N}$ is a solid lattice of sets, then $\mathfrak{R} \vdash \mathfrak{N} := \{A \mid A \text{ a set}, \exists B \in \mathfrak{R} \text{ with } A \triangle B \in \mathfrak{N}\}$ is a ring of sets such that $\mathfrak{R} \cup \mathfrak{N} \subset \mathfrak{R} \vdash \mathfrak{N}$.

(β) If $\mathfrak{N}$ is a solid lattice of sets and $\mathfrak{N} \cap \mathfrak{R}(\mu) \subset \mathfrak{N}(X(\mathfrak{R}), \mu)$, then there is exactly one positive measure, $\mu_{\mathfrak{N}}$, on $\mathfrak{R} \vdash \mathfrak{N}$ such that $\mu_{\mathfrak{N}}|_{\mathfrak{R}} = \mu$ and $\mu_{\mathfrak{N}}|_{\mathfrak{N}} = 0$.

(γ) If $(X, \mathfrak{R}, \mu)$ is a closed positive measure space and $\mathfrak{N}$ a solid σ-ring such that $\mathfrak{N} \cap \mathfrak{R} \subset \mathfrak{N}(X, \mu)$, then $(X, \mathfrak{R} \vdash \mathfrak{N}, \mu_{\mathfrak{N}})$ is an admissible extension of $(X, \mathfrak{R}, \mu)$.

(δ) Let $(X, \mathfrak{R}', \mu')$ be a closed extension of the positive measure space $(X, \mathfrak{R}, \mu)$. Define $\mathfrak{N} := \{A \in \mathfrak{R}' \mid \mu'(A) = 0\}$. Let Ψ_a be as in Ex. 5.2.8. Then $(X, \mathfrak{R}', \mu') \in \Psi_a$ iff $(X, \mathfrak{R}', \mu') = (X, \mathfrak{R}(\mu) \vdash \mathfrak{N}, (\bar{\mu})_{\mathfrak{N}})$.

These statements correspond to those formulated for functionals in Section 4.1. Now let $(X, \mathfrak{R}, \mu)$ be a positive measure space such that $\bigcup_{n \in \mathbb{N}} A_n \in \mathfrak{R}$ for every increasing sequence in $\mathfrak{R}$ with $\sup_{n \in \mathbb{N}} \mu(A_n) < \infty$. Set $\mathfrak{N} := \{A \subset X \mid \exists C \in \mathfrak{R},\ A \subset C,\ \mu(C) = 0\}$ and $\mathfrak{N}' := \{A \subset X \mid \forall B \in \mathfrak{R},\ \exists C \in \mathfrak{R},\ A \cap B \subset C,\ \mu(C) = 0\}$. Prove the next statements:

(ε) $(X, \mathfrak{R}(\mu), \bar{\mu}) = (X, \mathfrak{R} \vdash \mathfrak{N}, \mu_{\mathfrak{N}})$.

(ζ) $\mathfrak{N}' = \mathfrak{N}(\mu)$ and $(X, \mathfrak{L}(\mu), \mu^X) = (X, \mathfrak{R} \vdash \mathfrak{N}', \mu_{\mathfrak{N}'})$.

5.2.11(E) Let $(X, \mathfrak{R}, \mu)$ be a positive measure space, and let $f, g \in \mathscr{L}^1(\mu)$. Prove the following statements:

(α) If $\int_A f\,d\mu \geq 0$ for every $A \in \mathfrak{R}$, then $f \geq 0$ μ-a.e.

(β) If $\int_A f\,d\mu = 0$ for every $A \in \mathfrak{R}$, then $f = 0$ μ-a.e..

(γ) $f \geq g$ μ-a.e. whenever $\int_A f\,d\mu \geq \int_A g\,d\mu$ for every $A \in \mathfrak{R}$.

(δ) $f = g$ μ-a.e. whenever $\int_A f\,d\mu = \int_A g\,d\mu$ for every $A \in \mathfrak{R}$.

5.2.12(E) We wish to show that the closure and the completion of a positive measure space can be constructed exactly as in the case of Daniell spaces. Let $(X, \mathfrak{R}, \mu)$ be a positive measure space. Define

$$\mathfrak{R}^{\uparrow} := \left\{ \bigcup_{n \in \mathbb{N}} A_n \,|\, (A_n)_{n \in \mathbb{N}} \text{ an increasing sequence from } \mathfrak{R} \right\}$$

$$\mathfrak{R}^{\downarrow} := \left\{ \bigcap_{n \in \mathbb{N}} A_n \,|\, (A_n)_{n \in \mathbb{N}} \text{ a decreasing sequence from } \mathfrak{R} \right\}$$

$$\mu^{\uparrow} : \mathfrak{R}^{\uparrow} \to \mathbb{R}, A \mapsto \sup\{ \mu(B) \,|\, B \in \mathfrak{R}, B \subset A\}$$

$$\mu^{\downarrow} : \mathfrak{R}^{\downarrow} \to \mathbb{R}, A \mapsto \inf\{ \mu(B) \,|\, B \in \mathfrak{R}, A \subset B\}$$

Prove the following propositions:

(α) Given $A \subset X$, $A \in \mathfrak{R}(\mu)$ iff for every $\varepsilon \in \mathbb{R}$, $\varepsilon > 0$ there are $B \in \mathfrak{R}^{\downarrow}$ and $C \in \mathfrak{R}^{\uparrow}$ such that $B \subset A \subset C$, $\mu^{\uparrow}(C) < \infty$ and $\mu^{\uparrow}(C) - \mu^{\downarrow}(B) < \varepsilon$.

(β) If $A \in \mathfrak{R}(\mu)$ then $\bar{\mu}(A) = \sup\{\mu^{\downarrow}(B) \,|\, B \in \mathfrak{R}^{\downarrow}, B \subset A\} = \inf\{\mu^{\uparrow}(C) \,|\, C \in \mathfrak{R}^{\uparrow}, A \subset C\}$.

The pair (B, C) in (α) is called an **ε-bracket** for A, completing the analogy with Section 3.1. Prove the following proposition:

(γ) If $A \in \mathfrak{R}(\mu)$, then $\bar{\mu}(A)$ is the uniquely determined real number with the property that $\mu^{\downarrow}(B) \leq \bar{\mu}(A) \leq \mu^{\uparrow}(C)$ for every ε-bracket (B, C) for A.

In Ex. 5.2.8 it was shown that $(X, \mathfrak{L}(\mu), \mu^X)$ can be constructed from $(X, \mathfrak{R}(\mu), \bar{\mu})$ in the same way as $(X, \mathscr{L}^1(\ell), \int_\ell)$ can be constructed from the Daniell space $(X, \bar{\mathscr{L}}(\ell), \bar{\ell})$.

5.2.13(E) **The Borel Extension of a Positive Measure Space.** Let $(X, \mathfrak{R}, \mu)$ be a positive measure space. Set $\mathfrak{R}_0 := \mathfrak{R}$. Let $\alpha > 0$ be an ordinal. Define

$$\tilde{\mathfrak{R}}^{\alpha} := \bigcup_{\beta < \alpha} \mathfrak{R}_{\beta}$$

$$\mathfrak{R}^{\alpha} := \left\{ \bigcup_{n \in \mathbb{N}} A_n \,\middle|\, \begin{array}{l} (A_n)_{n \in \mathbb{N}} \text{ is an increasing sequence} \\ \text{in } \tilde{\mathfrak{R}}^{\alpha}, \sup \bar{\mu}(A_n) < \infty \end{array} \right\}$$

$$\mathfrak{R}_{\alpha} := \left\{ \bigcap_{n \in \mathbb{N}} A_n \,|\, (A_n)_{n \in \mathbb{N}} \text{ is a decreasing sequence in } \mathfrak{R}^{\alpha} \right\}$$

Prove the following propositions, where ω_1 denotes the first uncountable ordinal:

(α) $\mathfrak{R}_\alpha \subset \mathfrak{R}_\beta$ whenever $\alpha \le \beta$.
(β) $\mathfrak{R}_\alpha \subset \mathfrak{R}(\mu)$ for any ordinal α.
(γ) $\mathfrak{R}_\alpha = \mathfrak{R}_{\omega_1}$ for every ordinal $\alpha \ge \omega_1$.
(δ) $\bigcup_{n\in\mathbb{N}} A_n \in \mathfrak{R}_{\omega_1}$ for every increasing sequence $(A_n)_{n\in\mathbb{N}}$ in $\mathfrak{R}_{\omega_1}$ with $\sup_{n\in\mathbb{N}}\bar{\mu}(A_n) < \infty$.

$(X, \mathfrak{R}_{\omega_1}, \bar{\mu}|\mathfrak{R}_{\omega_1})$ is called the **Borel extension** of $(X, \mathfrak{R}, \mu)$. Prove the next proposition:

(ε) $\mathfrak{R}_{\omega_1}$ is a δ-ring. It is the smallest δ-ring $\mathfrak{S} \subset \mathfrak{R}(\mu)$ containing $\mathfrak{R}$ with the property that $\bigcup_{n\in\mathbb{N}} A_n \in \mathfrak{S}$ for every increasing sequence $(A_n)_{n\in\mathbb{N}}$ from $\mathfrak{S}$ with $\sup_{n\in\mathbb{N}}\bar{\mu}(A_n) < \infty$.

It is easy to show that, in general, $(X, \mathfrak{R}_{\omega_1}, \bar{\mu}|\mathfrak{R}_{\omega_1}) \ne (X, \mathfrak{R}(\mu), \bar{\mu})$. Let the set X consist of more than two points. Define $\mathfrak{R} := \{X, \varnothing\}$ and $\mu := 0$. Then $(X, \mathfrak{R}, \mu)$ has the desired property.

5.2.14(E) Let $(X, \mathfrak{R}, \mu)$ be a positive measure space. Prove the following propositions.

(α) For every $A \subset X$ the following statements are equivalent:

(α1) A is a μ-null set
(α2) For every $B \in \mathfrak{R}$ and for every $\varepsilon > 0$ there exists a sequence $(A_n)_{n\in\mathbb{N}}$ from $\mathfrak{R}$ such that $A \cap B \subset \bigcup_{n\in\mathbb{N}} A_n$ and $\sum_{n\in\mathbb{N}}\mu(A_n) < \varepsilon$.

(β) Let $\mathfrak{A}$ be a subset of $\mathfrak{R}$ such that $\mathfrak{R} \subset \mathfrak{A}_\sigma$. Then, if B is a subset of X such that $A \cap B \in \mathfrak{R}(\mu)$ for every $A \in \mathfrak{A}$, B is a μ-null set.

5.2.15(E) Let $(X, \mathfrak{R}, \mu)$ be a positive measure space and $f \in \overline{\mathbb{R}}^X$ a function with the property that $(f \vee \alpha e_X) \wedge \beta e_X \in \mathcal{L}^1(\mu)$ for all $\alpha, \beta \in \mathbb{R}$, $\alpha \le 0$, $\beta \ge 0$. Let $\gamma \in \mathbb{R}$. Prove that the following statements are equivalent:

(α) $f \in \mathcal{L}^1(\mu)$, $\int f\,d\mu = \gamma$.
(β) $\lim_{\substack{\beta\to\infty\\ \beta\ge 0}} \int((f \vee \alpha e_X) \wedge \beta e_X)\,d\mu$ exists whenever $\alpha \le 0$, and

$$\gamma = \lim_{\substack{\alpha\to-\infty\\ \alpha\le 0}} \lim_{\substack{\beta\to\infty\\ \beta\ge 0}} \left(\int((f \vee \alpha e_X) \wedge \beta e_X)\,d\mu\right).$$

(γ) $\lim_{\substack{\alpha\to-\infty\\ \alpha\le 0}} \int((f \vee \alpha e_X) \wedge \beta e_X)\,d\mu$ exists whenever $\beta \ge 0$, and

$$\gamma = \lim_{\substack{\beta\to\infty\\ \beta\ge 0}} \lim_{\substack{\alpha\to-\infty\\ \alpha\le 0}} \left(\int((f \vee \alpha e_X) \wedge \beta e_X)\,d\mu\right).$$

(δ) $\lim_{\substack{(\alpha,\beta)\to(-\infty,\infty)\\ \alpha\le 0,\ \beta\ge 0}} \int((f \vee \alpha e_X) \wedge \beta e_X)\,d\mu = \gamma$ (i.e., for every $\varepsilon \in \mathbb{R}$,

$\varepsilon > 0$ there exist $\alpha_0 \in \mathbb{R}$, $\alpha_0 \le 0$ and $\beta_0 \in \mathbb{R}$, $\beta_0 \ge 0$ such that

$$\left| \int ((f \vee \alpha e_X) \wedge \beta e_X) \, d\mu - \gamma \right| < \varepsilon$$

whenever $\alpha, \beta \in \mathbb{R}$, $\alpha \le \alpha_0$, $\beta \ge \beta_0$).

5.2.16(C) Completion of Regular Measure Spaces. The notion of regularity was introduced in Ex. 2.2.12. We continue now with the discussion of the completion of such measure spaces.

Let $(X, \mathfrak{S}, \mu)$ be a positive measure space and $\mathfrak{G}$ a lattice of sets. We call μ **$\mathfrak{G}$-regular** iff $\mathfrak{G} \subset \mathfrak{S}$ and μ is $\mathfrak{G}$-regular in the sense of Ex. 2.2.12. $(X, \mathfrak{S}, \mu)$ is called **$\mathfrak{G}$-regular** iff μ is.

Assume now that $\mathfrak{R}$ is a $\cap$-semi-lattice, $\mathfrak{G}$ is a lattice of sets, $\mathfrak{G} \subset \mathfrak{R}$, and $\varnothing \in \mathbb{G}$. Let $\nu : \mathfrak{R} \to \mathbb{R}$ be an alternating completely increasing null-continuous, $\mathfrak{G}$-regular function. In Ex. 2.2.12 (α) we proved the existence of a unique positive measure μ on $\mathfrak{R}_r$ with the property that $\mu|_{\mathfrak{R}} = \nu$. Prove now the following statements for some set $X \supset X(\mathfrak{R})$:

(α) $(X, \mathfrak{R}(\mu), \bar{\mu})$ is a $\mathfrak{G}_\delta$-regular positive measure space.

(β) $(X, \mathfrak{L}(\mu), \mu^X)$ is a $\mathfrak{G}_\delta$-regular positive measure space.

(γ) Denote by Φ the set of all $\mathfrak{G}_\delta$-regular positive measure spaces $(X, \mathfrak{R}', \mu')$ extending $(X, \mathfrak{R}, \nu)$. Then $(X, \mathfrak{L}(X), \mu^X)$ is the greatest element of Φ with respect to $\preccurlyeq$.

(δ) Assume that $\bigcap_{n \in \mathbb{N}} A_n \in \mathfrak{G}$ for every sequence $(A_n)_{n \in \mathbb{N}}$ from $\mathfrak{G}$. Then the following hold:

($\delta1$) $(X, \mathfrak{R}(\mu), \bar{\mu})$ is a $\mathfrak{G}$-regular positive measure space.

($\delta2$) $(X, \mathfrak{L}(\mu), \mu^X)$ is the greatest element with respect to $\preccurlyeq$ in the set of all $\mathfrak{G}$-regular positive measure spaces $(X, \mathfrak{R}', \mu')$ extending $(X, \mathfrak{R}, \nu)$.

Let $(X, \mathfrak{R}, \mu)$ be a positive measure space. Prove the next two statements:

(ε) $(X, \mathfrak{R}(\mu), \bar{\mu})$ is $\mathfrak{R}_\delta$-regular.

(ζ) $(X, \mathfrak{L}(\mu), \mu^X)$ is the greatest element with respect to $\preccurlyeq$ in the set of all $\mathfrak{R}_\delta$-regular positive measure spaces $(X, \mathfrak{R}', \mu')$ extending $(X, \mathfrak{R}, \mu)$.

(ζ) gives a new characterization of the completion of a positive measure space. This statement is a further justification of the definition of these objects.

We give some hints for the proof of (γ). Assume that $(X, \mathfrak{R}', \mu')$ is a $\mathfrak{G}_\delta$-regular extension of $(X, \mathfrak{R}, \mu)$. Then $(X, \mathfrak{R}(\mu'), \overline{\mu'})$ is also a $\mathfrak{G}_\delta$-regular extension of $(X, \mathfrak{R}, \mu)$. Take $A \in \mathfrak{R}(\mu')$. Then there is a $B \in \mathfrak{R}(\mu)$ with $B \subset A$ and $\overline{\mu'}(A) = \bar{\mu}(B)$. Let $C \in \mathfrak{R}(\mu)$ be arbitrary. We find $D \in \mathfrak{R}(\mu)$ such that $D \subset C \setminus (A \setminus B)$ and $\bar{\mu}(D) = \bar{\mu}(C)$. We conclude that $C \setminus D \in \mathfrak{R}(\mu)$; hence $C \cap (A \setminus B) \in \mathfrak{R}(\mu)$. But then $A \setminus B \in \mathfrak{R}(\mu)$ since C

was arbitrary. Finally, we get $A = B \cup (A \setminus B) \in \mathfrak{L}(\mu)$ and conclude that

$$(X, \mathfrak{R}', \mu') \preccurlyeq (X, \mathfrak{L}(\mu), \mu^X).$$

5.2.17(C) Completion of Positive Measures on Hausdorff Spaces. Let X be a Hausdorff space, and denote by $\mathfrak{K}$ the set of all compact subsets of X.

Let $\nu : \mathfrak{K} \to \mathbb{R}_+$ be a function with the following properties:

(i) $\nu(A \cup B) \leq \nu(A) + \nu(B)$ whenever $A, B \in \mathfrak{K}$.

(ii) $\nu(A \cup B) = \nu(A) + \nu(B)$ whenever $A, B \in \mathfrak{K}$, $A \cap B = \varnothing$.

(iii) $\nu(\bigcap_{A \in \mathfrak{A}} A) = \inf_{A \in \mathfrak{A}} \nu(A)$ for every nonempty downward directed subset $\mathfrak{A}$ of $\mathfrak{K}$.

In Ex. 2.2.13 we proved the existence of a unique positive measure μ on $\mathfrak{K}_r$ with the property that $\mu|_{\mathfrak{K}} = \nu$. We also proved that μ is $\mathfrak{K}$-regular. Prove now these statements:

(α) $(X, \mathfrak{L}(X, \mu), \mu^X)$ is the greatest element with respect to $\preccurlyeq$ in the set of all $\mathfrak{K}$-regular positive measure spaces $(X, \mathfrak{R}', \mu')$ extending $(X, \mathfrak{K}, \nu)$.

(β) $A \in \mathfrak{N}(X, \mu)$ iff $A \subset X$ and $A \cap K \in \mathfrak{N}(X, \mu)$ for every $K \in \mathfrak{K}$.

(γ) Let $\mathfrak{A}$ be the set of all open sets in $\mathfrak{N}(X, \mu)$. Then $\bigcup_{A \in \mathfrak{A}} A \in \mathfrak{N}(X, \mu)$.

(δ) $A \in \mathfrak{L}(X, \mu)$ iff $A \subset X$, $A \cap K \in \mathfrak{L}(X, \mu)$ for every $K \in \mathfrak{K}$ and $\sup_{K \in \mathfrak{K}} \mu^X(A \cap K) < \infty$.

(ε) For every $A \in \mathfrak{L}(X, \mu)$, $\mu^X(A) = \sup_{K \in \mathfrak{K}} \mu^X(A \cap K)$.

A subset A of X is called **σ-compact** iff there is a sequence $(K_n)_{n \in \mathbb{N}}$ from $\mathfrak{K}$ such that $A = \bigcup_{n \in \mathbb{N}} K_n$. Prove the following statement:

(ζ) For every set $B \in \mathfrak{L}(X, \mu)$ there is a σ-compact set A such that $A \subset B$ and $B \setminus A \in \mathfrak{N}(X, \mu)$.

5.2.18(C) The Riesz Representation Theorem. This exercise is the completion of Ex. 2.5.4 (γ). Let X be a locally compact space. Let $\mathfrak{K}$ denote the set of all compact subsets of X, and $\mathscr{K}$ the set of all continuous real functions on X with compact support. The following proposition is known as the Riesz Representation Theorem (F. Riesz, 1909):

For every additive positive functional ℓ on $\mathscr{K}$ there is exactly one positive measure, μ^ℓ, on $\mathfrak{K}_r$ such that $\mu^\ell(K) = \inf\{\ell(f) \mid f \in \mathscr{K}, f \geq e_K\}$ for each $K \in \mathfrak{K}$. μ^ℓ is $\mathfrak{K}$-regular, and $(X, \mathscr{K}, \ell) \preccurlyeq (X, \mathfrak{L}(\mu^\ell), (\mu^\ell)^X)$.

5.2.19(E) Completion of Restrictions, Images, and Products. Restrictions, images, and products of positive measure spaces were introduced earlier (Exs. 2.2.7, 2.4.6, and 5.1.8). We could now simply apply the completion procedure to these objects. But that is not in the spirit of this book. Admissibility in the case of products or images cannot be the same as in the general case, since we have a richer structure. Admissible extensions of such objects should have some properties connected with the supplementary structure. Hence we need to adjust the concept of admissibility in these cases. We do not solve this problem in this exercise. We only want to discuss some phenomena in this connection. An extensive discussion follows in Volume 3.

(a) **Restrictions.** Let $(X, \mathfrak{R}, \mu)$ be a positive measure space and Y a subset of X. Prove the following statements:

(α) $(Y, \mathfrak{R}(\mu)|_Y, \bar{\mu}|_Y) \supset (Y, \mathfrak{R}(\mu|_Y), \overline{(\mu|_Y)})$.

(β) There exists an example for which the inclusion in (α) is strict.

(γ) Prove that $(Y, \mathfrak{R}(\mu)|_Y, \bar{\mu}|_Y)$ is, in general, not an admissible extension of $(Y, \mathfrak{R}|_Y, \mu|_Y)$.

(δ) Prove that $(Y, \mathfrak{L}(\mu)|_Y, \mu^X|_Y)$ is, in general, not complete.

We give some hints. Let $\mathfrak{R}$ be the ring of all interval forms on $\mathbb{R}$, and denote by λ the Lebesgue measure on $\mathfrak{R}$. Let Y be a Cantor set. Then $\mathfrak{R}|_Y = \{\varnothing\}$ and $\lambda|_Y = 0$. So we have

$$\left(Y, \mathfrak{R}(\lambda|_Y), \overline{(\lambda|_Y)}\right) = (Y, \{\varnothing\}, 0).$$

On the other hand, $Y \in \mathfrak{R}(\lambda)|_Y$, and this proves ($\beta$) and ($\gamma$).

For the proof of (δ), consider the ring of sets $\mathfrak{F}(X) \cup \{A \subset X \mid X \setminus A \in \mathfrak{F}(X)\}$ for an uncountable set X. Define $\mu(A) := 0$ if $A \in \mathfrak{F}(X)$ and $\mu(A) := 1$ if $X \setminus A \in \mathfrak{F}(X)$, and let Y be an uncountable subset of X with the property that $X \setminus Y$ is uncountable too.

(b) **Images.** Let $(X, \mathfrak{R}, \mu)$ be a positive measure space, Y a set, and $\varphi\colon X \to Y$ a mapping. Prove the following statements:

(α) $(Y, \varphi(\mathfrak{R}(\mu)), \varphi(\bar{\mu})) \supset (Y, \mathfrak{R}(\varphi(\mu)), \overline{(\varphi(\mu))})$.

(β) There exists an example for which the inclusion in (α) is strict.

(γ) $(Y, \varphi(\mathfrak{R}(\mu)), \varphi(\bar{\mu}))$ is, in general, not an admissible extension of $(Y, \varphi(\mathfrak{R}), \varphi(\mu))$.

(δ) The space $(Y, \varphi(\mathfrak{L}(\mu)), \varphi(\mu^X))$ is, in general, not complete.

We give some hints. Take $X :=]0,1[$ and $Y := [0,1]$. Denote by $\mathfrak{R}$ the set of all interval forms on $]0,1[$, and let μ be the Lebesgue measure on $\mathfrak{R}$. Define

$$\varphi\colon X \to Y, \qquad x \mapsto \begin{cases} 1 & \text{if } x \text{ is rational} \\ 0 & \text{if } x \text{ is not rational} \end{cases}.$$

Then we have $\varphi(\mathfrak{R}) = \{\varnothing\}$ and $\varphi(\mu) = 0$. On the other hand, we have $Y \in \varphi(\mathfrak{R}(\mu))$, and this proves ($\beta$) and ($\gamma$).

For the proof of (δ), take the example just considered but denote by μ the Stieltjes measure on $\mathfrak{R}$ induced by a function g that is not bounded on $]0,1[$. Then we have $Y \in \mathfrak{L}(\varphi(\mu^X))$, but $Y \notin \varphi(\mathfrak{L}(\mu))$.

(c) **Products.** Take the product $(X \times Y, \mathfrak{R} \otimes \mathfrak{S}, \mu \otimes \nu)$ of two positive measure spaces $(X, \mathfrak{R}, \mu)$ and $(Y, \mathfrak{S}, \nu)$ in the sense of Ex. 2.4.6. For every function $f \in \overline{\mathbb{R}}^{X \times Y}$ define

$$f(\cdot, y) : X \to \mathbb{R}, x \mapsto f(x, y) \quad \text{whenever } y \in Y$$

$$f(x, \cdot) : Y \to \mathbb{R}, y \mapsto f(x, y) \quad \text{whenever } x \in X$$

$$\int f(x,y)\,d\nu(y) := \begin{cases} \int f(x,y)\,d\nu(y) & \text{if } f(x,\cdot) \in \mathscr{L}^1(Y,\nu) \\ \text{arbitrary} & \text{otherwise} \end{cases}$$

whenever $x \in X$

$$\int f(x,y)\,d\mu(x) := \begin{cases} \int f(x,y)\,d\mu(x) & \text{if } f(\cdot,y) \in \mathscr{L}^1(X,\mu) \\ \text{arbitrary} & \text{otherwise} \end{cases}$$

whenever $y \in Y$.

Prove the following statements:

(α) For every function $f \in \bar{\mathscr{L}}(\ell_{\mu \otimes \nu})$ the following are true:

(α1) $\{x \in X \mid f(x,\cdot) \notin \mathscr{L}^1(Y,\nu)\} \in \mathfrak{N}(X,\mu)$.
(α2) $\{y \in Y \mid f(\cdot,y) \notin \mathscr{L}^1(X,\mu)\} \in \mathfrak{N}(Y,\nu)$.
(α3) $\int f(\cdot,y)\,d\nu(y) \in \mathscr{L}^1(X,\mu)$.
(α4) $\int f(x,\cdot)\,d\mu(x) \in \mathscr{L}^1(Y,\nu)$.
(α5) $\int f\,d(\mu \otimes \nu) = \int(\int f(x,y)\,d\mu(x))\,d\nu(y)$
$= \int(\int f(x,y)\,d\nu(y))\,d\mu(x)$.

This is one form of the **Theorem of Fubini** about integration on product spaces (Fubini, 1907).

(β) Prove that (α) is not true, in general, for $f \in \mathscr{L}^1(\mu \otimes \nu)$.

For the proof of (α) use the Induction Principle.

For the proof of (β), take $(X, \mathfrak{R}, \mu) := ([0,1], \mathfrak{F}([0,1]), \chi)$, and denote by ν the Lebesgue measure on the ring $\mathfrak{S}$ of all interval forms on $[0,1]$. Set $A := \{(x,y) \in [0,1]^2 \mid x = y\}$ and $f := e_A$. Then (α) is not true for f.

5.2.20(E) Let X be a set, $\mathfrak{R}$ a ring of subsets of X and μ, ν positive measures on $\mathfrak{R}$. Prove

(α) $\overline{\mathfrak{R}}(\mu + \nu) = \overline{\mathfrak{R}}(\mu) \cap \overline{\mathfrak{R}}(\nu)$
(β) $\mathfrak{L}(X, \mu + \nu) = \mathfrak{L}(X,\mu) \cap \mathfrak{L}(X,\nu)$
(γ) $\overline{\mu + \nu}\,(A) = \bar{\mu}(A) + \bar{\nu}(A)$ for any $A \in \overline{\mathfrak{R}}(\mu + \nu)$
(δ) $(\mu + \nu)^X(A) = \mu^X(A) + \nu^X(A)$ for any $A \in \mathfrak{L}(X, \mu + \nu)$.

5.2.21(E) Let $(X, \mathfrak{R}, \mu)$ be a positive measure space and φ: $X \to X$ a mapping such that $\varphi^{-1}(A) \in \mathfrak{R}$ and $\mu(\varphi^{-1}(A)) = \mu(A)$ for every $A \in \mathfrak{R}$. Prove that $f \circ \varphi \in \mathscr{L}^1(\mu)$ and $\int f \circ \varphi\,d\mu = \int f\,d\mu$ for every $f \in \mathscr{L}^1(\mu)$.

5.3. MEASURABLE SPACES AND MEASURABILITY

The notion of measurability was introduced by Lebesgue as a means of defining his integral. Today as well the notion of measurability is often used to construct the integral associated with a given measure. The Daniell construction, which we have used in this book, does not require the notion of measurability, which is the reason we can introduce measurability so late in the book.

Measurability plays several roles in the theory of integration. First, it yields an integrability criterion, in fact the most important integrability criterion, which we treat in Section 5.4. It turns out that measurability, when properly defined, provides a necessary, though not sufficient, condition for integrability. Thus the set of measurable functions is larger than the set of integrable functions. Measurable functions, it further turns out, form a class that is large enough to permit all appropriate limit operations as well as algebraic operations, a property that makes this class very important in the theory of integration. One can place in this framework, for instance, the entire theory of $\mathscr{L}^p$-spaces (which will be treated in Chapter 7). Measurable functions also play a specific role in the theory of products of measures.

The sections at hand treat measurability for extended-real functions. Measurability for general mappings will be introduced later.

Sections 5.3 and 5.4 discuss Lebesgue's notion of measurability in a form that is attractive for our purposes. Section 5.3 treats measurability of sets and functions relative to a δ-ring. Section 5.4 then investigates these objects in the context of measures and measure-space integrals.

Definition 5.3.1. *For $\mathfrak{R}$ a δ-ring on a set X, we define*

$$\mathfrak{M}(X, \mathfrak{R}) := \{A \subset X \mid A \cap B \in \mathfrak{R}, \forall B \in \mathfrak{R}\}$$

$$\mathscr{M}(X, \mathfrak{R}) := \{f \in \overline{\mathbb{R}}^X \mid \{f < \alpha\} \in \mathfrak{M}(X, \mathfrak{R}), \forall \alpha \in \mathbb{R}\}$$

*and we say that X is a **measurable space** with δ-ring $\mathfrak{R}$ or that $(X, \mathfrak{R})$ is a **δ-measurable space**. Sets belonging to $\mathfrak{M}(X, \mathfrak{R})$ are called **$\mathfrak{R}$-measurable subsets** of X. Functions belonging to $\mathscr{M}(X, \mathfrak{R})$ are called **$\mathfrak{R}$-measurable functions** on X.* □

Proposition 5.3.2. *For every measurable space X with δ-ring $\mathfrak{R}$, the set $\mathfrak{M}(X, \mathfrak{R})$ contains $\mathfrak{R}$ and is a σ-algebra on X. If X belongs to $\mathfrak{R}$, then $\mathfrak{M}(X, \mathfrak{R}) = \mathfrak{R}$.*

Proof. That $\mathfrak{R} \subset \mathfrak{M}(X, \mathfrak{R})$ is evident, as is the opposite inclusion in case X itself belongs to $\mathfrak{R}$. To show that $\mathfrak{M}(X, \mathfrak{R})$ is a σ-algebra on X, we use the characterization from Proposition 5.2.16 (c) $\Rightarrow$ (a). The empty set certainly belongs to $\mathfrak{M}(X, \mathfrak{R})$. If A belongs to $\mathfrak{M}(X, \mathfrak{R})$, then the identity

$$(X \setminus A) \cap B = B \setminus (A \cap B)$$

shows that $(X \setminus A) \cap B$ belongs to $\mathfrak{R}$ for every B in $\mathfrak{R}$, that is, that $X \setminus A$ is in $\mathfrak{M}(X, \mathfrak{R})$. Finally, let $(A_n)_{n \in \mathbb{N}}$ be a sequence from $\mathfrak{M}(X, \mathfrak{R})$. Note for $B \in \mathfrak{R}$ that

$$\left(\bigcup_{n\in\mathbb{N}} A_n\right) \cap B = \bigcup_{n\in\mathbb{N}} (A_n \cap B) \subset B.$$

Using Proposition 2.1.5, we conclude that $\bigcup_{n\in\mathbb{N}} A_n$ belongs to $\mathfrak{M}(X, \mathfrak{R})$. By Proposition 5.2.16 (c) $\Rightarrow$ (a), $\mathfrak{M}(X, \mathfrak{R})$ is a σ-algebra on X. □

Proposition 5.3.3. *The following assertions hold, for every measurable space X with δ-ring $\mathfrak{R}$.*

(*a*) *For $A \subset X$, A is an $\mathfrak{R}$-measurable subset of X iff e_A is an $\mathfrak{R}$-measurable function on X.*

(*b*) *Constant extended-real functions on X are $\mathfrak{R}$-measurable.*

(*c*) *For every $\mathfrak{R}$-measurable function f on X, the sets*

$$\{f = \infty\} \qquad \{f = -\infty\} \qquad \{x \in X \mid f(x) \text{ is real}\}$$

are $\mathfrak{R}$-measurable.

(*d*) *If $(A_\iota)_{\iota \in I}$ is a disjoint countable family of $\mathfrak{R}$-measurable subsets of X whose union is X, and if $(f_\iota)_{\iota \in I}$ is a family of $\mathfrak{R}$-measurable functions on X, then the function*

$$f: X \to \overline{\mathbb{R}}, \qquad x \mapsto f_\iota(x) \qquad (x \in A_\iota, \quad \iota \in I)$$

is $\mathfrak{R}$-measurable.

(*e*) *For every $\mathfrak{R}$-measurable function f on X, the function*

$$h: X \to \mathbb{R}, \qquad x \mapsto \begin{cases} f(x) & \text{if } f(x) \in \mathbb{R} \\ 0 & \text{if } f(x) \notin \mathbb{R} \end{cases}$$

is $\mathfrak{R}$-measurable.

(*f*) *For every real number α, and for all $\mathfrak{R}$-measurable functions f and g on X, the sets*

$$\{f < g + \alpha\} \qquad \{f \leq g + \alpha\} \qquad \{f = g + \alpha\} \qquad \{f \neq g + \alpha\}$$

are $\mathfrak{R}$-measurable.

Proof. (a) For $\alpha \in \mathbb{R}$

$$\{e_A < \alpha\} = \begin{cases} X & \text{if } 1 < \alpha \\ X \setminus A & \text{if } 0 < \alpha \leq 1 \\ \varnothing & \text{if } \alpha \leq 0. \end{cases}$$

In view of Proposition 2 the assertion follows.

(b) For every extended-real number β and every real number α, the set $\{\beta e_X < \alpha\}$ is either X or the empty set. Since X and $\varnothing$ are both $\mathfrak{R}$-measurable, the assertion follows.

(c) Since

$$\{f = \infty\} = X \setminus \bigcup_{n \in \mathbb{N}} \{f < n\}$$

$$\{f = -\infty\} = \bigcap_{n \in \mathbb{N}} \{f < -n\}$$

$$\{x \in X \mid f(x) \text{ is real}\} = X \setminus (\{f = \infty\} \cup \{f = -\infty\})$$

for every $f \in \overline{\mathbb{R}}^X$, (c) follows from the fact that $\mathfrak{M}(X, \mathfrak{R})$ is a σ-algebra on X.

(d) For each real number α,

$$\{f < \alpha\} = \bigcup_{\iota \in I} (\{f_\iota < \alpha\} \cap A_\iota).$$

Since $\mathfrak{M}(X, \mathfrak{R})$ is a σ-algebra, the assertion follows.

(e) Assertion (e) follows from (b), (c), and (d).

(f) To verify that $\{f < g + \alpha\}$ is $\mathfrak{R}$-measurable, it suffices to note, in view of Proposition 2, that

$$\{f < g + \alpha\} = \bigcup_{\beta \in \mathbb{Q}} (\{f < \beta\} \setminus \{g < \beta - \alpha\}).$$

After all, $f(x) < g(x) + \alpha$ iff there exists a rational number β such that $f(x) < \beta \leq g(x) + \alpha$. The $\mathfrak{R}$-measurability of the other three sets now follows, again in view of Proposition 2, since

$$\{f \leq g + \alpha\} = X \setminus \{g < f - \alpha\}$$

$$\{f = g + \alpha\} = \{f \leq g + \alpha\} \setminus \{f < g + \alpha\}$$

$$\{f \neq g + \alpha\} = X \setminus \{f = g + \alpha\}.$$

□

Now we can describe the structure of the set $\mathscr{M}(X, \mathfrak{R})$.

Theorem 5.3.4. *The following assertions hold, for every measurable space X with δ-ring $\mathfrak{R}$.*

(*a*) *If f is an $\mathfrak{R}$-measurable function on X, then so is αf, for every extended-real number α.*

(*b*) *If f and g are $\mathfrak{R}$-measurable functions on X, and if $f + g$ is defined, then $f + g$ is $\mathfrak{R}$-measurable.*

(*c*) *For every countable family $(f_\iota)_{\iota \in I}$ of $\mathfrak{R}$-measurable functions on X, the functions*

$$\bigwedge_{\iota \in I} f_\iota \quad \textit{and} \quad \bigvee_{\iota \in I} f_\iota$$

are both $\mathfrak{R}$-measurable.

(*d*) *For $f \in \overline{\mathbb{R}}^X$, f is $\mathfrak{R}$-measurable iff both f^+ and f^- are $\mathfrak{R}$-measurable.*

(*e*) *If f is an $\mathfrak{R}$-measurable function on X, then so is $|f|$.*

(*f*) *If f and g are $\mathfrak{R}$-measurable functions on X, then so is fg.*

(*g*) *If f is an $\mathfrak{R}$-measurable function on X and*

$$g: X \to \mathbb{R}, \qquad x \mapsto \begin{cases} \dfrac{1}{f(x)} & \textit{if } f(x) \in \mathbb{R} \setminus \{0\} \\ 0 & \textit{otherwise} \end{cases}$$

then g is $\mathfrak{R}$-measurable.

(*h*) *For every sequence $(f_n)_{n \in \mathbb{N}}$ of $\mathfrak{R}$-measurable functions on X, the functions*

$$\limsup_{n \to \infty} f_n \quad \textit{and} \quad \liminf_{n \to \infty} f_n$$

are both $\mathfrak{R}$-measurable.

(*i*) *If $(f_n)_{n \in \mathbb{N}}$ is a sequence of $\mathfrak{R}$-measurable functions on X that order-converges in $\overline{\mathbb{R}}^X$, then the function $\lim_{n \to \infty} f_n$ is $\mathfrak{R}$-measurable.*

(*j*) *For every countable family $(f_\iota)_{\iota \in I}$ of $\mathfrak{R}$-measurable functions on X, the function $\sum^*_{\iota \in I} f_\iota$ is $\mathfrak{R}$-measurable.*

(*k*) *Every $\mathfrak{M}(X, \mathfrak{R})$-step function on X is $\mathfrak{R}$-measurable.*

Proof. (a) If $\alpha = 0$, then αf is a constant function and therefore $\mathfrak{R}$-measurable. If α is a nonzero real number, then for every real number β we have

$$\{\alpha f < \beta\} = \begin{cases} \left\{ f < \dfrac{\beta}{\alpha} \right\} & \text{if } \alpha > 0 \\ \left\{ f > \dfrac{\beta}{\alpha} \right\} & \text{if } \alpha < 0 \end{cases}$$

and Proposition 3 (b), (f) imply αf $\mathfrak{R}$-measurable. For each real number β,

$$\{\infty f < \beta\} = \begin{cases} \{f \leq 0\} & \text{if } \beta > 0 \\ \{f < 0\} & \text{if } \beta \leq 0. \end{cases}$$

In view of Proposition 3 (b), (f), ∞f is $\mathfrak{R}$-measurable. Finally,

$$\{-\infty f < \beta\} = \{-\beta < \infty f\}$$

for every real number β, so $-\infty f$ is also $\mathfrak{R}$-measurable.

(b) If $f + g$ is defined, then

$$\{f + g < \alpha\} = \{f < -g + \alpha\}$$

for every real number α. Thus (b) follows from (a) by Proposition 3 (b), (f).

(c) For every real number α

$$\left\{\bigwedge_{\iota \in I} f_\iota < \alpha\right\} = \bigcup_{\iota \in I} \{f_\iota < \alpha\}.$$

Since $\mathfrak{M}(X, \mathfrak{R})$ is a σ-algebra on X, we conclude that $\bigwedge_{\iota \in I} f_\iota$ is $\mathfrak{R}$-measurable. Since

$$\bigvee_{\iota \in I} f_\iota = -\bigwedge_{\iota \in I} (-f_\iota)$$

it follows from (a) that $\bigvee_{\iota \in I} f_\iota$ is also $\mathfrak{R}$-measurable.

(d), (e) These assertions follow from (a)–(c) [Theorem 1.2.7(p)–(r)].

(f) Suppose first that $f, g \geq 0$, and let $\alpha \in \mathbb{R}$. If $\alpha \leq 0$, then $\{fg < \alpha\}$ is empty; so suppose that $\alpha > 0$. We claim that

$$\{fg < \alpha\} = \{f = 0\} \cup \left[\bigcup_{\substack{\beta \in \mathbb{Q} \\ \beta > 0}} \left(\{g < \beta\} \cap \left\{f < \frac{\alpha}{\beta}\right\}\right)\right]. \tag{1}$$

That $\supset$ holds is evident. Suppose that x fails to belong to the set on the right-hand side. Then $f(x) > 0$, and for every strictly positive rational number β we have

$$x \notin \{g < \beta\} \cap \left\{f < \frac{\alpha}{\beta}\right\}.$$

One possibility is that $g(x) \geq \beta$ for every rational number β, in which case $g(x) = \infty = f(x)g(x)$ and $x \notin \{fg < \alpha\}$. Otherwise, $g(x) < \beta$ for some strictly positive rational number β. Then $f(x) \geq \alpha/\beta$ and $\beta f(x) \geq \alpha$. In fact, if β is any strictly positive rational number for which $g(x) < \beta$, then $\beta f(x) \geq$

α. Since

$$f(x)g(x) = \inf\{\beta f(x) \mid \beta \in \mathbb{Q}, \beta > g(x)\}$$

it follows that $f(x)g(x) \geq \alpha$ and $x \notin \{fg < \alpha\}$. We have established (1). It follows that fg is $\mathfrak{R}$-measurable.

Now let f and g be arbitrary $\mathfrak{R}$-measurable functions on X. Set

$$A := \{f \geq 0\} \qquad B := \{f < 0\}$$

$$C := \{g \geq 0\} \qquad D := \{g < 0\}.$$

The sets $A \cap C, A \cap D, B \cap C, B \cap D$ are pairwise disjoint, and their union is X. According to what we have already established, the functions $f^+g^+, -f^+g^-, -f^-g^+, f^-g^-$ are all $\mathfrak{R}$-measurable. Since

$$fg = \begin{cases} f^+g^+ & \text{on } A \cap C \\ -f^+g^- & \text{on } A \cap D \\ -f^-g^+ & \text{on } B \cap C \\ f^-g^- & \text{on } B \cap D \end{cases}$$

it follows from Proposition 3 (d) that fg is $\mathfrak{R}$-measurable.

(g) For $\alpha \in \mathbb{R}$,

$$\{g < \alpha\} = \begin{cases} \{f \leq 0\} \cup \left\{\frac{1}{\alpha} < f\right\} & \text{if } \alpha > 0 \\ \{f < 0\} \setminus \{f = -\infty\} & \text{if } \alpha = 0 \\ \{f < 0\} \cap \left\{\frac{1}{\alpha} < f\right\} & \text{if } \alpha < 0 \end{cases}$$

so (g) follows from Proposition 3 (b), (f).

(h) Assertion (h) follows from (c).

(i) Assertion (i) follows from (h) [Proposition 1.1.20 (b)].

(j) As usual we write Σ instead of Σ^* whenever the family in question is readily seen to be summable. If I is empty, then $\Sigma_{\iota \in I} f_\iota$ is a constant function and therefore $\mathfrak{R}$-measurable, so assume that $I \neq \emptyset$. Set

$$A := \left\{x \in X \,\middle|\, \sum_{\iota \in I}^{*} f_\iota(x) \in \mathbb{R}\right\}$$

$$B := \left\{\sum_{\iota \in I}^{*} f_\iota = \infty\right\}$$

$$C := \left\{\sum_{\iota \in I}^{*} f_\iota = -\infty\right\}.$$

For the case where I is finite and every f_ι is real valued, the $\mathfrak{R}$-measurability of $\sum_{\iota \in I} f_\iota$ follows from (b) by complete induction on the number of elements in I. Suppose that I is finite but the f_ι are not necessarily real valued. The identities

$$A = \bigcap_{\iota \in I} \bigcup_{n \in \mathbb{N}} \{ |f_\iota| < n \}$$

$$B = \bigcup_{\iota \in I} \bigcap_{n \in \mathbb{N}} \{ n < f_\iota \}$$

$$C = X \setminus (A \cup B)$$

show that A, B, and C are all $\mathfrak{R}$-measurable. Since they are also disjoint and since

$$\sum_{\iota \in I}{}^{*} f_\iota = \sum_{\iota \in I} f_\iota e_A + \infty e_B + (-\infty e_C)$$

we conclude that $\sum^{*}_{\iota \in I} f_\iota$ is $\mathfrak{R}$-measurable.

Suppose, finally, that I is countably infinite. Let

$$\mathfrak{F}(I) := \{ J \mid J \text{ is a finite subset of } I \}$$

and note that $\mathfrak{F}(I)$ is countable. For $J \in \mathfrak{F}(I)$ and $n \in \mathbb{N}$, set

$$A_{J,n} := \left\{ \left| \sum_{\iota \in J}{}^{*} f_\iota \right| < \frac{1}{n} \right\}$$

$$B_{J,n} := \left\{ n < \sum_{\iota \in J}{}^{*} f_\iota \right\}.$$

The sets $A_{J,n}$ and $B_{J,n}$ are $\mathfrak{R}$-measurable, and Theorem 3.4.20 shows that

$$A = \bigcap_{n \in \mathbb{N}} \bigcup_{J \in \mathfrak{F}(I)} \bigcap_{\substack{K \in \mathfrak{F}(I) \\ J \cap K = \varnothing}} A_{J,n}$$

$$B = \bigcap_{n \in \mathbb{N}} \bigcup_{J \in \mathfrak{F}(I)} B_{J,n}.$$

Since $\mathfrak{M}(X, \mathfrak{R})$ is a σ-algebra on X, it follows that A, B, and C are $\mathfrak{R}$-measurable. Moreover

$$\sum_{\iota \in I}{}^{*} f_\iota = \bigvee_{J \in \mathfrak{F}(I)} \bigwedge_{\substack{K \in \mathfrak{F}(I) \\ J \subset K}} \sum_{\iota \in K} f_\iota e_A + \infty e_B + (-\infty e_C)$$

(Corollary 3.4.19). We conclude that $\sum^{*}_{\iota \in I} f_\iota$ is $\mathfrak{R}$-measurable.

(k) Assertion (k) follows from (a) and (j), together with Proposition 3 (a). □

In $\mathcal{M}(X, \mathfrak{R})$ we have exactly the properties we want: all important operations and limit processes can be carried out inside $\mathcal{M}(X, \mathfrak{R})$. To avoid any misunderstanding, we stress that $\mathcal{M}(X, \mathfrak{R})$ is not, in general, a Riesz lattice. To wit, the constant function ∞e_X is always $\mathfrak{R}$-measurable, so $\mathcal{M}(X, \mathfrak{R})$ cannot be a Riesz lattice unless $\mathcal{M}(X, \mathfrak{R}) = \overline{\mathbb{R}}^X$, which is seldom the case.

Theorem 4 has the following useful corollary.

Corollary 5.3.5. *For every measurable space X with δ-ring $\mathfrak{R}$, the conditions*

$$f \in \mathcal{M}(X, \mathfrak{R}) \qquad h \in \overline{\mathbb{R}}^X \qquad fh \in \mathcal{M}(X, \mathfrak{R})$$

$$\{h \neq 0\} \subset \{f \neq 0\}$$

imply that h also belongs to $\mathcal{M}(X, \mathfrak{R})$.

Proof. With g defined as in Theorem 4 (g), we have $h = g(fh)$, and the assertion follows from Theorem 4 (f), (g). □

We conclude this section with a theorem concerning approximation of measurable functions by step functions.

Theorem 5.3.6. *Let X be a measurable space, with δ-ring $\mathfrak{R}$. Then the following assertions hold, for every (positive) $\mathfrak{R}$-measurable function f on X.*

(*a*) *There exists a sequence (an increasing sequence) $(f_n)_{n \in \mathbb{N}}$ of (positive) $\mathfrak{M}(X, \mathfrak{R})$-step functions on X such that $\lim_{n \to \infty} f_n = f$ and $|f_n| \leq |f|$ for every $n \in \mathbb{N}$.*

(*b*) *If A is a subset of X for which*

$$\sup_{x \in A} |f(x)| < \infty$$

then the sequence in (a) can be chosen so that the convergence to f is uniform on A.

(*c*) *If $B \in \mathfrak{R}_\sigma$, then corresponding to the function fe_B, there exists a sequence $(f_n)_{n \in \mathbb{N}}$ of $\mathfrak{R}$-step functions on X with the properties described in (a).*

Proof.

(a), (b) Assertions (a) and (b) follow from Proposition 5.1.3 in case f belongs to $\mathcal{M}(X, \mathfrak{R})_+$. If f is an arbitrary $\mathfrak{R}$-measurable function on X, then f^+ and f^- belong to $\mathcal{M}(X, \mathfrak{R})_+$. Using Proposition 5.1.3, choose increasing sequences $(g_n)_{n \in \mathbb{N}}$ and $(h_n)_{n \in \mathbb{N}}$ of positive $\mathfrak{M}(X, \mathfrak{R})$-step functions whose suprema are f^+ and f^-, respectively. Set $f_n := g_n - h_n$. Then the sequence $(f_n)_{n \in \mathbb{N}}$ has the properties specified in (a). If for some $A \subset X$ the sequences $(g_n)_{n \in \mathbb{N}}$ and $(h_n)_{n \in \mathbb{N}}$ converge uniformly on A to f^+ and f^-, respectively, then the convergence of $(g_h - h_n)_{n\in\mathbb{N}}$ to f is also uniform on A. Thus (b) holds as well for arbitrary $\mathfrak{R}$-measurable f.

(c) Since $\mathfrak{R}$ is a δ-ring, we can choose an increasing sequence $(B_n)_{n\in\mathbb{N}}$ from $\mathfrak{R}$ such that $B = \bigcup_{n\in\mathbb{N}} B_n$ (Proposition 2.1.11). Let $(f_n)_{n\in\mathbb{N}}$ be a sequence with the properties described in (a). Then $(f_n e_{B_n})_{n\in\mathbb{N}}$ is the desired sequence. □

EXERCISES

5.3.1(E) Let $(X, \mathfrak{R})$ be a measurable space. Prove the following:

(α) If $(A_n)_{n\in\mathbb{N}}$ is a sequence from $\mathfrak{M}(X,\mathfrak{R})$, then ${}^{\mathfrak{P}(X)}\limsup_{n\to\infty} A_n \in \mathfrak{M}(X,\mathfrak{R})$ and ${}^{\mathfrak{P}(X)}\liminf_{n\to\infty} A_n \in \mathfrak{M}(X,\mathfrak{R})$.

(β) If $(A_n)_{n\in\mathbb{N}}$ is an order-convergent sequence from $\mathfrak{M}(X,\mathfrak{R})$, then ${}^{\mathfrak{P}(X)}\lim_{n\to\infty} A_n \in \mathfrak{M}(X,\mathfrak{R})$.

5.3.2(E) **The Generation of Algebras and σ-Algebras.** Criteria for the generation of algebras and σ-algebras can be deduced directly from the propositions formulated in the exercises from Section 2.1. for the generation of rings of sets, δ-rings, and σ-rings on X. We wish to collect a few of these here.

Take $\mathfrak{S} \subset \mathfrak{P}(X)$. We use the notation introduced in Ex. 2.1.2. Prove the following:

(α) If $\mathfrak{S}$ is a $\cap$-semi-lattice with $X \in \mathfrak{S}$, then $\mathfrak{S}_{udv}$ is the algebra on X generated bv $\mathfrak{S}$.

(β) If $\mathfrak{S}$ is a semi-ring with $X \in \mathfrak{S}$, then $\mathfrak{S}_v$ is the algebra on X generated by $\mathfrak{S}$.

(γ) If $\mathfrak{S}$ is an algebra on X, then the σ-algebra on X generated by $\mathfrak{S}$, the monotone set generated by $\mathfrak{S}$, and the δ-ring generated by $\mathfrak{S}$ all coincide.

(δ) The algebra on X generated by $\mathfrak{S}$ is the set of all sets $A \subset X$, for which there is a finite subset $\mathfrak{S}' \subset \mathfrak{S}$ such that A belongs to the algebra on X generated by $\mathfrak{S}'$. An analogous statement, with "finite" replaced by "countable," holds for σ-algebras.

(ε) If $\mathfrak{S}$ is an algebra on X, set $\mathfrak{R}_0 := \mathfrak{S}$ and for each ordinal $\alpha > 0$, $\tilde{\mathfrak{R}}_\alpha := \bigcup_{\beta<\alpha} \mathfrak{R}_\beta$ and $\mathfrak{R}_\alpha := \{{}^{\mathfrak{P}(X)}\lim_{n\to\infty} A_n \mid (A_n)_{n\in\mathbb{N}}$ is a convergent sequence of elements of $\tilde{\mathfrak{R}}_\alpha\}$. Then $\mathfrak{R}_{\omega_1}$ is the σ-algebra on X generated by $\mathfrak{S}$ (where ω_1 is the first uncountable ordinal).

(ζ) If $\mathfrak{S}$ is a ring of sets, $\mathfrak{S} \subset \mathfrak{P}(X)$, then $\mathfrak{A}(\mathfrak{S}) := \{A \subset X \mid A \in \mathfrak{S}$ or $(X \setminus A) \in \mathfrak{S}\}$ is the algebra on X generated by $\mathfrak{S}$.

(η) If $\mathfrak{S}$ is a σ-ring, then $\mathfrak{A}(\mathfrak{S})$ is the σ-algebra on X generated by $\mathfrak{S}$, where $\mathfrak{A}(\mathfrak{S})$ is as in (ζ).

To prove (ζ), notice that $\mathfrak{A}(\mathfrak{S})$ is clearly contained in the algebra on X generated by $\mathfrak{S}$. We show that $\mathfrak{A}(\mathfrak{S})$ is itself an algebra. Clearly $X \in \mathfrak{A}(\mathfrak{S})$ and $X \setminus A \in \mathfrak{A}(\mathfrak{S})$ for any $A \in \mathfrak{A}(\mathfrak{S})$. There are three cases to consider in order to prove that $A \cup B \in \mathfrak{A}(\mathfrak{S})$ for $A, B \in \mathfrak{A}(\mathfrak{S})$:

CASE 1: $A, B \in \mathfrak{S}$. Then $A \cup B \in \mathfrak{S} \subset \mathfrak{A}(\mathfrak{S})$.

CASE 2: $X \setminus A, X \setminus B \in \mathfrak{S}$. Then $(X \setminus (A \cup B)) = (X \setminus A) \cap (X \setminus B) \in \mathfrak{S}$, so that $A \cup B \in \mathfrak{A}(\mathfrak{S})$.

CASE 3: $A \in \mathfrak{S}$, $X \setminus B \in \mathfrak{S}$. Then $(X \setminus (A \cup B)) = (X \setminus B) \setminus A \in \mathfrak{S}$. Thus $A \cup B \in \mathfrak{A}(\mathfrak{S})$.

Hence $A \cup B \in \mathfrak{A}(\mathfrak{S})$ whenever $A, B \in \mathfrak{A}(\mathfrak{S})$.

5.3.3(E) Dynkin Systems. Let $\mathfrak{R} \subset \mathfrak{P}(X)$ be a Dynkin system (Ex. 2.1.12). Prove the following statements:

(α) Every σ-algebra on X is a Dynkin system on X.

(β) Let X be a finite set with $2n$ elements ($n \in \mathbb{N}$). Set $\mathfrak{R} := \{A \subset X \mid A$ has an even number of elements$\}$. Then $\mathfrak{R}$ is a Dynkin system on X. If $n > 1$, then $\mathfrak{R}$ is not an algebra on X.

(γ) Let $\mathfrak{R}$ be a Dynkin system on X. Then $\mathfrak{R}$ is a σ-algebra on X iff $A \cap B \in \mathfrak{R}$ whenever $A, B \in \mathfrak{R}$.

(δ) If $\mathfrak{R} \subset \mathfrak{P}(X)$ is a $\cap$-semi-lattice, then the σ-algebra on X generated by $\mathfrak{R}$ coincides with the Dynkin system on X generated by $\mathfrak{R}$.

(ε) Formulate and prove analogous statements for conditionally Dynkin systems (Ex. 2.1.12).

To prove (δ), let $\mathfrak{R}_{X,\sigma}$ denote the σ-algebra on X generated by $\mathfrak{R}$, and $\mathfrak{R}_{X,d}$ the Dynkin system. Clearly $\mathfrak{R}_{X,d} \subset \mathfrak{R}_{X,\sigma}$. To establish the reverse inclusion, it suffices to show that $A \cap B \in \mathfrak{R}_{X,d}$ whenever $A, B \in \mathfrak{R}_{X,d}$ [cf., (γ)]. For $A \in \mathfrak{R}_{X,d}$ set $\mathfrak{R}_A := \{B \subset X \mid A \cap B \in \mathfrak{R}_{X,d}\}$. Then $\mathfrak{R}_A$ is a Dynkin system for each A. Thus $\mathfrak{R}_{X,d} \subset \mathfrak{R}_A$ for each $A \in \mathfrak{R}$. Hence for each $B \in \mathfrak{R}_{X,d}$ and $A \in \mathfrak{R}$, $A \cap B \in \mathfrak{R}_{X,d}$. Thus $\mathfrak{R} \subset \mathfrak{R}_B$. Hence $\mathfrak{R}_{X,d} \subset \mathfrak{R}_B$.

5.3.4(E) Measurable Mappings. Let $(X, \mathfrak{R})$ and $(Y, \mathfrak{S})$ be measurable spaces. A mapping $\varphi: X \to Y$ is called a **morphism** iff $\varphi^{-1}(A) \in \mathfrak{R}$ for every $A \in \mathfrak{S}$. φ is called **measurable** iff $\varphi^{-1}(A) \in \mathfrak{M}(X, \mathfrak{R})$ for every $A \in \mathfrak{M}(Y, \mathfrak{S})$. Prove the following assertions:

(α) Let $(X, \mathfrak{R})$, $(Y, \mathfrak{S})$ and $(Z, \mathfrak{T})$ be measurable spaces and $\varphi: X \to Y$ and $\psi: Y \to Z$ morphisms. Then $\psi \circ \varphi$ is a morphism.

(β) Let $(X, \mathfrak{R})$ be a measurable space. Then the identity map id_X is a morphism.

(γ) Let X, Y be sets, $\mathfrak{R} \subset \mathfrak{P}(X)$ and $\mathfrak{S} \subset \mathfrak{P}(Y)$. Assume that $\varphi: X \to Y$ is a mapping with $\varphi^{-1}(A) \in \mathfrak{R}$ for every $A \in \mathfrak{S}$. Then the following assertions hold:

(γ1) $\varphi^{-1}(A) \in \mathfrak{R}_r$ for every $A \in \mathfrak{S}_r$.
(γ2) $\varphi^{-1}(A) \in \mathfrak{R}_\delta$ for every $A \in \mathfrak{S}_\delta$.
(γ3) $\varphi^{-1}(A) \in \mathfrak{R}_\sigma$ for every $A \in \mathfrak{S}_\sigma$.
(γ4) If $\mathfrak{R}_{X,a}$ and $\mathfrak{S}_{Y,a}$ denote the algebras generated by $\mathfrak{R}$ on X, respectively, by $\mathfrak{S}$ on Y, then $\varphi^{-1}(A) \in \mathfrak{R}_{X,a}$ for every $A \in \mathfrak{S}_{Y,a}$.

($\gamma 5$) If $\mathfrak{R}_{X,\sigma}$ and $\mathfrak{S}_{Y,\sigma}$ denote the σ-algebras generated by $\mathfrak{R}$ on X, respectively, by $\mathfrak{S}$ on Y, then $\varphi^{-1}(A) \in \mathfrak{R}_{X,\sigma}$ for every $A \in \mathfrak{S}_{Y,\sigma}$.

(δ) Let $(X, \mathfrak{R})$ and $(Y, \mathfrak{S})$ be measurable spaces and φ: $X \to Y$ a morphism. If $\mathfrak{R} = \{\varphi^{-1}(A) \mid A \in \mathfrak{S}\}$, then these assertions hold:

($\delta 1$) φ is measurable.

($\delta 2$) $f \circ \varphi \in \mathscr{M}(X, \mathfrak{R})$ for every $f \in \mathscr{M}(Y, \mathfrak{S})$.

Consider now a topological space $(X, \mathfrak{T})$. Every such space is provided with a canonical structure of a measurable space, the σ-algebra of Borel sets. $A \subset X$ is called a **Borel set** on X with respect to $\mathfrak{T}$ iff $A \in \mathfrak{T}_\sigma$. Denote by $\mathfrak{B}(X, \mathfrak{T})$ the set of all Borel sets on X with respect to $\mathfrak{T}$.

Let $(X, \mathfrak{T})$ and $(Y, \mathfrak{S})$ be topological spaces. A mapping φ: $X \to Y$ is called **Borel measurable** iff φ is measurable with respect to $(X, \mathfrak{B}(X, \mathfrak{T}))$ and $(Y, \mathfrak{B}(Y, \mathfrak{S}))$. Prove the next assertions:

(ε) Every continuous mapping φ: $X \to Y$ is Borel measurable.

(ζ) Let $(X, \mathfrak{R})$ be a measurable space and $(f_k)_{k \in \mathbb{N}_n}$ a family of measurable functions on X. Let φ: $\mathbb{R}^n \to \mathbb{R}$ be Borel measurable (with respect to the natural topologies on the spaces $\mathbb{R}^n, \mathbb{R}$). Define

$$h: X \to \mathbb{R}, \qquad x \mapsto \varphi(f_1(x), f_2(x), \ldots, f_n(x)).$$

Then $h \in \mathscr{M}(X, \mathfrak{R})$.

Finally, let $(X, \mathfrak{R})$ be a measurable space, and denote by $\mathfrak{B}$ the set of all Borel sets on $\mathbb{R}$.

(η) For any $f \in \overline{\mathbb{R}}^X$, the following statements are equivalent:

($\eta 1$) f is $\mathfrak{R}$-measurable.

($\eta 2$) f is a measurable map with respect to $(X, \mathfrak{R})$ and $(\mathbb{R}, \mathfrak{B})$.

5.3.5(E) Let X be a set. Let $\mathfrak{R} := \{A \subset X \mid A$ is countable or $X \setminus A$ is countable$\}$. Prove the following:

(α) $\mathfrak{R}$ is the σ-algebra generated by $\{\{x\} \mid x \in X\}$.

(β) If X is uncountable, then there are subsets of X that are not $\mathfrak{R}$-measurable.

(γ) If X is uncountable, then there are functions $f \in \overline{\mathbb{R}}^X$ such that f is not $\mathfrak{R}$-measurable but $|f|$ is.

5.3.6(E) Show that if X is a set, then any σ-algebra on X is either finite or uncountable.

5.3.7(E) Let X be a nonempty set and $\mathfrak{R}$ a finite algebra of sets on X. Prove the following:

(α) $\mathfrak{R}$ is a σ-algebra on X.

(β) $\mathfrak{R}$ contains exactly 2^n elements for some $n \in \mathbb{N}$.

(γ) There is a surjection φ: $X \to \mathbb{N}_n$, where n is as in (β), such that $\mathfrak{R} = \{\varphi^{-1}(A) \mid A \subset \mathbb{N}_n\}$.

5.3.8(E) Let X be a set and $\mathfrak{S} \subset \mathfrak{P}(X)$. Let $\aleph_0$ denote the cardinality of $\mathbb{N}$, $\mathfrak{c}$ that of the continuum, and $|Y|$ that of Y for any set Y. Prove the following:

(α) If $|\mathfrak{S}| \geq \aleph_0$, then $\mathfrak{S}$ and the algebra on X generated by $\mathfrak{S}$ have the same cardinality.
(β) If $|\mathfrak{S}| \geq \aleph_0$, then the cardinality of the σ-algebra on X generated by $\mathfrak{S}$ is at least $\mathfrak{c}$.
(γ) If $|\mathfrak{S}| \leq \mathfrak{c}$, then the cardinality of the σ-algebra on X generated by $\mathfrak{S}$ does not exceed $\mathfrak{c}$.

Let X be a topological space with countable base. Prove that if $\mathfrak{B}$ denotes the set of Borel sets of X, then the following hold:

(δ) $|\mathfrak{B}| \leq \mathfrak{c}$.
(ε) The set of all Borel sets of $\mathbb{R}^n$ $(n \in \mathbb{N})$ has cardinality $\mathfrak{c}$.

5.3.9(E) **Borel Sets and Baire Sets.** Let $(X, \mathfrak{T})$ be a topological space. Take $A \subset X$. A is a **Borel set** iff $A \in \mathfrak{T}_\sigma$. A is a **Baire set** iff A belongs to the least σ-algebra on X with respect to which all continuous real functions on X are measurable. A is an **F_σ-set** if $A = \bigcup_{n \in \mathbb{N}} A_n$, where $(A_n)_{n \in \mathbb{N}}$ is a sequence of closed sets. A is a **G_δ-set** if $A = \bigcap_{n \in \mathbb{N}} A_n$, where $(A_n)_{n \in \mathbb{N}}$ is a sequence of open sets. Denote by $\mathfrak{B}$ the set of all Borel sets and by $\mathfrak{B}_0$ the set of all Baire sets of X. The functions in $\mathscr{M}(X, \mathfrak{B})$ are called **Borel measurable** and those in $\mathscr{M}(X, \mathfrak{B}_0)$, **Baire measurable**.

(a) Prove the following propositions:

(α) $\mathfrak{B}$ is a σ-algebra on X.
(β) $\mathfrak{B}$ is the σ-algebra on X generated by the closed subsets of X. (Hence $\mathfrak{B}$ contains, in particular, all closed and open subsets of X.)
(γ) Every continuous real function on X is Borel measurable.
(δ) $\mathfrak{B}_0 \subset \mathfrak{B}$.

(b) Let $\mathfrak{S}$ denote the set of all F_σ-sets. Prove the following:

(α) $\mathfrak{S} \subset \mathfrak{B}$
(β) $\bigcap_{\iota \in I} A_\iota \in \mathfrak{S}$ for every nonempty finite family $(A_\iota)_{\iota \in I}$ in $\mathfrak{S}$.
(γ) $\bigcup_{\iota \in I} A_\iota \in \mathfrak{S}$ for every countable family $(A_\iota)_{\iota \in I}$ in $\mathfrak{S}$.

(c) Let $\mathfrak{D}$ denote the set of all G_δ-sets. Prove the following:

(α) $\mathfrak{T} \subset \mathfrak{D} \subset \mathfrak{B}$.
(β) $\bigcup_{\iota \in I} A_\iota \in \mathfrak{D}$ for every finite family $(A_\iota)_{\iota \in I}$ in $\mathfrak{D}$.
(γ) $\bigcap_{\iota \in I} A_\iota \in \mathfrak{D}$ for every nonempty countable family $(A_\iota)_{\iota \in I}$ in $\mathfrak{D}$.
(δ) If f is a continuous real function on X and $\alpha \in \mathbb{R}$, then $f^{-1}(\{\alpha\}) \in \mathfrak{D} \cap \mathfrak{B}_0$.

(d) **Baire Functions**. Denote by $\mathscr{C}$ the set of all continuous real functions on X. Define $\mathscr{B}_\alpha$ for each ordinal α, by $\mathscr{B}_0 := \mathscr{C}$, $\mathscr{B}_\alpha := \{f \in \overline{\mathbb{R}}^X \mid \exists (f_n)_{n \in \mathbb{N}}$ in $\bigcup_{\beta < \alpha} \mathscr{B}_\beta$ with $f = \lim_{n \to \infty} f_n\}$ for $\alpha > 0$. The functions in $\mathscr{B}_{\omega_1}$ are called

Baire functions. Prove the following:

(α) If $\omega_1 \leq \alpha$, then $\mathscr{B}_\alpha = \mathscr{B}_{\omega_1}$, where ω_1 is the first uncountable ordinal.
(β) The Baire functions are precisely the Baire measurable ones.

(e) Let $(X, \mathfrak{T})$ be a normal space. Prove the following:

(α) For every closed G_δ-set A, there is a continuous function on X, f, with $f^{-1}(\{0\}) = A$.
(β) $\mathfrak{B}_0$ is the σ-algebra on X generated by the set of closed G_δ-sets.
(γ) $\mathfrak{B}_0$ is the σ-algebra on X generated by the set of open F_σ-sets.

(f) **Perfectly Normal Spaces.** $(X, \mathfrak{T})$ is **perfectly normal** iff $(X, \mathfrak{T})$ is normal and every closed set is a G_δ-set. Prove the following:

(α) Every metrizable space $(X, \mathfrak{T})$ is perfectly normal.
(β) If $(X, \mathfrak{T})$ is perfectly normal, then $\mathfrak{B} = \mathfrak{B}_0$.

(g) Let $(X, \mathfrak{T})$ be an arbitrary topological space. Denote by Φ the set of all subsets $\mathfrak{R}$ of $\mathfrak{P}(X)$ with the following properties:

(i) $\mathfrak{R}$ contains all open and all closed subsets of X.
(ii) $\bigcap_{n \in \mathbb{N}} A_n \in \mathfrak{R}$ for every sequence $(A_n)_{n \in \mathbb{N}}$ in $\mathfrak{R}$.
(iii) $\bigcup_{n \in \mathbb{N}} A_n \in \mathfrak{R}$ for every disjoint sequence $(A_n)_{n \in \mathbb{N}}$ in $\mathfrak{R}$.

Prove that $\mathfrak{B}$ is the least element of Φ (with respect to $\subset$). Moreover, if X is metrizable, then (i) can be replaced by

(i′) $\mathfrak{T} \subset \mathfrak{R}$.

5.3.10(C) Measurability on Baire Spaces. Let $(X, \mathfrak{T})$ be a topological space. Denote for any $A \subset X$ by A° the interior of A and by $\overline{A}$ the closure of A with respect to $\mathfrak{T}$.

(α) Prove the equivalence of the following conditions:

(α1) If the subset A of X is of the first category, then $X \setminus A$ is a dense subset of X.
(α2) There is no nonempty open subset of X of the first category.
(α3) If $(A_n)_{n \in \mathbb{N}}$ is a sequence of closed subsets of X such that $A_n^\circ = \emptyset$ for every $n \in \mathbb{N}$, then $\bigcup_{n \in \mathbb{N}} A_n$ has no interior points.
(α4) If $(A_n)_{n \in \mathbb{N}}$ is a sequence of open subsets of X with $\overline{A_n} = X$ for every $n \in \mathbb{N}$, then $\overline{\bigcap_{n \in \mathbb{N}} A_n} = X$.

(For the notion of a set of the first category, see Ex. 2.1.4.) A **Baire space** is a topological space satisfying one (hence all) of the conditions (α1)–(α4). Prove the following statements:

(β) If (X, d) is a complete metric space, then the associated topological space is a Baire space.
(γ) Every locally compact space is a Baire space.

We give some hints for the proof of (β). Let (X, d) be a complete metric space. Let A be a subset of X of the first category, and denote by $(A_n)_{n \in \mathbb{N}}$

a sequence of nowhere dense subsets of X with $A = \bigcup_{n\in\mathbb{N}} A_n$. We want to show that $B \cap (X \setminus A) \neq \varnothing$ for any nonempty open subset B of X. To prove this, construct a sequence $(x_n)_{n\in\mathbb{N}}$ from X and a sequence $(\varepsilon_n)_{n\in\mathbb{N}}$ of strictly positive numbers with the following properties: (For $x \in X$ and $\varepsilon > 0$ we set $B(x, \varepsilon) := \{y \in X \mid d(x, y) < \varepsilon\}$.)

(i) $\lim_{n\to\infty} \varepsilon_n = 0$.
(ii) $B(x_{n+1}, \varepsilon_{n+1}) \subset B(x_n, \varepsilon_n)$ for every $n \in \mathbb{N}$.
(iii) $B(x_n, \varepsilon_n) \cap A_n = \varnothing$ for every $n \in \mathbb{N}$.
(iv) $\overline{B(x_1, \varepsilon_1)} \subset B$.

Then $(x_n)_{n\in\mathbb{N}}$ is a Cauchy sequence converging to an element x of $B \cap (X \setminus A)$.

The proof for the case of a locally compact space is similar.

Now prove the following result:

(δ) Let $(X, \mathfrak{T})$ be a Baire space. Then, for any Borel-measurable function $f \in \mathbb{R}^X$ there is a dense G_δ-subset A of X such that $f|_A$ is continuous.

Hint: Let $\mathfrak{B}$ be a countable base for the natural topology on $\mathbb{R}$. For every $V \in \mathfrak{B}$ there is an open subset U_V of X such that $f^{-1}(V) \triangle U_V$ is of the first category [see Ex. 2.1.4 (ϑ)]. Hence for every $V \in \mathfrak{B}$ there is a sequence $(F_n^V)_{n\in\mathbb{N}}$ of closed nowhere dense subsets of X such that

$$f^{-1}(V) \triangle U_V \subset \bigcup_{n\in\mathbb{N}} F_n^V.$$

Thus

$$A := X \setminus \bigcup_{(n,V)\in\mathbb{N}\times\mathfrak{B}} F_n^V$$

is a G_δ-subset of X, and it is dense because X is a Baire space. Now let W be an open subset of $\mathbb{R}$ and denote by $\mathfrak{W}$ the set of all $V \in \mathfrak{B}$ with $V \subset W$. Then, $W = \bigcup_{V\in\mathfrak{W}} V$ and

$$(f|_A)^{-1}(W) = \left(\bigcup_{V\in\mathfrak{W}} f^{-1}(V)\right) \cap A = \left(\bigcup_{V\in\mathfrak{W}} U_V\right) \cap A.$$

We conclude that $(f|_A)^{-1}(W)$ is an open subset of A and $f|_A$ is continuous.

5.3.11(C) As a prerequisite for this exercise, the reader should study Exs. 2.1.4 and 5.3.10.

(a) The first step of this exercise is a generalization of Ex. 5.3.10 (δ). Consider a Baire space X and a topological space Y with a countable base. Let f: $X \to Y$ be a mapping such that $f^{-1}(U)$ is approximable for any open subset U of Y. Prove that there is a dense G_δ-subset A of X such that $f|_A$ is continuous.

The idea of the proof is the same as in Ex. 5.3.10 (δ).

(b) Take the discrete topology on $\{0,1\}$ and the product topology on $\{0,1\}^{\mathbb{N}}$. Let $\mathfrak{T}$ be the uniquely determined topology on $\mathfrak{P}(\mathbb{N})$ for which the mapping

$$\{0,1\}^{\mathbb{N}} \to \mathfrak{P}(\mathbb{N}), \qquad f \mapsto \{f=1\}$$

is a homeomorphism. Prove the following statements.

(α) $(\mathfrak{P}(\mathbb{N}), \mathfrak{T})$ is a compact metrizable space.

(β) Let $\mathfrak{A}$ be a dense G_δ-subset of $\mathfrak{P}(\mathbb{N})$ with respect to $\mathfrak{T}$. Then there are sets $A, B, C \in \mathfrak{A}$ such that $A \cap B = \varnothing$, $A \cup B = C$, and

$$\{C \triangle K \mid K \subset \mathbb{N},\ K \text{ finite}\} \subset \mathfrak{A}.$$

Hint for (β): Let $(\mathfrak{F}_n)_{n \in \mathbb{N}}$ be a sequence of closed nowhere dense subsets of $\mathfrak{P}(\mathbb{N})$ such that $\mathfrak{A} = \mathfrak{P}(\mathbb{N}) \setminus \bigcup_{n \in \mathbb{N}} \mathfrak{F}_n$. Define

$$\mathfrak{B} := \mathfrak{P}(\mathbb{N}) \setminus \bigcup_{n \in \mathbb{N}} \{D \triangle K \mid D \in \mathfrak{F}_n, K \subset \mathbb{N},\ K \text{ finite}\}.$$

$\mathfrak{B}$ is a dense G_δ-subset of $\mathfrak{P}(\mathbb{N})$. Now set

$$\varphi : \mathfrak{P}(\mathbb{N}) \times \mathfrak{P}(\mathbb{N}) \to \mathfrak{P}(\mathbb{N}), \qquad (D, E) \mapsto D \cup E$$

$$\psi : \mathfrak{P}(\mathbb{N}) \times \mathfrak{P}(\mathbb{N}) \to \mathfrak{P}(\mathbb{N}), \qquad (D, E) \mapsto D \setminus E.$$

φ and ψ are continuous open mappings. We conclude that $\varphi^{-1}(\mathfrak{B})$ and $\psi^{-1}(\mathfrak{B})$ are dense G_δ-sets. But then there exists

$$(D, A) \in (\mathfrak{B} \times \mathfrak{B}) \cap \varphi^{-1}(\mathfrak{B}) \cap \psi^{-1}(\mathfrak{B}).$$

Define $B := \psi(D, A)$ and $C := \varphi(D, A)$. Then $A, B, C \in \mathfrak{B} \subset \mathfrak{A}$, $A \cap B = \varnothing$, $A \cup B = C$, and

$$\{C \triangle K \mid K \subset \mathbb{N},\ K \text{ finite}\} \subset \mathfrak{B} \subset \mathfrak{A}.$$

(c) Denote by $\mathfrak{T}$ the topology on $\mathfrak{P}(\mathbb{N})$ defined in (b). Prove the following assertion:

(α) Let $\mu : \mathfrak{P}(\mathbb{N}) \to \mathbb{R}$ be a positive additive function such that $\mu^{-1}(U)$ is approximable for every open subset U of $\mathbb{R}$. Then μ is a positive measure on $\mathfrak{P}(\mathbb{N})$.

We give some hints. Define

$$\nu : \mathfrak{P}(\mathbb{N}) \to \mathbb{R}, M \mapsto \sum_{n \in M} \mu(\{n\}), \qquad \lambda := \mu - \nu.$$

ν is a positive measure on $\mathfrak{P}(\mathbb{N})$ and λ is a positive additive function on $\mathfrak{P}(\mathbb{N})$ such that $\lambda^{-1}(U)$ is approximable for any open subset U of $\mathbb{R}$. (a) implies the existence of a dense G_δ-subset $\mathfrak{A}$ of $\mathfrak{P}(\mathbb{N})$ such that $\lambda|_{\mathfrak{A}}$ is continuous. Because of (b), (β) there are sets $A, B, C \in \mathfrak{A}$ such that $A \cap B = \varnothing$, $A \cup B = C$, and

$$\{C \triangle K \mid K \subset \mathbb{N},\ K \text{ finite}\} \subset \mathfrak{A}.$$

For any finite subset K of $\mathbb{N}$, $\lambda(C) = \lambda(C \triangle K)$. Hence λ is constant on $\{C \triangle K \mid K \subset \mathbb{N},\ K \text{ finite}\}$. But these sets are dense in $\mathfrak{A}$, and we conclude that $\lambda|_{\mathfrak{A}}$ is constant. From

$$\lambda(C) = \lambda(A) + \lambda(B) = 2\lambda(C)$$

it follows that $\lambda(C) = 0$. Because $\{\mathbb{N} \setminus D \mid D \in \mathfrak{A}\}$ is also a dense G_δ-subset of $\mathfrak{P}(\mathbb{N})$, there is $E \in \mathfrak{A} \cap \{\mathbb{N} \setminus D \mid D \in \mathfrak{A}\}$. We have $E \in \mathfrak{A}$, $\mathbb{N} \setminus E \in \mathfrak{A}$, and it follows that

$$\lambda(\mathbb{N}) = \lambda(E) + \lambda(\mathbb{N} \setminus E) = 0.$$

Hence $\lambda = 0$, and $\mu = \nu$ is a positive measure.

Assertion (α) can be reformulated as follows. Denote by $\mathfrak{R}$ the σ-algebra of all approximable subsets of $\mathfrak{P}(\mathbb{N})$ and let $\mathfrak{B}$ be the set of all Borel subsets of $\mathbb{R}$. Consider now the positive additive function $\mu: \mathfrak{P}(\mathbb{N}) \to \mathbb{R}$, and prove the final assertion:

(β) If μ is a measurable map with respect to $(\mathfrak{P}(\mathbb{N}), \mathfrak{R})$ and $(\mathbb{R}, \mathfrak{B})$, then μ is a positive measure on $\mathfrak{P}(\mathbb{N})$.

In this way null continuity of μ can be described by a measurability property.

5.3.12(C) Boolean Algebras, Stone Theorem.

(a) Let X be a set and $\mathfrak{R}$ an algebra on X. Prove the following propositions about $\mathfrak{R}$ with regard to the order relation $\subset$:

(α) $\mathfrak{R}$ is a lattice.

(β) Given $A, B, C \in \mathfrak{R}$, $A \cap (B \cup C) = (A \cap B) \cup (A \cap C)$, and $A \cup (B \cap C) = (A \cup B) \cap (A \cup C)$.

(γ) X is the greatest element in $\mathfrak{R}$ and $\varnothing$ the least.

(δ) For each $A \in \mathfrak{R}$ there is an $A' \in \mathfrak{R}$ with $A \cap A' = \varnothing$ and $A \cup A' = X$.

The concept of a Boolean algebra is abstracted from these properties and is of great importance in different branches of mathematics.

A **Boolean algebra** is a lattice, X, with the following properties:

(i) For $x, y, z \in X$, $x \wedge (y \vee z) = (x \wedge y) \vee (x \wedge z)$ and $x \vee (y \wedge z) = (x \vee y) \wedge (x \vee z)$.

(ii) X has a greatest element, denoted by 1, and a least element, denoted by 0.

(iii) For each $x \in X$ there is an $x' \in X$ such that $x \vee x' = 1$ and $x \wedge x' = 0$.

The Boolean algebra X is **σ-complete** (resp. **complete**) iff X is σ-complete (resp. complete) as a lattice. A lattice that satisfies (i) is called **distributive**.

(ε) Prove that for the lattice X the following are equivalent:

(ε1) X is distributive.
(ε2) $x \wedge (y \vee z) = (x \wedge y) \vee (x \wedge z)$ for all $x, y, z \in X$.
(ε3) $x \vee (y \wedge z) = (x \vee y) \wedge (x \vee z)$ for all $x, y, z \in X$.

(ζ) Let X be a Boolean algebra. Suppose that $x, y, z \in X$ satisfy $x \vee y = x \vee z = 1$ and $x \wedge y = x \wedge z = 0$. Prove that $y = z$.

(ζ) shows that the element x' posited in (iii) is uniquely determined by x. x' is called the **complement** of x, and is denoted by $x^{\perp}$. Let X be a Boolean algebra. Prove the following:

(η) $x^{\perp\perp} = x$ for each $x \in X$.
(ϑ) $x^{\perp} \leq y^{\perp}$ for each $x, y \in X$ with $y \leq x$.
(ι) If $(x_\iota)_{\iota \in I}$ is a family in X, then $\bigvee_{\iota \in I} x_\iota$ exists iff $\bigwedge_{\iota \in I} x_\iota^{\perp}$ exists. In that case $(\bigvee_{\iota \in I} x_\iota)^{\perp} = (\bigwedge_{\iota \in I} x_\iota^{\perp})$.
(κ) The following are equivalent:

(κ1) X is (σ-) complete.
(κ2) For every (countable) family $(x_\iota)_{\iota \in I}$ in X, $\bigvee_{\iota \in I} x_\iota$ exists.
(κ3) For every (countable) family $(x_\iota)_{\iota \in I}$ in X, $\bigwedge_{\iota \in I} x_\iota$ exists.

(b) We investigate the relation between Boolean algebras and algebras of subsets of a set X. From the introduction to the definition of Boolean algebras, we have the following statement:

(α) If X is a set and $\mathfrak{R}$ an algebra on X, then $\mathfrak{R}$ is a Boolean algebra with respect to $\subset$.

The converse asserts that every Boolean algebra can be regarded as (i.e., is isomorphic to) an algebra of sets. This is a deep theorem, due to M. H. Stone (1937). We turn to an account of the considerations that lead to the conclusion of the theorem.

Let X be a Boolean algebra. Let $\mathfrak{A}$ be the set of all subsets A of X with the following properties:

(i) $0 \in A$ and $1 \notin A$.
(ii) If $x \in A$ and $y \in X$ with $y \leq x$, then $y \in A$.
(iii) If $x, y \in A$, then $x \vee y \in A$.

Prove the next statement:

(β) If A is a maximal element of $\mathfrak{A}$ with respect to $\subset$, then $x \wedge y \in X \setminus A$ whenever $x, y \in X \setminus A$.

To prove (β), take $x \in X \setminus A$, and set $B := \{y \in X \mid \exists z \in A,\ y \leq x^{\perp} \vee z\}$. Then $B \in \mathfrak{A}$ since clearly $0 \in B$ and B trivially satisfies (ii). If $1 \in B$, then $\exists z \in A$ with $x^{\perp} \vee z = 1$. Thus $x \leq z$ so that $x \in A$, contradicting the choice of x. It is left to the reader to show that B satisfies (iii). It follows that $x^{\perp} \in A$ for each $x \in X \setminus A$. Furthermore, if $x, y \in X \setminus A$, then $(x \wedge y)^{\perp} = x^{\perp} \vee y^{\perp} \in A$ so that $x \wedge y \in X \setminus A$.

Let X and Y be Boolean algebras. The mapping φ: $X \to Y$ is a **homomorphism of Boolean algebras** iff φ is a lattice homomorphism [i.e., $\varphi(x \vee y) = \varphi(x) \vee \varphi(y)$ and $\varphi(x \wedge y) = \varphi(x) \wedge \varphi(y)$ for $x, y \in X$], and $\varphi(x^{\perp}) = (\varphi(x))^{\perp}$ for each $x \in X$. φ is an **isomorphism of Boolean algebras** iff φ is bijective and both φ and φ^{-1} are homomorphisms of Boolean algebras.

(γ) Let X and Y be Boolean algebras, and φ: $X \to Y$ a mapping. Then the following are equivalent:

(γ1) φ is a homomorphism of Boolean algebras.
(γ2) Given $x, y \in X$, $\varphi(x \vee y) = \varphi(x) \vee \varphi(y)$ and $\varphi(x^{\perp}) = (\varphi(x))^{\perp}$.
(γ3) Given $x, y \in X$, $\varphi(x \wedge y) = \varphi(x) \wedge \varphi(y)$ and $\varphi(x^{\perp}) = (\varphi(x))^{\perp}$.

(δ) If φ: $X \to Y$ is a homomorphism of Boolean algebras, then $\varphi(1) = 1$ and $\varphi(0) = 0$.

(ε) A homomorphism of Boolean algebras is an isomorphism of Boolean algebras iff it is bijective.

We are now in a position to formulate the **Stone Representation Theorem**.

(ζ) Let X be a Boolean algebra. Let $\mathfrak{A}$ denote the set of all subsets, A, of X with the following properties:

(i) $0 \in A$, $1 \notin A$.
(ii) If $x \in A$ then $\{y \in X \mid y \leq x\} \subset A$.
(iii) If $x, y \in A$, then $x \vee y \in A$.
(iv) If $x, y \in X \setminus A$, then $x \wedge y \in X \setminus A$.

Define φ: $X \to \mathfrak{P}(\mathfrak{A})$, $x \mapsto \{A \in \mathfrak{A} \mid x \notin A\}$ and $\mathfrak{R} := \varphi(X)$. Then $\mathfrak{R}$ is an algebra of sets on $\mathfrak{A}$ and φ: $X \to \mathfrak{R}$ is an isomorphism of Boolean algebras.

The proof can be broken up into several steps:

STEP 1: $\varphi(x \vee y) = \varphi(x) \cup \varphi(y)$ for $x, y \in X$: Take $A \in \varphi(x \vee y)$. Then $x \vee y \notin A$ so that $x \notin A$ or $y \notin A$ [by (iii)]. Thus $A \in \varphi(x) \cup \varphi(y)$. Conversely, if $A \in \varphi(x) \cup \varphi(y)$, then without loss of generality, $x \notin A$; thus $x \vee y \notin A$ so that $A \in \varphi(x \vee y)$.

STEP 2: $\varphi(x^{\perp}) = \mathfrak{A} \setminus \varphi(x)$ for $x \in X$: Take $A \in \varphi(x^{\perp})$. Then $x^{\perp} \notin A$ so that $x \in A$ [using (iv) and (i)]. Thus $A \in \mathfrak{A} \setminus \varphi(x)$. Conversely, if $A \in \mathfrak{A} \setminus \varphi(x)$, then $A \notin \varphi(x)$ so that $x \in A$. Thus $x^{\perp} \notin A$, and so $A \in \varphi(x^{\perp})$.

STEP 3: $\mathfrak{R}$ is an algebra of sets on $\mathfrak{A}$, and φ is a homomorphism of Boolean algebras: Take $\mathfrak{B}, \mathfrak{B}' \in \mathfrak{R}$. Then $\mathfrak{B} = \varphi(x)$ and $\mathfrak{B}' = \varphi(y)$ for suitable x and $y \in X$. Then by Step 1, $\mathfrak{B} \cup \mathfrak{B}' = \varphi(x \vee y) \in \mathfrak{R}$. By Step 2, $\mathfrak{A} \setminus \mathfrak{B} = \varphi(x^{\perp}) \in \mathfrak{R}$. Finally, $\mathfrak{A} = \varphi(1) \in \mathfrak{R}$.

STEP 4: $\varphi(x) \neq \varnothing$ if $x \in X \setminus \{0\}$: Set $\mathfrak{S} := \{A \subset X \mid A$ satisfies (i), (ii), (iii)$\}$. Then $\{y \in X \mid y \leq x^{\perp}\} \in \mathfrak{S}$. By the Zorn Lemma, there is a maximal $B \in \mathfrak{S}$ such that $\{y \in X \mid y \leq x^{\perp}\} \subset B$. By (β), $B \in \mathfrak{A}$. Now $x^{\perp} \in B$ and $1 \notin B$. Thus $x \notin B$, and so $B \in \varphi(x)$.

STEP 5: φ is an isomorphism of Boolean algebras: It remains to show that φ is injective: Take $x, y \in X$ with $\varphi(x) = \varphi(y)$. By Step 3, $\varphi(x \wedge y^{\perp}) = \varphi(x) \setminus \varphi(y) = \varnothing$. Thus $x \wedge y^{\perp} = 0$. Now $y = y \vee 0 = y \vee (x \wedge y^{\perp}) = (y \vee x) \wedge (y \vee y^{\perp}) = y \vee x \geq x$. Similarly $y \leq x$. Thus $y = x$.

5.3.13(C) In this exercise we investigate an algebraic aspect of the notion of an algebra of sets. We assume that the reader is familiar with some fundamentals of the theory of rings.

We consider a ring E in the algebraic sense with identity 1 and with the property that $x^2 = x$ for every $x \in E$. Such rings are called **Boolean rings**. Prove the following statements:

(α) $x + x = 0$ for every $x \in E$.

(β) E is commutative.

(γ) Let F be a prime ideal of E. Then E/F is isomorphic to the field $\{0,1\}$.

(δ) An ideal of E is maximal iff it is a prime ideal.

(ε) Let F be a maximal ideal of E. Then for every $x \in E \setminus F$ and $y \in E \setminus F$, $x + y \in F$ and $x - y \in F$ hold.

For (α), note for every $x \in X$, that

$$x + x = (x + x)^2 = (x + x)(x + x) = x^2 + x^2 + x^2 + x^2$$
$$= (x + x) + (x + x).$$

For (β), take $x, y \in E$. Then, using (α), we get

$$x - y = (x - y)(x - y) = (x - y)(x + y) = x^2 + xy - yx - y^2$$
$$= (x - y) + (xy - yx).$$

For (γ), since F is a prime ideal, E/F has no zero divisor. For every $x \in E$ denote by $\dot{x}$ the equivalence class with respect to F, containing x. Then, for every $x \in E$,

$$\dot{0} = \dot{x}^2 - \dot{x} = \dot{x}(\dot{x} - \dot{1})$$

and we conclude, that $\dot{x} = \dot{0}$ or $\dot{x} = \dot{1}$.

For (δ), because E is a ring with identity, every maximal ideal of E is also a prime ideal. The reverse follows from (γ).

For (ε), let $x \in E \setminus F$ and $y \in E \setminus F$. From (γ) it follows that $\dot{x} = \dot{y}$. But then there is a $z \in F$ such that $x = y + z$, and we find $x - y = z \in F$.

The second assertion now follows from (α).

Important examples of Boolean rings arise from the notion of an algebra of sets. Let X be a set and $\mathfrak{R}$ an algebra of sets on X. Define

$$+ : \mathfrak{R} \times \mathfrak{R} \to \mathfrak{R}, \qquad (A, B) \mapsto A \triangle B$$

and

$$\cdot : \mathfrak{R} \times \mathfrak{R} \to \mathfrak{R}, \qquad (A, B) \mapsto A \cap B.$$

Prove the next statement:

(ζ) $\mathfrak{R}$ is a Boolean ring with respect to the addition $+$ and the multiplication $\cdot$.

We show now, that every Boolean ring is in essence of this kind.

Define X to be the set of all maximal ideals of E. For every $x \in E$ define

$$f(x) := \{ F \in X \,|\, x \notin F \}.$$

Prove the following statements:

(η) $f\colon E \to \mathfrak{P}(X)$ is injective.

(ϑ) $f(x+y) = f(x) \triangle f(y)$ whenever $x, y \in E$, and $f(xy) = f(x) \cap f(y)$ whenever $x, y \in E$.

(ι) $f(E)$ is an algebra of sets on X, and $E \to f(E)$, $x \mapsto f(x)$ is an isomorphism of rings.

For (η), take $x, y \in E$, $x \neq y$. Using the lemma of Zorn, we find a maximal ideal F of E such that $x + y \notin F$. Indeed, take F maximal under the additional conditions $x + y \notin F$ and $1 - (x + y) \in F$. Then F must be maximal, period. To see this, let G be a maximal ideal of E such that $F \subset G$. Because $1 \notin G$ and $1 - (x + y) \in G$, $x + y \notin G$. The maximality of F under these conditions then implies $F = G$.

Now, by (ε), we have either $x \in F$ and $y \notin F$ or $y \in F$ and $x \notin F$, and we conclude that $f(x) \neq f(y)$.

For (ϑ), take $x, y \in E$ and $F \in f(x + y)$. Then, by (ε), either $x \in F$ and $y \notin F$ or $y \in F$ and $x \notin F$, and this implies $F \in f(x) \triangle f(y)$. If, on the other hand, $F \in f(x) \triangle f(y)$, we conclude that either $x \in F$ and $y \notin F$ or $y \in F$ and $x \notin F$. But then, $x + y \notin F$.

The second assertion can be left to the reader.

Assertion (ι) follows from (η) and (ϑ).

5.3.14(C) Once again, we concern ourselves with the representation of algebras of sets. Let X be a set and $\mathfrak{R}$ an algebra of sets on X. We call $\mathfrak{R}$ **separating** iff, for every $x, y \in X$, $x \neq y$, there are sets $A, B \in \mathfrak{R}$ such that $x \in A$, $y \in B$ and $A \cap B = \varnothing$.

For every $x, y \in X$ define

$$x \sim y : \Leftrightarrow (x \in A \Leftrightarrow y \in A \text{ whenever } A \in \mathfrak{R}).$$

It is easy to show that $\sim$ is an equivalence relation on X. For every $x \in X$ denote by $\dot{x}$ the equivalence class containing x, and for every subset A of X define $\dot{A} := \{\dot{x} \mid x \in A\}$. Finally, define $\dot{\mathfrak{R}} := \{\dot{A} \mid A \in \mathfrak{R}\}$. Prove the statements that follow:

(α) $\dot{\mathfrak{R}}$ is a separating algebra of sets on $\dot{X}$.

(β) The map $\omega: \mathfrak{R} \to \dot{\mathfrak{R}}$, $A \mapsto \dot{A}$ is an isomorphism of algebras of sets.

Now let $\mathfrak{R}$ be a separating algebra of sets on X. Denote by $\mathfrak{T}$ the topology on X generated by $\mathfrak{R}$. Prove these statements:

(γ) Every set $A \in \mathfrak{R}$ is open and closed.

(δ) e_A is a continuous function whenever $A \in \mathfrak{R}$.

(ε) $(X, \mathfrak{T})$ is a completely regular Hausdorff space.

For every set $A \in \mathfrak{R}$ define

$$d_A : X \times X \to \mathbb{R}, \qquad (x, y) \mapsto |e_A(x) - e_A(y)|$$

and

$$U(A, \varepsilon) := \{(x, y) \in X \times Y \mid d_A(x, y) < \varepsilon\} \text{ for } \varepsilon \in \mathbb{R},\ \varepsilon > 0.$$

Finally, denote by $\mathfrak{U}$ the uniformity on X generated by $\{U(A, \varepsilon) \mid A \in \mathfrak{R}, \varepsilon \in \mathbb{R}, \varepsilon > 0\}$. Prove the next three statements:

(ζ) The uniformity $\mathfrak{U}$ is precompact.

(η) Let $\mathfrak{F}$ be a Cauchy filter with respect to $\mathfrak{U}$. Then for every $A \in \mathfrak{R}$, either $A \in \mathfrak{F}$ or $X \setminus A \in \mathfrak{F}$.

(ϑ) The topology on X generated by $\mathfrak{U}$ is identical with $\mathfrak{T}$.

Denote now by $(\overline{X}, \overline{\mathfrak{T}})$ the completion of $(X, \mathfrak{T})$ with respect to $\mathfrak{U}$. For every $A \in \mathfrak{R}$ let $\overline{A}$ be the closure of A with respect to $\overline{\mathfrak{T}}$. Define

$$\varphi: \mathfrak{R} \to \mathfrak{P}(\overline{X}), \qquad A \mapsto \overline{A}$$

and prove the following statements:

(ι) $\varphi(A \cup B) = \varphi(A) \cup \varphi(B)$ whenever $A, B \in \mathfrak{R}$.

(κ) $\varphi(A \cap B) = \varphi(A) \cap \varphi(B)$ whenever $A, B \in \mathfrak{R}$.

(λ) $\varphi(A \setminus B) = \varphi(A) \setminus \varphi(B)$ whenever $A, B \in \mathfrak{R}$.

(μ) $\varphi(\varnothing) = \varnothing$ and $\varphi(X) = \overline{X}$.

(ν) $(\overline{X}, \overline{\mathfrak{T}})$ is a compact Hausdorff space.

(ξ) $\varphi(A)$ is compact and open whenever $A \in \mathfrak{R}$.

(o) φ is an isomorphism of the algebra of sets $\mathfrak{R}$ onto the algebra of all compact-open subsets of $\overline{X}$.

Combining (o) with (ζ) of Ex. 5.3.10, we get the following result:

(π) Every Boolean algebra is isomorphic to the Boolean algebra of the compact-open sets of a compact space.

5.3.15(C) There are direct ways for constructing isomorphisms between Boolean algebras and systems of compact-open subsets of a compact Hausdorff space, as considered in Ex. 5.3.14 (π). We now describe such a way.

Let X be a Boolean algebra. A **filter** of X is a subset F of X with the following properties:

(i) $0 \notin F, 1 \in F$.

(ii) $x \in F,\ y \in X,\ x \leq y \Rightarrow y \in F$.

(iii) $x,\ y \in F \Rightarrow x \wedge y \in F$.

Maximal filters of X are called **ultrafilters** of X. Let X be a set and consider $\mathfrak{P}(X)$ as a Boolean algebra. Then the filters of $\mathfrak{P}(X)$ are exactly the filters $\mathfrak{F} \neq \mathfrak{P}(X)$ on X. Use this analogy to get the ideas for the proof of the following propositions. Let X be a Boolean algebra.

(α) For every filter F of X there is an ultrafilter G of X such that $F \subset G$.

(β) For every filter F of X, the following assertions are equivalent:

(β1) F is an ultrafilter of X.
(β2) If $x \in X$ such that $x \wedge y \neq 0$ for all $y \in F$, then $x \in F$.
(β3) If $(x_\iota)_{\iota \in I}$ is a finite family from X such that $\bigvee_{\iota \in I} x_\iota \in F$, then there is $\lambda \in I$ with $x_\lambda \in F$.
(β4) For every $x \in X$, either $x \in F$ or $x^\perp \in F$.

Let Y be the set of all ultrafilters of the Boolean algebra X. Define

$$f: X \to \mathfrak{P}(Y), \qquad x \mapsto \{F \in Y \mid x \in F\}$$

and

$$\mathfrak{R} := f(X).$$

Denote by $\mathfrak{T}$ the topology on Y induced by $\mathfrak{R}$. Prove the following:

(γ) $\mathfrak{R}$ is an algebra of sets on Y and the mapping f is an isomorphism of Boolean algebras.

(δ) Y is compact with respect to $\mathfrak{T}$, and $\mathfrak{R}$ is a base of $\mathfrak{T}$.

(ε) $\mathfrak{R}$ is the set of all subsets of Y that are both open and compact.

(ζ) For every family $(x_\iota)_{\iota \in I}$ from X the following assertions are equivalent.

(ζ1) $\bigvee_{\iota \in I} x_\iota$ exists.
(ζ2) $\overline{\bigcup_{\iota \in I} f(x_\iota)}$ is open.

If one of these conditions is fulfilled, then

$$f\Big(\bigvee_{\iota \in I} x_\iota\Big) = \overline{\bigcup_{\iota \in I} f(x_\iota)}.$$

(η) X is a complete lattice iff the closed hull of every open subset of Y is open.

(ϑ) Let Y' be a compact space. Let $\mathfrak{R}'$ be the set of all compact-open subsets of Y' and let f': $X \to \mathfrak{R}'$ be an isomorphism of Boolean algebras. Then if $\mathfrak{R}'$ is a base for the topology of Y', there is a homeomorphism φ: $Y \to Y'$ such that $f'(x) = \varphi(f(x))$ for all $x \in X$.

5.4. MEASURABILITY VERSUS INTEGRABILITY

In the preceding section, measurability relative to a δ-ring was defined and investigated. It was exactly the defining property of δ-rings that implied the important properties formulated in Theorem 5.3.4: measurability relative to a ring of sets would have been too weak. Given a positive measure space $(X, \mathfrak{R}, \mu)$, then, it won't be particularly useful to speak of measurability relative to $\mathfrak{R}$. Instead, we shall use either the δ-ring generated by $\mathfrak{R}$ or the completion of $(X, \mathfrak{R}, \mu)$ discussed in Section 5.2. That is, we shall use either $\mathfrak{L}(X, \mu)$-measurability (which we refer to as μ-measurability) or $\mathfrak{R}_\delta$-measurability.

The first part of the present section discusses μ-measurable objects and their relation to μ-integrable objects. Upper and lower μ-integrals, and inner and outer μ-measures, which are introduced in Definition 4, provide useful tools for describing the relationship between measurability and integrability.

The discussion of $\mathfrak{R}_\delta$-measurability includes various results concerning the possibility of approximating by $\mathfrak{R}_\delta$-measurable objects. The section concludes with the Egoroff Theorem: μ-a.e.-convergent sequences of μ-measurable μ-a.e. real functions are "nearly" uniformly convergent.

Proposition 5.4.1. *For every positive measure space $(X, \mathfrak{R}, \mu)$, X is a measurable space with δ-ring $\mathfrak{L}(X, \mu)$.*

Proof. The set rings for closed positive measure spaces are always δ-rings [Theorem 5.1.8(b)]. □

The terminology "$\mathfrak{L}(X, \mu)$-measurable" is too cumbersome.

Definition 5.4.2. *Let $(X, \mathfrak{R}, \mu)$ be a positive measure space. Then*

$$\mathfrak{M}(X, \mu) := \mathfrak{M}(X, \mathfrak{L}(X, \mu))$$

$$\mathcal{M}(X, \mu) := \mathcal{M}(X, \mathfrak{L}(X, \mu)).$$

Sets belonging to $\mathfrak{M}(X, \mu)$ are called **μ-measurable** *in X. Functions belonging to $\mathcal{M}(X, \mu)$ are called* **μ-measurable** *on X.* □

Proposition 5.4.3. *The following assertions hold, for every positive measure space* $(X, \mathfrak{R}, \mu)$.

(*a*) $\mathfrak{N}(X, \mu) \subset \mathfrak{L}(X, \mu) \subset \hat{\mathfrak{R}}(\mathscr{L}^1(X, \mu) \subset \mathfrak{M}(X, \mu)$.

(*b*) *If* $A \in \mathfrak{L}(X, \mu)$, $B \in \mathfrak{M}(X, \mu)$, *and* $B \subset A$, *then* B *belongs to* $\mathfrak{L}(X, \mu)$.

(*c*) *If* B *is a subset of* X *for which*

$$A \triangle B \in \mathfrak{N}(X, \mu)$$

for some $A \in \mathfrak{M}(X, \mu)$, *then* B *belongs to* $\mathfrak{M}(X, \mu)$.

(*d*) $\mathscr{N}(X, \mu) \subset \mathscr{L}^1(X, \mu) \subset \mathscr{M}(X, \mu)$.

(*e*) *If* g *is an extended-real function on* X *that is* μ-*a.e. equal to a* μ-*measurable function* f *on* X, *then* g *is* μ-*measurable.*

(*f*) *The conditions* $f, g \in \mathscr{M}(X, \mu)$, $h \in \overline{\mathbb{R}}^X$,

$$h(x) = f(x) + g(x) \ \mu\text{-a.e.}$$

imply that h *belongs to* $\mathscr{M}(X, \mu)$.

(*g*) *If* $f \in \mathscr{L}^1(X, \mu)_+$, *then the sets*

$$\{f > \alpha\} \qquad \{f \geq \alpha\}$$

belong to $\mathfrak{L}(X, \mu)$ *for every real number* $\alpha > 0$, *and the set*

$$\{\alpha \in]0, \infty[\mid \mu(\{f > \alpha\}) \neq \mu(\{f \geq \alpha\})\}$$

is countable.

Proof. (a) The first two inclusions are already known. According to Proposition 5.1.4(d), every set belonging to $\hat{\mathfrak{R}}(\mathscr{L}^1(X, \mu))$ is the union of a sequence from $\mathfrak{R}(\mathscr{L}^1(X, \mu))$, i.e. from $\mathfrak{L}(X, \mu)$ and therefore from $\mathfrak{M}(X, \mu)$. Since $\mathfrak{M}(X, \mu)$ is closed under the formation of countable unions (Proposition 5.3.2), the third inclusion also holds.

(b) Assertion (b) is evident.

(c) The hypothesis implies that both $B \setminus A$ and $A \setminus B$ are μ-measurable. Since

$$B = (A \setminus (A \setminus B)) \cup (B \setminus A)$$

B is also μ-measurable (Proposition 5.3.2).

(d) The first inclusion is already known. For the second, it suffices, since f is μ-measurable if f^+ and f^- are [Theorem 5.3.4(d)], to show that

$$\mathscr{L}^1(X, \mu)_+ \subset \mathscr{M}(X, \mu). \tag{1}$$

By Proposition 5.1.4(c), every function in $\mathscr{L}^1(X, \mu)_+$ is the supremum of a sequence of $\mathfrak{L}(X, \mu)$-step functions, which implies that (1) holds [Theorem 5.3.4(a), (b), (c), (k)].

(e) With f and g as hypothesized, set

$$A := \{f \neq g\}.$$

The sets A and $X \setminus A$ both belong to $\mathfrak{M}(X, \mu)$, and the function $fe_{X \setminus A}$ belongs to $\mathscr{M}(X, \mu)$ [Proposition 5.3.3(a), Theorem 5.3.4(f)]. Since ge_A is μ-null, it belongs to $\mathscr{L}^1(X, \mu)$ and therefore to $\mathscr{M}(X, \mu)$, by (d). The relation

$$g = fe_{X \setminus A} + ge_A$$

shows that g belongs to $\mathscr{M}(X, \mu)$ [Theorem 5.3.4(b)].

(f) Assertion (f) follows from (e) and Theorem 5.3.4(b).

(g) Let $f \in \mathscr{L}^1(X, \mu)_+$. For $\alpha > 0$ the sets $\{f > \alpha\}$ and $\{f \geq \alpha\}$ are μ-measurable subsets [Proposition 5.3.3(f)] of the μ-integrable set $\{(1/\alpha)f \neq 0\}$. Hence the first assertion follows from (b). Since summable families of real numbers can never have more than countably many nonzero members [Theorem 3.4.8(a) ⇒ (f)], we can prove the second assertion by showing, for each rational number $\beta > 0$, that the family

$$\big(\mu(\{f \geq \alpha\}) - \mu(\{f > \alpha\})\big)_{\alpha \in]\beta, \infty[}$$

is summable. Let $\beta > 0$. Let $(\alpha_n)_{n \in \mathbb{N}_m}$, for some m in $\mathbb{N}$, be a finite family from $]\beta, \infty[$. Assume, with (as we shall see) no loss of generality, that the family $(\alpha_n)_{n \in \mathbb{N}_m}$ is strictly increasing. Setting $\alpha_0 := \beta$, we have

$$\begin{aligned}\sum_{n=1}^{m} [\mu(\{f \geq \alpha_n\}) - \mu(\{f > \alpha_n\})] &\leq \sum_{n=1}^{m} [\mu(\{f > \alpha_{n-1}\}) - \mu(\{f > \alpha_n\})] \\ &= \mu(\{f > \beta\}) - \mu(\{f > \alpha_m\}) \\ &\leq \mu(\{f > \beta\}).\end{aligned}$$

Consequently

$$\begin{aligned}&\sup\Big\{\sum_{\alpha \in F} [\mu(\{f \geq \alpha\}) - \mu(\{f > \alpha\})] \,\Big|\, F \text{ is a finite subset of }]\beta, \infty[\Big\} \\ &\quad \leq \mu(\{f > \beta\}) < \infty.\end{aligned}$$

According to Theorem 3.4.8(c) ⇒ (a), we have established the desired summability and therefore the assertion. □

Having embedded μ-integrable objects in the larger collection of μ-measurable objects, we want to describe that embedding. In particular, we seek usable criteria for distinguishing which μ-measurable sets and functions are also μ-integrable. Some preparation, having to do with upper and lower closure and the related notions of outer and inner measure, is required.

Definition 5.4.4. *Let* $(X, \mathfrak{R}, \mu)$ *be a positive measure space. For* $f \in \overline{\mathbb{R}}^X$, *we define the* ***upper and lower μ-integral*** *of* f *by*

$$\int_X^* f\,d\mu := \left(\int_{X,\mu}\right)^*(f)$$

$$\int_{*X} f\,d\mu := \left(\int_{X,\mu}\right)_*(f)$$

respectively. For $A \subset X$, *we define the* ***outer and inner μ-measure*** *of* A *to be, respectively,*

$$(\mu^X)^*(A) := \int_X^* e_A\,d\mu$$

$$(\mu^X)_*(A) := \int_{*X} e_A\,d\mu. \qquad \square$$

The rules governing upper and lower integrals were already discussed in Section 3.3. It goes without saying that analogous rules hold for inner and outer measures. It will prove useful to display the most important of these results.

Proposition 5.4.5. *The following assertions hold, for every positive measure space* $(X, \mathfrak{R}, \mu)$.

(*a*) *For every subset* A *of* X,

$$0 \le \mu_*(A) \le \mu^*(A).$$

(*b*) *If* $A \subset B \subset X$, *then*

$$\mu_*(A) \le \mu_*(B)$$

$$\mu^*(A) \le \mu^*(B).$$

(*c*) *For every nonempty, countable, directed upward family* $(A_\iota)_{\iota \in I}$ *from* $\mathfrak{P}(X)$,

$$\mu^*\left(\bigcup_{\iota \in I} A_\iota\right) = \sup_{\iota \in I} \mu^*(A_\iota).$$

(*d*) *For every countable family* $(A_\iota)_{\iota \in I}$ *from* $\mathfrak{P}(X)$,

$$\mu^*\left(\bigcup_{\iota \in I} A_\iota\right) \le \sum_{\iota \in I}{}^* \mu^*(A_\iota).$$

(*e*) *For every disjoint family* $(A_\iota)_{\iota \in I}$ *from* $\mathfrak{P}(X)$,

$$\mu_*\left(\bigcup_{\iota \in I} A_\iota\right) \ge \sum_{\iota \in I}{}^* \mu_*(A_\iota).$$

(*f*) *A subset A of X belongs to* $\mathfrak{L}(X, \mu)$ *iff* $\mu_*(A)$ *and* $\mu^*(A)$ *are real and equal. If A belongs to* $\mathfrak{L}(X, \mu)$, *then*

$$\mu^X(A) = \mu_*(A) = \mu^*(A).$$

(*g*) $\mathfrak{N}(X, \mu) = \{A \subset X \mid \mu^*(A) = 0\}$.

Proof. For (a), (b), and (c), use Proposition 3.3.6, Propositions 3.3.7(d) and 3.3.3(a) and Theorem 3.3.4(a), respectively. Assertions (d) and (e) follow from Proposition 3.5.5. Assertions (f) and (g) follow from Proposition 3.3.10. □

Next we prove the important result that measurable sets have identical inner and outer measure.

Theorem 5.4.6. *For every positive measure space* $(X, \mathfrak{R}, \mu)$, *if A belongs to* $\mathfrak{M}(X, \mu)$, *then*

$$\mu_*(A) = \mu^*(A). \tag{2}$$

Proof. Since $\mu_*(A) \le \mu^*(A)$, there is nothing to prove if $\mu_*(A) = \infty$. Assume therefore that

$$\mu_*(A) < \infty.$$

In this case we claim that the μ-measurable set A must actually be μ-integrable and therefore satisfy (2). The argument that A belongs to $\mathfrak{L}(X, \mu)$ follows a familiar pattern. Using an extremal sequence, we construct a subset B of A that belongs to $\mathfrak{L}(X, \mu)$ and is somehow maximal. Then we show that the complementary set $A \setminus B$ is μ-exceptional and therefore μ-integrable.

Let

$$\alpha := \sup\{\mu^X(B') \mid B' \in \mathfrak{L}(X, \mu), B' \subset A\}$$

and note that α is real [Proposition 5(b), (f)]. Choose an increasing sequence

$(B_n)_{n \in \mathbb{N}}$ from $\mathfrak{L}(X, \mu)$ such that

$$B_n \subset A \qquad (n \in \mathbb{N})$$

$$\sup_{n \in \mathbb{N}} \mu^X(B_n) = \alpha.$$

Set

$$B := \bigcup_{n \in \mathbb{N}} B_n$$

$$C := A \setminus B.$$

Evidently $B \subset A$. According to Theorem 5.1.8(f), B belongs to $\mathfrak{L}(X, \mu)$ and $\mu^X(B) = \alpha$. Hence C belongs to $\mathfrak{M}(X, \mu)$. To show that C belongs to $\mathfrak{N}(X, \mu)$ we use Theorem 5.2.6(d). Let $D \in \mathfrak{L}(X, \mu)$. Then $C \cap D$ belongs to $\mathfrak{L}(X, \mu)$ and

$$\alpha = \mu^X(B) \leq \mu^X(B) + \mu^X(C \cap D) = \mu^X(B \cup (C \cap D)) \leq \alpha$$

so $\mu^X(C \cap D) = 0$ and $C \cap D$ belongs to $\mathfrak{N}(X, \mu)$. Since D was an arbitrary set in $\mathfrak{L}(X, \mu)$, we conclude that C belongs to $\mathfrak{N}(X, \mu)$ and therefore to $\mathfrak{L}(X, \mu)$. As was claimed, A belongs to $\mathfrak{L}(X, \mu)$ and (2) holds. □

Corollary 5.4.7. *For every positive measure space $(X, \mathfrak{R}, \mu)$, if $f \in \mathfrak{M}(X, \mu)_+$, then*

$$\int_{*X} f \, d\mu = \int_X^* f \, d\mu. \tag{3}$$

Proof. The validity of (3) follows immediately from Theorem 6 if f is the characteristic function of a μ-measurable set.

Suppose next that f is a positive $\mathfrak{M}(X, \mu)$-step function, say

$$f = \sum_{\iota \in I} \alpha_\iota e_{A_\iota}$$

for some finite family $(A_\iota)_{\iota \in I}$ from $\mathfrak{M}(X, \mu)$ and some family $(\alpha_\iota)_{\iota \in I}$ from $\mathbb{R}_+$. Using Proposition 3.5.5, we have

$$\int_X^* f \, d\mu \leq \sideset{}{^*}\sum_{\iota \in I} \alpha_\iota \int_X^* e_{A_\iota} \, d\mu = \sideset{}{^*}\sum_{\iota \in I} \alpha_\iota \int_{*X} e_{A_\iota} \, d\mu \leq \int_{*X} f \, d\mu.$$

Since the opposite inequality always holds, (3) holds.

Finally, let f be an arbitrary positive μ-measurable function on X. According to Theorem 5.3.6, there exists an increasing sequence $(f_n)_{n \in \mathbb{N}}$ of positive

$\mathfrak{M}(X, \mu)$-step functions whose supremum is f. Using Theorem 3.3.4(a) and Proposition 3.3.7(d), we have

$$\int_X^* f\, d\mu = \sup_{n \in \mathbb{N}} \int_X^* f_n\, d\mu = \sup_{n \in \mathbb{N}} \int_{*X} f_n\, d\mu \leq \int_{*X} f\, d\mu.$$

Once again, (3) holds. □

The characterization of μ-integrable objects, within the class of μ-measurable objects, takes the following form.

Theorem 5.4.8. *Let $(X, \mathfrak{R}, \mu)$ be a positive measure space. Then the following assertions hold, for every set $A \subset X$ and for every function $f \in \overline{\mathbb{R}}^X$.*

(*a*) $A \in \mathfrak{L}(X, \mu)$ *iff* $A \in \mathfrak{M}(X, \mu)$ *and* $\mu^*(A) < \infty$.

(*b*) $f \in \mathscr{L}^1(X, \mu)$ *iff* $f \in \mathscr{M}(X, \mu)$ *and* $\int^*_X |f|\, d\mu < \infty$.

Proof. (a) We have already established that μ-integrability implies μ-measurability [Proposition 3 (a), (d)]. The rest of (a) restates Proposition 5 (f) and Theorem 6.

(b) If f is μ-integrable, then so is $|f|$ and the characterization in Proposition 3.3.10 shows that

$$\int_X^* |f|\, d\mu < \infty. \tag{4}$$

Suppose, conversely, that f is μ-measurable and satisfies (4). Then both f^+ and f^- are μ-measurable [Theorem 5.3.4(d)] and

$$\int_X^* f^+\, d\mu < \infty$$

$$\int_X^* f^-\, d\mu < \infty$$

[Proposition 3.3.3(a)]. Combining Corollary 7 with Proposition 3.3.10, we conclude that f^+ and f^- belong to $\mathscr{L}^1(X, \mu)$. Hence f belongs to $\mathscr{L}^1(X, \mu)$. □

Corollary 5.4.9. *Let $(X, \mathfrak{R}, \mu)$ be a positive measure space, and denote by $(X, \mathscr{L}, \ell)$ its associated Daniell space. Then the following assertions hold.*

(*a*) *If $f \in \mathscr{L}^1(X, \mu)$, $g \in \mathscr{M}(X, \mu)$, and*

$$|g| \leq |f| \ \mu\text{-a.e.}$$

then g belongs to $\mathscr{L}^1(X, \mu)$.

(*b*) *If* $f \in \bar{\mathscr{L}}(\ell)$, $g \in \mathscr{M}(X, \mu)$, *and*

$$|g| \leq |f| \quad \bar{\mathscr{L}}(\ell)\text{-a.e.}$$

then g *belongs to* $\bar{\mathscr{L}}(\ell)$.

(*c*) *The product of a bounded* μ*-measurable function with a* μ*-integrable function is* μ*-integrable. In fact, if* f *belongs to* $\mathscr{M}(X, \mu)$ *and satisfies*

$$\alpha e_X \leq f \leq \beta e_X \ \mu\text{-a.e.}$$

with $\alpha, \beta \in \mathbb{R}$, *then* fg *belongs to* $\mathscr{L}^1(X, \mu)$ *for every* $g \in \mathscr{L}^1(X, \mu)$ *and satisfies*

$$\alpha \int_X g \, d\mu \leq \int_X fg \, d\mu \leq \beta \int_X g \, d\mu \tag{5}$$

for every $g \in \mathscr{L}^1(X, \mu)_+$.

Proof. (a) The hypotheses imply that g is μ-measurable and has finite upper integral, so (a) follows from Theorem 8(b).

(b) By (a), g belongs to $\mathscr{L}^1(X, \mu)$. The set $\{f \neq 0\}$ belongs to $\hat{\mathfrak{R}}(\bar{\mathscr{L}}(\ell))$. By Theorem 5.2.6 (e), $ge_{\{f \neq 0\}}$ belongs to $\bar{\mathscr{L}}(\ell)$. But $|g| \leq |f|$ $\bar{\mathscr{L}}(\ell)$-a.e. implies that $g = ge_{\{f \neq 0\}}$ $\bar{\mathscr{L}}(\ell)$-a.e., so g belongs to $\bar{\mathscr{L}}(\ell)$.

(c) Suppose that $f \in \mathscr{M}(X, \mu)$, $g \in \mathscr{L}^1(X, \mu)$, and $\alpha \leq f \leq \beta$ μ-a.e. As a product of μ-measurable functions, fg is μ-measurable. Since

$$|fg| \leq (|\alpha| + |\beta|)|g| \ \mu\text{-a.e.}$$

it follows from (a) that fg is μ-integrable. The validity of (5) in case g is positive is a trivial consequence of μ-integral monotonicity. □

We present some equivalent characterizations of μ-measurable objects.

Proposition 5.4.10. *Let* $(X, \mathfrak{R}, \mu)$ *be a positive measure space. Then the following assertions are equivalent, for every subset* A *of* X.

(*a*) $A \in \mathfrak{M}(X, \mu)$.

(*b*) $fe_A \in \mathscr{M}(X, \mu)$ *for every* $f \in \mathscr{M}(X, \mu)$.

(*c*) $A \cap B \in \mathfrak{L}(X, \mu)$ *for every* $B \in \mathfrak{R}$.

(*d*) $fe_A \in \mathscr{L}^1(X, \mu)$ *for every* $f \in \mathscr{L}^1(X, \mu)$.

Proof. (a) ⇒ (b). A is μ-measurable iff e_A is and the product of two μ-measurable functions on X is μ-measurable.

(b) ⇒ (c). Let $B \in \mathfrak{R}$. Then B belongs to $\mathfrak{L}(X, \mu)$, hence to $\mathfrak{M}(X, \mu)$, so the function e_B is μ-measurable. Since

$$e_{A \cap B} = e_A e_B$$

the function $e_{A \cap B}$ belongs to $\mathscr{M}(X, \mu)$ by hypothesis. Therefore $A \cap B$ belongs to $\mathfrak{M}(X, \mu)$. Since

$$A \cap B = (A \cap B) \cap B$$

we conclude that $A \cap B$ belongs to $\mathfrak{L}(X, \mu)$.

(c) $\Rightarrow$ (d). We use the Induction Principle [Theorems 5.2.6(b), 4.1.18]. Let

$$\mathscr{F} := \{ f \in \mathscr{L}^1(X, \mu) \mid f e_A \in \mathscr{L}^1(X, \mu) \}.$$

The identity

$$e_B e_A = e_{B \cap A}$$

shows that e_B belongs to $\mathscr{F}$ if B belongs to $\mathfrak{R}$. Consequently $\mathscr{F}$ contains all $\mathfrak{R}$-step functions on X. Let $(g_n)_{n \in \mathbb{N}}$ be a μ-sequence from $\mathscr{F}$ and g an extended-real function on X such that

$$\lim_{n \to \infty} g_n = g \ \mu\text{-a.e.}$$

Note that g must belong to $\mathscr{L}^1(X, \mu)$. Since

$$-(|g_1| + |g|) \leq g_n e_A \leq |g_1| + |g| \ \ \mu\text{-a.e.}$$

for every n, $(g_n e_A)_{n \in \mathbb{N}}$ is a μ-sequence from $\mathscr{L}^1(X, \mu)$. Moreover

$$\lim_{n \to \infty} g_n e_A = g e_A \ \mu\text{-a.e.}$$

We conclude that $g e_A$ belongs to $\mathscr{L}^1(X, \mu)$. In other words, g belongs to $\mathscr{F}$. By the Induction Principle, $\mathscr{F} = \mathscr{L}^1(X, \mu)$.

(d) $\Rightarrow$ (a). If B belongs to $\mathfrak{L}(X, \mu)$, then e_B belongs to $\mathscr{L}^1(X, \mu)$, so $e_{A \cap B}$ belongs to $\mathscr{L}^1(X, \mu)$, so $A \cap B$ belongs to $\mathfrak{L}(X, \mu)$. Thus A is μ-measurable.

□

Proposition 5.4.11. *Let $(X, \mathfrak{R}, \mu)$ be a positive measure space. Then the following assertions are equivalent, for every function $f \in \overline{\mathbb{R}}_+^X$.*

(*a*) $f \in \mathscr{M}(X, \mu)$.

(*b*) $f e_A \in \mathscr{M}(X, \mu)$ *for every* $A \in \mathfrak{R}$.

(*c*) $f \wedge \alpha e_A \in \mathscr{L}^1(X, \mu)$ *for every* $A \in \mathfrak{R}$ *and every* $\alpha \in \mathbb{R}_+$.

(*d*) $f \wedge n e_A \in \mathscr{L}^1(X, \mu)$ *for every* $A \in \mathfrak{R}$ *and every* $n \in \mathbb{N}$.

Proof. (b) $\Rightarrow$ (a). Let $\alpha \in \mathbb{R}$. For every A in $\mathfrak{R}$, the hypothesis implies, since

$$\{f < \alpha\} \cap A = \{fe_A < \alpha\} \cap A$$

that $\{f < \alpha\} \cap A$ belongs to $\mathfrak{L}(X, \mu)$. According to Proposition 10(c) $\Rightarrow$ (a), $\{f < \alpha\}$ is μ-measurable. Since α was arbitrary, (a) follows.

(a) $\Rightarrow$ (c). For $A \in \mathfrak{R}$ and $\alpha \in \mathbb{R}_+$, the function αe_A is μ-measurable, and so is $f \wedge \alpha e_A$. Moreover

$$\int^* (f \wedge \alpha e_A)\, d\mu \leq \int^* \alpha e_A\, d\mu = \alpha\mu(A) < \infty.$$

According to the criterion in Theorem 8(b), $f \wedge \alpha e_A$ belongs to $\mathscr{L}^1(X, \mu)$.

(c) $\Rightarrow$ (d). This implication is trivial.

(d) $\Rightarrow$ (b). Let $A \in \mathfrak{R}$. Hypothesis (d) implies that $f \wedge ne_A$ is μ-measurable for every natural number n. Since

$$fe_A = \bigvee_{n \in \mathbb{N}} (f \wedge ne_A)$$

it follows that fe_A is μ-measurable. □

Suppose $A \subset X$ and $f \in \overline{\mathbb{R}}^X$. With part (d) of Proposition 10 available, the obvious definition of $\int_A f\, d\mu$ makes sense, as long as A is μ-measurable and f is μ-integrable on X. We give the relevant definition in more general form.

Definition 5.4.12. *Let $(X, \mathfrak{R}, \mu)$ be a positive measure space. For f an arbitrary extended-real function on X and A an arbitrary subset of X,*

$$\int_A^* f\, d\mu := \int_X^* fe_A\, d\mu$$

$$\int_{*A} f\, d\mu := \int_{*X} fe_A\, d\mu$$

and if $fe_A \in \mathscr{L}^1(X, \mu)$, then

$$\int_A f\, d\mu := \int_X fe_A\, d\mu.$$ □

The dependence on X of the sets $\mathfrak{M}(X, \mu)$ and $\mathscr{M}(X, \mu)$ is fundamentally the same as that of the sets $\mathscr{L}^1(X, \mu)$ and $\mathfrak{L}(X, \mu)$. We have the following analogue to Proposition 5.2.7(d), (e).

Proposition 5.4.13. *Let μ be a positive measure on a set-ring $\mathfrak{R}$, and let X, Y be sets such that*

$$X \supset Y \supset X(\mathfrak{R}).$$

Then the following assertions hold.

(*a*) $\mathfrak{M}(X, \mu) = \{A \subset X \mid A \cap Y \in \mathfrak{M}(Y, \mu)\}$.
(*b*) $\mathcal{M}(X, \mu) = \{f \in \overline{\mathbb{R}}^X \mid f\mid_Y \in \mathcal{M}(Y, \mu)\}$.

Proof. (a) Using Proposition 10(a) $\Leftrightarrow$ (c) and Proposition 5.2.7(d), we have

$$\begin{aligned}\mathfrak{M}(X, \mu) &= \{A \subset X \mid A \cap B \in \mathfrak{L}(X, \mu), \forall B \in \mathfrak{R}\} \\ &= \{A \subset X \mid (A \cap B) \cap Y \in \mathfrak{L}(Y, \mu), \forall B \in \mathfrak{R}\} \\ &= \{A \subset X \mid A \cap Y \in \mathfrak{M}(Y, \mu)\}.\end{aligned}$$

(b) Since

$$\{f\mid_Y < \alpha\} = \{f < \alpha\} \cap Y$$

for every f in $\overline{\mathbb{R}}^X$ and every real number α, (b) follows from (a). □

Also of interest is the question of when two positive measure spaces generate the same measurable objects.

Proposition 5.4.14. *Let $(X, \mathfrak{R}, \mu)$ and $(X, \mathfrak{S}, \nu)$ be positive measure spaces. If $(X, \mathcal{L}^1(X, \nu), \int_{X,\nu})$ is an admissible extension of the Daniell space associated with $(X, \mathfrak{R}, \mu)$, then*

$$\mathfrak{M}(X, \mu) \subset \mathfrak{M}(X, \nu)$$

and

$$\mathcal{M}(X, \mu) \subset \mathcal{M}(X, \nu).$$

Proof. Let $A \in \mathfrak{M}(X, \mu)$. Thus $A \cap B$ belongs to $\mathfrak{L}(X, \mu)$ whenever B does. We want to show that $A \cap C$ belongs to $\mathfrak{L}(X, \nu)$ whenever C does, so let $C \in \mathfrak{L}(X, \nu)$. Then e_C belongs to $\mathcal{L}^1(X, \nu)$, and the admissibility hypothesis ensures the existence of a set B, belonging to $\mathfrak{L}(X, \mu)$, such that

$$C \triangle B \in \mathfrak{N}(X, \nu).$$

Since subsets of ν-null sets are ν-null and therefore ν-integrable, and since

$$A \cap C = [(A \cap B) \setminus (A \cap (B \setminus C))] \cup [A \cap (C \setminus B)]$$

we conclude that $A \cap C$ belongs to the ring of sets $\mathfrak{L}(X, \nu)$. Now A was arbitrary, so the first inclusion holds. The second inclusion follows from the first. □

For additional criteria the reader is referred to Proposition 5.2.8 and Corollaries 5.2.10, 5.2.12.

One aim of the discussion begun in Section 5.3 was to expand the collections of μ-integrable sets and functions in order to obtain better properties. With μ-measurable objects, upper and lower μ-integrals, and inner and outer μ-measures, that objective is achieved. Theorems 5.3.4 and 5.4.8 enunciate the desired properties.

The story continues, however: formulating measurability in terms of the measurable space $(X, \mathfrak{L}(X, \mu))$ is not always satisfactory. Difficulties arise, in the first place, when we work with different measures. Beyond that, one often needs relationships to the set-ring used to determine measurability, and in this respect $\mathfrak{L}(X, \mu)$ is too large. In these two respects, measurability relative to the measurable space $(X, \mathfrak{R}_\delta)$ is sometimes very useful. Accordingly, we investigate $\mathfrak{R}_\delta$-measurability yet, studying in particular its relation to μ-measurability.

Proposition 5.4.15. *The following assertions hold, for every positive measure space* $(X, \mathfrak{R}, \mu)$.

(*a*) $\mathfrak{M}(X, \mathfrak{R}_\delta) \subset \mathfrak{M}(X, \mu)$.

(*b*) $\mathcal{M}(X, \mathfrak{R}_\delta) \subset \mathcal{M}(X, \mu)$.

Proof. (a) Since $\mathfrak{R} \subset \mathfrak{R}_\delta \subset \mathfrak{L}(X, \mu)$, (a) follows from the characterization

$$\mathfrak{M}(X, \mu) = \{A \subset X \mid A \cap B \in \mathfrak{L}(X, \mu), \forall B \in \mathfrak{R}\}$$

established in Proposition 10.

(b) Assertion (b) is a consequence of (a). □

We investigate the extent to which various kinds of functions from $\overline{\mathbb{R}}^X$ can be approximated by $\mathfrak{R}_\delta$-measurable functions.

Theorem 5.4.16. *Let* $(X, \mathfrak{R}, \mu)$ *be a positive measure space, and denote by* $(X, \mathcal{L}, \ell)$ *its associated Daniell space. Then the following assertions hold.*

(*a*) *For each* $f \in \bar{\mathcal{L}}(\ell)$ *there exist*

$$g, h \in \mathcal{M}(X, \mathfrak{R}_\delta) \cap \bar{\mathcal{L}}(\ell)$$

such that

$$g \le f \le h$$

$$g = h \ \mu\text{-}a.e.$$

(*b*) *For each* $f \in \mathscr{L}^1(X, \mu)$ *there exists*

$$g \in \mathscr{M}(X, \mathfrak{R}_\delta) \cap \bar{\mathscr{L}}(\ell) \cap \mathbb{R}^X$$

such that

$$f = g \ \mu\text{-}a.e.$$

(*c*) *For each* $f \in \overline{\mathbb{R}}^X$ *there exists* $g \in \mathscr{M}(X, \mathfrak{R}_\delta)$ *such that*

$$g \leq f \ \mu\text{-}a.e.$$

$$\int_{*X} f \, d\mu = \int_{*X} g \, d\mu.$$

If $\int_{*X} f \, d\mu$ *is real, then g can be chosen from*

$$\mathscr{M}(X, \mathfrak{R}_\delta) \cap \bar{\mathscr{L}}(\ell) \cap \mathbb{R}^X.$$

If there exists a countable family from $\mathfrak{R}$ *whose union contains* $\{f < 0\}$, *then g can be chosen so that* $g \leq f$.

(*d*) *For each* $f \in \overline{\mathbb{R}}^X$ *there exists* $h \in \mathscr{M}(X, \mathfrak{R}_\delta)$ *such that*

$$f \leq h \ \mu\text{-}a.e.$$

$$\int_X^* f \, d\mu = \int_X^* h \, d\mu.$$

If $\int_X^* f \, d\mu$ *is real, then h can be chosen from*

$$\mathscr{M}(X, \mathfrak{R}_\delta) \cap \bar{\mathscr{L}}(\ell) \cap \mathbb{R}^X.$$

If there exists a countable family from $\mathfrak{R}$ *whose union contains* $\{f > 0\}$, *then h can be chosen so that* $f \leq h$.

Proof. (a) Let $f \in \bar{\mathscr{L}}(\ell)$. According to Proposition 3.1.16(a) ⇒ (c), there exist an increasing sequence $(g_n)_{n \in \mathbb{N}}$ from $\mathscr{L}^\downarrow$ and a decreasing sequence $(h_n)_{n \in \mathbb{N}}$ from $\mathscr{L}^\uparrow$ such that the sequences $(\ell^\downarrow(g_n))_{n \in \mathbb{N}}$ and $(\ell^\uparrow(h_n))_{n \in \mathbb{N}}$ are real,

$$\bigvee_{n \in \mathbb{N}} g_n \leq f \leq \bigwedge_{n \in \mathbb{N}} h_n$$

and

$$\sup_{n \in \mathbb{N}} \ell^\downarrow(g_n) = \inf_{n \in \mathbb{N}} \ell^\uparrow(h_n) = \bar{\ell}(f).$$

Set

$$g := \bigvee_{n \in \mathbb{N}} g_n, \qquad h := \bigwedge_{n \in \mathbb{N}} h_n.$$

Functions belonging to $\mathscr{L}$, that is to say $\mathfrak{R}$-step functions on X, are $\mathfrak{R}_\delta$-measurable (after all they are also $\mathfrak{R}_\delta$-step functions on X). Moreover infima and suprema of sequences of $\mathfrak{R}_\delta$-measurable functions are still $\mathfrak{R}_\delta$-measurable. Thus g and h must belong to $\mathscr{M}(X, \mathfrak{R}_\delta)$. According to Proposition 3.2.1, g and h also belong to $\bar{\mathscr{L}}(\ell)$ and

$$\bar{\ell}(g) = \bar{\ell}(f) = \bar{\ell}(h).$$

Obviously $g \le f \le h$, so $g = h$ μ-a.e. (Proposition 1.4.6).

(b) Let $f \in \mathscr{L}^1(X, \mu)$. According to Theorem 5.2.6(e), there exist disjoint sets

$$A \in \hat{\mathfrak{R}}(\bar{\mathscr{L}}(\ell)), \qquad B \in \mathfrak{N}(X, \mu)$$

so that

$$f = fe_A + fe_B, \qquad fe_A \in \bar{\mathscr{L}}(\ell).$$

By (a), there exists

$$g' \in \mathscr{M}(X, \mathfrak{R}_\delta) \cap \bar{\mathscr{L}}(\ell)$$

with

$$g' = fe_A \ \mu\text{-a.e.}$$

Define

$$g: X \to \mathbb{R}, \qquad x \to \begin{cases} g'(x) & \text{if } g'(x) \in \mathbb{R} \\ 0 & \text{if } g'(x) \notin \mathbb{R}. \end{cases}$$

The function g is real valued and $\mathfrak{R}_\delta$-measurable [Proposition 5.3.3(e)]. Moreover $g = g'$ $\bar{\mathscr{L}}(\ell)$-a.e., so

$$g = f \ \mu\text{-a.e.}$$

and g belongs to $\bar{\mathscr{L}}(\ell)$.

(c) Let $f \in \overline{\mathbb{R}}^X$. If $\int_{*X} f\,d\mu = -\infty$, the constant function

$$g := -\infty e_X$$

satisfies $g \le f$, $\int_{*X} g\,d\mu = \int_{*X} f\,d\mu$ [Proposition 3.3.7(d)] and is $\mathfrak{R}_\delta$-measurable [Proposition 5.3.3(b)]. Assume therefore that

$$\int_{*X} f\,d\mu > -\infty. \tag{6}$$

Set

$$\mathscr{H} := \left\{ h \in \mathscr{L}^1(X,\mu)^{\downarrow} \mid h \le f \right\}.$$

By (6), $\mathscr{H}$ is nonempty. Choose an increasing sequence $(h_n)_{n\in\mathbb{N}}$ from $\mathscr{H}$ such that $h_n \le f$ for every n, and

$$\sup_{n\in\mathbb{N}} \int_{*X} h_n\,d\mu = \int_{*X} f\,d\mu.$$

[Note that $\int_{*X} h_n\,d\mu = \int_{X,\mu}^{\downarrow}(h_n)$, by Proposition 3.3.7(b).] Assume, without loss of generality, that $\int_{*X} h_n\,d\mu$ is real for every n. Then each h_n must belong to $\mathscr{L}^1(X,\mu)$ [Proposition 3.3.7(c)]. According to (b), there exists a sequence $(g_n)_{n\in\mathbb{N}}$ from $\mathscr{M}(X,\Re_\delta)\cap\bar{\mathscr{L}}(\ell)\cap\mathbb{R}^X$ such that

$$g_n = h_n \ \mu\text{-a.e.}$$

If $\int_{*X} f\,d\mu = \infty$, we can take

$$g := \bigvee_{n\in\mathbb{N}} g_n$$

[Theorem 5.3.4(c), Proposition 3.3.7(e)]. Otherwise, $\int_{*X} f\,d\mu$ is real, and we have

$$\sup_{n\in\mathbb{N}} \bar{\ell}(g_n) \le \int_{*X} f\,d\mu < \infty.$$

In this case Theorem 1.4.12(a) implies that the function $\bigvee_{n\in\mathbb{N}} g_n$ belongs to $\bar{\mathscr{L}}(\ell)$ and therefore to $\mathscr{L}^1(X,\mu)$. Applying (b) to $\bigvee_{n\in\mathbb{N}} g_n$, we obtain the required function g.

Now assume that $\{f<0\}\subset\bigcup_{\iota\in I}A_\iota$, for some countable family $(A_\iota)_{\iota\in I}$ from $\Re$. For each $n\in\mathbb{N}$ choose a sequence $(B_k)_{k\in\mathbb{N}}$ from $\Re$ such that $\{h_n\ne 0\}\setminus\bigcup_{k\in\mathbb{N}}B_k$ is μ-exceptional. Set

$$C_n := \left(\bigcup_{\iota\in I} A_\iota\right)\cup\left(\bigcup_{k\in\mathbb{N}} B_k\right)$$

for each n. By Proposition 4.1.15(b), each C_n belongs to $\hat{\Re}(\bar{\mathscr{L}}(\ell))$. We have $h_n e_{C_n} = h_n$ μ-a.e., $h_n e_{C_n} \le f$, and, by Theorem 5.2.6(e), $h_n e_{C_n}$ belongs to $\bar{\mathscr{L}}(\ell)$. By (a), there exists for each n a function g_n' belonging to

$\mathcal{M}(X, \mathfrak{R}_\delta) \cap \bar{\mathcal{L}}(\ell)$ such that $g'_n \leq h_n e_{C_n}$ and $g'_n = h_n e_{C_n}$ μ-a.e. The function $g := \bigvee_{n \in \mathbb{N}} g'_n$ possesses the required properties.

(d) The argument for (d) is analogous to that for (c). □

Using Theorem 16, we can describe the connection between the closure $(X, \mathfrak{R}(\mu), \bar{\mu})$ of a positive measure space $(X, \mathfrak{R}, \mu)$ and the inner measure μ_* associated with μ. Note that $\mathfrak{R} \subset \mathfrak{R}_\delta \subset \mathfrak{R}(\mu)$.

Corollary 5.4.17. *For every positive measure space $(X, \mathfrak{R}, \mu)$, and for every subset A of X,*

$$\mu_*(A) = \sup\{\bar{\mu}(B) \mid B \in \mathfrak{R}_\delta, B \subset A\}.$$

Proof. Denote by $(X, \mathcal{L}, \ell)$ the Daniell space associated with the positive measure space $(X, \mathfrak{R}, \mu)$. Let $A \subset X$. Theorem 16 justifies the key last step in the following computation:

$$\mu_*(A) = \int_{*X} e_A \, d\mu = \sup\left\{ \int_{X,\mu}^{\downarrow} (g) \mid g \in \mathcal{L}^1(X,\mu)^{\downarrow}, g \leq e_A \right\}$$

$$= \sup\left\{ \int_{X,\mu}^{\downarrow} (g) \mid g \in \left(\mathcal{L}^1(X,\mu)^{\downarrow}\right)_+, g \leq e_A \right\}$$

$$= \sup\left\{ \int_X g \, d\mu \mid g \in \mathcal{L}^1(X,\mu)_+, g \leq e_A \right\}$$

$$= \sup\{\bar{\ell}(g) \mid g \in \bar{\mathcal{L}}(\ell)_+, g \leq e_A\}$$

$$= \sup\{\bar{\ell}(g) \mid g \in \mathcal{M}(X, \mathfrak{R}_\delta) \cap \bar{\mathcal{L}}(\ell)_+, g \leq e_A\}$$

[Proposition 3.3.7(c), (d), Theorem 5.2.6(e), Theorem 16(a), (b)]. By the formula of Theorem 5.2.6(i), if g belongs to $\bar{\mathcal{L}}(\ell)_+$ and satisfies $g \leq e_A$, then

$$\bar{\ell}(g) = \sup_{B \in \mathfrak{R}} \bar{\ell}(g e_B) \leq \sup_{B \in \mathfrak{R}} \bar{\ell}(e_{B \cap \{g \neq 0\}}) = \sup_{B \in \mathfrak{R}} \bar{\mu}(B \cap \{g \neq 0\}).$$

If g belongs to $\mathcal{M}(X, \mathfrak{R}_\delta)$, then $B \cap \{g \neq 0\}$ belongs to $\mathfrak{R}_\delta$ for every B in $\mathfrak{R}$. We conclude that

$$\mu_*(A) \leq \sup\{\bar{\mu}(B) \mid B \in \mathfrak{R}_\delta, B \subset A\} \leq \mu_*(A).$$

□

Approximation by $\mathfrak{R}_\delta$-measurable sets closely parallels the approximation by $\mathfrak{R}_\delta$-measurable functions described in Theorem 16.

Corollary 5.4.18. *Let $(X, \mathfrak{R}, \mu)$ be a positive measure space, and denote by $(X, \mathscr{L}, \ell)$ its associated Daniell space. Then the following assertions hold.*

(*a*) *For each $A \in \mathfrak{R}(\mu)$ there exist sets*

$$B, C \in \mathfrak{M}(X, \mathfrak{R}_\delta) \cap \mathfrak{R}(\mu)$$

such that

$$B \subset A \subset C, \quad C \setminus B \in \mathfrak{N}(X, \mu).$$

(*b*) *For each $A \in \mathfrak{L}(X, \mu)$ there exists a set*

$$B \in \mathfrak{M}(X, \mathfrak{R}_\delta) \cap \mathfrak{R}(\mu)$$

such that

$$B \subset A, \quad A \setminus B \in \mathfrak{N}(X, \mu).$$

(*c*) *For each $A \in \mathfrak{P}(X)$ there exists a set $B \in \mathfrak{M}(X, \mathfrak{R}_\delta)$ such that*

$$B \subset A, \qquad \mu_*(A) = \mu_*(B).$$

If $\mu_*(A)$ is finite, then B can be chosen from

$$\mathfrak{M}(X, \mathfrak{R}_\delta) \cap \mathfrak{R}(\mu).$$

(*d*) *For each $A \in \mathfrak{P}(X)$ there exists $C \in \mathfrak{M}(X, \mathfrak{R}_\delta)$ such that*

$$A \setminus C \in \mathfrak{N}(X, \mu)$$

$$\mu^*(A) = \mu^*(C).$$

If $\mu^*(A)$ is finite, then C can be chosen from

$$\mathfrak{M}(X, \mathfrak{R}_\delta) \cap \mathfrak{R}(\mu).$$

If there exists a countable family from $\mathfrak{R}$ whose union contains A, then C can be chosen to contain A.

Proof. (a) Let $A \in \mathfrak{R}(\mu)$. For A to belong to $\mathfrak{R}(\mu)$ means that e_A belongs to $\bar{\mathscr{L}}(\ell)$. By Theorem 16(a) there exist functions g, h such that

$$g, h \in \mathscr{M}(X, \mathfrak{R}_\delta) \cap \bar{\mathscr{L}}(\ell)$$

$$g \leq e_A \leq h, \qquad g = h \ \mu\text{-a.e.}$$

Let

$$B := \{g > 0\} \qquad C := \{h \geq 1\}.$$

The sets B and C belong to $\mathfrak{M}(X, \mathfrak{R}_\delta)$ and $B \subset A \subset C$. Since $C \setminus B$ is contained in $\{g \neq h\}$, $C \setminus B$ is a μ-null set. To see that B and C belong to $\mathfrak{R}(\mu)$, we can argue as follows. Notice that

$$g = e_A = h \ \mu\text{-a.e.}$$

All three of these functions belong to $\bar{\mathscr{L}}(\ell)$, so

$$g = e_A = h \ \bar{\mathscr{L}}(\ell)\text{-a.e.}$$

(Proposition 4.1.4). It follows that

$$e_B = e_A = e_C \ \bar{\mathscr{L}}(\ell)\text{-a.e.}$$

Thus e_B and e_C are each $\bar{\mathscr{L}}(\ell)$-a.e. equal to a function in $\bar{\mathscr{L}}(\ell)$. Therefore e_B and e_C belong to $\bar{\mathscr{L}}(\ell)$ (Proposition 1.2.30). In other words, B and C belong to $\mathfrak{R}(\mu)$.

(b) Let $A \in \mathfrak{L}(X, \mu)$. According to Theorem 5.2.6(f), there exist sets C, D such that

$$A = C \cup D \qquad C \in \mathfrak{R}(\mu) \qquad D \in \mathfrak{N}(X, \mu).$$

In view of (a) there exists a set B such that

$$B \in \mathfrak{M}(X, \mathfrak{R}_\delta) \cap \mathfrak{R}(\mu)$$

$$B \subset C \qquad C \setminus B \in \mathfrak{N}(X, \mu).$$

Since

$$A \setminus B = (C \setminus B) \cup D$$

the assertion follows.

(c) Let $A \subset X$. By Corollary 17, there exists an increasing sequence $(B_n)_{n \in \mathbb{N}}$ from $\mathfrak{R}_\delta$ such that

$$B_n \subset A \qquad (n \in \mathbb{N})$$

$$\sup_{n \in \mathbb{N}} \bar{\mu}(B_n) = \mu_*(A).$$

Set $B := \bigcup_{n \in \mathbb{N}} B_n$, and note that B satisfies the stated conditions. In particular, if $\mu_*(A)$ is finite, then B must belong to $\mathfrak{R}(\mu)$ since $\mu_*(A)$ finite implies that $(e_{B_n})_{n \in \mathbb{N}}$ is an $\bar{\ell}$-sequence from $\bar{\mathscr{L}}(\ell)$.

(d) Let $A \subset X$. According to Theorem 16(d), we can choose an $\mathfrak{R}_\delta$-measurable function h such that

$$e_A \leq h \ \mu\text{-a.e.}$$

$$\mu^*(A) = \int_X^* h \, d\mu.$$

Moreover we can assume that $h \geq 0$ (everywhere!), and if $\mu^*(A)$ is finite, we can assume that h belongs to $\bar{\mathscr{L}}(\ell)$. Set

$$C := \{h \geq 1\}.$$

Then C is $\mathfrak{R}_\delta$-measurable [Proposition 5.3.3(f)], hence μ-measurable (Proposition 15) and

$$e_A \leq e_C \leq h \ \mu\text{-a.e.}$$

As a subset of $\{h < e_A\}$, the set $A \setminus C$ is μ-null, and we have

$$\mu^*(A) = \mu^*(C).$$

Finally, if $\mu^*(A)$ is finite, we have $e_C \leq h$, e_C is μ-measurable, and h belongs to $\bar{\mathscr{L}}(\ell)$. It follows from Corollary 9 (b) that e_C belongs to $\bar{\mathscr{L}}(\ell)$; that is, C belongs to $\mathfrak{R}(\mu)$. If there exists a countable family from $\mathfrak{R}$ whose union contains A, then we can choose h so that $e_A \leq h$ and therefore $A \subset C$. □

Some remarks are in order. The approximation theorems 5.4.16 and 5.4.18 also describe the relation of $\mathfrak{R}_\delta$-measurability to μ-measurability. Evidently connections become weaker as one goes further from $\mathfrak{R}$. The reason for this is that μ-exceptional objects need not be $\mathfrak{R}_\delta$-measurable.

Certain of our assertions could be even further sharpened. There is good reason, however, to forgo these further "sharpenings." To form the approximating objects from the corresponding objects in $\mathscr{L}$ and $\mathfrak{R}$, we have already passed to the limit twice. Measurability should serve, not least of all, to avoid passing to the limit.

Finally, it is worth stressing that in case $\mathfrak{R}$ itself is already a δ-ring, the approximation theorems hold true with $\mathfrak{R}_\delta$ replaced throughout by $\mathfrak{R}$.

One consequence of Theorem 16(d) is the following.

Corollary 5.4.19. *Let $(X, \mathfrak{R}, \mu)$ be a positive measure space. Then for every $f \in \overline{\mathbb{R}}_+^X$, and for every countable disjoint family $(A_\iota)_{\iota \in I}$ from $\mathfrak{M}(X, \mu)$,*

$$\int_{\bigcup_{\iota \in I} A_\iota}^* f \, d\mu = \sum_{\iota \in I}^* \int_{A_\iota}^* f \, d\mu \tag{7}$$

Proof. Let

$$A := \bigcup_{\iota \in I} A_\iota$$

and let $(X, \mathscr{L}, \ell)$ be the Daniell space associated with $(X, \Re, \mu)$. By Proposition 3.5.5,

$$\int_A^* f \, d\mu \le \sum_{\iota \in I}^* \int_{A_\iota}^* f \, d\mu .$$

Assume that

$$\int_A^* f \, d\mu < \infty$$

since otherwise there is nothing more to prove. Since $f \ge 0$, we then have $\int_A^* f \, d\mu$ real. By Theorem 16(d), there exists a function h such that

$$h \in \mathscr{M}(X, \Re_\delta) \cap \bar{\mathscr{L}}(\ell) \cap \mathbb{R}^X$$

$$0 \le f e_A \le h \ \mu\text{-a.e.}$$

$$\int_A^* f \, d\mu = \int_X h \, d\mu .$$

We may also assume that $h \ge 0$ (everywhere). By Corollary 9(b), the function $h e_A$ belongs to $\bar{\mathscr{L}}(\ell)$, as do all of the functions $h e_{A_\iota}$. We have

$$f e_A \le h e_A \ \mu\text{-a.e.}$$

and

$$f e_{A_\iota} \le h e_{A_\iota} \ \mu\text{-a.e.}$$

for every ι. It follows that

$$\int_A^* f \, d\mu \le \int_A^* h \, d\mu = \int_A h \, d\mu \le \int_X h \, d\mu = \int_A^* f \, d\mu$$

so

$$\int_A^* f \, d\mu = \int_A h \, d\mu .$$

Using Theorem 3.5.2(b) ⇒ (f), we have

$$\int_A^* f\,d\mu = \int_A h\,d\mu$$

$$= \sum_{\iota \in I} \int_{A_\iota} h\,d\mu$$

$$= \sum_{\iota \in I} \int_{A_\iota}^* h\,d\mu$$

$$\geq \sum_{\iota \in I}^* \int_{A_\iota} f\,d\mu$$

and (7) follows. □

Corollary 5.4.20. *For every positive measure space* $(X, \mathfrak{R}, \mu)$ *and for every positive extended-real-valued function* f *on* X,

$$\int_X^* f\,d\mu = \sup_{A \in \mathfrak{R}} \int_A^* f\,d\mu. \tag{8}$$

Proof. Set

$$\alpha := \sup_{A \in \mathfrak{R}} \int_A^* f\,d\mu$$

We distinguish three cases:

CASE 1. If $\alpha = \infty$, then (8) follows from the monotonicity of $\int^*$ [Proposition 3.3.3(a)].

CASE 2. Suppose $\alpha = 0$. Then

$$\int_A^* f\,d\mu = 0$$

for every A in $\mathfrak{R}$. By Proposition 3.3.10(c), fe_A belongs to $\mathcal{N}(X, \mu)$ for every A in $\mathfrak{R}$. By Theorem 5.2.6(c), f belongs to $\mathcal{N}(X, \mu)$. Therefore

$$\int_X^* f\,d\mu = 0 = \alpha.$$

CASE 3. Suppose that $0 < \alpha < \infty$. Choose from $\mathfrak{R}$ an increasing sequence $(A_n)_{n \in \mathbb{N}}$ such that

$$\sup_{n \in \mathbb{N}} \int_{A_n}^* f \, d\mu = \alpha$$

and set

$$B := X \setminus \bigcup_{n \in \mathbb{N}} A_n.$$

We claim that fe_B is μ-null. Let $C \in \mathfrak{R}$. For every n, Corollary 19 yields

$$\alpha \geq \int_{C \cup A_n}^* f \, d\mu = \int_{C \setminus A_n}^* f \, d\mu + \int_{A_n}^* f \, d\mu$$

$$\geq \int_{C \cap B}^* f \, d\mu + \int_{A_n}^* f \, d\mu.$$

It follows that

$$\alpha \geq \int_{C \cap B}^* f \, d\mu + \alpha$$

so

$$\int_{C \cap B}^* f \, d\mu = 0.$$

Since C was an arbitrary element of $\mathfrak{R}$, the same argument used in Case 2 shows that fe_B belongs to $\mathscr{N}(X, \mu)$, as claimed. Using Theorem 3.3.4(a), we have

$$\int_X^* f \, d\mu = \int_B^* f \, d\mu + \int_{\bigcup_{n \in \mathbb{N}} A_n}^* f \, d\mu = \sup_{n \in \mathbb{N}} \int_{A_n}^* f \, d\mu = \alpha. \qquad \square$$

Corollary 5.4.21. *For every positive measure space $(X, \mathfrak{R}, \mu)$ and for every subset A of X,*

$$\mu^*(A) = \sup_{B \in \mathfrak{R}} \mu^*(A \cap B). \qquad \square$$

We conclude this section with an important theorem about the convergence of sequences of μ-measurable functions. In preparation, we distinguish the collection of μ-measurable functions on X that are μ-a.e. finite, a subset of $\mathscr{M}(X, \mu)$ that will also play an important role in Volume 2.

Definition 5.4.22. *Let $(X, \mathfrak{R}, \mu)$ be a positive measure space. Then*

$$\mathscr{L}^0(X, \mu) := \{ f \in \mathscr{M}(X, \mu) \mid \{ |f| = \infty \} \in \mathfrak{N}(X, \mu) \}. \qquad \square$$

It has already been noted that $\mathscr{M}(X, \mu)$ is not a Riesz lattice. In contrast, $\mathscr{L}^0(X, \mu)$ is a Riesz lattice with additional properties besides. These properties are described in the next proposition, which can be verified by the reader.

Proposition 5.4.23. *The following assertions hold, for every positive measure space $(X, \mathfrak{R}, \mu)$.*

(*a*) *$\mathscr{L}^0(X, \mu)$ is a Stone lattice that is conditionally σ-completely embedded in $\overline{\mathbb{R}}^X$.*

(*b*) *The conditions*

$$f \in \mathscr{L}^0(X, \mu) \qquad g \in \mathscr{M}(X, \mu) \qquad |g| \leq |f| \ \mu\text{-a.e.}$$

imply that g belongs to $\mathscr{L}^0(X, \mu)$.

(*c*) *$\mathscr{L}^1(X, \mu) \subset \mathscr{L}^0(X, \mu)$.*

(*d*) *$\mathscr{N}(\mathscr{L}^0(X, \mu)) = \mathscr{N}(X, \mu)$*
$\mathfrak{N}(\mathscr{L}^0(X, \mu)) = \mathfrak{N}(X, \mu)$.

(*e*) *$\mathfrak{R}(\mathscr{L}^0(X, \mu)) = \hat{\mathfrak{R}}(\mathscr{L}^0(X, \mu)) = \mathfrak{M}(X, \mu)$.* $\square$

The following theorem, then, describes order convergence in $\mathscr{L}^0(X, \mu)$.

Theorem 5.4.24 (Egoroff, 1911). *Let $(X, \mathfrak{R}, \mu)$ be a positive measure space. Let $f \in \mathscr{L}^0(X, \mu)$, and suppose that $(f_n)_{n \in \mathbb{N}}$ is a sequence from $\mathscr{L}^0(X, \mu)$ such that*

$$f(x) = \lim_{n \to \infty} f_n(x) \ \mu\text{-a.e.}$$

Then for every set $A \in \mathfrak{L}(X, \mu)$ and for every real number $\varepsilon > 0$, there exists an $\mathfrak{R}_\delta$-measurable set $B \subset A$ such that

$$\mu^X(A \setminus B) < \varepsilon$$

and the sequence $(f_n)_{n \in \mathbb{N}}$ converges to f uniformly on B.

Proof. Set

$$C := \left\{ x \in X \;\middle|\; \begin{array}{c} f(x) \in \mathbb{R} \\ f_n(x) \in \mathbb{R}, \forall n \in \mathbb{N} \\ f(x) = \lim\limits_{n \to \infty} f_n(x) \end{array} \right\}.$$

By hypothesis, $X \setminus C$ belongs to $\mathfrak{N}(X, \mu)$.

Let $A \in \mathfrak{L}(X, \mu)$. For all $m, n \in \mathbb{N}$, set

$$A_{m,n} := A \cap \left\{ x \in C \mid |f(x) - f_n(x)| \geq \frac{1}{m} \right\}$$

$$B_{m,n} := \bigcup_{k \geq n} A_{m,k}.$$

For each m, $(B_{m,n})_{n \in \mathbb{N}}$ is a decreasing sequence from $\mathfrak{L}(X, \mu)$. We claim that

$$\bigcap_{n \in \mathbb{N}} B_{m,n} = \varnothing \tag{9}$$

for every m. Suppose that, for some m in $\mathbb{N}$, x belongs to $\bigcap_{n \in \mathbb{N}} B_{m,n}$. Then for every n in $\mathbb{N}$ there exists an integer $k \geq n$ such that

$$|f(x) - f_k(x)| \geq \frac{1}{m}.$$

On the other hand, x must belong to C so

$$\lim_{k \to \infty} f_k(x) = f(x).$$

This contradiction verifies (9). By null continuity we conclude that

$$\inf_{n \in \mathbb{N}} \mu^X(B_{m,n}) = 0$$

for every m.

Now let $\varepsilon > 0$ be given. For each m in $\mathbb{N}$ choose n_m in $\mathbb{N}$ such that

$$\mu^X(B_{m,n_m}) \leq \frac{\varepsilon}{2^m}.$$

Set

$$B' := (A \cap C) \setminus \left(\bigcup_{m \in \mathbb{N}} B_{m,n_m} \right).$$

Observe that B' belongs to $\mathfrak{L}(X, \mu)$. Since

$$A \setminus B' = (A \setminus C) \cup \left(\bigcup_{m \in \mathbb{N}} B_{m,n_m} \right)$$

the set $A \setminus B'$ also belongs to $\mathfrak{L}(X, \mu)$, and

$$\mu^X(A \setminus B') \le \mu^X(A \setminus C) + \mu^X\Big(\bigcup_{m \in \mathbb{N}} B_{m,n_m}\Big) \le \sum_{m \in \mathbb{N}} \mu^X(B_{m,n_m}) \le \varepsilon$$

(Theorem 3.5.1). We want to show that $(f_n)_{n \in \mathbb{N}}$ converges uniformly on B' to f. Given $\varepsilon' > 0$, choose $m \in \mathbb{N}$ such that $1/m < \varepsilon'$. If x belongs to B', then x belongs to $(A \cap C) \setminus B_{m,n_m}$ so

$$|f(x) - f_n(x)| < \frac{1}{m} < \varepsilon'$$

whenever $n \ge n_m$. We have established uniform convergence on B'.

Finally, Theorem 18(b) allows us to choose a set B such that

$$B \subset B'$$

$$B \in \mathfrak{M}(X, \mathfrak{R}_\delta)$$

$$B' \setminus B \in \mathfrak{N}(X, \mu).$$

This set meets all the requirements of the theorem. □

EXERCISES

5.4.1(E) **The Dirac measure**. This notion was first introduced in Ex. 2.2.1. Let $(X, \mathfrak{R}, \delta_x)$ be a positive measure space such that δ_x is the Dirac measure concentrated in $x \in X(\mathfrak{R})$. Prove the following statements:

(α) $\mathscr{L}^1(\delta_x) = \{f \in \overline{\mathbb{R}}^X \mid f(x) \in \mathbb{R}\}$.
(β) $\int f\, d\delta_x = f(x)$ whenever $f \in \mathscr{L}^1(\delta_x)$.
(γ) $\mathscr{N}(\delta_x) = \{f \in \overline{\mathbb{R}}^X \mid f(x) = 0\}$.
(δ) $\mathfrak{N}(\delta_x) = \{A \subset X \mid x \notin A\}$.
(ε) $\mathscr{M}(\delta_x) = \overline{\mathbb{R}}^X$, $\mathfrak{M}(\delta_x) = \mathfrak{L}(\delta_x) = \mathfrak{P}(X)$.
(ζ) $\int^* f\, d\delta_x = \int_* f\, d\delta_x = f(x)$ whenever $f \in \overline{\mathbb{R}}^X$.

5.4.2(E) The considerations of Ex. 5.4.1 can be generalized in the following direction. Let X be a set, $\mathfrak{R}$ a ring of subsets of X and $A \subset X(\mathfrak{R})$ a countable set. Let $(\alpha_x)_{x \in A}$ be a summable family of strictly positive numbers, and define

$$\mu: \mathfrak{R} \to \mathbb{R}, \qquad B \mapsto \sum_{x \in A} \alpha_x e_B(x).$$

Prove the following statements:

(α) $(X, \mathfrak{R}, \mu)$ is a positive measure space.

(β) $(X, \mathfrak{R}, \mu)$ is bounded.

(γ) $\mathscr{L}^1(\mu) = \{f \in \overline{\mathbb{R}}^X \mid (\alpha_x f(x))_{x \in A}$ is summable$\}$.

(δ) $\int f\,d\mu = \sum_{x \in A} \alpha_x f(x)$ whenever $f \in \mathscr{L}^1(\mu)$.

(ε) $\mathscr{N}(\mu) = \{f \in \overline{\mathbb{R}}^X \mid f(A) = \{0\}\}$.

(ζ) $\mathfrak{N}(\mu) = \{B \subset X \mid B \cap A = \varnothing\}$.

(η) $\mathscr{M}(\mu) = \overline{\mathbb{R}}^X$, $\mathfrak{M}(\mu) = \mathfrak{L}(\mu) = \mathfrak{P}(X)$.

(ϑ) $\int^* f\,d\mu = \sum^*_{x \in A} \alpha_x f(x)$ for each $f \in \overline{\mathbb{R}}^X$.

(ι) $\mu^*(B) = \mu^X(B) = \sum_{x \in A} \alpha_x e_B(x)$ for every $B \subset X$.

5.4.3(E) Let X be a set. Consider χ_g on $\mathfrak{F}(X)$ for $g \in \mathbb{R}^X_+$. This positive measure was first introduced in Ex. 2.2.3. Prove the following assertions.

(α) $\mathscr{L}^1(\chi_g) = \ell^1_g(X)$ and $\int f\,d\chi_g = \sum_{x \in X} f(x)g(x)$ whenever $f \in \mathscr{L}^1(\chi_g)$.

(β) $\mathfrak{L}(\chi_g) = \mathfrak{R}_g$ and $\chi^X_g(A) = \sum_{x \in A} g(x)$ for every $A \in \mathfrak{L}(\chi_g)$.

(γ) $\mathscr{M}(\chi_g) = \overline{\mathbb{R}}^X$.

(δ) $\mathfrak{M}(\chi_g) = \mathfrak{P}(X)$.

(ε) $\int^* f\,d\chi_g = \sum^*_{x \in X} f(x)g(x)$ for every $f \in \overline{\mathbb{R}}^X$.

(ζ) $\chi^*_g(A) = \sum^*_{x \in A} g(x)$ for every $A \subset X$.

5.4.4(E) Let $\mathfrak{F}$ be a δ-stable ultrafilter on the set X and define

$$\mu: \mathfrak{P}(X) \to \mathbb{R}, \qquad A \mapsto \begin{cases} 1 & \text{if } A \in \mathfrak{F} \\ 0 & \text{if } A \notin \mathfrak{F} \end{cases}.$$

We know that $(X, \mathfrak{P}(X), \mu)$ is a positive measure space (see Ex. 2.2.1). Prove the following statements:

(α) $\mathscr{M}(\mu) = \overline{\mathbb{R}}^X$.

(β) $\int^* f\,d\mu = \int_* f\,d\mu = \lim_{\mathfrak{F}} f$ whenever $f \in \overline{\mathbb{R}}^X$.

(γ) $\mathscr{L}^1(\mu) = \{f \in \overline{\mathbb{R}}^X \mid \lim_{\mathfrak{F}} f \in \mathbb{R}\}$.

(δ) $\int f\,d\mu = \lim_{\mathfrak{F}} f$ whenever $f \in \mathscr{L}^1(\mu)$.

5.4.5(E) Let X be an uncountable set, and define

$$\mathfrak{R} := \{A \subset X \mid A \text{ countable}\} \cup \{A \subset X \mid X \setminus A \text{ countable}\}$$

$$\mu: \mathfrak{R} \to \mathbb{R}, \qquad A \mapsto \begin{cases} 0 & \text{if } A \text{ countable} \\ 1 & \text{if } X \setminus A \text{ countable} \end{cases}.$$

$(X, \mathfrak{R}, \mu)$ is a positive measure space. Prove the following statements:

(α) $\mathfrak{M}(\mu) = \mathfrak{L}(\mu) = \mathfrak{R}(\mu) = \mathfrak{R}$.

(β) $\mathscr{M}(\mu) = \bigcup\{\langle \alpha e_X \dot{+} g\rangle \mid \alpha \in \overline{\mathbb{R}},\ g \in \overline{\mathbb{R}}^X,\ \{g \neq 0\} \text{ countable}\}$.

(γ) $\int^* f\,d\mu = \int_* f\,d\mu = \alpha$ whenever $f \in \langle \alpha e_X \dot{+} g\rangle \subset \mathscr{M}(\mu)$.

5.4.6(E) Prove the following assertions:

(α) If $\mathfrak{R}$ is a σ-algebra on the set X, then $\mathfrak{M}(\mathfrak{R}) = \mathfrak{R}$.

(β) Let $(X, \mathfrak{R}, \mu)$ be a bounded positive measure space. Then the following assertions hold:

($\beta 1$) $\mathfrak{M}(\mu) = \mathfrak{L}(\mu)$.
($\beta 2$) $\mathscr{M}(\mu)_+ = \{f \in \overline{\mathbb{R}}_+^X \mid f \wedge n \in \mathscr{L}^1(\mu)$ whenever $n \in \mathbb{N}\}$.
($\beta 3$) $\mathscr{M}(\mu) = \{f \in \overline{\mathbb{R}}^X \mid (f \wedge n) \vee (-n) \in \mathscr{L}^1(\mu)$ whenever $n \in \mathbb{N}\}$.

5.4.7(E) Let $(X, \mathfrak{R}, \mu)$ be a positive measure space, $Y \subset X$ and $\mathfrak{S} := \{A \cap Y \mid A \in \mathfrak{R}\}$. Prove the following assertions:

(α) $\mathfrak{S}$ is a ring of sets.

(β) $\nu: \mathfrak{S} \to \mathbb{R}_+$, $A \mapsto \mu^*(A)$ is a positive measure on $\mathfrak{S}$.

(γ) $(Y, \mathfrak{R}|_Y, \mu|_Y)$ and $(Y, \mathfrak{S}, \nu)$ need not generate the same integral.

5.4.8(E) Set $X := \{0,1\}$, $Y := \{0\}$ and $\mathfrak{R} := \{\varnothing, X\}$. Let μ be a positive measure on $\mathfrak{R}$ with $\mu(X) = 1$. Define

$$f: X \to \mathbb{R}, \qquad x \mapsto \begin{cases} 1 & \text{if } x = 0 \\ -1 & \text{if } x = 1 \end{cases}.$$

Prove the following propositions:

(α) $f \notin \mathscr{L}^1(\mu)$.

(β) $f|_Y \in \mathscr{L}^1(\mu|_Y)$.

(γ) $\int^* f|_Y \, d\mu|_Y = \int f|_Y \, d\mu|_Y = 0$, $\int^* f \, d\mu = 1$.

(δ) $\int^* f e_Y^X \, d\mu = 1$ and $\int_* f e_Y^X = 0$.

(ε) $\int^*((-f)|_Y) \, d\mu|_Y = 0$.

(ζ) $\int_*(-f) e_Y^X \, d\mu = -1$.

(η) Y is a $\mu|_Y$-null set, but not a μ-null set.

5.4.9(E) Let $(X, \mathfrak{R}, \mu)$ be a σ-bounded positive measure space. Denote by $(A_n)_{n \in \mathbb{N}}$ an increasing sequence from $\mathfrak{R}$ such that $X \setminus \bigcup_{n \in \mathbb{N}} A_n \in \mathfrak{N}(\mu)$. (See Proposition 5.2.14.) Prove the following:

(α) $\mathfrak{M}(\mu)$ is the σ-algebra generated by $\mathfrak{L}(\mu)$ on X.

(β) For every set $B \in \mathfrak{M}(\mu)$ there is an increasing sequence $(B_n)_{n \in \mathbb{N}}$ from $\mathfrak{L}(\mu)$ such that $B = \bigcup_{n \in \mathbb{N}} B_n$.

For every $f \in \overline{\mathbb{R}}^X$ and for all natural numbers m, n define

$$f_{mn}: X \to \overline{\mathbb{R}}, \qquad x \mapsto \begin{cases} f(x) \wedge m & \text{if } x \in A_n \\ 0 & \text{if } x \in X \setminus A_n \end{cases}$$

and

$$\bar{f}_{mn}: X \to \overline{\mathbb{R}}, \qquad x \mapsto \begin{cases} (f(x) \wedge m) \vee (-m) & \text{if } x \in A_n \\ 0 & \text{if } x \in X \setminus A_n \end{cases}.$$

Prove the following assertions for $f \in \overline{\mathbb{R}}^X$:

(γ) $f \in \mathcal{M}(\mu)_+$ iff $f \geq 0$ and $f_{mn} \in \mathcal{L}^1(\mu)$ whenever $m, n \in \mathbb{N}$.

(δ) $f \in \mathcal{M}(\mu)$ iff $\bar{f}_{mn} \in \mathcal{L}^1(\mu)$ whenever $m, n \in \mathbb{N}$.

5.4.10(E) Let $(X, \mathfrak{R}, \mu)$ be a positive measure space. Prove the following statements:

(α) If $f \in \mathcal{L}^1(\mu)$, then $(\mu^*(\{f \geq n\}))_{n \in \mathbb{N}}$ is summable.

(β) Let $(X, \mathfrak{R}, \mu)$ be bounded. Then the following assertions are equivalent:

($\beta 1$) $f \in \mathcal{M}(\mu)$ and $(\mu^*(\{|f| \geq n\}))_{n \in \mathbb{N}}$ is summable.

($\beta 2$) $f \in \mathcal{L}^1(\mu)$.

5.4.11(E) Let $(X, \mathfrak{R}, \mu)$ be a positive measure space. Take $f \in \mathcal{M}(\mu)_+$, and for each $n \in \mathbb{N}$ define

$$f_n: X \to \mathbb{R}, \qquad x \mapsto \begin{cases} f(x) & \text{if } f(x) \leq n \\ 0 & \text{if } f(x) > n \end{cases}$$

Prove the following:

(α) $(f_n)_{n \in \mathbb{N}}$ is increasing and

$$\sup_{n \in \mathbb{N}} f_n(x) = \begin{cases} f(x) & \text{if } f(x) < \infty \\ 0 & \text{if } f(x) = \infty \end{cases}$$

(β) $\int^* f \, d\mu = \lim_{n \to \infty} \int^* f_n \, d\mu$ if $\{f = \infty\} \in \mathfrak{N}(\mu)$.

5.4.12(E) Let $(X, \mathfrak{R}, \mu)$ be a positive measure space. Let $f \in \overline{\mathbb{R}}^X$ be μ-measurable. Take an increasing family of real numbers $(\alpha_n)_{n \in \mathbb{Z}}$ such that $\lim_{n \to \infty} \alpha_n = \infty$, $\lim_{n \to -\infty} \alpha_n = -\infty$, and $\sup_{n \in \mathbb{Z}} (\alpha_{n+1} - \alpha_n) < \infty$. Prove the following under the assumption that $(X, \mathfrak{R}, \mu)$ is bounded:

(α) $f \in \mathcal{L}^1(\mu)$ iff $(\alpha_n \mu^*(f^{-1}([\alpha_n, \alpha_{n+1}[)))_{n \in \mathbb{Z}}$ is summable.

(β) For every $k \in \mathbb{N}$ take a family $(\alpha_n^k)_{n \in \mathbb{Z}}$ as described but with the additional property that $\sup_{n \in \mathbb{Z}} (\alpha_{n+1}^k - \alpha_n^k) \leq 1/k$. Now, if f is μ-integrable, then

$$\int f \, d\mu = \lim_{k \to \infty} \sum_{n \in \mathbb{Z}} \alpha_n^k \mu^*\left(f^{-1}\left(\left[\alpha_n^k, \alpha_{n+1}^k\right[\right)\right).$$

Show that the boundedness of $(X, \mathfrak{R}, \mu)$ is a necessary condition.

5.4.13(E) Let $(X, \mathfrak{R}, \mu)$ be a positive measure space. Take $f, g \in \overline{\mathbb{R}}^X$, and prove the following statements:

(α) If $\int^* f\,d\mu + \int_* g\,d\mu$ is defined, then

$$\int^* (f \dot{+} g)\,d\mu \geq \int^* f\,d\mu + \int_* g\,d\mu.$$

(β) If $\int^* f\,d\mu + \int_* g\,d\mu$ is defined and $\langle f \dot{+} g \rangle \subset \mathscr{L}^1(\mu)$, then

$$\int (f \dot{+} g)\,d\mu = \int^* f\,d\mu + \int_* g\,d\mu.$$

(γ) If $g \in \mathscr{L}^1(\mu)$, then

$$\int^* (f \dot{+} g)\,d\mu = \int^* f\,d\mu + \int g\,d\mu$$

$$\int_* (f \dot{+} g)\,d\mu = \int_* f\,d\mu + \int g\,d\mu.$$

(δ) If $\int^* f\,d\mu < \infty$ and $\int^* g\,d\mu < \infty$, then $\int^* (f \vee g)\,d\mu < \infty$.

5.4.14(E) Let $(X, \mathfrak{R}, \mu)$ be a positive measure space and $(X, \mathscr{L}, \ell)$ the associated Daniell space. Prove the following statements for $f \in \overline{\mathbb{R}}^X$:

(α) $\int^* f\,d\mu = \inf\{\ell^{\uparrow}(g) \mid g \in \mathscr{L}^{\uparrow}, g \geq f\ \ \mu\text{-a.e.}\} = \inf\{\ell^{\uparrow}(g) \mid g \in \mathscr{L}^{\uparrow}, g \geq f\ \mu\text{-a.e.}, g \geq f \wedge 0\}$.

(β) If there is a countable family $(A_\iota)_{\iota \in I}$ from $\mathfrak{R}$ with $\{f > 0\} \subset \bigcup_{\iota \in I} A_\iota$, then $\int^* f\,d\mu = \inf\{\ell^{\uparrow}(g) \mid g \in \mathscr{L}^{\uparrow},\ g \geq f\}$.

(γ) If $f \geq 0$, then the following statements are equivalent:

($\gamma 1$) $\int^* f\,d\mu = 0$.

($\gamma 2$) $f = 0$ μ-a.e.

(δ) If $f \geq 0$, then $\int^* \infty f\,d\mu = \infty \int^* f\,d\mu$.

(ε) If $\int^* f\,d\mu < \infty$, then $f < \infty$ μ-a.e., and there is an increasing sequence $(A_n)_{n \in \mathbb{N}}$ from $\mathfrak{R}$ such that $\{f > 0\} \setminus \bigcup_{n \in \mathbb{N}} A_n \in \mathfrak{N}(\mu)$.

(ζ) $\int_* f\,d\mu = \sup\{\ell^{\downarrow}(g) \mid g \in \mathscr{L}^{\downarrow},\ g \leq f\ \mu\text{-a.e.}\}$.

(η) $\int^* f\,d\mu = \int_* f\,d\mu$ whenever $f \in \mathscr{L}^{\uparrow} \cup \mathscr{L}^{\downarrow}$.

(ϑ) If $f \geq 0$, then $\int_* f\,d\mu = \sup\{\ell^{\downarrow}(g) \mid g \in \mathscr{L}^{\downarrow}, 0 \leq g \leq f\}$.

5.4.15(E) Prove the equivalence of the following propositions for a positive measure space $(X, \mathfrak{R}, \mu)$:

(α) $\mu = 0$.

(β) $X \in \mathfrak{N}(\mu)$.

(γ) $\int^* f\,d\mu = 0$ whenever $f \in \overline{\mathbb{R}}^X$.

5.4.16(E) Define

$$\mathfrak{R} := \{ A \subset \mathfrak{P}(\mathbb{R}) \mid A \text{ countable or } \mathfrak{P}(\mathbb{R}) \setminus A \text{ countable} \}$$

$$\mathfrak{S} := \{ B \subset \mathbb{R} \mid B \text{ countable or } \mathbb{R} \setminus B \text{ countable} \}$$

$$\mu: \mathfrak{S} \to \mathbb{R}, \qquad B \mapsto \begin{cases} 0 & \text{if } B \text{ countable} \\ 1 & \text{if } \mathbb{R} \setminus B \text{ countable} \end{cases}$$

$$\mathfrak{T} := \{ A \times \{ y \} \mid A \in \mathfrak{R}, y \in \mathbb{R} \} \cup \{ \{ x \} \times B \mid x \in \mathfrak{P}(\mathbb{R}), B \in \mathfrak{S} \}.$$

Prove the following:

(α) $\mathfrak{T}_r$ is a δ-ring.

(β) There is a unique positive measure ν on $\mathfrak{T}_r$ such that

$$\nu(A \times \{ y \}) = 0 \text{ whenever } A \in \mathfrak{R} \text{ and } y \in \mathbb{R}$$

$$\nu(\{ x \} \times B) = \mu(B) \text{ whenever } B \in \mathfrak{S} \text{ and } x \in \mathfrak{P}(\mathbb{R}).$$

Let $(A_y)_{y \in \mathbb{R}}$ be a disjoint family of uncountable subsets of $\mathfrak{P}(\mathbb{R})$, and define $C := \bigcup_{y \in \mathbb{R}} (A_y \times \{ y \})$. Take a $\mathfrak{T}_\delta$-measurable subset D of $\mathfrak{P}(\mathbb{R}) \times \mathbb{R}$ such that $C \subset D$, and prove the following:

(γ) $C \in \mathfrak{N}(\nu)$.

(δ) $D \notin \mathfrak{N}(\nu)$.

Hint for (δ): For each $y \in \mathbb{R}$ there is exactly one $A'_y \in \mathfrak{R}$ such that

$$A'_y \times \{ y \} = D \cap (\mathfrak{P}(\mathbb{R}) \times \{ y \}).$$

$A_y \subset A'_y$ implies that $A''_y := \mathfrak{P}(\mathbb{R}) \setminus A'_y$ is countable for every $y \in \mathbb{R}$. Because $\bigcup_{y \in \mathbb{R}} A''_y$ has at least the same cardinality as $\mathbb{R}$, there is a point $x \in \mathfrak{P}(\mathbb{R}) \setminus \bigcup_{y \in \mathbb{R}} A''_y$. We conclude that $\{ x \} \times \mathbb{R} \subset D$; hence $D \notin \mathfrak{N}(\nu)$.
Remark: This example shows that in Corollary 18(d) the assertion $A \setminus C \in \mathfrak{N}(\mu)$ cannot be replaced by the stronger assertion $A \subset C$.

5.4.17(E) Set $\mathfrak{R} := \{0, \mathbb{N}\}$, and $\mu := 0$. $(\mathbb{N}, \mathfrak{R}, \mu)$ is a positive measure space. Define

$$f: \mathbb{N} \to \mathbb{R}, \qquad n \mapsto (-1)^n n$$

and prove the following:

(α) $f \in \mathscr{N}(\mu)$.

(β) If $g, h \in \mathscr{M}(\mathbb{N}, \mathfrak{R})$ such that $g \leq f \leq h$, then $g = -\infty e_{\mathbb{N}}$ and $h = \infty e_{\mathbb{N}}$.

Remark: This example shows that in Theorem 16(a), g and h need not be finite, even if f is finite. Compare it with Theorem 16(c) and (d)!

5.4.18(E) Consider Theorem 16. Prove the following:

(α) In Theorem 16(c), g can be chosen such that $g \le f \vee 0$.
(β) In Theorem 16(d), h can be chosen such that $f \wedge 0 \le h$.

5.4.19(C) **Caratheodory Method.** The following characterization of μ-measurability is essentially due to C. Caratheodory (1914).

(a) Take a positive measure space $(X, \mathfrak{R}, \mu)$. Prove the following proposition:

(α) If $A \subset X$, then the following are equivalent:

(α1) A is μ-measurable.
(α2) $\mu^*(B) = \mu^*(B \cap A) + \mu^*(B \setminus A)$ for every $B \subset X$.
(α3) $\mu(B) = \mu^*(B \cap A) + \mu^*(B \setminus A)$ for every $B \in \mathfrak{R}$.

For (α1) $\Rightarrow$ (α2), observe that $\mu^*(B) \le \mu^*(B \cap A) + \mu^*(B \setminus A)$ whenever $B \subset X$. Thus (α2) is trivial if $\mu^*(B) = \infty$. So suppose that $\mu^*(B) < \infty$. Then $\{C \in \mathfrak{L}(\mu) \mid B \subset C\} \neq \varnothing$. Take any such C. Then,

$$C \setminus A \in \mathfrak{L}(\mu),\ C \cap A \in \mathfrak{L}(\mu),\ B \cap A \subset C \cap A,\ B \setminus A \subset C \setminus A$$

and

$$\mu^X(C) = \mu^X(C \cap A) + \mu^X(C \setminus A) \ge \mu^*(B \cap A) + \mu^*(B \setminus A).$$

Hence

$$\mu^*(B) \ge \mu^*(B \cap A) + \mu^*(B \setminus A).$$

For (α3) $\Rightarrow$ (α1), take $B \in \mathfrak{R}$. Suppose $C \in \mathfrak{L}(\mu)$ with $B \setminus A \subset C$. Then

$$\mu(B) = \mu^X(B \cap C) + \mu^X(B \setminus C) \le \mu^X(B \cap C) + \mu_*(B \cap A).$$

Hence

$$\mu(B) \le \mu^*(B \setminus A) + \mu_*(B \cap A) \le \mu^*(B \setminus A) + \mu^*(B \cap A) = \mu(B).$$

Hence $\mu_*(B \cap A) = \mu^*(B \cap A)$ for each $B \in \mathfrak{R}$ so that A is μ-measurable [5.4.10(c) $\Rightarrow$ (a)].

Caratheodory applied the characterization of measurability just presented to the construction of extensions of positive measure spaces. We pursue these ideas.

(b) Let X be a set. An **outer measure** on X is a mapping $\mu^*\colon \mathfrak{P}(X) \to \overline{\mathbb{R}}_+$ with the following properties:

(i) $\mu^*(\varnothing) = 0$.
(ii) $\mu^*(A) \le \mu^*(B)$ whenever $A \subset B \subset X$.
(iii) $\mu^*(\bigcup_{n \in \mathbb{N}} B_n) \le \sum^*_{n \in \mathbb{N}} \mu^*(B_n)$ for any sequence $(B_n)_{n \in \mathbb{N}}$ from $\mathfrak{P}(X)$.

Every positive measure μ on a ring of sets $\mathfrak{R} \subset \mathfrak{P}(X)$ generates an outer measure μ^* on X.

Let μ^* be an outer measure on X. $A \subset X$ is called μ^*-**measurable** iff $\mu^*(B) = \mu^*(B \cap A) + \mu^*(B \setminus A)$ for every $B \in \mathfrak{P}(X)$. Prove the following:

(α) The set $\mathfrak{M}$ of μ^*-measurable subsets of X is a σ-algebra on X.

(β) $\mu^*(\bigcup_{n \in \mathbb{N}} A_n) = \sum^*_{n \in \mathbb{N}} \mu^*(A_n)$ for every disjoint sequence $(A_n)_{n \in \mathbb{N}}$ from $\mathfrak{M}$.

(γ) $\mathfrak{R} := \{A \in \mathfrak{M} \mid \mu^*(A) < \infty\}$ is a δ-ring.

(δ) Define $\mu := \mu^*|_{\mathfrak{R}}$. Then $(X, \mathfrak{R}, \mu)$ is a closed positive measure space.

For (α), (β), it is clear that $\varnothing$, $X \in \mathfrak{M}$ and $X \setminus A \in \mathfrak{M}$ whenever $A \in \mathfrak{M}$. So take $A, B \in \mathfrak{M}$ and $C \subset X$. Then

$$\mu^*(C \cap A) = \mu^*(C \cap A \cap B) + \mu^*((C \cap A) \setminus B)$$

and

$$\mu^*(C) = \mu^*(C \cap A) + \mu^*(C \setminus A).$$

Put $D := C \setminus (A \cap B)$. Then $D \cap A = (C \cap A) \setminus B$ and $D \setminus A = C \setminus A$. Now $A \in \mathfrak{M}$. Thus

$$\mu^*(D) = \mu^*(D \cap A) + \mu^*(D \setminus A)$$

or, in other words,

$$\mu^*(C \setminus (A \cap B)) = \mu^*((C \cap A) \setminus B) + \mu^*(C \setminus A).$$

Hence

$$\mu^*(C) = \mu^*(C \cap A \cap B) + \mu^*(C \setminus (A \cap B))$$

so that $A \cap B \in \mathfrak{M}$. Hence $\mathfrak{M}$ is an algebra on X. Now take $A, B \in \mathfrak{M}$ with $A \cap B = \varnothing$, and $C \subset X$. Then

$$\mu^*(C \cap (A \cup B)) = \mu^*(C \cap A) + \mu^*(C \cap B).$$

If $C = X$, then

$$\mu^*(A \cup B) = \mu^*(A) + \mu^*(B).$$

Take a disjoint sequence $(A_n)_{n \in \mathbb{N}}$ from $\mathfrak{M}$ and set $A := \bigcup_{n \in \mathbb{N}} A_n$ and $B_n := \bigcup_{k=1}^n A_k$ for every $n \in \mathbb{N}$. Then $B_n \in \mathfrak{M}$ and

$$\mu^*(B_n) = \sum_{k=1}^{n}{}^* \mu^*(A_k)$$

for every $n \in \mathbb{N}$. Furthermore, if $C \subset X$, then

$$\mu^*(C) = \mu^*(C \cap B_n) + \mu^*(C \setminus B_n) = \sum_{k=1}^{n}{}^* \mu^*(C \cap A_k) + \mu^*(C \setminus B_n).$$

But $B_n \subset A$ so that $C \setminus A \subset C \setminus B_n$. Thus for $n \to \infty$,

$$\sum_{n \in \mathbb{N}}{}^* \mu^*(C \cap A_n) + \mu^*(C \setminus A) \leq \mu^*(C).$$

On the other hand,

$$\mu^*(C \cap A) \leq \sum_{n \in \mathbb{N}}{}^* \mu^*(C \cap A_n) \text{ and } \mu^*(C) \leq \mu^*(C \cap A) + \mu^*(C \setminus A).$$

Hence for any $C \subset X$

$$\mu^*(C) = \mu^*(C \cap A) + \mu^*(C \setminus A) = \sum_{n \in \mathbb{N}}{}^* \mu^*(C \cap A_n) + \mu^*(C \setminus A).$$

Hence $A \in \mathfrak{M}$, and taking $C = A$,

$$\mu^*(A) = \sum_{n \in \mathbb{N}}{}^* \mu^*(A_n).$$

Hence outer measures generate positive measure spaces, which provides an avenue for constructing positive measure spaces.

Let μ^* be an outer measure on X, and denote by $(X, \mathfrak{R}, \mu)$ the positive measure space generated by μ^* in the sense of (δ). μ^* is called **complete** iff $(X, \mathfrak{R}, \mu)$ is complete. Prove the following:

(ε) If $A \subset X$ and $\mu^*(A) = 0$, then $A \in \mathfrak{R}$ and $\mu(A) = 0$. (The reverse is trivially true.)

(ζ) The following statements are equivalent:

(ζ1) μ^* is complete.
(ζ2) If $A \subset X$ such that $\mu^*(A \cap B) = 0$ whenever $B \subset X$ with $\mu^*(B) < \infty$, then $\mu^*(A) = 0$.
(ζ3) μ^* is the outer measure generated by a positive measure space.
(ζ4) μ^* is the outer measure generated by $(X, \mathfrak{R}, \mu)$.

Give an example of an outer measure that is not complete. Prove:

(η) If there exists a sequence $(A_n)_{n \in \mathbb{N}}$ in $\mathfrak{P}(X)$ such that $\mu^*(A_n) < \infty$ for every $n \in \mathbb{N}$, and if $\bigcup_{n \in \mathbb{N}} A_n = X$, then μ^* is complete.

Assume now that μ^* is complete, and prove the following:

(ϑ) $\mathfrak{M}(\mu) = \{A \subset X \mid A\ \mu^*\text{-measurable}\}$.

5.4.20(C) In this exercise we consider an application of the Caratheodory construction discussed in Ex. 5.4.19.

Let X be a metric space. Let μ^* be an outer measure on X with the property that $\mu^*(A \cup B) = \mu^*(A) + \mu^*(B)$ whenever $A, B \in \mathfrak{P}(X)$ such that $\inf\{d(x,y) \mid (x,y) \in A \times B\} > 0$. Prove the following:

(α) Every open subset of X is μ^*-measurable.
(β) Every Borel set of X is μ^*-measurable.

Hint for (α): Let U be an open subset of X, $U \neq X$. Take $A \subset U$ such that $\mu^*(A) < \infty$. For every $n \in \mathbb{N}$ define

$$U_n := \{x \in U \mid d(x, X \setminus U) > 1/n\}$$

and

$$A_n := (U_n \setminus U_{n-1}) \cap A \qquad (\text{set } U_0 := \varnothing).$$

Then

$$\lim_{n \to \infty} \mu^*(U_n \cap A) \leq \mu^*(A).$$

For every $n \in \mathbb{N}$,

$$\mu^*(U_{n+1} \cap A) \geq \mu^*((U_{n-1} \cap A) \cup A_{n+1})$$
$$= \mu^*(U_{n-1} \cap A) + \mu^*(A_{n+1})$$

From this, it can be concluded that $(\mu^*(A_n))_{n \in \mathbb{N}}$ is summable. From

$$\mu^*(A) \leq \mu^*(U_n \cap A) + \sum_{m \geq n+1} \mu^*(A_m)$$

it follows that

$$\mu^*(A) = \lim_{n \to \infty} \mu^*(U_n \cap A).$$

For an arbitrary $A \subset X$,

$$\mu^*(A) \geq \mu^*(A \cap U_n) + \mu^*(A \setminus U)$$

and, using the preceding considerations,

$$\mu^*(A) \geq \mu^*(A \cap U) + \mu^*(A \setminus U).$$

We now give an important example. Let $h\colon \mathbb{R}_+ \to \mathbb{R}_+$ be a function. Let X be a metric space, and for every $x \in X$ and every $\varepsilon > 0$ define $B(x, \varepsilon) := \{y \in X \mid d(x,y) < \varepsilon\}$. Take $n \in \mathbb{N}$. For every $A \subset X$ and for every $\varepsilon > 0$

define

$$\mu_\varepsilon^*(A) := \inf\left\{ \sum_{\iota\in I}^* h(\alpha_\iota) \mid A \subset \bigcup_{\iota\in I} B(x_\iota, \alpha_\iota),\ \alpha_\iota < \varepsilon \right\}.$$

Finally, define for every $A \subset X$

$$\mu^*(A) := \lim_{\substack{\varepsilon\to 0\\ \varepsilon>0}} \mu_\varepsilon^*(A).$$

Prove:

(γ) μ^* is an outer measure with the property formulated before.

The positive measures generated by an outer measure of this kind are called **Hausdorff measures** on X. For $h(t) = t^\alpha$ ($t \in \mathbb{R}_+$, $\alpha \in]0, \infty[$), we get the **α-dimensional measures** on X. Now consider $X = \mathbb{R}^n$ with respect to the natural metric. Prove the following:

(δ) For every set $A \subset \mathbb{R}^n$ there exists $\alpha_0 \in [0, n]$ such that for every $\alpha \in]0, \alpha_0[$ the α-dimensional measure is ∞ and for every $\alpha \in]\alpha_0, n]$ the α-dimensional measure is 0.

The number α_0 is called the **dimension** of A.

(ε) Let $\alpha_0 := \log 2/\log 3$. Show that the dimension of the Cantor set is α_0 and its α_0-dimensional measure is 1.

(ζ) Show that for every $n \in \mathbb{N}$, $\alpha \in]0, n]$ and $\beta \in \mathbb{R}_+$ there exists a compact set of $\mathbb{R}^n$ whose dimension is α and whose α-dimensional measure is β.

Hint for (ζ): Use constructions similar to those for Cantor sets.

5.4.21(C) The Theorem of Lusin. Let $(X, \mathfrak{T})$ be a Hausdorff space. Denote by $\mathfrak{K}$ the set of all compact subsets of X. Finally let μ be a $\mathfrak{K}$-regular positive measure on $\mathfrak{K}_r$. The completions of such measures were investigated in Ex. 5.2.15.

There is an interesting connection between μ-measurability and continuity of functions on X due to Vitali and Lusin (1905, 1912). We want to prove the corresponding theorem in this exercise. First prove the following proposition:

(α) For any subset A of X, the following assertions are equivalent:

(α1) A is μ-measurable.
(α2) For every $\varepsilon > 0$ and for every $K \in \mathfrak{K}$ there are sets K' and K'' in $\mathfrak{K}$ such that $K' \subset K \cap A$, $K'' \subset K \setminus A$ and $\mu(K \setminus (K' \cup K''))$ $< \varepsilon$.

(α1) $\Rightarrow$ (α2). $K \cap A$ and $K \setminus A$ are μ-integrable. The $\mathfrak{K}$-regularity of μ^X implies the existence of increasing sequences $(K_n')_{n\in\mathbb{N}}$ and $(K_n'')_{n\in\mathbb{N}}$

from $\mathfrak{K}$ such that

$$\bigcup_{n \in \mathbb{N}} K_n' \subset K \cap A, \qquad K \cap A \setminus \bigcup_{n \in \mathbb{N}} K_n' \in \mathfrak{N}(\mu)$$

and

$$\bigcup_{n \in \mathbb{N}} K_n'' \subset K \setminus A, \qquad (K \setminus A) \setminus \bigcup_{n \in \mathbb{N}} K_n'' \in \mathfrak{N}(\mu).$$

From this $(\alpha 2)$ follows.

$(\alpha 2) \Rightarrow (\alpha 1)$. Take $K \in \mathfrak{K}$. Then for any $n \in \mathbb{N}$ there are sets K_n' and K_n'' in $\mathfrak{K}$ such that $K_n' \subset K \cap A$, $K_n'' \subset K \setminus A$ and $\mu(K \setminus (K_n' \cup K_n'')) < 1/n$. Define $K' := \bigcup_{n \in \mathbb{N}} K_n'$ and $K'' := \bigcup_{n \in \mathbb{N}} K_n''$. Then we find $K \setminus (K' \cup K'') \in \mathfrak{N}(\mu)$ and

$$K \cap A = K' \cup \big((K \cap A) \setminus K'\big) \in \mathfrak{L}(\mu).$$

Because K is arbitrary, $(\alpha 1)$ follows.

We now formulate the important **Theorem of Lusin**:

(β) For every $f \in \overline{\mathbb{R}}^X$ the following assertions are equivalent:

$(\beta 1)$ f is μ-measurable.
$(\beta 2)$ For every $\varepsilon > 0$ and for every $K \in \mathfrak{K}$ there is a set $L \in \mathfrak{K}$ such that $L \subset K$, $\mu(K \setminus L) < \varepsilon$ and so that $f|_L$ is continuous.
$(\beta 3)$ For every $K \in \mathfrak{K}$ there is a disjoint sequence $(K_n)_{n \in \mathbb{N}}$ from $\mathfrak{K}$ such that $\bigcup_{n \in \mathbb{N}} K_n \subset K$, $K \setminus \bigcup_{n \in \mathbb{N}} K_n$ is a μ-null set and $f|_{K_n}$ is continuous for every $n \in \mathbb{N}$.

$(\beta 1) \Rightarrow (\beta 2)$. First, assume that there is $A \subset X$ such that $f = e_A$. Let $K \in \mathfrak{K}$ and take K', K'' as in $(\alpha 2)$ for an $\varepsilon > 0$. Then $L := K' \cup K''$ has the desired properties.

Second, assume that there is a finite family $(A_\iota)_{\iota \in I}$ of μ-measurable subsets of X such that $f = \sum_{\iota \in I} e_{A_\iota}$. Apply the result just proved.

Third, assume that $0 \le f \le 1$. For every $n \in \mathbb{N}$ define

$$f_n := \frac{1}{2^n} \sum_{m=1}^{2^n} e_{\{f \ge m/2^n\}}.$$

By the second step there is for every $n \in \mathbb{N}$ a set $K_n \in \mathfrak{K}$ such that $K_n \subset K$, $\mu(K \setminus K_n) < \varepsilon/2^n$, and $f_n|_{K_n}$ is continuous. Define $L := \bigcap_{n \in \mathbb{N}} K_n$. Then L has the desired properties. The reader should note that $(f_n)_{n \in \mathbb{N}}$ converges uniformly on L to f.

Fourth, let f be an arbitrary function. Define

$$g := \frac{1}{\pi}\left(\arctan f + \frac{\pi}{2}\right)$$

Apply the third step.

($\beta 2$) $\Rightarrow$ ($\beta 3$) This proof is left to the reader.

($\beta 3$) $\Rightarrow$ ($\beta 1$). Consider first the case $f \geq 0$. Take $A \in \mathfrak{R}_\delta$ and $\alpha \in \mathbb{R}_+$. There is an increasing sequence $(K_n)_{n \in \mathbb{N}}$ from $\mathfrak{R}$ such that $\bigcup_{n \in \mathbb{N}} K_n \subset A$ and $(A \setminus \bigcup_{n \in \mathbb{N}} K_n) \in \mathfrak{N}(\mu)$. Take $n \in \mathbb{N}$. There is a disjoint sequence $(L_m)_{m \in \mathbb{N}}$ from $\mathfrak{R}$ such that $\bigcup_{m \in \mathbb{N}} L_m \subset K_n$, $(K_n \setminus \bigcup_{m \in \mathbb{N}} L_m) \in \mathfrak{N}(\mu)$, and $f|_{L_m}$ is continuous for every $m \in \mathbb{N}$. For every $m \in \mathbb{N}$ we have $f \wedge \alpha e_{L_m} \in \mathcal{L}^1(\mu)$, and we get

$$f \wedge \alpha e_{K_n} = \sum_{m \in \mathbb{N}} (f \wedge \alpha e_{L_m}) \ \mu\text{-a.e.}$$

From

$$\sum_{m \in \mathbb{N}} \int (f \wedge \alpha e_{L_m}) \, d\mu \leq \alpha \mu(K_n) < \infty$$

it follows that $f \wedge \alpha e_{K_n} \in \mathcal{L}^1(\mu)$. $(f \wedge \alpha e_{K_n})_{n \in \mathbb{N}}$ is an increasing sequence from $\mathcal{L}^1(\mu)$ and $\bigvee_{n \in \mathbb{N}} (f \wedge \alpha e_{K_n}) = f \wedge \alpha e_A$ μ-a.e. From

$$\bigvee_{n \in \mathbb{N}} (f \wedge \alpha e_{K_n}) \leq \alpha e_A$$

it follows that $f \wedge \alpha e_A \in \mathcal{L}^1(\mu)$. From this and from the fact that A and α are arbitrary, we conclude that f is μ-measurable. For an arbitrary function f our assertion now follows from $f = f^+ - f^-$.

5.4.22(E) Let X be a set, $\mathfrak{R}$ a ring of subsets of X, and μ, ν positive measures on $\mathfrak{R}$. Prove:

(α) $(\mu + \nu)^* = \mu^* + \nu^*$

(β) $(\mu + \nu)_* = \mu_* + \nu_*$

5.4.23(E) Let $(X, \mathfrak{R}, \mu)$ be a positive measure space. Set

$$\mathfrak{S} := \{ A \subset X \mid A \cap B \in \mathfrak{R} \text{ for all } B \in \mathfrak{R} \}$$

and

$$\mathfrak{T} := \left\{ A \in \mathfrak{S} \mid \sup_{B \in \mathfrak{R}} \mu(A \cap B) < \infty \right\}.$$

Prove the following propositions:

(α) $\mathfrak{S}$ is an algebra of sets on X and any set $A \in \mathfrak{S}$ is $\mathfrak{R}_\delta$-measurable.

(β) $\mathfrak{T}$ is a ring of sets. It is an algebra of sets iff μ is bounded.

(γ) $\mathfrak{T} \subset \mathfrak{L}(\mu)$.

(δ) $\mu^X(A) = \sup_{B \in \mathfrak{R}} \mu(A \cap B)$ for every $A \in \mathfrak{T}$.

5.4.24(E) Let $(X, \mathfrak{R}, \mu)$ and $(X, \mathfrak{S}, \nu)$ be positive measure spaces. Prove that the following assertions are equivalent:

(α) For every $f \in \mathscr{L}^1(\mu)$, $f \in \mathscr{L}^1(\nu)$ and $\int f\,d\mu = \int f\,d\nu$.

(β) For every $f \in \overline{\mathbb{R}}^X$, $\int_* f\,d\mu \le \int_* f\,d\nu \le \int^* f\,d\nu \le \int^* f\,d\mu$.

(γ) For every $A \subset X$, $\mu_*(A) \le \nu_*(A) \le \nu^*(A) \le \mu^*(A)$.

5.4.25(E) Let $(X, \mathfrak{R}, \mu)$ and $(X, \mathfrak{S}, \nu)$ be positive measure spaces such that $(X, \mathfrak{S}, \nu)$ is an admissible extension of $(X, \mathfrak{R}, \mu)$. Prove that every μ-measurable set is ν-measurable. The converse does not hold in general.

5.5. STIELTJES MEASURES

NOTATION FOR SECTION 5.5:
The same as for Sections 2.5 and 3.6

This section continues our study of the Stieltjes functionals. Although the integral that comes from extending such a functional has already been thoroughly described, we want to study the measure-theoretic aspects of these functionals. The measures in question are of interest in their own right, and they also provide a nontrivial example for the theory of the preceding sections.

Corollary 5.5.1. *Let ℓ be a positive, linear, nullcontinuous functional on $\mathscr{K}(A)$. Then the positive measure space $(A, \mathfrak{R}(\mathscr{L}^1(\ell)), \mu^{\int_\ell})$ induced by the integral $(A, \mathscr{L}^1(\ell), \int_\ell)$ of the Daniell space $(A, \mathscr{K}(A), \ell)$ is a closed positive measure space. The pair $(A, \mathfrak{R}(\mathscr{L}^1(\ell)))$ is a σ-finite measurable space.*

Proof. The first assertion means that $(A, \mathscr{L}^1(\ell), \int_\ell)$ is a closed Daniell space and $\mathscr{L}^1(\ell)$ is a Stone lattice. Since $\mathscr{K}(A)$ is a Stone lattice, so is $\mathscr{L}^1(\ell)$ [Proposition 5.1.2 (b)]. Second, since intervals from A that are compact in A belong to $\mathfrak{R}(\mathscr{L}^1(\ell))$, it follows that the pair $(A, \mathfrak{R}(\mathscr{L}^1(\ell)))$ is σ-finite. That $\mathfrak{R}(\mathscr{L}^1(\ell))$ is a δ-ring has already been noted (Theorem 5.1.8). □

In Section 2.5 we associated to each increasing real-valued function g on A a positive, linear, nullcontinuous functional ℓ_g defined on $\mathscr{K}(A)$. The closure of the Daniell space $(A, \mathscr{K}(A), \ell_g)$ was described in Corollary 3.6.7, and in Section 4.2 we showed that the integral $(A, \mathscr{L}^1(\ell_g), \int_{\ell_g})$ is just that same closure (Proposition 4.2.15, Example 4.2.16). Now, in the manner described in Corollary 1, we can associate to each such g a positive measure, which we call μ_g, as well as a closed positive measure space.

Definition 5.5.2. *Let g be an increasing real-valued function on A. Then the integral for the Daniell space $(A, \mathscr{K}(A), \ell_g)$, which we denote as usual by $(A, \mathscr{L}^1(\ell_g), \int_{\ell_g})$, is called the* **Stieltjes integral** *associated with g. The positive measure induced on $\mathfrak{R}(\mathscr{L}^1(\ell_g))$ by $\int_{\ell_g}$ shall be denoted by μ_g. The closed positive measure space $(A, \mathfrak{R}(\mathscr{L}^1(\ell_g)), \mu_g)$ induced by the Stieltjes integral associated with g is called the* **Stieltjes measure space** *associated with g. The measure μ_g is called the* **Stieltjes measure** *on A associated with g. For $f \in \mathscr{L}^1(\mu_g)$, we write*

$$\int_A f\,dg := \int_A f(x)\,dg(x) := \int_A f\,d\mu_g.$$

Quite generally, a positive measure μ on a ring of subsets of A is said to be a **Stieltjes measure** *on A iff $\mu = \mu_g$ for some increasing real-valued function g on A.*

□

According to this definition, the Daniell-space integral called the Stieltjes integral associated with g is the same as the measure-space integral of the Stieltjes measure space associated with g; that is,

$$\left(A, \mathfrak{R}\left(\mathscr{L}^1(\ell_g)\right), \mu_g\right) = \left(A, \mathfrak{L}(A, \mu_g), \mu_g\right).$$

In particular,

$$\mathfrak{L}(A, \mu_g) = \mathfrak{R}\left(\mathscr{L}^1(\ell_g)\right).$$

As was already remarked, the Stieltjes integral associated with g is also the same as the closure of the Daniell space associated with g; that is,

$$\left(A, \mathscr{L}^1(\ell_g), \int_{\ell_g}\right) = \left(A, \bar{\mathscr{L}}(\ell_g), \bar{\ell}_g\right).$$

Stieltjes measures require closer investigation. We want to know what kind of sets belong to $\mathfrak{L}(A, \mu_g)$ and what values μ_g takes on these sets. Some of the properties to be established here will be encountered again in the treatment of measures on Hausdorff spaces. Indeed, A is a Hausdorff space, and the Stieltjes measures on A provide an example for the general theory developed later. The reader should keep in mind, however, that A is also an ordered set, and we began with this order property. The important object, after all, is the increasing function g. It might be mentioned that we could have used other function spaces instead of $\mathscr{K}(A)$, for instance a space of step functions.

In preparation for describing what kind of sets belong to $\mathfrak{L}(A, \mu_g)$ we want to define Borel sets, K_σ-sets, and G_δ-sets, and to examine a few of their properties.

Definition 5.5.3. *$\mathfrak{B}(A)$ shall denote the δ-ring generated by the subsets of A that are closed in A. Sets belonging to $\mathfrak{B}(A)$ are called the* **Borel sets** *in A. We use "**Borel measurable**" as a synonym for "$\mathfrak{B}(A)$-measurable" (provided A is evident from the context).*

$\mathfrak{B}_b(A)$ shall denote the set consisting of all Borel sets in A that are also bounded in A.

A subset C of A is called a **K_σ-set** *in A (**F_σ-set** in A) iff it can be written as the union of a countable family of sets that are compact (closed) in A. A subset C of A is called a* **G_δ-set** *in A iff it can be written as the intersection of a countable family of sets that are open in A.* □

The reader can readily verify the following proposition.

Proposition 5.5.4.

(*a*) *Intervals from A are K_σ-sets in A.*

(*b*) *The K_σ-sets in A, the F_σ-sets in A, and the G_δ-sets in A are all Borel sets in A.* □

Proposition 5.5.5.

(*a*) *The Borel sets in A form a σ-algebra on A.*

(*b*) *$\mathfrak{B}(A)$ is the σ-algebra on A generated by the sets that are compact in A.*

(*c*) *$\mathfrak{B}(A)$ is the σ-algebra on A generated by the closed intervals from A.*

(*d*) *The Borel sets in A that are bounded in A form a δ-ring. In fact, $\mathfrak{B}_b(A)$ is the δ-ring generated by the closed intervals from A.*

(*e*) *For $B \subset A$, B is Borel-measurable iff B is $\mathfrak{B}_b(A)$-measurable.*

Proof. (a) Since A is closed in A, A is a Borel set in A. As a δ-ring on A that contains A, $\mathfrak{B}(A)$ is a σ-algebra on A (Proposition 5.2.16).

(b) Let $\mathfrak{S}$ denote the σ-algebra on A generated by the sets that are compact in A. Since compact sets in A are also closed in A, $\mathfrak{S} \subset \mathfrak{B}(A)$, by (a). Sets that are open in A are countable unions of intervals from A (Proposition 2.5.4). By Proposition 4 (a), sets that are open in A are K_σ-sets in A and therefore belong to $\mathfrak{S}$. We conclude that all sets closed in A must also belong to $\mathfrak{S}$ and $\mathfrak{B}(A) \subset \mathfrak{S}$.

(d) Denote by $\mathfrak{R}$ the specified δ-ring. Evidently

$$\mathfrak{R} \subset \mathfrak{B}_b(A).$$

For D a closed interval from A, set

$$\mathfrak{S}(D) := \{B \in \mathfrak{B}_b(A) \mid B \cap D \in \mathfrak{R}\}.$$

Using the fact that δ-rings are conditionally σ-complete (Proposition 2.1.5), we conclude that $\mathfrak{S}(D)$ is a σ-ring. Obviously $\mathfrak{S}(D)$ contains all closed intervals from A. But then $\mathfrak{S}(D)$ also contains all subsets of A that are open in A, since each such set is the union of countably many closed intervals from A. In particular, A itself belongs to $\mathfrak{S}(D)$, so $\mathfrak{S}(D)$ is a σ-algebra. Thus $A \setminus C$ belongs to $\mathfrak{S}(D)$ whenever C does, so $\mathfrak{S}(D)$ contains all subsets of A that are closed in A. It follows that

$$\mathfrak{B}(A) \subset \mathfrak{S}(D)$$

for every closed interval D from A.

Now let $B \in \mathfrak{B}_b(A)$. Choose in A a closed interval D that contains B. Then B belongs to $\mathfrak{S}(D)$, so $B \cap D$, which is B itself, belongs to $\mathfrak{R}$. We have shown that

$$\mathfrak{B}_b(A) \subset \mathfrak{R}.$$

(c) Assertion (c) follows from (a) and (d).

(e) The inclusion $\mathfrak{M}(A, \mathfrak{B}(A)) \subset \mathfrak{M}(A, \mathfrak{B}_b(A))$ is evident; so let $C \in \mathfrak{M}(A, \mathfrak{B}_b(A))$. Since $\mathfrak{B}(A)$ is a σ-algebra on A, we have

$$\mathfrak{M}(A, \mathfrak{B}(A)) = \mathfrak{B}(A).$$

Choose an increasing sequence $(A_n)_{n \in \mathbb{N}}$ of intervals from A that are compact in A and whose union is A. For each n, the set $C \cap A_n$ belongs to $\mathfrak{B}_b(A)$ and therefore to $\mathfrak{B}(A)$. Since $\mathfrak{B}(A)$ is a σ-algebra and C can be written as the union of the sequence $(C \cap A_n)_{n \in \mathbb{N}}$, it follows that C belongs to $\mathfrak{B}(A)$, that is, to $\mathfrak{M}(A, \mathfrak{B}(A))$. □

We can now characterize μ_g-integrable sets.

Theorem 5.5.6. *Let μ be a Stieltjes measure on A. Then the following assertions hold.*

(a) Every Borel set in A is μ-measurable.

(b) Every Borel set in A that is bounded in A is μ-integrable.

Moreover, the following assertions are equivalent, for every subset D of A.

(c) D is μ-integrable.

(d) For every real number $\varepsilon > 0$ there exist a set B that is open in A and a set C that is compact in A, both belonging to $\mathfrak{L}(A, \mu)$ such that

$$C \subset D \subset B \quad \text{and} \quad \mu(B) - \mu(C) < \varepsilon.$$

(e) There exist a G_δ-set B and a K_σ-set C, both belonging to $\mathfrak{L}(A, \mu)$ such that

$$C \subset D \subset B \quad \text{and} \quad B \setminus C \in \mathfrak{N}(A, \mu).$$

(f) D can be written as the disjoint union of a μ-integrable K_σ-set in A and a μ-null set in A.

Proof. (a) Let C be a closed interval from A. It is possible to construct a function f belonging to $\mathscr{K}(A)_+$ such that

$$C = \{f = 1\}.$$

Such an f is μ-integrable and therefore μ-measurable. It follows that C is μ-measurable [Proposition 5.3.3 (f)]. Since $\mathfrak{M}(A, \mu)$ is a σ-algebra (Proposition 5.3.2), Proposition 5(c) implies that

$$\mathfrak{B}(A) \subset \mathfrak{M}(A, \mu).$$

(b) Let $B \in \mathfrak{B}_b(A)$. There exists a closed interval C from A such that $B \subset C$. Taking f as in the proof of (a), we have $e_B \le f$. Since e_B is μ-measurable and f is μ-integrable, it follows that e_B is μ-integrable [Theorem 5.4.8 (a)]. Thus B belongs to $\mathfrak{L}(A, \mu)$.

(c) $\Rightarrow$ (d). Let g be an increasing real-valued function on A such that $\mu = \mu_g$, and let $\varepsilon > 0$ be given. For D to be μ-integrable means that e_D belongs to $\mathscr{L}^1(A, \ell_g)$ and therefore to $\overline{\mathscr{L}}(\ell_g)$. According to Corollary 3.6.7, there exist an upper semicontinuous function f' and a lower semicontinuous function f'' such that

$$f' \le e_D \le f''$$

$$\forall \alpha \in \mathbb{R}, \alpha > 0, \{f' \ge \alpha\} \text{ is bounded in } A$$

$$\overline{\ell}_g(f'') - \overline{\ell}_g(f') < \varepsilon/2.$$

The sequence

$$\left(e_{\{f' \ge 1/n\}}\right)_{n \in \mathbb{N}}$$

is an increasing μ-sequence whose supremum lies between f' and e_D. Therefore

$$\overline{\ell}_g(f') \le \sup_{n \in \mathbb{N}} \overline{\ell}_g(e_{\{f' \ge 1/n\}})$$

and there exists $m \in \mathbb{N}$ such that

$$\overline{\ell}_g(e_{\{f' \ge 1/m\}}) > \overline{\ell}_g(f') - \frac{\varepsilon}{2}.$$

Set

$$B := \{f'' > 1\}$$

$$C := \left\{f' \ge \frac{1}{m}\right\}.$$

By Proposition 3.6.5, B is open in A and C is closed in A. The choice of f' guarantees that C is also bounded in A and therefore compact in A. By (b), C belongs to $\mathfrak{L}(A, \mu)$. By (a), B is μ-measurable, and the inequality $e_B \leq f''$ guarantees

$$\mu^*(B) \leq \int_X^* f''\, d\mu = \overline{\ell}_g(f'') < \infty$$

so B also belongs to $\mathfrak{L}(A, \mu)$ (Theorem 5.4.8). By construction, $C \subset D \subset B$ and

$$\begin{aligned} \mu(B) - \mu(C) &= \overline{\ell}_g(e_B) - \overline{\ell}_g(e_C) \\ &\leq \overline{\ell}_g(f'') - \overline{\ell}_g(f') + \frac{\varepsilon}{2} < \varepsilon. \end{aligned}$$

(d) ⇒ (e). Using (d), we can choose from $\mathfrak{L}(A, \mu)$ a sequence $(C_n)_{n \in \mathbb{N}}$ of sets that are compact in A and a sequence $(B_n)_{n \in \mathbb{N}}$ of sets that are open in A, such that for every n,

$$C_n \subset D \subset B_n \quad \text{and} \quad \mu(B_n) - \mu(C_n) < \frac{1}{n}.$$

Let

$$B := \bigcap_{n \in \mathbb{N}} B_n \qquad C := \bigcup_{n \in \mathbb{N}} C_n.$$

Noting that $\mu(B \setminus C) = 0$, we see that B and C have all the required properties.

(e) ⇒ (f) ⇒ (c). These implications are all trivial. □

Corollary 5.5.7. *Let μ be a Stieltjes measure on A. Then, for every set $D \in \mathfrak{L}(A, \mu)$,*

$$\begin{aligned} \mu(D) &= \inf\{\mu(B) \mid B \in \mathfrak{L}(A, \mu), D \subset B,\ B \text{ is open in } A\} \\ &= \sup\{\mu(C) \mid C \in \mathfrak{L}(A, \mu), C \subset D,\ C \text{ is compact in } A\}. \end{aligned}$$ □

Corollary 5.5.8. *Every lower semicontinuous extended-real-valued function on A is Borel measurable and $\mathfrak{B}_b(A)$-measurable, and it is μ-measurable for every Stieltjes measure μ on A. The same assertions held for every upper semicontinuous extended-real-valued function on A.*

Proof. Note that for each real α, the set $\{f < \alpha\}$ is open in A if f is upper semicontinuous and is an F_σ-set in A if f is lower semicontinuous (Proposition 3.6.5). Now apply the definitions and use Theorem 6 (a) as needed. □

Proposition 5.5.9. *Let μ be a Stieltjes measure on A. Let $\mathfrak{R}$ be a set-ring contained in $\mathfrak{L}(A,\mu)$ such that $\mathfrak{R}_\delta$ contains every interval from A that is bounded in A. Set*

$$\nu := \mu|_{\mathfrak{R}}.$$

Then ν is a positive measure on $\mathfrak{R}$, such that the closure and the completion of $(A,\mathfrak{R},\nu)$ are identical and coincide with the given Stieltjes measure space:

$$(\mathfrak{L}(A,\mu),\mu) = (\mathfrak{L}(A,\nu),\nu^A) = (\mathfrak{R}(\nu),\bar{\nu}).$$

Proof. By Proposition 2.1.10,

$$\mathfrak{R}_\delta \subset \{B \subset A \mid B \subset C \text{ for some } C \in \mathfrak{R}\}.$$

Using the hypothesis on intervals from A, we conclude that A can be written as a countable union of sets belonging to $\mathfrak{R}$. In other words, the pair $(A,\mathfrak{R})$ is σ-finite. The closure and the completion of $(A,\mathfrak{R},\nu)$ are therefore identical (Proposition 5.2.13). Since $(A,\mathfrak{L}(A,\mu),\mu)$ is a closed positive measure space extending $(A,\mathfrak{R},\nu)$, we conclude that

$$(\mathfrak{L}(A,\nu),\nu^A) = (\mathfrak{R}(\nu),\bar{\nu}) \preccurlyeq (\mathfrak{L}(A,\mu),\mu)$$

(5.2.2). We only need to show that

$$\mathfrak{L}(A,\mu) \subset \mathfrak{L}(A,\nu).$$

Let

$$\mathfrak{S} = \{B \in \mathfrak{L}(A,\mu) \mid B \in \mathfrak{L}(A,\nu) \quad \text{and} \quad \nu^A(B) = \mu(B)\}.$$

We claim that all of the μ-integrable G_δ-sets from A and all of the μ-integrable K_σ-sets from A belong to $\mathfrak{S}$.

Let $B \in \mathfrak{L}(A,\mu)$, and suppose that B is open in A. Then B can be written as a countable disjoint union of intervals open in A. Each interval open in A can be written as a countable disjoint union of intervals (half-open, half-closed) that are bounded in A. These latter intervals belong to $\mathfrak{S}$. Countable additivity (Theorem 3.5.1) together with Theorem 5.1.8 (h) imply that B belongs to $\mathfrak{S}$. Since $\mathfrak{L}(A,\mu) \cap \mathfrak{L}(A,\nu)$ is a δ-ring and μ and ν^A are nullcontinuous, it follows that μ-integrable G_δ-sets in A must all belong to $\mathfrak{S}$.

Now let $B \in \mathfrak{L}(A,\mu)$, and suppose that B is compact in A. Choose a bounded interval C from A such that $B \subset C$ and C is open in A. Then C belongs to $\mathfrak{L}(A,\mu)$ [Theorem 6 (b)], $C \setminus B$ is open in A, and both C and $C \setminus B$ must belong to $\mathfrak{S}$. Writing B as $C \setminus (C \setminus B)$, we see that B belongs to $\mathfrak{S}$. Each K_σ-set from A can be written as the union of a countable, directed upward family of sets that are compact in A. Using Theorem 5.1.8 (f), we conclude that μ-integrable K_σ-sets in A belong to $\mathfrak{S}$.

Having verified our claim, we need only apply Theorem 6 (c) ⇒ (e) to see that $\mathfrak{L}(A, \mu) \subset \mathfrak{L}(A, \nu)$. Indeed, let $B \in \mathfrak{L}(A, \mu)$. By Theorem 6 (c) ⇒ (e), there exist a μ-integrable G_δ-set D and a μ-integrable K_σ-set C such that

$$C \subset B \subset D$$

and $D \setminus C$ is μ-null. Both C and D belong to $\mathfrak{S}$, so $D \setminus C$ is ν-null, so $B \setminus C$ is ν-null. As the union of the ν-integrable set C and the ν-null set $B \setminus C$, B is ν-integrable. □

Proposition 9 is very useful. It enables us to specify simple set systems for which the accompanying collection of measures is in one-to-one correspondence with the set of Stieltjes measures on A. This possibility, which we pursue more thoroughly in Volume 2 after real Stieltjes measures have been defined, makes the rather complicated Stieltjes measures easier to manage.

In preparation for stating the values assigned by Stieltjes measures to the intervals from A, we make a straightforward observation, which the reader can readily verify.

Proposition 5.5.10. *The following assertions hold, for every increasing real-valued function g on A.*

(*a*) *For every $x \in A \setminus \{a\}$, $g(x-)$ exists and*

$$g(x-) = \sup\{ g(y) \mid y \in]a, x[\}.$$

(*b*) *For every $x \in A \setminus \{b\}$, $g(x+)$ exists and*

$$g(x+) = \inf\{ g(y) \mid y \in]x, b[\}.$$ □

Theorem 5.5.11. *Let g be an increasing real-valued function on A. If $x, y \in A$ and $x \leq y$, then*

$$\mu_g([x, y]) = g(y+) - g(x-).$$

If $x, y \in A$ and $x < y$, then

$$\mu_g(]x, y]) = g(y+) - g(x+)$$

$$\mu_g([x, y[) = g(y-) - g(x-)$$

$$\mu_g(]x, y[) = g(y-) - g(x+).$$

For each $x \in A$, $\{x\}$ is a μ_g-null set iff g is continuous at x.

Proof. Let x and y belong to A with $x \leq y$. First, choose from A a sequence $(x_n)_{n \in \mathbb{N}}$ that is either strictly increasing (if $x \neq a$) or constant (if $x = a$) but in any case has its supremum equal to x. Second, choose from A a sequence $(y_n)_{n \in \mathbb{N}}$ that is either strictly decreasing (if $x \neq b$) or constant (if $x = b$) but in any case has its infimum equal to y. For n in $\mathbb{N}$ define

$$f_n\colon A \to \mathbb{R}, \qquad z \mapsto \begin{cases} 1 & \text{if } z \in [x, y] \\ 0 & \text{if } z \in A \setminus [x_n, y_n] \\ \dfrac{z - x_n}{x - x_n} & \text{if } z \in [x_n, x[\\ \dfrac{y_n - z}{y_n - y} & \text{if } z \in]y, y_n]. \end{cases}$$

Note that $(f_n)_{n \in \mathbb{N}}$ is a decreasing sequence from $\mathscr{K}(A)$ whose infimum is $e_{[x,y]}$. Hence

$$\mu_g([x, y]) = \inf_{n \in \mathbb{N}} \ell_g(f_n).$$

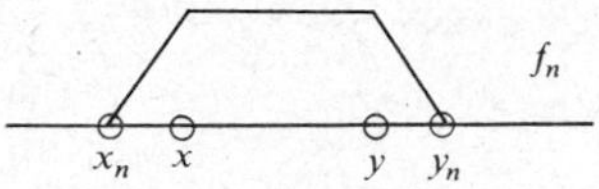

Figure 1

Consider the partition

$$p := \left(x_n, x - \frac{1}{n}(x - x_n), x, y, y + \frac{1}{n}(y_n - y), y_n\right)$$

of the interval $[x_n, y_n]$. Estimating $\ell_g(f_n)$ from above and below by $\varphi^*(f_n, g; p)$ and $\varphi_*(f_n, g; p)$, respectively, we have

$$\begin{aligned} \left(1 - \frac{1}{n}\right)&\left[g(x) - g\left(x - \frac{1}{n}(x - x_n)\right)\right] + [g(y) - g(x)] \\ &+ \left(1 - \frac{1}{n}\right)\left[g\left(y + \frac{1}{n}(y_n - y)\right) - g(y)\right] \\ &\leq \ell_g(f_n) \\ &\leq \left(1 - \frac{1}{n}\right)\left[g\left(x - \frac{1}{n}(x - x_n)\right) - g(x_n)\right] \\ &+ \left[g(x) - g\left(x - \frac{1}{n}(x - x_n)\right)\right] + [g(y) - g(x)] \\ &+ \left[g\left(y + \frac{1}{n}(y_n - y)\right) - g(y)\right] \\ &+ \left(1 - \frac{1}{n}\right)\left[g(y_n) - g\left(y + \frac{1}{n}(y_n - y)\right)\right]. \end{aligned}$$

By Proposition 10,

$$\lim_{n\to\infty} g\left(x - \frac{1}{n}(x - x_n)\right) = g(x-)$$

$$\lim_{n\to\infty} g\left(y + \frac{1}{n}(y_n - y)\right) = g(y+)$$

so

$$\lim_{n\to\infty}\left[g\left(x - \frac{1}{n}(x - x_n)\right) - g(x_n)\right] = \lim_{n\to\infty}\left[g(y_n) - g\left(y + \frac{1}{n}(y_n - y)\right)\right]$$

$$= 0.$$

Let n tend to ∞ in the estimate for $\ell_g(f_n)$ to obtain

$$g(y+) - g(x-) \leq \inf_{n\in\mathbb{N}} \ell_g(f_n) \leq g(y+) - g(x-).$$

Consequently

$$\mu_g([x, y]) = g(y+) - g(x-).$$

Now suppose that $x, y \in A$ and $x < y$. Since

$$[x, y] =]x, y] \cup [x, x] = [x, y[\cup [y, y]$$

$$[x, y[=]x, y[\cup [x, x]$$

and these are all disjoint unions, the remaining formulas follow from the formula already established.

From what has already been proved, $\mu_g(\{x\}) = g(x+) - g(x-)$ for every x in A. Thus $\mu_g(\{x\}) = 0$ iff g is continuous at x. □

Corollary 5.5.12. *For $A =]a, b[$, the following assertions hold, for every $x \in A$ and for every increasing real function g on A.*

(*a*) $]a, x[\in \mathfrak{L}(\mu_g) \Leftrightarrow]a, x] \in \mathfrak{L}(\mu_g) \Leftrightarrow g(a+) \neq -\infty$, *and in this case*

$$\mu_g(]a, x[) = g(x-) - g(a+)$$

$$\mu_g(]a, x]) = g(x+) - g(a+).$$

(*b*) $]x, b[\in \mathfrak{L}(\mu_g) \Leftrightarrow [x, b[\in \mathfrak{L}(\mu_g) \Leftrightarrow g(b-) \neq +\infty$, *and in this case*

$$\mu_g(]x, b[) = g(b-) - g(x+)$$

$$\mu_g([x, b[) = g(b-) - g(x-).$$

(*c*) $A \in \mathfrak{L}(\mu_g) \Leftrightarrow g$ *is bounded* $\Leftrightarrow \mu_g$ *is bounded, and in this case*

$$\mu_g(A) = g(b-) - g(a+).$$

(*d*) $\mu_g = 0 \Leftrightarrow g$ *is constant.*
(*e*) *The sets*

$$\{y \in A \mid \exists z \in]y, b[, g(z) = g(y-)\}$$

$$\{y \in A \mid \exists z \in]a, y[, g(z) = g(y+)\}$$

are Borel sets that are μ_g-null.

Proof. (a)–(d) follow immediately from Theorem 11.
In (e) each of the sets can be written as the union of a countable family of intervals from A that are μ_g-null. □

Corollary 5.5.13. *The following assertions are equivalent, for every increasing real-valued function g on A.*

(*a*) *The function g is continuous on A.*
(*b*) *Every countable subset of A is μ_g-null.*
(*c*) *For every $x \in A$, the set $\{x\}$ is μ_g-null.* □

Corollary 5.5.14. *Two positive Stieltjes measures on A coincide iff they coincide on the closed intervals from A.* □

EXERCISES

5.5.1(E) Let A be an interval of $\mathbb{R}$ satisfying the prescriptions of this section. Prove the following:

(α) Take $g \in \mathbb{R}^A$ increasing. For each $x \in A$ define $\alpha_x := g(x+) - g(x-)$. Then $\alpha_x \geq 0$ for each $x \in A$, and g is continuous at x iff $\alpha_x = 0$.
(β) If $[\alpha, \beta] \subset A$, then $(\alpha_x)_{x \in [\alpha, \beta]}$ is summable, and $\Sigma_{x \in [\alpha, \beta]} \alpha_x \leq g(\beta) - g(\alpha)$.
(γ) The set of points at which g is discontinuous is countable.

(δ) If g_1 and g_2 are increasing functions on A with the same points of continuity such that $g_1(x) = g_2(x)$ in each point x of continuity, then $\mu_{g_1} = \mu_{g_2}$.

(ε) If g is an increasing function on A, then there is exactly one left-continuous increasing function $\tilde{g}$ on A such that $g(x) = \tilde{g}(x)$ in every point x of continuity of g.

(ζ) μ_g restricted to the ring of sets of the interval forms of A is identical with $\mu_{\tilde{g}}$ in the sense of Ex. 2.2.10.

(η) If g is an increasing function on A, then $(A, \mathfrak{R}(\mathscr{L}^1(\ell_g)), \mu_g) = (A, \mathfrak{L}(\mu_{\tilde{g}}), \mu_{\tilde{g}}^A)$.

Hence we have available two different paths to a positive Stieltjes measure on A. The path via the functionals on $\mathscr{K}(A)$ leads to the same thing as does the path via the measures on the interval forms. The definition using the positive linear functionals on $\mathscr{K}(A)$ is somewhat more general in that it makes no assumptions about the increasing function g. The construction in Ex. 2.2.10 used essentially the left continuity of g. (δ) and (ε) show that this is no real restriction on g.

5.5.2(E) The Cantor Set. Consider the interval $[0,1]$. The Cantor set $C \subset [0,1]$ was introduced in Ex. 1.1.12, and its topological properties were discussed there. We now turn to some properties of C that are important for measure theory. For $n \in \mathbb{N}$ we denote by I_{ni} ($i \in \mathbb{N}_{2^{n-1}}$) the nth distinguished open complementary intervals (as in Ex. 1.1.12). Prove the following:

(α) There is exactly one increasing function g on $[0,1]$ such that g is equal to $(2i-1)/2^n$ on I_{ni} for each $n \in \mathbb{N}$ and $i \in \mathbb{N}_{2^{n-1}}$.

(β) The function g defined in (α) is continuous, and $g(0) = 0$, $g(1) = 1$.

(γ) $\mu_g(C) = 1$, but C is a Lebesgue null-set.

(δ) For each $\alpha \in [0,1]$ there is an increasing function h on $[0,1]$ such that $\mu_h(C) = \alpha$ and $h(0) = 0$, $h(1) = 1$.

(ε) Define

$$f: [0,1] \to \mathbb{R}, \qquad x \mapsto \begin{cases} 0 & \text{for } x \in C \\ n & \text{for } x \in I_{ni} \end{cases}$$

Then $\int f\, d\lambda = 3$, where λ is the Lebesgue measure on $[0,1]$, and $\int f\, d\mu_g = 0$ where g is the function described in (α).

(ζ) Set $\mathfrak{R} := \{[\alpha, \beta[\setminus C \mid \alpha, \beta \in [0,1]\}$ and

$$\nu: \mathfrak{R} \to \mathbb{R}, \qquad [\alpha, \beta[\setminus C \mapsto g(\beta) - g(\alpha).$$

Then the following are true:

($\zeta 1$) $\mathfrak{R}$ is a semi-ring.

($\zeta 2$) ν is an additive positive content such that $\inf_{n \in \mathbb{N}} \nu(A_n) = 0$ for any decreasing sequence $(A_n)_{n \in \mathbb{N}}$ from $\mathfrak{R}$ with $\bigcap_{n \in \mathbb{N}} A_n = \varnothing$.

($\zeta 3$) ν is not σ-additive.

(η) If μ is a positive additive real function on the semi-ring $\{[x, y[\,|\, x, y \in \mathbb{R}\}$, then μ is σ-additive iff it is nullcontinuous.

To prove (δ), take g as in (α), and define $h(x) := \alpha g(x) + (1 - \alpha)x$ for every $x \in [0,1]$. (ε) reduces to calculating $\Sigma_{n \in \mathbb{N}} n2^{n-1}/3^n$. Observe that $\Sigma_{n \in \mathbb{N}} x^n = 1/(1 - x)$ for $|x| < 1$. Differentiate to find $\Sigma_{n \in \mathbb{N}} nx^{n-1} = 1/(1 - x)^2$. Thus

$$\sum_{n \in \mathbb{N}} \frac{n2^{n-1}}{3^n} = \frac{1}{3} \sum_{n \in \mathbb{N}} n\left(\frac{2}{3}\right)^{n-1} = 3.$$

5.5.3(E) Let A be an interval satisfying the general prescriptions of this section. Prove the propositions that follow:

(α) Let g be an increasing function on A. Then every nonempty perfect subset of A contains a nonempty nowhere dense perfect set of μ_g-measure 0.

(β) For every increasing function g on A there is a dense K_σ-set of μ_g-measure 0 with the cardinality $\mathfrak{c}$ of the continuum.

For (α), the construction is essentially the same as in the case of the Cantor set. For the proof of (β) use the result of (α). Prove the following consequence:

(γ) Given an increasing function g on A, there is a μ_g-measurable set B that is not a Borel set. B can be chosen to be a μ_g-null-set.

A simple counting argument suffices for (γ). The set of all Borel subsets of A has the cardinality $\mathfrak{c}$. On the other hand, there is a μ_g-null-set which has also the cardinality $\mathfrak{c}$. Every subset of this null-set is also a null-set, and there are $2^{\mathfrak{c}} > \mathfrak{c}$ of them.

5.5.4(E) Let A be an interval as prescribed at the beginning of this section. Let $g \in \mathbb{R}^A$ be increasing. Consider μ_g on $\mathfrak{K}$, the set of all compact subsets of A. Prove the following:

(α) $\mu_g(K \cup L) \le \mu_g(K) + \mu_g(L)$ for any $K, L \in \mathfrak{K}$.

(β) $\mu_g(K \cup L) = \mu_g(K) + \mu_g(L)$ for any $K, L \in \mathfrak{K}$, $K \cap L = \emptyset$.

(γ) $\mu_g(\bigcap_{\iota \in I} K_\iota) = \inf_{\iota \in I} \mu_g(K_\iota)$ for any nonempty downward directed family $(K_\iota)_{\iota \in I}$ from $\mathfrak{K}$.

Observe that μ_g is $\mathfrak{K}$-regular in the sense of Ex. 5.2.16. Conclude the following result:

(δ) $(A, \mathfrak{L}(\mu_g), \mu_g)$ is the greatest $\mathfrak{K}$-regular positive measure space extending $(A, \mathfrak{K}, \mu_g|_{\mathfrak{K}})$.

5.5.5(E) **Parameter Integrals.** Let $(X, \mathfrak{R}, \mu)$ be a positive measure space throughout this exercise.

(a) Let (Y, d) be a metric space. Let $f\colon X \times Y \to \mathbb{R}$ have the following properties:

(i) $f(\cdot, y)\colon X \to \mathbb{R}$, $x \mapsto f(x, y)$ is μ-measurable for each $y \in Y$.
(ii) There is a $g \in \mathscr{L}^1(\mu)$ such that $|f(\cdot, y)| \leq g$ for every $y \in Y$.
(iii) The functions $f(x, \cdot)\colon Y \to \mathbb{R}$, $y \mapsto f(x, y)$ are continuous on Y for μ-almost all $x \in X$.

Prove the following statements:

(α) $f(\cdot, y)$ is μ-integrable for each $y \in Y$.
(β) $h\colon Y \to \mathbb{R}$, $y \mapsto \int f(x, y)\, d\mu(x)$ is continuous on Y.

(b) Suppose that X is a compact space and that $\mathfrak{R}$ contains all compact subsets of X. Let Y be a compact space and f a continuous real function on $X \times Y$. Prove the following:

(α) $f(\cdot, y) \in \mathscr{L}^1(\mu)$ for every $y \in Y$.
(β) The function $Y \to \mathbb{R}$, $y \mapsto \int f(x, y)\, d\mu(x)$ is continuous on Y.

Hint: This is trivial if there are finite families $(f_\iota)_{\iota \in I}$ of continuous functions on X and $(g_\iota)_{\iota \in I}$ of continuous functions on Y such that $f(x, y) = \sum_{\iota \in I} f_\iota(x) g_\iota(y)$ for all $(x, y) \in X \times Y$. For the proof of the general case use the Theorem of Weierstrass and Stone (see Ex. 2.5.2).

(c) Let A be an open subset of $\mathbb{R}$. Let $f\colon X \times A \to \mathbb{R}$ have the following properties:

(i) $f(\cdot, y)$ is μ-measurable for each $y \in A$.
(ii) There is a $z \in A$ such that $f(\cdot, z) \in \mathscr{L}^1(\mu)$.
(iii) There is an $N \in \mathfrak{N}(\mu)$ such that $\partial f(x, y)/\partial y$ exists for each $y \in A$ and each $x \in X \setminus N$.
(iv) There is a $g \in \mathscr{L}^1(\mu)$ such that for each $x \in X \setminus N$ (N as in (iii)) and each $y \in A$, $|\partial f(x, y)/\partial y| \leq g(x)$.

Now set $\partial f(x, y)/\partial y := 0$ for each $x \in N$ and $y \in A$. Prove the following propositions:

(α) $\partial f(\cdot, y)/\partial y \in \mathscr{L}^1(\mu)$ for each $y \in A$.
(β) $f(\cdot, y) \in \mathscr{L}^1(\mu)$ for each $y \in A$.
(γ) Define $h\colon A \to \mathbb{R}$, $y \mapsto \int f(x, y)\, d\mu(x)$. Then h is differentiable with respect to y on A and

$$\frac{dh}{dy}(y) = \int \frac{\partial}{\partial y} f(x, y)\, d\mu(x).$$

(d) We now turn to a special case. Let A and B be intervals of $\mathbb{R}$ with nonempty interior. Let $f\colon B \times A \to \mathbb{R}$ be continuous, and let μ be a Stieltjes measure on B. Prove the following:

(α) $f(\cdot, y) \in \mathscr{L}^1(\mu)$ for every $y \in A$.

(β) $h\colon A \to \mathbb{R},\ y \mapsto \int f(x, y)\, d\mu(x)$ is continuous on A.

(γ) If $f(x, \cdot)$ is differentiable on A for each $x \in B$ and if there is a $k \in \mathscr{L}^1(\mu)$ such that $|\partial f(x, y)/\partial y| \le k(x)$ for each $(x, y) \in B \times A$, then h is differentiable and

$$\frac{dh}{dy}(y) = \int \frac{\partial}{\partial y} f(x, y)\, d\mu(x).$$

5.5.6(E) The Theorem of Lusin for Stieltjes Measures. We refer to Ex. 5.4.21. We discussed there the important connection between measurability and continuity of functions on a Hausdorff space. The reader may prove the following form of the theorem of Lusin for a positive Stieltjes measure μ on the interval A: For any $f \in \overline{\mathbb{R}}^A$, the following statements are equivalent:

(α) f is μ-measurable.

(β) For every real number $\varepsilon > 0$ there is a continuous function $g\colon A \to \mathbb{R}$ such that $\mu^A(\{f \neq g\}) < \varepsilon$.

5.5.7(C) Baire Functions on $\mathbb{R}$. This exercise is devoted to sets of Baire functions on closed intervals of $\mathbb{R}$. For the definitions, we refer the reader to Ex. 5.3.9. We know from there that in our special case the Baire functions coincide with the Borel measurable functions.

The sets of Baire functions, $\mathscr{B}_\alpha$, were defined by a transfinite procedure, which stabilized at $\alpha = \omega_1$ so that $\mathscr{B}_\alpha = \mathscr{B}_{\omega_1}$ for any $\alpha \ge \omega_1$. The question is: Which stages α ($\alpha < \omega_1$) are really necessary? The answer is: All of them are, in the sense that $\mathscr{B}_\alpha = \mathscr{B}_\beta$ for $\alpha, \beta \le \omega_1$ iff $\alpha = \beta$. To prove this, we need some preparation.

(a) Let $[a, b] \subset \mathbb{R}$ be fixed. Let α be an ordinal. Prove the following:

(α) $\mathscr{B}_\alpha \cap \mathbb{R}^{[a,b]}$ is a real Riesz lattice closed under the pointwise multiplication of functions.

(β) If $(f_n)_{n \in \mathbb{N}}$ is a sequence from $\mathscr{B}_\alpha$ that converges uniformly on $[a, b]$ to $f \in \mathbb{R}^{[a,b]}$, then $f \in \mathscr{B}_\alpha$.

We shall also need the following properties:

(γ) If $f \in \mathbb{R}^{[a,b]}$ is discontinuous at only finitely many places, then $f \in \mathscr{B}_1$.

(δ) Take $x \in [0,1]$. Express $x \neq 1$ as a decimal $0.\ \alpha_1(x)\alpha_2(x)\ldots$ in such a way that for each $n \in \mathbb{N}$ there is an $m \in \mathbb{N}$, $m \ge n$ such that $\alpha_m(x) \neq 9$. Set $\alpha_n(1) = 9$ for every $n \in \mathbb{N}$. Then $[0,1] \to \mathbb{R}$, $x \mapsto \alpha_n(x)$ belongs to $\mathscr{B}_1$ for each $n \in \mathbb{N}$.

We also need the notion of universal function for $\mathscr{F} \subset \overline{\mathbb{R}}^{[0,1]}$. Call $g\colon [0,1] \times [0,1] \to \overline{\mathbb{R}}$ a **universal function** for $\mathscr{F}$ if for each $f \in \mathscr{F}$ there is a $y \in [0,1]$ such that $f = g(\cdot, y)$. [Define $g(\cdot, y)\colon [0,1] \to \mathbb{R}$, $x \mapsto g(x, y)$.]

Now let g be a universal function for $\mathscr{F} \subset \overline{\mathbb{R}}^{[0,1]}$, and set

$$\mathscr{F}' := \left\{ \lim_{n\to\infty} f_n \mid (f_n)_{n\in\mathbb{N}} \text{ a convergent sequence from } \mathscr{F} \right\}.$$

We wish to show that there is also an universal function for $\mathscr{F}'$. To this end take $y \in [0,1]$. Write $y = 0.\ \alpha_1(y)\alpha_2(y)\ldots$ as in (δ).

For each $n \in \mathbb{N}$ define

$$h_n: [0,1] \to \overline{\mathbb{R}}, \qquad y \mapsto \sum_{k\in\mathbb{N}} 10^{-k}\alpha_{2^{n-1}(2k-1)}(y)$$

and

$$h: [0,1]\times[0,1] \to \overline{\mathbb{R}}, \qquad (x,y) \mapsto \limsup_{n\to\infty} g(x, h_n(y)).$$

Prove the following:

(ε) h is a universal function for $\mathscr{F}'$.

(ζ) $h_n \in \mathscr{B}_1$ for each $n \in \mathbb{N}$.

(η) h is a Baire function on $[0,1]\times[0,1]$ whenever g is.

To prove (ε), take $f \in \mathscr{F}'$. Then there is a sequence $(f_n)_{n\in\mathbb{N}}$ from $\mathscr{F}$ such that $f = \lim_{n\to\infty} f_n$. But for each $n \in \mathbb{N}$ there is a $y_n \in [0,1]$ such that $f_n = g(\cdot, y_n)$. Now choose $y \in [0,1]$ such that $h_n(y) = y_n$. Then

$$h(x,y) = \limsup_{n\to\infty} g(x, y_n) = \limsup_{n\to\infty} f_n(x) = f(x)$$

for any $x \in [0,1]$.

For (ζ), notice that for each $n \in \mathbb{N}$

$$h_n(y) = \lim_{m\to\infty} \sum_{k=1}^{m} 10^{-k}\alpha_{2^{n-1}(2k-1)}(y)$$

is a uniform limit on $[0,1]$. Now apply (γ).

Lebesgue showed that there is a universal function for each set $\mathscr{B}_\alpha$ which is itself a Baire function on the unit square. It follows that each set $\mathscr{B}_\alpha$ contains functions that belong to no $\mathscr{B}_\beta$ where $\beta < \alpha$ is countable. This is the content of the next two propositions:

(ϑ) Let α be a countable ordinal. Then there is a Baire function g_α: $[0,1]\times[0,1] \to \overline{\mathbb{R}}$ that is a universal function for $\bigcup_{\beta<\alpha} \mathscr{B}_\beta$.

(ι) Let α be a countable ordinal. Then $\mathscr{B}_\alpha \setminus \bigcup_{\beta<\alpha} \mathscr{B}_\beta \neq \emptyset$.

To prove (ϑ), let $\mathscr{P}_{\mathbb{Q}}$ denote the set of all polynomial functions on $[0,1]$ with rational coefficients. $\mathscr{P}_{\mathbb{Q}}$ is countable so that there is a bijection

$\varphi\colon \mathbb{N} \to \mathscr{P}_{\mathbb{Q}}$. Write p_n for $\varphi(n)$, and for each $n \in \mathbb{N}$ define

$$g_n\colon [0,1] \to \mathbb{R}, \qquad x \mapsto \begin{cases} 1 & \text{if } x = 1/n \\ 0 & \text{if } x \neq 1/n \end{cases}$$

and

$$g\colon [0,1] \times [0,1] \to \mathbb{R}, \qquad (x, y) \mapsto \sum_{n \in \mathbb{N}} p_n(x) g_n(y).$$

g is a Baire function on $[0,1] \times [0,1]$, and for each $n \in \mathbb{N}$, $p_n(x) = g(x, 1/n)$ for each $x \in [0,1]$ so that g is a universal function for $\mathscr{P}_{\mathbb{Q}}$.

Recall now that by definition $\mathscr{B}_0$ consists of all continuous functions on $[0,1]$. With help of the Weierstrass Approximation Theorem, we find that every function in $\mathscr{B}_0$ is the limit of some sequence from $\mathscr{P}_{\mathbb{Q}}$. Using (ε), conclude that there is a universal function for $\mathscr{B}_0$ of the required kind.

Now use transfinite induction to prove the existence of Baire universal functions g_α for each $\bigcup_{\beta < \alpha} \mathscr{B}_\beta$ (α countable). Assume that the proposition has been proved for all $\beta < \alpha$, and consider the two cases possible:

CASE 1: $\alpha = \beta + 1$ for some $\beta < \alpha$. Then (ε) yields the existence of g directly.

CASE 2: α is a limit ordinal. Then there is a sequence $(\beta_n)_{n \in \mathbb{N}}$ of ordinals $\beta_n < \alpha$ with $\alpha = \bigvee_{n \in \mathbb{N}} \beta_n$. Set

$$g_\alpha(x, y) := \limsup_{n \to \infty} g_{\beta_n}(x, h_n(y))$$

for $(x, y) \in [0,1] \times [0,1]$ with h_n defined as in (ζ). Take $f \in \mathscr{B}_\beta$, $\beta < \alpha$. Then for some $m \in \mathbb{N}$, $f \in \mathscr{B}_{\beta_n}$ for every $n \geq m$, and for each such n there is, consequently, some $y_n \in [0,1]$ such that $f = g_{\beta_n}(\cdot, y_n)$. Now choose $y \in [0,1]$ with $h_n(y) = y_n$ for each $n \geq m$. Then

$$g_\alpha(x, y) = \limsup_{\substack{n \to \infty \\ n \geq m}} g_{\beta_n}(x, y_n) = f(x)$$

for each $x \in [0,1]$.

(ι) can be proved indirectly by assuming first that $\mathscr{B}_\alpha \setminus \bigcup_{\beta < \alpha} \mathscr{B}_\beta = \varnothing$ for some countable ordinal α, and then defining

$$\varphi\colon [0,1] \times [0,1] \to \mathbb{R}, \qquad (x, y) \mapsto \begin{cases} 1 & \text{if } g_\alpha(x, y) > 1 \\ g_\alpha(x, y) & \text{if } 0 \leq g_\alpha(x, y) \leq 1 \\ 0 & \text{if } g_\alpha(x, y) < 0. \end{cases}$$

If f is a Baire function on $[0,1]$ with $f([0,1]) \subset [0,1]$, then there is a

$y \in [0,1]$ such that $f = \varphi(\cdot, y)$. Now define

$$\psi: [0,1] \times [0,1] \to \mathbb{R}, \qquad (x,y) \mapsto \lim_{n\to\infty} \frac{n\varphi(x,y)}{1 + n\varphi(x,y)}$$

Then ψ is a Baire function whose range is $\{0,1\}$. If f is a Baire function on $[0,1]$ whose range is $\{0,1\}$, then there is a $y \in [0,1]$ such that $f = \psi(\cdot, y)$. In particular consider $f: [0,1] \to \mathbb{R}$, $x \mapsto 1 - \psi(x,x)$. So $f = \psi(\cdot, y)$ for some $y \in [0,1]$. Now set $x = y$. Then $1 - \psi(y,y) = \psi(y,y)$ so that $\psi(y,y) = 1/2$ contradicting the properties of ψ.

(b) Some of the functions in $\mathscr{B}_1$ are of particular importance. We investigate this now, beginning with the following theorem of Lebesgue:

(α) Take $[a,b] \subset \mathbb{R}$ and $f \in \overline{\mathbb{R}}^{[a,b]}$. Then the following are equivalent:

(α1) $f \in \mathscr{B}_1$.
(α2) $\{f > \gamma\}$ and $\{f < \gamma\}$ are F_σ-sets for each $\gamma \in \mathbb{R}$.
(α3) $\{f \geq \gamma\}$ and $\{f \leq \gamma\}$ are G_δ-sets for each $\gamma \in \mathbb{R}$.

For (α1) $\Rightarrow$ (α2), take $f \in \mathscr{B}_1$. Then f is the limit of a sequence $(f_n)_{n \in \mathbb{N}}$ of continuous functions on $[a,b]$. Take $\gamma \in \mathbb{R}$. Then

$$\{f < \gamma\} = \bigcup_{k\in\mathbb{N}} \bigcup_{m\in\mathbb{N}} \bigcap_{n \geq m} \left\{ f_n \leq \gamma - \frac{1}{k} \right\}$$

and thus an F_σ-set. A similar argument applies to $\{f > \gamma\}$.

For (α2) $\Rightarrow$ (α1), suppose first that f is bounded. Choose $\delta_*, \delta^* \in \mathbb{R}$ such that $\{\delta_* < f < \delta^*\} = [a,b]$. For each $n \in \mathbb{N}$ choose elements $\delta_0^{(n)}, \delta_1^{(n)}, \ldots, \delta_n^{(n)}$ of $[\delta_*, \delta^*]$ such that $\delta_0^{(n)} = \delta_*$, $\delta_n^{(n)} = \delta^*$ and $\delta_{k-1}^{(n)} < \delta_k^{(n)}$, $\delta_k^{(n)} - \delta_{k-1}^{(n)} = (\delta^* - \delta_*)/n$ whenever $k \in \mathbb{N}_n$.

For every $k \in \mathbb{N}_n \cup \{0\}$ define

$$A_k^{(n)} := \begin{cases} f^{-1}(]\delta_{n-1}^{(n)}, \delta^*[) & \text{if } k = n \\ f^{-1}(]\delta_{k-1}^{(n)}, \delta_{k+1}^{(n)}[) & \text{if } 0 < k < n \\ f^{-1}(]\delta_*, \delta_1^{(n)}[) & \text{if } k = 0. \end{cases}$$

Each $A_j^{(n)}$ is an F_σ-set. For each $n \in \mathbb{N}$ there is a disjoint family $(B_j^{(n)})_{j \in \mathbb{N}_n \cup \{0\}}$ of F_σ-sets such that $[a,b] = \bigcup_{j=0}^{n} B_j^{(n)}$ and $B_j^{(n)} \subset A_j^{(n)}$ for each $j \in \mathbb{N}_n \cup \{0\}$. Next define for each $n \in \mathbb{N}$

$$f_n: [a,b] \to \mathbb{R}, \qquad x \mapsto \delta_j^{(n)} \text{ where } x \in B_j^{(n)}.$$

Then $(f_n)_{n \in \mathbb{N}}$ is a sequence of functions from $\mathscr{B}_1$ that converges uniformly on $[a,b]$ to f. Thus $f \in \mathscr{B}_1$.

Now consider the general case, in which f need not be bounded. Set $g := \arctan \circ f$. Then $g \in \mathscr{B}_1$, and it is left to the reader to verify that $f = \tan \circ g \in \mathscr{B}_1$.

There is a characterization of $\mathscr{B}_1$ originating from Baire. To present this, we need some preliminaries:

Let (X, d) be a metric space. We repeat some notions. $A \subset X$ is called a set of the first category iff A is the union of a countable family of nowhere dense subsets of X. $A \subset X$ is called a set of the second category iff it is not of the first category.

The following theorem is also of importance in other contexts.

(β) **Baire Theorem.** Every nonempty complete metric space is of the second category (cf., Ex. 5.3.10).

We also need the **Cantor-Bernstein Theorem:**

(γ) Let (X, d) be a separable metric space. Then

(γ1) Every set of open (resp. closed) subsets of X that is well ordered with respect to $\subset$ is countable.
(γ2) Every set of open (resp. closed) subsets of X that is well ordered with respect to $\supset$ is countable.

Let $\mathfrak{R}$ be a set of open subsets of X well ordered with respect to $\subset$. Choose a countable dense subset Y of X, and define

$$\mathfrak{S} := \{ B(y, 1/n) \mid y \in Y, n \in \mathbb{N} \}$$

with $B(y, \varepsilon) := \{ x \in X \mid d(x, y) < \varepsilon \}$ for every $y \in X$ and every $\varepsilon > 0$. Then $\mathfrak{S}$ is countable, so that there is an injection $\varphi\colon \mathfrak{S} \to \mathbb{N}$. For each $A \in \mathfrak{R}$ define

$$N_A := \{ \varphi(B(y, 1/n)) \mid y \in Y, B(y, 1/n) \subset A \}.$$

Then $\psi\colon \mathfrak{R} \to \mathfrak{P}(\mathbb{N})$, $A \mapsto N_A$ is an increasing injection. Thus $\{ N_A \mid A \in \mathfrak{R} \}$ is a subset of $\mathfrak{P}(\mathbb{N})$ well ordered with respect to $\subset$.

The case of open subsets of X well ordered by $\supset$ is dealt with completely analogously.

For closed sets, notice that the family $(F_n)_{n \in \mathbb{N}}$ is well ordered by $\subset$ (resp. $\supset$) iff $(X \setminus F_n)_{n \in \mathbb{N}}$ is a family of open subsets well ordered by $\supset$ (resp. $\subset$).

We now are in a position to formulate and prove the characterization of $\mathscr{B}_1$ due to Baire:

(δ) Take $[a, b] \subset \mathbb{R}$ and $f \in \mathbb{R}^{[a,b]}$. Then $f \in \mathscr{B}_1$ iff every nonempty closed subset A of $[a, b]$ contains a point x such that $f|_A$ is continuous at x.

Suppose that $f \in \mathscr{B}_1$, and that A is a closed subset of $[a, b]$. If x is any isolated point of A, then $f|_A$ is continuous at x. So assume that $A \neq \varnothing$ has no isolated points. Then A is perfect. The aim now is to show that for every $[\alpha, \beta] \subset [a, b]$ with $]\alpha, \beta[\cap A \neq \varnothing$ and for each $\varepsilon > 0$ there is a $[\gamma, \delta]$

$\subset]\alpha, \beta[$ such that

$$]\gamma, \delta[\cap A \neq \varnothing \quad \text{and} \quad \sup_{x \in [\gamma, \delta] \cap A} f(x) - \inf_{x \in [\gamma, \delta] \cap A} f(x) < \varepsilon.$$

To this end, choose a closed interval $C \subset [\alpha, \beta]$ with $A \cap C$ perfect and nonempty. There is a sequence of continuous functions on $[a, b]$, $(f_n)_{n \in \mathbb{N}}$, such that $f = \lim_{n \to \infty} f_n$. For each $n \in \mathbb{N}$ put

$$B_n := \bigcap_{m \in \mathbb{N}} (C \cap \{ |f_n - f_{n+m}| \leq \varepsilon/4 \}).$$

Then $A \cap C = \bigcup_{n \in \mathbb{N}} (A \cap B_n)$. Thus, by (β), there exist an open interval $B \subset [a, b]$ and an $n \in \mathbb{N}$ such that $\varnothing \neq (A \cap C) \cap B \subset A \cap B_n$. For each $x \in A \cap C \cap B$ and each $m \in \mathbb{N}$, $|f_n(x) - f_{n+m}(x)| \leq \varepsilon/4$. Let $m \to \infty$. Then $|f_n(x) - f(x)| \leq \varepsilon/4$. Since $A \cap C$ is perfect, $A \cap C \cap B$ cannot be finite. Choose $z \in A \cap C \cap B$ such that z is an interior point of C, and seek $[\gamma, \delta] \subset [a, b]$ with the following properties:

(i) $z \in [\gamma, \delta] \subset C$.
(ii) $[\gamma, \delta] \subset B$.
(iii) $\sup_{x \in [\gamma, \delta]} f_n(x) - \inf_{x \in [\gamma, \delta]} f_n(x) < \varepsilon/3$.

Given then $x, y \in A \cap [\gamma, \delta]$ arbitrarily chosen,

$$|f(x) - f(y)| \leq |f(x) - f_n(x)| + |f_n(x) - f_n(y)| + |f_n(y) - f(y)|$$

$$\leq 5\varepsilon/6$$

which gives the result sought. So construct by recursion a decreasing sequence $([\gamma_n, \delta_n])_{n \in \mathbb{N}}$ of closed intervals in $[a, b]$ with the following properties:

(iv) $]\gamma_n, \delta_n[\cap A \neq \varnothing$.
(v) $\delta_n - \gamma_n < 1/n$.
(vi) $\sup_{x \in [\gamma_n, \delta_n] \cap A} f(x) - \inf_{x \in [\gamma_n, \delta_n] \cap A} f(x) < 1/n$.

Then $\{x\} := \bigcap_{n \in \mathbb{N}} [\gamma_n, \delta_n] \subset A$ and $f|_A$ is continuous at x.

Conversely, simply check that for each $\gamma \in \mathbb{R}$, both $\{f < \gamma\}$ and $\{f > \gamma\}$ are F_σ-sets and apply (α). For this, take $\delta_1, \delta_2 \in \mathbb{R}$, $\delta_1 < \delta_2$. Put $D_1 := \{f > \delta_1\}$ and $D_2 := \{f < \delta_2\}$. Clearly $D_1 \cup D_2 = [a, b]$. Take $A \subset [a, b]$, A nonempty and closed. Then there is a nonempty closed subset A' of $[a, b]$ such that $A \setminus A' \neq \varnothing$, and $A \setminus A' \subset D_1$ or $A \setminus A' \subset D_2$. This is easy to see, since, by hypothesis, there is an $x \in A$ such that $f|_A$ is continuous at x. Either $f(x) > \delta_1$ or $f(x) < \delta_2$. Hence there is an open interval B with $x \in B$ and $f(y) > \delta_1$ for each $y \in A \cap B$ (if $f(x) > \delta_1$) or $f(y) < \delta_2$ for each $y \in A \cap B$ (if $f(x) < \delta_2$). Put $A' := A \setminus B$.

Now put $A_0 := [a, b]$, and construct a closed subset A_α of A for each ordinal α satisfying the following conditions:

(vii) For each α, $A_{\alpha+1} \subset A_\alpha$ and either $A_\alpha \setminus A_{\alpha+1} \subset D_1$ or $A_\alpha \setminus A_{\alpha+1} \subset D_2$.

(viii) $A_\alpha \setminus A_{\alpha+1} \neq \varnothing$ and $A_{\alpha+1} \neq \varnothing$ if $A_\alpha \neq \varnothing$.

(ix) $A_\alpha = \bigcap_{\beta < \alpha} A_\alpha$ if α is a limit ordinal.

Applying (γ), there is a countable ordinal β such that $A_\beta = \varnothing$, so that $A_{\beta'} = \varnothing$ for any $\beta' > \beta$. Thus $[a, b] = \bigcup_{\alpha < \beta}(A_\alpha \setminus A_{\alpha+1})$.

Define $P_k := \{\alpha < \beta \mid A_\alpha \setminus A_{\alpha+1} \subset D_k\}$ $(k = 1, 2)$. Then

$$[a, b] = \bigcup_{\alpha \in P_1} (A_\alpha \setminus A_{\alpha+1}) \cup \bigcup_{\alpha \in P_2} (A_\alpha \setminus A_{\alpha+1})$$

and

$$\bigcup_{\alpha \in P_1} (A_\alpha \setminus A_{\alpha+1}) \cap \bigcup_{\alpha \in P_2} (A_\alpha \setminus A_{\alpha+1}) = \varnothing.$$

Putting all of this together, if $(\delta_1, \delta_2) \in \mathbb{R}^2$ and $\delta_1 < \delta_2$, then there are disjoint F_σ-sets C_1 and C_2 such that $C_1 \subset D_1$, $C_2 \subset D_2$ and $C_1 \cup C_2 = [a, b]$. Now choose $\gamma \in \mathbb{R}$. Let $(\gamma_n)_{n \in \mathbb{N}}$ be a decreasing sequence from $\mathbb{R} \setminus \{\gamma\}$ such that $\gamma = \inf_{n \in \mathbb{N}} \gamma_n$. Denote by C_{1n} and C_{2n} the corresponding F_σ-sets. Verify that $\{f < \gamma\} = \bigcup_{n \in \mathbb{N}} C_{1n}$. Thus $\{f < \gamma\}$ is an F_σ-set. Argue similarly for $\{f > \gamma\}$.

(c) Finally we present some illustrative examples.

(α) If $f \in \mathbb{R}^{[a,b]}$ is continuous at all but at most countably many points, then $f \in \mathscr{B}_1$.

(β) The Dirichlet function $e_{\mathbb{Q}}|_{[a,b]}$ is an element of $\mathscr{B}_2 \setminus \mathscr{B}_1$.

(γ) Let $C \subset [a, b]$ be a Cantor set. Then $e_C|_{[a,b]} \in \mathscr{B}_1$. But define

$$g\colon [a, b] \to \mathbb{R}, \qquad x \mapsto \begin{cases} 1 & \text{if } x \in C \text{ and } x \text{ is not an end-point of a complementary interval} \\ 0 & \text{if } x \in [a, b] \setminus C \text{ or if } x \text{ is an endpoint of a complementary interval.} \end{cases}$$

Then $g \notin \mathscr{B}_1$.

(δ) Every upper semicontinuous function and every lower semicontinuous function belong to $\mathscr{B}_1$ (see Section 3.6).

It should be mentioned that $[a, b]$ can be replaced throughout the entire exercise by an interval A satisfying the hypotheses of this section.

5.6. LEBESGUE MEASURE

NOTATION FOR SECTION 5.6:

A denotes an open, half-open, or closed interval from $\overline{\mathbb{R}}$, containing neither ∞ nor $-\infty$, and having a nonempty interior.

The Lebesgue measure is by far the most important example of a Stieltjes measure. It has already been mentioned that modern integration theory developed out of this example. The Lebesgue measure distinguishes itself from other Stieltjes measures by its translation invariance, its special link with the algebraic structure of $\mathbb{R}$. This aspect has led to the development of a general theory of measures on topological groups, which is a topic we shall study in Volume 3.

Definition 5.6.1. *The Stieltjes measure μ_g associated with the identity function*

$$g: A \to A, \qquad x \mapsto x$$

is called the ***Lebesgue measure*** *on A. We denote the Lebesgue measure on A by λ_A. The integral associated with λ_A is called the* ***Lebesgue integral*** *on A. We use "**Lebesgue measurable** on or in A", "**Lebesgue integrable** on or in A", and "**Lebesgue null**" as synonyms for "λ_A-measurable", "λ_A-integrable", and "λ_A-null" respectively.*

For $f \in \overline{\mathbb{R}}^A$ and $x, y \in A$, if $x \leq y$ and $fe_{]x,y[} \in \mathscr{L}^1(\lambda_A)$, we define

$$\int_x^y f(t)\,dt := \int_x^y f\,dt := \int fe_{]x,y[}\,d\lambda_A$$

$$\int_y^x f(t)\,dt := \int_y^x f\,dt := -\int_x^y f(t)\,dt. \qquad \square$$

Corollary 5.6.2. *Every countable subset of A is Lebesgue null. The Lebesgue measure on A is bounded iff the interval A is bounded in $\mathbb{R}$.*

Proof. Merely apply the results of the previous section, specifically Corollaries 5.5.12 (c) and 5.5.13. $\square$

Theorem 5.6.3. (Translation invariance of Lebesgue measure on $\mathbb{R}$). *Denote by λ the Lebesgue measure on $\mathbb{R}$. Let f be an extended–real-valued function on*

$\mathbb{R}$, *let* B *be a subset of* $\mathbb{R}$, *and for* $\gamma \in \mathbb{R}$ *define*

$$B_\gamma := \{x + \gamma \mid x \in B\}$$

$$f_\gamma:\quad \mathbb{R} \to \overline{\mathbb{R}},\quad x \mapsto f(x - \gamma).$$

Then the following assertions hold, for every real number γ.

(*a*) f *is Lebesgue integrable on* $\mathbb{R}$ *iff* f_γ *is Lebesgue integrable on* $\mathbb{R}$, *and in this case*

$$\int_{\mathbb{R}} f\,d\lambda = \int_{\mathbb{R}} f_\gamma\,d\lambda.$$

(*b*) B *is Lebesgue integrable in* $\mathbb{R}$ *iff* B_γ *is Lebesgue integrable in* $\mathbb{R}$, *and in this case*

$$\lambda(B) = \lambda(B_\gamma).$$

(*c*) f *is Lebesgue measurable on* $\mathbb{R}$ *iff* f_γ *is.*

(*d*) B *is Lebesgue measurable in* $\mathbb{R}$ *iff* B_γ *is.*

Proof. (a) We shall use Proposition 5.5.9 and the first form of the Induction Principle (Theorem 3.2.7).

Let $\mathfrak{J}_r$ denote the set-ring generated by bounded intervals from $\mathbb{R}$. Denote by ν the restriction of λ to $\mathfrak{J}_r$, and by $(\mathbb{R}, \mathscr{L}, \ell)$ the Daniell space associated with the positive measure space $(\mathbb{R}, \mathfrak{J}_r, \nu)$. In particular, $\mathscr{L}$ consists of the $\mathfrak{J}_r$-step functions on $\mathbb{R}$.

We use Proposition 5.5.9 to show that

$$(\mathbb{R}, \bar{\mathscr{L}}(\ell), \bar{\ell}) = \left(\mathbb{R}, \mathscr{L}^1(\mathbb{R}, \lambda), \int_{\mathbb{R},\lambda}\right). \tag{1}$$

To apply Proposition 5.5.9, we need only check that

$$\mathfrak{J}_r \subset \mathfrak{L}(\mathbb{R}, \lambda).$$

But this inclusion certainly holds, since intervals bounded in $\mathbb{R}$ are all Lebesgue integrable [Theorem 5.5.6 (b)]. We conclude that

$$(\mathfrak{R}(\nu), \bar{\nu}) = (\mathfrak{L}(\mathbb{R}, \lambda), \lambda).$$

The closed positive measure space $(\mathbb{R}, \mathfrak{R}(\nu), \bar{\nu})$ is induced by $(\mathbb{R}, \bar{\mathscr{L}}(\ell), \bar{\ell})$, whereas $(\mathbb{R}, \mathfrak{L}(\mathbb{R}, \lambda), \lambda)$ is induced by $(\mathbb{R}, \mathscr{L}^1(\mathbb{R}, \lambda), \int_{\mathbb{R},\lambda})$. Since a given closed positive measure space is induced by a uniquely determined closed Daniell space [Theorem 5.1.8 (a)], (1) follows.

Now let

$$\mathscr{F} := \left\{ f \in \mathscr{L}^1(\mathbb{R}, \lambda) \;\middle|\; \begin{array}{l} \forall \gamma \in \mathbb{R}: \\ f_\gamma \in \mathscr{L}^1(\mathbb{R}, \lambda) \\ \int_{\mathbb{R}} f_\gamma \, d\lambda = \int_{\mathbb{R}} f \, d\lambda \end{array} \right\}.$$

We use Theorem 3.2.7—Induction Principle of the first form, applied to $(\mathbb{R}, \mathscr{L}, \ell)$ and $(\mathbb{R}, \bar{\mathscr{L}}(\ell), \bar{\ell})$ and taking (1) into account—to show that

$$\mathscr{F} = \mathscr{L}^1(\mathbb{R}, \lambda).$$

There are three hypotheses to be checked.

Hypothesis 1: The formula for Stieltjes measure of an interval shows that e_B belongs to $\mathscr{F}$ for every bounded interval B from $\mathbb{R}$. It follows that $\mathfrak{J}_r$-step functions all belong to $\mathscr{F}$; that is, $\mathscr{L} \subset \mathscr{F}$.

Hypothesis 2: Let $(f_n)_{n \in \mathbb{N}}$ be a λ-sequence from $\mathscr{F}$. For each real γ, $(f_{n\gamma})_{n \in \mathbb{N}}$ is a λ-sequence from $\mathscr{F}$. Evidently

$$\left(\lim_{n \to \infty} f_n \right)_\gamma = \lim_{n \to \infty} f_{n\gamma}.$$

It follows that $\lim_{n \to \infty} f_n$ belongs to $\mathscr{L}^1(\mathbb{R}, \lambda)$, $\lim_{n \to \infty} f_{n\gamma}$ belongs to $\mathscr{L}^1(\mathbb{R}, \lambda)$ for every real γ, and

$$\int_{\mathbb{R}} \lim_{n \to \infty} f_{n\gamma} \, d\lambda = \lim_{n \to \infty} \int_{\mathbb{R}} f_{n\gamma} \, d\lambda = \lim_{n \to \infty} \int_{\mathbb{R}} f_n \, d\lambda = \int_{\mathbb{R}} \lim_{n \to \infty} f_n \, d\lambda.$$

Thus $\lim_{n \to \infty} f_n$ belongs to $\mathscr{F}$.

Hypothesis 3: Finally, let $f, g \in \mathscr{F}$, let $h \in \mathscr{L}^1(\mathbb{R}, \lambda)$, and suppose that

$$f \le h \le g, \qquad \int_{\mathbb{R}} f \, d\lambda = \int_{\mathbb{R}} g \, d\lambda.$$

For every real γ we have

$$f_\gamma \le h_\gamma \le g_\gamma$$

so

$$\int_{\mathbb{R}} f_\gamma \, d\lambda \le \int_{\mathbb{R}} h_\gamma \, d\lambda \le \int_{\mathbb{R}} g_\gamma \, d\lambda$$

so

$$\int_{\mathbb{R}} h_\gamma \, d\lambda = \int_{\mathbb{R}} h \, d\lambda.$$

Thus h belongs to $\mathscr{F}$.

In view of (1), the Induction Principle implies that $\mathscr{F} = \mathscr{L}^1(\mathbb{R}, \lambda)$.

(b) Note that $(e_B)_\gamma = e_{(B_\gamma)}$, and apply (a).

(c) Suppose $f \in \mathscr{M}(\mathbb{R}, \lambda)_+$ and $\gamma \in \mathbb{R}$. By (b), $A_{(-\gamma)}$ belongs to $\mathfrak{L}(\mathbb{R}, \lambda)$ for every Lebesgue-integrable set A. Applying Proposition 5.4.11 (a) $\Rightarrow$ (c) to the measure space $(\mathbb{R}, \mathfrak{L}(\mathbb{R}, \lambda), \lambda)$, we conclude that $f \wedge \alpha e_{A_{(-\gamma)}}$ belongs to $\mathscr{L}^1(\mathbb{R}, \lambda)$ for every positive real number α and every Lebesgue-integrable set A. By (a), $f_\gamma \wedge \alpha e_A$ belongs to $\mathscr{L}^1(\mathbb{R}, \lambda)$ for every positive real number α and every Lebesgue-integrable set A. Now Proposition 5.4.11 (c) $\Rightarrow$ (a) shows that f_γ is Lebesgue measurable.

For arbitrary Lebesgue-measurable f on $\mathbb{R}$, note that $f = (f^+) - (f^-)$, and apply what has already been proved to the functions f^+, f^-.

(d) This assertion follows from (c). □

An analogous theorem obviously holds for Lebesgue measure on an arbitrary interval A, provided one restricts γ so that $\lambda_A(B_\gamma)$ and $\int f_\gamma \, d\lambda_A$ are defined. We obtain this analogue later, in an essentially more general setting, as a consequence of theorems concerning image measures and restrictions.

For every positive measure μ on a set-ring $\mathfrak{R}$, and for every set X containing $X(\mathfrak{R})$, we have defined the collection $\mathfrak{M}(X, \mu)$ of μ-measurable sets in X. An interesting question arises: Under what conditions is every subset of X μ-measurable? This question is in fact an important one in the foundations of mathematics. Although we cannot investigate this question very thoroughly in this book, we at least want to verify that not all subsets of $\mathbb{R}$ are Lebesgue measurable:

$$\mathfrak{M}(\mathbb{R}, \lambda_{\mathbb{R}}) \neq \mathfrak{P}(\mathbb{R}).$$

Eventually we shall obtain some sharper, more general results.

The following example is due to Vitali (1905).

Theorem 5.6.4. *Every Lebesgue-integrable set in $\mathbb{R}$, the Lebesgue measure of which is strictly positive, has a subset that is not Lebesgue measurable in $\mathbb{R}$.*

Proof. Let $B \in \mathfrak{L}(\mathbb{R}, \lambda_{\mathbb{R}})$ with $\lambda_{\mathbb{R}}(B) > 0$. Without loss of generality assume, for some m in $\mathbb{N}$, that

$$B \subset [-m, m].$$

Define an equivalence relation on B as follows:

$$x \sim y :\Leftrightarrow x - y \in \mathbb{Q}.$$

Choose one element (Axiom of Choice!) from each equivalence class, and let C be the set consisting of the chosen elements of B. We claim that the subset C is not $\lambda_{\mathbb{R}}$-measurable.

Let

$$\varphi: \mathbb{N} \to [-2m, 2m] \cap \mathbb{Q}$$

be bijective. For n in $\mathbb{N}$ define

$$C_n := \{x + \varphi(n) \mid x \in C\}.$$

Then $(C_n)_{n \in \mathbb{N}}$ is a disjoint sequence, and

$$B \subset \bigcup_{n \in \mathbb{N}} C_n \subset [-3m, 3m]. \tag{2}$$

As a subset of a $\lambda_{\mathbb{R}}$-integrable set, if C were $\lambda_{\mathbb{R}}$-measurable, it would also be $\lambda_{\mathbb{R}}$-integrable [Proposition 5.4.3 (b)]. Since $\lambda_{\mathbb{R}}$ is translation invariant, each C_n would also be $\lambda_{\mathbb{R}}$-integrable and satisfy

$$\lambda_{\mathbb{R}}(C_n) = \lambda_{\mathbb{R}}(C).$$

In view of (2) we would have

$$0 < \lambda_{\mathbb{R}}(B) \le \sum_{n \in \mathbb{N}}{}^{*} \lambda_{\mathbb{R}}(C_n) \le 6m$$

hence

$$0 < n\lambda_{\mathbb{R}}(C) \le 6m$$

for every n in $\mathbb{N}$. This last inequality is not possible. Thus C cannot belong to $\mathfrak{M}(\mathbb{R}, \lambda_{\mathbb{R}})$. □

We conclude this section by showing how the integral of a given function relative to an arbitrary positive measure μ can be represented as a Lebesgue integral. It was noted earlier that for positive μ-integrable functions f, the sets $\{f > \alpha\}, \{f \ge \alpha\}$ are μ-integrable if $\alpha > 0$ [Proposition 5.4.3 (g)]. The functions f' and f'' specified in this representation theorem are therefore defined.

Theorem 5.6.5. *Let $(X, \mathfrak{R}, \mu)$ be a positive measure space. Denote by λ the Lebesgue measure on $]0, \infty[$. To each $f \in \mathscr{L}^1(X, \mu)_+$, associate functions f', f''*

defined as follows:

$$f' :]0, \infty[\to \mathbb{R}, \qquad \alpha \mapsto \mu(\{f > \alpha\})$$

$$f'' :]0, \infty[\to \mathbb{R}, \qquad \alpha \mapsto \mu(\{f \geq \alpha\}).$$

Then for every $f \in \mathscr{L}^1(X, \mu)_+$ the functions f', f'' are λ-integrable and

$$\int_X f\,d\mu = \int_{]0,\infty[} f'\,d\lambda = \int_{]0,\infty[} f''\,d\lambda. \tag{3}$$

Proof. For $f \in \mathscr{L}^1(X, \mu)_+$, the set $\{f' \neq f''\}$ is countable [Proposition 5.4.3 (g)], hence λ-null. Thus for $f \in \mathscr{L}^1(X, \mu)_+$, f' is λ-integrable iff f'' is, and in this case

$$\int_{]0,\infty[} f'\,d\lambda = \int_{]0,\infty[} f''\,d\lambda.$$

Let

$$\mathscr{F} := \{g \in \mathscr{L}^1(X, \mu) \mid (3) \text{ holds for } f := g^+\}$$

and denote by $(X, \mathscr{L}, \ell)$ the Daniell space associated with $(X, \mathfrak{R}, \mu)$. We use the Induction Principle for Admissible Extensions (Proposition 4.1.18), applied to $(X, \mathscr{L}, \ell)$ and $(X, \mathscr{L}^1(X, \mu), \int_{X,\mu})$.

Let $g \in \mathscr{L}$; that is, let g be an $\mathfrak{R}$-step function on X. Then g^+ is a positive $\mathfrak{R}$-step function on X. As such, it has a disjoint $\mathfrak{R}$-representation with positive coefficients:

$$g^+ = \sum_{\iota \in I} \beta_\iota e_{B_\iota}$$

for some finite disjoint family $(B_\iota)_{\iota \in I}$ from $\mathfrak{R}$ and some family $(\beta_\iota)_{\iota \in I}$ from $\mathbb{R}_+$. For every real $\alpha > 0$,

$$\{g^+ \geq \alpha\} = \bigcup_{\substack{\iota \in I \\ \beta_\iota \geq \alpha}} B_\iota$$

so

$$(g^+)''(\alpha) = \mu(\{g^+ \geq \alpha\}) = \sum_{\substack{\iota \in I \\ \beta_\iota \geq \alpha}} \mu(B_\iota) = \sum_{\iota \in I} \mu(B_\iota) e_{]0,\beta_\iota]}(\alpha).$$

In other words,

$$(g^+)'' = \sum_{\iota \in I} \mu(B_\iota) e_{]0, \beta_\iota]}.$$

We conclude from this representation that $(g^+)''$ belongs to $\mathscr{L}^1(]0, \infty[, \lambda)$ and

$$\int_X g^+ \, d\mu = \sum_{\iota \in I} \beta_\iota \mu(B_\iota) = \int_{]0, \infty[} (g^+)'' \, d\lambda.$$

In view of the remarks at the beginning of the proof, it follows that $\mathscr{L} \subset \mathscr{F}$.

Now let $(g_n)_{n \in \mathbb{N}}$ be a μ-sequence from $\mathscr{F}$, and let h be an extended-real function on X such that

$$h = \lim_{n \to \infty} g_n \ \mu\text{-a.e.}$$

Suppose first that $(g_n)_{n \in \mathbb{N}}$ increases, and set

$$g := \bigvee_{n \in \mathbb{N}} g_n.$$

For $0 < \alpha < \infty$, the sequence $(\{g_n^+ > \alpha\})_{n \in \mathbb{N}}$ increases, and its union is $\{g^+ > \alpha\}$. Therefore $((g_n^+)')_{n \in \mathbb{N}}$ is an increasing sequence from $\mathscr{L}^1(]0, \infty[, \lambda)$ whose supremum is $(g^+)'$. In fact

$$\sup_{n \in \mathbb{N}} \int_{]0, \infty[} (g_n^+)' \, d\lambda = \sup_{n \in \mathbb{N}} \int_X g_n^+ \, d\mu = \int_X g^+ \, d\mu < \infty$$

so $((g_n^+)')_{n \in \mathbb{N}}$ is a λ-sequence. It follows that $(g^+)'$ belongs to $\mathscr{L}^1(]0, \infty[, \lambda)$ and

$$\int_{]0, \infty[} (g^+)' \, d\lambda = \sup_{n \in \mathbb{N}} \int_{]0, \infty[} (g_n^+)' \, d\lambda = \int_X g^+ \, d\mu.$$

We conclude that $\lim_{n \to \infty} g_n$ belongs to $\mathscr{F}$ in case the μ-sequence $(g_n)_{n \in \mathbb{N}}$ increases. If $(g_n)_{n \in \mathbb{N}}$ decreases, set

$$g := \bigwedge_{n \in \mathbb{N}} g_n.$$

Carrying out a similar argument, but using $(g_n^+)''$ and $(g^+)''$ instead of $(g_n^+)'$ and $(g^+)'$, we conclude that $(g^+)''$ belongs to $\mathscr{L}^1(]0, \infty[, \lambda)$ and

$$\int_{]0, \infty[} (g^+)'' \, d\lambda = \int_X g^+ \, d\mu$$

so g belongs to $\mathscr{F}$. Thus $\lim_{n\to\infty} g_n$ belongs to $\mathscr{F}$ in either case. Since

$$h = \lim_{n\to\infty} g_n \ \mu\text{-a.e.}$$

we have

$$\mu(\{h^+ > \alpha\}) = \mu\left(\left\{\left(\lim_{n\to\infty} g_n\right)^+ > \alpha\right\}\right)$$

$$\mu(\{h^+ \geq \alpha\}) = \mu\left(\left\{\left(\lim_{n\to\infty} g_n\right)^+ \geq \alpha\right\}\right)$$

for $0 < \alpha < \infty$, so h must belong to $\mathscr{F}$. By the Induction Principle of Proposition 4.1.18,

$$\mathscr{F} = \mathscr{L}^1(X, \mu).$$

The theorem follows. □

EXERCISES

5.6.1(C) Approximable subsets of $\mathbb{R}$. For further information about the objects considered here see Ex. 2.1.4.

(α) Let A be an approximable subset of $\mathbb{R}$. For any $A \subset \mathbb{R}$ and $x \in \mathbb{R}$ define $A + x := \{y + x \mid y \in A\}$. Assume that there is a sequence $(x_n)_{n\in\mathbb{N}}$ from $\mathbb{R}$ with $\lim_{n\to\infty} x_n = 0$ such that $A \cap (A + x_n)$ is of the first category for every $n \in \mathbb{N}$. Then A is of the first category itself.

Hint: Let U be an open subset of $\mathbb{R}$ such that $A \triangle U$ is of the first category. Prove that $U \cap (U + x_n) = \varnothing$ for every $n \in \mathbb{N}$. Hence $U = \varnothing$, and it follows that A is of the first category.

(β) For any $x, y \in \mathbb{R}$ define $x \sim y :\Leftrightarrow x - y \in \mathbb{Q}$. The symbol $\sim$ defines an equivalence relation. From every equivalence class choose an element, and denote by A the set of these chosen elements. Prove that for any subset B of $\mathbb{R}$ not of the first category, there is an $x \in \mathbb{Q}$ such that $(A + x) \cap B$ is not approximable. In particular, A is not approximable.

Hint: For any $x, y \in \mathbb{Q}$, $x \neq y$, we have

$$B = \bigcup_{z\in Q} (A + z) \cap B \quad \text{and} \quad (A + x) \cap (A + y) = \varnothing$$

If $(A + x) \cap B$ is approximable for any $x \in \mathbb{Q}$, then $(A + x) \cap B$ is of the first category. This follows from (a). But B is not of the first category. Hence there must be an $x \in \mathbb{Q}$ such that $(A + x) \cap B$ is not approximable.

Now prove the following statements:

(γ) There is an approximable subset of $\mathbb{R}$ that is not Lebesgue measurable. This set can be chosen to be nowhere dense.

(δ) There is a Lebesgue-measurable subset of $\mathbb{R}$ that is not approximable. This set can be chosen to be of Lebesgue measure zero.

(ε) Consider (β). Prove that the set A constructed there cannot be Lebesgue measurable.

Hint for (γ): Choose a Cantor set with strictly positive measure. Apply 5.6.4.

Hint for (δ): Take a Lebesgue null set B that is not of the first category, and apply (β).

(γ) and (δ) show that there is no simple connection between Lebesgue measurability and approximability of subsets of $\mathbb{R}$.

5.6.2(C) Lebesgue Null Sets and Sets of the First Category. In several exercises we dealt with sets of the first category and with null sets with respect to a positive measure. Any of these sets can be regarded as small in some sense. Indeed, nowhere dense sets are small in the geometric sense of being perforated with holes. Sets of the first category are countable unions of nowhere dense sets. In general, they do not have holes, but they always have a dense set of gaps. On the other hand, null sets are small in a metric sense. They have no content. It is natural to ask about the connection between these notions. That is our goal in this exercise. We consider the case of $\mathbb{R}$ with the Lebesgue measure λ.

(a) Prove the following statements.

(α) There are subsets A and B of $\mathbb{R}$ with the following properties.

(i) A is a Lebesgue null set and a dense G_δ-set.
(ii) B is a K_σ-set of the first category.
(iii) $A \cap B = \varnothing$, and $A \cup B = \mathbb{R}$.

(β) Every subset of $\mathbb{R}$ can be represented as the disjoint union of a Lebesgue null set and a set of the first category.

(γ) There is a Lebesgue null set that is a dense G_δ-set and therefore not of the first category.

(δ) There is a subset of $\mathbb{R}$ of the first category that is a K_σ-set and not a Lebesgue null set.

We give some hints for (α). The other statements are easy consequences. Let $\varphi\colon \mathbb{N} \to \mathbb{Q}$ be a bijective map, and define $\alpha_n := \varphi(n)$ for each $n \in \mathbb{N}$. For any $i, j \in \mathbb{N}$ let I_{ij} be the open interval centered in α_i with length $1/2^{i+j}$. For any $j \in \mathbb{N}$ set $G_j := \bigcup_{i \in \mathbb{N}} I_{ij}$, and define $A := \bigcap_{j \in \mathbb{N}} G_j$. Then

$$\lambda(G_j) \leq \sum_{i \in \mathbb{N}} \frac{1}{2^{i+j}} = \frac{1}{2^j}$$

and so $\lambda(A) = 0$. We set $B := \mathbb{R} \setminus A$.

(b) **The Theorem of Erdös (1943).** Accepting the Continuum Axiom, (that is, $2^{\aleph_0} = \aleph_1$) Erdös proved a very interesting result about the connection between Lebesgue null sets and subsets of $\mathbb{R}$ of the first category. For the proof, we need some preliminaries from set theory.

(α) Let X be a set of cardinality $\aleph_1$ and $\mathfrak{R}$ a set of subsets of X with the following properties:

(i) $\mathfrak{R}$ is a solid σ-ring (solid means that $A \in \mathfrak{R}$ whenever $B \in \mathfrak{R}$ exists with $A \subset B$).
(ii) $\bigcup_{A \in \mathfrak{R}} A = X$.
(iii) There is a subset $\mathfrak{S}$ of $\mathfrak{R}$ with cardinality $\aleph_1$ such that for any $A \in \mathfrak{R}$ there is a $B \in \mathfrak{S}$ with $A \subset B$.
(iv) The complement of each $A \in \mathfrak{R}$ contains a set of cardinality $\aleph_1$ that belongs to $\mathfrak{R}$.

Then there is a disjoint family $(A_\iota)_{\iota \in I}$ of subsets of X with $\bigcup_{\iota \in I} A_\iota = X$ such that I and each A_ι have the cardinality $\aleph_1$ and so that for any $A \subset X$, $A \in \mathfrak{R}$ iff there is a countable set $J \subset I$ with $A \subset \bigcup_{\iota \in J} A_\iota$.

Let ω_1 be the first uncountable ordinal, and define $\Phi := \{\alpha \mid \alpha \text{ ordinal}, \alpha < \omega_1\}$. Φ has the cardinality $\aleph_1$, and there is a mapping $\alpha \mapsto C_\alpha$ of Φ onto $\mathfrak{S}$. For each $\alpha \in \Phi$ define

$$H_\alpha := \bigcup_{\beta < \alpha} C_\beta \quad \text{and} \quad K_\alpha := H_\alpha \setminus \bigcup_{\beta < \alpha} H_\beta .$$

Set $\Psi := \{\alpha \in \Phi \mid K_\alpha \text{ is uncountable}\}$. Properties (i), (iii), and (iv) imply that Ψ has no upper bound in Φ. Therefore there exists an order isomorphism φ of Φ onto Ψ. For each $\alpha \in \Phi$ define

$$A_\alpha := H_{\varphi(\alpha)} \setminus \bigcup_{\beta < \alpha} H_{\varphi(\beta)} .$$

$(A_\alpha)_{\alpha \in \Phi}$ is a disjoint family from $\mathfrak{R}$, and each A_α has cardinality $\aleph_1$. For any $\beta \in \Phi$ we have $\beta < \varphi(\alpha)$ for some $\alpha \in \Phi$, and therefore

$$C_\beta \subset H_{\varphi(\alpha)} = \bigcup_{\gamma \le \alpha} A_\gamma .$$

Hence, by (iii), each member of $\mathfrak{R}$ is contained in a countable union of sets A_α. Using (ii), it follows that

$$X = \bigcup_{A \in \mathfrak{R}} A \subset \bigcup_{\alpha \in \Phi} A_\alpha .$$

(β) Let X be a set of cardinality $\aleph_1$. Let $\mathfrak{R}$ and $\mathfrak{R}'$ be systems of subsets of X each of which has properties (i)–(iv) of (α). Suppose further that X is the union of two disjoint sets A and B, $A \in \mathfrak{R}$, $B \in \mathfrak{R}'$. Then there is a bijective mapping φ of X onto itself such that $\varphi = \varphi^{-1}$ and such that $\varphi(C) \in \mathfrak{R}'$ iff $C \in \mathfrak{R}$.

Let $(A_\alpha)_{\alpha\in\Phi}$ be a decomposition of X in the sense of (α) with respect to $\mathfrak{N}$. Let $\mathfrak{S}$ be the set described in (iii) with respect to $\mathfrak{N}$. We may assume that $A \in \mathfrak{S}$ and $A = A_0$. Similarly let $(B_\alpha)_{\alpha\in\Phi}$ be the decomposition of X with respect to $\mathfrak{N}'$, and assume that $B \in \mathfrak{S}'$ and $B = B_0$ whereby $\mathfrak{S}'$ denotes the set described in (iii) with respect to $\mathfrak{N}'$. Then

$$A = \bigcup_{0<\alpha<\omega_1} B_\alpha \quad \text{and} \quad B = \bigcup_{0<\alpha<\omega_1} A_\alpha. \tag{1}$$

For each $\alpha \in \Phi$, $\alpha \neq 0$ let φ_α be a bijective mapping of A_α onto B_α. Define

$$\varphi\colon X \to X, \qquad x \mapsto \begin{cases} \varphi_\alpha(x) & \text{if } x \in A_\alpha \quad (0 < \alpha < \omega_1) \\ \varphi_\alpha^{-1}(x) & \text{if } x \in B_\alpha \quad (0 < \alpha < \omega_1). \end{cases}$$

Then φ is a bijective mapping of X onto itself, $\varphi = \varphi^{-1}$, and $\varphi(A_\alpha) = B_\alpha$ whenever $\alpha \in \Phi$, $\alpha \neq 0$. From (1) it follows that $\varphi(A_0) = \varphi(B_0)$. From the properties stated in (α), it follows that $f(C) \in \mathfrak{N}'$ iff $C \in \mathfrak{N}$.

(γ) Let $\mathfrak{N}$ be the set of all Lebesgue null sets $A \subset \mathbb{R}$. Denote by $\mathfrak{S}$ the set of all G_δ-sets $A \in \mathfrak{N}$. Assume that the Continuum Axiom holds. Then $\mathfrak{N}$ and $\mathfrak{S}$ have the properties (i)–(iv) with respect to $X := \mathbb{R}$.

(δ) Let $\mathfrak{N}$ be the set of all subsets of $\mathbb{R}$ of the first category, and denote by $\mathfrak{S}$ the set of all K_σ-sets $A \in \mathfrak{N}$. Then, if the Continuum Axiom is accepted, $\mathfrak{N}$ and $\mathfrak{S}$ have the properties (i)–(iv) with respect to $X := \mathbb{R}$.

The proof is left to the reader.

It is now easy to prove the theorem of Erdös.

(ε) If we accept the Continuum Axiom, there is a bijective mapping φ of $\mathbb{R}$ onto $\mathbb{R}$ such that $\varphi(A)$ is a Lebesgue null set if and only if A is a set of the first category.

Denote by $\mathfrak{N}$ the set of all Lebesgue null-sets and by $\mathfrak{N}'$ the set of all subsets of $\mathbb{R}$ of the first category. Apply (α)–(δ).

5.6.3(C) In this exercise our goal is the construction of interesting representations of Stieltjes measures.

First, we have to introduce some notation. Let $\mathfrak{J}$ be the set of all interval forms on $\mathbb{R}$. Denote by $(k_n)_{n\in\mathbb{N}}$ a sequence from $\mathbb{N}$, and define

$$X_n := \{\, i \in \mathbb{N} \cup \{0\} \mid i < k_n \,\}$$

and

$$\mathfrak{B}_n := \{\, \{x\} \mid x \in X_n \,\}$$

whenever $n \in \mathbb{N}$. For any $n \in \mathbb{N}$ provide X_n with the discrete topology and with the order relation inherited from $\mathbb{N} \cup \{0\}$. Further define

$$X := \prod_{n \in \mathbb{N}} X_n$$

$$X_r := \{(i_n)_{n \in \mathbb{N}} \in X \mid \exists m \in \mathbb{N}, (\forall n \in \mathbb{N}, n \geq m \Rightarrow i_n = 0)\}$$

$$X_l := \{(i_n)_{n \in \mathbb{N}} \in X \mid \exists m \in \mathbb{N}, (\forall n \in \mathbb{N}, n \geq m \Rightarrow i_n = k_n - 1)\}$$

and

$$\mathfrak{B} := \left\{ \prod_{n \in \mathbb{N}} A_n \,\middle|\, \begin{array}{l} \exists m \in \mathbb{N}, (\forall n \in \mathbb{N}, n < m \Rightarrow A_n \in \mathfrak{B}_n) \\ \text{and } (\forall n \in \mathbb{N}, n \geq m \Rightarrow A_n = X_n) \end{array} \right\}.$$

Provide X with the product topology and with the lexicographical order, defined by

$$(i_n)_{n \in \mathbb{N}} < (j_n)_{n \in \mathbb{N}} :\Leftrightarrow (i_m < j_m \quad \text{where} \quad m := \inf\{n \in \mathbb{N} \mid i_n \neq j_n\}).$$

We now formulate several subexercises:

(a) Prove the following assertions:

(α) The following properties are equivalent:

(α1) X is finite.
(α2) $\{n \in \mathbb{N} \mid k_n > 1\}$ is finite.
(α3) $X = X_r = X_l$.
(α4) $X_r \cap X_l \neq \varnothing$.

(β) If $\{n \in \mathbb{N} \mid k_n > 1\}$ is not finite, then X has the cardinality $\mathfrak{c}$ of the continuum.

(γ) X is a totally ordered complete lattice.

(δ) The following properties are equivalent for any $x \in X$:

(δ1) $x \in X_r$.
(δ2) $\{y \in X \mid y < x\}$ is either empty or has a largest element. This largest element is an element of X_l.

(ε) The following properties are equivalent for any $x \in X$:

(ε1) $x \in X_l$.
(ε2) $\{y \in X \mid y > x\}$ is either empty or has a smallest element. This smallest element is an element of X_r.

(ζ) The following properties are equivalent for any $x, y \in X$ with $x < y$:

(ζ1) $x \in X_l$, $y \in X_r$, and $]x, y[= \varnothing$.
(ζ2) $]x, y[= \varnothing$.
(ζ3) $]x, y[\cap X_r = \varnothing$.
(ζ4) $]x, y[\cap X_l = \varnothing$.
(ζ5) If $x = (i_n)_{n \in \mathbb{N}}$ and $y = (j_n)_{n \in \mathbb{N}}$, then there is $m \in \mathbb{N}$ such

that the following hold:

(i) $i_n = j_n$ whenever $n \in \mathbb{N}$, $n < m$.
(ii) $i_m + 1 = j_m$.
(iii) $i_n = k_n - 1$, $j_n = 0$ whenever $n \in \mathbb{N}$, $n > m$.

(η) For any $V \in \mathfrak{B}$,

$$\bigwedge_{x \in V} x \in X_r, \qquad \bigvee_{x \in V} x \in X_l, \qquad V = \left[\bigwedge_{x \in V} x, \bigvee_{x \in V} x\right].$$

(b) Prove the following assertions:

(α) X is compact and metrizable.

(β) If X is not finite, then X has no isolated points.

(γ) A sequence $(x_n)_{n \in \mathbb{N}}$ from X converges to $x \in X$ iff it is order convergent to x.

(δ) $]x, y[$ is open and $[x, y]$ compact whenever $x, y \in X$.

(ε) $[x, y[$ is open and $[y, x[$ compact whenever $x \in X_r$ and $y \in X$.

(ζ) $]y, x]$ is open and $]x, y]$ compact whenever $x \in X_l$ and $y \in X$.

(η) $[x, y]$ is open and $]y, x[$ compact whenever $x \in X_r$ and $y \in X_l$.

(ϑ) Every $V \in \mathfrak{B}$ is open and compact.

(ι) $\mathfrak{B}$ is a countable base of X.

(κ) X_l and X_r are dense subsets of X.

(λ) X is totally disconnected.

(c) Let f be a real function on X. Let $x_0 \in X$, and define f to be left-continuous at x_0 iff $\lim_{n \to \infty} f(x_n) = f(x_0)$ for any increasing sequence $(x_n)_{n \in \mathbb{N}}$ from X with $\bigvee_{n \in \mathbb{N}} x_n = x_0$. f is called right-continuous at x_0 iff $\lim_{n \to \infty} f(x_n) = f(x_0)$ for any decreasing sequence $(x_n)_{n \in \mathbb{N}}$ from X with $\bigwedge_{n \in \mathbb{N}} x_n = x_0$. f is called left-continuous (resp. right-continuous) iff f is left-continuous (resp. right-continuous) at every point of X. Prove the following:

(α) f is continuous at x_0 iff f is left-continuous and right-continuous at x_0.

(β) f is continuous at $x_0 \in X_r$ iff f is right-continuous at x_0.

(γ) f is continuous at $x_0 \in X_l$ iff f is left-continuous at x_0.

(δ) Denote by Z the set of all $x \in X$ at which f is not continuous. Assume that f is increasing. Then Z is countable.

(ε) Assume that f is increasing and $f(X)$ an interval of $\mathbb{R}$. Then the following assertions hold:

(ε1) f is continuous.
(ε2) $f(V)$ is an interval of $\mathbb{R}$ for any $V \in \mathfrak{B}$.
(ε3) If $x \in X_l$, $y \in X_r$, $x < y$ and $]x, y[= \varnothing$, then $f(x) = f(y)$.

(ζ) Let x_0 (resp. x_1) be the smallest (resp. greatest) element of X and let f be increasing and not constant. Then the following assertions are equivalent:

($\zeta 1$) For every $x \in X \setminus (X_r \cup X_l)$, $f([x_0, x[)$ and $f(]x, x_1])$ are open sets of $[f(x_0), f(x_1)]$.
($\zeta 2$) $f(X) = [f(x_0), f(x_1)]$ and f is strictly increasing on $X \setminus (X_r \cup X_l)$.
($\zeta 3$) For every $V \in \mathfrak{B}$, $f(V)$ is a nondegenerate interval of $\mathbb{R}$.

These conditions imply:

($\zeta 4$) For every $a \in [f(x_0), f(x_1)]$ the following holds: Either $f^{-1}(a)$ consists of exactly one point which belongs to $X \setminus (X_r \cup X_l)$ or $f^{-1}(a)$ consists of two points x, y with $x \in X_l$, $y \in X_r$, $x < y$ and $]x, y[= \varnothing$.

(d) Define

$$f: X \to \mathbb{R}, \qquad (i_n)_{n \in \mathbb{N}} \mapsto \sum_{n \in \mathbb{N}} \frac{i_n}{\left(\prod_{m=1}^{n} k_m\right)}$$

$$g: X \to \mathbb{R}, \qquad (i_n)_{n \in \mathbb{N}} \mapsto \sum_{n \in \mathbb{N}} \frac{2i_n}{\left(\prod_{m=1}^{n} (2k_m - 1)\right)}.$$

Suppose that the set $\{n \in \mathbb{N} \mid k_n > 1\}$ is not finite, and prove the following:

(α) f and g are increasing.

(β) $f(X) = [0,1]$ and for every $V \in \mathfrak{B}$, $f(V)$ is a nondegenerate interval of $\mathbb{R}$.

(γ) f is continuous and its restriction to $X \setminus X_r$ (to $X \setminus X_l$) is strictly increasing.

(δ) Let x_0(resp. x_1) be the smallest (resp. greatest) element of X. Then for every $x \in X \setminus (X_r \cup X_l)$, $f([x_0, x[)$ and $f(]x, x_1])$ are open sets of $[0,1]$.

(ε) For every $a \in [0,1]$ the following holds: Either $f^{-1}(a)$ consists of exactly one point, which belongs to $X \setminus (X_r \cup X_l)$, or $f^{-1}(a)$ consists of two point x, y such that $x \in X_l$, $y \in X_r, x < y,]x, y[= \varnothing$ (in this case a is rational).

(ζ) g is continuous and the mapping $X \to g(X)$, $x \mapsto g(x)$ is a homeomorphism.

(η) There is a unique increasing continuous mapping h: $[0,1] \to [0,1]$ such that $f \circ g^{-1}(x) = h(x)$ for each $x \in g(X)$. h is constant for any interval of $[0,1] \setminus g(X)$.

(ϑ) If $k_n = 2$ for each $n \in \mathbb{N}$, then $g(X)$ is the Cantor set.

(ι) If $k_n = k \in \mathbb{N}$ for each $n \in \mathbb{N}$ then the dimension of $g(X)$ in the sense of Ex. 5.4. 20 (δ) is equal to $\log k/\log(2k - 1)$ and its measure in this dimension is 1. In particular $g(X)$ is a Legesgue null set. (The Lebesgue measure on $\mathbb{R}$ is exactly the 1-dimensional measure on $\mathbb{R}$ in the sense of Ex. 5.4.20 (δ).)

(κ) For any $\alpha \in]\log 2/\log 3, 1[$ and any $\beta \in \overline{\mathbb{R}}_+$ there exists a sequence $(k_n)_{n\in\mathbb{N}}$ in $\mathbb{N}$ such that the dimension of $g(X)$ is α and the measure of $g(X)$ in this dimension is β.

(e) We set

$$\mu: \mathfrak{B}_r \to \mathbb{R}, \qquad \prod_{n\in\mathbb{N}} A_n \mapsto \prod_{\substack{n\in\mathbb{N}\\ A_n\neq X_n}} |A_n|/k_n.$$

($|A_n|$ denotes the cardinality of A_n.)

Let $\mathfrak{J}$ be the ring of sets of all interval forms on $\mathbb{R}$, f an increasing real function on X,

$$f(\mu): \mathfrak{J} \to \mathbb{R}, \qquad A \mapsto \sup\{\mu(V)\,|\,V\in\mathfrak{B}_r,\, V\subset f^{-1}(A)\}$$

and $\mathcal{N}$ the set of all positive measures ν on $\mathfrak{J}$ such that there is an $A \in \mathfrak{J}$ with

$$\sup_{B\in\mathfrak{J}} \nu(B) = \nu(A) = 1.$$

Prove:

(α) $f(\mu) \in \mathcal{N}$.

(β) There exists a unique increasing, left continuous real function g on $\mathbb{R}$ such that

$$(f(\mu))([\alpha,\beta[) = g(\beta) - g(\alpha)$$

for all $\alpha, \beta \in \mathbb{R}$ with $\alpha < \beta$ and $\lim_{\alpha\to-\infty} g(\alpha) = 0$. One has $\lim_{\alpha\to\infty} g(\alpha) = 1$.

(γ) g is continuous (or equivalently: $f(\mu)$ is atomfree) iff for all $V \in \mathfrak{B}$, $f(V)$ contains at least two points.

(δ) The following assertions are equivalent:

(δ1) $f(X)$ is a nondegenerate interval of $\mathbb{R}$.
(δ2) There exists an interval $A \in \mathfrak{J}$ such that $(f(\mu))(A) = 1$ and such that the restriction of g to A is strictly increasing.
(δ3) The support of $f(\mu)$ is an interval of $\mathbb{R}$.

(ε) The following assertions are equivalent:

(ε1) For every $V \in \mathfrak{B}$, $f(V)$ is a nondegenerate interval of $\mathbb{R}$.
(ε2) g is continuous and (δ2) holds.
(ε3) $f(\mu)$ is atomfree and its support is an interval of $\mathbb{R}$.

(ζ) If f is the mapping of (d) then $f(\mu)$ is the Lebesgue measure on $[0,1]$ in the following sense:

$$(\forall A \in \mathfrak{J}) \qquad (f(\mu))(A) = \text{the length of } A \cap [0,1].$$

(η) For every $\nu \in \mathcal{N}$ there exists a unique left continuous (resp. right continuous) f, such that $f(\mu) = \nu$.

5.6.4(C) **Bernstein Sets.** In this exercise we construct subsets of $\mathbb{R}$ that are neither approximable nor measurable with respect to the Lebesgue measure λ. The construction presented here is due to F. Bernstein (1908). First prove the following statement:

(α) There is a subset B of $\mathbb{R}$ such that both B and $\mathbb{R} \setminus B$ meet every uncountable closed subset of $\mathbb{R}$.

We give some hints. Let $\mathfrak{F}$ be the set of all closed uncountable subsets of $\mathbb{R}$. Note that every set $F \in \mathfrak{F}$ has cardinality $\mathfrak{c}$. Let A be the set of all ordinals of cardinality strictly smaller than $\mathfrak{c}$. A has cardinality $\mathfrak{c}$. Hence there is a bijective map φ: $A \to \mathfrak{F}$. Let $\preccurlyeq$ be a well-ordering on $\mathbb{R}$. Then every set $F \in \mathfrak{F}$ is well ordered by $\preccurlyeq\restriction_F$. Denote by x_0 and y_0 the first two elements of $\varphi(0)$. Suppose that x_β and y_β have been defined for each ordinal β less than $\alpha \in A$. Then let x_α and y_α be the first two elements of $\varphi(\alpha) \setminus \bigcup_{\beta<\alpha}\{x_\beta, y_\beta\}$. Finally, define $B := \{x_\alpha \mid \alpha \in A\}$. B has the desired property.

Now prove the following consequences:

(β) Let B be a Bernstein set. Then every λ-measurable subset of B is a λ-null-set. Hence $\lambda_*(B) = 0$, $\lambda_*(\mathbb{R} \setminus B) = 0$, and $\lambda^*(B) = \infty$.

(γ) Every approximable subset of a Bernstein set is of the first category.

(δ) If B is a Bernstein set then B is neither λ-measurable nor approximable.

APPENDIX

HISTORICAL REMARKS ON THE DEFINITION OF THE INTEGRAL

It is difficult to pinpoint when the (definite) integral first appears in the mathematical literature, as the notion of integral arose slowly, and in small steps. The roots of integration theory can already be found in antiquity, in Eudoxos or Archimedes, for example, or later in J. Kepler's book *Nova stereometria doliorum vinariorum*, which appeared in Linz in 1615. *Nova stereometria* was dedicated to determining the volumes of solids of revolution, and about 90 such solids are treated in the book. By the way, traces can be found in Kepler's book not only of integration theory but also of the calculus of variations (the parallelepiped of greatest volume that can be inscribed in a sphere is a cube) and of differential calculus (functions change only slowly at extreme points).

Substantially greater recognition as birthplaces of integration theory must be accorded three works of a famous student of Galileo Galilei, namely, Bonaventura Cavalieri: *Geometria indivisibilibus continuorum nova quadam ratione promota* (Bononiae, Bologna, 1635, 2nd edition 1653), *Centuria di varii problemi per dimonstrare l'uso e la facilità de' logaritmi nella gnomonica, astronomia, geograffia, altimetria, pianimetria, stereometria & aritmetica prattica* (Monti e Zenero, Bologna 1639), and *Exercitationes geometricae sex* (Bononiae, Bologna, 1647). For there, in contrast to the work of Cavalieri's predecessors, area and volume are obtained not by artificial devices (and sometimes extraordinary at that!) but by a well-laid-out method of computation. For example, the method appears principally in the fact that the exhaustion is always accomplished by parallel slices. This method rests on a clearly formulated theorem, which asserts that if two solids (plane regions) have equal altitudes, and if cross sections made by planes parallel to the bases and at equal distances from them are always in a given ratio, then the volumes (areas) of the solids (plane regions) are also in this ratio. It cannot be said, though, that this theorem was actually proved.

To give a first, very vague idea of what can be found of the beginning of integration theory in these works of Cavalieri, let us say that he proved theorems about area and volume that are distantly analogous to the theorem

$$\int_0^a x^n\,dx = \frac{a^{n+1}}{n+1}$$

for $n \in \{1,2\}$ in the first work and for $n \in \{3,\dots,9\}$ in the second. In *Centuria* Cavalieri even claims, although he never proved it, that this formula holds for all natural numbers. In *Exercitationes* Cavalieri reports how he came to this discovery. An exercise from Kepler's *Doliometrie* enabled him to handle the case $n = 4$, where he found the denominator 5. He remembered that in the cases $n = 1$, $n = 2$, the denominators 2 and 3 had arisen. In order to leave no gaps, he investigated the case $n = 3$ and found, as expected, the denominator 4. He was astounded that this rule extended to all natural numbers.

In the preceding description of Cavalieri's accomplishments we have been rather generous in two senses. First, what Cavalieri actually proved reads:

$$\sum \overline{AB}^n = \frac{1}{n+1}\sum \overline{AC}^n$$

where $\overline{AC}$ varies over the set of line segments that lie inside a parallelogram and are parallel to one side (and therefore have constant length) and $\overline{AB}$ signifies the portions of these segments $\overline{AC}$ that lie on one side of a diagonal. Second, with such a vaguely stated formula one cannot speak of a proof in the strong sense of the word. Concerning *Geometria indivisibilibus*, Maximilien Marie wrote in his book *Histoire des sciences mathematiques et physiques* (Volume IV, *De Descartes à Huygens*, Paris, 1884, p. 90): "If prizes for obscurity were awarded, it would, without a doubt, have won first prize. One simply cannot read it; one is continually forced to guess." Finally, one must remark that the proofs were somewhat long and complicated (for each n the already proven results for smaller n were used) and that the proofs for various n were not unified. The proofs are purely geometric with no trace of analytic geometry, which was then only beginning to emerge.

Arithmetic proofs for the formula

$$\int_0^a x^\alpha\,dx = \frac{a^{\alpha+1}}{\alpha+1} \tag{1}$$

(for α a positive rational number) were given in the years following by P. Fermat, E. Toricelli, B. Pascal, and G. P. de Roberval, all of them, unfortunately, either unpublished or published only after great delay. The same is true for the quadrature of various trigonometric functions, of the cycloid, and of other curves studied by the aforementioned mathematicians. In his book *Arithmetica Infinitorum*, printed in 1655, John Wallis asserts without proof that the formula (1) holds for all positive real numbers α.

In the seventeenth century came the discovery of differential and integral calculus, one of the most important discoveries ever made in the field of mathematics. Fermat, R. Descartes, Toricelli, G. de Saint-Vincent, Wallis, de Roberval, Pascal, C. Huygens, J. Gregory, and I. Barrow took part as forerunners to this achievement; the discovery itself came independently from I. Newton and G. W. Leibniz; thus the names are of the first rank. (The terminology "integral" appears for the first time in 1690 with Jakob Bernoulli; the symbol $\int$ was introduced by Leibniz in 1675.) Infinitesimal calculus points out three fundamental concepts: derivative, antiderivative, and definite integral, with the definite integral being computed by means of the antiderivative. The integral of contemporary integration theory is a generalization of this definite integral. In this sense our interest here is primarily in the third concept. From such a standpoint infinitesimal calculus appears primarily as a fantastic method for calculating definite integrals, a method that outshadows all earlier results.

Although the eighteenth century knows such names as L. Euler and J. L. Lagrange, and although at that time the infinitesimal calculus developed enormously, having an extraordinary influence, both on other branches of mathematics and on the natural sciences, the first definition of definite integral is not found until the nineteenth century, in 1823, in a book of A. -L. Cauchy. This fact is not surprising, since many mathematical theories were developed at that time without the fundamental concepts being rigorously defined. Besides, one must remember that the notion of function was much more narrowly conceived then, and the existence of the integral was assured by the existence of an antiderivative. The latter was, to be sure, not assured but was nevertheless known for a large class of functions.

At the beginning of the nineteenth century the limits of the notion of function, as defined, for example, by Johann Bernoulli in 1718 and fixed in the famous encyclopedia of J. d'Alembert and D. Diderot (*Encyclopédie on dictionnaire raisonné des sciences, des arts et de métiers*, Paris, 1751–1765) were exceeded in the works of Euler and J. B. Fourier. A definition of definite integral became an urgent necessity. Cauchy gave this definition under a certain didactical pressure. He was active not only in research but also in instruction, which was not at all a forgone conclusion in those days. He carried out both tasks with great enthusiasm, and also published his lectures. The first part (*Cours d'analyse de l'École Polytechnique*, *Analyse algébrique*, Debure, Paris) appeared in 1821; the lectures on differential and integral calculus, in which we encounter the definition of definite integral, were published in abbreviated form in 1823 (*Résumé des leçons à l'École Polytechnique données sur le calcul infinitésimal*, Debure, Paris).

The integral was defined as follows. Let f be a continuous real function on an interval $[x_0, X]$ of the real axis. Partition the interval $[x_0, X]$ into small subintervals using points x_i $(i = 0, 1, \ldots, n)$, where $x_n = X$, and form the sums

$$\sum_{i=1}^{n} f(x_{i-1})(x_i - x_{i-1}).$$

These sums converge toward a value when the partition of the interval is refined in such a way that the length of the largest interval in the partition converges to zero. This limiting value is called the integral of f. Cauchy proposed three notations for the integral, giving preference to the notation $\int_{x_0}^{X} f(x)\,dx$ introduced by Fourier. Fourier's notation has persisted.

The continuity that appears in this definition was defined by Cauchy in his first book, that is, in 1821. This definition reads: "The function f will be said to be continuous between two fixed values of the variable x if, for each value of x between the specified limits, the numerical value of the difference $f(x+\alpha)-f(x)$ decreases indefinitely with that of α." Unfortunately the expression "decreases indefinitely" is not further defined. From a related description it is seen that this expression stands for "converges to zero." That impairs of course the actual definition of continuity. A further aggravation is the fact that continuity is defined on an interval and not at a point. Continuity at a point is not mentioned at all by Cauchy. He explains only the notion "continuous in the neighborhood of a point," that is, "continuous on an interval containing the point," a notion of no mathematical significance. By "discontinuous at a point" Cauchy understands, in contrast to our notion, "not continuous in the neighborhood of the point," again a useless notion in mathematics.

Here it is appropriate to mention the name of B. Bolzano. Bolzano found, already before Cauchy, a better definition of continuity, the very first definition of continuity, even ["Rein analytischer Beweis des Lehrsatzes, dass zwischen je zwey Werthen, die ein entgegengesetztes Resultat gewähren, wenigstens eine reele Wurzel der Gleichung liege," *Abh. König. Böhm. Gesell. Wiss., Phys.-Math. Theil* 5 (1817) 1–60]. The definition reads: "By a proper explanation one namely understands by the expression that a function f varies continuously for all values of x lying inside or outside certain limits simply that, if x be any such value, then the difference $f(x+\omega)-f(x)$ can be made smaller than every given quantity if one can take ω to be as small as one wants."

Concerning the proof for the convergence of the sums mentioned previously with regard to the Cauchy definition of integral, it must be said that Cauchy's proof is also unsatisfactory. Cauchy assumes without proof that the function is uniformly continuous. He very carefully estimates the difference between two sums and shows this difference to be smaller than the length $X - x_0$ multiplied by the greatest variation of the function over the intervals of the partition. Since the function is continuous—the proof proceeds—one can choose the intervals so small that the variations can be made arbitrarily small. That does hold, but one must first show that the function is uniformly continuous, which Cauchy, as has been said, did not do. One can assume it never occurred to Cauchy that two different concepts are involved here. Presumably he confused continuity with uniform continuity.

In all likelihood P. G. Lejeune-Dirichlet provided the first proper proof of the aforementioned convergence. Since this proof is needed if the definition of integral is to make sense, Dirichlet's name must also be associated with this definition. Exactly when Dirichlet first produced a proof that every continuous real function on a closed interval is uniformly continuous (the result known today as the Heine Theorem), and then reported Cauchy's proof, is not known. In his famous paper "Über die Darstellung ganz willkürlicher Funktionen durch Sinus- und Cosinusreihen," which he published in 1837 in *Repertoiren der Physik*, Dirichlet asserted, in conjunction with the definition of integral, that the appropriate convergence can easily be proved rigorously. Since Dirichlet was famous for the rigor of his proofs,* the adverb "rigorously" provides food for thought. It can be assumed that Dirichlet carried out the proof in question in some of his lectures in Berlin and Göttingen. Fortunately indications of two such lectures remain intact. One can be found in the book of G. Arendt, *G. Lejeune-Dirichlets Vorlesungen über die Lehre von den einfachen und mehrfachen bestimmten Integralen*, which appeared in 1904 in Braunschweig. In his preface, Arendt writes: "The book which I here present to the public has as its basis an elaboration, prepared by me with great care, usually on the day of the lecture itself and therefore to a large degree reliable, with no gaps, of the lectures held by Dirichlet, for four hours a week in the summer of 1854 at the local university, on the study of the definite integral, and of the accompanying one-hour *Publikum*, in which were shown some applications of the one-dimensional definite integral treated in the first part of the principal lecture. ... I took the greatest pains in the writing up of my book to reproduce the lecture of Dirichlet in its entire originality, without abbreviations or alterations, but without any of my own or any other additions."

On page 4 of this book, under the heading "Fundamental Property of Continuous Functions," we read the following text:

> Let $y = f(x)$ be a function of x continuous in the finite interval from a to b, and by a subinterval let one understand the difference between two arbitrary values of x, thus each arbitrary piece of the abscissa-axis between a and b. Then the possibility always exists for an absolute quantity σ chosen arbitrarily small to find a second proportionally small quantity ρ of such a kind that in every subinterval that is $\leq \sigma$ the function y varies by no more than at most ρ.

Connected to this is a rather long and detailed proof. The definition of the integral follows on page 8. The proof of the convergence of the Cauchy sums is based on the uniform continuity of the function proved earlier.

In many problems functions appear that have discontinuities or even become infinite but are to be integrated. Cauchy as well as Dirichlet therefore extended the integral in such cases by first calculating the integral on intervals where the function is continuous and then passing to a limit. The result is a patchwork with quite limited applicability.

*"He alone, not I, not Cauchy, not Gauß, knows what a completely rigorous mathematical proof is, instead we learn it first from him. If Gauß says he has proven something it seems probable to me, if Cauchy says it, then just as much to be wagered for as against, if Dirichlet says it, it is certain" (from a letter dated December 21, 1846, from C. G. J. Jacobi to Alexander von Humboldt).

The well-known definition of the Riemann integral is found in B. Riemann's *Habilitationsschrift* of 1853, "Über die Darstellbarkeit einer Function durch eine trigonometrische Reihe." This work is dedicated to Fourier series and consists of two parts. In the first part the history of Fourier series is reported in detail. The second part presents new results in this field. Between these two parts, in four pages, we find Riemann's definition of the integral as well as his criterion for integrability. Except for the investigations appearing on these four pages, Riemann never occupied himself with integration theory. He also never published his *Habilitationsschrift*. It was R. Dedekind who in 1867, after Riemann's death, published the paper [in *Abh. K. Gesell. Wiss. Göttingen, Math. Classe* 13 (1866–1867) 87–132].

Before Riemann presented the new definition, he wanted, for the very reason of its newness, somehow to justify it. Indeed, he wrote:

> However great our uncertainty is about how material forces and positions vary infinitesimally in time and space, we can certainly assume, that the functions to which Dirichlet's investigations do not apply do not appear in nature. Notwithstanding that, these cases that were not disposed of by Dirichlet deserve attention for a twofold reason. First, as Dirichlet himself remarked at the conclusion of his paper, this object is immediately related to infinitesimal calculus and can thereby serve to bring its principles into greater clarity and precision. In this connection its treatment is of immediate interest in itself. Secondly, however, the application of Fourier series is not restricted to physical investigations; it has now also been applied with success in a field of pure mathematics, the theory of numbers, and here exactly those functions whose representation by trigonometric series Dirichlet did not investigate appear to be of importance.

He defines the integral as Cauchy does, yet with the following differences: (a) the function f, although bounded on $[x_0, X]$, need not be continuous; (b) the "Cauchy sums" $\sum_{i=1}^{n} f(x_{i-1})(x_i - x_{i-1})$ are replaced by "Riemann sums" $\sum_{i=1}^{n} f(\xi_i)(x_i - x_{i-1})$ where $\xi_i \in [x_{i-1}, x_i]$ for all i; (c) it is, of course, no longer claimed that these sums converge as the partitions of the interval are refined; rather, this property is required, and the function f is then declared to be integrable (today we call f "Riemann integrable" and the value of the limit is called the integral of f). Riemann's criterion for integrability reads: for every $\alpha > 0$ there exists a partition of the interval so that the sum of the lengths of the intervals on which the variation of the function is greater than α is arbitrarily small. Finally, Riemann gives an example of an integrable function with a dense set of discontinuities. The upper and lower integrals of a function, as they are presented in books and lectures on the Riemann integral, were not defined by Riemann but were first introduced in 1875 by G. Darboux, K. J. Thomae, G. Ascoli, and H. J. S. Smith, independently of each other; the corresponding symbols $\overline{\int}, \underline{\int}$ were invented by V. Volterra in 1881.

The publication of the Riemann definition of integral in 1867 influenced the mathematicians of the last third of the past century in great measure, and numerous papers of that period concern themselves with this concept. We report only on a few that were especially important in smoothing the way for the Lebesgue integral.

First of all, two works of G. Peano from 1883 and 1887 deserve mention. ["Sulla integrabilitàdelle funzioni," *Atti Acc. Sci. Torino* 18 (1883) 439–446; *Applicazione geometriche del calcolo infinitesimale*, Bocca, Torino, 1887]. Inner and outer measure of a set are defined in these papers. Peano presents the theory separately for the dimensions 1, 2, and 3; in the following only dimension 2 is considered. Let A be a bounded subset of the plane. The outer measure of A is the infimum of the areas of polygonal figures containing A, and the inner measure of A is the supremum of the areas of polygonal figures contained in A. If the two measures of A coincide, one says that A has an area that is equal to the inner or outer measure of A. In contemporary terminology, which we use from here on, such sets are said to be Jordan measurable, and the corresponding area is called the Jordan measure. Peano proves the following theorems: (a) a set is Jordan measurable iff its boundary has outer measure zero; (b) if two disjoint sets are Jordan measurable, then so is their union, and the Jordan measure of the union is equal to the sum of the Jordan measures of the two sets (this property is called the additivity of Jordan measure); (c) if f is a bounded positive function on an interval $[a, b]$ and if E is the set $\{(x, y) \mid a \leq x \leq b, 0 \leq y \leq f(x)\}$, then the lower Riemann integral of f is equal to the inner measure of E, and the upper Riemann integral of f is equal to the outer measure of E; in particular, f is Riemann integrable iff E is Jordan measurable, and in this case the Riemann integral of f is equal to the Jordan measure of E.

C. Jordan in 1892 extended Peano's work, without citing Peano ["Remarques sur les intégrales définies," *J. Math. Pures et Appl.* (4) 8 (1892) 69–99]. He developed the theory of Jordan-measurable sets and of the Riemann integral directly in Euclidean space. Actually Jordan repeated the Riemann definition of integral, yet two important differences must be noted. For Riemann the function to be integrated is always defined on an interval, so no difficulties arise from the nature of the domain. In contrast, for Jordan the domain of definition of the function is an arbitrary bounded Jordan-measurable set in Euclidean space. To define the integral, Jordan partitions this set into finitely many Jordan-measurable subsets, whereas Riemann partitions the interval into finitely many subintervals. To be sure, Jordan's new method leads to no extension, for the one-dimensional case, of the collection of integrable functions, but it already contains the basic idea of the Lebesgue method. The method of Lebesgue differs from that of Jordan merely by the fact that the sets in the partition need no longer be Jordan measurable, only Lebesgue measurable. Thus it is measure theory that first provides the means for a usable definition of integral.

E. Borel took a half step in this direction. He developed, in his 1898 book *Leçons sur la théorie des functions* (Gauthier-Villars, Paris), the theory of "Borel sets" and "Borel measures." The Borel sets (in $\mathbb{R}$) are defined recursively. One begins with the open sets and allows two operations, namely, forming the union of countably many sets and taking the complement of a set. These operations may be performed arbitrarily many times, even transfinitely many. In this way one obtains the σ-algebra of Borel sets. Borel further asserts, without providing clear proofs, that to each bounded Borel set one can associate a positive real number, its Borel measure, which tells something about the length of the set and which, in the case of an interval, gives exactly the length of the interval. Very important here is the fact that the Borel measure is σ-additive; in other words, that if $(A_n)_{n \in \mathbb{N}}$ is a sequence of pairwise disjoint Borel sets whose union is bounded, then the Borel measure of $\bigcup_{n \in \mathbb{N}} A_n$ is exactly the sum of the Borel measures of the A_n. Jordan measure has this property (which was not remarked) only under the very strong hypothesis that $\bigcup_{n \in \mathbb{N}} A_n$ is Jordan measurable, which is not always the case. In particular, countable sets are always Borel sets and in fact have Borel measure zero. Thus there exist Borel sets that are not Jordan measurable (an example is the set of rational numbers from a bounded interval), yet there also exist Jordan-measurable sets that are not Borel sets. This may be one of the reasons why Borel made not the slightest attempt to connect his theory with integration theory, thereby missing a great discovery.

The correct definition of the integral, which at the same time signifies the birth of modern integration theory, was achieved by H. Lebesgue in 1901 ["Sur une généralisation de l'intégrale définie," *C. R. Acad. Sci. Paris* 132 (1901) 1025–1028]. This definition relies on what is today called Lebesgue measure. Let A be a bounded set in $\mathbb{R}$. Consider the coverings of A by means of countably many intervals; the infimum of the sums of the lengths of these intervals is called the measure of A (today, the outer Lebesgue measure of A). A is said to be measurable (we say Lebesgue measurable) if the sum of the measures of A with those of the complementary sets of A relative to an interval $[a, b]$ containing A is equal to the length $b - a$ of the interval $[a, b]$. Now let y be a bounded real function on an interval $[a, b]$ with values between m and M. Further let

$$m = m_0 < m_1 < m_2 < \cdots < m_{p-1} = M < m_p$$

be a partition of the interval $[m, M]$ and

$$E_0 := \{x \in [a, b] \mid y(x) = m\}$$

$$E_i := \{x \in [a, b] \mid m_{i-1} < y(x) \leq m_i\}.$$

Lebesgue assumes that these sets are measurable and considers the sums

$$m_0\lambda_0 + \sum m_i\lambda_i \qquad m_0\lambda_0 + \sum m_{i-1}\lambda_i$$

where, for each i, λ_i denotes the measure of the set E_i. These two sums converge toward the same value as the partitions of the interval $[m, M]$ are refined. This value Lebesgue called the integral of y (we call it the Lebesgue integral of y). Those functions for which the preceding sets E_i are measurable and for which the Lebesgue integral can be defined Lebesgue called summable (today one says Lebesgue measurable). Lebesgue further remarked (without proving so) that every bounded Riemann-integrable function is integrable in his sense and that the sum and the product of two (bounded) Lebesgue-integrable functions as well as the limit of a uniformly bounded sequence of Lebesgue-integrable functions are Lebesgue integrable as well. The last assertion is a special case of the Lebesgue Convergence Theorem and does not hold in general for the Riemann integral, but instead only under additional hypotheses such as that of uniform convergence. As a special case Lebesgue showed that every bounded derivative is Lebesgue integrable and that the definite Lebesgue integral provides an antiderivative.

One year later Lebesgue published his doctoral dissertation ["Intégrale, longuer, aire," *Ann. Mat. Pura Appl.* (3) 7 (1902) 231–359], in which he develops the ideas described here. Measure theory is formulated for bounded sets of arbitrary dimension, although only carried out for one and two dimensions. There is a certain inclination to anchor the theory axiomatically, which would carry with it a kind of uniqueness. Unfortunately this intention is not consistently carried out since axioms are given only for the measure and not for its domain of definition, thus calling the uniqueness into question. It is proved (in contemporary terminology) that the bounded Lebesgue-measurable sets form a δ-ring and that Lebesgue measure is σ-additive. In addition the translation invariance is shown. Every Borel-measurable set is Lebesgue measurable, as is every Jordan-measurable set, and there exist sets that are Lebesgue measurable but neither Jordan nor Borel measurable. A bounded set E is Lebesgue measurable iff there exist Borel measurable sets E_1, E_2 such that $E_1 \subset E \subset E_2$ and the Lebesgue measure of $E_2 \setminus E_1$ is zero. The question of whether sets exist that are not Lebesgue-measurable is, to be sure, formulated, but remains open. (G. Vitali answered this question, constructing, by means of the Axiom of Choice, a set that is not Lebesgue measurable, in *Sul problema della misura dei gruppi de punti di una retta*, Gamberini e Parmeggiani, Bologna, 1905.) The integral of a bounded real function on a bounded Lebesgue-measurable set is viewed geometrically as the measure of the corresponding set in a higher dimension. It is shown that this definition coincides with the analytic one from the Note mentioned before. Lebesgue now remarks, however, that the analytic definition can also be applied to unbounded real functions, in fact to extended-real functions that are infinite at finitely many points.

In the year 1902–1903 Lebesgue held 20 lectures at the Collège de France on integration theory, which he then brought out as a book: *Leçons sur l'intégration et la recherche des fonctions primitives* (Gauthier-Villars, Paris, 1904). In the course of discussing the Riemann definition of integral, Lebesgue shows that the criterion of Riemann can be formulated (in the new terminology) as follows: a bounded real function on an interval is Riemann integrable iff its set of points of discontinuity has Lebesgue measure zero. Apart from the beauty of this criterion, it strikes one that for integrability in the sense of Riemann what is decisive is not the nature of the discontinuities but their extent. This has the consequence that the property of being Riemann integrable is preserved under many operations. New in Lebesgue's work is an attempt to define the integral axiomatically, as was attempted for measures in his doctoral thesis.

A didactical problem in the construction of the Lebesgue integral arises from the fact that one must first construct Lebesgue measure on the corresponding δ-ring. In order to circumvent this difficulty, F. Riesz developed a new method of construction using simple functions as the starting point for the integral; in this connection a function is called simple if it has only finitely many points of discontinuity and is constant on the neighboring intervals ["Sur quelques points de la théorie des fonctions sommables," *C. R. Acad. Sci. Paris* 154 (1912) 641–643 and "Sur l'intégrale de Lebesgue," *Acta Math.* 42 (1920) 191–205]. This method of F. Riesz deserves mention insofar as one may view it as the first construction of an integral where the measure is defined on a ring of sets and not on a δ-ring. In this connection we also mention a work of C. Caratheodory ["Über das lineare Maβ von Punktmengen—eine Verallgemeinerung des Längenbegriffs," *Nachr. König. Gesell. Wiss. Göttingen, Math. -Phys. Klasse* (1914) 404–426] in which the theory of outer measure for Euclidean space is developed and which leads to the completion of measures. In particular, one obtains from this paper, in special cases, the extension of a positive measure from a ring of sets to the generated δ-ring.

In the foregoing description we skipped over the Stieltjes integral. This integral was introduced by T.-J. Stieltjes in his paper "Recherches sur les fractions continues" of 1894 (*Ann. Sci. Fac., Toulouse, Sci. Math. Sci. Phys.* 8 J1–J122). As the title of the paper suggests, the problem with which Stieltjes concerned himself had nothing to do directly with the theory of integration, but various investigations led him to expound the "Stieltjes integral" for continuous functions. It is the Cauchy definition of integral with the difference that the length of intervals of the form $[x, y]$ is replaced by $g(y) - g(x)$, where g is an increasing real function. Stieltjes himself notes that one can also define the integral for a noncontinuous function, but he sees no value that such a definition could bring.

The Stieltjes integral was ignored by the mathematical world until 1909, when it was rescued from oblivion by a paper of F. Riesz: "Sur les opérations fonctionnelles linéaires," *C. R. Acad. Sci. Paris* 149 (1909) 974–977. If $[a, b]$ is an interval of $\mathbb{R}$ and $\mathscr{C}$ is the vector lattice of continuous real functions on $[a, b]$, then Riesz shows that for every positive linear form φ on $\mathscr{C}$ there exists an increasing real function g on $[a, b]$ such that

$$\varphi(f) = \int_a^b f(x)\, dg(x)$$

for every $f \in \mathscr{C}$. This question was posed and somewhat unsatisfactorily answered by J. Hadamard ["Sur les opérations fonctionnelles," *C. R. Acad. Sci. Paris* 136 (1903) 351–354]. Hadamard's paper is also important because continuous linear forms first appear there. The fact that one can construct a definite integral in the sense of Lebesgue but for the Stieltjes measure, an integral that includes both the Stieltjes and the Lebesgue integrals, was remarked by J. Radon ["Theorie und Anwendungen der absolut additiven Mengenfunktionen," *Sitzungsberichte der math. naturwiss. Klasse der Akad. der Wiss.*, (Wien) 122 Abt. IIa (1913) 1295–1438]. Radon goes essentially further, however, insofar as he develops a general theory of measure and corresponding theory of integration and proves for arbitrary dimensions the theorem of Riesz mentioned here.

The first step away from Euclidean space was taken by M. Fréchet ["Sur l'intégrale d'une fonctionnelle étendue à un ensemble abstrait," *Bull. Soc. Math. France* 43 (1915) 248–265]. He defines the integral on a measure space $(X, \mathfrak{R}, \mu)$, where $\mathfrak{R}$ is a σ-algebra on X. A more general standpoint was developed by P. J. Daniell ["A general form of integral," *Annals Math.* 19 (1917–1918) 279–294], when he introduced the closure of a Daniell space. The method of Daniell also has the advantage that it can be applied directly to measures that are defined on rings of sets. The Carathéodory method of first completing a measure and only then defining the integral is thereby circumvented. Moreover the results of Carathéodory are consequences of the Daniell construction. Daniell himself, however, does not seem to have noticed this fact, which first appears in the papers of H. H. Goldstine ["Linear functionals and integrals in abstract spaces," *Bull. Amer. Math. Soc.* 47 (1941) 615–620] and M. H. Stone ["Notes on Integration II," *Proc. Nat. Acad. USA* 34 (1948) 447–455].

In developing the theory of integration on locally compact spaces, N. Bourbaki encountered several difficulties that forced the definition of a second kind of integral, which he called the essential integral (*Intégration*, Ch. 5, Sec. 2, Hermann, Paris, 1956). The difference from the (ordinary) integral consists in using locally null sets (sets for which every point of the locally compact space has a neighborhood whose intersection with the given set is a null set) instead of null sets for the construction of the integral. It turns out that the essential integral has better properties than the integral. When the theory of integration was developed in the late 1960s on Hausdorff spaces, a definition of the integral was introduced that in the locally compact case corresponds to the definition of the essential integral. I. E. Segal and R. A. Kunze suggested in their book (*Integrals and Operators*, Sec. 3.7, McGraw-Hill, New York, 1968) that one use locally null sets in defining the integral in the abstract case. A set would be viewed as locally null if its intersection with every set of the domain of definition of the measure is a null set (in the locally compact case these are exactly the locally null sets described earlier). Although they stress the usefulness of such a definition, Segal and Kunze nevertheless worked with the old definition in their book. The integral introduced in the present book is exactly that proposed by Segal and Kunze. One advantage of this integral is that the definition of integrals on Hausdorff spaces becomes, by virture of this definition, a special case of the general definition (a fact not mentioned by Segal and Kunze); but measures on Hausdorff spaces possess an additional property (regularity) that describes the compatibility of the measures with the topology and that is reflected in additional properties of the integral. The definition used here achieves the long-desired goal of embedding in abstract integration theory the theory of integration on Hausdorff spaces.

IMPORTANT EXERCISE TOPICS

LIST OF SYMBOLS AND NOTATION

The following list contains notation introduced in the main text. Notation used throughout the text but assumed familiar to the reader is listed in Notation and Terminology but not included here. The reader should also note the comment on page 5 concerning our use of notation.

INDEX

Boldface numbers indicate definitions appearing in the main text. *Italicized* numbers indicate definitions appearing in the exercises.